HYDROGEOLOGY OF AN ARID REGION: THE ARABIAN GULF AND ADJOINING AREAS

HYDROGEOLOGY OF AN ARID REGION: THE ARABIAN GULF AND ADJOINING AREAS

A.S. ALSHARHAN
Faculty of Science, United Arab Emirates University
Al-Ain, United Arab Emirates

Z.A. RIZK
Previous Address: Faculty of Science, United Arab Emirates University
Al-Ain, United Arab Emirates

Present Address: Department of Geology, Menoufia University
Egypt

A.E.M. NAIRN
Earth Sciences & Resources Institute, University of South Carolina
Columbia, SC 29208, U.S.A.

D.W. BAKHIT
Previous Address: Ministry of Electricity & Water
Dubai, United Arab Emirates

Present Address: Department of Civil Aviation
Abu Dhabi, United Arab Emirates

S.A. ALHAJARI
Department of Geology, University of Qatar
Doha, Qatar

2001

ELSEVIER
Amsterdam - London - New York - Oxford - Paris - Tokyo

ELSEVIER SCIENCE B.V.
Sara Burgerhartstraat 25
P.O. Box 211, 1000 AE Amsterdam, The Netherlands

First edition 2001

British Library Cataloguing in Publication Data
Hydrogeology of an arid region : the Arabian Gulf and
adjoining areas
1.Hydrogeology - Persian Gulf Region
I. Alsharhan, A. S.
551.4'9'09536

ISBN 0444502254

Library of Congress Cataloging in Publication Data
A catalog record from the Library of Congress has been applied for.

ISBN: 0-444-50225-4

∞ The paper used in this publication meets the requirements of ANSI/NISO Z39.48-1992 (Permanence of Paper).
Printed in The Netherlands.

PREFACE

The Arabian Peninsula is an arid to semi-arid region, with a low rainfall and high temperatures most of the year, but with a high humidity in the coastal areas during the summer months. Water resources are limited, yet the availability of a sufficient supply of good quality water is the major requirement for the social, industrial, agricultural and economic development of the region. The increased demand for water arises from the improved standard of living, population growth and development arising from the oil revenues.

The countries of the Arabian Peninsula have made great efforts, to remedy the water shortage, by providing the financial and technical backing, for water desalination, treatment of wastewater and improved management and conservation techniques. The various water ministries, universities and research centres have supported scientific research, and applied the most recent technologies, in the search for new and alternative water supplies. Laws have been promulgated and economic and public relation campaigns have been developed, to promote and encourage the practice of efficient water use and the conservation of this scarce commodity.

In this book we have tried to provide the most important source of information for senior undergraduate and graduate students and researchers of the Gulf area, and more generally of arid regions, in order to comprehend the nature of the problems and how they interact with all aspects of life. In an area with a water deficiency, these interactions are more clearly defined than in water rich environments. For this reason sections on water laws and management, not usually found in regular hydrology was appropriately placed. The first part of the book is of a general character, it provides a geographic and geologic setting and emphasizing the climatic parameters, followed by a discussion on the aquifers and water chemistry. The second part of the book is devoted to the legal and management aspects of water resources, the more detailed studies of individual areas follows, and the book ends with the application of computer modeling of water flow and aquifers. Obviously the coverage cannot be complete, but a substantial bibliography provides a key to more detailed study.

ACKNOWLEDGEMENTS AND COPYRIGHT PERMISSIONS

The authors of this book would like to thank His Highness Sheikh Nahyan Mubarak Al Nahyan, Minister of Higher Education and Scientific Research and Chancellor of the United Arab Emirates University for his inspiration, encouragement and support. Without his support this publication would not have been possible.

Thanks are given to those authors and publishers who kindly allowed figures and tables from their publications to be reproduced in this volume. Every reasonable effort has been made to contact copyright holders in these regards. To any whose rights have unintentionally been infringed we offer our unreserved apologies.

We greatly appreciated permission from:

- Dr. P.G. Macumber, for figures 5.2b; 5.14; 8.94a,b; 8.95a,b; 8.96; 8.97; 8.98; 8.10 and 8.102.
- Dr. Moujahed Husseini (Editor-in-Chief of GeoArabia), for figures 8.92, 8.93 and Table 3.1.
- Geological Society, London (Quarterly Journal of Engineering Geology), for figures 8.14, 8.15, 8.16, 8.17 and Tables 8.5, 8.6.
- Prof. Peter Rogers, for figures 4.1, 4.2 and 4.3 and tables 10.2 and 10.3.
- Prof. Walid Abderrahman, Editor, (The Arabian Journal for Science and Engineering), for figures 4.5, 4.17, 5.10, 5.11, 5.12, 8.24, 8.33, 8.34, 8.35, 8.36 and Tables 5.5 and 8.10.
- Prof. Ali A. Alshamlan (Kuwait Foundation for the Advancement of Sciences), for figures 8.2, 8.5, 8.6, 8.10, 11.16, 11.17, 11.18, 11.19, 11.20, 11.21, 11.22, 11.23 and Tables 11.9, 11.10, 11.11, 11.12, 11.13.
- Dr. Ian Clark, for figures 8.99, 8.100, 8.103, 8.104, 8.105.
- Prof. Peter H. Gleick, for Tables 2.4, 2.5 and 2.6.
- Springer-Verlags, for figures 2.23, 2.51, 8.21 and Tables 8.7, 8.8 and 8.9.

We greatly appreciate the effort of Mr. M. Shahid who assisted us in more ways than could be imagined, he processed the chapters for this volume from inception to final completion, incorporated the author's changes and handled all correspondences between the authors. A mammoth task in this project is the figures. We would like to express our thanks to Mr. Hamdi Kandil for drafting all the figures and arranged them in proper position in this book and produced the final camera-ready copy of this volume.

We would like to thank Prof. Andrew Goudie (University of Oxford, UK), Dr. Anthony Lomando (Chevron) and Dr. Richard Ives (US Bureau of Reclamation), who read critically initial rough drafts of chapters 2, 3 and 10 respectively, and their comments improved the final text. Also to Prof. H. Edgell who provide us with many of his papers on the water resources of Saudi Arabia. In attempting to synthesize such field as water resources and management in the Arabian Peninsula, we have undoubtedly missed many references and under-represented a part of the field of study.

We thank Drs. Femke Wallien of Elsevier for her patience and encouragement for the inception of this book to its completion.

We dedicate this publication for geoscientists of water resources in the Middle East and comparable areas around the world.

TABLE OF CONTENTS

Chapter 6: Traditional Water Resources: Springs and Falajes

Chapter 7: Non-Traditional Water Resources: Desalination and Treated Wastewater

Chapter 8: Case Studies on the Hydrogeology of the Cenozoic Aquifer Systems in the Arabian Peninsula

Chapter 9: The Legal Basis for Groundwater Protection in the Gulf States

Chapter 1

AN INTRODUCTION TO WATER RESOURCES IN THE ARABIAN PENINSULA

INTRODUCTION

The Arabian Peninsula, located in southwest Asia with a population of 49 million, occupies approximately 3,000,000 km^2. It includes the political units of Kuwait, Saudi Arabia, Bahrain, the United Arab Emirates (UAE), Qatar and Oman. It lies within an arid-semi-arid zone lacking renewable surface water; the only surface waters are those of the Tigris-Euphrates river system which become saline upon entering the Arabian Gulf. The deserts of Arabia, the Rub al Khali, and Hijaz deserts, pass into a marginal zone of pasture, which has been subjected to over-grazing and cutting of the few trees for fuel. One result of overgrazing is the replacement of edible plants by inedible thorny perennial species depriving livestock of inexpensive fodder. It increases the process of desertification amplified by the current loss of soil through wind erosion and through channel erosion during the infrequent rainstorms. The groundwater resources and their conservation are essential for the entire region for both present and future generations. Rain-supported agriculture exists only in southwestern Saudi Arabia and in Oman where the mountains receive relatively higher rains than other parts of Arabia, but soil salinity has increased as a result of more saline water being drawn up by capillary action.

The local population adapted to the arid environment, the population was small and restricted to oases and better watered upland areas which could support cattle and crops. Because of the rapid development and rise in population consequent upon the discovery and exploitation of the rich hydrocarbon resources a large volume of groundwater is required depleting the aquifers in the Gulf area, the Gulf States face a real water shortage problem. As neither the amount or quality can satisfy the ever-increasing demands for water, the number of desalination plants is increasing. The high cost of production restricts its use in agriculture which is only partially alleviated by using treated water.

The countries of the Arabian Peninsula lack permanent and renewable surface water resources such as streams and lakes because they lie within the arid belt of the earth. The high temperatures, sand storms and low surface rainfall (annual average ~100 mm) causes high evaporation rates (annual average ~3,500 mm). These factors increase the severity of arid climate, enhance erosion and accelerate desertification.

The countries depend on groundwater (from both shallow and deep aquifers), and a small number of springs and falajes. The two latter resources are being seriously depleted at present as a result misuse, excessive pumping and poor maintenance. Because of the large volume needed for agriculture, groundwater is being depleted the Gulf States are facing a real water shortage problem. In the meantime, the high cost of producing desalinated water restricts the possibility of its use for agricultural purposes.

The Gulf States depend on several water-bearing formations (aquifers) for their groundwater resources. There are approximately 30 aquifers composed mainly of limestone and sandstone. The names of these aquifers vary from one country to another; but the same name may describe a specific aquifer in several neighbouring countries. The most important deep aquifers in the Gulf region are the Wajid, Saq, Minjur, Wasia, Umm er Radhuma, Dammam, and the Neogene. Most of these aquifers exist in Saudi Arabia, while some of them exist in other Gulf States. For example, the Umm er Radhuma, Dammam and Neogene aquifers also exist in Kuwait, Bahrain and Qatar. Other aquifers also exist in the United Arab Emirates and Qatar. Table 1.1 shows the most important features of these aquifers.

Because neither the amount nor the quality of groundwater produced in the Gulf States satisfies the ever-increasing demands for water, these countries started desalination of saline water in the 1970's. Coastal desalination plants draw raw water from the Arabian Gulf or the Gulf of Oman while the inland plants use brackish and saline groundwaters. During rainy seasons, some rain and flood waters are retained behind dams and recharge shallow aquifers. Despite their limited uses, sewage-treated water also represents an additional source of water in the Gulf region.

The exponential rise of water demands in the Gulf States began in 1980. The water resources deficit was met by water desalination. However,

Table 1.1. The most important aquifers in the Gulf States (compiled from Al-Mogren, 1995; Dabbagh and Abderrahman, 1997).

Aquifer	Thickness (m)	Total dissolved Solids (mg/l)	Depth from ground surface (m)	Country
Wajid	300-400	500-1,000	15-1,110	Saudi Arabia
Saq	500-600	500-1,500	100-1,500	Saudi Arabia
Tabuk	1,000	500-3,500	10-1400	Saudi Arabia
Minjur	360	400-1600	1400	Saudi Arabia
Wasia	200-230	1,000-3,000	230-1,200	Saudi Arabia
Umm er Radhuma	500	300-1,000	250-600	UAE, Bahrain, Oman
Dammam	200	1,000-6,000	100-500	Bahrain, Qatar, Kuwait
Neogene	30-100	100-4,000	10-150	UAE, Bahrain, Oman

desalinated water can only meet the increasing domestic needs and is still not economically feasible for agricultural purposes.

The most important water-related problems in these countries are the depletion of aquifers in several areas, saline-water intrusion problems, and water quality problems such as those associated with oil industry or agricultural activities.

Because agriculture consumes between 75 to 85% of water resources in the Gulf States, management and conservation measures target this particular sector. The effort spent in water conservation and management in the United Arab Emirates is evident.

Improvement of the present water management can lead to water conservation, maintain better water quality, and restore deteriorated aquifer systems in many areas of the Gulf States. The use of advanced irrigation technologies, construction of recharge dams, and growing salt-tolerant crops are proper agricultural approaches. Development of human resources is a priority and helps training national experts in water-related fields. Establishment of data banks and application of advanced groundwater modelling techniques represent powerful management tools.

1. Water Losses

The Gulf States are characterized by high evaporation rates and scarce rainfall. However, conservation of each drop of water is needed. Water loss can occur from drinking water, irrigation water and rain and flood.

A) Drinking water losses

The loss of drinking water is the difference between amount of water produced and the amount recorded by water meters. Water loss can occur through one or more of the following:

i) Water loss from the network itself which can reach 30% of water production.
ii) Loss associated with poor network maintenance.
iii) Water loss resulting from the improper equipment such as counters, floats and pumps.
iv) Loss as result of misuse, flooding of tanks or error in their construction.

B) Irrigation water losses

The loss of irrigation water is the difference between amount of water produced and the amount actually used by plants or crops. Water is usually lost through evaporation or seepage from water-transport channels. Traditional irrigation techniques lead to the loss of huge amounts of water and are economically unfeasible in the Gulf States. The irrigation water losses occur through:

i) Water loss from transport channels through natural evaporation and seepage.
ii) Traditional flood irrigation leads to large evaporation losses and waste of water.
iii) Growth of weeds and unwanted plants, which consume additional amounts of water.
iv) Excess of irrigation water as a result of lack of experience or negligence of some farmers.

C) Rain and flood water losses

As rainwater reaches the ground surface, a considerable part of it is lost through evaporation and infiltration. In coastal areas, a part of rainwater can be lost to the sea. Runoff water is the part of rainfall that can be properly managed. Dams are constructed to utilize runoff water by retention or diversion to recharge groundwater.

2. Dams for Water Conservation and Protection

Hydrogeologic investigations indicate that annual runoff volume varies from 206 Mm^3 in Oman to 270 Mm^3 in United Arab Emirates to 250 Mm^3 in Saudi Arabia. More than 200 dams of various designs and capacities were constructed in 1995 in the three countries for water conservation and flood protection.

A) Dam Construction Measures

To make the best use of runoff water and dam construction, the following measures must be taken into account:

i) Reduction of the velocity of runoff water to move as slowly as possible.
ii) Construction of dams to retain floodwater for direct use or to divert it to recharge groundwater.
iii) The topography, gradient and area of drainage basins must be taken into account during design of either retention or recharge dams.
iv) The geology, rock type, dominant soil and geologic structures control the velocity of runoff water and infiltration rate.
v) The water retained behind dams is directly used for irrigation and domestic purposes. Part of this water recharge underlying aquifers.

B) Types of Dams

Types of dams vary according to the nature of basins in which they are built, the purpose of dam construction and the geologic setting of the site. The major dam types in the Gulf States are:

i) Concrete dams

This type of dams is constructed in mountainous areas, especially where the cross-sectional area of the stream channel is narrow. These dams tolerate climatic conditions and speedy-moving runoff water. Costs of construction of this type of dam are usually high. Several dams of this type were constructed in Saudi Arabia and United Arab Emirates.

ii) Stones dams

Stones available in the site and sand are used to fill the dam and compact its body. Concrete and hard stones prevent water seepage and dam protection against severe climatic conditions line both sides of the dam. Dams of this type exist in the Saudi Arabia and Oman.

iii) Earth dams

These dams are constructed in plain areas where construction materials are usually available. However, construction of these dams may need the removal of a huge amount of surficial material to reach the solid bedrock where the foundation of the dam must be placed. Earth dams are usually constructed to recharge underlying and surrounding aquifers. Dams belonging to this type are common in the Saudi Arabia, United Arab Emirates and Oman.

iv) Subsurface dams

Because of the prevailing arid climate and the extremely high evaporation rates (3,500 mm/yr) compared to very low rainfall (average 100 mm/yr), subsurface dams represent good alternatives. These dams are constructed in the subsurface such as Al Taif dam in Saudi Arabia. The advantages of this type of dam are the absence of evaporation losses and siltation problems. However, construction of these dams needs advanced technology, proper site selection and high costs. The storage capacities of existing and planned dams in Saudi Arabia, Oman and United Arab Emirates are 850 Mm^3 (from 190 dams), 67 Mm^3 (from 15 dams) and 18.5 Mm^3 (from 11 dams) respectively.

The wadi beds in Saudi Arabia and Oman represent good aquifers and their recharge through dams depends on the amount and intensity of the annual rain. The runoff water usually carries huge amounts of silt, which is deposited on the upstream sides of dams. Despite the high fertility of this type of soil, they greatly reduce the infiltration capacity of sediments on the upstream sides of groundwater recharge dams.

In the United Arab Emirates dams are mainly constructed to recharge aquifers and natural springs. The heights of these dams vary between 3 and 33m, their storage capacity was 18.5 Mm^3 in 1995 but about 75 Mm^3 in 2000. In Saudi Arabia, earth dykes of 1.5m are constructed to slow down the velocity of floodwater, increase infiltration volumes and protect surrounding farms.

3. Water Resources

A) Water resources in Saudi Arabia

The agriculture in Saudi Arabia depends mainly on groundwater for rain-supported agriculture is limited to parts of the southwestern part of the country. The water sources in Saudi Arabia summarized in the following:

The rainfall in Saudi Arabia exhibits a wide variation in space and time. Occasional heavy, short rainstorms cause floods in soil-rich wadi channels. To control floodwater the Ministry of Agriculture and Water has constructed more than 190 dams of variable sizes and storage capacities. The total storage capacity of dams in Saudi Arabia is 850 Mm^3. These dams are intended to retain floodwater for irrigation and recharging aquifers. After proper treatment, floodwater can be also used for domestic and drinking purposes.

Spring waters are used for irrigation in areas such as Al Hofuf, Al Qatif and Al Aflaj. A small number of springs exist in the western region of Saudi Arabia and their water is mainly used for drinking.

Both shallow (5 to 50 m) and deep (50-2,000 m) aquifers are utilized in Saudi Arabia. Groundwater in the shallow aquifers seems to be renewable as parts of rainwater and occasional floods may recharge them. Groundwater satisfies about 70% of water needs in Saudi Arabia and the number of drilled wells has reached over 78,000 in 1995.

The Saudi Arabia is the largest producer of desalinated water in the world. This is attributed to the steadily rising demands for water in the country as a result of population growth and rising standard of living. The industrial and urban developments also need additional water resources. Several recent desalination plants were constructed and pipelines from these plants were extended to areas of use. Twenty-three desalination plants built by 1995 supply the water needs of 40 city and village along the eastern and western coasts of Saudi Arabia. Desalination plants produced 2.2 Mm^3/d, 57.4% of it served the towns of the eastern coast, whereas 42.6% of it served the towns of the western coast. Four desalination plants of an approximate capacity of 380,000 m^3/d are under construction of present. Upon completion of these projects the daily water production in Saudi Arabia is predicted to reach 3 Mm^3/d. Fifteen additional projects for desalination plants are also being evaluated. Desalinated water is used mainly for domestic purposes. In some areas, desalinated water is mixed with groundwater to improve its quality.

Sewage-treated water is used for irrigation of some farms in Riyadh city. The sewage treatment plant produces over 220,000 m^3/d. Treated water is transported via pipelines to nearby farms.

B) Water resources in Oman

Oman realized the importance of water and initiated the Ministry of Water Resources in 1994. The ministry responsibilities include research studies, evaluation and quality of water in Oman and the producing aquifers. Groundwater is the main source of water used for irrigation, domestic purposes and drinking in Oman.

The total number of wells tapping both shallow and deep groundwater in Oman is more than 167,000. These wells produce about 56% of water used for irrigation.

Falajes represent one of the oldest irrigation technologies developed by Omani people hundreds of years ago. The falaj waters meet 40% of the irrigation needs. The individuals who have constructed them or their families own the falajes. The falaj water is distributed among owners on an accurate time-share basis. The Ministry of Agriculture and Fisheries fix and maintain falaj systems all over the country. The falajes of Oman are classified into two main types; Daudi and Gheli. The Daudi falajes represent 80% of the falajes used for irrigation in Oman. These falajes are subsurface tunnels constructed to transfer groundwater from the foothills of mountains, where the water table is usually shallow, to farms further away from the mountains. The falajes are designed to have vertical shafts for aeration and maintenance. As the falaj tunnel intersects the ground surface, water is distributed to different farms via a system of cement-lined small channels. According to a definite time-share, falaj water is directed through these channels to different farms. Because groundwater is the main source of recharge for the Daudi falajes, they maintain discharge throughout the year. On the other hand, the Gheli falajes, which represent 20% of the falaj systems in Oman are fed directly from the base flow of natural wadi channels. In contrast to the Daudi falajes, the Gheli falajes are small open canals in which water freely flows under gravity. The discharge of the Gheli falajes is highly variable, depending mainly on the amount and intensity of annual rains. The falaj length varies from 0.1 to 12 km. The total number of falajes in Oman is about 4,200, while the presently active ones about 3,045 falajes.

Retention dams are very important in Oman. Dams are constructed to retain rainwater before it reaches the Gulf of Oman. Oman constructed 4 dams of a total storage capacity of 46 Mm^3. Oman is the least dependent on desalinated water of the Gulf States. The production of the water desalination plants in Oman reached 5 Mm^3 in 1995. Sewage-treated water is used for irrigation of green areas, gardens, parks and roundabouts. The sewage treatment plants in Oman produce 60,000 gallons/day.

C) Water resources in United Arab Emirates

The mean annual runoff on the main wadis in United Arab Emirates is 125 Mm^3. A large volume of runoff water is now harvested by 35 recharge dams with a total storage capacity of 75 Mm^3. A few dams are under construction at present and several others are planned in the future.

Permanent springs provide about 3.0 Mm^3 of water per year. Spring discharges range from 0.06 Mm^3/yr to 2.50 Mm^3/yr, with little change over the years. Discharge of some springs is directly related to rainfall, whereas the discharge of others is not directly related to rainfall. During the 1984-1991 period, spring salinity has increased by 10% (e.g., Khatt South in Ras Al Khaimah) to 50% (e.g., Bu Sukhnah in Al-Ain) as a result of low rainfall and heavy groundwater pumping in the recharge areas.

Despite their limited discharge, falaj water is a renewable resource which is directly related to rainfall. During 1978-1995, the total falaj discharge in United Arab Emirates varied between 9.0×10^6 m^3/yr in 1994 and 31.2×10^6 m^3/yr in 1982, which represents 2.8 to 9.7% of the total water use in the country.

The annual recharge for groundwater in United Arab Emirates as 120 Mm^3 was estimated by Khalifa (1995). The current annual groundwater extraction averages 880 Mm^3, reflecting a highly unbalanced situation resulting in aquifer depletion in many areas such as Al Ain and Al Dhaid, dryness of many shallow wells, and saline water-intrusion problems. Due to excessive groundwater pumping, cones-of-depression ranging from 50 to 100 km in diameter now exist in the Al Dhaid, Hatta, Al Ain and Liwa areas.

The volume of desalinated water has increased from 7 Mm^3 in 1973 to 694 Mm^3 in 2000. In 1985, the desalination plants in the United Arab Emirates produced 204 Mm^3 of water, which represents 60% of the domestic water needs. In 1998, the production of desalinated water reached 526.6 Mm^3, which is 76% of the water used for domestic purposes. In 1997, the United Arab Emirates production of desalinated water was 57% in Abu Dhabi, 35% in Dubai, 5% in Sharjah, and 3% in the northern Emirates.

The sewage water discharge in the United Arab Emirates increased from 1.5 Mm^3 in 1973 to 142 Mm^3 in 1994 and reached 175 Mm^3 in 2000. There is about 10% annual increase in sewage water production in the United Arab Emirates as a result of increasing population, increasing per capita water use, and extension of sewage network to serve about 70% of the population.

D) Water resources in Qatar

The water resources in Qatar include groundwater, mostly in Tertiary aquifer systems, desalinated water and sewage treated water. In 1995, Qatar had two desalination plants, which produced 130 Mm^3. There were also two sewage treatment plants producing 30 Mm^3 of water. Treated water was used for irrigation of animal-forage crops, green areas and public parks.

E) Water resources in Kuwait

Water resources in Kuwait include ground-water, rainwater, desalinated water and sewage treated water. The desalinated water represents 62% of the total water resources in Kuwait, groundwater represent 20% and sewage-treated water represent 18%.

F) Water resources in Bahrain

Bahrain used to depend mainly on water of natural fresh water springs. The discharge of these springs decreased over the time until most of them have disappeared at present. Water wells penetrating the Dammam aquifer are the main source of groundwater on the island, while desalinated water is now used for drinking and domestic purposes. Desalinated water is produced from 4 desalination plants producing 40 Mm^3 of water. The volume of sewage-treated water reached 8 Mm^3. This water is reused in agriculture.

4. Water consumption

Because of serious deficit of water resources, the Gulf States rely on desalinated water to meet the increasing demands for water. The desalination plants numbered 56 in 1995 mainly located along Arabian Gulf and the Gulf of Oman, and producing 1,552 Mm^3/yr. After being mixed with groundwater, desalinated water is used for domestic and drinking purposes. Additional desalination plants are operated by oil companies and other industrial companies.

Despite the fact that the agricultural activities consume between 75 and 80% of groundwater pumped in the Gulf States, water needs for certain specific irrigation activities are met by treated-wastewater. This water is used for irrigation of public parks, animal-feeding crops, and certain trees. The volume of produced treated wastewater is about 2Mm^3/day, however, 700,000 m^3/day are only used.

The water need of the agricultural sector is steadily increasing in Saudi Arabia, United Arab Emirates and Kuwait. The volume of water used in agriculture was estimated at 16,000 Mm^3 in 1988. About 87% of this amount was used in Saudi Arabia.

The water resources in Gulf States are subject to a great depletion, especially by the agricultural sector. The excessive use of water devoted for domestic and drinking purposes represent an additional stress. Statistics show that the per capita water consumption in the Gulf States exceeds 300 liters per day, value that exceeds the individual share in some industrial countries. The high investment by the government of the Gulf States to meet the increasing needs for water has to be recognized by individuals through water conservation and proper management.

SCOPE OF THE VOLUME

The intent of this book is to provide the researchers in the Gulf region with an integrated approach to the problems of water, technical, economic and social. The book provides a geographic and geological setting, emphasizing the climatic parameters. This is followed by a discussion of the aquifers and of the water geochemistry. The final chapters are devoted to the legal and management aspects of water resources. The recognition of water as an economic good with competition not only from domestic, but industrial and agricultural users for a scarce commodity, forces a re-evaluation of water. It ceases being a low cost commodity to one with a distinct value. Competition for a scarce commodity raises the question of allocation with charging as an economic tool which affects demand through conservation and the efficient use. The ultimate aim is full cost recovery.

The change in the view of water has obvious social and political importance. The traditional water laws existed before the onset of development. Nevertheless the Gulf Sates are bound by their constitutions to honor Islamic Law. So a new code has to be devised which, while honoring Islamic Law, is nevertheless appropriate to modern times. This is achieved through Water Resource Management policies which integrate all elements involved, production, distribution, and the appropriate social and legal aspects. The book ends with a number of case studies to illustrate some of the problems in more detail, and numerical modelling of certain aquifer systems.

This book therefore deals with several issues, not all directly related to water, but to its effects upon society, effects which must be integrated into a successful water resource management problem. Listed below is an outline of these topics:

A) Water Resources. In a semi-arid to arid region where rainfall is insufficient to supply the needs of a growing population and a higher standard of living, the deficit is normally made up by extracting groundwater. Groundwater which is not being recharged under present climatic conditions. The result is a falling groundwater level, changes in the water geochemistry with increasing total dissolved solids and the uprise of saline water from deeper horizons and water deteriorating in quality and quantity. The water currently being withdrawn is fossil water emplaced during the pluvial epochs, of the last ice age.

   Attempts at conservation and improving supplies by the construction of retention dams to retard the run-off from infrequent storms, while laudable is not a solution to the shortage problem, and treated water is insufficient in quantity to meet agricultural needs. The construction of many desalination plants (32 in Saudi Arabia prior to 1995) while providing for domestic supply is too expensive to maintain a major agricultural program. The adoption of modern irrigation techniques will require major financial support. The only rain fed agriculture is in the mountainous areas made possible by the use of the traditional falaj system.

B) Aquifer Systems. The main aquifer system extends from central Arabia towards the Arabian Gulf to the north and east, with an eastward groundwater flow. The system is made of sedimentary formations extending from early Cretaceous to Quaternary time. There are three main hydrogeological units hydraulically connected. A secondary aquifer system is in discontinuous unconsolidated sands and gravels where the fresh water may be floating on top of highly saline artesian groundwater.

C) Water Types. Three chemically distinct water types are recognized, bicarbonate, sulphate, and chloride which reflect the nature of the rock through which the water passes and residence time. The groundwater usually changes from bicarbonate to sulphate to chloride as the water moves away from the recharge area to the discharge area. Bicarbonate water is generally characteristic of low salinity groundwater, renewable groundwater resources and low residence time. Sulphate waters predominate in groundwater passing through gypsum and anhydrite aquifers, and is usually associated with intermediate salinity in unconfined aquifers. Chloride groundwater is dominant in the discharge areas in high salinity springs and chloride rich sabkha deposits.

D) Social, Legal and Economic constraints. In the modern complex society of the Gulf, the States have taken over the ownership and distribution of water supplies to meet the steadily rising demand for water from industrial and urban projects in addition to domestic and agricultural demands. The competition for a limited resource requires some form of allocation, an assessment and prioritization based upon current needs bearing in mind planning for future needs. It requires an integration of water supplies from all sources, groundwater, treated water, and desalinated water and a corpus of laws to provide the basis for agreements and for the resolution of disagreements which involves all facets of society.

Chapter 2

PHYSICAL GEOGRAPHY OF THE ARABIAN PENINSULA

GEOMORPHOLOGY

Geographic Setting

The Arabian Peninsula lies between latitudes 13° and 32° N and longitudes 35° and 60°E. It forms a part of the great desert belt which stretches from the Atlantic Ocean, near the coast of northwestern Africa, to the Thar Desert of northwestern India. It has an area of approximately three million square kilometers, about 10.5% of the Earth's surface and supports a population of about 49 million. This is only 8% of the global population. Included within the Arabian Peninsula are the relatively small states of Kuwait, Bahrain, Qatar and the United Arab Emirates as well as the proportionately larger ones of Oman, Yemen and Saudi Arabia (Table 2.1; Fig. 2.1). The Arabian Peninsula, a southwestern projection of Asia, is separated from Africa by the Red Sea, from Iran by the Arabian Gulf and the Gulf of Oman, and is bounded on the south by the Arabian Sea and Gulf of Aden (Fig. 2.2). It is divided into three main divisions: the Arabian Shield, Arabian Shelf and Mountains belts (Powers et al., 1966; Alsharhan and Nairn, 1986).

Table 2.1. Total area and population distribution of countries of the Arabian Peninsula as of 1999.

Country	Area (km^2)	Population (million)
Saudi Arabia	2,149,690	22.25
Kuwait	17,818	2.25
Bahrain	00,652	0.70
Qatar	11,610	0.80
United Arab Emirates	77,700	2.50
Oman	312,000	2.50
Yemen	528,000	18.00
TOTAL	**3,097,470**	**49.00**

To the northeast the Arabian Peninsula meets the alluvial deposits of the Tigris-Euphrates river system draining the mountains to the north and east. Sediments from the mountains form the Tigris-Euphrates delta which is prograding into and gradually filling the Arabian Gulf. The peninsula's eastern flank contains a major part of the world's known hydrocarbon resources and a disproportionately large number of the world's giant and super-giant oil and gas fields. The Arabian Gulf gradually passes into shallow, submerged areas with average depths of only 60 m, increasing in the deepest part to 100m in the southeast. Bathymetric charts show a depth asymmetry, whereby the deeper parts lie closer to the Iranian side. At its southeastern end the Arabian Gulf narrows, forming the strait of Hormuz where the Musandam Peninsula projects northwards towards the Iranian shore. Eastwards, beyond the strait, a profound geological change occurs, whereas the Arabian Gulf is floored by continental crust, the Gulf of Oman and the Gulf of Aden are oceanic in character. The Red Sea in the west is long and narrow with the Strait of Bab al Mandab marking its boundary with the Gulf of Aden. The bathymetry of the Red Sea varies, and the maximum depth recorded is about 2,850m.

Topography

Geomorphology, climate and the availability of water, have influenced human settlement and communications in the Arabian Peninsula. The whole region lies within the arid subtropical zone. During summer the main track of the jet stream controlling the passage of atmospheric depressions, lies north of the Pontic Mountains in Turkey. During winter this track moves southwards and covers the northern Arabian Gulf. The few depressions pass south of 30° N latitude, increase the region's relatively low rainfall regime of approximately 300 mm/yr. The lower the precipitation, the greater its variability. For example, Bahrain with an average of 76 mm/yr, may receive from 10 to 170 mm. Only the Arabian Sea coast benefits to a limited extent from the passage of the monsoon. Settlements in the Arabian Peninsula are restricted to areas of permanent springs and oases to areas where irrigation is possible. In the deserts, a few nomads continue to eke out a precarious existence grazing livestock. Since prehistoric time the greater part of the population has been found in the "fertile crescent" along the Tigris-Euphrates River system and in small coastal pockets in Kuwait, eastern Saudi Arabia, Bahrain, Qatar, United Arab Emirates and Oman. Prior to the discovery of oil, pearl fishing and coastal transport provided subsistence for many among the coastal populations.

The Arabian Peninsula has a varied relief, combining extensive disserted plateau with rugged mountains along the western and southeastern rim. In the west, the Al Hijaz Mountains stretch from the Gulf of Aqaba into Yemen, gradually increasing in height southwards from about 1,500m near Mecca to Yemen, where the highest peaks in the vicinity of the Yemen's capital Sana'a reach 3,660m with an average elevation range from 1,800 to 2,400m. As a result of faulting associated with the late Tertiary separation of the Arabian Peninsula from Africa, there is a precipitous drop in elevation from the mountains to a narrow coastal plain bordering the Red Sea (Fig. 2.3). Elevations are less conspicuous along the southeastern Yemen-Oman coast. The terrain is less rough and elevations decrease to about 900m. These mountains are separated by a low saddle from the higher peaks in the northern Oman Mountains. In the main Oman Mountains chain, facing the Arabian Sea and the Gulf of Oman, peaks may reach heights of around 3,000m. The mountains slope steeply both to the east and west. In the west, the mountains disappear under the great sand sea of the Rub Al Khali. The central core of the mountain chain projects northwards as the Musandam Peninsula, which reaches the Strait of

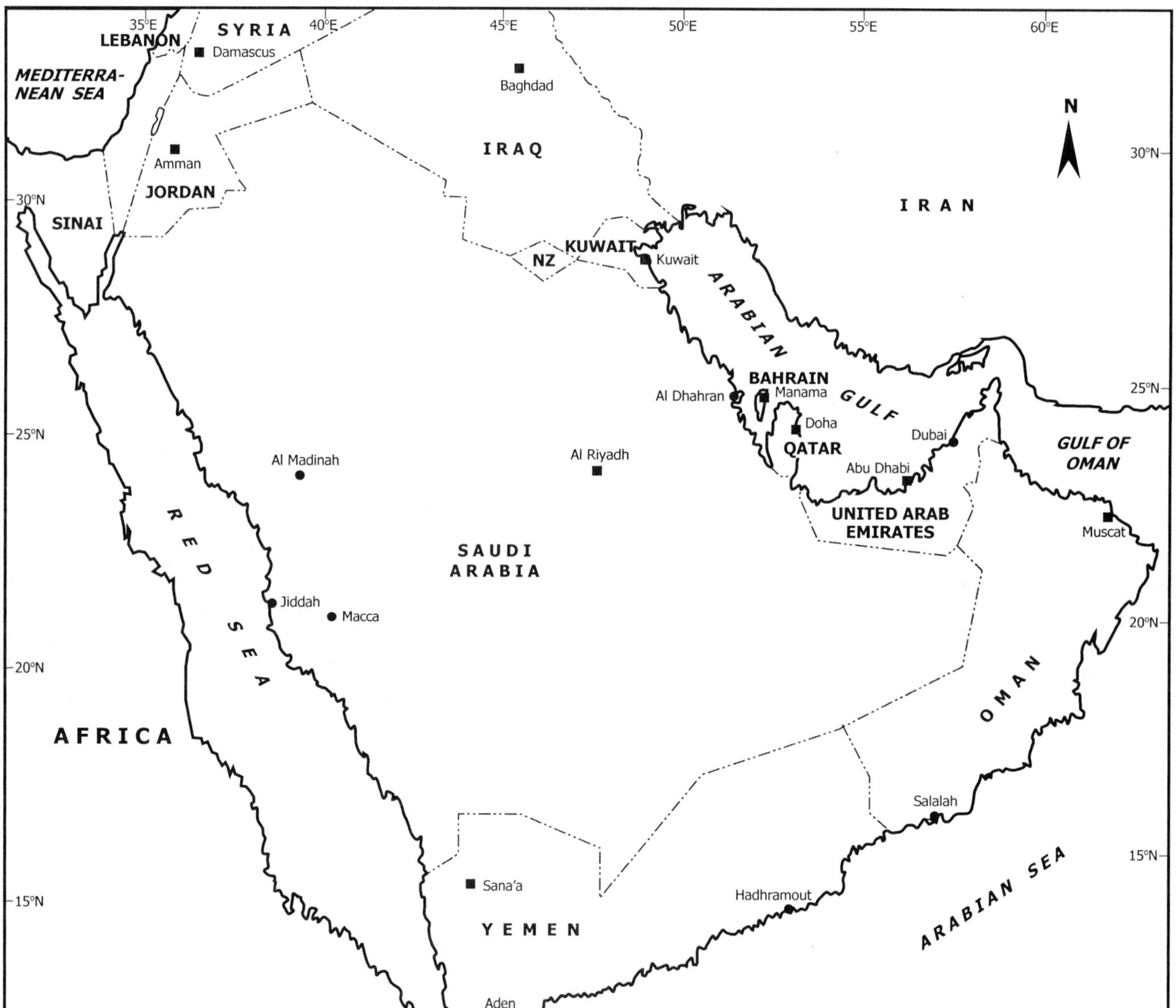

Fig. 2.1. Geographic location map of the Arabian Peninsula countries and adjacent areas.

Hormuz. In the east, the mountains decline gradually near the Gulf of Oman, and in some areas leave only a narrow coastal plain bordering the sea.

The heavier rainfall associated with the Al Hijaz Mountains is responsible for maintaining the variety of crops in southern Saudi Arabia and Yemeni sections of the coastal plain. Similarly, in Oman, precipitation on Jebel Akhdar supports some agriculture on the Batinah coastal plain and to lesser extent around Salalah. However, much of the coastal region is barren and sandy, with many spits, bars and low-lying salt flats (sabkhas). In the north of Arabia, the coastal region is dominated by sand-covered plains, which pass westward into flat gravel plains bordering an ancient drainage system in Wadi Al Batin (Fig. 2.4). This valley crosses into Kuwait, and at one time drained an area from the Al Hijaz Mountains to the channels of the Tigris-Euphrates river system. Over the greater part of this arid region the ground cover consists of dunes and sandy (erg) or stony (hamada) deserts with little or no vegetation (See Figs. 2.5 and 2.6).

In central Arabia, west of the sandy area, is a series of west-facing escarpments, where the Mesozoic and older sedimentary rocks form long ridges with steep west facing scarps and shallow,

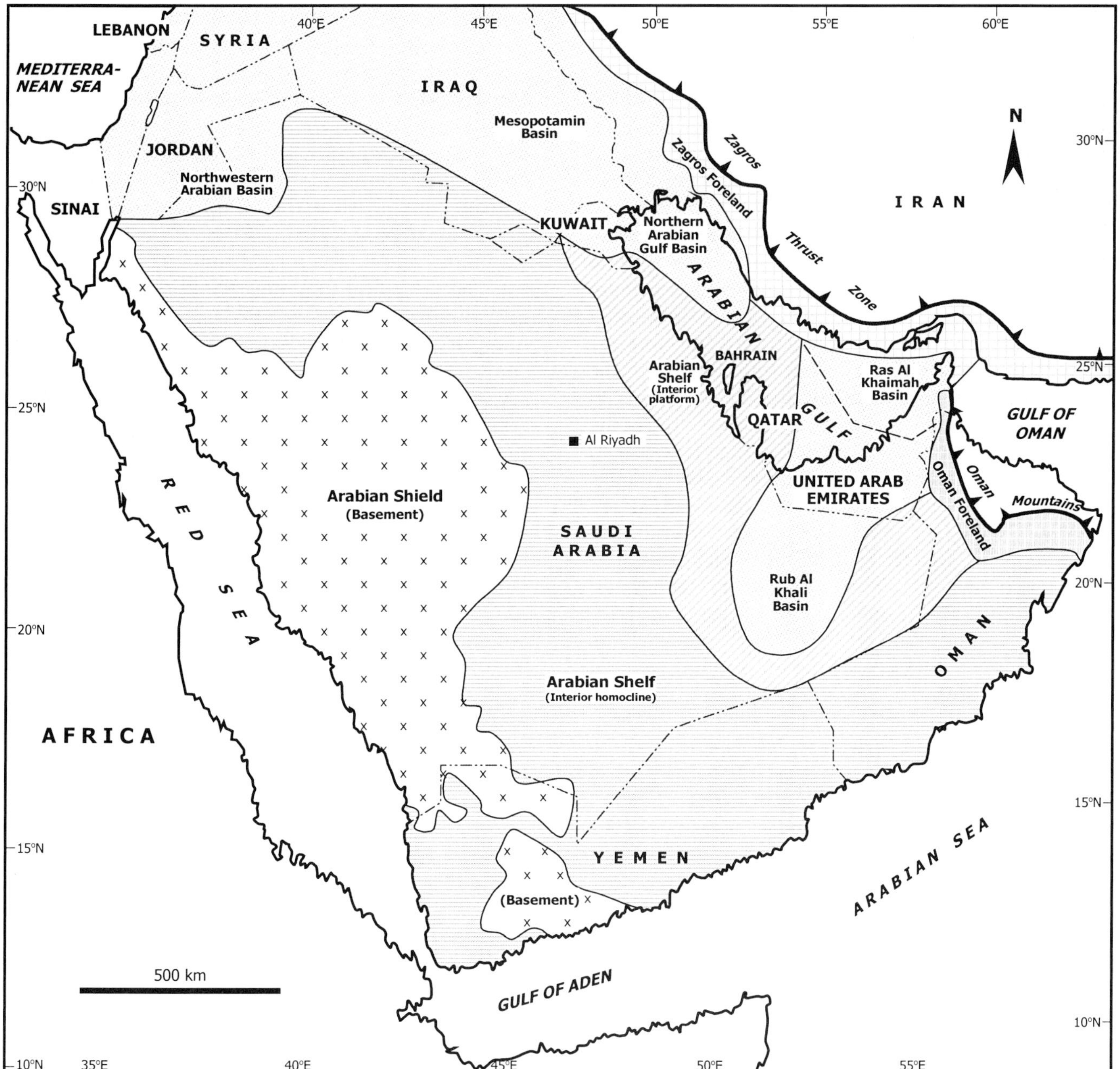

Fig. 2.2. Main geologic subdivision of the Arabian Peninsula (modified after Powers et al., 1966; Alsharhan and Nairn, 1986).

easterly dipping back slopes. Of these, the Tuwaiq limestone scarp is the most prominent, reaching an elevation of 240m above mean sea level, and 100m above the surrounding terrain. Several major wadis cut across the strike of the Tuwaiq escarpment.

These wadis provide access to the central parts of Saudi Arabia except during infrequent rain events (Figs. 2.3; 2.4). West of the escarpment, the land continues towards the Al Hijaz Mountains, forming a rugged and extensive plateau of igneous and metamorphic rocks, with elevations ranging between 1,200m and 1,800m. These Precambrian igneous and metamorphic rocks are overlain by recent lava flows (harratts), and gradually merge with the coastal mountains

Geologic Setting

Geologically the Arabian Peninsula is bounded by the Owen Fracture Zone and the Gulf of Aden rifting to the south, the rift system of the Red Sea/Gulf of Aqaba to the west and the Oman Mountains to the east.

The Arabian Peninsula is divided geologically into the western Arabian Shield, part of a Precambrian crustal plate, and the Arabian Shelf, which consists of an eastward thickening sedimentary wedge separated into an interior homocline and interior platform (Fig.2.2) (See Powers, et al., 1966; Alsharhan and Nairn, 1997). In general, sedimentary strata dip away from the shield at a very low angle, from less than a degree in the older beds, to a third of a degree in the younger beds. In the interior homocline of Arabia, beds have been subjected to minor folding and faulting, and some tectonic activity is evident along structural axes, such as the Ha'il-Jauf-Rutbah- Khleissia-Mosul, the Central Arabian, and the Qatar - south Fars, the Hadhramout, and the Huqf arches. Dips remain low in the interior platform, but several major north-south anticlinal axes rise above the level of the platform, exemplified by the Ghawar, Burgan and Dukhan Highs. The latter are believed to be related to basement ridges and are superposed lineaments.

The source of the sediments is the peneplaned Arabian Shield, which has been subjected to mild epeirogenic uplift. The sediments through the Phanerozoic were deposited in shallow to deep shelf seas, giving an alternation of continental and marine deposits, punctuated by evaporitic events. The total thickness of the Phanerozoic deposits increases from 5,500m in Central Arabia to about 7,500m along the Arabian Gulf.

Geomorphological Zones

1. The Coastal Zone

The flat desert landscape which characterizes the coastal regions of the Arabian Gulf and stretches from Kuwait to Oman, has few distinctive features. There are some positive topographic features, such as the Dammam dome, and the Abqaiq and Dukhan anticlines, interpreted as developing over salt plugs, though they rise only a few tens of meters above the desert surface. The most marked feature is the presence of inland and coastal sabkhas, particularly in the United Arab Emirates and Saudi Arabia, and the presence of a large number of collapse structures at Qatar and in parts of Saudi Arabia, in the vicinity of Riyadh.

Sea-level changes in the recent history of the region are reflected in the development of offshore terraces, widespread flat inland surfaces, and rock pavements. Inland, this zone passes into gravel and stony plains, sometimes covered by sand dunes. In contrast, in the northern and northwestern parts of Kuwait, the gravel surfaces are replaced by fluvial and estuarine deposits, associated with the Tigris-Euphrates and Karun fluvial complex.

The shoreline of the Arabian Gulf is irregular and dominated by supratidal sabkhas, sand spits and carbonate sands. Bordering this zone to the south, is a 30 to 120 km wide zone of active dunes, often resting directly on a gravel surface. In Saudi Arabia, however, there is a dissected limestone plateau, 80 to 250 km wide, that narrows to the south until it loses its identity under the sands of the Rub Al Khali, which intervene between the coastal strip and the Ad Dhahna sands.

Tidal flats along the Arabian Gulf coast of the United Arab Emirates, as far as Kuwait, are made up of sandy, silt-sized carbonate sediments with anhydrite and halite resting on calcareous beds. Solution of these calcareous beds, in some areas such as Qatar, normally leads to the development of extensive depressions, forming a modified karst topography, mantled by fine-grained sediments. Low eroded ridges are the topographic expression of small anticlines, and salt piercement structures. The importance of the collapse structures, is their function as groundwater recharge and discharge areas. Sabkha Matti in western United Arab Emirates, is thought to be the largest coastal sabkha in the Arabian Gulf (Glennie, 1970). It extends 40-60 km east-west, and up to 120 km north-south. Most of the sabkha consists of partly cemented dune sand, and is undergoing slow deflation. The whole sabkha surface is salt encrusted, because the water table coincides with the surface of the gently sloping plain.

The major inland sabkhas in Qatar occur at Sauda Nathil and Jaww As Salama, in the south. These sabkhas occupy depressions which lie close to sea level, and are even lower locally. The origin of these depressions seems to be related to the dissolution of fractured limestone rock in the presence of abundant groundwater of relatively low salinity. Sabkhat Sauda Nathil in southern Qatar, is about 8 km long and 3 km wide, and has an area of 22.5 km^2. The land surface is 1 m below sea level in some areas, and in general does not exceed 1 m above sea level. Sabkhat Jaww As Salama, west of sabkhat Sauda Nathil, has an area in excess of 18 km^2. It occurs mostly at or below sea level. Several wadis discharge into the sabkha, and lower-salinity water gathers at the sabkha surface. Most of the collapse structures in Qatar date from post-Miocene time, as there are no examples of older Eocene land surface depressions filled with Miocene sediments (Cavalier et al., 1970). Reference is made to surface or mantled karst, in accordance with whether the limestone is exposed or not. Historically, the dissolution depressions in Qatar are important, in that they contained both fresh and brackish water. Conditions are similar in Bahrain. They are all arid, and traditionally the resident population has relied on these depressions for water.

The Umm-as-Samim sabkha basin, at the western borders of inner central Oman, is adjacent to the eastern limit of the Rub Al Khali sand desert.

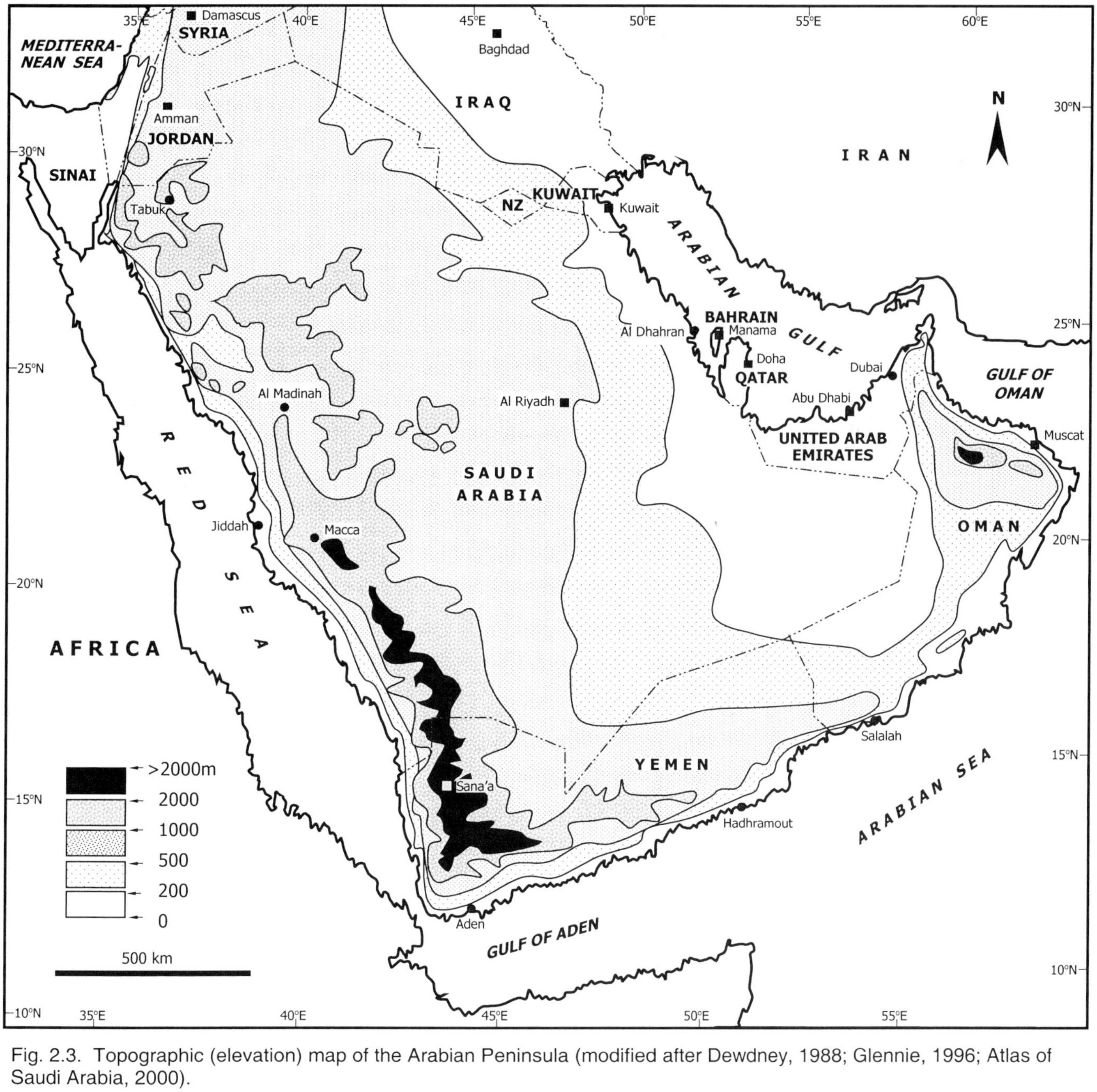

Fig. 2.3. Topographic (elevation) map of the Arabian Peninsula (modified after Dewdney, 1988; Glennie, 1996; Atlas of Saudi Arabia, 2000).

It covers an area of 2,500 to 3,000 km^2, extending 100 km from northwest to southeast, and is 30 km wide. The sabkha runs parallel to the strike of the mountain range, and lies 200 to 300 km from the nearest coast, at an average elevation of 60m above sea-level. The Umm-as-Samim sabkha is a salt-encrusted playa, which may have developed in a natural basin or deflation hollow, where the groundwater table is very close to, or reaches the surface. In this situation, efflorescence or capillarity evaporation causes crystallization of evaporites from groundwater. Alternatively, the evaporites may have developed in an area where a former lake dried out, as a result of increasing aridity.

Beydoun (1980) believes that the Umm-as-Samim was a lake during late Pleistocene pluvials, receiving its water via backslope drainage from the Oman Mountains. With the onset of Holocene aridity, the lake progressively dried up and inland sabkha formation commenced at about 4,000-5,000 years BP as described by Kinsman (1969). Glennie's (1970) hypothesis, that the Umm-as-Samim sabkha was originally a relict arm of the sea, needs further investigations, in view of the long distance between the sabkha and the nearest coast (200-300 km). In general, in this region of vast undulating plains, with a Tertiary sediment cover, there is a network of drainage channels radiating across the plain, from

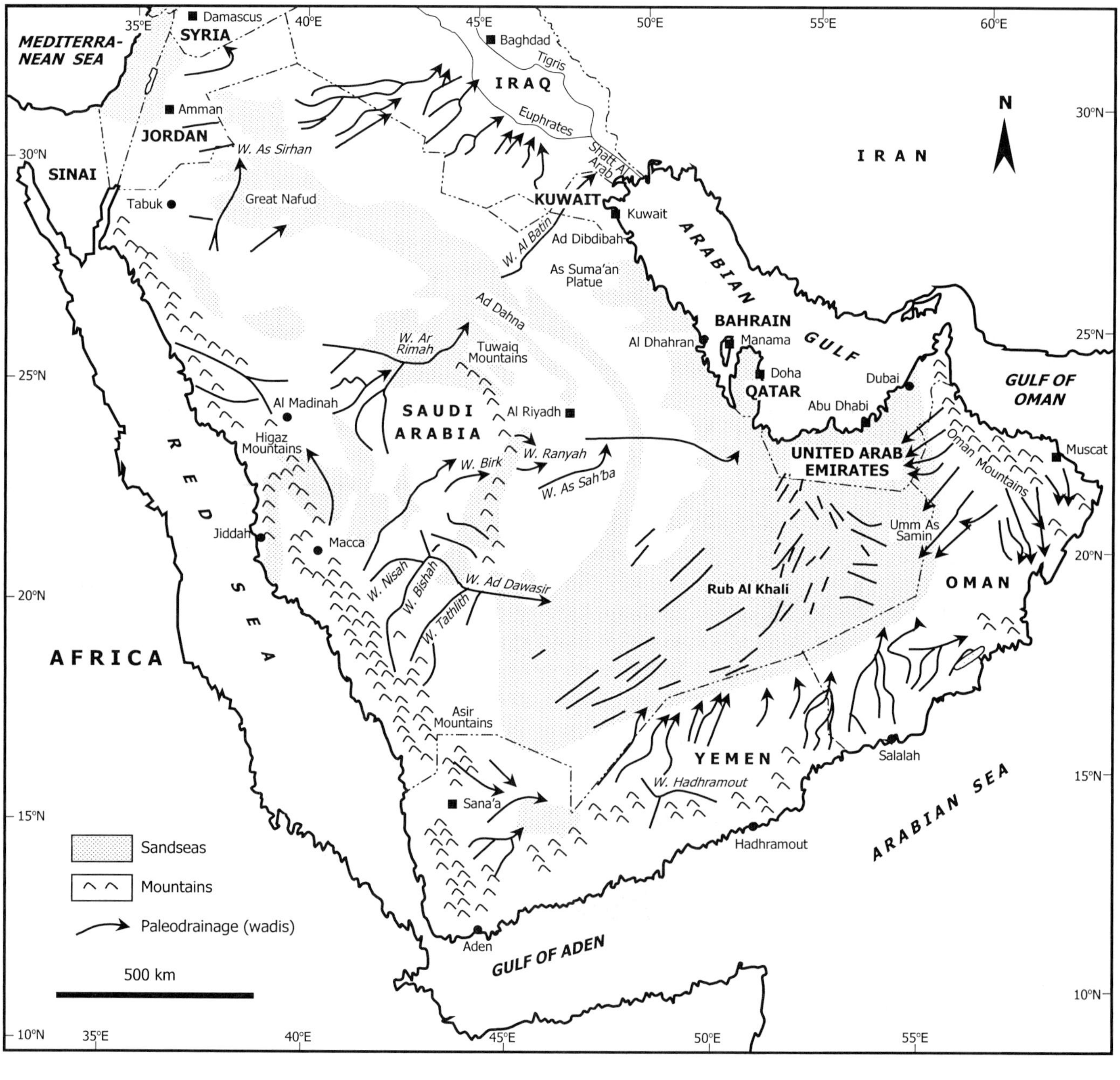

Fig. 2.4. Geomorphology of the Arabian Peninsula showing the mountaneous region, the sand seas and the ancient Paleodrainage wadis (compiled from Holm, 1960; Beydoun, 1980; Glennie et al., 1994).

higher elevations in the bordering areas, which terminate or dissipate at the margins of depressions, and coincide with the sabkhas. During the last 25,000 years BP, the surface has been alternately exposed to weathering, or submerged and accumulating sediment.

In eastern Arabia, the coastal strip of Oman provides good agricultural land, watered by the rainfall trapped by the Oman Mountains. Cultivation along the Batinah coast, and a smaller area around Salalah, provide a variety of tropical fruits that include dates, coconuts, bananas, pineapples and papayas. In the areas between, sand covers the embayments, separating promontories projecting into the Gulf of Oman. On the eastern coast of the United Arab Emirates, the sand flats and wadi fans coalesce to an almost continuous littoral strip between the mountains and the sea, and they retain some of the fresh water draining from the main wadis. On the other hand, the northern slopes of the mountains do not receive much rain and remain dry and arid.

2. The Gravel and Dune Zone

Inland from the coastal zone lies an extensive area covered by gravel and sand dunes. In Saudi Arabia this zone is separated from the coastal zone by the Summan uplift, a dissected limestone plateau, 80 to 250 km wide, covered by dikakah (small bushes and bunch grass). To the north and south, the Summan uplift grades into gravel plains, where wind ablation has produced an almost flat to gently undulating surface, readily traversable in any direction when dry. The sand covered area is known as Ad Dhahna, and is one of the most distinctive geomorphic features in the country. It is a belt of reddish sand, about 1300 km long and 25 km wide, extending between the Great Nafud in the north and

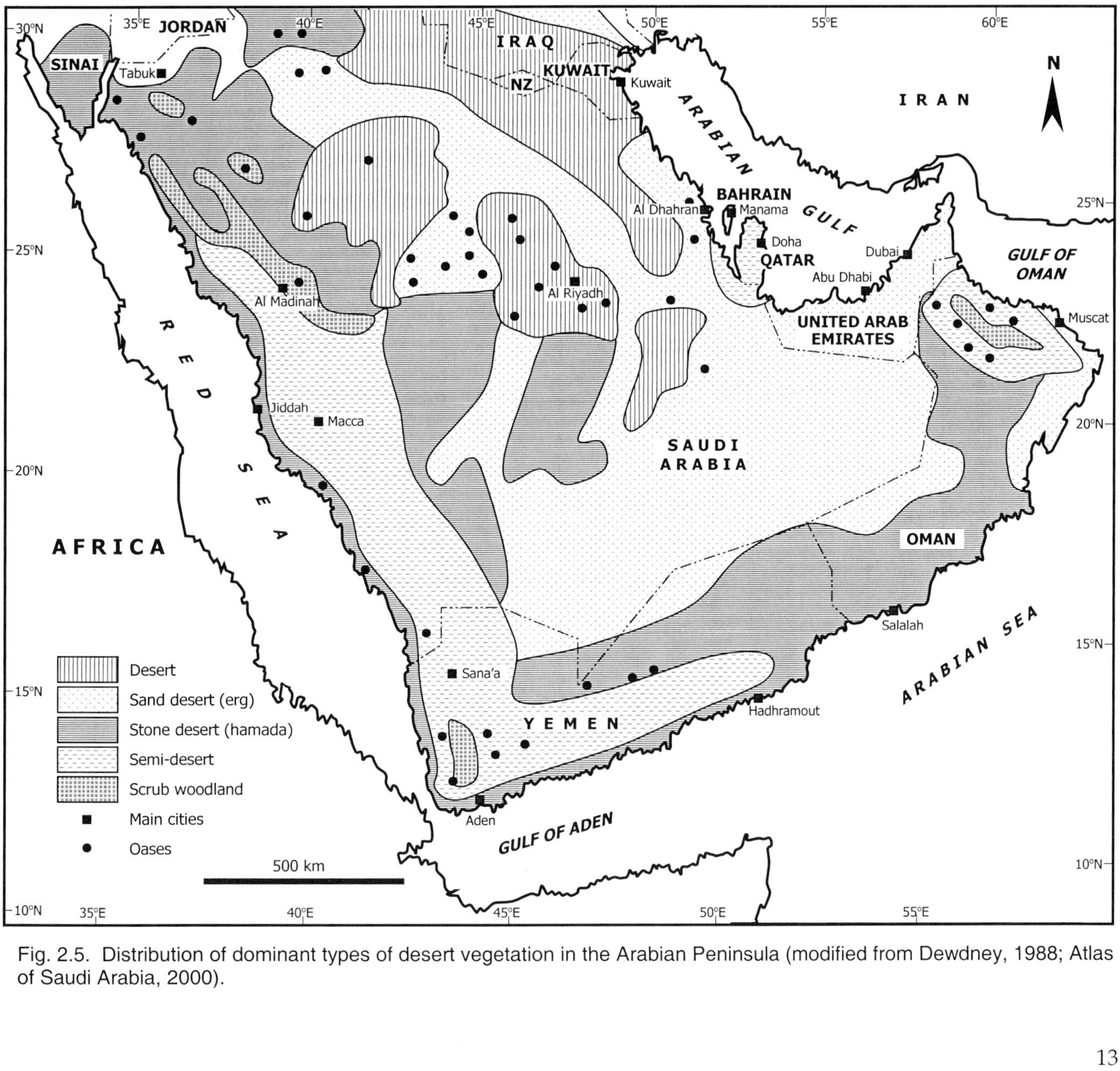

Fig. 2.5. Distribution of dominant types of desert vegetation in the Arabian Peninsula (modified from Dewdney, 1988; Atlas of Saudi Arabia, 2000).

the Rub Al Khali sand sea in the south. It is bordered to the west by the west facing ridges of the Aruma and Tuwaiq Mountains. The Great Nafud is an elliptical shaped sand body covering about 57,000 km^2.

The Rub Al Khali originated during the Late Quaternary, and is the largest contiguous sand desert in the world, having an area of about 640,000 km^2. During the Miocene and Pliocene, pluvial and humid climates prevailed, as indicated by fossils and shallow marine water deposits (Whybrow and McClure, 1981). Very large alluvial fans were formed at the end of the Pliocene and Early Quaternary. These are composed of conglomerate and sand deposits, where major periodic streams or wadis debouched into the Rub Al Khali (Edgell, 1990). The climate of the Rub Al Khali was not uniform during the Pleistocene to Holocene, with much sand movement, occasional rainy years, and several wetter intervals as shown in Table 2.2. Different types of dunes have been formed in the Rub Al Khali. The greater part of this sand desert is covered by linear dunes, including draa dunes, seif dunes, sigmoidal dunes, fishhook dunes, feather dunes and divergent dunes (Edgell, 1990). Some of these gigantic linear sand dunes are up to 260 km long, are spaced from 2 to 6 km apart, and have an average trend of N 60° E. These sand dunes and sand sheets are believed to have their provenance from the crystalline Precambrian Arabian Shield,

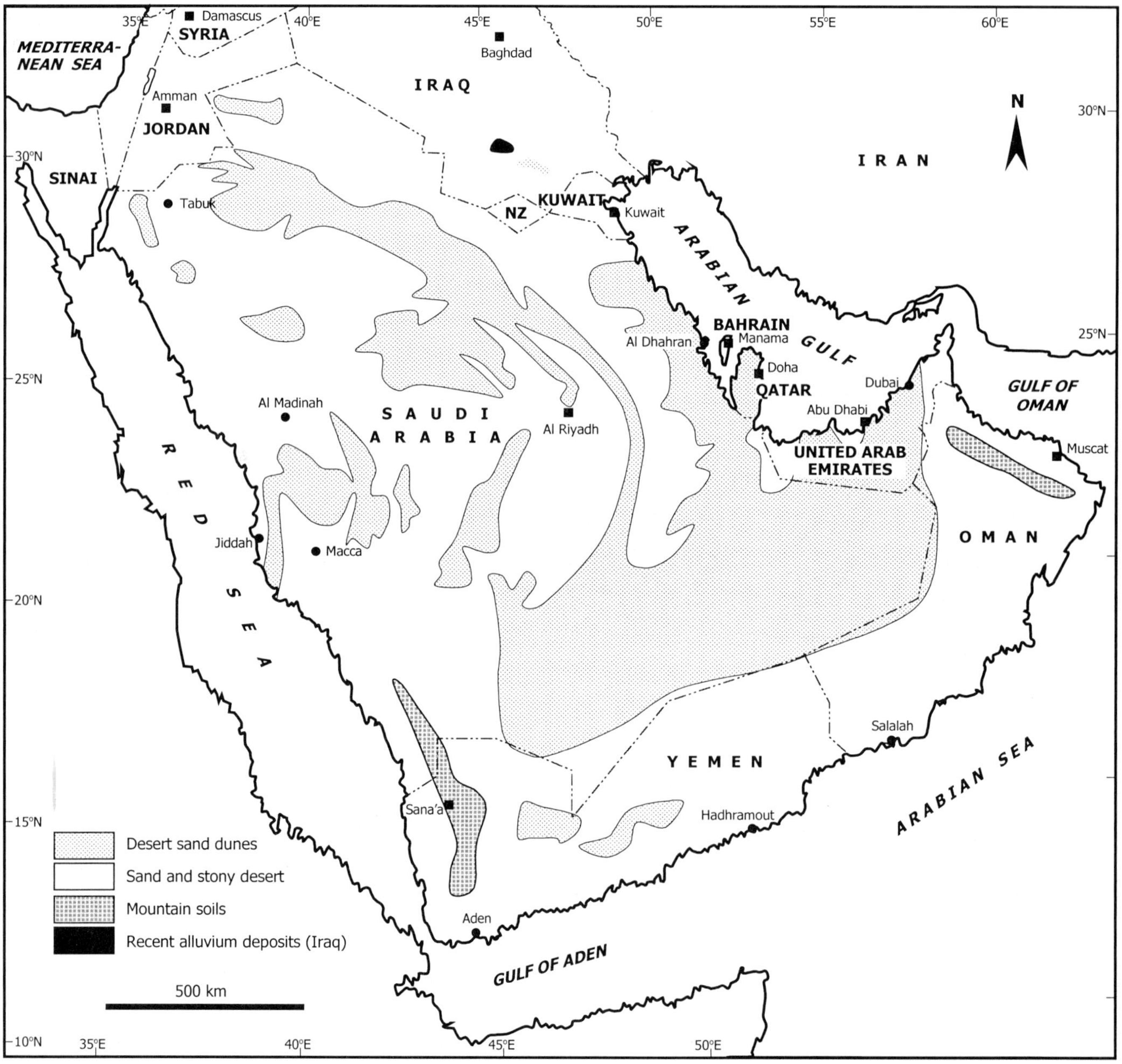

Fig. 2.6. Variation of soil types within the Arabian Peninsula (modified from Dewdney, 1988; Atlas of Saudi Arabia, 2000).

Neogene clastic formations such as the Hadrukh and Hofuf, the Cambro-Ordovician Saq and Wajid formations, the Lower Cretaceous sandstones of the Buwaib and Biyadh formations, Hadhramont Arch and from the high Oman Mountains. Many wadis draining from the sand dune seas (Fig. 2.3) are able to supply large volumes of sediment to the Rub Al Khali.

The desert plains in the United Arab Emirates are extensive with gravel plains skirting the mountains giving way to dunes, which cover 74% of the country. The desert plains occupy a triangular area with its east side along the coast and its apex at Ras al Khaimah in the north. The gravel plains are best developed at the outlets of the main wadis that dissect the Oman Mountains. Volume of incoming sediment controls the shape and size of these plains. The gravel plains effectively occupy the northern end of the Rub Al Khali sand sea. The dunes, which cover most of the area, increase in height from a few meters in the north, to more than 200m in the south. Several dune types are recognized, their shape being controlled by sand supply, climatic conditions, and to a lesser extent by the underlying sediments. Linear, barchan, barchanoid, transverse and star dunes, have all been described from this region (Embabi, 1991). The central and interior parts of Oman are also covered by gravel desert plains and sanddunes (Wahiba Sands).

A large portion of Kuwait also lies within the zone of low relief, sand and gravel desert. Sand dunes occur only in limited areas in northeastern Kuwait, where barchans with heights up to 25m have been reported. In northernmost and northeastern Kuwait recent deposits from the Tigris-Euphrates and Karun rivers have been reported. The gently undulating sand and gravel desert is known as Dibdibbah with a maximum elevation of 300m lies in the southwestern part of the country. It is crossed by the only major depression in the region, the southwest-northeast striking Wadi Al Batin, which has an average width of 6-8 km, with its lowest elevation defining a valley lying 50m below the general ground level. This feature runs parallel to the Jal el Zor escarpment, which lies along the northern shore of Kuwait Bay. The escarpment, which could have originated through faulting, ranges in elevation between 120 and 150m.

3. The Mountain Belt Zone

West of the Tuwaiq escarpment lies the central plateau of Saudi Arabia, with elevations in the range of 1,150-1,350m. Metamorphic and igneous rocks of the basement Arabian Shield are exposed in western Arabia. They grade towards the mountains, which form the platform edge, and have been the source of the clastic sediments laid down to the east. The mountainous belt ranges from 40 to 140 km wide and rises to the east to the lip of the Hijaz plateau. In the south, ridges and deep canyons extend from the foothills to the lip. In this area wadis are deeply incised. Further north the height and ruggedness decrease.

Table 2.2. A provisional chronology of Quaternary climate and events in the Rub'al Khali (after Edgell, 1990).

Geological Epoch	Chronology in Years (BP)	Climatic Phase	Events
Holocene	0 - 700	Hyperarid	Continued movement of high crested dunes
	700 - 1,300	Slightly moist	Hofuf river
	1,300 - 1,400	Arid	Dune movement
	1,400 - 2,100	Slightly moist	Sabean Kingdom flourished and also Kingdom of Kinda and Qaryat Al Fau
	2,100 - 5,000	Hyperarid	Dune movement
	5,000 - 5,500	Slightly moist	Neolithic camp site in SW Rub Al Khali 5120 years BP
	5,500 - 6,000	Hyperarid	High crested dunes; 'Irqs and interdune corridors
	6,000 - 10,000	Wet (Pluvial)	"Neolithic wet phase" lakes in SW Rub' Al Khali (C^{14} dating of organic remains and sinter)
Late Pleistocene	10,000 - 17,000	Hyperarid	Dune topography and longitudinal dunes extended
	17,000 - 36,000	Wet (Pluvial)	Lakes in the SW Rub Al Khali; Arabian Gulf Gulf dry, due to lowered sea-level of the last great ice age (C^{14} dating of organic remains and sinter)
	36,000 - 70,000	Arid	Main movement of sand from old wadis in the shrunken Arabian Gulf
	70,000 - 270,000	Moist	Early phase of glacial and interglacial (U/Th isotope dating)
	270,000 - 325,000	Arid	Summan Plateau caves dry
Middle Pleistocene	325,000 - 560,000	Wet	Active karstification and cave formation in Summan Plateau (U/Th isotope dating)
	560,000 - 700,000	Arid	Beginning of low dunes (δO^{18} isotope evidence of warmer climate)
Early Pleistocene	700,000 - 1,610,000 + (possibly to 2,500,000)	Wet humid (Pluvial)	Early Quaternary drainage systems in the Rub Al Khali. Large alluvial fans formed (δO^{18} isotope evidence of cooler climate)

The main topographic high areas of the Arabian Gulf region are the Oman Mountains, which stretch from the Musandam Peninsula in the north to central Oman in the south, extending over a distance of 700 km. The chain continues into the United Arab Emirates where the Ru'us al Jibal in the north is separated by the Dibba zone from the northern Oman Mountains to the south. The Ru'us Al Jibal Massif is primarily a carbonate sedimentary sequence, with units ranging in age from Late Paleozoic to Mesozoic. This sequence displays broad folding, block faulting and local thrusting. The Dibba zone is a northwest-southeast trending depression separating the Ru'us Al Jibal massive shelf from the Semail Ophiolite nappe, within which the ophiolite sequence is repeated by low angle, internal thrust faults. The mountains enclose a number of small basins on both sides of the watershed, the largest having an area of 5,000 km^2 and the smallest covering only 5 km^2.

Vegetation and Water

Throughout the Arabian Peninsula, vegetation is extremely sparse and in many areas non-existent. The basic soil cover consists of red desert soil which changes to sierozems or gray desert soil in the southwest and northwest. In the north, reddish prairie soils develop and within the neighboring mountains chernozem or chestnut soils may be found. The natural vegetation is characteristic of deserts or semi-deserts, with scrub woodland at higher elevations and steppe in the extreme north. Due to the scarcity of water, the growing season is affected by temperature, rainfall and elevation, and hence cultivation is restricted mainly to flood plains. Variations in soil and vegetation are also influenced by the steepness of slope, exposure, drainage conditions and geology. Along the low, flat and sandy shoreline, salt flats or sabkhas have formed in shallow depressions. Due to the high rate of evaporation, salt crusts develop which have been locally exploited where the salt is relatively free from sand. Under storm conditions, these low-lying areas, may be flooded by the sea, which may temporarily extend many miles inland. Under other conditions, sand dunes bury the sabkhas.

In the virtual absence of vegetation, maps of surficial sediments can be drawn. The principal ground cover is desert sand, often in the form of dunes or stony deserts (hamadas). The only significant vegetation type is scrub woodland found at higher elevations in Saudi Arabia, Yemen and Oman. There is also a very narrow coastal strip of sparsely vegetated dunes, whose water supply is maintained by dew condensed at night from the humid air developed over the sea, and carried inland by local onshore winds, during late afternoon (see Fig. 2.5).

Among all the parameters affecting growth and agriculture, the availability of water is the most critical. Precipitation abruptly declines inland where the largest area receives an average rainfall below 100 mm. Closer to the mountains in Yemen, rainfall may exceed 500 mm and in the Jabal Akhdar of Oman as much as 350 mm, has been recorded. This uneven distribution of precipitation has a major influence on the agricultural potential of the host countries, and on the distribution of cultivatable land. In recent decades the amount of land under irrigation throughout the region has increased dramatically, as a direct response to the ever increasing demand for agricultural products (Fig. 2.7).

Temperature and rainfall affect the length of the growing season and the availability of moisture. The latter is the dominant influence, and varies with latitude and distance from the sea. The major part of the Arabian Peninsula arable land is irrigated, except the uplands of Yemen and Oman.

The climatic water balance of the region indicates that precipitation exceeds potential evapotranspiration from January to April, and again from October to December. January to April are the months of soil moisture and water surplus, during which there is sufficient water available to support the growth of many cultivated plants. In May and June, evapotranspiration exceeds precipitation, but plants can still draw moisture stored in the soil. By July, soil moisture is exhausted, and potential evapotranspiration is far greater than precipitation. July to September are the months of soil moisture deficiency, when further plant growth can occur only with the aid of irrigation.

The fresh water supply is increasingly critical, and already the scarce resources are being stretched, by the rapidly growing populations, and by expanding agriculture and industries. Saudi Arabia has a program to build many dams of various sizes, all on seasonal water courses. The United Arab Emirates has already built 35 groundwater-recharge dams, with a total storage capacity of 75 Mm^3.

The use of groundwater (springs and wells) as a freshwater source has been practised for thousands of years in the eastern Arabian Peninsula. Similarly in Oman and United Arab Emirates, there are subterranean canal systems known as Qanats or Falajes. Because many states rely heavily on groundwater extensively used for irrigation, several water-related problems have now surfaced. The most serious of these are lack of aquifer recharge, over-pumping, aquifer depletion and continuously rising groundwater salinity.

There is clearly a need for other sources of fresh water. Desalination of sea water is one reasonable solution. However, the large investments needed and high production cost, limit its use to domestic purposes. Treated sewage for garden irrigation, and irrigation of some crops, is being tried in the Arabian Gulf region. Considerable caution has to be exercised to avoid the environmental consequences of the transmission of disease. There is no single solution, to the fresh water supply shortage in the Arabian Peninsula. Careful management of available sources, desalinization, practical recycling and conservation throughout the region are required to prevent severe shortages and socio-economic dislocation.

The irrigation schemes in Arabia have had only limited success, and because they depend upon groundwater, which has only limited possibilities for recharge, or fossil water there is a limit to the extent of development. Agriculture remains an important aspect of the economy of many countries in the region, not only providing food and export revenues, but as a source of employment. For environmental and technological reasons, crop yields are generally low, and crop variety is restricted. Oil revenues have meant that a progressively greater percentage of food requirements has been met through imports. The area of total cultivated land has changed with time, due to population growth, and increased food

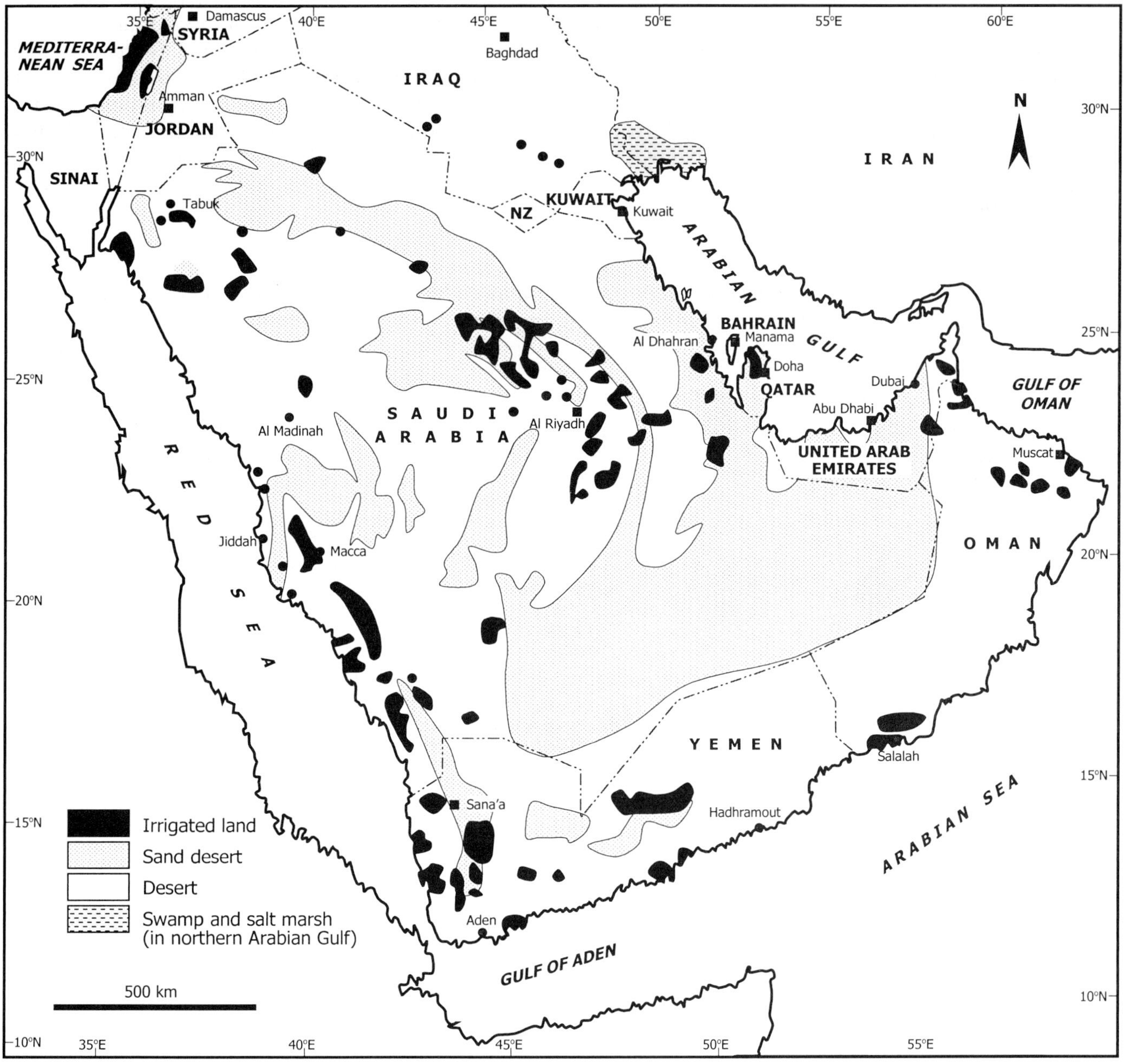

Fig. 2.7. Generalized landscape of the Arabian Peninsula, showing the irrigated land (modified after Dewdney, 1988; Atlas of Saudi Arabia, 2000).

demand. More than 80% of the cultivated area of the Arabian Peninsula is under irrigation, and rangeland is widespread (Table 2.3).

Since water resources and water management are important in all countries, the United Nations Water Research Council has adopted the Mar del Plata Action Plan (1977). This plan recommended that each country formulate a national policy for the use, management and conservation of fresh water. It also included research activities, and appropriate institutional structures and laws for development and administration of water resources (Gleick, 1993).

The principal alternative source for fresh water is desalinization, but this is still expensive because of the energy required. Gleick (1993) compiled data related to alternative power sources ranging from wind to solar energy (Table 2.4) and to water demand (Table 2.5). The availability of local freshwater resources, and water which can be transported to its place of use, vary greatly throughout the Arabian Peninsula. These are summarized in Table 2.6 from Gleick (1993). Stream, rainfall and groundwater availability is shown in Table 2.7.

Climate

The Arabian Peninsula lies within one of the world's great desert belts which are characterized by high temperatures and semi-arid to extremely arid conditions (Fig. 2.8). During summer the main track of the jet stream controlling the passage of depressions, passes north of the Pontic Mountains. During winter, the track of the jet stream moves southwards and covers the northern Arabian Gulf. However, few depressions pass south of 30°N. The effects of the Red Sea, Arabian Sea, Arabian Gulf and the Gulf of Oman on the regional climatic patterns appear to be minor. Detailed climatic measurements can be gleaned from the annual meteorological reports at the principal international airports and meteorological stations in the Arabian Gulf countries (Fig. 2.9). From these data isothermal and isohyet maps can be generated as well as a map of climatic zones (see Fig. 2.8).

The climate of the Arabian Gulf region features high temperatures, high relative humidities, seasonal rainfall and predominantly "shamal" winds. These features, either singly or together,

Table 2.3. Land use distribution in the Arabian Peninsula (compiled with modification from Kharin et al., 1999).

Country	Country Surface (x 10^3 km^2)	Permanent Cultivation (%)	Annual Crops (%)	Irrigated Area		Forest (%)	Rangeland (km^2)
				Surface (x10^3 km^2)	Percent		
Saudi Arabia	2,150.00	0.95	15.30	16.08	100.0	0.87	764.40
Yemen	528.00	2.00	8.50	3.80	45.7	20.00	158.40
Kuwait	18.00	0.05	0.04	0.05	100.0	0.01	1.34
Oman	313.00	0.43	0.18	0.62	100.0	0.02	10.60
Bahrain	0.65	0.02	0.02	0.03	100.0	0.01	0.11
Qatar	11.00	0.02	0.06	0.13	100.0	0.01	0.50
United Arab Emirates	76.00	0.33	0.20	0.67	100.0	0.04	1.52

Table 2.4. Wind and solar desalination plants with a capacity greater than 10 m^3/day in the Arabian Peninsula (compiled from Gleick, 1993).

Country	Date of operation	Capacity (m^3/day)	Process	Water supply	Energy source
Kuwait	1984	22	MSF	Seawater	--
	1988	45	RO	Brackish	Parabolic collector
Qatar	1986	20	MSF	Seawater	--
	1986	20	MSF	Seawater	--
Saudi Arabia	1987	210		Seawater	Point focus
	1987	250	RO	Seawater	Line focus
	1988	14	ME	Seawater	Heliostat
	1988	20	RO	Seawater	Heliostat
United Arab Emirates	1985	80	ME	Seawater	--
	1985	80	ME	Seawater	--

ME: multiple effect distillation; MSF: multi-stage flash distillation; RO: Reverse osmosis.

Table 2.5. Water demand, water resources and use in the Arabian Gulf countries (compiled with modification from Gleick, 1993).

Country	Water demand (10^9 m^3/yr)	Water resources (10^6 m^3/yr)			Water use (10^6 m^3/yr)				
		Surface	Groundwater recharge	Total renewable	Surface	Ground	Desalination	Treated	Total
Bahrain	0.112	-	90	90	-	153	16.5	0.5	170
Kuwait	0.804	-	160	160	-	283	404	80	767
Oman	0.512	1,470	564	2,034	1.30	400	15	8.6	424
Qatar	0.135	-	55	55	-	90	90	20	200
Saudi Arabia	3.530	3,208	2,338	5,546	450	3,000	903	217	4,570
United Arab Emirates	1.012	365	387	752	30	300	276	0.8	577

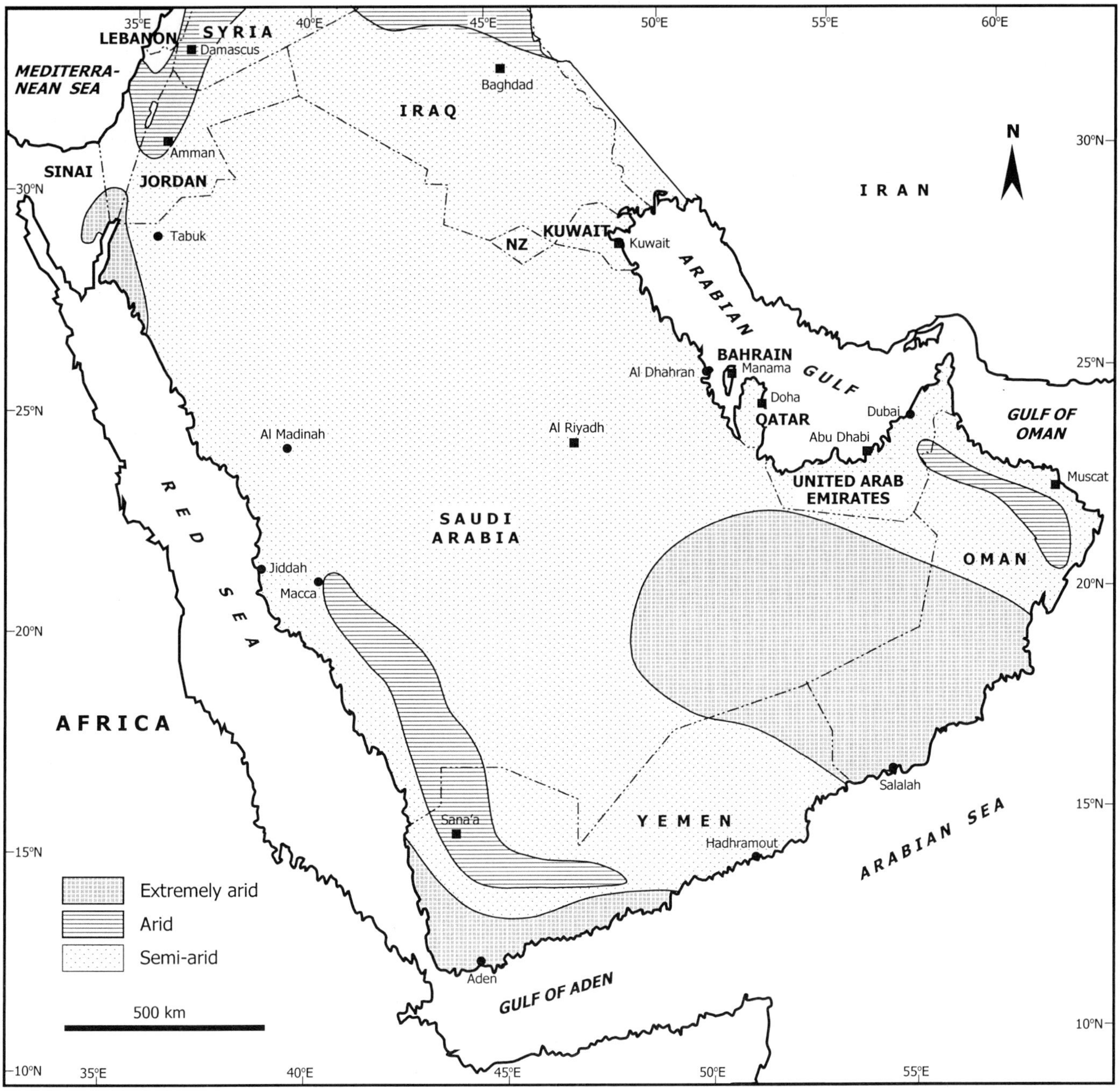

Fig. 2.8. The main climatic zones in the Arabian Peninsula (modified from Dewdney, 1988; Atlas of Saudi Arabia, 2000).

Table 2.6. Freshwater withdrawal in the Arabian Peninsula (compiled from Gleick, 1993).

Country	Year	Annual renewable water resources (km^3/yr)	Total freshwater withdrawal (km^3/yr)	Fraction of annual renewable water resources with-drawn (%)	Per capita withdrawal (m^3/person/yr)	Domestic use (%)	Industrial use (%)	Agricultural use (%)
Bahrain	1975	<1	0.31	>100	609	60	36	4
Kuwait	1974	<1	0.50	>100	238	64	32	4
Oman	1975	2.0	0.48	24	325	3	3	94
Qatar	1975	<1	0.15	663	415	36	26	38
Saudi Arabia	1975	2.2	3.60	164	255	45	8	47
United Arab Emirates	1980	<1	0.90	299	565	11	9	80
Yemen	1975	2.5	3.40	276	1,167	9	4	94

Table 2.7. Stream flow, groundwater, and rainfall resources in the Arabian Peninsula (compiled with modification from Gleick, 1993).

Country	Stream flow (km^3/yr)	Ground water (km^3/yr)	Rainfall	
			(km^3/yr)	(mm/yr)
Bahrain	No Stream flow	0.10	0.1	80
Kuwait	No stream flow	0.16	2.5	135
Oman	1.4	0.60	15.0	71
Qatar	No stream flow	0.05	0.8	73
Saudi Arabia	2.20	2.35	126.5	60
Yemen	3.80	1.40	68	310
United Arab Emirates	1.0	0.50	10.0	60

Table 2.8. Meterological summary of Bahrain (Source from Bahrain International Airport in Doornkamp et al., 1980).

Elements	January	February	March	April	May	June	July	August	September	October	November	December	Period Years
Mean daily maximum temperature °C	19.9	20.9	24.6	29.0	33.3	35.7	37.3	37.7	35.9	32.4	27.4	27.4	29
Mean daily minimum temperature °C	14.3	15.1	18.1	21.6	26.0	29.0	30.5	30.7	28.8	25.5	21.6	16.6	29
Highest maximum recorded	31.7	34.7	38.0	41.7	46.7	44.8	44.0	45.0	42.2	40.6	35.0	29.4	29
Lowest minimum recorded	2.8	7.2	7.8	13.9	18.7	22.7	25.3	27.4	24.4	20.6	13.5	6.4	29
Mean daily maximum wet bulb temperature	15.8	16.5	18.8	21.6	25.0	27.3	28.8	30.1	28.5	26.1	21.7	17.6	13
Highest wet bulb temperature recorded	21.1	22.2	23.9	27.6	30.6	32.1	32.2	34.0	31.8	30.0	28.5	23.9	13
Mean daily maximum relative humidity	88%	87%	85%	83%	80%	80%	80%	84%	86%	90%	84%	88%	16
Mean daily minimum relative humidity	59%	55%	50%	45%	39%	40%	41%	44%	45%	46%	52%	58%	16
Mean daily MSL pressure (mb)	1018.4	1017.1	1013.8	1010.8	1006.8	1000.9	997.9	999.0	1004.6	1011.8	1016.4	1018.2	44
Mean daily vapour pressure	14.8	15.2	16.9	20.3	25.2	27.9	31.8	34.7	32.1	27.2	21.7	17.1	44
Mean daily hours of sunshine	7.2	8.2	8.2	8.5	10.3	11.8	11.3	11.0	10.7	10.1	8.9	7.7	7
Mean monthly rainfall (mm)	17.8	12.2	9.7	8.9	1.6	00	0.05	0.0	0.05	0.3	3.2	16.3	29
Mean no. of rain days (1 mm or more)	2.0	1.9	1.8	1.3	0.9	Nil	Nil	Nil	Nil	0.05	0.9	1.8	29
Highest rainfall in one day	54.8	29.4	18.6	64.0	8.0	0.0	0.05	0.0	0.05	8.9	13.5	42.7	29
Mean no. of days per month with:													
a) Fog (Vis. 1 km or less)	1.7	0.8	0.5	0.1	0.2	0.2	0.2	0.1	0.05	0.7	0.8	1.0	29
b) Thick haze (Vis. 1 km or less)	0.1	0.3	0.6	0.7	0.5	1.6	1.6	0.2	0.3	0.1	0.0	0.2	29
c) Thunder	1.2	0.9	1.8	2.0	1.0	0.0	0.0	0.0	0.0	0.1	1.2	0.8	29

have an influence on landscape, soil and vegetation growth. Table 2.8 depicts statistics for current climatic conditions in Bahrain, which are fairly typical of other countries of the Arabian Gulf.

The areas of semi-desert are restricted to the higher ground in Saudi Arabia and Yemen. There are no permanent streams, and the number of oases is small. Storm water in flash floods, funneled through existing wadis, may reach the sea but these floods seldom flow for more than a few days. Through the construction of groundwater recharge dams, attempts have been made to reduce the velocity of flood water, and hence increase the amount of infiltration, as well as protect downstream features such as roads and bridges. It follows that the agricultural requirements have to be met primarily from groundwater. Hence the study of climate, porosity and hydraulic conductivity of surficial sediments is important in investigating the geohydrology of the region.

Temperature

Temperature, rainfall and local dew are the principal climatic parameters which affect the geohydrology of the region. During the summer months (June-August) over much of the central part of Arabia, shade temperatures frequently exceed

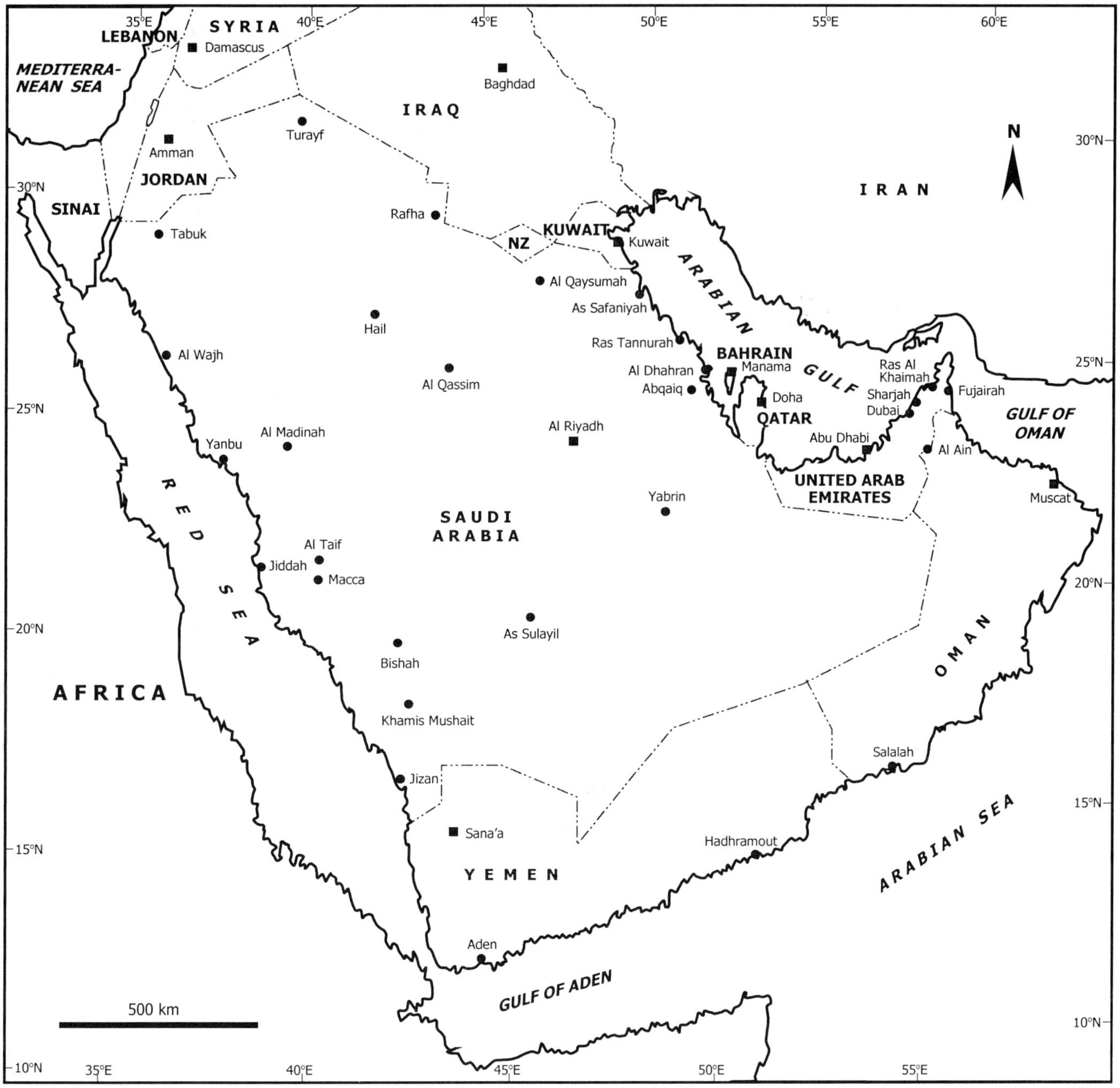

Fig. 2.9. The main meteorological stations in the Arabian Peninsula.

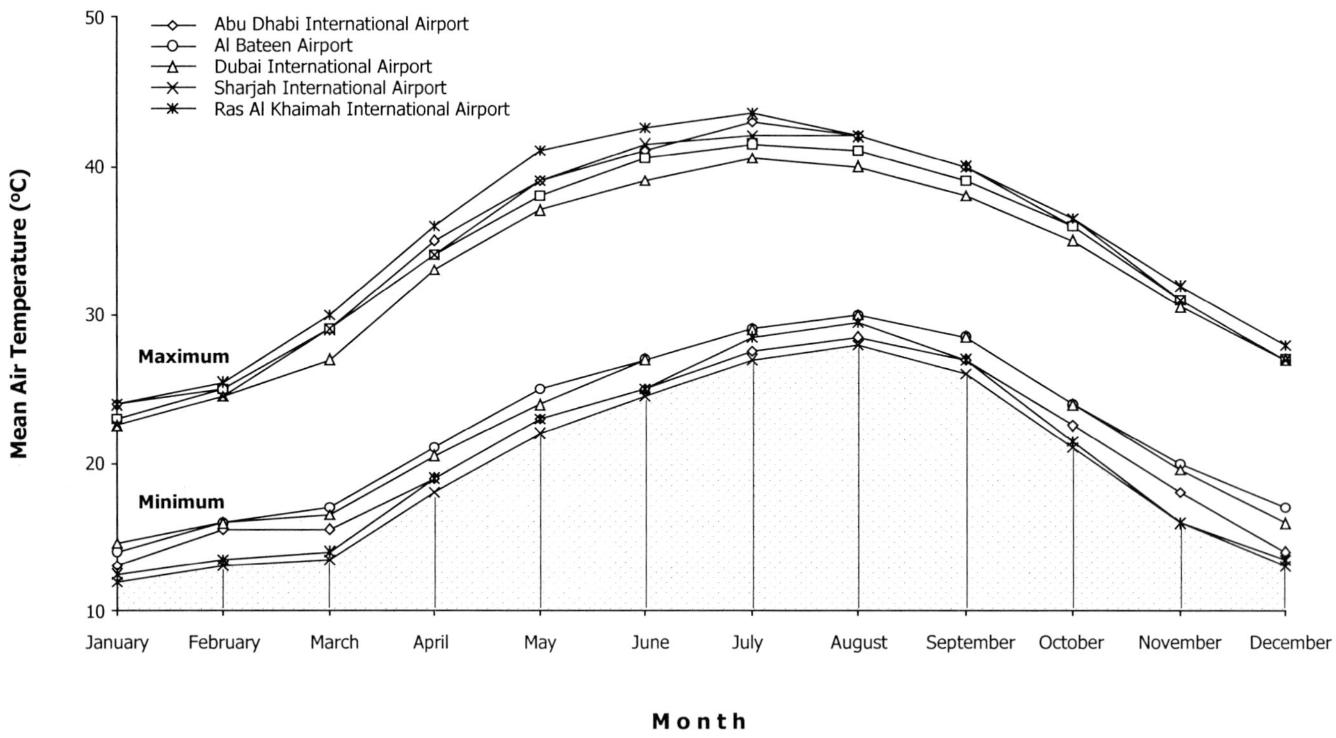

Fig. 2.10. Mean values of monthly maximum and minimum temperatures (°C) in United Arab Emirates for the year of 1995 from different airports (Compiled from Department of Civil Aviation annual reports).

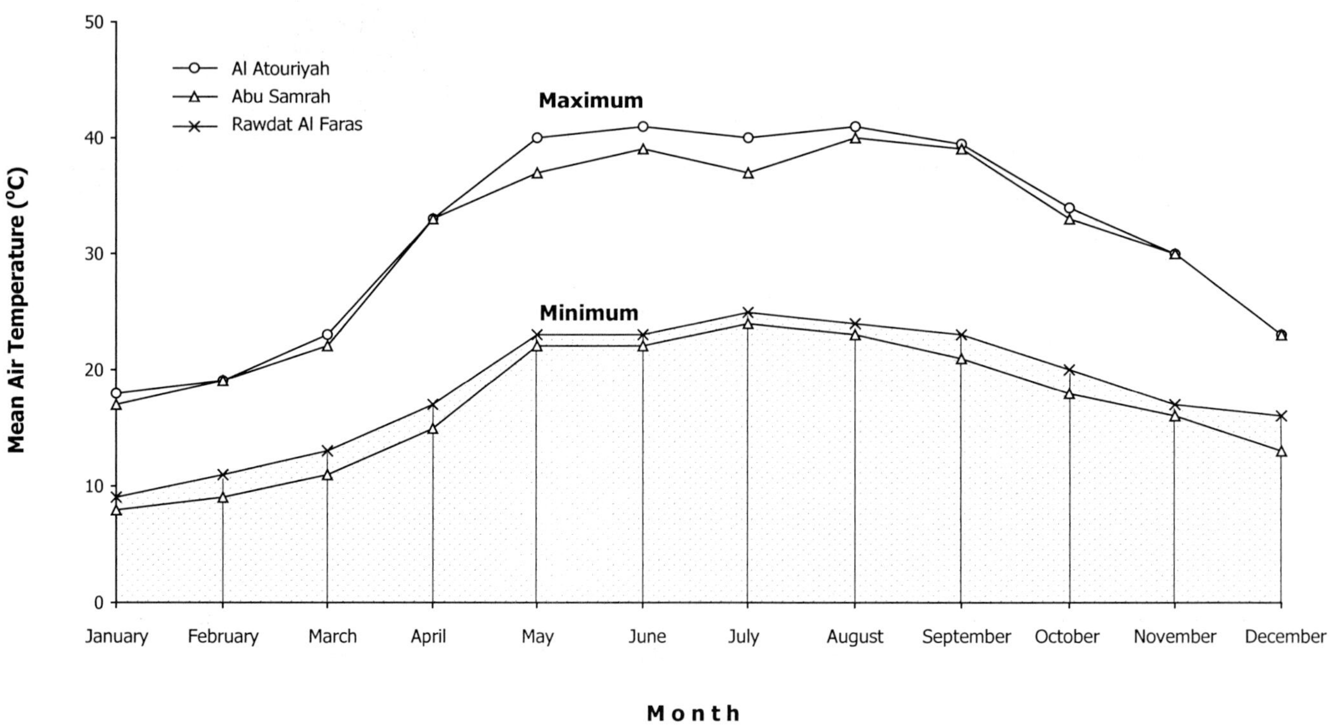

Fig. 2.11. Mean values of monthly maximum and minimum temperatures (°C) in Qatar for the year of 1992 from records of three meteorological stations (compiled from Ministry of Municipal Affairs and Agriculture, Annual Report, 1995).

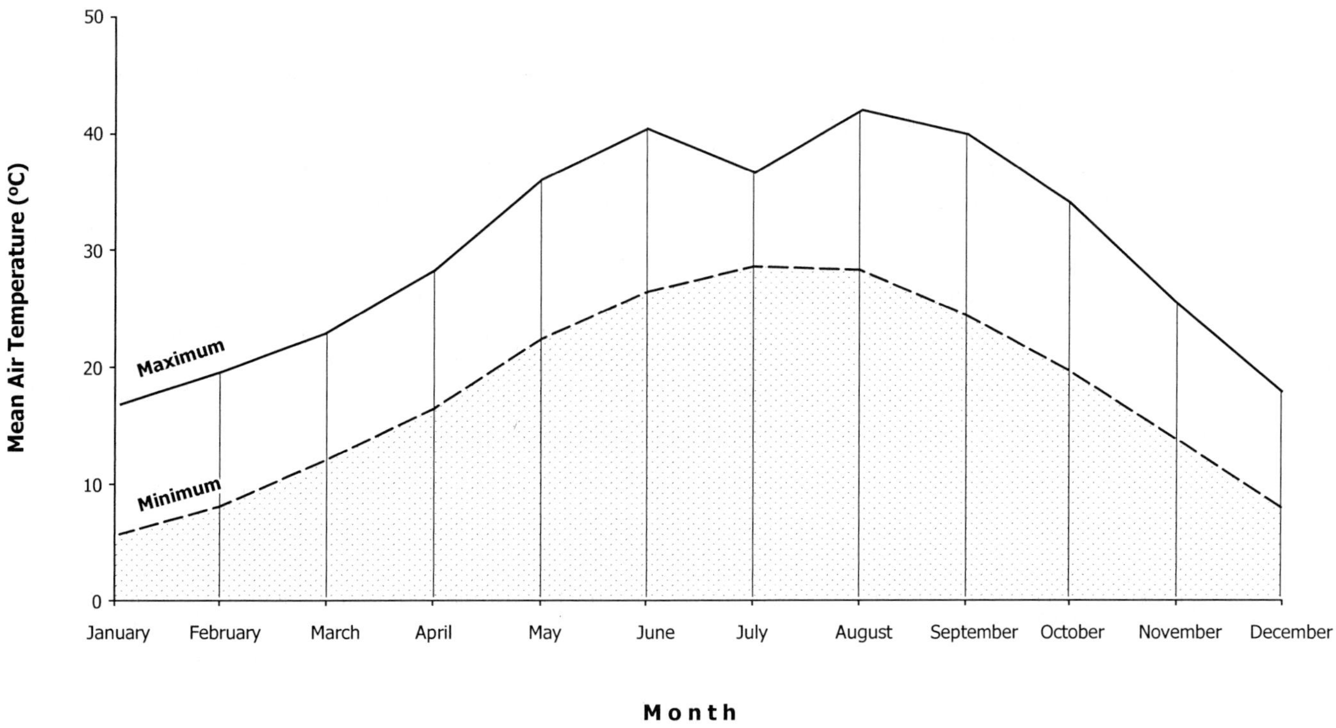

Fig. 2.12. Mean values of monthly maximum and minimum temperatures (°C) in Kuwait from eight meteorological stations (modified from Sharaf, 1980).

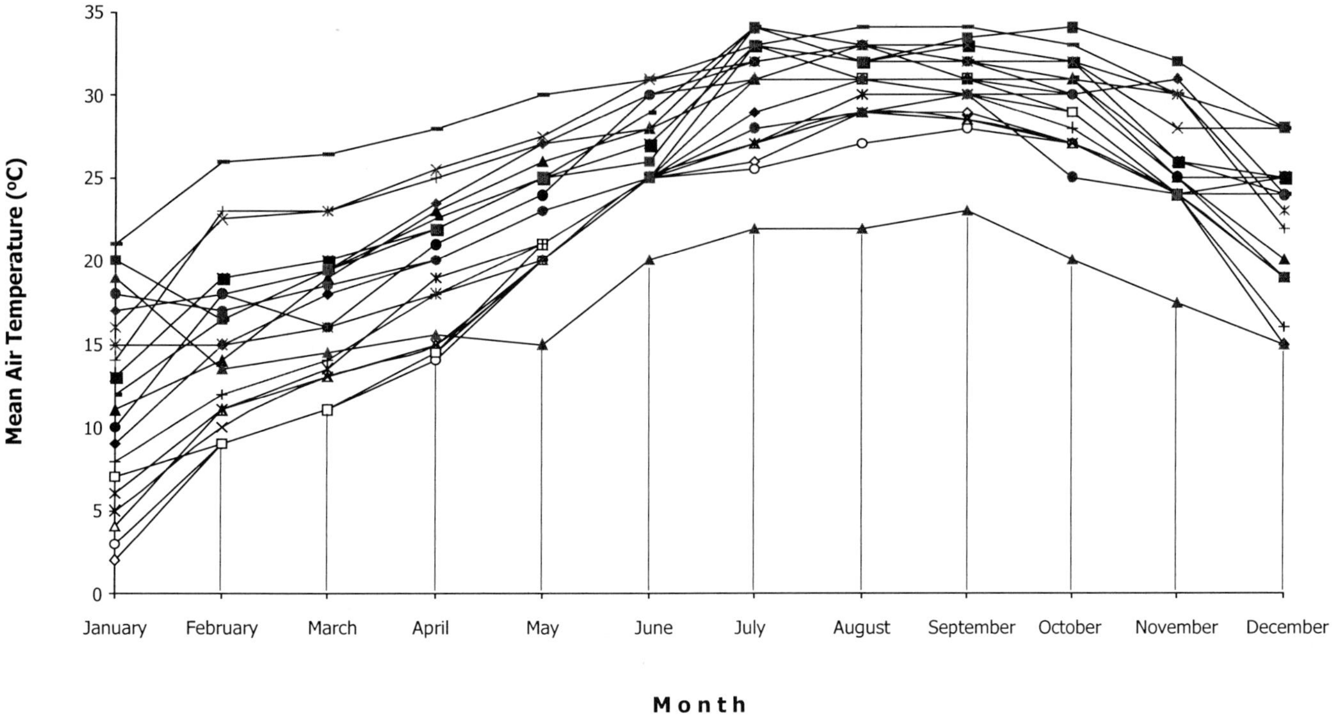

Fig. 2.13. Mean values of monthly average temperatures (°C) in Saudi Arabia from records of twenty meteorological stations (modified from Ahmed, 1993).

48°C, though the mean monthly temperatures are lower. Depending upon elevation, inland stations may record temperatures ranging from 28° to 37°C. Red Sea coastal regions have lower temperatures (from 28° to 34°C), whereas those in the Arabian Gulf tend to be higher (31° to 37°C). Data from the main airports show maximum daily values exceeding 40°C during the summer months (Figs. 2.10-2.13). The stability of these temperatures is emphasized by Figure (2.14) which shows that the daily temperature seldom deviates from the mean values in the coastal region by more than one degree. It should be noted that the deviations in the daily minimum temperature are somewhat greater. The departure from the mean monthly maximum and minimum temperatures in Saudi Arabia and Kuwait ranges from 10° to 15°C. Deviation in the monthly minimum temperatures is likewise greater (Figs. 2.15 and 2.16). Very few locations receive less than 80% of the potential sunshine hours (Fig. 2.17). For example, the mean of daily sunshine hours for Qatar in 1992 is 9.5 hours. Between May and September, the sunshine hours reach the maximum of 11.5 hours, whereas in December the average can drop to 7.5 hours (Fig. 2.18). Mean daily sunshine hours in Kuwait vary between 7 in January and 11 in August (Fig. 2.19).

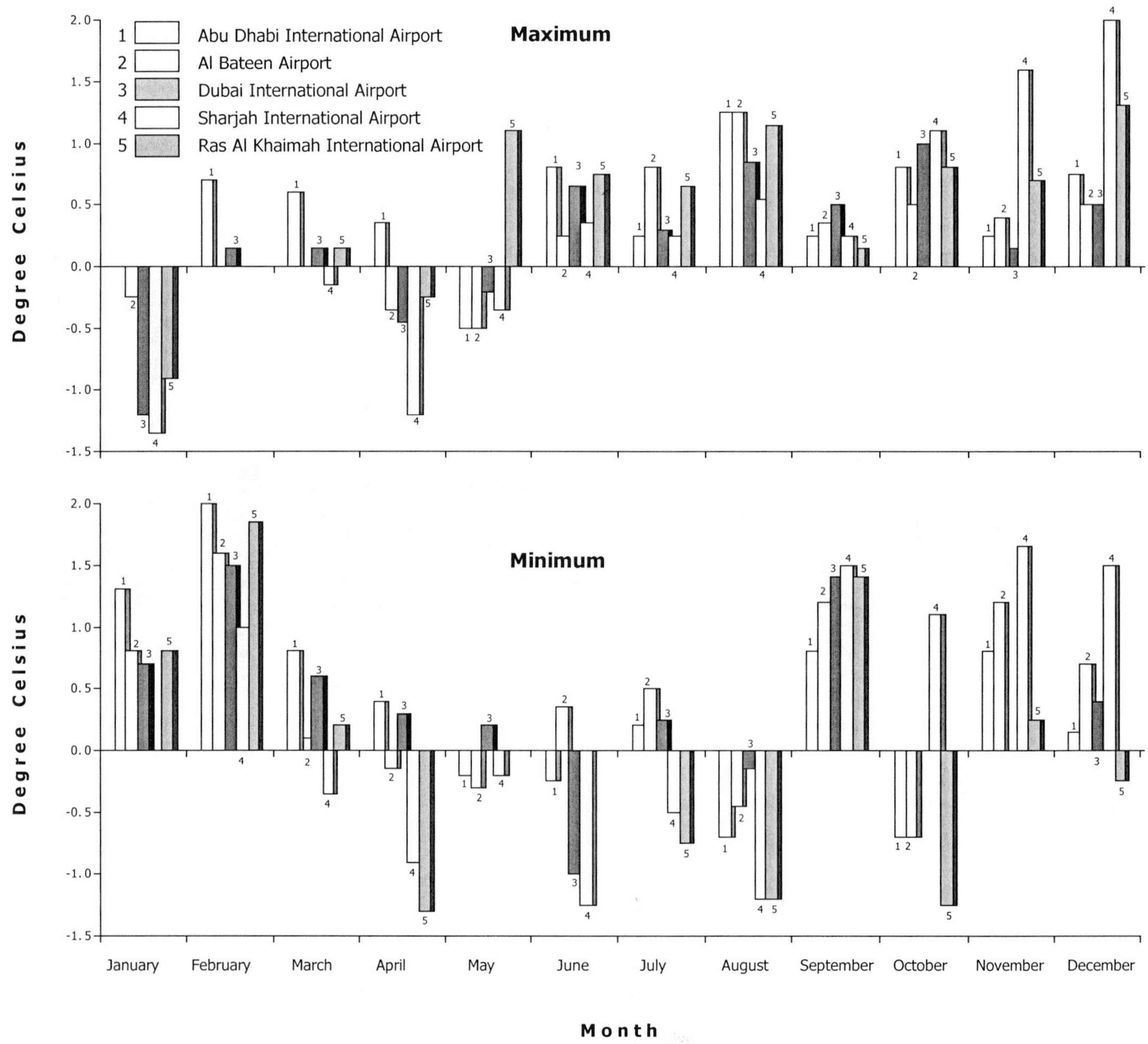

Fig. 2.14. Mean departure from daily maximum and minimum temperatures (°C) for the year 1993 in the United Arab Emirates (data from UAE Department of Civil Aviation, Annual Report, 1993).

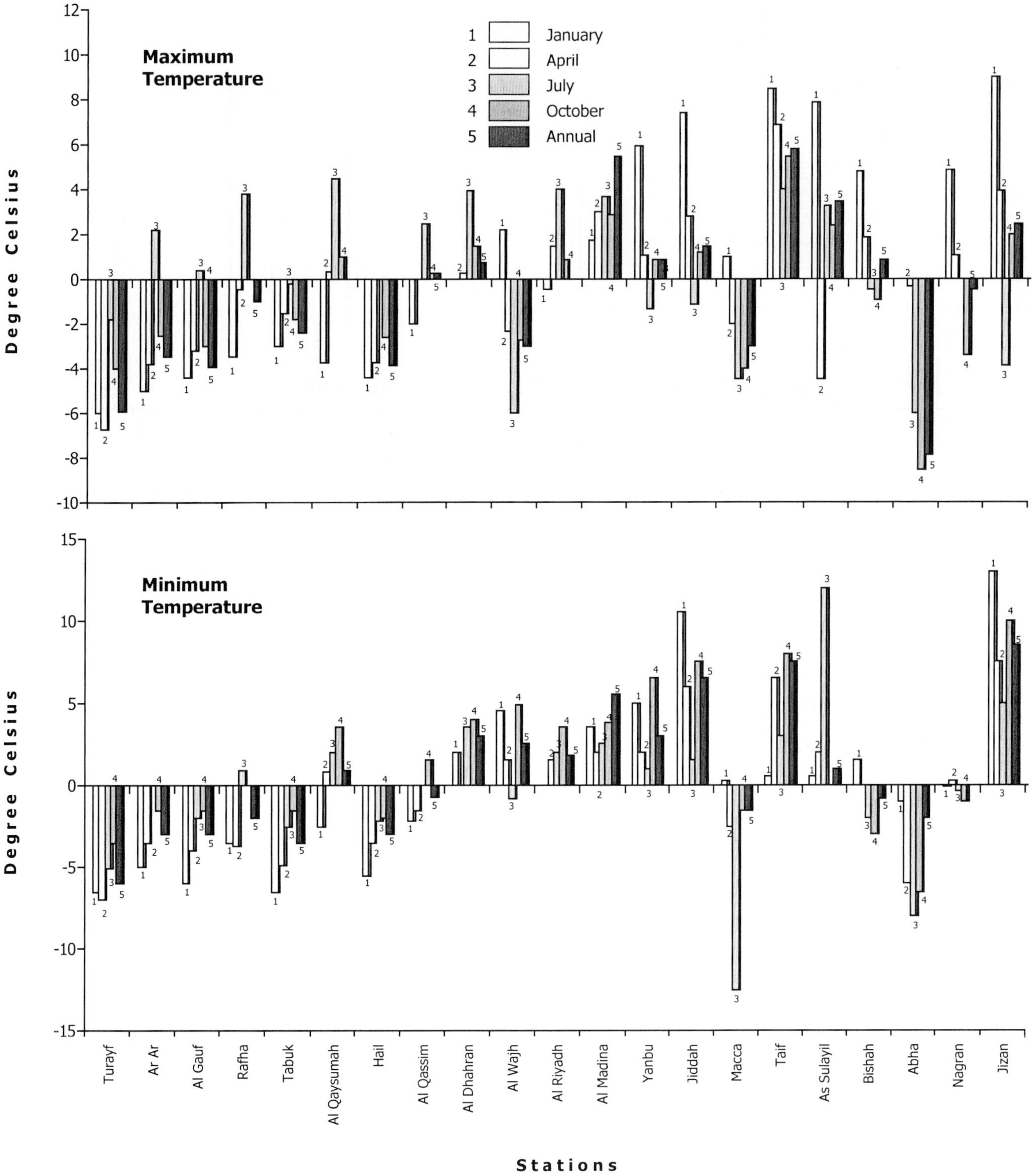

Fig. 2.15. Saudi Arabia mean departure from monthly maximum and minimum temperatures (°C) for the period 1967-1987 (modified after Ahmed, 1993).

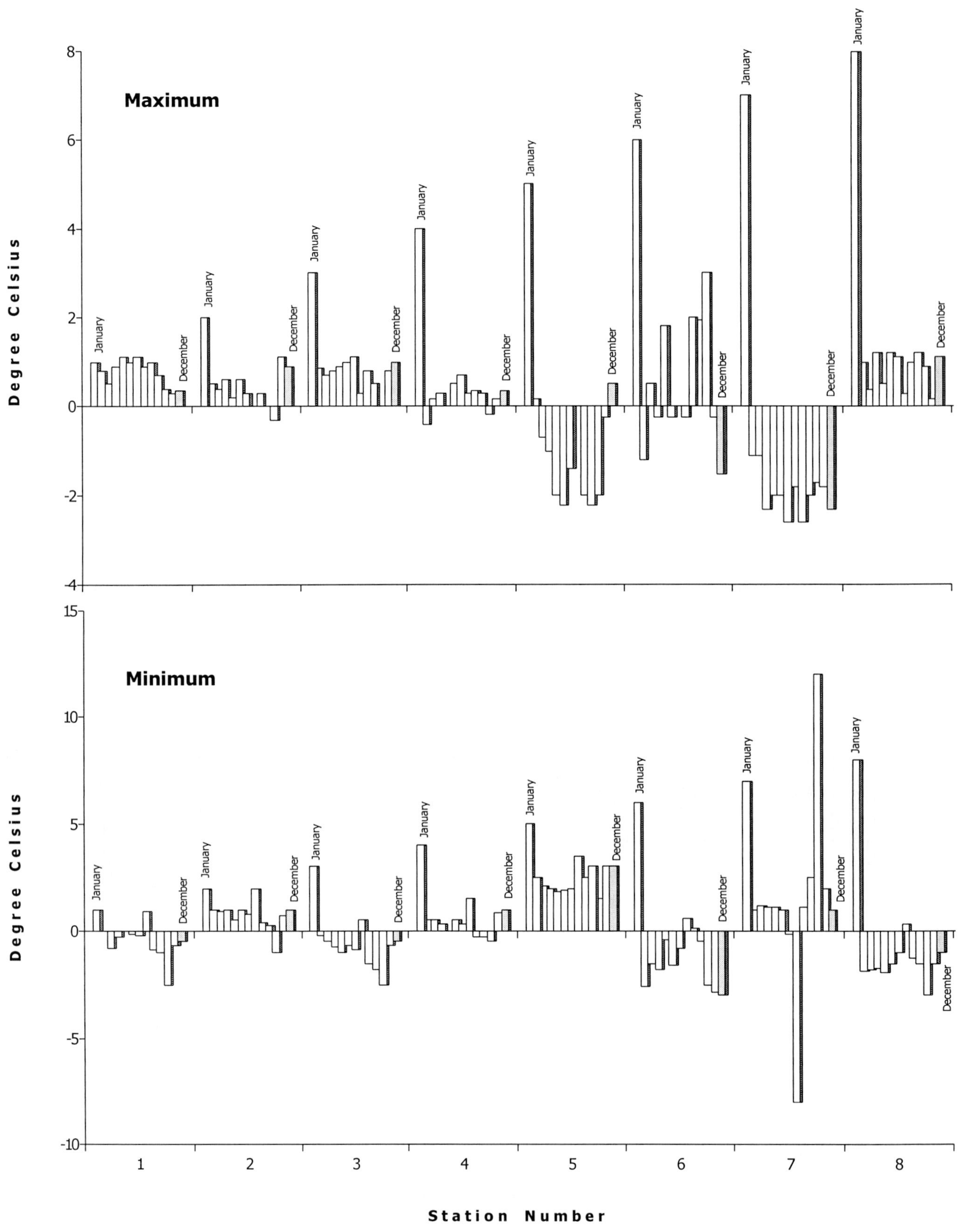

Fig. 2.16. Kuwait mean temperature from monthly maximum and minimum temperatures for the period 1957-1976 (modified after Sharaf, 1980).

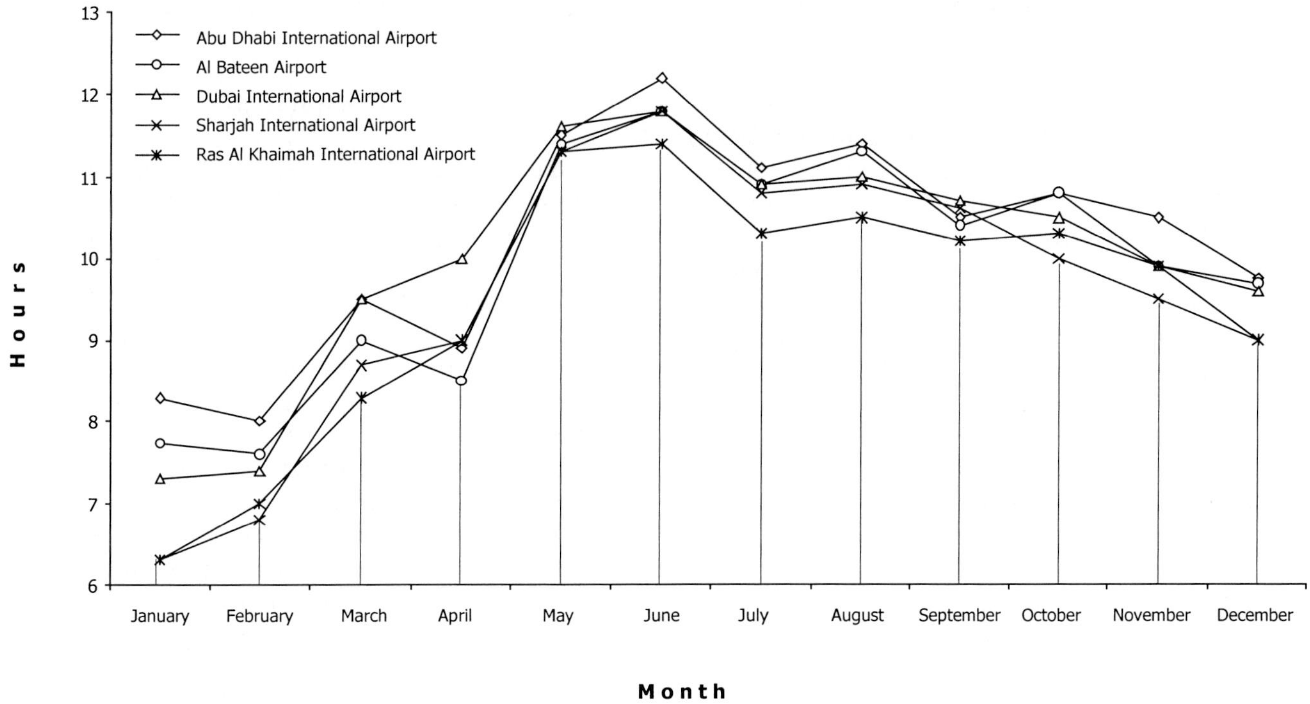

Fig. 2.17. Mean daily sunshine hours for the year 1993 from major airports in the United Arab Emirates (data from UAE Department of Civil Aviation, Annual Reports).

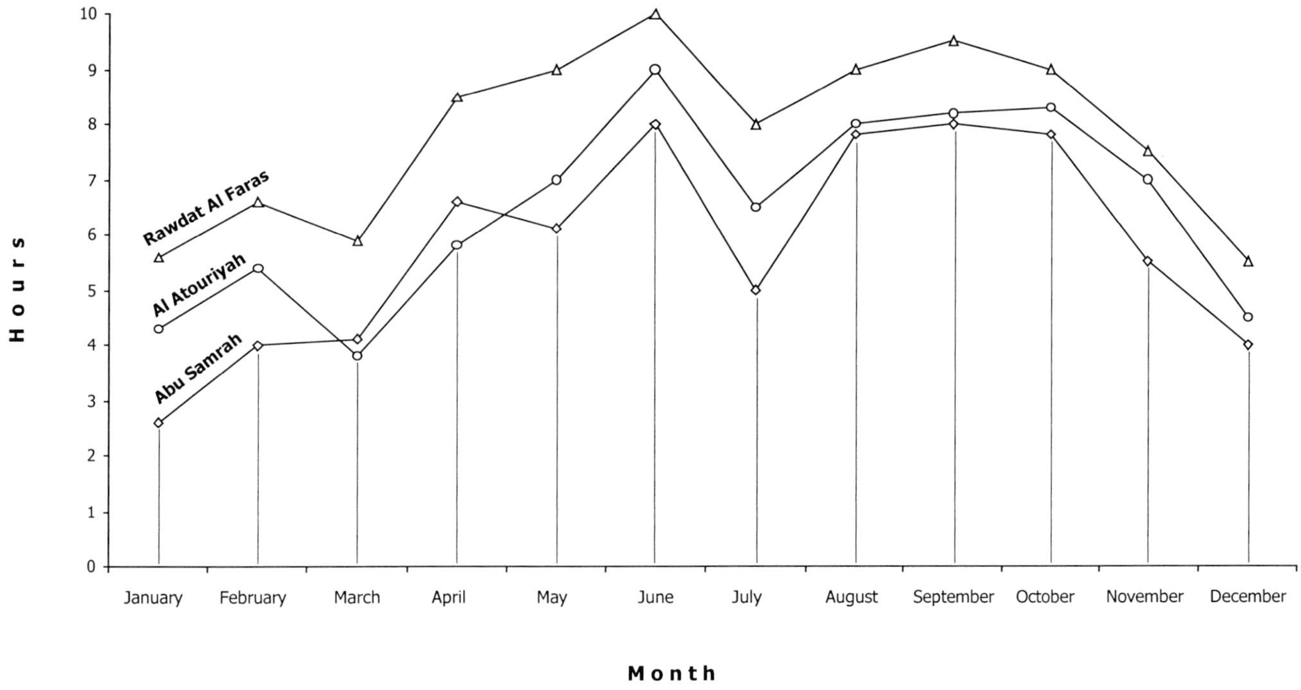

Fig. 2.18. Mean daily sunshine hours for 1992 from three meteorological stations in Qatar (data compiled from Ministry of Municipal Affairs and Agriculture, Annual Reports, 1995).

Temperatures during the winter months (December-February) often drop below freezing in central and northernmost Saudi Arabia. They generally range from 5° to 20°C, whereas, a narrower range of 8° to 17°C is common at most inland stations. As might be expected, temperatures along the coastal strips are somewhat higher, 19° to 27°C on the Red Sea coast and 11° to 22°C on the Arabian Gulf coast. During winter, temperatures below freezing often occur in central and northernmost Arabia, but snow and ice are unusual except at the highest elevations. Sub-zero temperatures are never found in the coastal areas. Figures 2.20 and 2.21 present the summer and winter isotherms (°C) in the Arabian Peninsula.

The most pleasant months are the transitional spring months (March-May), and the autumn months (September-November), with potential for precipitation during March and November. The days tend to be sunny and warm, and the night-time temperatures are relatively cool. Transition from summer to winter is relatively rapid. For example, in Bahrain the mean daily minimum temperature is 15°C in January/February, and 40°C in July/August. The highest temperature recorded in summer is 48°C, whereas in winter the lowest temperature is about 3°C (Fig. 2.22). The temperature contrast in the continental interior of Arabia from summer to winter is 8° to 17°C during winter, and 5° to 37°C during summer. This can be compared to with a seasonal change of around 10°C in the coastal regions, slightly greater in the Arabian Gulf and slightly lower than in the Red Sea.

Precipitation

The Arabian Peninsula, in keeping with its location in one of the great desert belts of the world, has a low and irregular rainfall, typically less than 150 mm/yr. It can be as low as 50 mm in the northern and central parts of the Arabian Gulf area (Fig. 2.23). As the region receives little benefit from barometric depressions during the summer months, it is not surprising that large areas have low rainfall, in the range of 100-300 mm/year. The lower the precipitation the greater its variability. In Bahrain, for example, the average annual rainfall is 76 mm, but ranges between 10 mm and 170 mm. Only the Arabian Sea coast benefits to even a limited extent from the passage of the monsoon. Climate is modified by regional factors including the distribution of land/water, the local topography, particulary the location of mountains. The Oman Mountains and the Al Hijaz Mountains of Yemen and Saudi Arabia provide a good example. Rainfall is less than 100 mm in Saudi Arabia, but sometimes exceeds 200 mm in Oman, where ice may be expected in the higher elevations during winter.

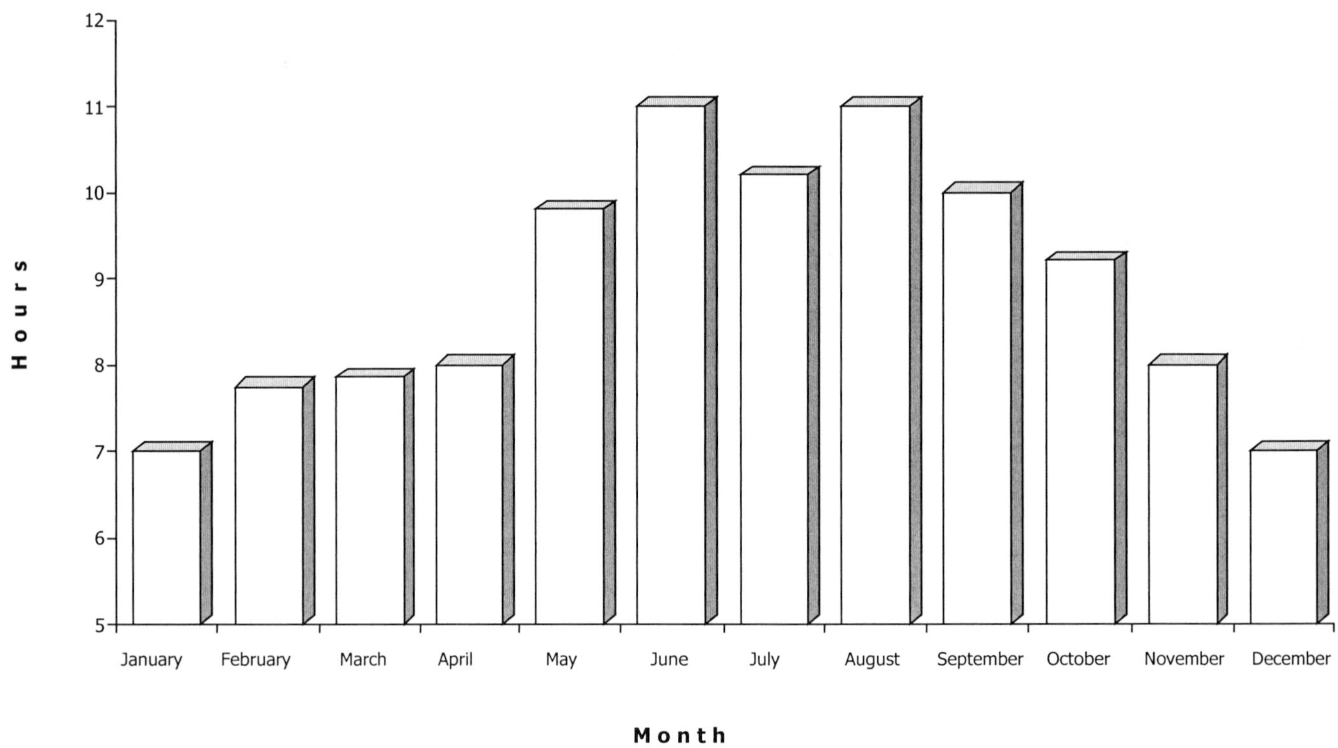

Fig. 2.19. Mean daily sunshine hours in Kuwait (modified after Sharaf, 1980).

Sporadic rainfall, sometimes with very heavy precipitation, may occur during short unpredictable intervals in limited areas, followed by long dry periods. Precipitation during such storms may equal the annual precipitation for the region. Precipitation of 109.2 mm was recorded during a twenty-four hour period in Sharjah where the average rainfall is 107 mm. In Bahrain, 71 mm fell during a twenty-four hour interval, in a country where the average annual precipitation is 74 mm. The principal rainfall occurs between November and March/April, with the greatest frequency of rain being during February and arch. The highest rainfall is in the mountainous regions in the Oman Mountains, and the Al Hijaz Mountains in Saudi Arabia and Yemen.

There are two sources of rain, the most important being winter cyclones which traverse the eastern Mediterranean. These are deflected south into the Arabian Gulf with northwesterly winds referred to as the "Shamal". When these cyclones fail to materialize the Arabian Peninsula may receive as little as 15 mm and the cyclonic track passes across the southern end of the Caspian Sea. The northern part of the United Arab Emirates, may receive a small amount of rain drawn from the Indian Ocean, that is, from the southeast into the Rub Al Khali low. Rain from both sources can be distinguished geochemically via the study of deuterium and oxygen isotopes.

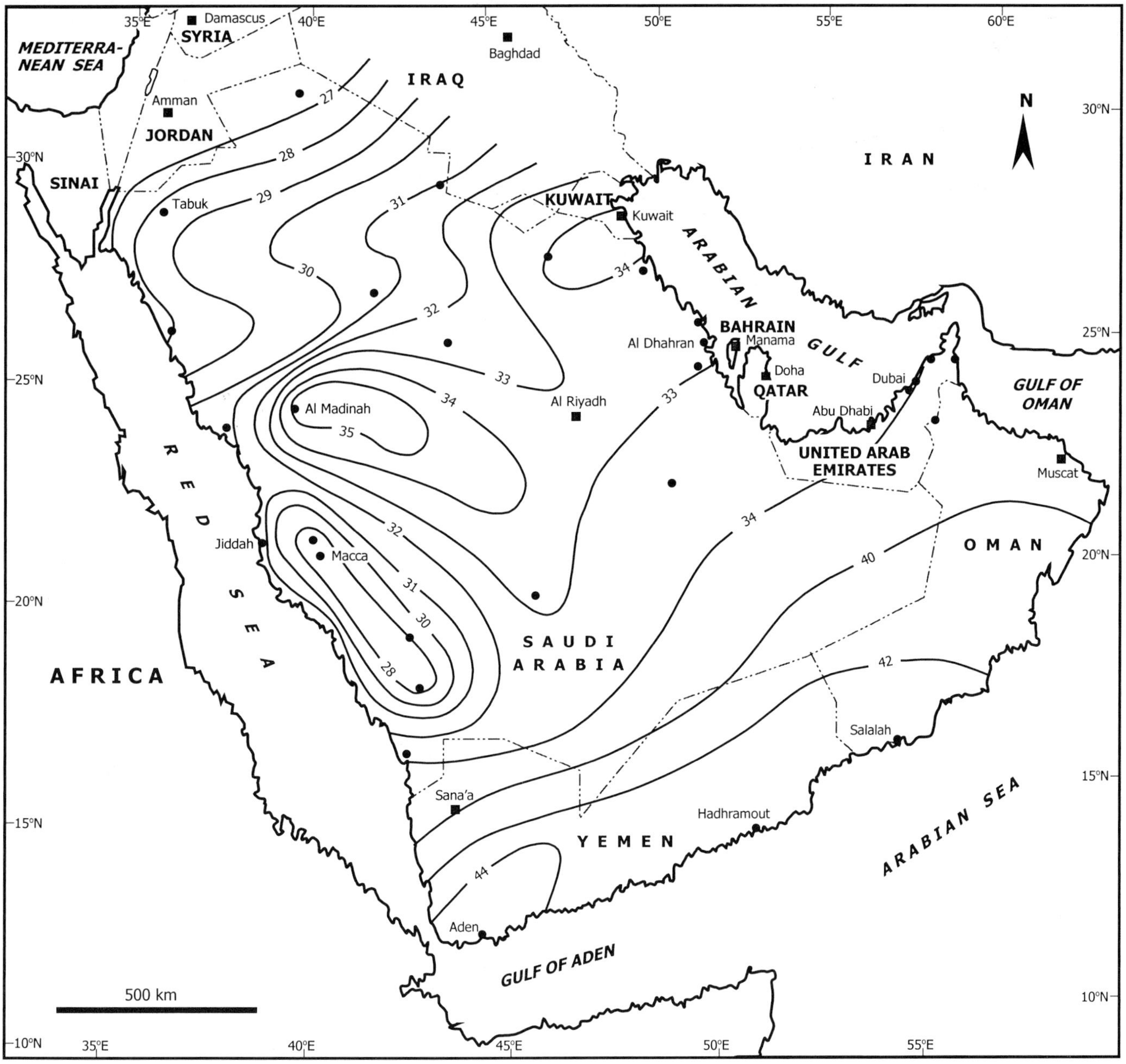

Fig. 2.20. Summer isotherms (°C), mean values for June, July and August, in the Arabian Peninsula (modified after Schyfsma, 1978; Atlas of Saudi Arabia, 2000).

There appears to be a cyclicity in the rainfall pattern evident in the United Arab Emirates, Bahrain and Qatar. Rainy years, which provide as much as 200 mm (as in the United Arab Emirates during 1982), are separated by four or five year periods during which the total precipitation may fall to as low as 15 mm (as in the United Arab Emirates during 1985) (Fig. 2.24).

In Bahrain, rainfall varies considerably. December and January have about 16.5-18.0 mm, while from February to May the amount of rain decreases steadily. July, September and October have only a trace of rain, and June and August are dry. As shown in Figure (2.25), rainfall in Bahrain is limited in amount and falls irregularly through the year.

Rainfall in Saudi Arabia varies with place and time (Figs. 2.26 and 2.27). The Aseer mountains in the southwest receive the highest recorded rainfall in the country. The highest average annual rainfall recorded in this area includes 577 mm at Jabal Fifa, and 357mm at Abha. Otherwise, the average annual rainfall in Saudi Arabia remains below 100 mm. In fact, the annual average of rainfall value in western Saudi Arabia varies between 20mm and 133mm (Ahmed, 1993).

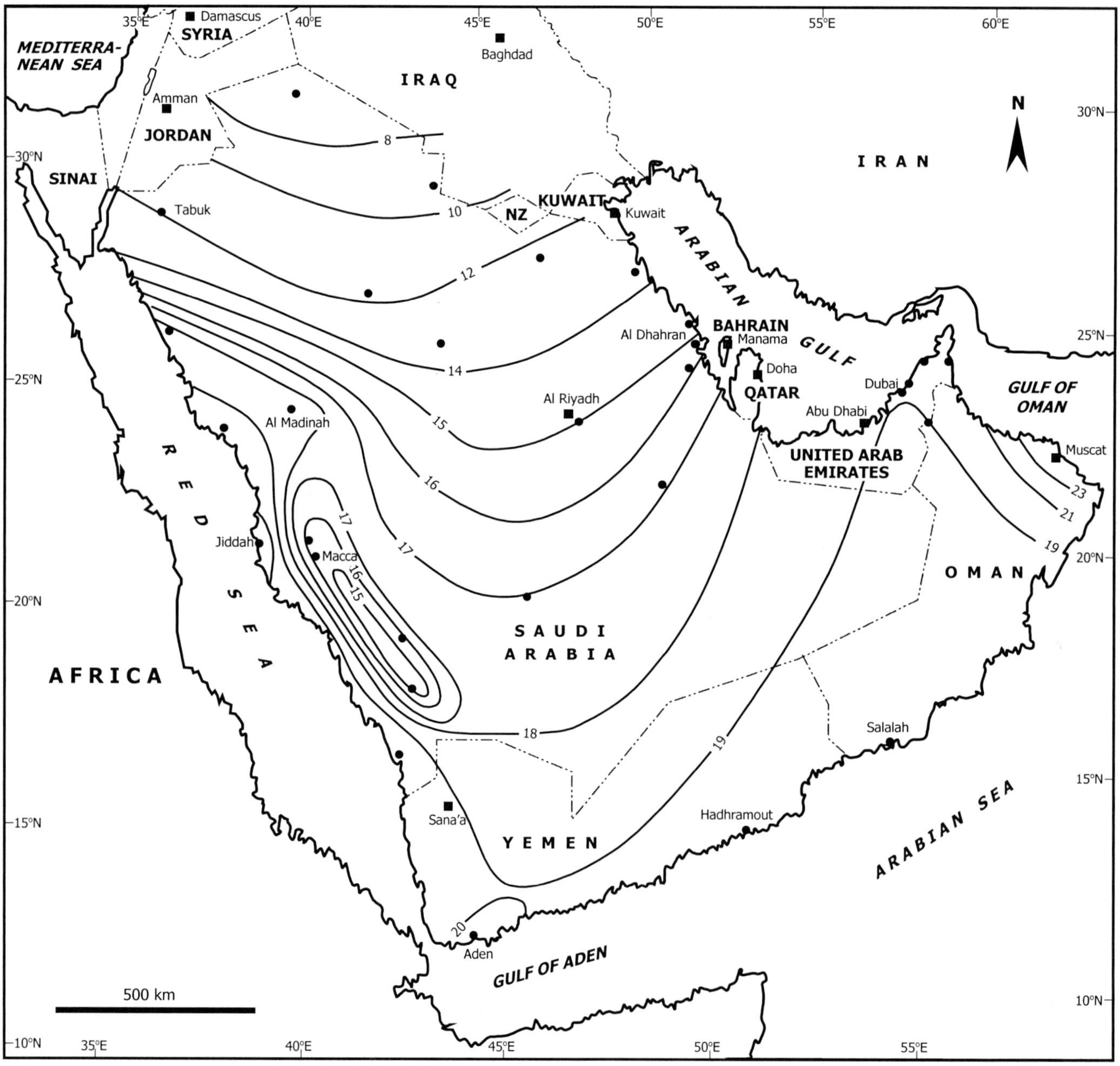

Fig. 2.21. Winter isotherms (°C), mean values for December, January and February, in the Arabian Peninsula (modified after Schyfsma, 1978; Atlas of Saudi Arabia, 2000).

Two types of rain are recognized in Qatar. The first is associated with the movement of weather fronts crossing the Arabian Gulf in a northwesterly - southeasterly direction. The second type is associated with convection (Fig. 2.28), heavy but local and occurs mostly in the central part of the country (Al Magid, 1988).

Rain in Kuwait is also variable with place and time, 28 mm in 1964 and 374 mm in 1972 (Fig. 2.29). Airport meteorological data provide a detailed information on rainfall patterns. During a wet year there may be as many as nine rainy days in a month, though six is more common. Since the maximum daily rainfall may constitute as much as 60% of the monthly total, the remaining rainy days may have only a few millimeters (yet potential daily evaporation is never less than 4 mm). Monthly and daily rainfall and the number of rainy days per month in 1986 for the United Arab Emirates is depicted in Figure (2.30).

There have been relatively few analyses of storm intensity, due to their infrequent occurrence. In the United Arab Emirates, the heaviest storms show a 50 year return frequency. The return frequency of one hour storms, during which 180 mm of rain falls, is only five years (Fig. 2.31). There is only a single analysis available at Sharjah of a 24 hour storm for period 1949-1970. A profile for this storm (Fig. 2.32) suggests that 75% of the total rain falls during the first ninety minutes. This affects runoff, since surficial sediments are rapidly saturated and the remainder of the flood water runs off with reduced chances of infiltration.

Wind Directions

The two principal systems affecting the climate of the Arabian Peninsula are the winter cyclonic depressions, which descend the Arabian Gulf from the north and northwest, and give rise to the cold northwesterly "Shamal" airflow; and the summer monsoonal low developed over the Rub Al Khali (Fig. 2.33). The latter is an extension of the Intertropical Convergence Zone which originates in East Africa, and results in a southeasterly airflow, as air is drawn into the low situated over the strongly heated Rub Al Khali. Wind direction and velocities are variable in the Arabian Peninsula. For example, the Shamal wind affects the northern part of the United Arab Emirates (wind rose for the years 1989 and 1993 in Fig. 2.34). Qatar near the northern limit of the area affected by the retreating monsoon, is impacted by the low frequency of south-southeasterly winds, (Fig. 2.35) and absence of rain.

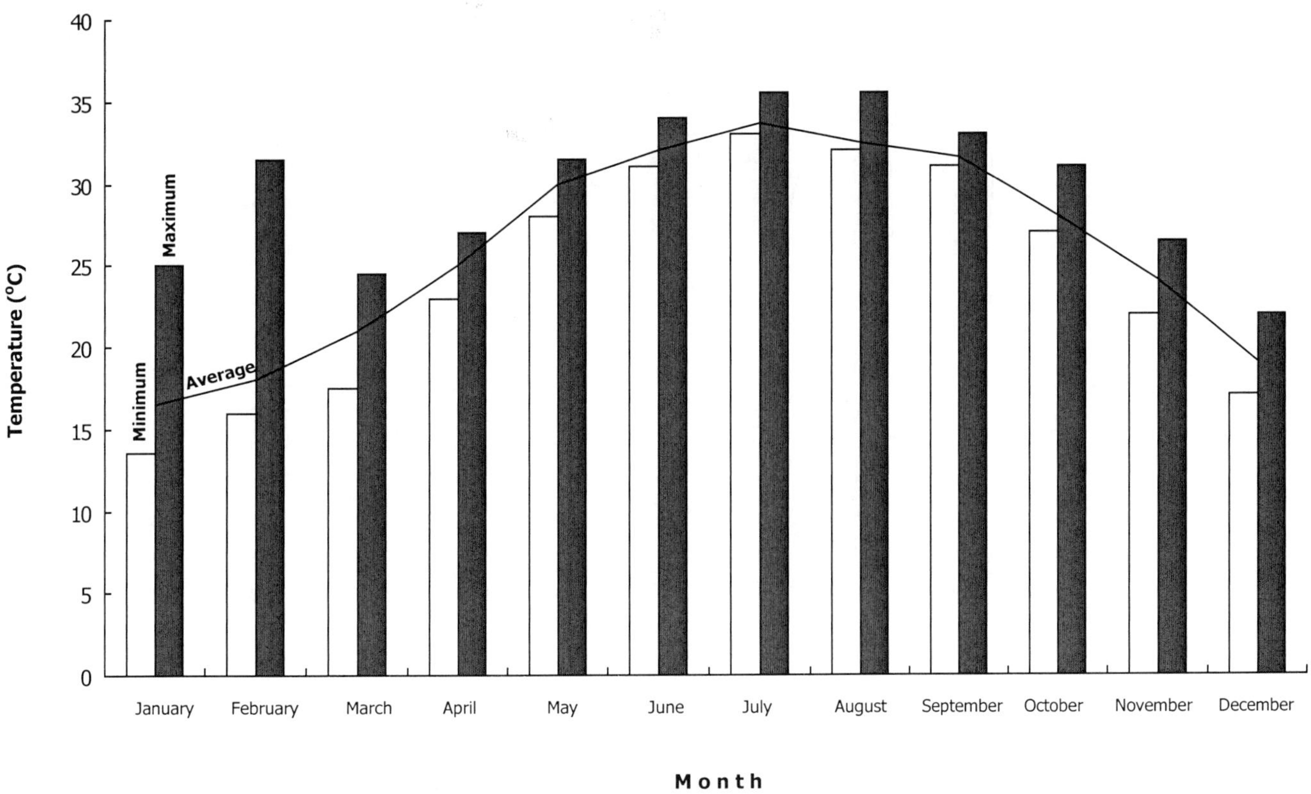

Fig. 2.22. Mean monthly temperature (maximum and minimum) in Bahrain during the period 1960-1989 (modified from Aba Hussain, 1992).

Ahmed (1993) depicts seven wind directions in Saudi Arabia (Fig. 2.36), with velocities which appear to be higher than those found in the southeastern Arabian Gulf, with steady north-northwesterly winds from 26 km/hr to 34 km/hr. Variability of wind direction is illustrated by the southeast trades in the eastern region of Saudi Arabia, the northwesterly winds in western Saudi Arabia and southerly dry winds in the southwest (the As Sumoom). The latter are responsible for causing dust storms during the summer months. The southeasterly winds dominate the northern, eastern and central parts, and a southwesterly wind crosses the western part of Saudi Arabia. The highest wind velocities, usually in excess of 13 km/hr, are in the coastal areas.

In Bahrain, the northwesterly air flow component is the predominant direction, and is also the direction from which the strongest winds blow (Fig. 2.37). Over 50% of the winds over Bahrain come from either the west-northwest, or north-west/north-northwest with speeds of 10(18), 12(22), and 12(22) knots (km/hr), respectively. The predominant dust-raising wind is the "Shamal", which sometimes brings airborne silts from mainland Arabia and the northern Arabian Gulf. Locally dust clouds are raised at wind speeds of 24-40 knots (46-47 km/hr), and can occur as gusts in any month of the year (Doornkamp et al., 1980).

The winter winds in Qatar and Kuwait are mainly from the northwest with velocities averaging 9 knots (16km/hr) but gusting to 45 knots

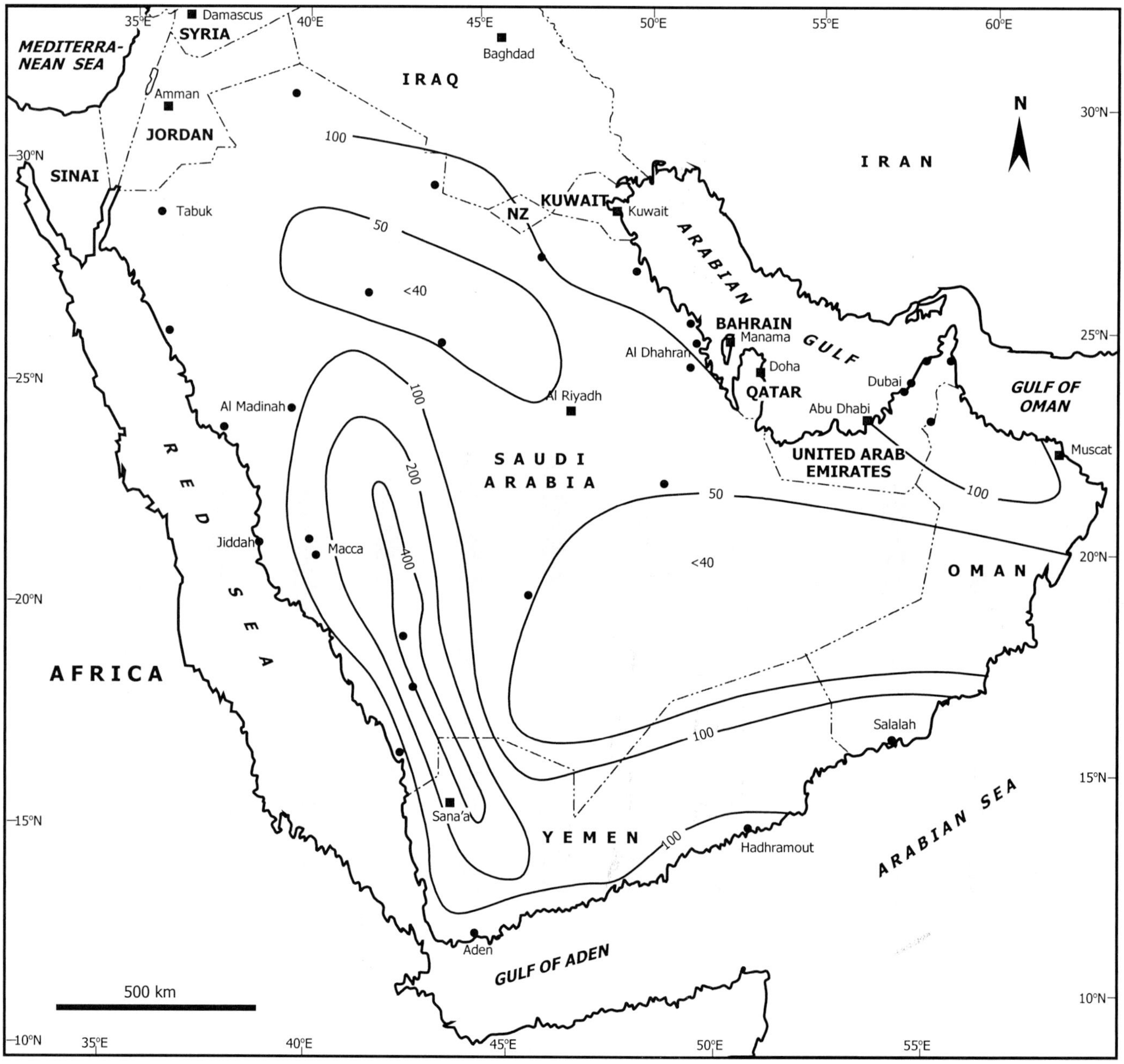

Fig. 2.23. The yearly average of the actual evaporation (mm) in the Arabian Peninsula (modified after Schyfsma, 1978).

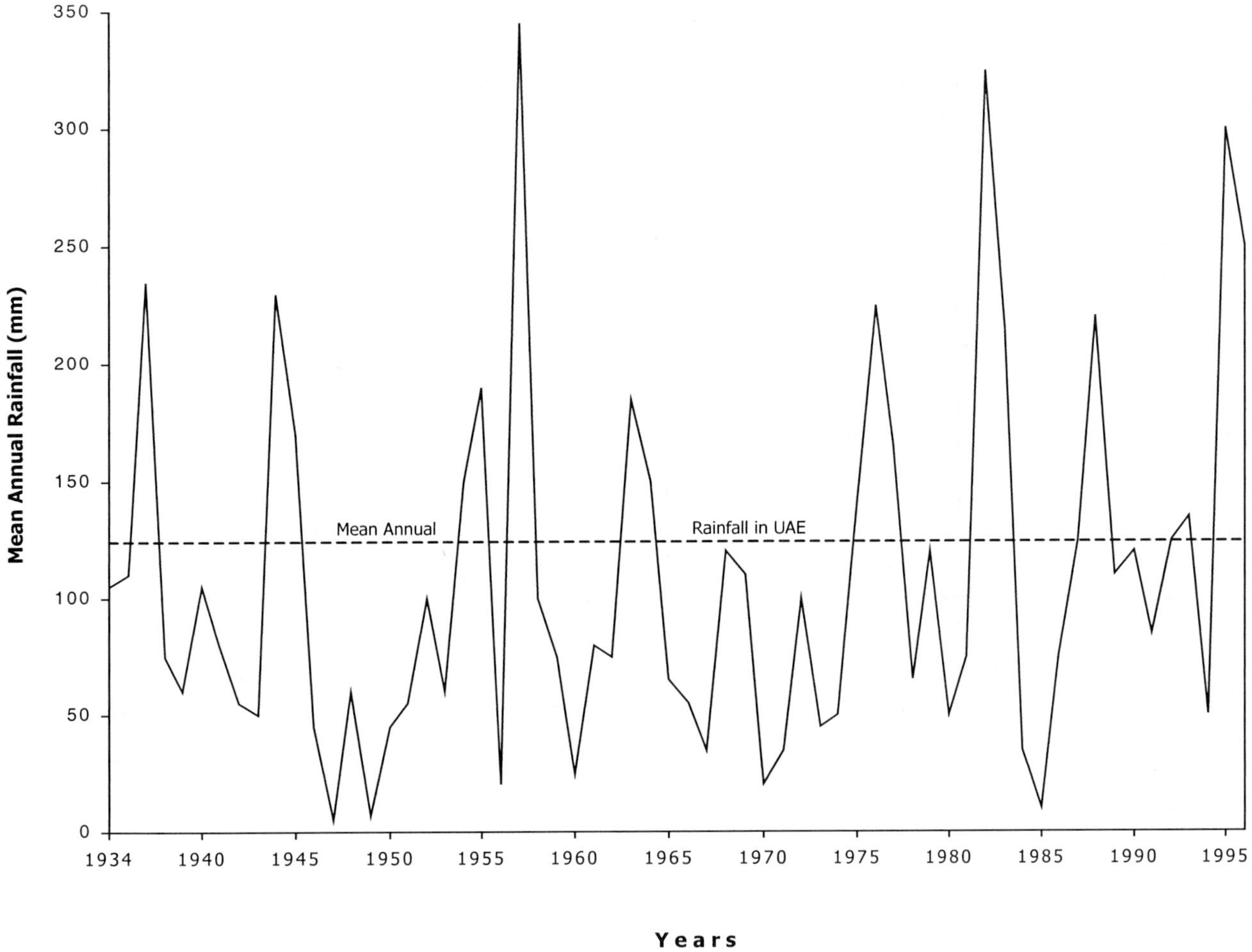

Fig. 2.24. Mean annual rainfall (mm) in the United Arab Emirates for the period 1934-1994

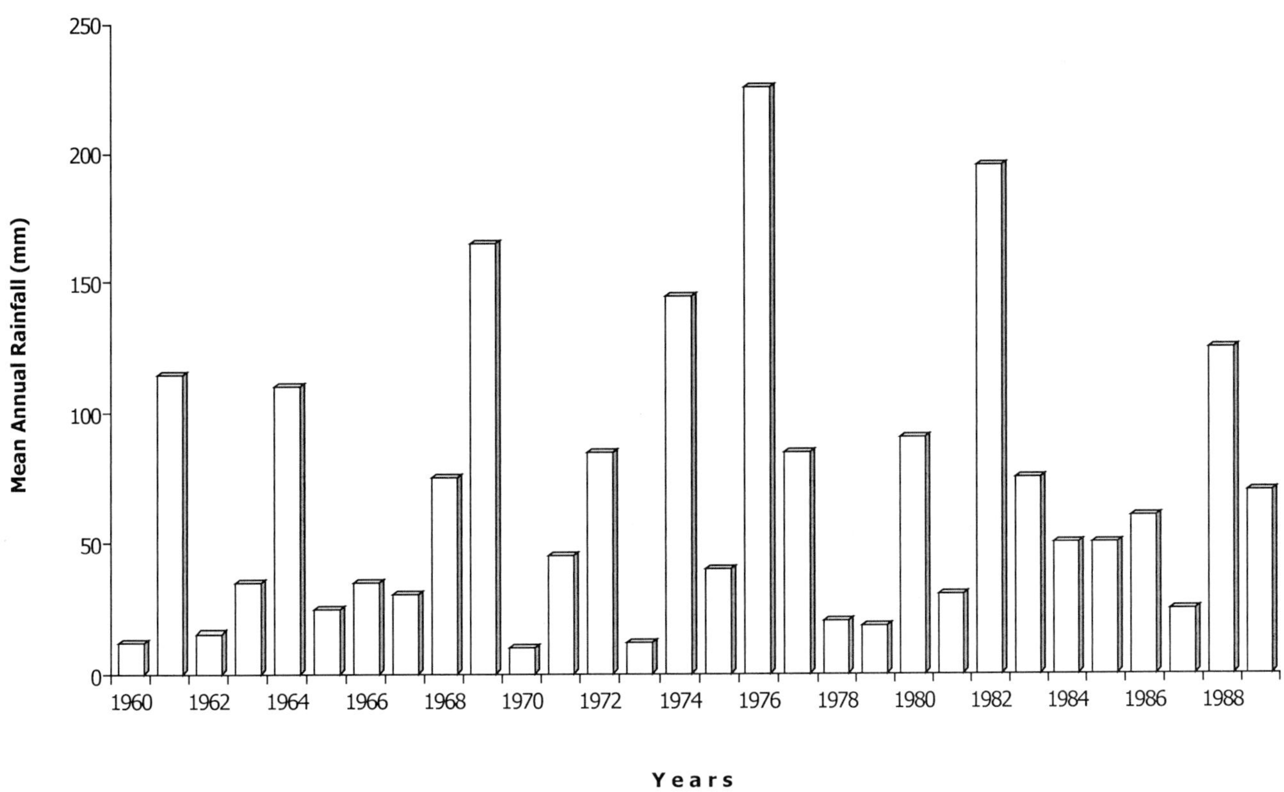

Fig. 2.25. Mean yearly rainfall in Bahrain during the period between 1960 and 1989 (modified after Aba Hussain, 1992).

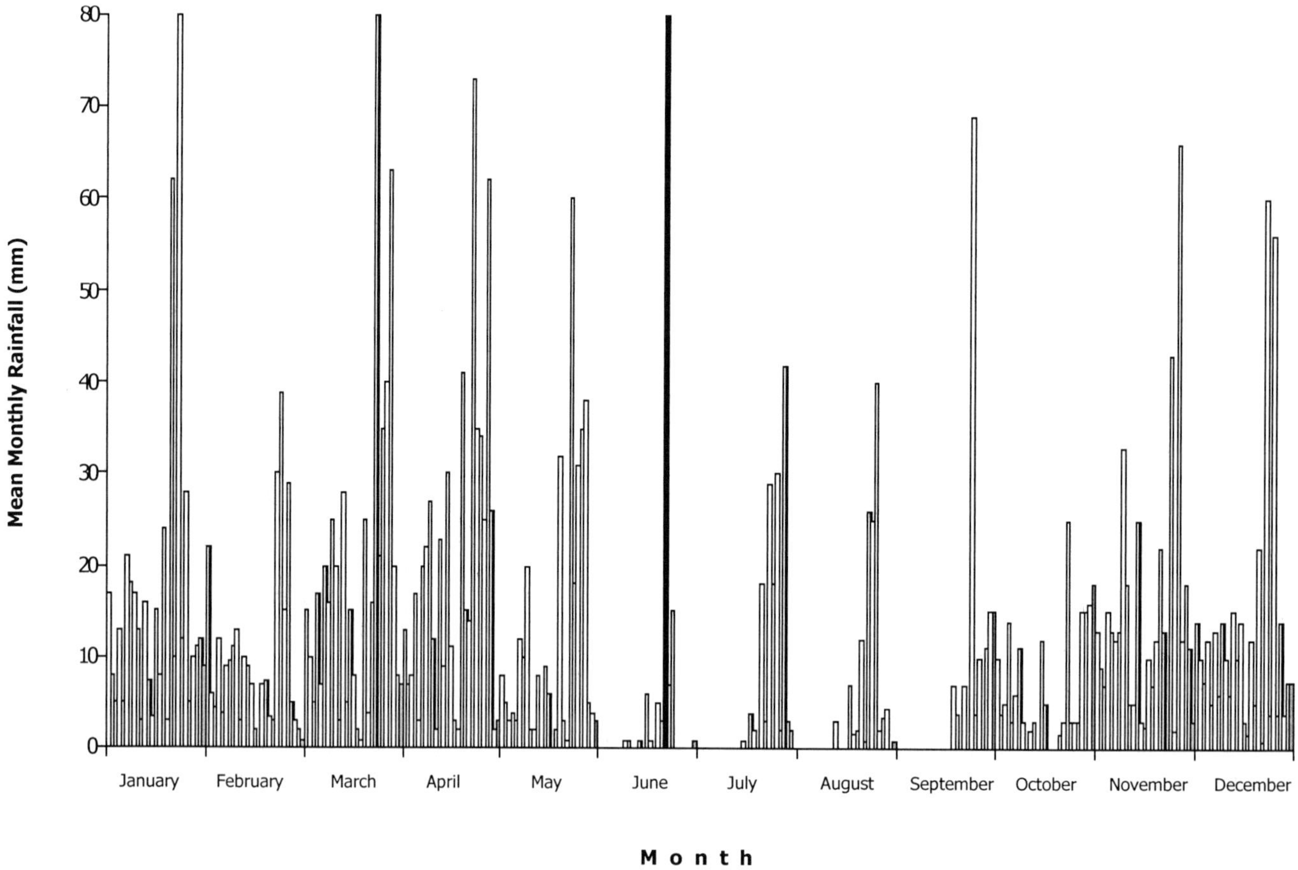

Fig. 2.26. Average monthly rainfall in Saudi Arabia (in mm) for the period 1967-1987 (modified after Ahmed, 1993).

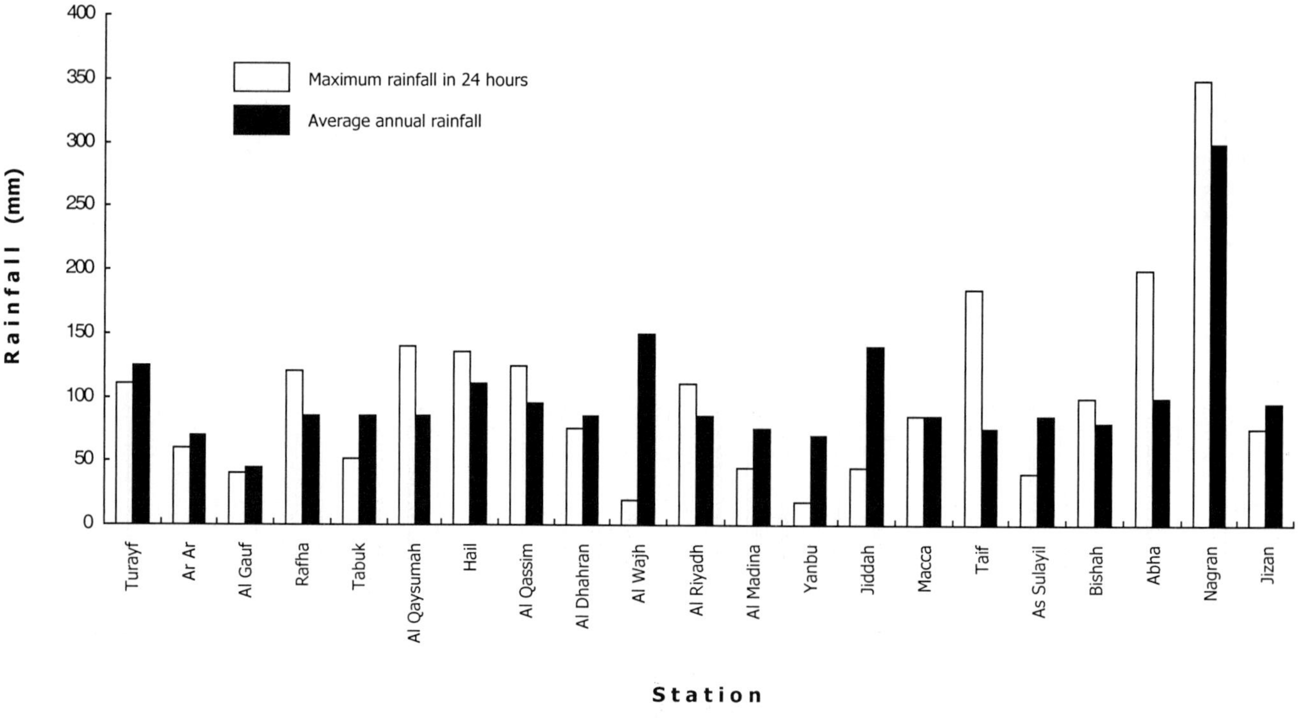

Fig. 2.27. Average annual rainfall and daily maximum at 22 meteorological stations in Saudi Arabia from 1960 to 1987 (modified after Ahmed, 1993).

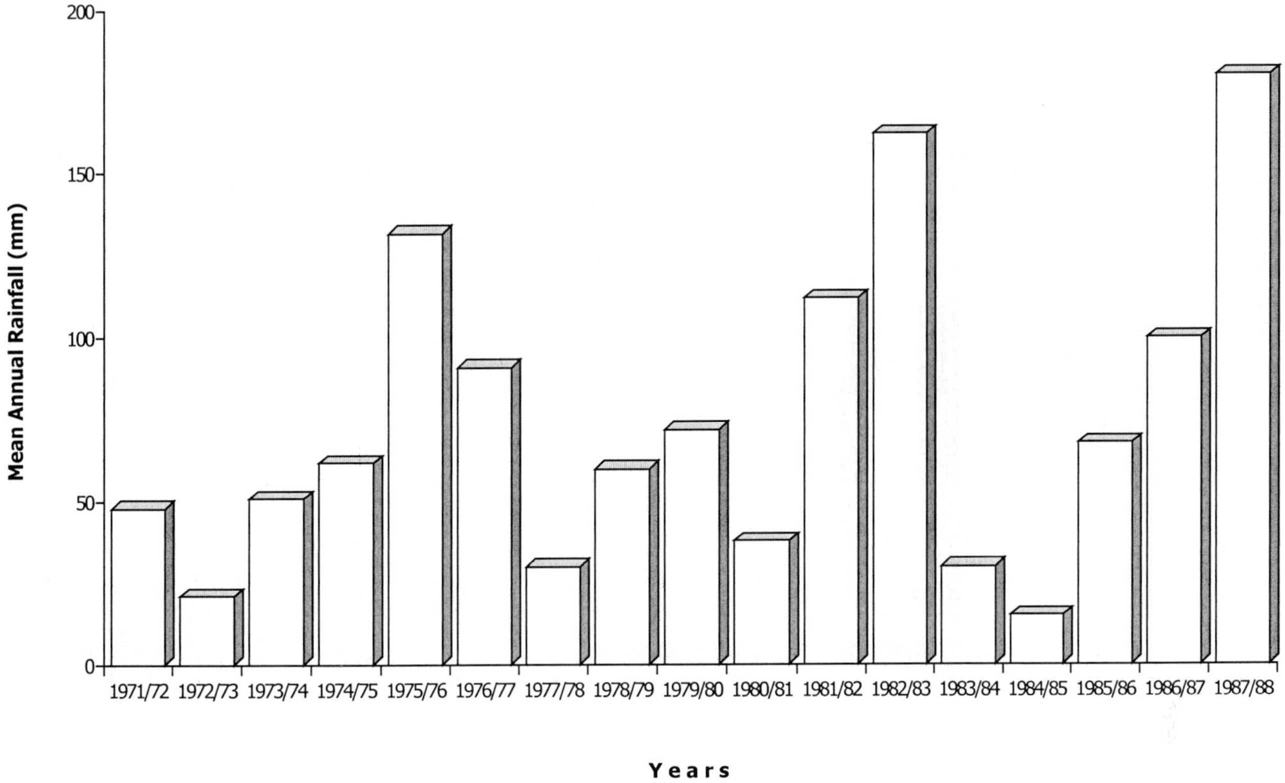

Fig. 2.28. Average annual rainfall (in mm) in Qatar for the period 1971-1988 (modified after Al-Magid, 1988).

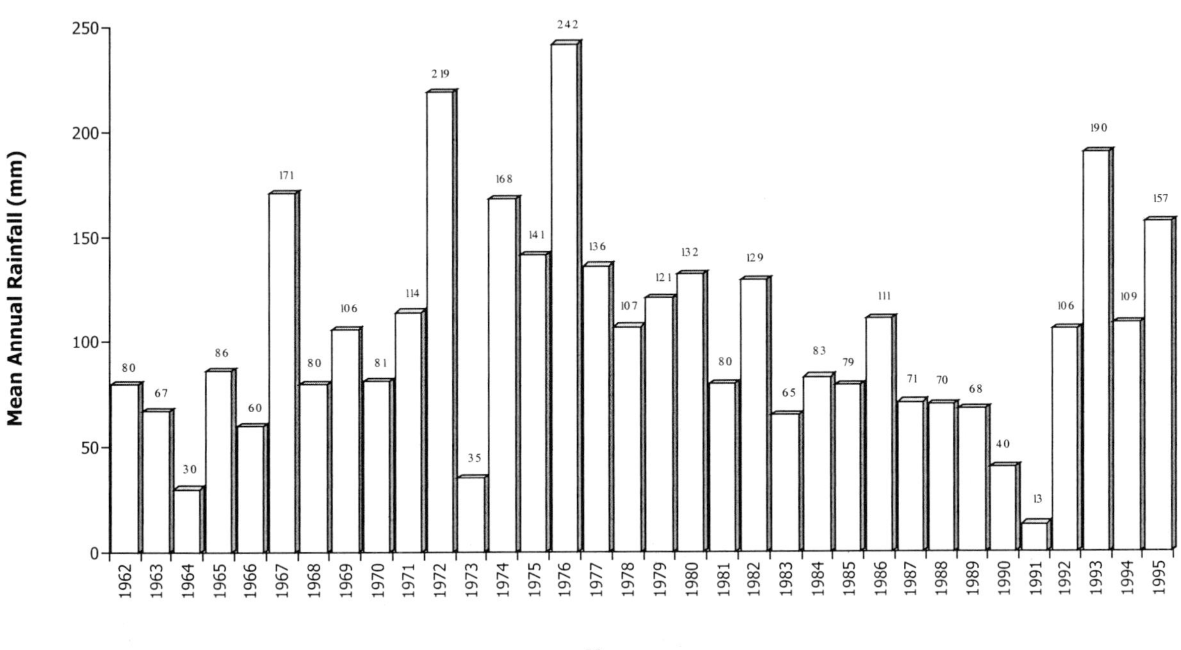

Fig. 2.29. Kuwait average yearly rainfall (modified after Sharaf, 1980).

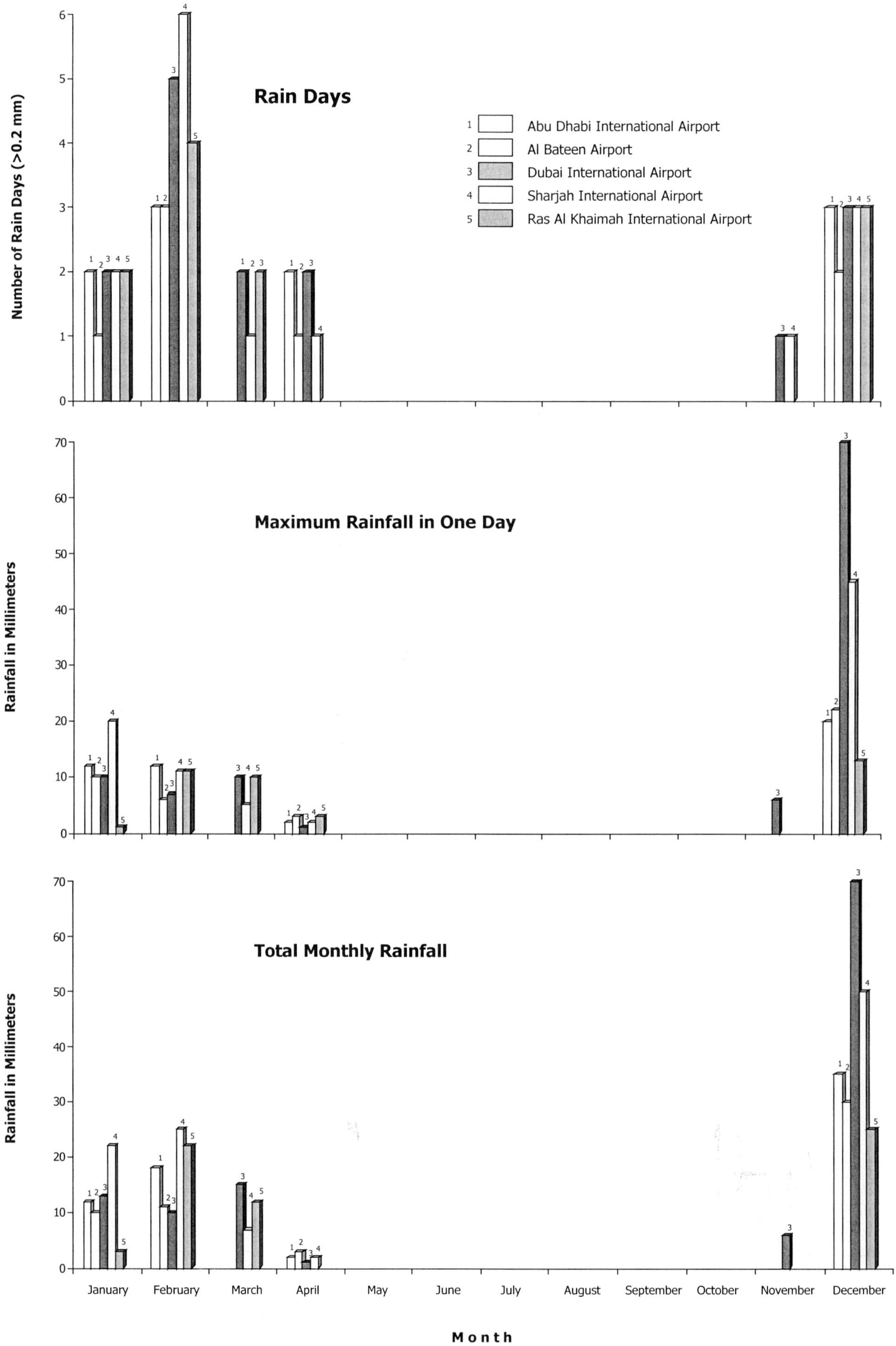

Fig. 2.30. Total monthly maximum single day rainfall per month and the number of rain days per month during the year 1986 in the United Arab Emirates (data from UAE Department of Civil Aviation annual report, 1986).

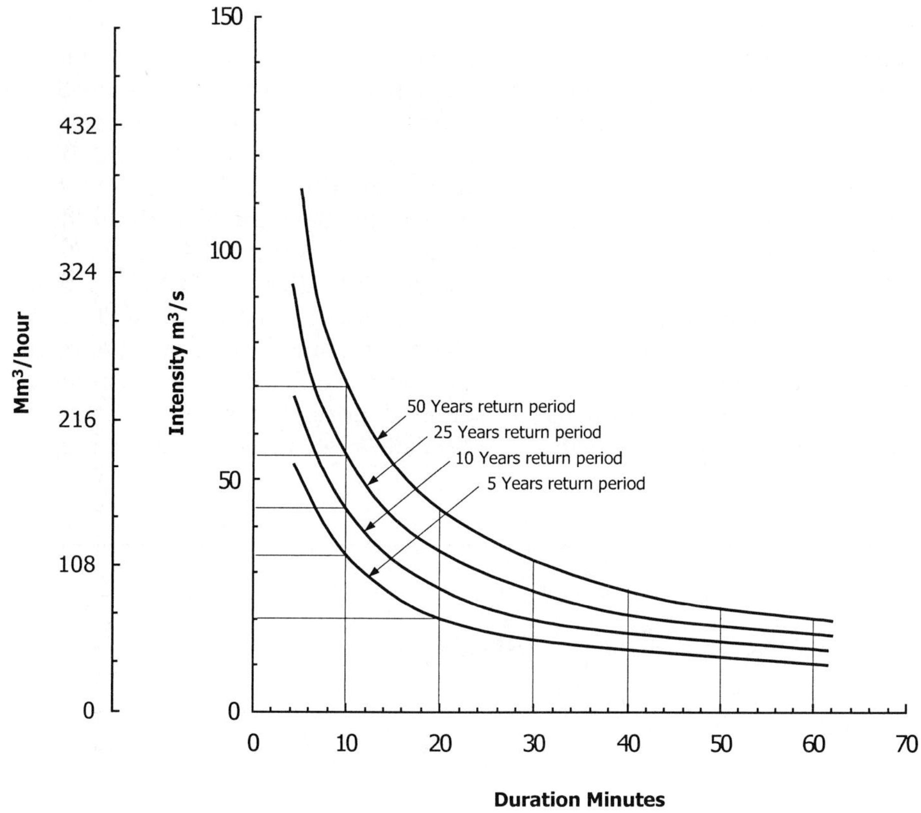

Fig. 2.31. Rainfall intensity and duration curves for Emirate of Sharjah (United Arab Emirates).

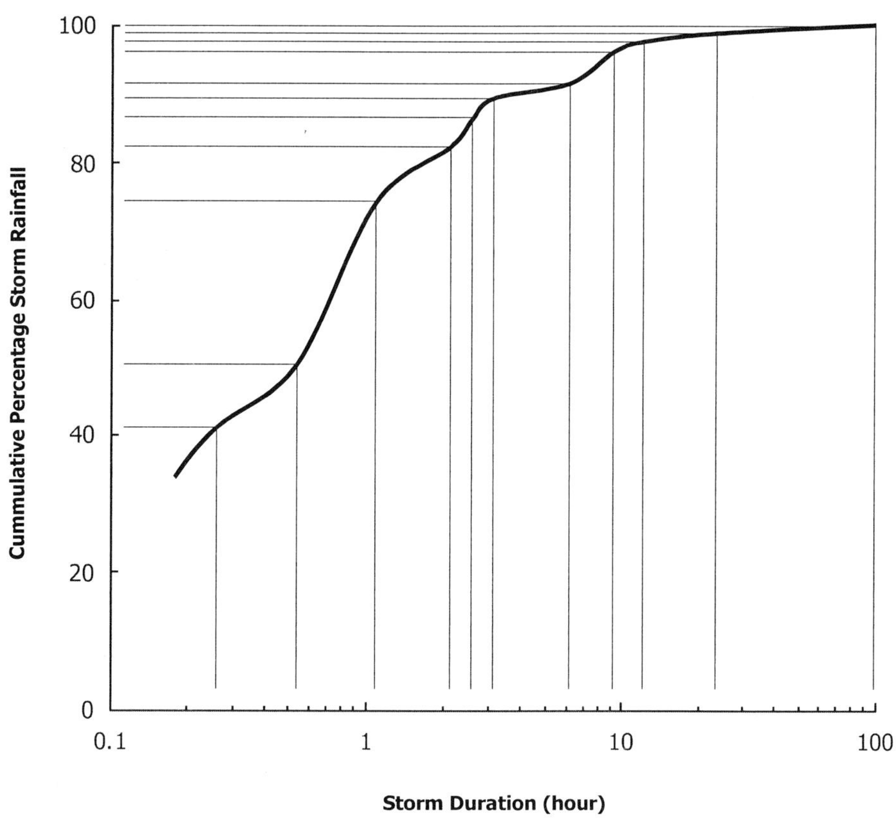

Fig. 2.32. Storm duration profile, for Emirate of Sharjah (United Arab Emirates).

(81km/hr). In spring and summer the westerly winds may reach 10 km/hr. In Kuwait the strong and dry northwestern wind is called As Sumoom ("the Poisonous") and is associated with dust storms. During spring and autumn, wind directions are more variable (Fig. 2.39).

Wind speeds in the southeastern Arabian Gulf are of the order of 5-7 knots (9-13km/hr) (Fig. 2.38). Airport data show small variations, lower velocities to be recorded in Ras Al Khaimah and higher ones at Abu Dhabi. In Qatar, wind speed ranges from 4.5km/hr in November to 20km/hr in July (Fig. 2.44). The magnitude of the variation, however, is only 1-2 knots (2-4km/hr). The maximum velocity over a ten minutes period is about double the average figures (14 knots about 25km/hr); gust velocities of 30 to 45 knots (54-81km/hr) may be recorded. Wind directions also depict regional differences. The shamal winds in Abu Dhabi are from the northwest, but a more westerly component is seen in Dubai and Sharjah. In the northern United Arab Emirates the winds have a more northeasterly component (the "Hashi" winds). Over the mountains in northern United Arab Emirates and Oman, cumulonimbus clouds can develop into thunderstorms, though the number of storms is not great. During the summer, when the general air circulation weakens, land/sea breezes are common

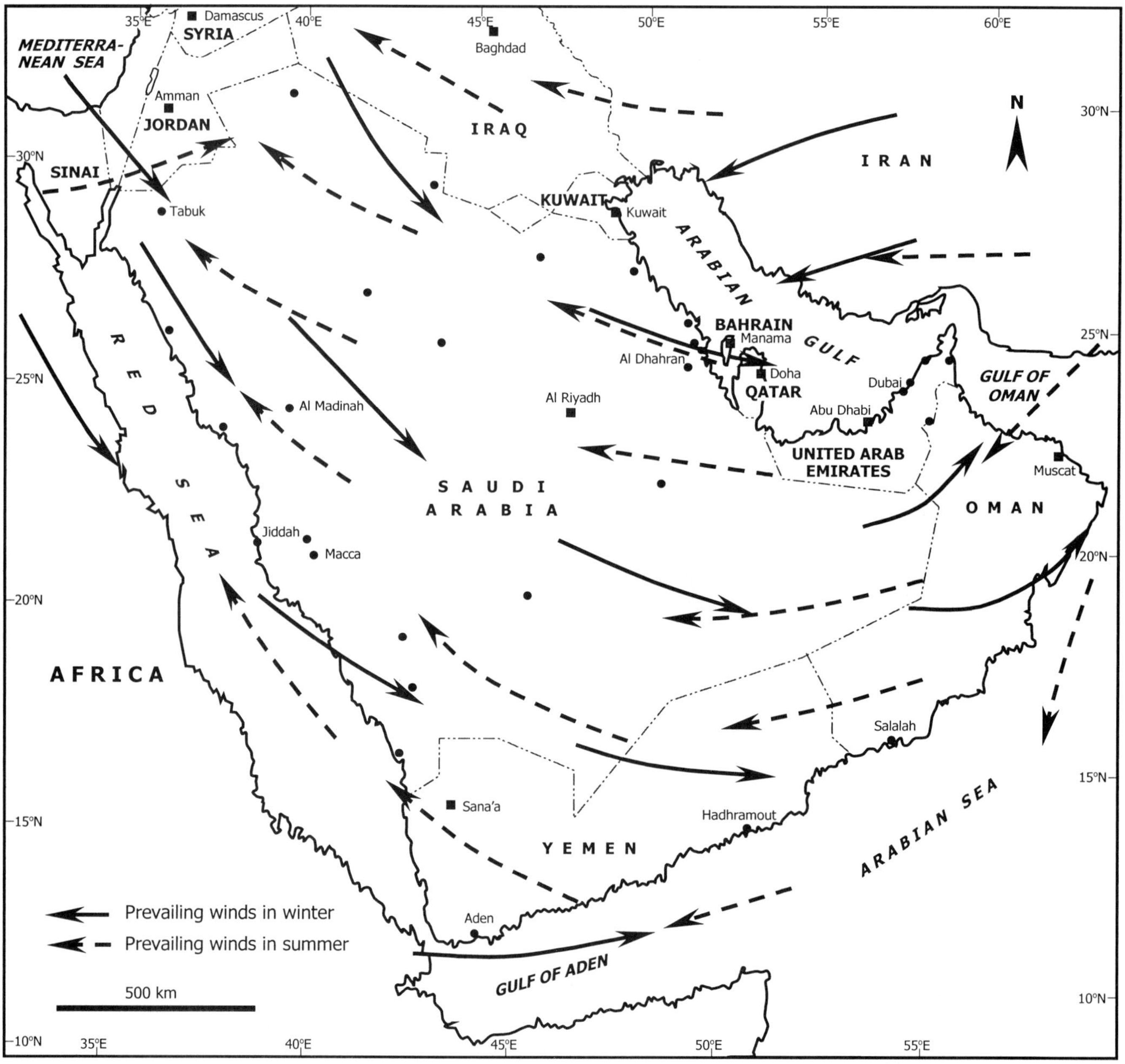

Fig. 2.33. Prevailing wind direction in the Arabian Peninsula (compiled from Dewdney, 1988; Atlas of Saudi Arabia, 2000).

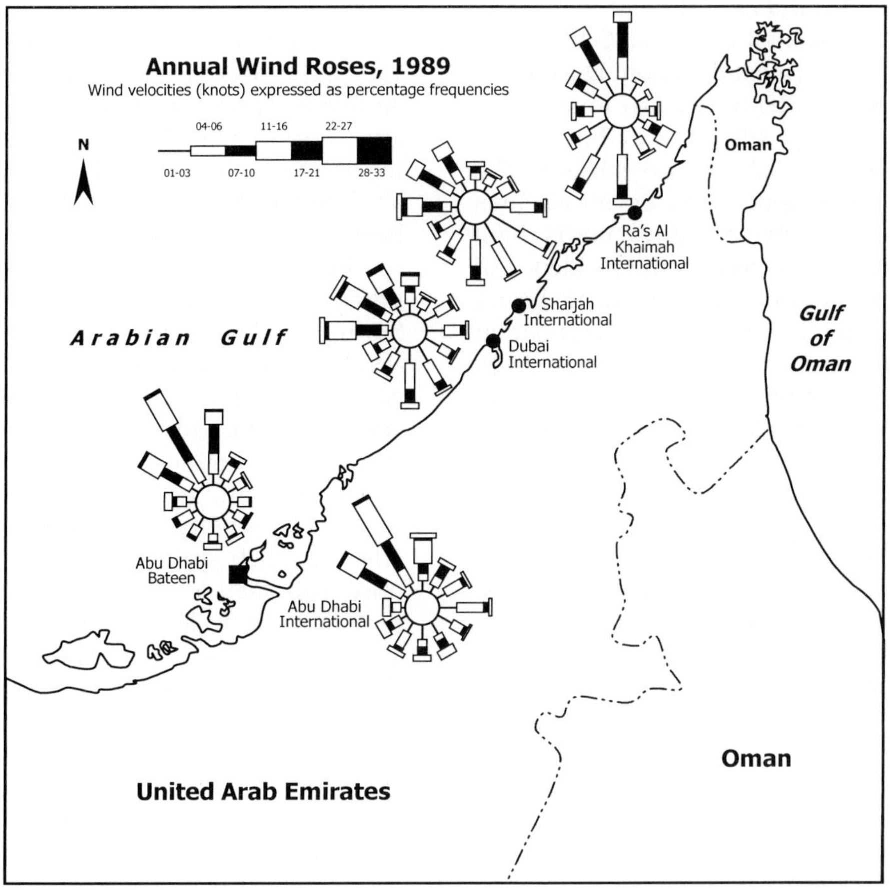

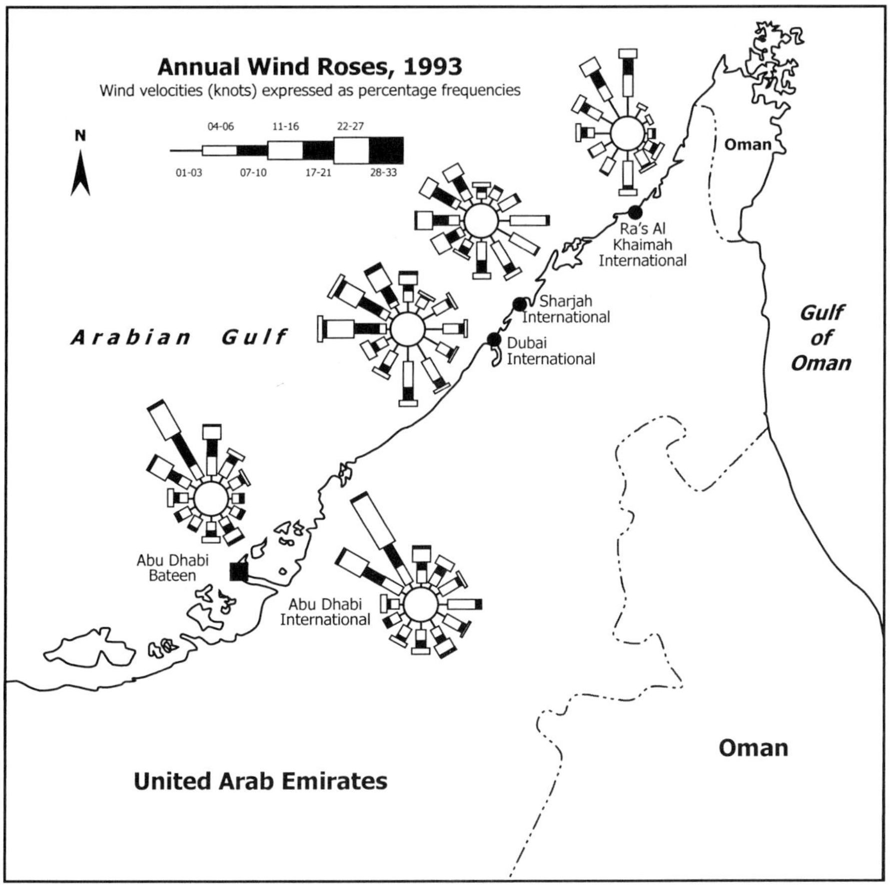

Fig. 2.34. Annual wind roses illustrating regional differences in wind direction for the years 1989 and 1993 as recorded at the five airports in the United Arab Emirates (UAE Department of Civil Aviation, Annual Reports, 1989 and 1993; Al Shamsi,1993).

in the coastal areas, associated with high humidity, with moisture derived from the Arabian Gulf. It is at this period when haze is most common. During the rest of the year, haze (visibility reduced to less than 8km) may occur on ten to fifteen days; but during the summer, it occurs on about two thirds of the days (Fig. 2.40). There appears to be no obvious correlation between haze and fog days, nor is there an obvious seasonal component to the incidence of fog, though it appears to be more frequent during September. In Kuwait, there is a good correlation between cloudy and foggy days (Figs. 2.41 and 2.42). Sandstorms, not common in the United Arab Emirates, mostly occur during winter months associated with strong depressions. The total number of sandstorms per year at a given location is seldom more than five.

The eastern regions of Saudi Arabia (e.g., Dhahran and Riyadh) are subject to dust storms and haze all year. This phenomena is less common on the Red Sea coast, as in Al Madinah and Al Taief. Dust storms occur throughout the year with the fewest in August. At Riyadh, 18 days of storms per year were recorded annualy during the period 1967-1987. The western mountainous region of Saudi Arabia has fewer dust storms, since the wind has to rise. Storms in 1967, 1968 and 1969, were unusual; for example in 1967, Riyadh recorded 157 days, Yanbou had 40 days, and Al Qaseem had 33 days of heavy storms (Ahmed, 1993).

In Qatar, dust storms are associated with northern and northwestern winds (Al Magid, 1988). Storm duration and season varies from one year to another but generally occurs during spring and summer (March-April and June-July). The spring storms can last for 18 days (1973), while summer storms may continue for 21 to 26 days (Fig. 2.43).

In Kuwait, active winds loaded with dust, decrease visibility to less than 1 km. Factors affecting such storms include the presence of vast unconsolidated sources of sand and dust, high solar radiation and low atmospheric pressure. The average of dusty days in June is 4.8 and in July is 4.5 (Sharaf, 1980) (Fig. 2.44).

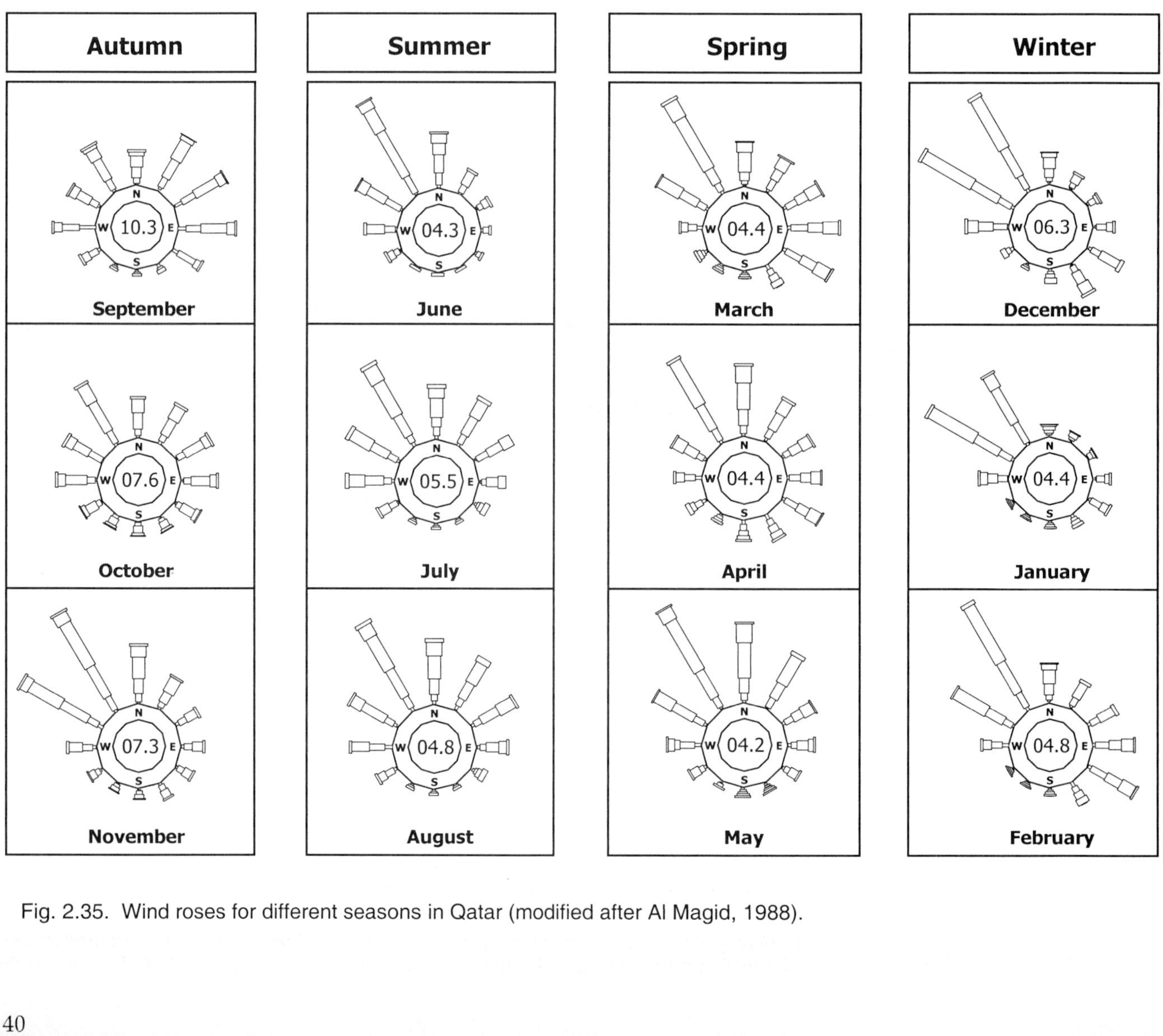

Fig. 2.35. Wind roses for different seasons in Qatar (modified after Al Magid, 1988).

Relative Humidity

Relative humidity in the United Arab Emirates is high in the coastal areas, decreasing sharply towards the interior. For example, 60% humidity or more in Abu Dhabi, 45% in Al Ain and 25% at Liwa in the northeastern corner of the Rub Al Khali desert. Mean maximum humidity varies from 76% during May to 89% during December, January and March. Mean minimum humidity varies, from 17% during May to 37% during December. The mean relative humidity varies from 46% during May, to 64% during December (Fig. 2.45). During the winter months, the mean relative humidity increases, by about 10 to 15%, from the interior to the coastal regions, while during the summer months, this variation is 20 to 25%. In general, the mean monthly relative humidity, for the United Arab Emirates is around 60% during winter and 50% during summer. The diurnal variation in relative humidity is more extreme, particularly during the summer months, when the daily variation in temperature is also high.

In Saudi Arabia, the coastal areas along the Red Sea coast and the Arabian Gulf have higher relative humidity than the interior parts. Relative humidity also declines from south to north, reaching a minimum of 24% in Al Madinah (Fig. 2.46). The winter months have a higher relative humidity than

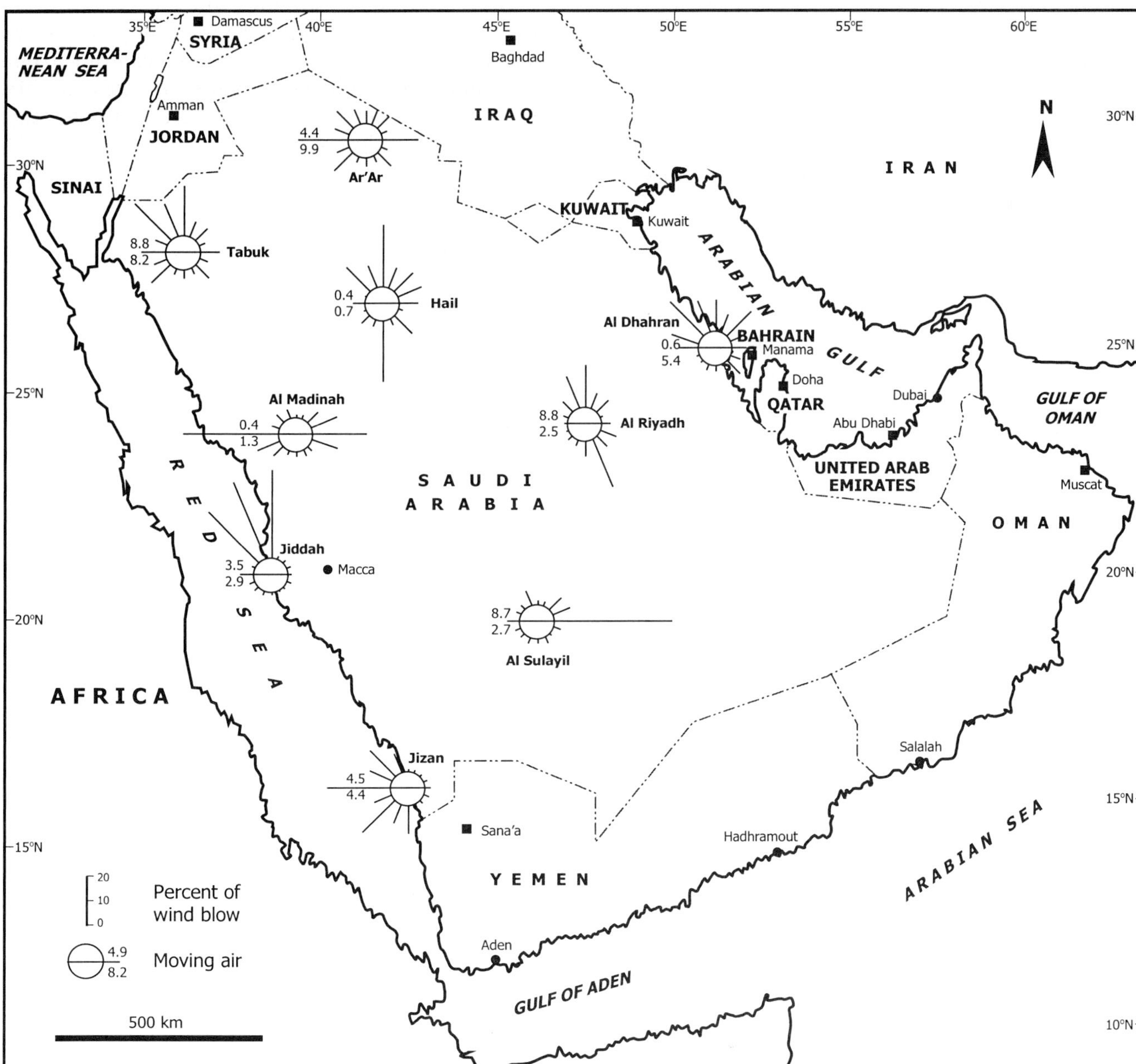

Fig. 2.36. Rates and wind directions in major meteorological stations in Saudi Arabia for the period 1967-1976 (modified after Ahmed, 1993; Atlas of Saudi Arabia, 2000).

summer, except in the coastal regions, which maintain high relative humidity all year long (Ahmed, 1993).

The relative humidity in Qatar is higher in winter than summer, with an average annual value of 50%, higher in coastal areas than in the interior. The relative humidity decreases from March to July and increases again from August (Fig. 2.47).

The mean maximum and minimum relative humidity values, at eight meteorological stations in Kuwait, show the lowest values occurring during July, and the highest during January (Figs. 2.48 and 2.49). In Bahrain, the mean daily maximum humidity is about 85% throughout the year, with the minimum daily humidity dropping to 40%. However, the absolute minimum relative humidity is in May, and the period from April to October, may also experience low minimum relative humidity (Fig. 2.50).

Evaporation

In the Arabian Peninsula, the significance of average annual precipitation is questionable. This due to the variability of rainfall, and the need to combine data from many stations. Only from those averaged over several numbers of years is it possible to produce an isohyet map with some level of confidence. Not surprisingly, the contours on the map showing actual evaporation (in mm) are similar (Fig. 2.51).

The potential for evaporation is much higher. In Bahrain, the evapotranspiration rate has been calculated using Thornthwaite's (1948) equation (see Zubari, 1987) as 1,850 mm/yr; for the United Arab Emirates, as 2,390 mm/yr in Dubai and 2,750 mm/yr in Al-Ain (Garamoon, 1996). These high negative deficits in water budget create an impossible condition for perennial surface water.

In the United Arab Emirates, evaporation is extremely high: The western coast has the lowest annual rate (7.5 to 8.0 mm/day) (Fig. 2.52) whereas the rate is 9.0 to 9.5 mm/day in the east because of higher wind speeds. In the mountains, gravel plains and desert foreland, evaporation ranges from 10.0 to 11.0 mm/day, while evaporation rates in the western and southwestern desert regions (e.g., Liwa), are the highest in the country, reaching 12.0 mm/day.

The average annual evaporation in Qatar is 8.80 mm/day, with the evaporation rate increasing from 3.82 mm/day in January to 12.81 mm/day in July.

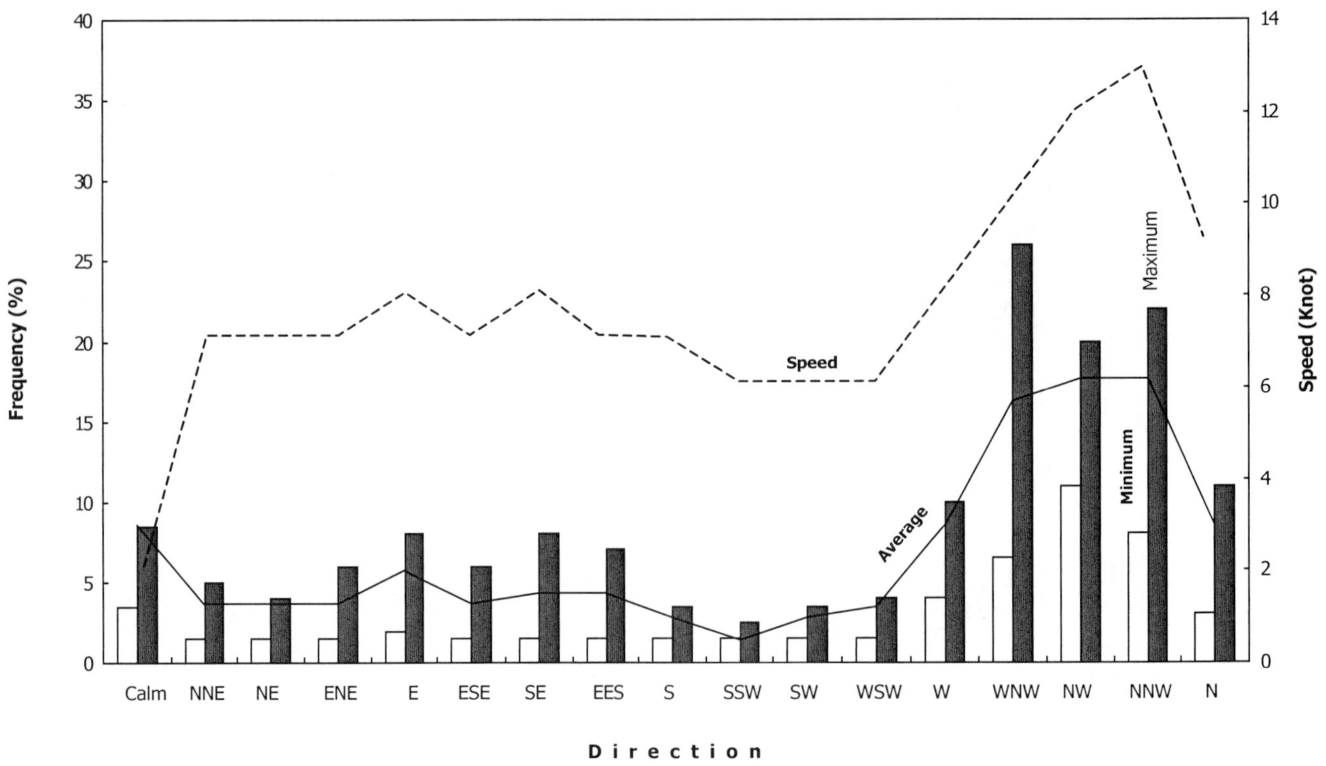

Fig. 2.37. Range and average frequency of wind directions and average wind speed in all directions for a period of fourteen years (from 1975-1989) in Bahrain (modified after Aba Hussain, 1992).
N = north, S = south, E = east, W = west, NNE = north-northeast, ENE = east-northeast, NE = northeast, ESE = east-southeast, SE = southeast, EES = east-eastsouth, SSW = south-southwest, SW = southwest, WSW = west-southwest, WNW = west-northwest, NW = northwest, NNW = north-northwest

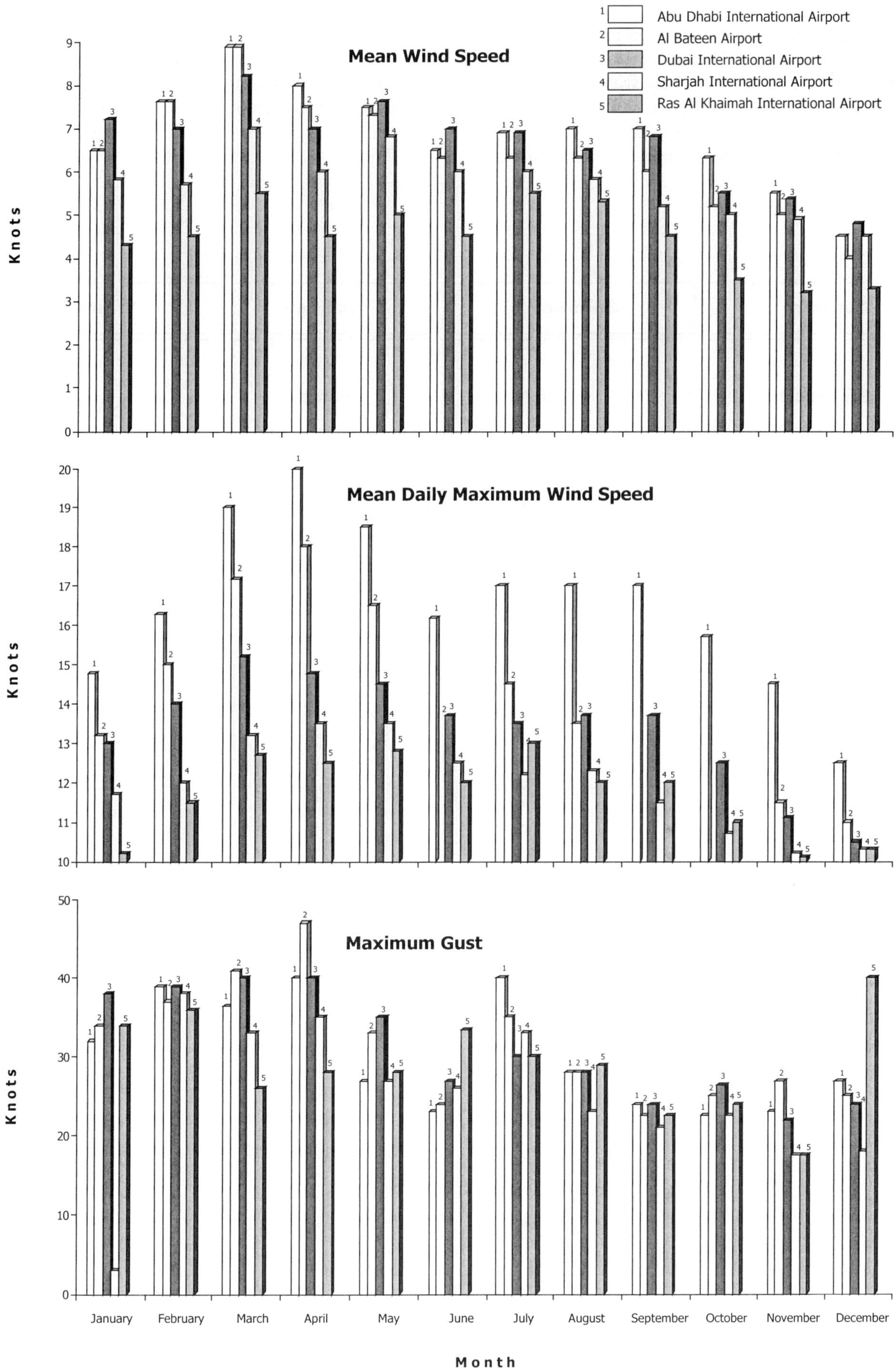

Fig. 2.38. United Arab Emirates: Mean wind speed, mean daily maximum wind speed and strength of gusts for the year 1993 (data from UAE Department of Civil Aviation, Annual Report, 1993).

The minimum evaporation rate (0.36 mm/day) was recorded in January 1976, while the maximum evaporation rate of 27.65 mm/day was recorded in July 1977 (Fig. 2.53). The pan evaporation at Kuwait Airport and Amriah meteorological station varies between 3 mm/day in January and 17.5 mm/day in July (Fig. 2.54). Potential evaporation in Bahrain varies from 2-5 mm/day in January/February, to 7-14 mm/day in June/July. Because of the moister conditions, evaporation in summer months may be a little higher, especially during the month of August (Fig. 2.55).

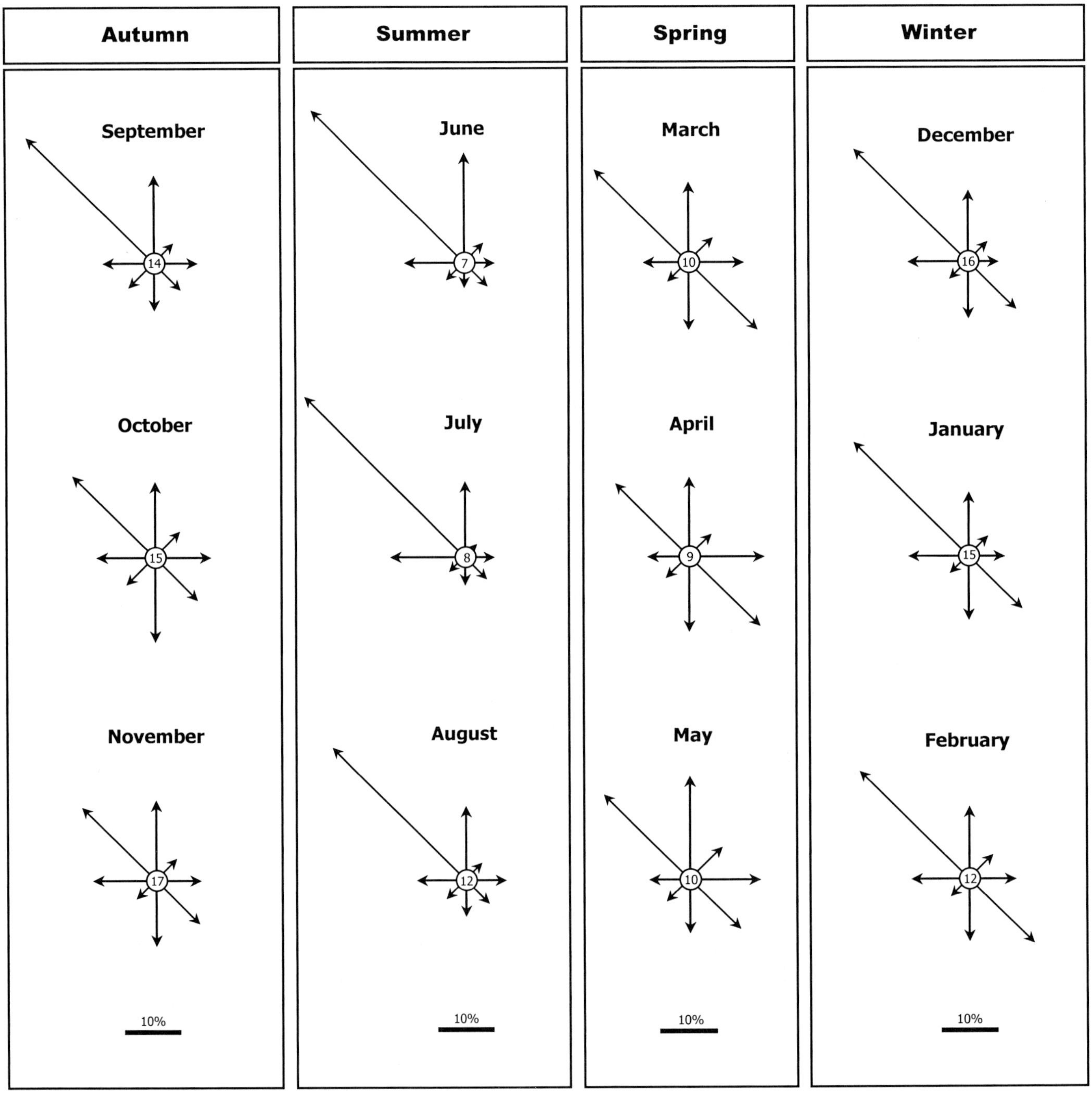

Fig. 2.39. Wind cycles in Kuwait airport during winter, spring, summer, and autumn (modified after Sharaf, 1980).

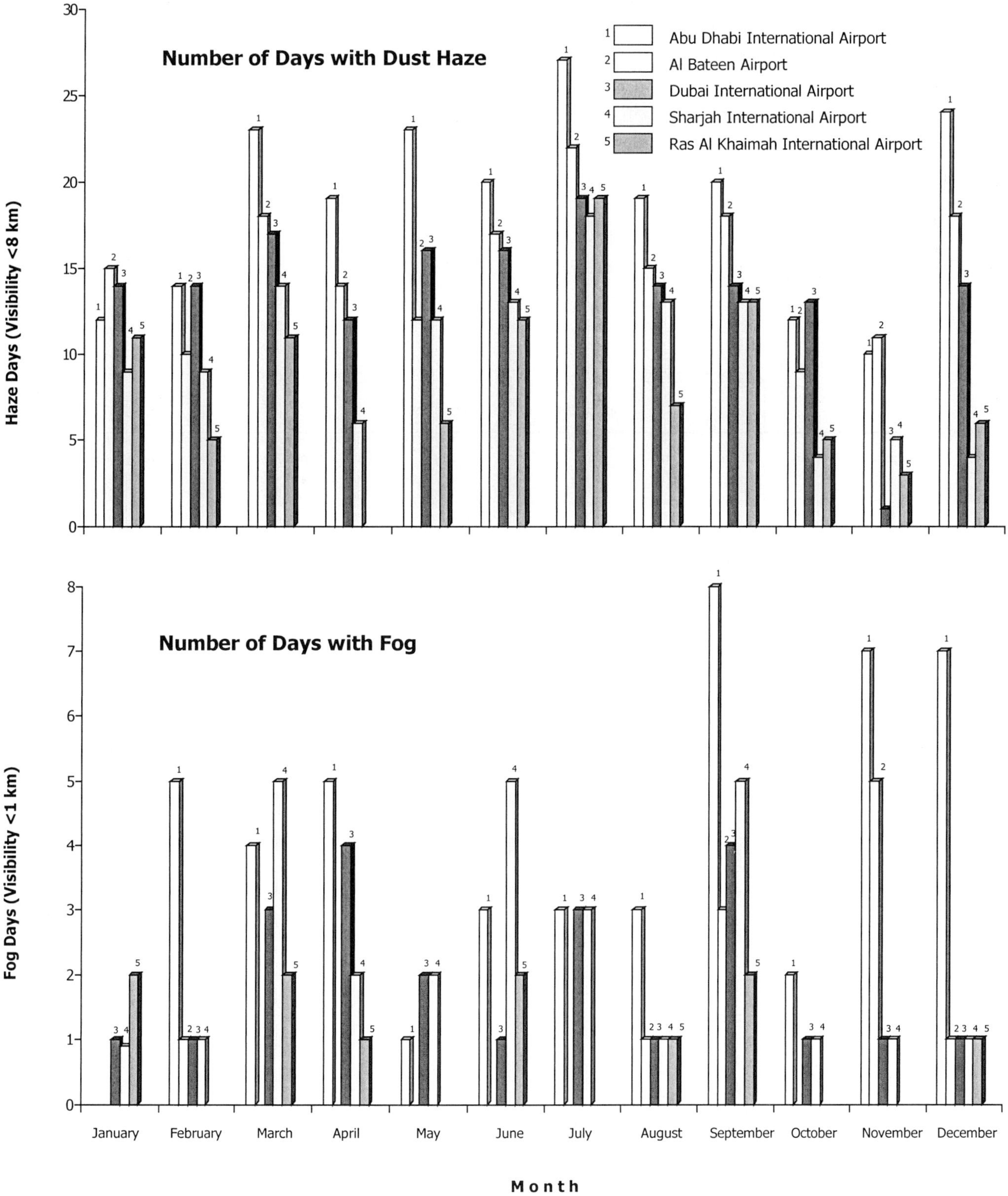

Fig. 2.40. Graphs illustrating the number of haze and fog days at the major airports in United Arab Emirates. Note the higher incidence of fog and haze in Abu Dhabi Emirate (data from UAE Department of Civil Aviation, annual report, 1993).

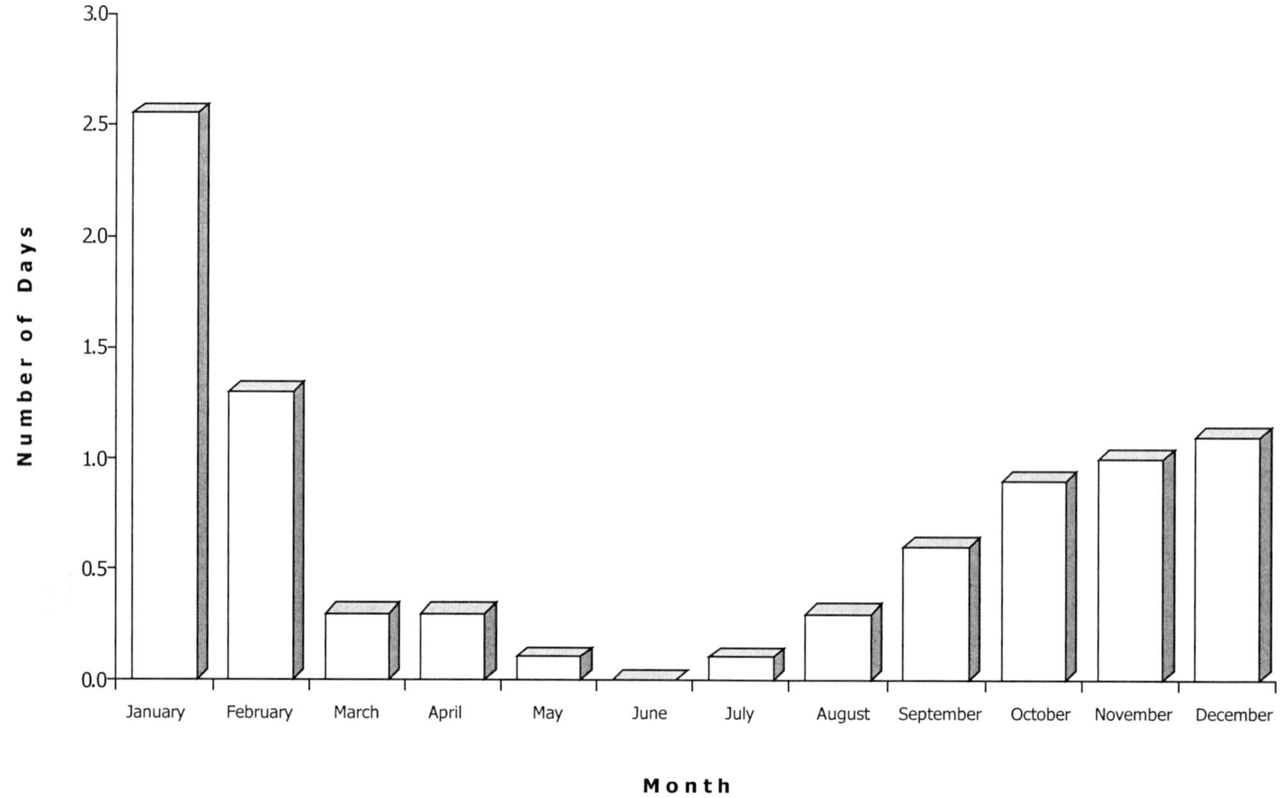

Fig. 2.41. Average monthly values of cloudy days in Kuwait (modified after Sharaf, 1980).

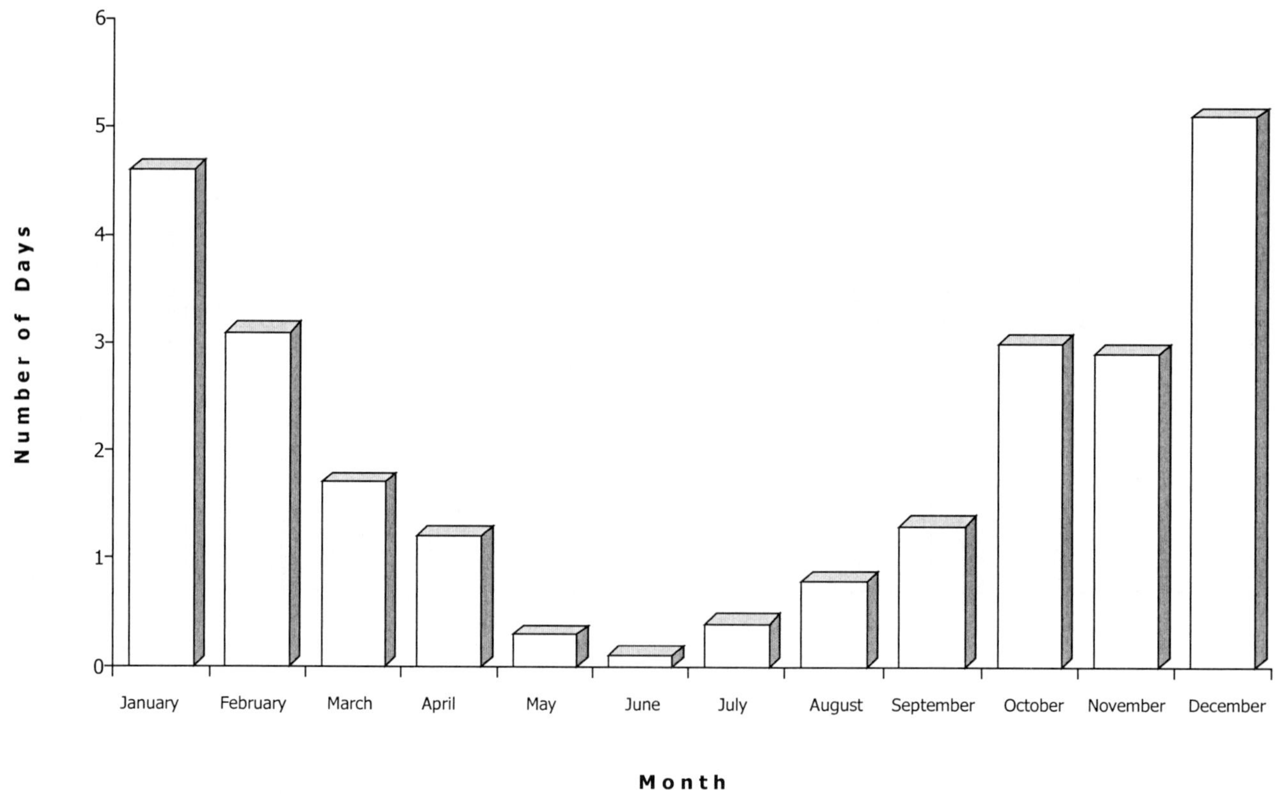

Fig. 2.42. Average monthly values of foggy days in Kuwait (modified after Sharaf, 1980).

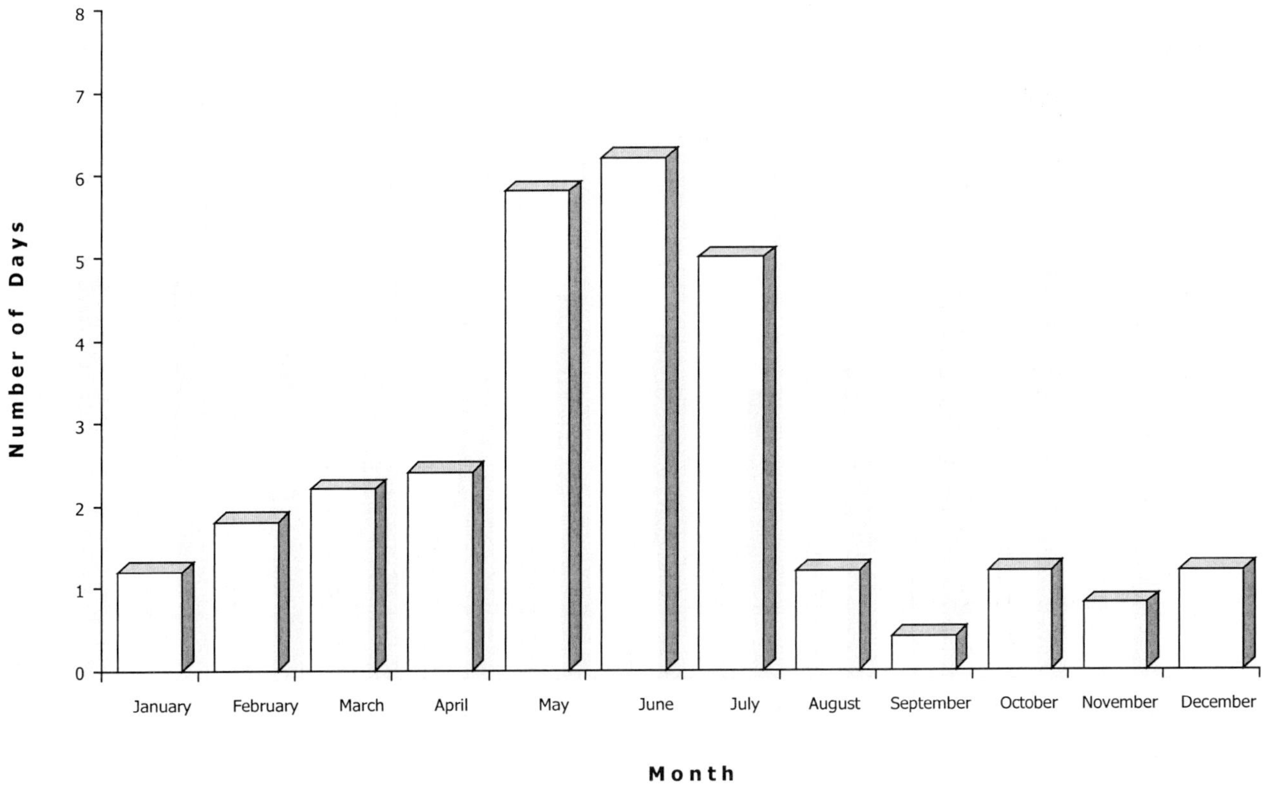

Fig. 2.43. Monthly mean of dust days in Qatar (modified after Al Majid, 1988).

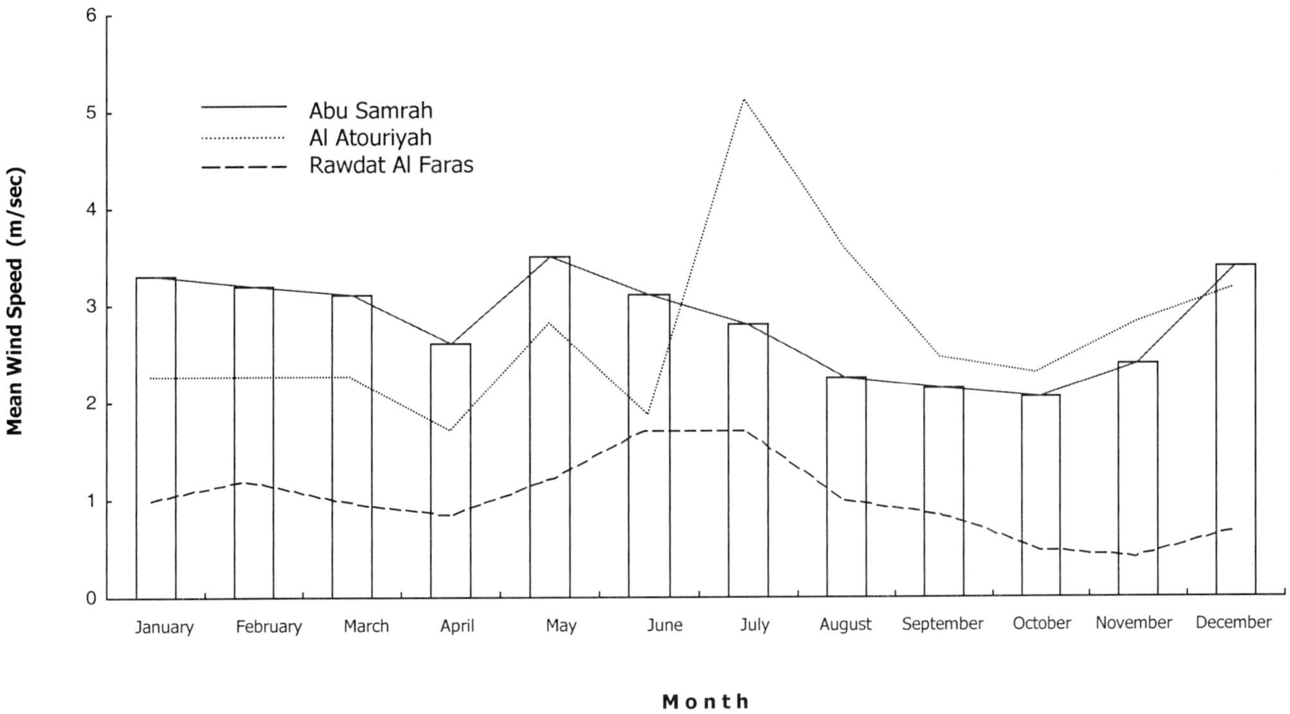

Fig. 2.44. Average monthly values of mean wind speed in Kuwait (modified after Sharaf, 1980).

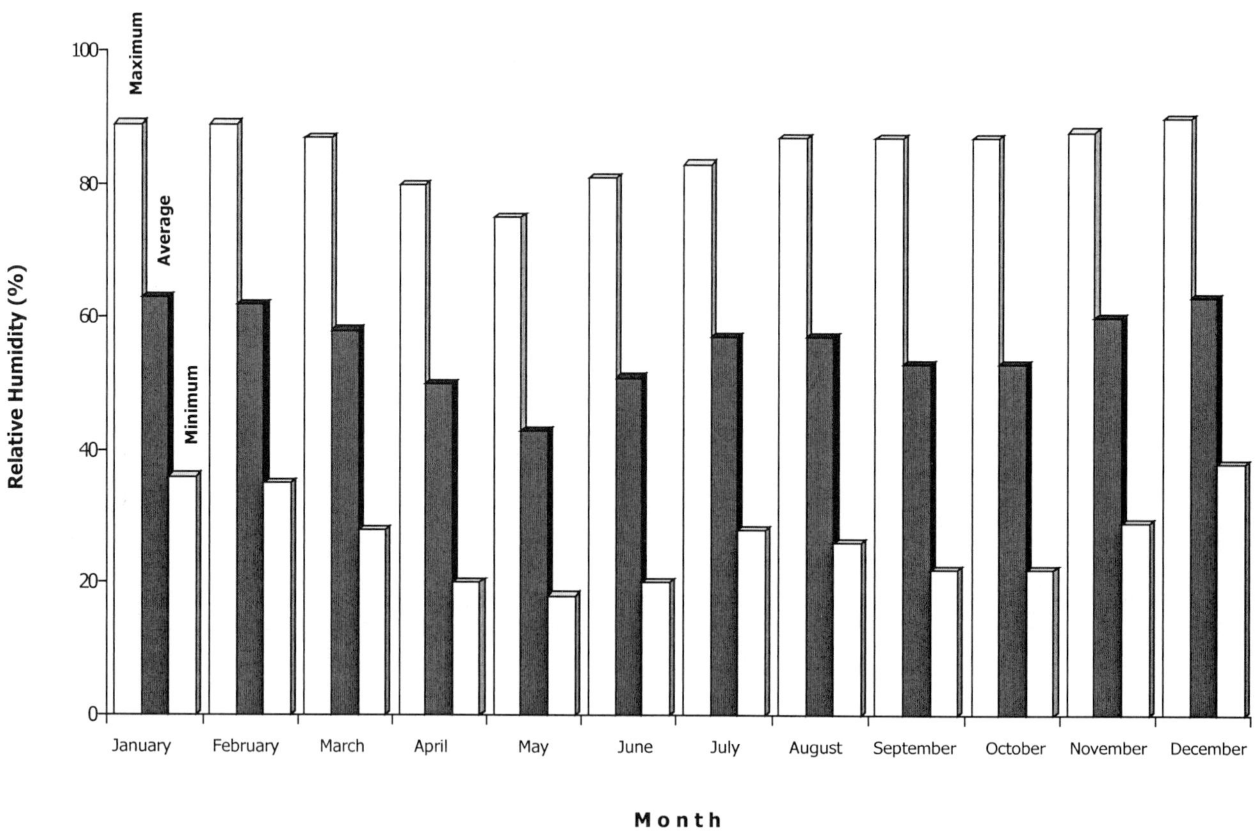

Fig. 2.45. Monthly average relative humidity (%) in the United Arab Emirates (data from UAE Ministry of Communications, Annual Report, 1996).

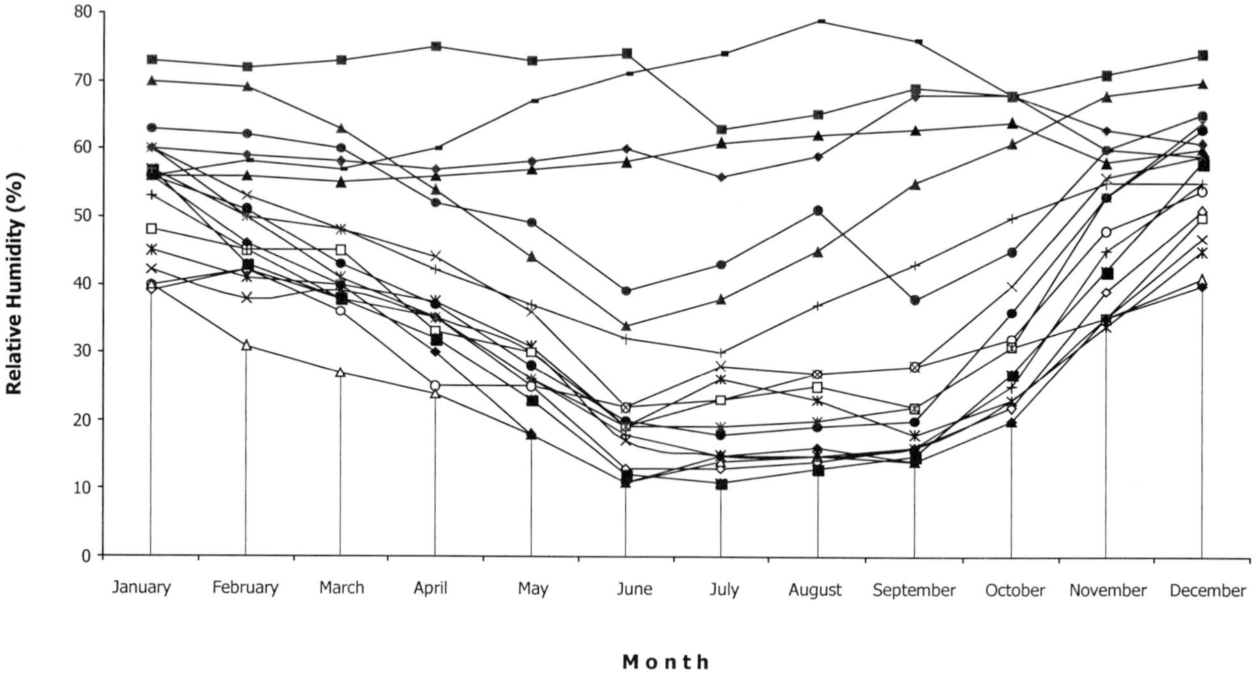

Fig. 2.46. Monthly average relative humidity (%) in twenty-one meteorological stations in Saudi Arabia for the period 1967-1987 (modified after Ahmed, 1993).

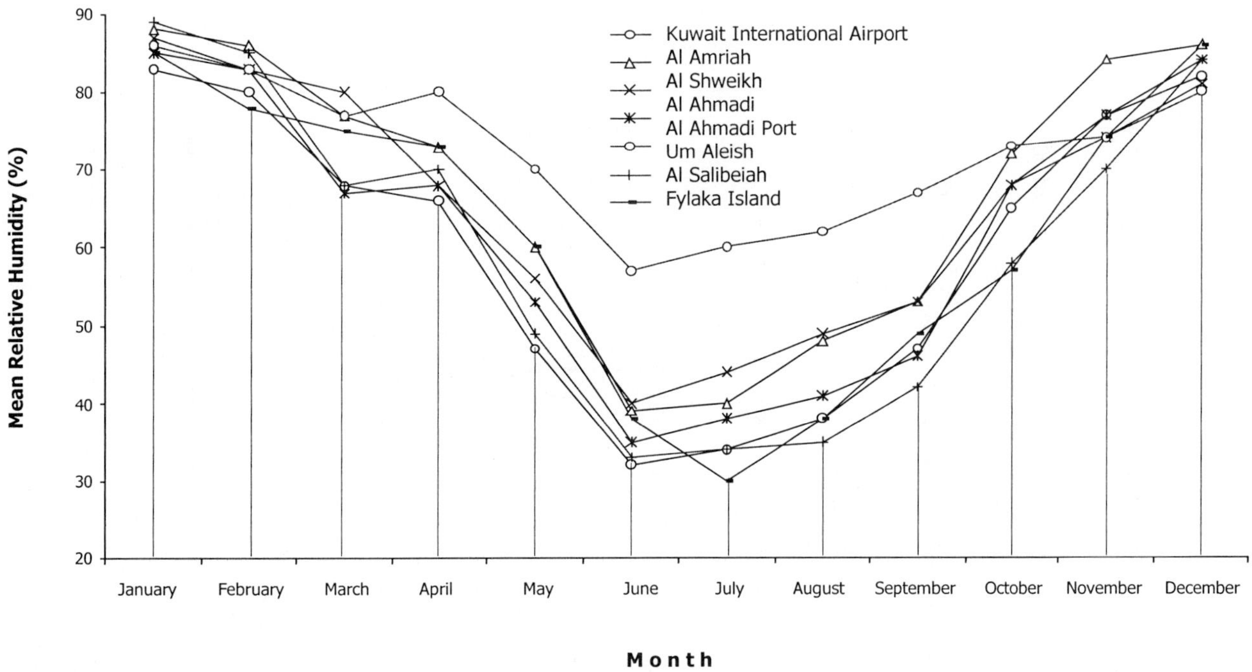

Fig. 2.47. Monthly average maximum relative humidity (%) in eight meteorological stations in Kuwait (modified after Sharaf, 1980).

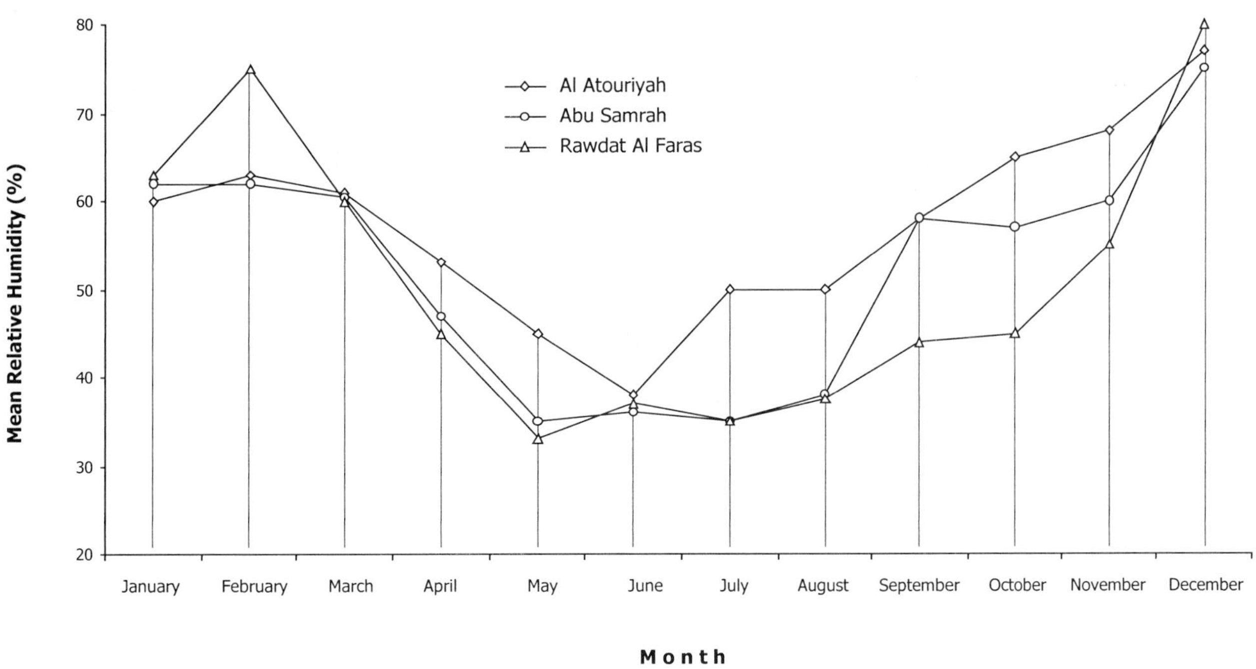

Fig. 2.48. Monthly average relative humidity (%) in three meteorological stations in Qatar for the year 1992 (compiled data from Ministry of Municipal Affairs and Agriculture, Annual Report, 1995).

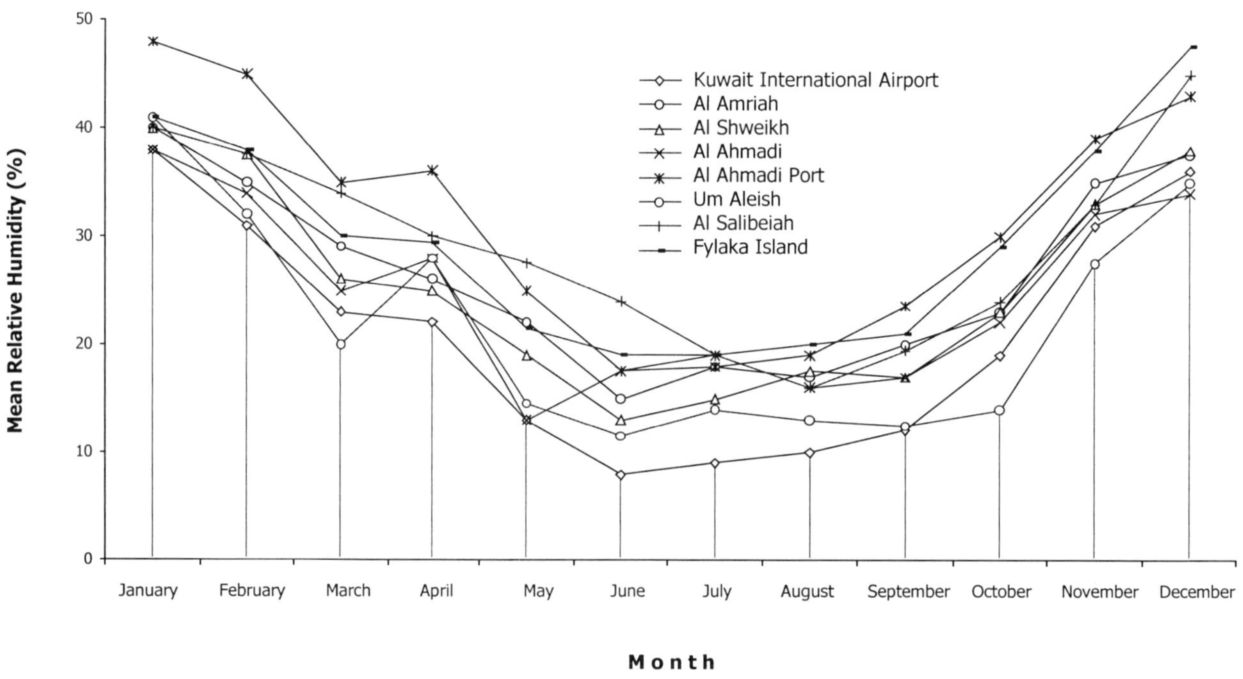

Fig. 2.49. Mean monthly relative humidity (%) in eight meteorological stations in Kuwait (modified after Sharaf, 1980).

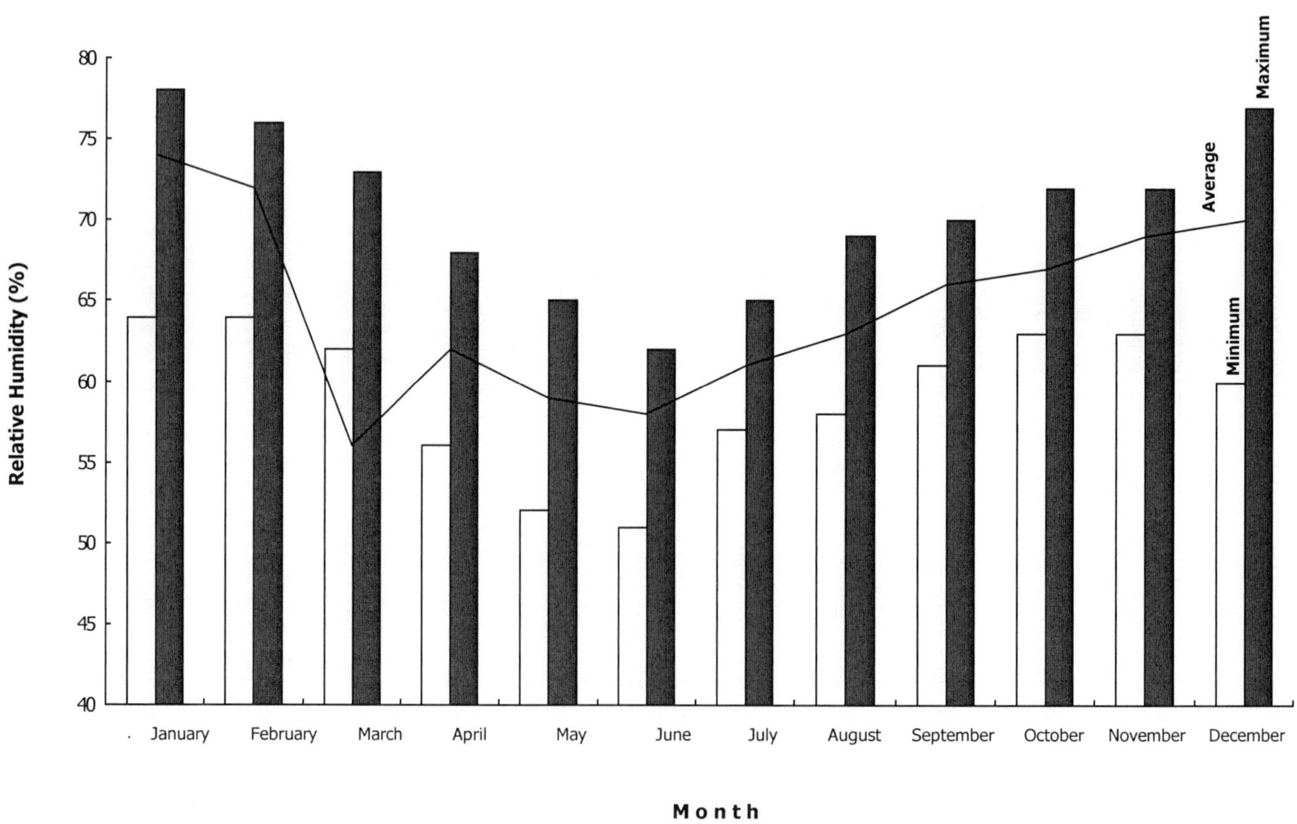

Fig. 2.50. Range and average monthly relative humidity in Bahrain from 1962-1989 (modified after Aba Hussain, 1992).

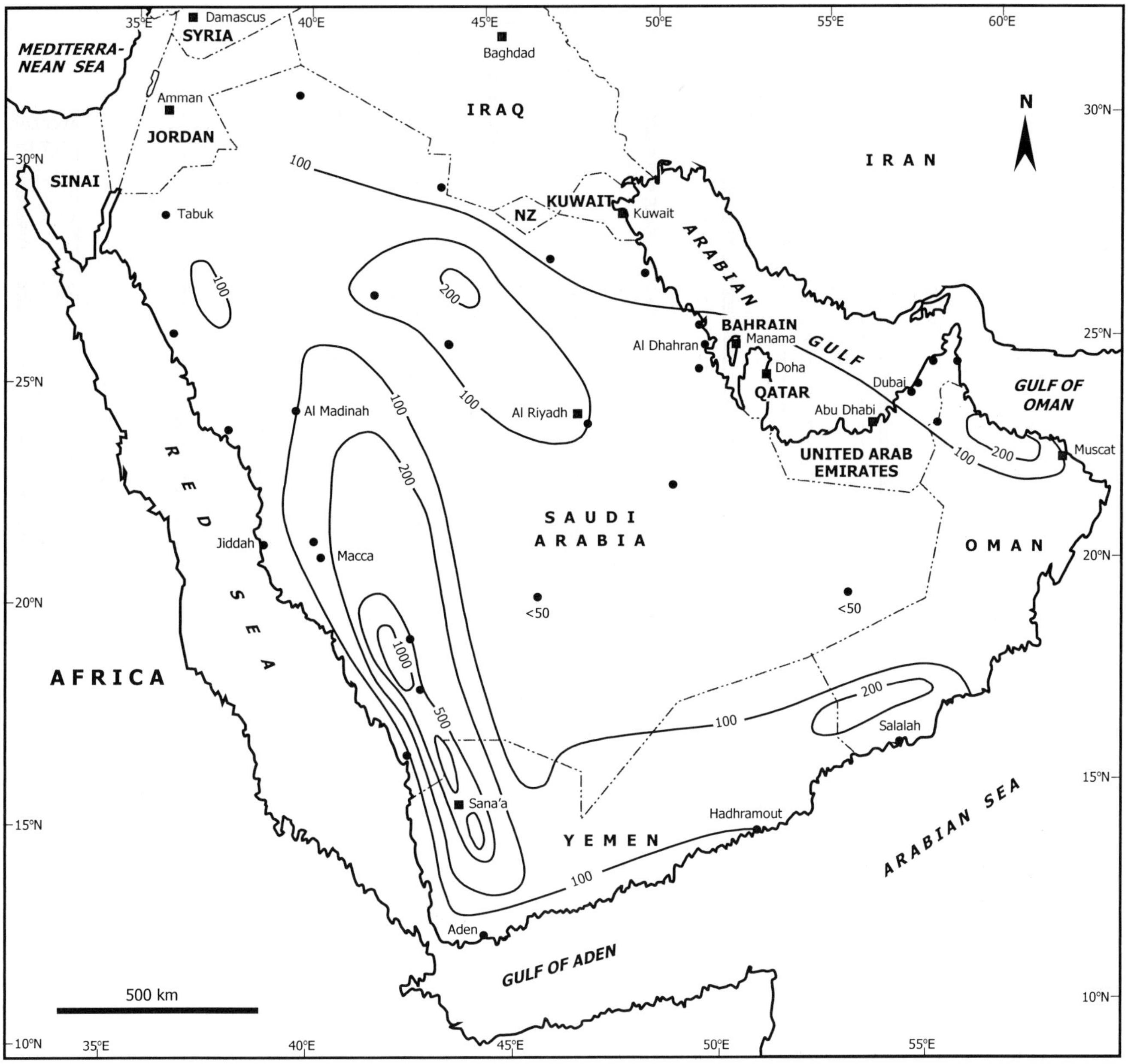

Fig. 2.51. Isohyets for average annual precipitation (mm) in the Arabian Peninsula (modified after Schyfsma, 1978; Atlas of Saudi Arabia, 2000).

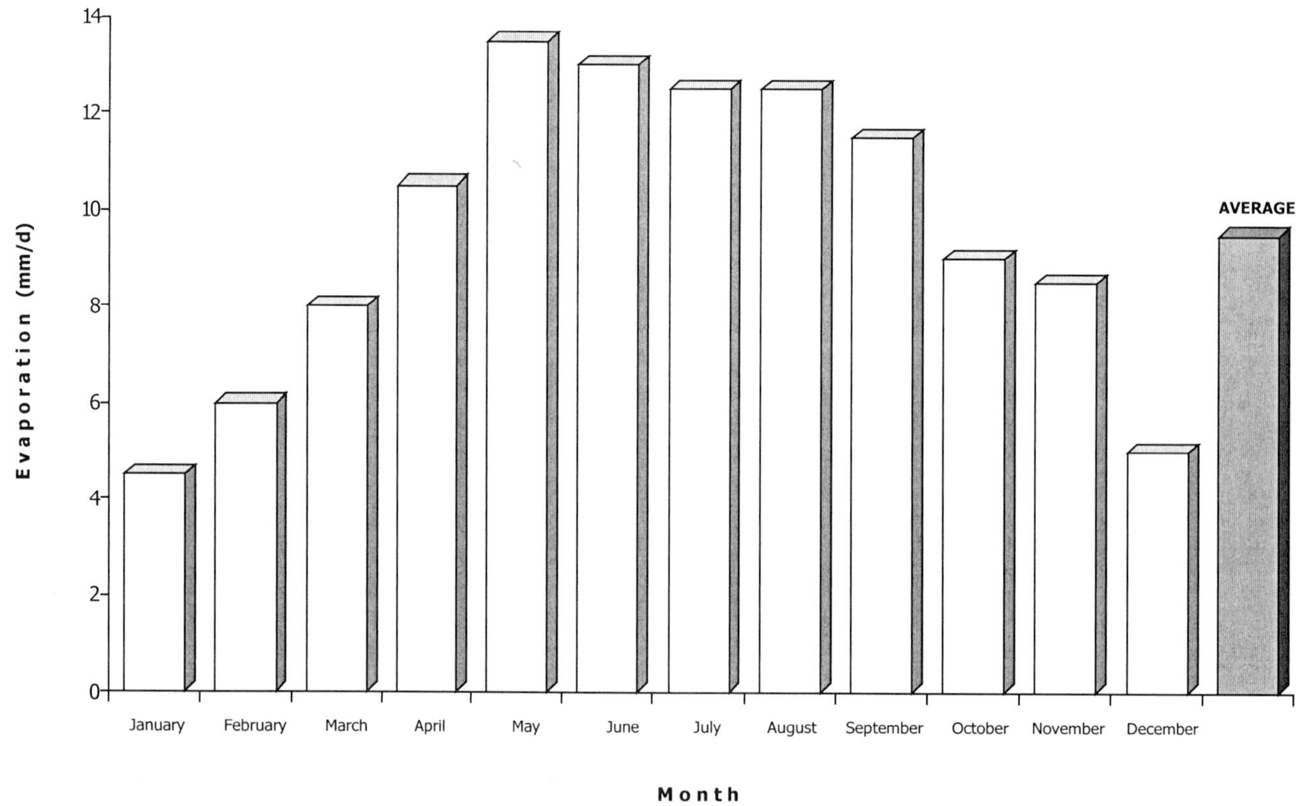

Fig. 2.52. Monthly pan evaporation in Abu Dhabi International Airport, United Arab Emirates (data from UAE Ministry of Communications, Annual Report, 1996).

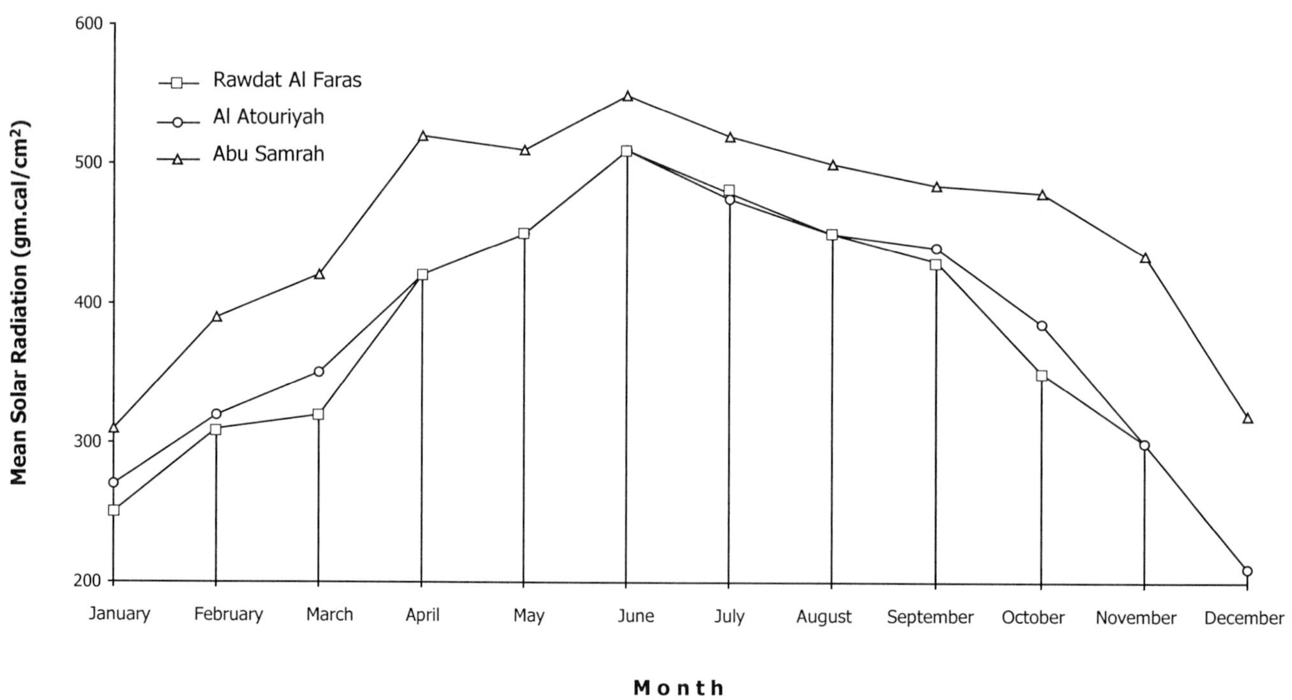

Fig. 2.53. Monthly pan evaporation in three meteorological stations in Qatar, 1992 (data from Ministry of Municipal Affairs and Agriculture, Annual Report, 1995).

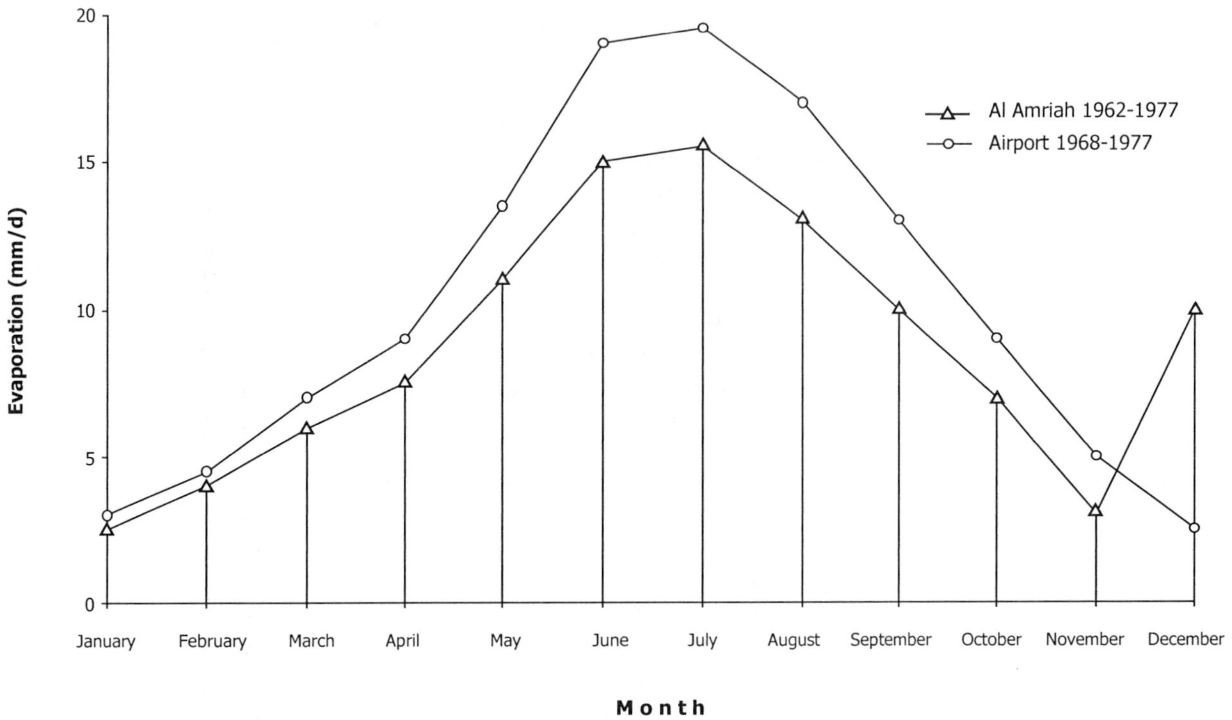

Fig. 2.54. Monthly pan evaporation in two meteorological stations in Kuwait from 1962-1977 (compiled data from Sharaf, 1980).

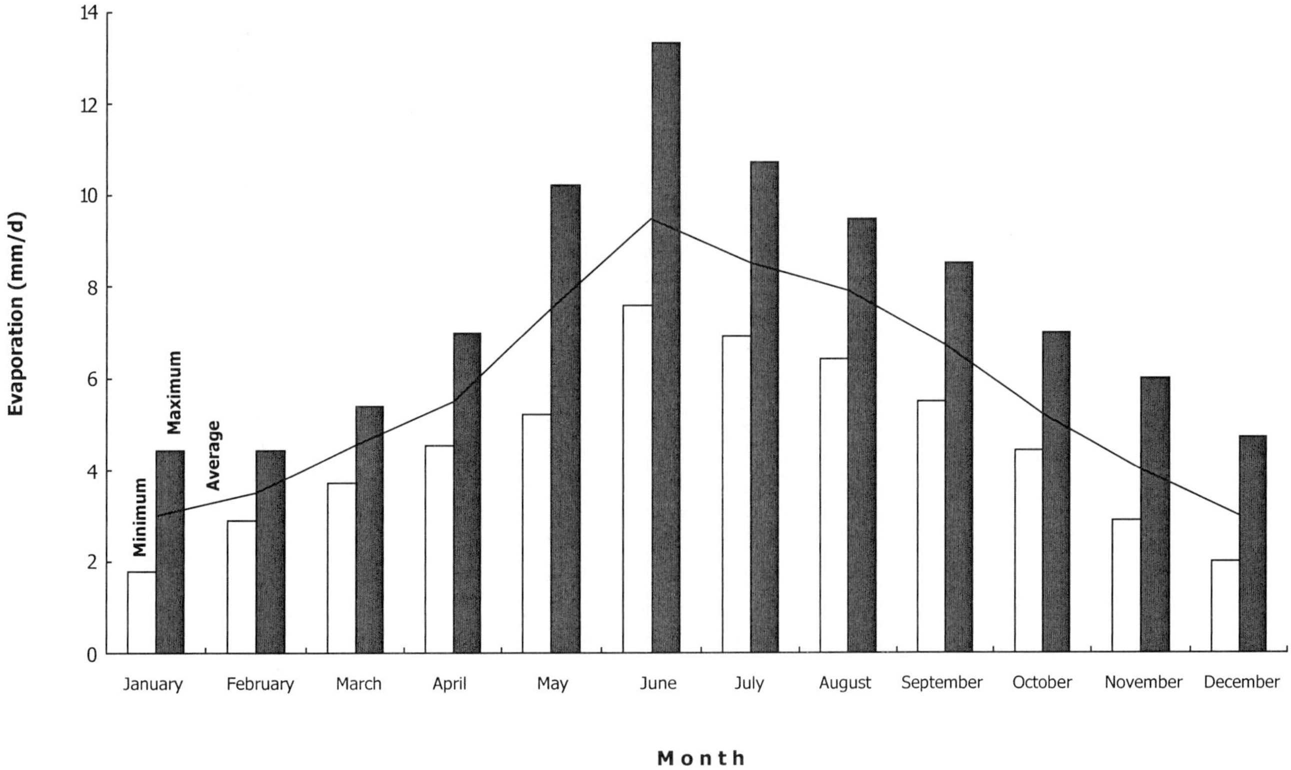

Fig. 2.55. Range and average monthly evaporation (mm/day) in Bahrain between 1983 and 1989 (modified after Aba Hussain, 1992).

Chapter 3

GEOLOGY OF THE ARABIAN PENINSULA AND GULF

INTRODUCTION

The Arabian Peninsula has an area of around 3,003,200 km^2 (1,159,500 mi^2). Its boundaries to the west, southwest and south are formed by tensional structures, the shear faults and grabens of the Gulf of Aden, the Red Sea and the Gulf of Aqaba - Dead Sea rifts. The boundaries to the north and northeast are defined by the compressional features associated with the Alpine Taurus, Zagros and Oman Mountains (Fig. 3.1). Henson (1951) and Powers et al. (1966) recognized four broad morphological

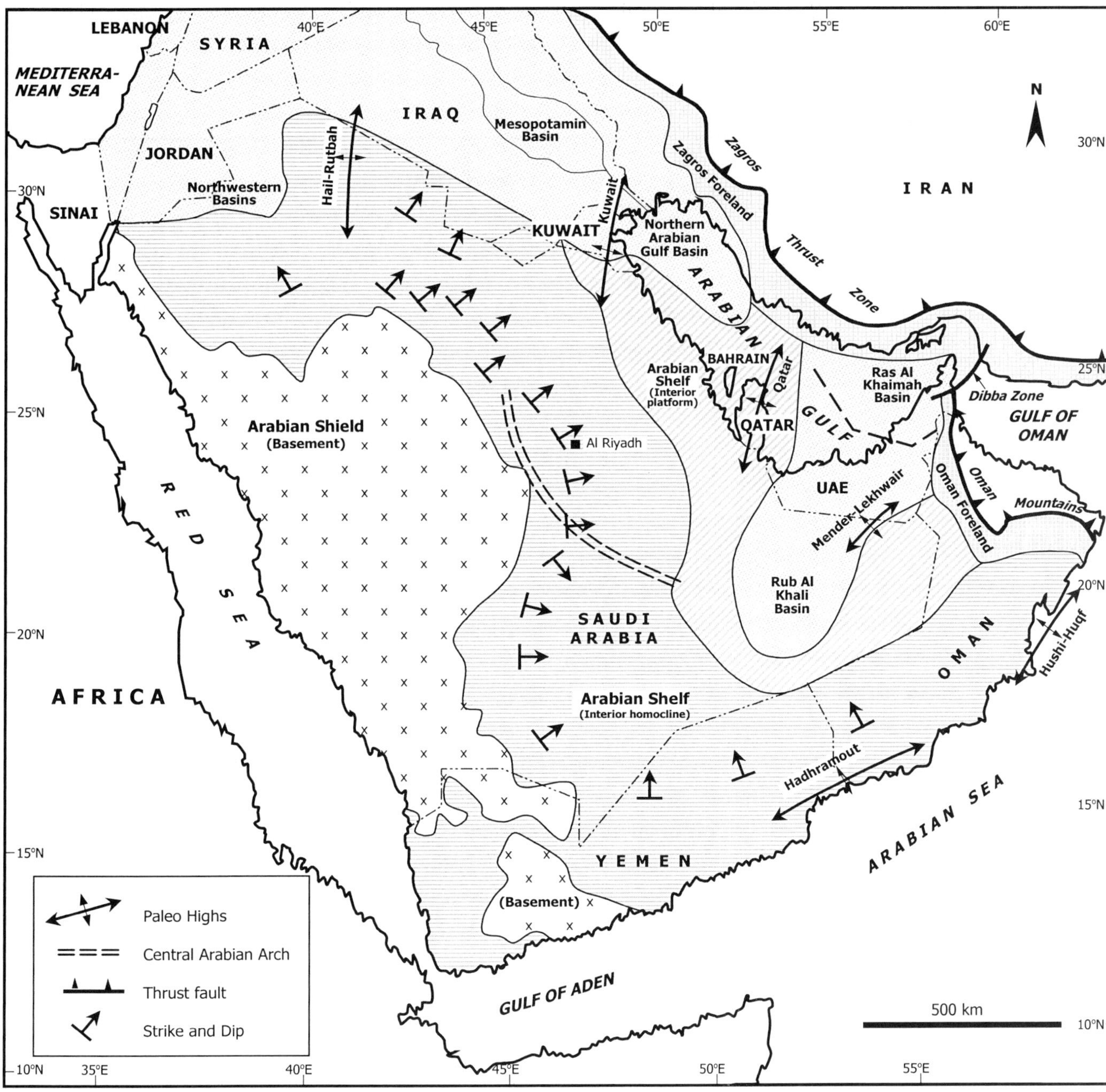

Fig. 3.1. Major tectonic elements in Arabian Peninsula (modified after Powers et al., 1966; Alsharhan and Nairn, 1986, 1997).

units, the Arabian-Nubian Shield, the interior homocline or the stable shelf, the outer or unstable shelf and the orogenic margins of the block, which include the Omani and Zagros fold belts. The geomorphological and geological characteristics of each of these units is the result of the interplay of several geological factors which in turn reflect the geological history of the area. This is considerably more complex than the simple morphological-geological units would suggest (Fig. 3.2). The Phanerozic sediments in the Arabian Peninsula reach more than 35,000 ft (about 10,670m) (Fig. 3.3), increasing gradually from the Arabian Shield toward the Arabian Gulf. The Paleozoic succession from Cambrian to mid-Permian is dominated by clastic sediments in which carbonates play a relatively minor role, whereas the late Permian to Triassic sequences consist of a mixed carbonate-clastic succession. During the Jurassic to mid-Cretaceous carbonates and evaporites were the dominant rocks whereas, following the Middle Cretaceous during the Late Cretaceous and Paleogene interval, clastics dominated the sequences. During the Neogene there was a more balanced clastic-carbonate sequence. This very broad generalized stratigraphic history however is a reflection of major structural and tectonic changes that affected the Arabian Peninsula. The aim here is to outline these changes and then focus on the progressively smaller events which make up the geology of the Arabian Peninsula and thereby integrate basin formation, the development of regional structures and relative sea-level changes into the context of regional geology.

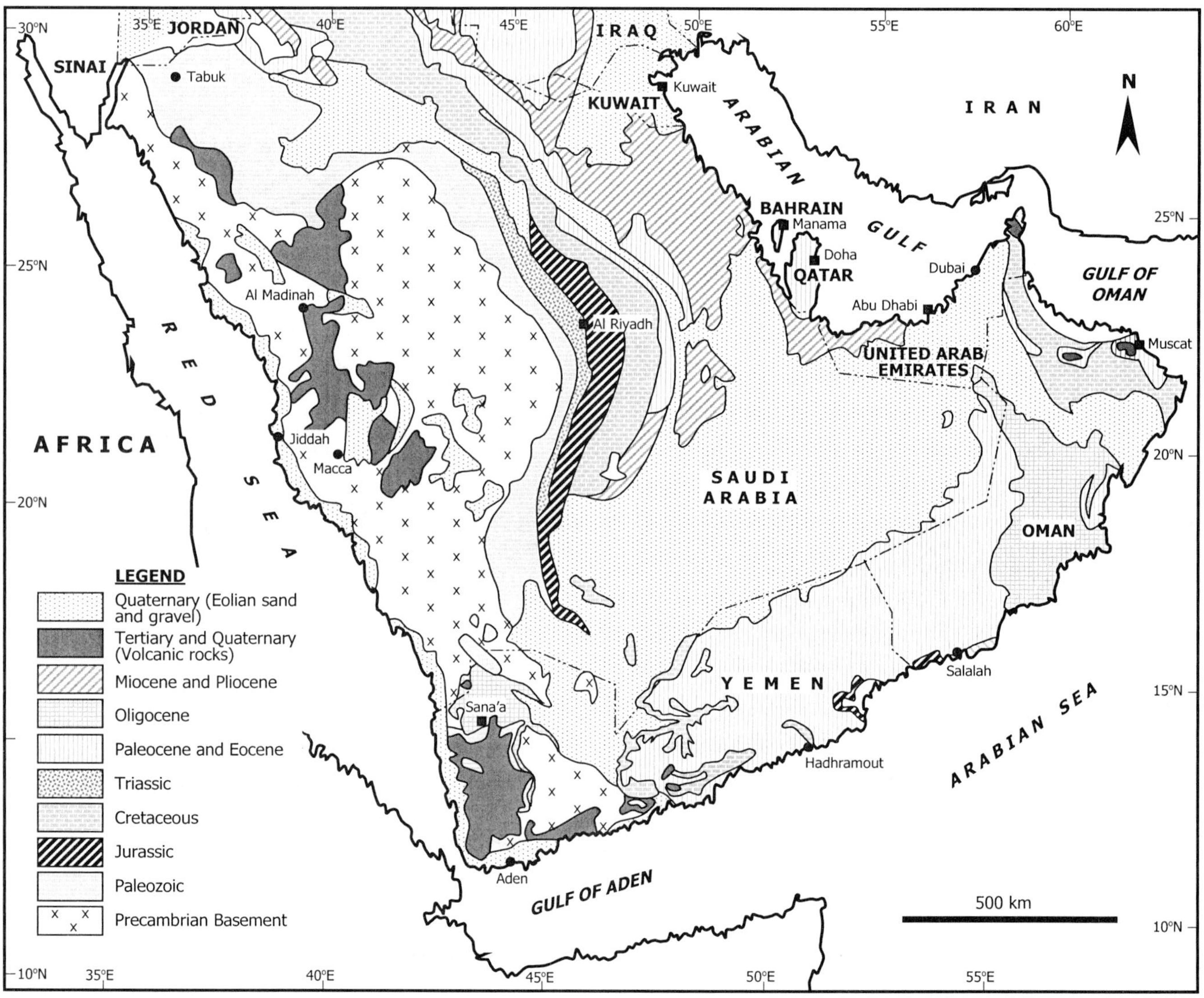

Fig. 3.2. Simplified geological map of the Arabian Peninsula (modified from Powers et al., 1966; Alsharhan and Nairn, 1997).

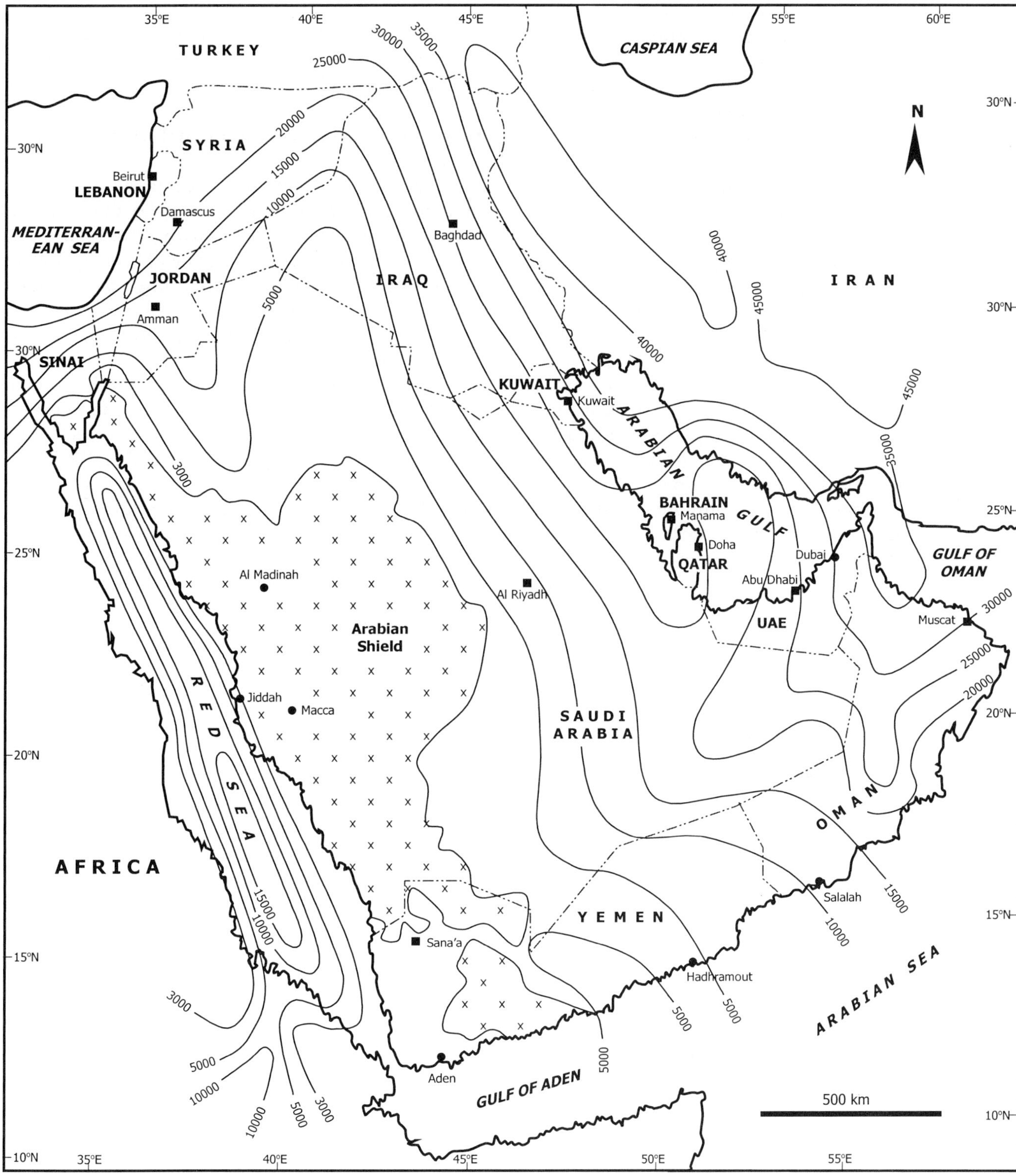

Fig. 3.3. Isopach map of the Phanerozoic sediments of the Arabian Peninsula (modified from Wilson and Peterson, 1986; Alsharhan and Nairn, 1997). Contours labeled thousand feet, and contour interval every five thousand feet.

The Succession of Tectonic Events

Phase 1. The consolidation of the Arabian Shield

The shield is an ancient polygonal mass of continental crust occupying some 770,000 km^2, about one third of the Arabian Peninsula (Fig. 3.1). It lies on the western side of the peninsula flanking the Red Sea with a relatively flat, slightly undulating surface which can be traced north, northwest and west. It dips below the thick sedimentary cover of the eastern side of the peninsula, and forms the basement of the Interior Homocline and Unstable Shelf (Fig. 3.4). Outcrops of these, Precambrian basement rocks have been described from southwest Yemen and a smaller outcrop also occurs in eastern Oman. The shield has remained as a stable mass persisting as a positive area throughout the Phanerozoic. It consists of gneiss, metasediments and volcanic rocks intruded by granitic plutons, which were peneplaned during the Late Precambrian time. These plutonic intrusions represent as much as 57% of the rocks of the shield. During the Tertiary they were, in part, covered by extensive lava flows (harrats) (Alsharhan and Nairn, 1997).

The cratonization of the shield resulted from the amalgamation of island arc volcanics and sedimentary rocks, accompanied by tectonic movement and metamorphism from 1200-520 Ma ago. The sequences are thick (9,000 to 15,000 m). The sutures formed by the sweeping together of the volcanic arcs recognized in Saudi Arabia, can be traced into the Sudan, and can probably be assigned to the Kibaran, Hijaz and Najd orogenies. The final disappearance of oceanic crust and the incorporation of the crustal sequences in Gondwana (end of the first megacycle), ended during the Najd orogeny and was marked by the faulting activities of the Najid Fault System, where granite is both cut by the faults or in some cases intrudes into the faults (Alsharhan and Nairn, 1997). The isotope studies of Fleck et al. (1980) suggest that the continental lithosphere evolved before 680-900 Ma. Unfortunately little is known of the basement geology of the Interior Homocline since it has yet to be drilled and consequently its relationship to the shelf remains open at least for the present. The Arabian Shield may dip gently under the homocline as generally assumed, but there could be a shear fault or suture between the two.

As a result of the consolidation, a source of unmetamorphosed sediments from Gondwana flooded the Infracambrian and early Paleozoic of the shield. In South Oman these sediments included clastic, carbonate and evaporite deposits which can be related to the Hormuz evaporites of the Arabian Gulf and Oman salt basins and to the deposits of the Salt Range in Pakistan.

Phase 2. The phase of tectonic stability

The tectonic events which occurred around the Arabian plate margins are related to the rifting of Gondwana and later collision with Asia. Figure (3.5) shows reconstruction of the plates (African, Indian, Iranian and Eurasian) surrounding the Arabian plate. The timing of the major tectonic events that affected Arabian plate during the Mesozoic-Tertiary time are listed in Table 3.1.

Following the accretion of the Arabo-Nubian massif to the margin of Gondwana, a long period of relative stability supervened during which time sediments from Gondwana spread over the shield during the Infracambrian and the early Paleozoic. The Infracambrian is marked by the development of braided stream and flood plain deposits over Saudi Arabia and supratidal to intertidal and lagoonal deposits in a trough in Oman and the Arabian Gulf forming major salt basins. In western Arabia, Cambrian-Ordovician (Sauk sequence) clastic sheets spread widely over an unconformable surface (the infra-Tassilian unconformity of French authors). At present these are found in the Wajid Basin south of the Central Arabian Arch and the Tabuk basin lying north of the arch. Since the sediment transport directions from both basins are consistently from the south, it is suggested that they may have formed a single basin which was subsequently divided by the uplift of the Central Arabian Arch. These sediments are strikingly similar to the early Paleozoic clastics of Algeria separated into two areas following the Hercynian uplift of the Hoggar Massif. If this inference holds true, then it dates a major structural change with the separation of the Rub Al Khali/Ras Al Khaimah basin to the south from the Zagros Basin to the north. Beyond the limits of the epicontinental shield area around the Gondwana margin Turkey, Syria, northern Iraq and southwest Iran, a thick wedge of marine sediments accumulated in intrashelf basins (geosynclinal deposits in the older terminology).

This early phase of tectonic stability marks a transition from the predominantly clastic depositional events of the early Paleozoic to a phase marked by upwarping and erosion. Close to the northern border of Gondwana, a series of east-west striking highs plunging eastwards developed in Algeria, Libya and Egypt as well as in the Arabian Peninsula. Enormous thicknesses of lower Paleozoic

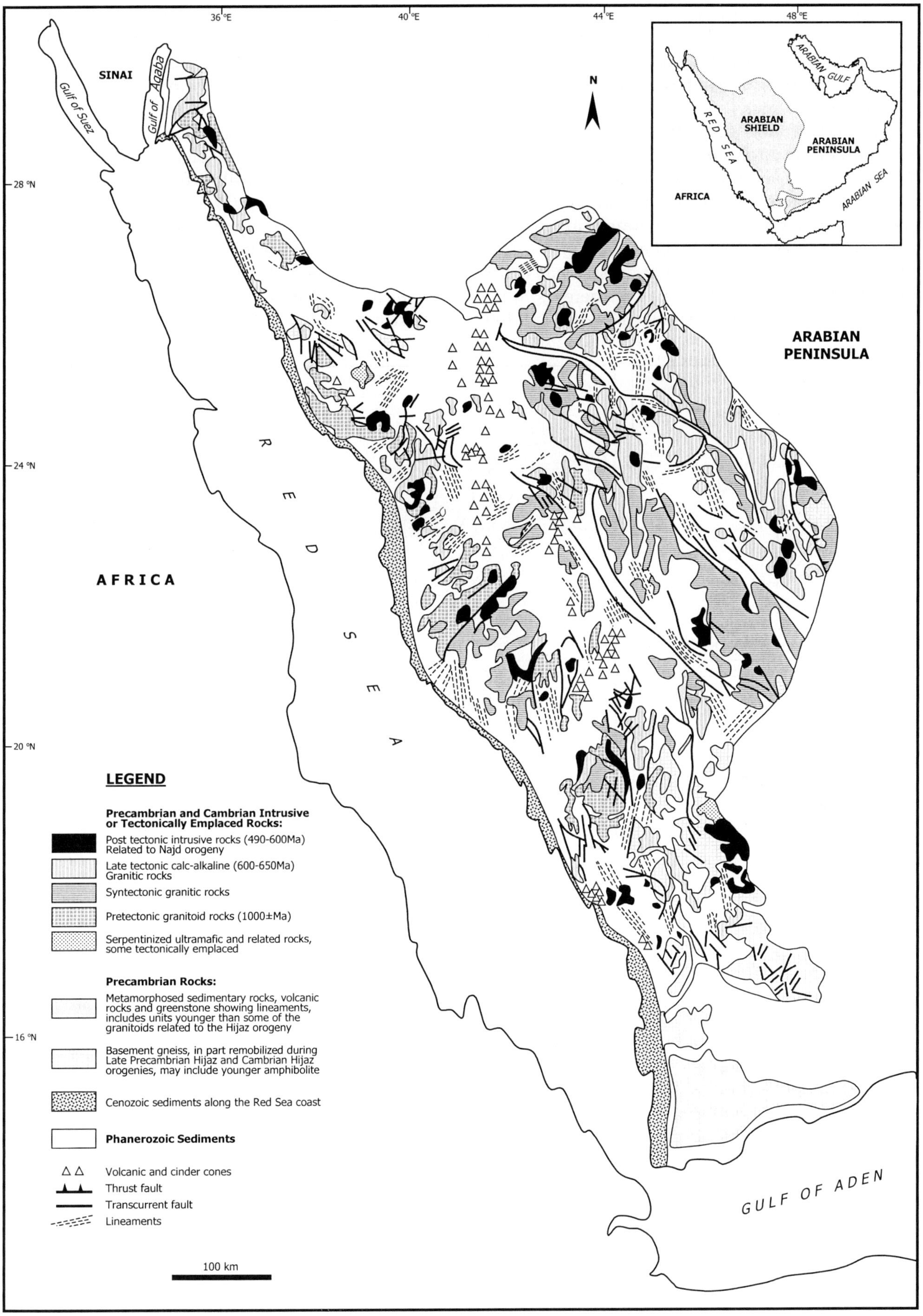

Fig. 3.4. Tectonic sketch map of the Arabian Shield, showing terranes and suture zones. Northwest-trending solid lines indicate fractures of the Najd Fault System (modified from Stoesser and Camp, 1985; Alsharhan and Nairn, 1997).

rocks were stripped from the upwarps such that, for example, the Late Permian may rest directly upon Precambrian over the Central Arabian Arch. Further to the east, drilling in Qatar revealed that Silurian and Devonian rocks are still present over the arch, and seismic data from the Rub Al Khali basin south of the Arch showed that thick sequences still exist. These are presumed to be Devonian or older, but they have not been drilled as yet except in Oman. Carboniferous rocks are found in most of Arabia north of the Central Arabian Arch.

While the timing of uplift and erosion has led it to be called the Hercynian event, there is little evidence of significant tectonic activity in the Arabian Peninsula, but in contrast it has dramatic effects on the geology of Asia. The important orogenic and collisional events in Laurasia, such as the closing of the Ural seaway and the collision of Baltica and Laurentia occurred at this time. The major effects in Arabia are the north-south trending, basement-involved horst blocks formed in central and eastern Arabia with which major oil and gas reserves are associated (Wender et al., 1998).

Two regional components may be recognized in the basement structure of the Arabian block. The first is represented by a north-south trend in the western Arabian Gulf region which changes into a northeast-southwest in the southeastern Arabian

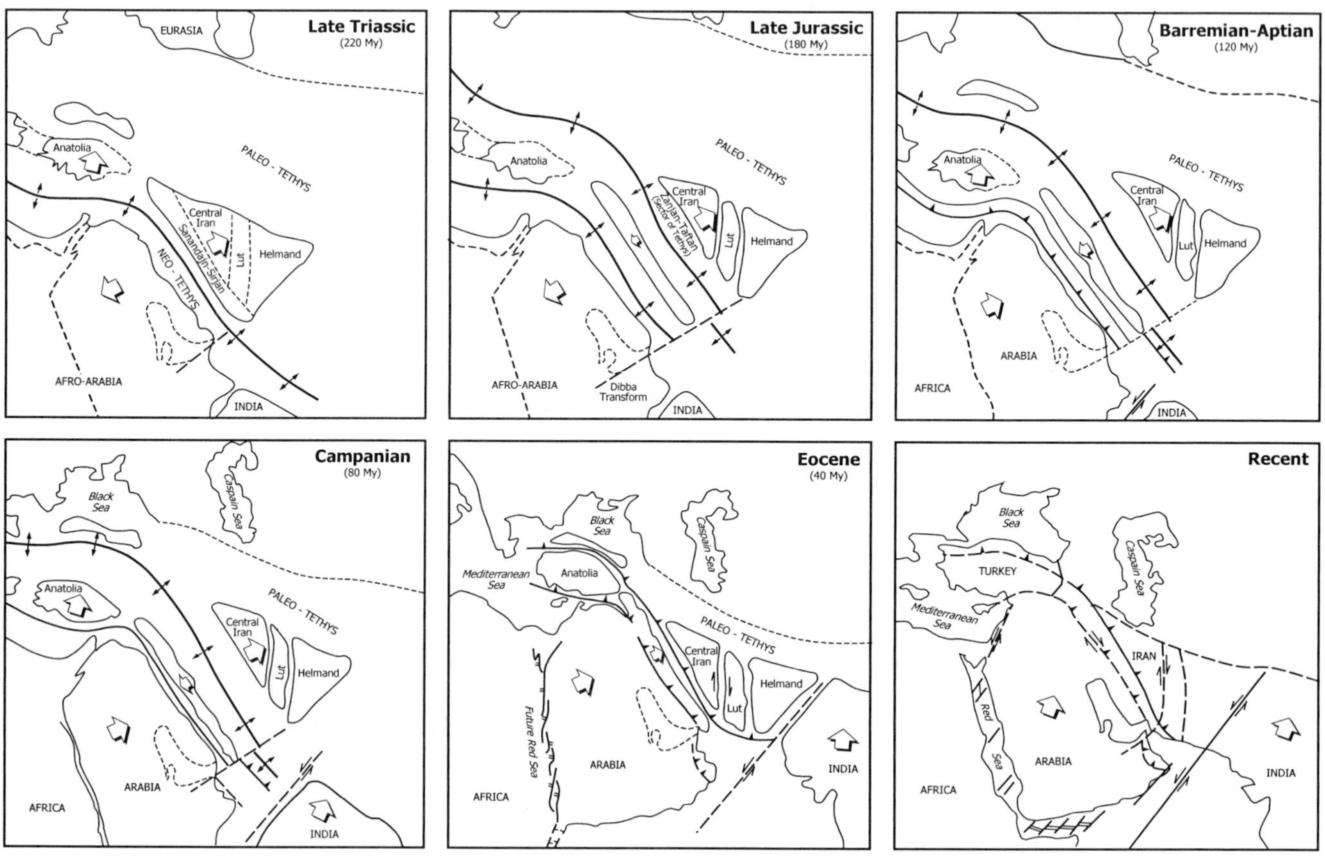

Fig. 3.5. Continental drift of the Arabian plate and adjacent plates in Late-Triassic to present (modified from Glennie, 1992).

Table 3.1. List of the major tectonic events that affected the Arabian Plate since the Late Paleozoic (after Grabowski and Norton, 1995).

Series/Stage	Tectonic Event	Effect on Arabian Plate
Permian	Hercynian orogeny ends	Quiescent period, rifting of Central Iranian blocks from Arabian Plate
Liassic	Break-up of Atlantic begins	Unconformity with the major erosion on margins of Arabian Plate
Oxfordian	Rifting of India from Africa begins	Rifting in south, plate titled down to north
Berriasian	Rifting of eastern Mediterranean begins	Uplift of western margin of Arabian Plate, Hail-Jawf arch raised in the north
Aptian	Opening of Mediterranean begins	Sub-basin form on Arabian Platform
Turonian	Ophiolitic obduction in eastern Arabia	Fault reactivation, warping of Arabian Plate with erosion and onlap
Eocene	Collision on northern Arabian margin	Mesopotamian foredeep of Zagros Foldbelt formed
Miocene	Collision begins on eastern Arabian margin	Zagros-Oman Foldbelt formed, clastic pile formed

Gulf region. These structural components are thought to have been initiated as early as the Middle Jurassic. The second, a northwest-southeast structural grain is superimposed on these trends in the crescentic structure of the Oman foredeep and orogen and in the Zagros fold belt. The northwest to southeast trending Zubair, west Qurna and Rumaila fields in southwestern Iraq indicate that the late Miocene to Pliocene Zagros tectonic influence extended westward well onto the Arabian platform. Similar relationships may have existed for structures in the west side of the Ras Al Khaimah Foredeep flanking the Musandam Peninsula in northern Oman.

The Infracambrian Hormuz salt supplied the structural impetus in the Arabian Gulf region that resulted in the development of many producing oil fields (Fig. 3.6). Most productive structures on the Arabian platform have been interpreted as growth structures which began movement in the Jurassic and accelerated in post-Cretaceous times until the middle Eocene. Examples include Ghawar in Saudi Arabia, Dukhan in Qatar and Burgan in Kuwait (Alsharhan and Kendall, 1986). The deep oil tests at Burgan show at least part of that structure moved

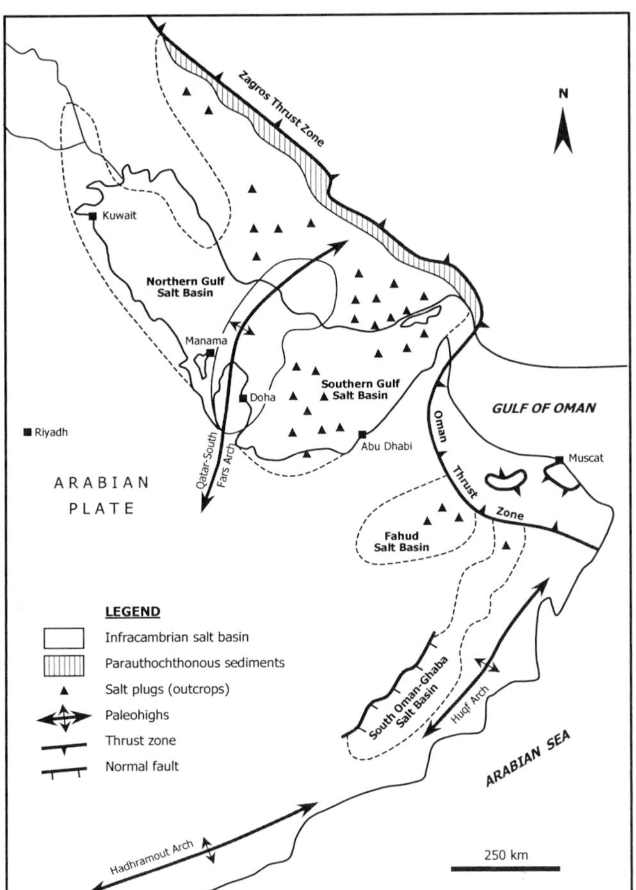

Fig. 3.6. Distribution of Infracambrian (Hormoz) salt basins in the Arabian Gulf region (modified from Murris, 1981; Alsharhan and Nairn,1997).

during the Triassic. Ibrahim et al., (1981) have shown that the first movement of the Harmaliyah structure, (southeast of Ghawar Field) in Saudi Arabia began during the Turonian (Late Cretaceous).

The other structures of the stable craton are the gently deformed intraplatform basins that were developed within the carbonate shelf. These include the kerogen-rich Late Jurassic Hanifa and Sargelu formations of the southwestern and northern parts of the Arabian Gulf region, respectively, and the Lower Cretaceous of the United Arab Emirates. Organic-rich source rocks have accumulated in these basins and charged the prolific Jurassic and Lower Cretaceous reservoirs of the region. The basins were probably formed by a gentle epeirogenic downwarp of the craton in response to locally accelerated rates of subsidence related to crustal cooling. Paleozoic sediments in these cratonal sags were probably deposited in water depths only a few tens (exceptionally hundreds) of meters deeper than the adjacent carbonate platforms. In basins or portions of basins where bottom waters were anoxic, some of the organic material survived in the sediment to become kerogen. The prolific oil fields of the southwestern Arabian Gulf resulted from having Jurassic basinal source beds, shelf carbonate reservoirs, juxtaposed with evaporite seals and convergent timing of migration.

Phase 3 - Paleotethys, Neotethys and the break-up of Gondwana

The first steps in the break-up of Gondwana was the development of tensional related basins and highs. The central Iranian segment separated from the Arabian plate along the line of the present Zagros Shear Zone and began drifting eastwards. The split can be traced westwards into the Mediterranean sea, its position marked by the development of extensive evaporitic facies, as for example the late Triassic-early Jurassic facies in both Algeria and Morocco. The closure of Paleotethys occurred as the Central Iranian segment collided with the Turan block north of the Kopet Dagh (Fig. 3.5). In the wake of the drifting fragment a small ocean basin, the Neotethys, opened up. In Oman this is referred to as the Hawasina Ocean by Glennie et al. (1974). The deepwater sediments accumulated in this basin are associated with the Haybi volcanics and pelagic carbonates suggesting the existence of oceanic islands and shallow carbonate platforms within the basin.

The change is also marked by a change in the sedimentation pattern with the cyclic development of extensive shelf limestones (lagoonal and tidal flats) grading to carbonate ramp and platforms in post-Toarcian time (Murris, 1981; Alsharhan and

Table 3.2. Time constraints and nature of crust in the evolution of the Red Sea, Gulf of Aden and Afar depressions (from Behre, 1986; Alsharhan and Nairn, 1997).

Area	Age of initiation of rifting	Initiation of Seafloor spreading	Name of crust
Gulf of Suez	a. Pre-Carboniferous b. after late Eocene (37-40 Ma)	No development of spreading axis	Underlain by continental crust
Gulf of Aqba	a. Early Miocene b. Post-early Miocene	No development of spreading axis	Continental crust
Red Sea	a. Late Cretaceous b. Late Oligocene (30 Ma) c. Late Oligocene (22.5 Ma)	Late Miocene (9 Ma)	In the northern part of the Red Sea: a. subdivided and attenuated continental crust b. subdivided and attenuated continental crust c. oceanic crust consisting of basalts interlayered evaporites In the central part of the Red Sea: a. oceanic crust from coast to coast b. limited oceanic crust
Gulf of Aden	a. Early Mesozoic b. Early Oligocene (30 Ma)	a. Early Oligocene (30 Ma) b. Late Oligocene (23.5 Ma) c. Late Miocene (10 Ma)	a. oceanic crust from coast to coast b. limited oceanic crust c. attenuation of continental crust below the coastal plain
Afar	a. Late Mesozoic (?) b .Middle-Late Eocene (41-34 Ma) c. Late Oligocene (ca. 25 Ma) d. Middle Miocene (9-14 Ma)	Development of axial volcanic range 1-2 Ma	Continental to transitional crust, except in the Eritaria spreading axis.

Nairn, 1997). The Qatar- South Fars Arch acted as a positive structure and all units thicken east of the arch. To the west and closer to the shield where the epicontinental sea shallowed, marginal clastic facies developed. The closure of Neotethys occurred during the Cenozoic. Several phases can be recognized in the subduction of the Afro-Arabian plate below the Eurasian plate (to which was now attached the central Iranian segment which closed off Paleotethys). The first phase is marked by the Late Cretaceous (80-90 Ma) emplacement of ophiolites in Oman, Iran (Kermanshah to Neyriz) and southeast Turkey associated with the collision and partial subduction of the eastern Arabian plate margin (Alavi, 1994). Traditionally the limit of the Zagros is set at the Sanandaj-Sirjan zone, however the Arabian plate continues further to the east to the line of the Zagros suture. The first deformed package dates from the middle to late Eocene collision of the Eurasian plate with the Arabian platform. The allochthon, as a package, was thrust over the autochthon during a later, post middle-Miocene convergence. The final phase was one of epeirogenic uplift of the central and southern part of the folded mountain belt.

These movements are synchronous with the Late Cretaceous-Paleogene uplift and doming in the Red Sea -Gulf of Aden and Arabian Sea. The uplift and rifting in the Gulf of Aden and Red Sea culminated near the end of the Eocene. Reactivation of the Hadhramout Arch is dated as early as Paleocene with the arch and the Rub Al Khali depression to the north both reaching their present form by the end of Eocene. Thus the separation of Arabia from formerly Somalia was the result of rifting of what had the been a local depression beginning in late Eocene. It was accompanied by widespread volcanism. Table (3.2) summarizes the rifting events in the Gulf of Aden - Red Sea region.

The rifting through the Gulf of Tadjura and Afar links up with Red Sea - Dead Sea rifts which end up against the Taurus Crush Zone in the northern part of the Arabian Plate.

The final structural phase to affect the eastern Arabian block was the Zagros orogenic phase of Late Miocene to Pliocene age. In northern Oman this involved thrusting in perhaps a similar manner to that of the folded Zagros Mountains. In the central and southern Oman parts of the orogen, this involved additional vertical epeirogenic uplift of the folded mountain belt with the concomitant development of folded and thrust gravity features to the west of the mountain belt in the present-day foreland. Latest Tertiary to Holocene events included the deposition of extensive alluvial fans of debris derived from the uplifted mountainous areas along the west side of the Oman orogen and the flooding of the Arabian Gulf basin during the late Pliocene or Pleistocene. Continued tectonic activity is manifested in the Makran coast of Iran and the northwestern part of the Gulf of Oman where relatively Recent marine sediments are being subducted and deformed under the fault structures rimming the coast.

Arches/Paleohighs and Basins/Depressions

East of the Arabian Shield and within the limits set by the Alpine orogenic belts (the Taurus-Zagros-

Oman Mountains), lies the Interior Homocline bordered to the east by the unstable shelf. In the subsurface the two zones are divided into a number of basins or sub-basins by a series basement corrugations, ridges or arches, which strike in two principal directions, north-south and northeast-southwest.

From their timing, the arches appear to be the only reflection in Arabia of two major tectonic events; the Caledonian and Hercynian. Some of these arches have affected subsequent sedimentation over long periods of time, but different stretches of a single arch may have been affected at different times. However in the absence of reliable data on the basement nothing can be said concerning their location and magnitude.

The longest arch in the Middle East, which extends from northern Saudi Arabia to northwestern Iraq, is the Ha'il-Rutbah-Ga'ara-Khleissia Arch. Traditionally the Ha'il-Rutbah segment is regarded as the margin of the Tabuk basin to the west; however it is covered by 4-6 km of early Paleozoic sediments which show no signs of thinning or facies change over the arch. Up-arching must therefore post date early Paleozoic but may pre-date Devonian since few Devonian sediments have been recorded west of the arch. Late Tertiary erosion however exposes some 900m of continental Permian sediments flanking the Ga'ara depression. The northernmost segment, the Khleissia high, was uplifted and peneplaned during the Late Jurassic and is marked by a reduced Mesozoic thickness. It remained above sea-level as an integral part of the Rutbah arch until separated from it by the Campanian-Maastrichtian Anah trough.

The Hail arch and its continuation in east Jordan and western Iraq as Rutbah- Ga'ara-Khleissia high is marked by a prominent positive gravity anomaly indicating the presence of basement at depth (around 6 kms.). Seismic evidence in northeastern Jordan points to a post-Paleozoic history of the Rutbah - Ga'ara Arch from early Cretaceous to Turonian and from Paleocene to middle Eocene (Anon 1986; Abu Jaber et al., 1989). The arch therefore shows evidence of sporadic activity during much of the post early Paleozoic with different segments active at different times.

The Central Arabian Arch is a relatively gentle structure but has a profound effect on the surface rock distribution in western Arabia. It trends in an east-northeasterly direction across the central part of the Arabian Platform from the easternmost point of the Arabian Shield to as far as Qatar where it probably links with the Qatar - South Fars Arch. It has a complex geologic history and separates the northeast dipping central segment of the stable shelf from its southern, east dipping segment (Al Khadi and Hancock, 1980). It could be a residual high caused by the sagging of regions on either side, as indicated by the tensional nature of the central Arabian graben. During the early Paleozoic, the arch was mildly positive as pre-Permian sediments seem to pinch out against it, but strong upwarping during post-early Devonian and pre-Late Permian stripped off the older sediments from the crest of the arch. As a result, Late Permian carbonates rest directly on basement rocks across the axis. Eastwards tilting has preserved some Devonian rocks found in deep drilling in Qatar. The Qatar-South Fars segment of the arch is more clearly a Hercynian, Late Paleozoic feature, which divided the Arabian Gulf basin into two sub-basins and hence influenced later sedimentation since all units thicken away from the high.

The Huqf-Haushi arch, a NE-SW trending broad, faulted, anticlinal structure in eastern Oman, appears to have behaved as a structural high since it was initiated during the Cambrian. As a result of epeirogenic movements younger sediments progressively onlap the flanks of the arch (Ries and Shackleton, 1990). The western flank dips gently into the South Oman - Ghaba salt basin whereas active faulting on the eastern flank has resulted in the accumulation of thick Cretaceous and Tertiary sequences.

In southern Arabia, the Hadhramout Arch is a younger structure which only began to develop during the Paleocene reaching its present emergent form by the end of the Eocene.

The Stratigraphic and Sedimentological Framework

As a broad generalization, clastic sedimentation dominated the Paleozoic section from the Cambrian until Middle Carboniferous with the occurrence of arenaceous deposits during the Cambro-Ordovician followed by argillaceous deposits accompanied by minor carbonates. From the Late Carboniferous until the Miocene carbonate sediments came to dominate a stable, wide continental platform that opened eastwards into the seaway of the Tethys Ocean, although not always continuously. From Permian to early Jurassic dessication of the carbonate platform led to the formation of evaporitic basins. From middle Jurassic to the Turonian the carbonate platform was covered with argillaceous or clastic sediments. From the late Turonian the effects of the Alpine orogeny became apparent with the accumulation of deep flysch and shale sediments from an easterly source. Ophiolite intrusion and thrusting over the Arabian platform began during late Campanian-Maastrichtian. To the west carbonate and shale deposition continued on the platform whereas, clastic sediments, derived from newly uplifted beds, accumulated on the eastern edge of the basin.

A summary of the stratigraphy, sedimentology and paleontology of Arabia is best illustrated by a series of paleogeographic maps by Murris (1981) and Alsharhan and Nairn (1997) and the reader can refer to these references for more detailed descriptions. Each map is a record relating stratigraphy, sedimentology and associated facies changes to the overall geography. They aim to define accurately each specific time-slice and identify depositional environments. The closer spaced the time slices, the better they present a continuous history of the region and reflect progressive changes in basin development. It is also possible to incorporate paleontological data as an indication of the faunal response to the changing environmental conditions.

Two basic patterns of carbonate deposition alternating through the Mesozoic-Cenozoic were delineated and defined by Murris (1981) and Alsharhan and Nairn (1997) as the carbonate ramp and the differentiated shelf facies. The ramp sequence is characterized by cyclic sequences of shale and carbonate units each of which can be correlated over hundreds of kilometers with little change in lithology or thickness. The carbonates consist of peloidal bioclastic wackestones or oolitic grainstones. The argillaceous units comprise pyritized pelletal limestones which pass horizontally to marls or shales. Ramp sedimentation was contemporaneous with periods of high clastic inflow which occurred during marine lowstands. In contrast, when sea level was high, clastic flow diminished and the shallow parts of the differentiated platform contained clean algal-foraminiferal wackestone/packstone and ooidal-peloidal packstone and grainstone. In deeper water, depositional rates were low and dominated by the deposition of marl and mudstone. Such carbonate cycles were less consistent in thickness and lithological character than the ramp carbonates.

Only a brief account of the stratigraphy is given here, for detailed descriptions the reader is referred to Alsharhan and Nairn (1997). The sequence has been divided into major time intervals due to the considerable data accumulated on the Mesozoic rocks during the last few decades because of their economic importance.

Infracambrian: Stratigraphy and Sedimentation

The first unmetamorphosed Precambrian and Cambrian sedimentary rocks deposited on the basement rocks of the Arabian Peninsula were clastics with some carbonates. Evaporites were found only in South Oman and the Arabian Gulf area (in the South Oman - Ghaba salt basin, Fahud salt basin and the Hormuz salts in the north and south Arabian Gulf basins). Although direct evidence of a connection between the Arabian Gulf and Oman basins is lacking since they were separated by a high between the Rub Al Khali and South Oman Basin, the wide distribution of evaporites and stromatolitic limestones and dolostones during the late Precambrian - early Cambrian suggests similar climatic and sedimentation regimes covered vast areas of the Gondwana continent.

In South Oman, the Infracambrian sediments were assigned to the Huqf Group which is divided into five formations (Fig. 3.7), and consist of clastic and carbonate sediments laid down in a shallow marine environment, ending in a thick halite (the Ara salt of Gorin et al., 1982; Alsharhan and Kendall 1986; Wright et al., 1990; Alsharhan and Nairn, 1997).

The Abu Mahara Formation contains the oldest sedimentary rocks found in Oman. It was deposited on metamorphosed basement rocks. The formation is made up of thin beds of dolomite overlying the basement rocks followed by a clastic sequence composed of alternations of cemented sandstone with laminae of dolomitized siltstone. The rocks of this formation are believed to have been deposited in a braided stream to tidal flat environment.

The Khufai Formation represents a facies change from the dolomitized siltstone of Abu Mahara Formation to dolomite typical of the lower part of the formation. The base of the Khufai Formation is characterized by the presence of collapse structures probably resulting from the dissolution of anhydrite, dolomite and halite. The formation is composed of dolomite with some oolitic limestone, stromatolites and chert. Some beds show indications of folding and slipping due to the dissolution of dolomite. The upper part is characterized by well bedded dolomite, stromatolites, pelletal, oncoidal and oolitic limestone, with thin bedded of current-bedded sandstone and siltstone alternating with dolomite. The rocks of this formation were laid down in supratidal and shallow intertidal environments.

The Shuram Formation is separated from the Khufai Formation by a disconformity surface. It is made up of two sedimentary units; the lower unit is composed of brown to red shaly calcareous siltstone with muscovite and biotite. The upper unit consists of alternations of siliciclastic rocks with oolitic, dolomitic and lime mudstone containing argillaceous and micaceous limestone. The nature of the lithologic components and the sedimentary structures such as cross-bedding, ripple marks and collapse structures, suggest that the formation was deposited in a moderate to high energy shallow water marine environment.

The Buah Formation also consists of two units. The lower unit is made up of alternating thin beds of lime mudstone, siltstone and aphanitic dolomite containing flat and domal stromatolites. The upper unit is formed of thick beds of chert rich dolomite containing dissolution pores associated with anhydrite at different levels. The unit contains also large dolomite crystals and traces of oolites and oncoids. The base of the unit is characterized by the presence of sandy dolomite. The Buah Formation was deposited in tidal flat environment.

The Ara Formation is composed of two units. The lower is made up of thick beds of salt, anhydrite, shale and siltstone with thin beds of limestone. The upper section consists of dolomite and limestone intercalated with anhydrite, shale, silty mud and sandstone. The formation was deposited in a very shallow marine lagoonal environment.

Paleozoic: Stratigraphy and Sedimentation

The Paleozoic basins are relatively simple and confined mainly to Oman in the east and Saudi Arabia, Jordan and Northwestern Iraq. In the south Arabian Peninsula, there is the relatively poorly known Wajid Basin. In the north lies the Tabuk basin extending into Jordan from Saudi Arabia. East of it, using the Ha'il-Ga'ara Arch as the limit, lies the Widyan Basin with its more complete sequence of Silurian, Devonian and Carboniferous rocks. The lower Paleozoic sequences are made up mainly of arenaceous beds with intervening shale horizons. The sands are medium-coarse, brown to tan in colour, friable quartzose sandstones with a thick basal conglomerate. In the north, the Saq Sandstone Formation covers 300,000 km^2 while in the south the Wajid sandstone equivalent has an area of 196,000 km^2. Although the two sub-basins are now separated by the Arabian Shield Precambrian rocks, in both basins lithologies are similar and sediment transport directions are consistently from the south, suggesting a single basin by analogy to the sediment transport in rocks of the same age in Algeria north and south of the Hoggar Massif. The comparison may be extended to the late Ordovician - early Silurian for the descriptions and timing of the glacial event show a striking parallelism, even to minor oil fields preserved in glacially filled valleys in both areas. During the Late Carboniferous, Arabia was affected by the Hercynian Orogeny that resulted in uplift followed by strong erosion which cut down through the Paleozoic, in some places to the Precambrian. Such was the case in the southern part of the Arabian Peninsula near the Yemeni border where Permian rocks rest directly upon the basement rocks.

The Hercynian Orogeny was followed by the deposition glacial deposits during the late Carboniferous-Early Permian (Murris 1981), found in south of Southern Arabia, Yemen and in the subsurface sections of South Oman (Dhofar). These deposits are oil-bearing in the South Oman oilfields and are equivalent to the Dwyka glacial deposits in Gondwana (Alsharhan et al., 1993). The glacial influence in Oman decreases northwards and sediments of the same age in Zagros Mountains in Iran and southeast Turkey are neritic carbonates and non-glacial deposits.

There is a dearth of information about much of the Devonian and Carboniferous since the rocks lie deep and there has been little incentive to drill them. With the recognition of the Late Carboniferous clastics and the development of the Late Permian carbonates, it seems that there is a drastic change from arenaceous dominated sequence to a carbonate platform. Arenaceous deposits did not cease, but they tend to spread eastwards over the platform area during periods of marine regression. The shoreline position did not change much implying that an uplift and tilting of the shield maintained the shoreline in a relatively stable position.

The deposition of the sedimentary rocks in the Arabian Peninsula during the Paleozoic (Fig. 3.7) started with sandstones and shale during the Cambrian and Lower Ordovician. In late Ordovician to early Devonian, deposition was dominated by a series of shale and mica-rich sandstone with minor limestones. By the end of the Permian there was a clear change in the nature of sediments, ending with the deposition of limestone, dolomite, anhydrite and gypsiferous shale. There are also several unconformity surfaces which resulted from epirogenic movement which affected the depositional continuity. The most prominent unconformity surfaces were formed during pre-late Cambrian, Pre-Devonian and Early Carboniferous. In general, the sandy component in the formation increases westward, and the unconformity surfaces become more prominent. As a result the Lower Paleozoic rocks in the southern edge of the basin were formed mainly of conglomerates and sandstones which were separated by unconformity surfaces and time gaps. The Pre-Permian rocks in the Arabian Peninsula represent wide areas and beds of clastic materials derived mainly from the neighboring positive areas such as the Arabian Shield, Hadhramout Arch and Haushi-Huqf Arch.

From field observations and from the few deep wells drilled in different parts of the Arabian Peninsula, it is possible to deduce a general framework of the succession of formations during the Paleozoic. In southeast Arabia the pre-Carboniferous rocks consist mainly of clastic rocks ending with a large unconformity surface reflecting the extent of the tectonic movement which effected those areas during the Paleozoic. This was followed by Permo-Carboniferous deposits, the lower were glacial sediments (fluvio-glacial, outwash and tillites), the upper were dominated by red-bed facies formed in valleys and river environments except in the vicinity of the Arabian Gulf where the section is dominated by quartizitic sandstone, shale, siltstone and thin beds of dolomite and anhydrite deposited in continental to coastal-marine setting.

During the Late Carboniferous, Arabia was affected by the Hercynian orogeny that resulted in folding and strong erosion which cut down through the Paleozoic rocks in some areas to the Precambrian. That was the case in the southern part of the Arabian Peninsula near the Yemeni border, where the Upper Permian rocks rest directly on the basement rocks. Most of the paleozoic sediments were removed due to this movement.

During the Late Permian, the sediments in the Arabian Peninsula were characterized by shallow water carbonates, dolomitic limestone and anhydrite. In the central part of Saudi Arabia, north of the latitude 23°N, the Late Permian is made up of an alternation of neritic limestone and shale. The limestone shows evidence of dolomitization and contains some gypsum and anhydrite beds. In the area south of the 23°N latitude and near latitude 19°N, clastic sediments start to appear gradually, with shale followed by sandstone. The clastic thickness increases at the expense of the carbonate rocks. At the latitude 19°N, the sediments become completely clastic. In the eastern part of the Arabian Peninsula the subsurface sections of Late Permian are made up of limestone containing beds of dolomite and anhydrite. In the south toward the Rub Al Khali basin, the line separating clastic and carbonate rocks lies near latitude 20°N. It seems that Hadhramaut and Dhofar arches were positive areas which prevented a marine advance towards the south. On the western side of the Oman Mountains, the early Permian starts with the deposition of sandstone followed unconformably by Late Permian carbonate rocks. This can be seen in the Jebel Akhdar and south of Saih Hatat, where the rocks are mainly dolostone. In NW Oman Mountains, and in the northern United Arab Emirates, the Upper Permian sediments are made up of neritic limestone resting conformably upon the Lower Permian.

The presence of anhydrite lenses and beds in the area extending from Rub Al Khali to central United Arab Emirates indicates that the water movement in this marine belt was restricted. This may be due to the presence of an elevated area formed before the Late Permian marine transgression. The presence of this high area may be supported by the occurrence of an unconformity surface between the Lower and Upper Permian rocks in the central and southern areas of the Arabian Peninsula. This unconformity surface is not reported from Yemen but is clearly evident in the southern and central Oman Mountains although not present in the north.

In some parts of the Arabian Peninsula such as central and south Oman, there is a time gap, where the Upper and Middle Jurassic rocks rest unconformably over the Lower Permian rocks. It is believed that the Upper Permian, Triassic and Lower Jurassic sediments were laid down but removed later by erosion before the Middle Jurassic. This is indicated by the presence of an elevated area formed before Late Permian. This high area was probably responsible for the presence of many unconformity surfaces to the west. Eastward where the area was covered with marine sediments, the Upper Permian is unconformable as in the western central Oman Mountains. In Central Arabia and in the Rub Al Khali basin there is no unconformity. The trend of this high is completely different from the trends of other structures in the area which extend east-west but is in accordance with the directions of structures in which oil is found within the Jurassic and Cretaceous rocks in the Arabian Gulf area.

Triassic: Stratigraphy and Sedimentation

The Triassic was a period of regression but one which was not uniform over the whole area. This can be seen from the presence of marine rocks indicating a marine transgression during the Middle Triassic over the central part of the Arabian Peninsula and from the large number of unconformity surfaces during the late Middle Triassic in Iraq. During the Late Triassic and Early Jurassic, orogenic movements affected the central and southern parts of the Arabian Peninsula, where Jurassic rocks are unconformably resting on the Triassic rocks.

The Lower Triassic is made up mainly of green and red shale beds probably of continental origin intercalated with siltstone, limestone and gypsum. In the north (latitude 27°N) and in the south (latitude 20°N), lenses of limestone also appear. In the subsurface sections to the east in the Arabian Gulf region the shale alternates with beds of dolomite and anhydrite. The sediments of the Lower Triassic rest conformably above the sandstone of the Permian in the south of the Arabian Peninsula, and above its limestone equivalent in northern Saudi Arabia. The upper contact of the Lower Triassic is delineated by the contact between the red shale of the Lower Triassic and sandstone rocks overlying the Middle Triassic (Fig. 3.7).

A significant facies change occured during the Middle Triassic. The thickness of the sediments increased northward (latitude 22°N). They were made up of sandstone, marine shale and limestone in the exposed areas of the central part of the Arabian Peninsula (latitude 27°N) to limestone in the north. Eastward, in the Arabian Gulf region, the Middle Triassic in the subsurface sections, is made up of interbedded dolomite and anhydrite. In the northern Arabian Gulf, the Middle Triassic is dominated in the lower part by alternations of anhydrite, dolomite and minor thin beds of shale, grading upward to salt layers with thin interbeds of dolomite, and ending by alternations of dolomite and shale and thin interbeds of anhydrite.

During the Late Triassic, lateral facies changes can be observed since the formation in Arabia is wholly continental origin in all the exposed areas. The continental origin of the sediments can be deduced from the presence of mud-cracks and fossil plants (Powers et al., 1966; Sharief, 1982). Eastward, in Qatar and in the offshore areas of the United Arab Emirates, it appears that the formation is absent, and there is an unconformity since the Jurassic section rests unconformably on Middle Triassic rocks. The southern part of the Arabian Peninsula, southeast Yemen and southern Oman were elevated areas during the Late Permian. In northern Arabian Gulf the Upper Triassic consists predominantly of shale and subordinate dolomite and minor sandstone.

No Triassic rocks occur in south Oman, while in the outcrop in central west Oman, the rocks of the Upper Permian and Triassic age are considered to be transported (exotic blocks) of thick neritic limestones. In subsurface west of the Oman Mountains, the Triassic sediments consist of dolomite and some beds of marly anhydrite followed unconformably by the Jurassic rocks. In the central Oman Mountains they consist of detrital, dolomitic limestone, while in the northern Oman Mountains and northern part of the United Arab Emirates the Triassic rocks consist of dolomitic limestone, followed by beds of sandstone, marl, shale, topped unconformably by the Lower Jurassic carbonates (see also Sharief, 1983).

Jurassic: Stratigraphy and Sedimentation

The Tethys marine transgression began during the Toarcian, and increased greatly in scale during the Bajocian and the Bathonian, and continued without apparent interruption during the Late Jurassic. This gradual transgression resulted in a laterally different facies, which were given different formation names in various parts of the Arabian Peninsula (Fig. 3.7).

Early Jurassic

The Lower Jurassic sediments in the Arabian Gulf area represent part of a large Jurassic sedimentary cycle. The sediments of this cycle range from red shale to fine-grained deep-water carbonates to shallow neritic carbonates which were deposited under relatively high energy conditions and ended with very shallow supratidal conditions. In the central part of the Arabian Peninsula, the Lower Jurassic sediments are made up of red lamellar mudstone with some beds of limestone, and thin intercalations of sandy clastic rocks increasingly common toward the Arabian Gulf. In the offshore areas of Qatar and United Arab Emirates, the Lower

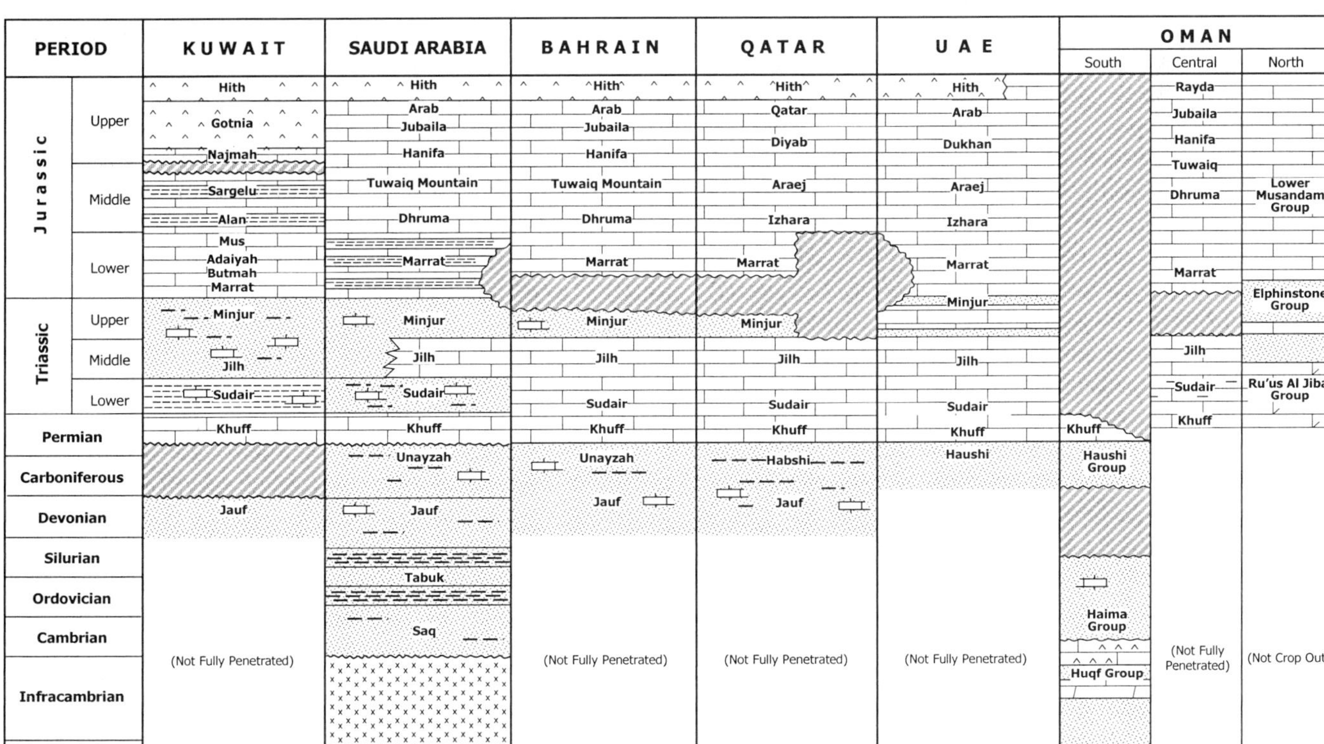

Fig. 3.7. Tentative stratigraphic Infracambrian-Paleozoic-Triassic and Jurassic rock units correlation chart across Arabian Gulf and adjacent areas (Alsharhan and Kendall, 1986).

Jurassic sequence consists of dolomite with anhydrite in the upper part, and a carbonate-shale in the lower part.

The alternation of Lower Jurassic sediments in the authochthonous mountainous area of Oman Begin with ferruginous sandstone or sandy limestone resting unconformably on the Triassic sediments. These are followed by a thick sequence of carbonate rocks in which the lower part is of early and middle Liassic age (Lower Jurassic). The carbonate rocks of the Lower Jurassic are made up of limestone deposited on a shallow water platform, containing dolomite and rare sand. The same sequence is encountered in the subsurface section in the oilfields of the west Oman Mountains.

The marine sediments of the Lower Jurassic are restricted to the north and the eastern borders of the central part of the Arabian Peninsula. During the Early Jurassic, the transgression was less widespread than during the Triassic. In the central part of the Arabian Peninsula, Upper Triassic and Lower Jurassic sediments are absent. This is attributed to the occurrence of an orogenic movement during the Late Triassic, which led to the uplift of the central part of the Arabian Peninsula. It seems that the northern part of the Arabian Peninsula was either slightly effected or was not influenced at all by this tectonic movement. As a result of this movement and the limited transgression of the Early Jurassic sea, the marine facies associated with the Tethys Seaway are found mainly along the eastern borders of SW Iran. In northeast and eastern area of Arabia, the marine carbonate sediments are mixed and intercalated with clastic sediments and evaporites.

Middle Jurassic

The transgression that covered wide areas of the Arabian Peninsula, began during the Toarcian and expanded during the Bajocian-Bathonian. The effect of this widespread transgression can be seen clearly in the southern part of the Arabian Peninsula.

In the central part of the Arabian Peninsula, the sediments of Bajocian-Bathonian age crop out along Tuwaiq Mountains from latitude 19°N to 27°N, and consist of shallow water limestone and shale. Toward the south, the sediments change from marine sandstone into continental sandstones. This change occurs in the lower beds at the latitude 24°N, and the rocks become totally clastic at latitude 21°N. This facies change may provide a measure of the extent of the marine advance to the south. To the north of Saudi Arabia and along the Tuwaiq Mountains, the change from limestone to shale and then to continental sandstone probably indicates the presence of shoreline to the west. In southwestern Saudi Arabia, the sequence is made up of clastic rocks including sandstone, conglomerate, shale and marl containing plant remains indicating a continental deposition in a slowly subsiding basin. At the end of the sequence, marine beds continued to be deposited during the Middle Jurassic in the area covered by the sea.

In Rub Al Khali Basin, and areas of South Oman, it is difficult to delineate the extent of the marine transgression. However, the marine transgression during the Toarcian was less extensive than that of the Middle Jurassic in the south of Rub Al-Khali. In the west, the Middle Jurassic sediments consist of marine sandstone, whereas, they are composed of marine limestone in the eastern part. In east central Oman, the Middle Jurassic sequence is formed of limestone, resting unconformably upon Upper Permian rocks. The lithofacies of this area are similar to those found in Oman Mountains but their thickness is less than in Haushi and Huqf area.

In the north Oman Mountains, the Toarcian, Bajocian and Bathonian sediments form part of the limestone sequence deposited without interruption on a shallow carbonate platform, from the Early Jurassic (Liassic) to the Late Jurassic and continuing into the late Lower Cretaceous. Their thickness increases from southwest to northeast.

In United Arab Emirates and Qatar areas, the sediments change laterally from shallow water carbonates to deeper water argillaceous and shaly limestone in the subsurface section. Many gaps are found in Qatar and along the Arabian Gulf area, where the Bajocian sediments rest unconformably above the Middle Triassic rocks. They are followed by lime mudstone grading upward into oolitic grainstone. In the Arabian Gulf at latitude 28° 13'N and longitude 48° 29' E, the Upper Toarcian sediments are missing, and the Bajocian sediments rest over the lower Toarcian sediments. The unconformity surface indicates the presence of a shallow, elevated high covered gradually during the transgression.

Along the northwestern margin of the southwestern Arabian Gulf (from Kuwait to Qatar) the absence of the Toarcian rocks, is attributed to the presence of an elevated area formed before the early Toarcian (and probably during the Late Triassic), but covered during the Middle Jurassic. This part of the Jurassic history was characterized by a widespread transgression from the Toarcian to the Bathonian which increased during the Late Jurassic. In the central and southern parts of Arabia, the shallow seas covered these areas to approximately latitude 22°N. In northern Arabia, the Toarcian represents a period of regression, between two periods of transgression, the Middle Liassic and Bajocian. The sediment distribution suggests that the Middle Jurassic transgression was larger in scale than that of the Toarcian.

In the central part of Arabia (the Tuwaiq Mountains region) and south of the Rub Al Khali basin, there was a change from marine to continental facies. The Hadhramaut area of Yemen was elevated. The sea also did not advance southward into southwest Yemen, but the area was low and continental deposits accumulated there. During the Late Jurassic, the area represented a passage between the Arabian and the African sea. Tectonically, the period was one of relative stability as indicated by bed conformity and continuous deposition in the north. The unconformity surface in the central and south of the western part of the Arabian Peninsula between the Toarcian or the Bajocian and the Late Triassic or older rocks was a result of orogenic tectonic movements which occurred during the Late Triassic and led to the elevation of these areas above the sea level during the Early Jurassic.

Late Jurassic

Late Jurassic rocks are exposed over wide areas in the Arabian Peninsula and surrounding areas. In general, these rocks are represented by shallow water carbonate rocks and the widespread occurrence of evaporite during the Late Kimmeridgian and Tithonian (Fig. 3.7). Probably the most important exposures of the Upper Jurassic rocks are those found in the central part of the Arabian Peninsula in the Tuwaiq Mountains where they extend about 1000 km from latitude 17° 30' N to 27° 30' N. During this period the sea covered large portions of the Arabian Shield, extending into Yemen in the south where it joined the East Africa sea.

The Callovian-Oxfordian which rests unconformably on the top of the Bajocian-Bathonian is made up of limestone deposited on a continental shelf and contains in situ corals. The Callovian-Oxfordian in northern Arabian platform is also carbonate. The lower Kimmeridgian is composed of carbonate rocks with high percentage of clastic materials. This is followed by middle Kimmeridgian to early Tithonian carbonates which are characterized by significant changes in the depositional conditions. These sediments are made up of alternations of neritic limestone and dolomite with anhydrite horizons. During the late Kimmeridgian and early Tithonian, evaporite deposition became dominant and was represented by the Tithonian anhydrites with rare intercalations of limestone and dolomite. In northern Arabian Gulf it is made up mainly of salt interbedded with anhydrite and carbonates.

In the eastern part of the Saudi Arabia, the Late Jurassic sediments are found in the subsurface sections but the limestone tends to be more argillaceous and finer-grained. Kimmeridgian evaporites are also present but they contain less anhydrite compared to the middle part. The Upper Jurassic sea covered the southern part of the Arabian Peninsula except south Oman. The wells drilled in the Rub Al Khali show that the Callovian, Oxfordian and lower Kimmeridgian rocks are formed of neritic limestones whereas the upper Kimmeridgian rocks are formed of evaporites. This sequence is similar to that found in the central part of the Arabian Peninsula. From Qatar to central United Arab Emirates area the deposition of the shallow neritic carbonates was continuous and similar to that in the wide area of the Arabian Peninsula with slight differences in the stratigraphic nomenclature (see Alsharhan and Nairn, 1997).

In Oman and the northern Emirates, the Callovian, Oxfordian and Lower Kimmeridigian rocks are conformable with the Bathonian rocks. They are made up of fossiliferous limestone deposited on a carbonate platform. There is an unconformity between the lower and upper parts of this sequence. It is indicated by the presence of breccia within the limestone, and it is believed that the upper Kimmeridgian sediments were missing. Above these sediments are dense limestones extending in age from the Tithonian to the Valanginian.

In southern part of the Arabian Peninsula, the Triassic to Middle Jurassic sediments are made up of fluvial clastic and lacustrine rocks. This succession is followed by the Callovian-Oxfordian sediments which consist of calcareous sandstone and sandy limestone of marine origin in the lower part of the sequence. Callovian to lower Kimmeridgian sediments are formed of neritic marine sediments in which detrital or oolitic limestone is interbedded with marl. The upper Kimmeridgian and lower Bathonian rocks are composed of intercalations of evaporite, shale and sandstone. The Upper Jurassic rocks in Somalia and Ethiopia are similar to those found in Yemen. This suggests that the seas which covered the Arabian Peninsula and east Africa were connected at that time (Beydoun, 1988).

A gradually subsiding deep depression trending from NNW to SSE formed in Iraq (one which already existed since the Middle Jurassic) and extended to the Zagros Mountains. In this basin deep marine sediments were laid down, and on the northeastern flank of this deep depression, a condensed sequence containing deep marine sediments was formed. The western flank was characterized by shallow marine sediments similar to those deposited at the central part of the Arabian Peninsula (Saint Marc, 1978; Buday, 1980).

In the central parts of the Arabian Peninsula, along the Arabian Gulf and in the Rub Al Khali and Oman, carbonate sediments were deposited in shallow seas, whereas clastic sediments were deposited in south Yemen. There is an unconformity between the Middle and Upper Jurassic in the central part of the Arabian Peninsula and from northeast of Qatar to Iraq, i.e. the whole area which bordered the depression zone. The unconformity probably was due to the gradual subsidence and the filling of available accommodation space rather than the orogenic movements. In other areas of Arabia, the Middle and Upper Jurassic sediments are conformable. The widespread transgression which occurred between the Callovian to the early Kimmeridgian was followed by a major regression during the upper Kimmeridgian, when evaporites were laid down in all areas, except Oman and the northwestern parts of the Arabian Peninsula.

During the Late Jurassic an orogenic movement effected Yemen. This was followed by another phase of movement during Late Jurassic and Early Cretaceous which effected the Arabian Peninsula and the surrounding region and led to the development of faults which can be observed in most regions except in a depression which extended from NNE to SSW. This tectonic movement was followed by strong erosion that affected high areas in southwestern and northwestern the Arabian Peninsula. These positive areas were the source which provided clastic sediments during the Lower Cretaceous. The other regions of Arabia were characterized by marine sediments which continued without interruption to the Lower Cretaceous.

Cretaceous: Stratigraphy and Sedimentation

In the Arabian Peninsula and the Gulf, it is a common practice to divide the Cretaceous succession into three subdivisions; the Lower, Middle and Upper Cretaceous. This division is based on the presence of regional unconformity surfaces at the end of the Lower Cretaceous during the late Aptian, during the late Cenomanian and early Turonian at the end of middle Cretaceous and again in late Upper Cretaceous during late Maastrichtian in most parts of the Arabian Peninsula (Harris et al., 1984; Alsharhan and Nairn, 1986) (Fig. 3.8). Carbonate sedimentation was interrupted by emergence and some clastic influx in the Jurassic and in the later part of the Early Cretaceous.

Regional changes in sedimentation patterns during the Jurassic and most of the Cretaceous appear to have been controlled more by fluctuation in eustatic sea-level than by tectonics. By Middle Cretaceous time, the epeiric seas had spread over an area of some 200,000 mi^2 of the eastern Arabian block, well onto the Interior Homocline rimming the eastern side of the shield. These extensive shelf

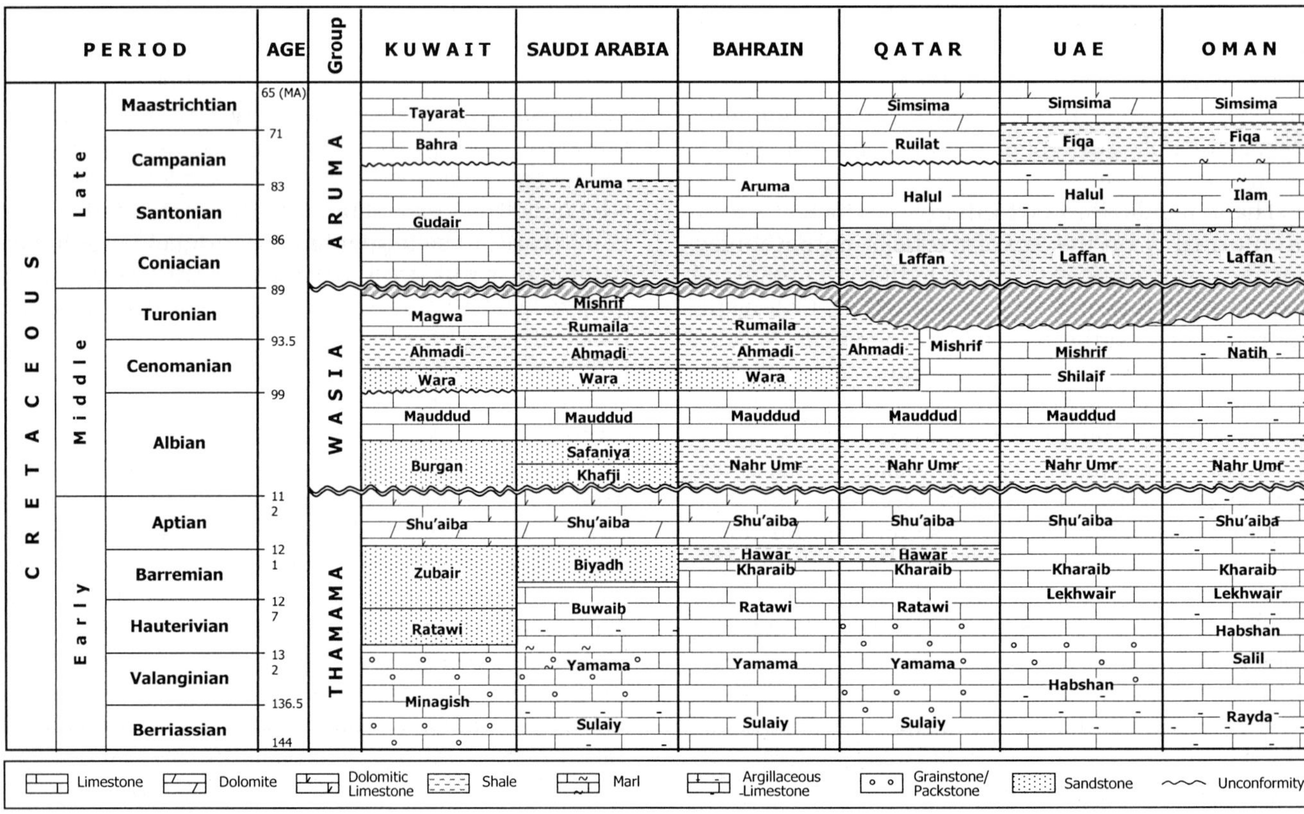

Fig. 3.8. Comparative Cretaceous stratigraphic columns, from Kuwait to Oman (compiled with modification after Alsharhan and Nairn, 1986, 1998 and 1990).

carbonates represent the final passive tectonic phase of the Arabian Shield area (Alsharhan and Nairn, 1988).

A major change in the tectonic and depositional regimes took place at the end of the Cenomanian or the early Turonian. This resulted mainly from the collision and partial subduction of a margin of the eastern Arabian sub-plate with a spreading ridge whose axis is thought to have been centered in the present-day Gulf of Oman. This structural event appears to be roughly coincident with a worldwide eustatic sea-level rise in the Late Cretaceous, both of which caused the change from shallow stable platform environments to deep water open marine, and in the case of the foredeep, bathyal and perhaps abyssal conditions (Alsharhan and Nairn, 1990). The Upper Cretaceous sequence, that developed on the stable craton in front of the Oman orogenic belt on the hingeline and in the foredeep trough, displays a two-part division which is readily seen in well-log cross sections. The lower part of the sequence is characterized by platform carbonates to the west and northwest of the hingeline and by a poorly calcareous or non-calcareous shale facies in the foredeep. In the central part of the foredeep trough, the shales grade into a clastic-limestone facies (Wilson, 1969; Glennie et al., 1974), and appear to coincide with the first phase of thrusting of nappes in the Omani orogen. The upper part of the Upper Cretaceous sequence, is typically represented by shelf carbonates which spread well onto the Interior Homocline portion of the Arabian block. In the foredeep trough the Upper Cretaceous of Oman is a sequence of turbidite sandstones, deep sea cherts, siliceous shales and olistostromes. These correspond to another phase of thrusting referred to the Hawasina nappe sequence and their emplacement in the late Campanian of the Semail Ophiolite. By the middle Maastrichtian, the foredeep had been filled with sediment and the Cretaceous foredeep sequence is capped by a widespread shelf carbonate. In the northern Oman Mountains the foredeep flanking the west side of the Musandam Peninsula or the northern part of the Oman orogen continued to subside through the latest Cretaceous into the early Tertiary.

Early Cretaceous

The Lower Cretaceous in Arabia was characterized by continental and shallow marine sediments reflected by marine transgressive cycles and encompasses the rocks extending from the top of the Tithonian anhydrite to the base of the Albian clastic sequence.

In the central part of the Arabian Peninsula the sediments indicate a marine transgression represented by the deposition of the Berriasian sediments followed by the lower Valanginian. They are made up of detrital limestones rich in shell fragments and foraminifera. The upper Valanginian was followed by the Hauterivian, which is formed of marine detrital limestone in the lower part changing upward into sandstone interbedded with argillaceous sediments.

Between the Barremian and the Albian, the sediments of central Arabia were continental clastics and can be divided among two main units: the Lower Unit is made up of cross-bedded sandstone with varicolored shale and conglomerate which extended from the Barremian to the Early Aptian. The Upper unit is separated from the lower unit by an unconformity surface representing the period between Aptian-Albian, and this is considered as the separating line between the Lower and Middle Cretaceous. Moshrif and Kelling (1984), suggested that these units were made to great extent of clastic facies with some carbonates. The clastics were coarse to fine-grained sandstone with channel-filling conglomerate, quartzite, siltstone, planar claystone and mudstone with shale. The carbonate units are formed of dolomite and oolitic limestone. The clastic materials indicate continued fluvial sedimentation, whereas the carbonate suggest tidal flat and shallow marine deposition.

In the eastern part of the Arabian Peninsula and the Arabian Gulf region, the Lower Cretaceous sediments change gradually to marine, and the sequence becomes complete with no gap or unconformity surface. Early Hauterivian rocks change into shale in the lower part and argillaceous limestone and carbonates in the upper part of the sequence. They are conformable with the Valanginian rocks.

The Barremian and early Aptian rocks are continental from longitude 40° E in the Rub Al Khali to northeastern Saudi Arabia. In the area bounded between 49° and 51°E, the sediments are formed of alternating of shale and limestone, and the area east of longitude 51°E is characterized by neritic limestone. During the early part of Barremian, the fluvial sediments deposited in central Saudi Arabia can also be traced over a wide area in Kuwait.

The Aptian rocks formed of porous, well-developed, dolomitic limestone, contain *Orbitolina* sp., algae and rudists. During the Aptian, the carbonate and clastic facies of the Aptian were deposited in the very shallow marine environment which existed between the Barremian and Albian. These sediments intertongue with the Albian exposed in the central part of the Arabian Peninsula. They are developed also in the subsurface sections of eastern Arabia and the Arabian Gulf region.

In the northern part of the Arabian Gulf, the Berriasian-lower Valanginian succession started with the deposition of argillaceous lime mudstone followed by algal-foraminiferal-oolitic limestone during the late Valanginian. The Hauterivian-upper Berriasian sediments are formed of dense, pyritic greenish/black shale. The lower part is intercalated with yellowish brown detrital, pyritic, fossiliferous, peloidal packstone. The Barremian-lower Aptian in northern Arabian Gulf is made up of clastic cycles starting from greenish black, fissile shale with zones of sandstone and siltstone at the base followed by very fine-grained sandstone with rare amounts of siltstone and hard fissile, greenish black shale. The cycle usually ends with finer-grained sandstone with minor quantities of shale and siltstone followed by shale with two-distinctive sandstone zones and lesser amounts of siltstone.

In southeastern Arabian Peninsula, the Hauterivian rocks rest conformably on the Valanginian rocks and consist of lime mudstone containing radiolaria. In South Oman the Barremian rocks are made up of shallow water limestone sometimes containing argillaeous material. The Aptian sediments are composed of massive Orbitolina-bearing limestones which cover the whole area. In the southwestern Oman Mountains and western United Arab Emirates, the Aptian becomes more differentiated and an intrashelf basin formed filled with argillaceous limestone and shale and on the edge of this shelf margin, shallow water carbonates contain algae and rudistids formed. The sediments of this epoch are among the most important petroleum reservoirs in the region.

In southern part of Arabia, the Early Cretaceous sediments are absent due either to non-deposition or later erosion. The Late/Middle Cretaceous rocks rest unconformably on the Jurassic rocks. They are formed of marine sediments in the east and continental sediments in the west. The sediments indicate the presence of an open marine environment in the east. Most parts of the west Arabian Peninsula were elevated and strongly eroded due to orogenic movements, which occurred during the Late Jurassic and Early Cretaceous. The erosion products formed the clastic materials of the Early Cretaceous which surround the western areas. The later orogenic movements activated the faults already present in the area and volcanic eruptions in other areas, which activity continued into the Albian. The sea transgressed the elevated areas from Barremian to Cenomanian broken only by a slight regression during the late Aptian.

The Lower Cretaceous sediments of the Arabian Peninsula are characterized by two carbonate cycles proposed by Alsharhan and Nairn (1986). The first cycle was a complete and nearly regular unit that covered most parts of the Arabian Peninsula except the high regions and ended in the early Hauterivian. At the same time, the clastic sediments began to invade the area from the Arabian Shield and extended nearly to Qatar. The Berriasian-Valanginian witnessed the deposition of the first stage of the carbonate rocks in the form of a narrow passageway of deeper water carbonate rocks in Oman, and shallow water carbonates in United Arab Emirates extending into Saudi Arabia. They are mixed with clastic materials near the margin of the Arabian Shield.

The second sedimentary cycle extends from the Hauterivian to early Barremian. This was distinguished by shallow marine environments near the Arabian Shield dipping toward the United Arab Emirates. The inclination increased with water depth in Oman due to the weak movement of Lekhwair-Mender High in Oman-Abu Dhabi. The north part of the Arabian Gulf is characterized by the presence of clastic rocks derived from the Arabian Shield in the form of valley-fill clastic rocks. At the base of the second carbonate cycle, there were some transported argillaceous materials in the western United Arab Emirates (Abu Dhabi region) but sedimentation continued into the mid-Aptian and ceased as a result of an interruption of sedimentation which prevailed all over the Arabian Peninsula.

During the late Barremian, the increase of the clastic supply pushed carbonates eastward in the direction of western United Arab Emirates, whereas during the Aptian at the end of the second cycle, a relatively shallow sea covered the area in which carbonates were deposited. The carbonate proportion increased and covered most of the Arabian Peninsula except near the shield area which was covered by fluvial and valley fill clastic sediments. In the southern Arabian Gulf, a relatively deep sedimentary basin was formed and characterized by fine-grained carbonates and shales, belonging to the mid-Aptian. Around this depression rudistids and algae flourished to subsequently form the petroleum reservoir rocks of many giant fields such as in Bu Hasa, Idd El Shargi, Yibal and Shaybah Fields in Abu Dhabi, Qatar, Oman, Saudi Arabia and United Arab Emirates, respectively. The Lower Cretaceous ended with a major erosional unconformity surface which extended from South Iraq to Oman. In the central part of the Arabian Peninsula, the Aptian sediments were eroded locally and the Middle Cretaceous beds rest directly on the Barremian sandstone.

Middle Cretaceous

The Middle Cretaceous sediments are exposed in the central part of the Arabian Peninsula for more than 140 km, and are formed of continental sandstone thought to be derived from the Arabian Shield. The rocks of Middle Cretaceous rest unconformably on the Lower Cretaceous.

The continental environment during the Albian resulted in the deposition of a series of quartzite sands extending from South Iraq to Qatar. In this belt, the sand component increased continuously southward but ended during a weak marine invasion to be replaced by shale which extended from east Qatar to Oman and argillaceous limestone in northern United Arab Emirates and Oman. The limestone beds of the late Albian form an excellent time stratigraphic marker in the area from south Iraq to Oman.

In Kuwait and Saudi Arabia the lower Cenomanian sequence started with shale and rare limestone beds which became sandy upward, but in other places, the sequence was represented mainly by shale and limestone. During the late Cenomanian to early Turonian, the deposition was an alternation of shale and limestone. In southern parts of the Arabian Gulf regions, the late Albian and lower Cenomanian deposition occurred in deep water and was characterized by argillaceous limestone and shale rich in organic matter, while the upper Cenomanian sediments were characterized by fossiliferous limestone deposited in a shallow marine environment.

The Albian sediments rest unconformably on the top of the Aptian and are made up of sandstones and minor shales. In northern Arabian Gulf they consist of about 90% sandstone and 10% shale. They represent a widespread accumulation of well-sorted beach sandstone probably deposited in fluvial-deltaic environment. The alternating shale beds have a fluvial character and contain large amounts of plant remains and traces of amber. The clastic sediment deposition was followed by shallow water carbonates, the thickness of which increases toward Oman. The late Albian carbonates were followed directly by early Cenomanian shale intercalated with carbonates, which grade into sandstone in central and eastern Arabia. In northeastern Arabian platform this was followed by shale with limestone intercalations, and the late Cenomanian to probably early Turonian is composed of, neritic carbonates and fine-grained, sub-basinal Oligosteginal limestone and shale.

In the southern regions of the Arabian Gulf, the deposition of the Albian shale started with rare sand lenses and glauconite. The percentage of carbonates increases in the direction of Oman. At the end of this sedimentary cycle, clastic deposition ceased and was replaced by a shallow carbonates. After that, an intrashelf basin formed in the southern Arabian Gulf, where fine-grained, argillaceous oligostegina limestone was deposited with some shale. After the basin was filled with limestone, an oolitic and pelloidal packstone and grainstone was deposited accompanied by rudists and algae within the Cenomanian. At the end of the deposition of these sediments a regional unconformity covering most of the area was established over which the Upper Cretaceous sediments were deposited.

Most of the western part of the Arabian Peninsula was elevated and subjected to extensive erosion due to the orogenic movement which started between the Late Jurassic and Early Cretaceous. The products of this erosion were the clastic rocks of the Lower and Mid-Cretaceous which surround the western area. The orogenic movements led to the activation of the faults present in the area, and to volcanic eruption which in some areas continued into the Albian. In eastern Iraq and in southeast Arabian Peninsula, the Cretaceous rocks are conformable with the Jurassic rocks, whereas in other areas they are unconformable resting on different levels of Jurassic rocks. The sea transgressed the elevated areas during Barremian-Cenomanian except for a minor regression during the late Aptian.

The sea transgressed over the continental areas twice; the first time during the Aptian and for a second time during the late Cenomanian. In the northwestern part of the Arabian Peninsula epeirogenic movement occurred during Cenomanian-Turonian which led to the formation of basins and elevated areas associated with deposition of different lithofacies. Turonian rocks are absent in most parts of the Arabian Peninsula whereas they are present in deep areas in Iraq. It was assumed that this was due to erosion which resulted from the orogenic movement occurring during the Coniacian, and led to an essential change in the submarine topography at that time. The mid-Cretaceous ended with an unconformity surface in most parts of the Arabian Peninsula.

Late Cretaceous

In the central part of the Arabian Peninsula and Rub Al Khali depressions, there are many time gaps and unconformity surfaces within the Upper Cretaceous rocks. These rocks were probably associated with orogenic movements, which started during the Turonian and continued till the Maastrichtian and affected different parts of Arabia.

The shale of the Coniacian extends from Qatar to Oman with the carbonate proportion increasing toward Oman Mountains. The shale is followed by the marly limestone of the lower Campanian-Santonian, whereas the upper Campanian sediments are represented by impure limestone and dolomite. Towards the United Arab Emirates the sediments change to limestone, marl, shale and some siliciclastics. During the Maastrichtian the sequence was dominated by algal-rudist-bearing limestones, dolomite, dolomitic limestone with some shale.

In northern part of the Arabian Gulf, the Coniacian to lower Campanian sediments consist of detrital limestone, glauconitic in some places. They also contain some shale horizons in the lower part of the section. The Campanian is made up of white to grey cryptocrystalline limestone containing, greyish nodules of chert. In the lower part of this section, there are some oolites and black shale occurs. The Maastrichtian is composed of dolomitic limestone, frequently anhydritic, granular or crystallized with whitish limestone, intercalated with thin layers of black, pyritic, bituminous shale.

The Upper Cretaceous ended with a large scale unconformity resulting from the regional changes in sea-level, initiated by the Alpine orogeny and other tectonic movement in the Oman and Zagros Mountains. During the early Upper Cretaceous most of the western regions of the central part of Saudi Arabia and extending into the eastern province near the coast were elevated above sea-level. There were also some highs along the Arabian Gulf, so the Turonian, Santonian and early Campanian rocks overlapped unconformably over different levels of the Cenomanian rocks in the basins situated between the elevated areas along the eastern area of the Saudi Arabia to central part of the Rub Al Khali basin.

A widespread transgression occurred during the Late Campanian and Maastrichtian, which led to the deposition of carbonate rocks, over wide areas in the central of the Arabian Peninsula. Those rocks in the subsurface sections of the eastern Arabia were formed of limestone and shale rich in *Globotruncana* spp. Limestones rich in benthonic foraminifera and rudistids were deposited in the central Rub Al Khali, and as a belt that extends from the latitude 24° N to 30°N. At latitude 30°N, the carbonate rocks intertongue with sandstone, indicating proximity to the old shoreline. South of latitude 22°N, the upper Campanian-Maastrichtian sediments become dominantly continental sandstone, which can be traced into southwest Arabia and west of Rub Al Khali. Most of the southern part of the west Arabian Peninsula was an elevated area during the Late Cretaceous and continental clastics were deposited in Yemen and southern Oman. Shallow water limestone containing foraminifera and rudistids were deposited during the Maastrichtian.

In southwest Arabia, volcanic activity began during the early Coniacian and continued to early Cenozoic, when basalt, andesite, rhyolite and trachyte rocks were formed. In extreme eastern Arabia in the Oman Mountains and the northern Emirates, the end of the Late Cretaceous was marked by important orogenic movement beginning during the Turonian and continuing through the Campanian culminating during the Maastrichtian.

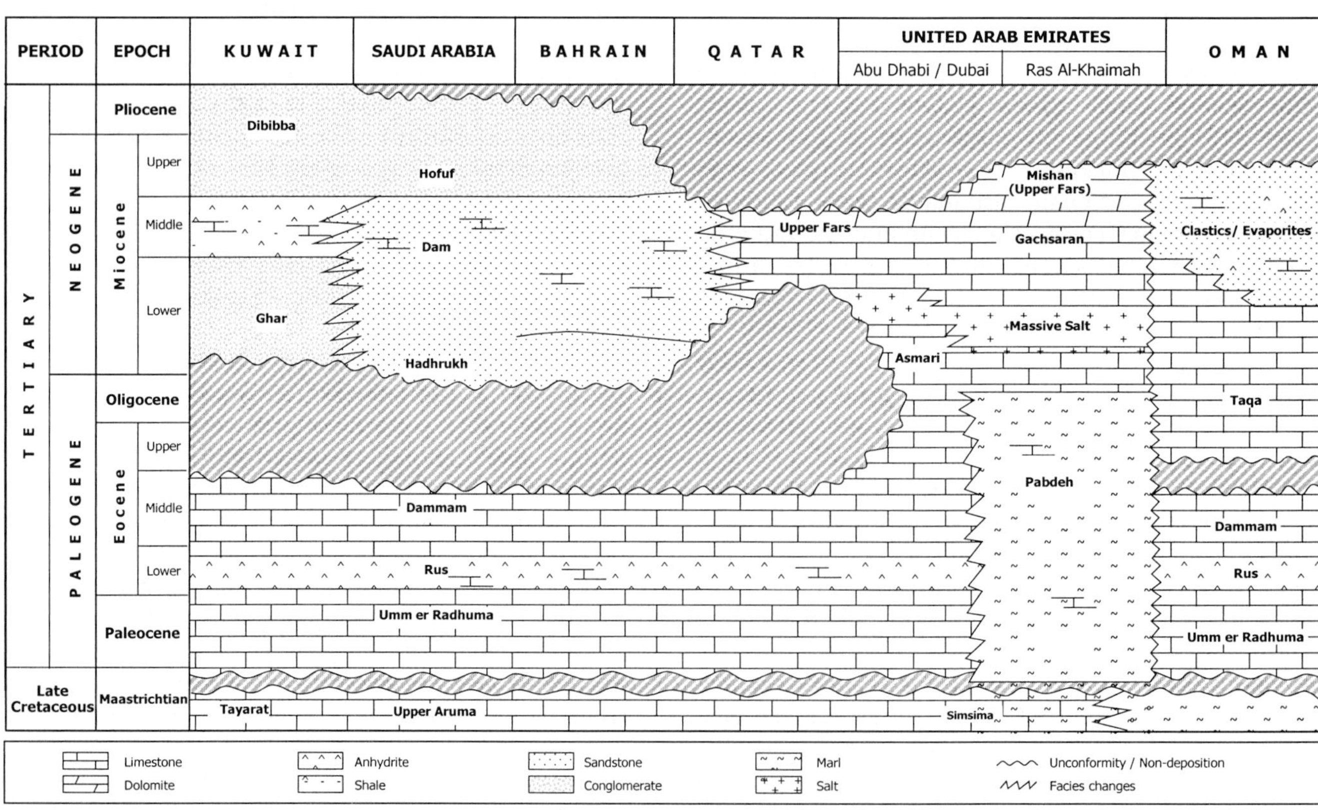

Fig. 3.9. Lithostratigraphic chart of the Paleogene-Neogene Formations in the Arabian Gulf and adjacent areas (Alsharhan and Nairn, 1995).

Tertiary: Stratigraphy and Sedimentation

The most important facies change during the Cenozoic was in eastern Arabia (the Ras Al Khaimah Basin) where a three-fold development of the basin occurs. The first was a Paleogene phase during which a Paleocene to Oligocene carbonate platform developed in the central and southern end of the Oman Foredeep. The second was marked by the flooding of the stable craton during late Paleocene time when a thick sequence of shelf carbonates and some evaporites were deposited over much of the platform extending to the Interior Homocline rim of the Arabian Shield. The sequence appears to be largely controlled by eustatic sea-level fluctuations, for this part of the craton was tectonically passive. The third development lay on the west side of the Musandam Peninsula orogen, where the northern end of the foredeep continued to subside, and receive a thick deposit of Paleocene through Oligocene calcareous flysch sediments. This was related to the continuation of thrusting from the Late Cretaceous orogenic phase. The axis of this early Tertiary foredeep was displaced westward from the axis of the Late Cretaceous Foredeep, and its continued subsidence resulted in a thick flysch deposits. By the early to middle Miocene, the subsidence rate in the Ras Al Khaimah trough decreased, to the point where a thick sequence of evaporites was deposited in a basin whose axis coincides with the southeastern part of the present-day Arabian Gulf. The great thickness of salt deposited in the early Miocene is overlain by middle Miocene mixed salt, anhydrite and carbonate. To the south in the area of the former central and southern Omani Foredeep, evaporitic sedimentation was less pronounced and much of the equivalent age strata consist of fine-grained clastic and carbonate rocks.

The lithostratigraphy of the Tertiary sediments (Fig. 3.9) indicate major lithofacies variations laterally and vertically, although some are too small or local to be shown. In the Arabian Gulf region and the Arabian Peninsula, the lower Tertiary sediments are made up of limestones, dolomites and evaporites. Although the Oligocene sediments were removed by erosion due to a worldwide drop in sea-level, some areas remained submerged and Oligocene sediments crop out in Abu Dhabi and Oman and in the subsurface in offshore United Arab Emirates.

Paleogene

The advancing Paleocene and lower Eocene seas deposited globigerinal marl in the central part of Iraq, and limestone with some shale in the area extending from Kuwait to the United Arab Emirates. In the northeastern Arabian platform, the continuation of the marine regression during the Paleocene led to the deposition of carbonates intercalated with evaporites (limestone, dolomitic and anhydritic limestone) with marl and shale in a shallow, locally restricted environment during Early Eocene. The upper part of the early Eocene witnessed the return of arid climatic conditions in the Arabian Gulf region, indicated by the deposition of evaporites, while the transgression of less saline water, during the middle Eocene, led to the deposition of dolomitic limestone and nummulitic limestone (Powers et al.,1966; Alsharhan and Nairn, 1994). This was followed by a slight elevation in the Arabian Platform at the end of the Middle Eocene, which led eventually to the regression of the remnant Tethys seaway, and the start of a long erosion process. In the northern part of the United Arab Emirates to the Iranian coast, a deep depositional depression was formed in which marl, argillaceous limestone, shale and some clastic rocks were deposited. Conditions in the eastern part of the Arabian Peninsula became continental with exceptional occurrence of small, intermittent marine flooding events which resulted in the deposition of the Miocene clastics (Powers et al., 1966; Alsharhan and Nairn, 1994).

Late Paleocene-lower Eocene sediments in the eastern region of the Saudi Arabia consist of a successive series of foraminifera-rich limestone, dolomitic limestone and chert. In Qatar the succession is made up of dolomite with intercalations of chert, marl, shale with some beds of anhydrite. Toward United Arab Emirates, a widespread transgression occurred during the Paleocene and thin beds of shale, bedded marl, followed by bioclastic bearing limestone with marly dolomite were deposited in some areas indicating shallow marine conditions.

The Paleocene-Eocene rocks in Oman Mountains are characterized by shallow neritic carbonates. In other areas early Eocene rocks are absent and an unconformity separates the Paleocene and middle Eocene. In the subsurface sections, west of the mountains, marine carbonate deposition continued during the upper Eocene-Oligocene with considerably increased thicknesses. During the Paleocene and Eocene in north Oman Mountains and the northern United Arab Emirates a deep depression (Ras Al Khaimah sub-basin which represent the continuation of Rub Al Khali basin) was formed in which argillaceous limestone and marl accumulated. On the eastern side of this basin, however, clastic sediments such as conglomerate, sandstone and shale with some beds of limestone were deposited. From the late Paleocene to early Eocene, shallow neritic carbonates were deposited in southwest Oman Mountains. The lower part of the sequence characterized by the presence of marl indicates low energy facies, whereas, the upper part contains nummulites-rich limestone with red algae and coral debris indicating a high energy environment.

In Kuwait, the continuation of the marine regression during the Paleocene led to the deposition of carbonate intercalated with evaporites (limestone, dolomitic and anhydritic limestone) with marl and shale in a shallow locally restricted environment. With the dominant regression at the beginning of the Eocene, carbonates were deposited in a quiet environment.

In eastern Arabia, the Ypresian sediments of Saudi Arabia are white crystalline anhydrite alternating with thin layers of green shale in the lower part, whereas the upper part is formed of grey marl alternating with shale and limestone. In Bahrain, the Ypressian consists of gypsum, anhydrite, dolomitic limestone with planar claystone, which also occurs in central Bahrain and decreases in thickness toward the upper part of the Bahrain fold and increases away from its axis. Toward Qatar, these rocks appear in the exposed sections wherever the overlying Middle Eocene has been removed by erosion. The thickness of the Early Eocene sediments is variable due to variations in the evaporite section caused by later leaching and dissolution, and consists of dolomitic limestone and limestone with thick horizons of gypsum and anhydrite.

In western United Arab Emirates (Abu Dhabi region), the Ypresian sediments were laid down on a wide evaporitic continental shelf, as alternations of evaporites and carbonates. Shallow, marly limestone prevail in the lower part of the formation their thickness increasing as their lithology changes into massive anhydrite representing supratidal and subaerial conditions.

Toward Oman, in the southwest Oman Mountains, similar sediments continued to be deposited during the Lower and Middle Eocene, whereas in the eastern part of the region the sequence is made up of impure sandy limestone, microporous limestone and shale. In the central and southern part of Oman, the Ypresian consists of a mixed facies of limestone, dolomite, shale, gypsum with anhydrite and evaporite the predominant lithology in most cases.

In eastern Arabia, the Middle Eocene sediments consist of coloured shale (brown to yellow, blue to bluish grey), *Alveolina*-rich limestone, dolomitic limestone, light to dark brown in colour with some marl and marly limestone. Toward northeastern Arabian Platform, the increasing regression during Lower and Middle Eocene led to the deposition of more evaporites with anhydrite, and mudstone. While the elevated areas witnessed erosion of the structures tops, the Ypresian rocks either were not deposited or were removed by later erosion.

Toward southern and southwestern Arabian Gulf as in Qatar, the Middle Eocene rests conformably on the Ypresian. The sediments consist mainly of shale with some thin argillaceous fossiliferous limestone. These are locally recrystallized, and in the north and northeastern areas, dolomitized. The upper Middle Eocene is widely distributed and rests conformably on the lower Middle Eocene. It consists of recrystallized, microporous limestone, with marl, fossiliferous shale and chert nodules. In the western United Arab Emirates the transgression of less saline marine water during the Eocene, led to the formation of a nearly normal, shallow marine shelf in which nummulite rich carbonates were laid down which form the Middle Eocene. The lower part of the formation was characterized by the presence of bedded calcareous marl and high energy barrier-type limestone, while the upper part is made up of subtidal limestone, dolomite with some marl.

During the Upper Eocene and the Oligocene the Arabian Peninsula was subjected to erosion due to uplift which had its maximum effect in Qatar and western part of United Arab Emirates. This led also to local erosion on the Eocene continental shelf followed by transgression and deposition of fossiliferous carbonates in shallow water during the Oligocene. These sediments are found in some exposures (Jabal Hafit in Abu Dhabi), and consist of coral and crinoidal limestone, dolomite with small calcite crystals, large foraminifera rich limestone (such as *Operculina* spp., and *Disocyclina* spp.) with algae in addition to a thin layer of greenish brown marl (Whittle et al., 1995). In exposures in the Oman Mountains, the Oligocene sediments found in the northeast mountain belt are made up of limestone, marl with coral patch reefs and evaporites. In northern region of the Arabian Gulf a transgression covered the whole area during the Lower Eocene and led to changes in the depositional environment and water depth. As a result, porous, dolomitic limestone and microporous limestone of the Middle Eocene, was deposited unconformably over the Ypresian, which is capped with very hard cap rocks, made up of siliceous cherty limestone. At the base, there is a brownish green bed of planar claystone with scattered dolomite and pyrite crystals, and sparse phosphatic nodules in some parts (Owen and Nasr, 1958; Alsharhan and Nairn, 1997).

Neogene

After the deposition of the Middle Eocene, regression was followed by uplift, that led to non-deposition and erosion of structure tops from upper

Eocene to Lower Miocene. This was followed by the deposition of a sandy facies in a shallow to very shallow environment from the Middle Miocene to the Pleistocene. This period was characterized by great floods leading to the deposition of river-mouth sediments. During the Middle Miocene, the sediments were made up of a succession of limestone, anhydrite, salt, red mudstone and bluish marl. These were followed by sandy limestone and fluvial sediments such as sandstone and conglomerate. The Pleistocene, was characterized by ungraded fluvial sediments made up mostly of cross-bedded sandstone and gravel with intercalations or lenses of sandy mudstone. The sediments were cemented locally by calcite or gypsum (Owen and Nasr, 1958; Fuchs, 1968). The most important uplift movements occurred during the early Oligocene and continued during the Miocene, leading to the erosion of most of the Oligocene sediments reducing the carbonate cover of the Upper Eocene and resulting in the local re-distribution of these sediments. With the continuation of these movements, the Oman Mountains and neighboring areas were uplifted, and compressional movements increased the folding and faulting, which formed the north Oman Mountains, but decreased in magnitude passing into asymetrical folds to the south and east.

It is probable that the Tertiary movements were activated, as early as the Late Cretaceous, as reflected by erosion in some areas. Among the prominent features which formed as a result of these movements, are the whale back structures, which appeared in the Tertiary which are similar in their position to those in the Zagros belt.

It seems that the climate and the paleogeography during the Paleocene affected to certain extent the geologic evolution of Oman Mountains and the neighboring areas during the Tertiary. The climate during the Miocene/Pliocene in Arabian Peninsula was nearly arid, so most of the known deposits of that period near the mountains were sandy valley deposits mixed with evaporites.

In summary, the sedimentary patterns of transgression and regression and the Paleogene stratigaphic sequence follow the global cycles suggested by Vail et al. (1977). The Lower Eocene sediments in the central and northern regions of Oman Mountains contain less evaporites than those found in the equivalent stratigraphic sequence in southwestern mountain belt and also in eastern Arabia and Qatar. It is also possible that this sequence may reflect past climatic effects (Glennie et al., 1974), since Oman was still in the tropical zone and in the same place which it occupied during the Late Cretaceous. However the climate was rainy due to the northern movement of the Indian plate across the old westerly wind zone.

During the Paleocene to the Early Eocene, most of the Arabian Peninsula was covered by a shallow evaporitic shelf. While in the northern United Arab Emirates to Zagros Mountains, the sediments were laid down in an open basin, in the central parts of this basin, the north Zagros Mountains, isolated carbonate islands were established, with the development of natural evaporitic depressions. The shallow nature of the basin was due to the development of Qatar-South Fars Arch, which exerted a positive regional influence. During the rest of the Eocene, the paleogeography remained the same but the climate was less arid, and consequently evaporites are missing. The open marine conditions contracted gradually. Late Eocene sediments are missing from the Arabian Platform due to non-deposition. In the Early and Middle Oligocene, there was a considerable subsidence in the sea-level.

At the end of the Oligocene and into the Early Miocene, the sedimentary basin contracted, while the evaporitic carbonate shelf remained trending NE/SW with the formation of depressions in the central part of the Gulf region. Some clastic debris from the east and west in northern part of the Gulf formed the Ahwaz delta. These sediments followed the Miocene and Pliocene carbonate and evaporates forming fluvial and valley-fill sediments of sandstone and conglomerate.

Chapter 4

AQUIFER AND AQUICLUDE SYSTEMS

INTRODUCTION

Groundwater in the Arabian Peninsula is found in thick, high yield aquifers within the large sedimentary basins to the north, east, and south of the Arabian Shield. Minor amounts of groundwater occur in fractured igneous and metamorphic rocks, which provide extensive, and permeable areas for surface runoff, and shallow wadi underflow.

Geological structures of the Arabian Plate (discussed in Chapter 3) control the geometry of the sedimentary basins and the hydraulic properties of major aquifers in the Arabian Gulf region. The area covered by sedimentary basins in the Gulf region is about $1.7x10^6$ km^2. These basins contain groundwater of different ages, at variable depths, and with a wide range of salinity, hydraulic properties, and recharge mechanisms. The sedimentary sequence (Fig. 4.1) has been classified and divided into many formations and members extending in age from Cambrian to Recent (see Powers, 1968; Alsharhan and Nairn, 1997). The water-bearing units in the Arabian Gulf can be divided into four aquifer groups:
(1) Precambrian-Paleozoic aquifers, which include, from older to younger, the Huqf, Haima, Saq, Tabuk, Wajid, Haushi and Khuff aquifers; (2) Triassic-Jurassic aquifers, which include Minjur, Dhruma, and Hanifa aquifers; (3) Cretaceous aquifers, which include Thamama (Sulaiy-Yamama-Buwaib), Wasia-Biyadh, Aruma and Simsima aquifers; and (4) Tertiary-Quaternary aquifers, which include Umm er Radhuma, Dammam, Rus and Neogene aquifers. The approximate areal extent of these aquifers in the Arabian Peninsula is shown in Figure (4.2), and hydrogeological groundwater flow direction is shown in Figure (4.3).

Whereas the aquifers in the north of the Arabian Peninsula consist of multiple, Lower Paleozoic, permeable clastic formations and confining layers, those of eastern Arabia include both karstified Tertiary carbonates and Mesozoic clastic/ carbonate formations. To the south of the Arabian Shield, a single thick Lower Paleozoic sandstone formation constitutes a high yield aquifer (Figs.4.1 and 4.3).

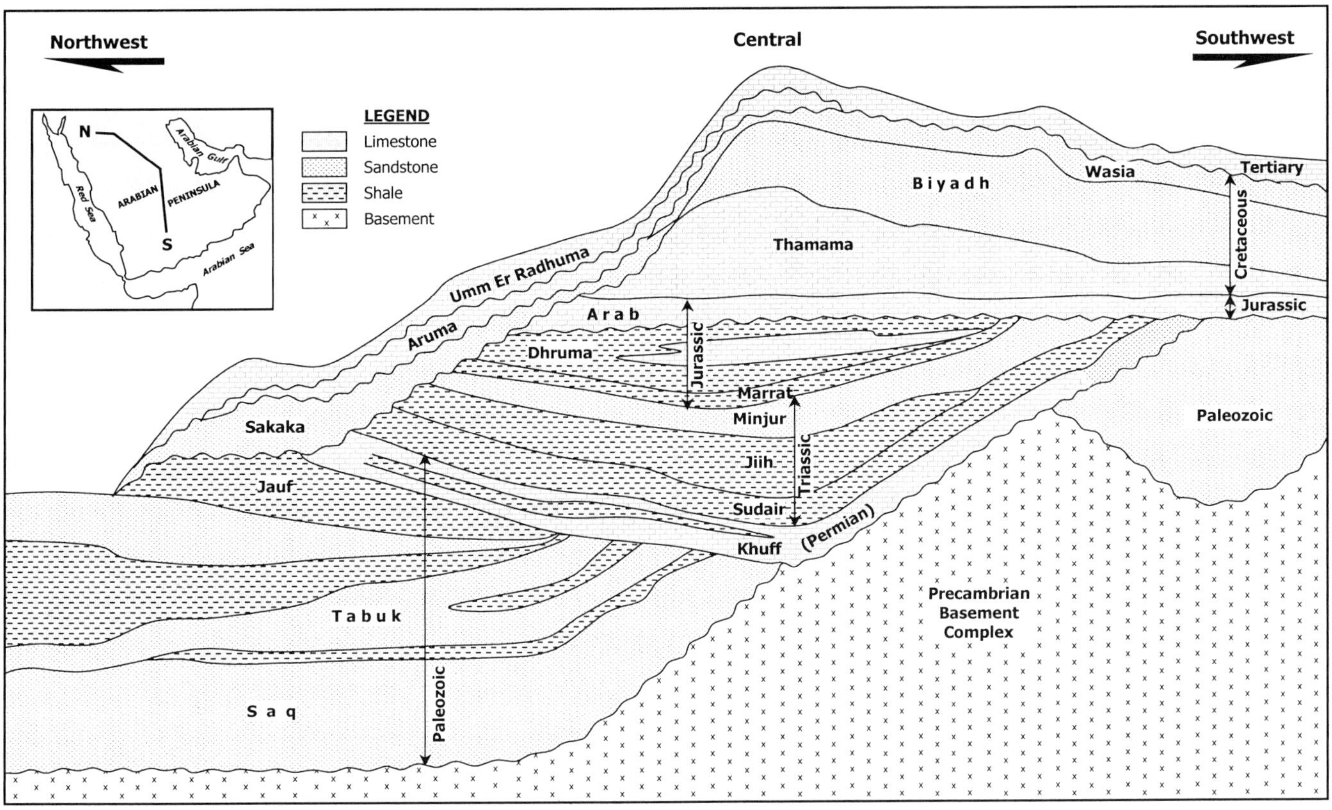

Fig. 4.1. Schematic stratigraphic-sedimentologic cross section in the Arabian Peninsula (modified from Powers, et al., 1966; Al Alawi and Abdulrazzak, 1994). These Arabian Shelf formations of mainly sandstone and limestone contains significant amounts of groundwater.

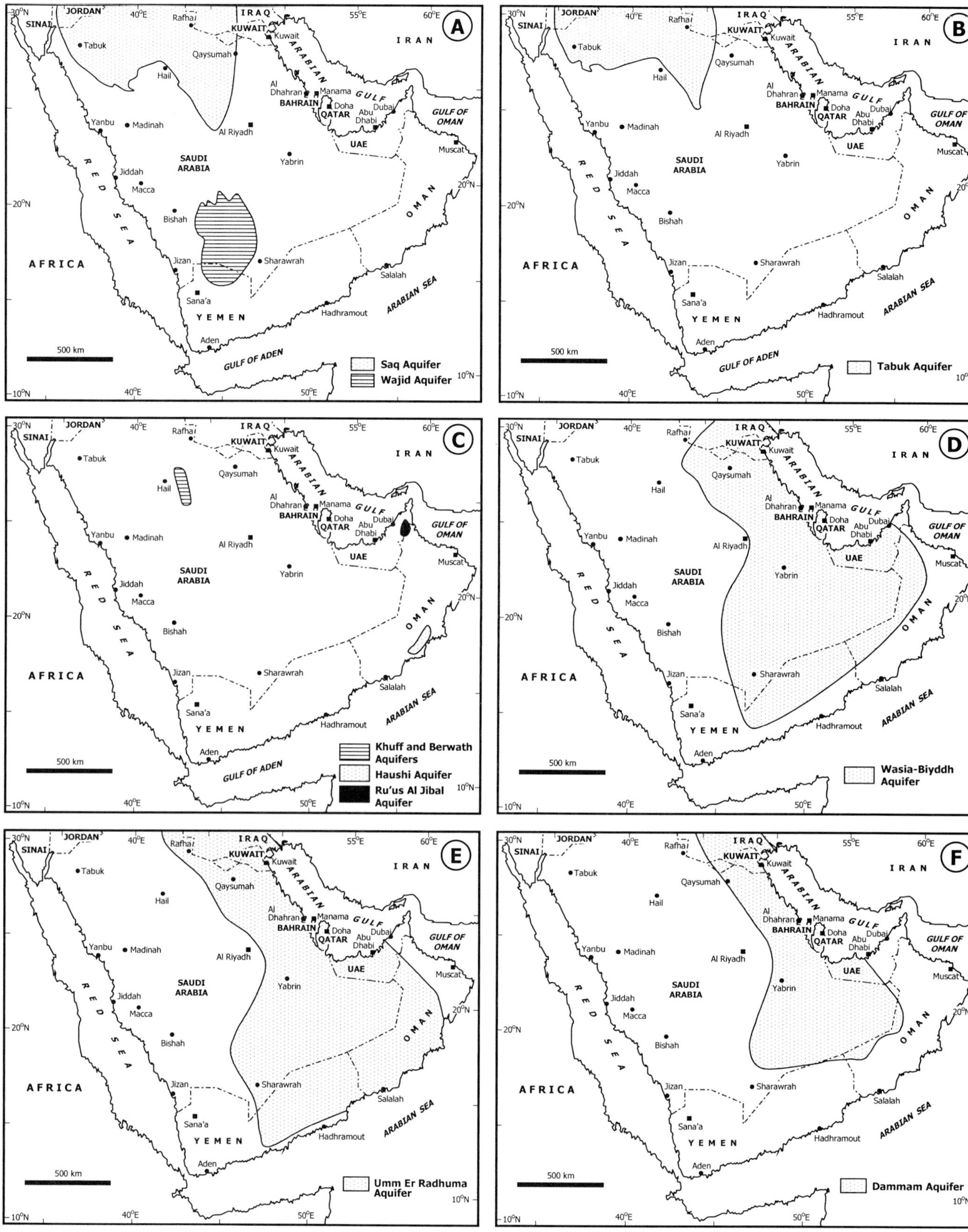

Fig. 4.2. Location map of the Arabian Peninsula showing distribution of major aquifers (modified from Saudi Arabian Ministry of Agriculture, 1984; Al Alawi and Abdulrazzak, 1994; Atlas of Saudi Arabia, 2000).

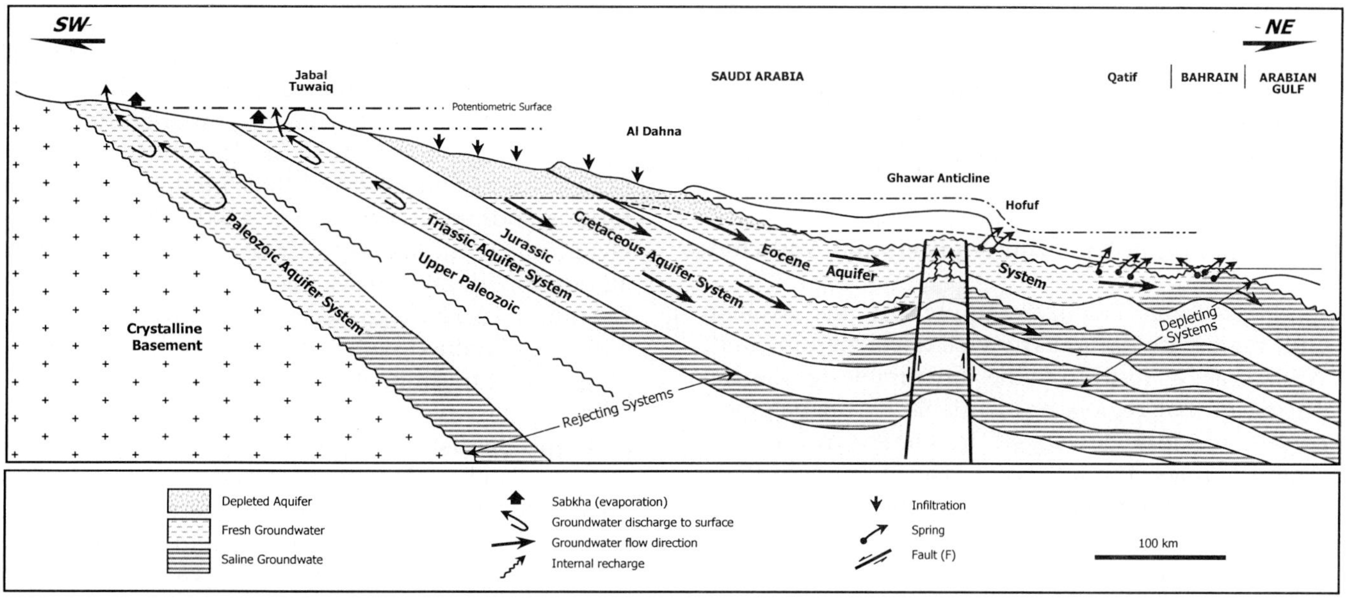

Fig. 4.3. Schematic hydrogeological cross section of the Eastern Arabian Peninsula (compiled with modification from Italconsult, 1969; FAO, 1981; Al Hajari, 1990; Al Alawi and Abdulrazzak, 1994).

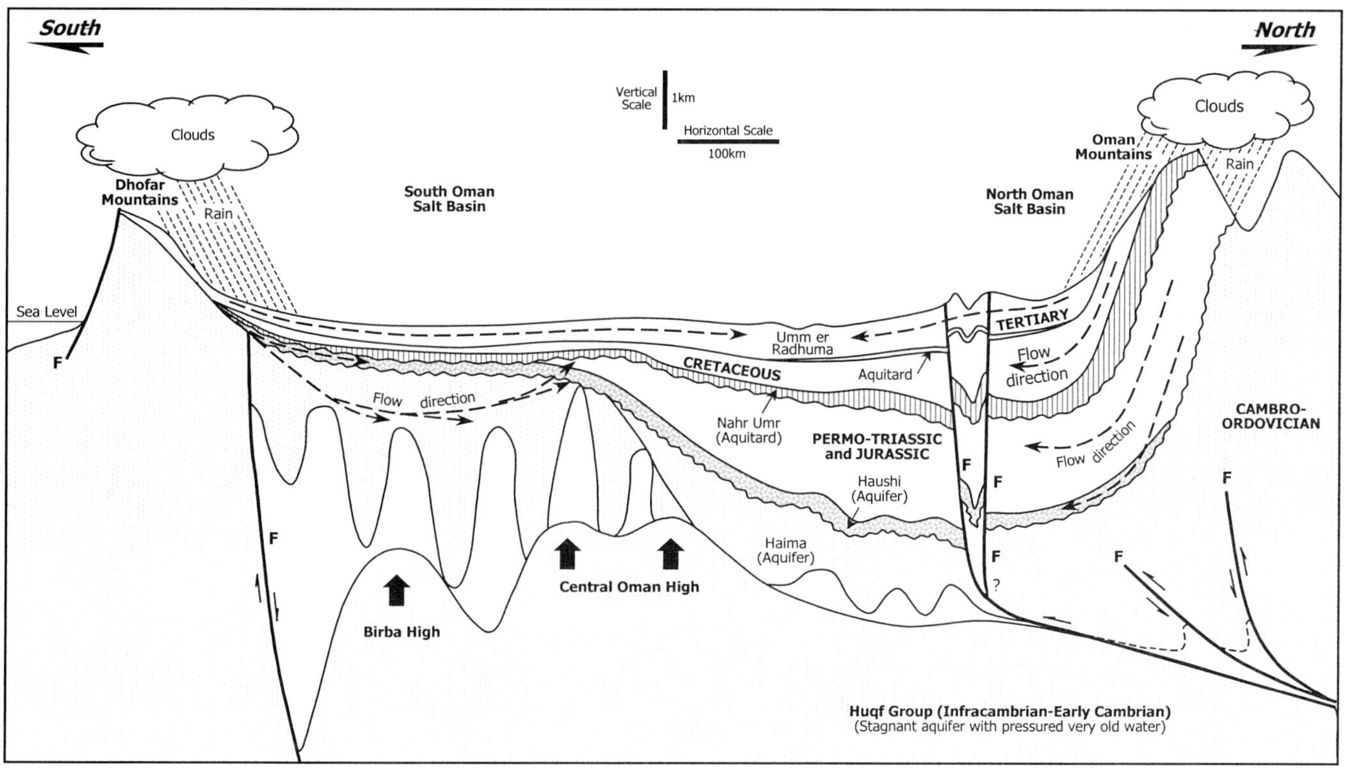

Fig. 4.4. A hydrogeological cross-section of Oman showing water flow regime (modified from Al Lamki and Terken, 1996).

The groundwater reserves in the numerous aquifers of the sedimentary basins of the Arabian Peninsula are fossil groundwater derived from earlier Quaternary intervals of much greater rainfall. Under the present arid to hyperarid conditions in the Arabian Gulf area very low rainfall conditions do not allow significant recharge to most aquifers in their exposed and unconfined parts.

Generally, groundwater is the most utilized water resource in the Arabian Peninsula, especially by the agricultural sector. Estimates indicate that 70 to 80% of groundwater is used for agricultural purposes as for example in United Arab Emirates, despite use of modern irrigation technologies which have increased the efficient use of water in agriculture. Nevertheless, more than 70% of groundwater consumption in agricultural projects.

In this chapter the most important aquifers and aquicludes in the Arabian Gulf area, and their hydraulic properties and chemical characteristics are described.

A. Precambrian-Paleozoic Aquifers and Aquicludes

Fresh-water Paleozoic aquifers in the Arabian Gulf region are particularly well developed in Saudi Arabia as a result of the favourable geological setting in that region, and suitable hydrogeological conditions, less developed in Oman and United Arab Emirates. The following is a summary of these aquifers and their hydraulic properties:

1. Huqf Aquifer

In Oman, the Huqf (Infracambrian-Early Cambrian) (Fig. 4.4), contains a stagnant water with salinities exceed 200,000 mg/l, presumably representing very old water with no significant flushing having occurred in the recent past (Al Lamki and Terken, 1996).

2. Saq Sandstone Aquifer

In northern Saudi Arabia, the major aquifer is the Cambro-Ordovician Saq Sandstone which rests directly on crystalline rocks of the Precambrian basement (Fig. 4.5). This aquifer ranges in thickness from 400 to 928m and consists of medium- to coarse-grained, brown to tan, friable, quartzose sandstone, with a basal conglomerate with moderate bedding thicknesses. Since the formation dips gradually towards the north-northeast under less permeable rocks (Fig. 4.5), it exists as a confined aquifer throughout most of the Tabuk Basin, and has an estimated area of almost 300,000 km^2. Effective porosity of the Saq Sandstone generally exceeds 15% (range: 10-20%) with transmissivity ranging from $9x10^{-3}$ to $3.8x10^{-2}$ m^2/sec, and storativity in the unconfined areas of $1.2x10^{-3}$ to $7x10^{2}$. In confined areas the Saq aquifer has storativity ranging from 10^{-4} to $2x10^{-3}$ (King Fahd University Staff, 1987).

The isotopic age of water in the Saq Sandstone indicates that it is a fossil groundwater, 22,000 to 28,000 years old. This precludes any significant recharge at the present time, and suggests that major recharge occurred during the Quaternary Ice Age pluvial intervals.

The Saq-Tabuk aquifer system in northern Saudi Arabia (Fig. 4.1) varies in thickness from 250 to 500m and is composed mainly of coarse-grained, well-sorted sandstone with thin silty interbeds. The isosalinity contour map indicates that the Total Dissolved Solids (TDS) of groundwater in the aquifer ranges from 1,300 to 3,000 mg/l. However, a high salinity of about 14,500 mg/l was measured in some wells, rendering the water unfit for any use before desalination and treatment (Khalifa and Rizk, 1994). The calculated transmissivity of the aquifer averages $0.01m^2/sec$. A trilinear plot of groundwater samples from these aquifers in the Al-Mustawi area (Al Qasim province) shows that the dominant water type is sodium chloride (Fig. 4.6a). A plot of the electrical conductivity (EC) versus sodium adsorption ratio (SAR) using the U.S. Salinity Laboratory Staff diagram proposed in 1954 indicates that the examined water samples are hazardous for irrigation of traditional crops (Fig. 4.6b).

3. Wajid Sandstone Aquifer

The Cambro-Ordovician Wajid sandstone covers an area of 196,000 km^2 in southwestern Saudi Arabia (Figs. 4.1 and 4.2). Within this sandstone is a confined aquifer with a thickness varying between 200 and 900m, occurs over an area of about 170,000 km^2 (Edgell, 1990). Effective porosity of this aquifer is in the order of 20%, and water stored in the aquifer is very conservatively estimated at 30 Bm^3 (Saudi Arabian Ministry of Agriculture and Water, 1984). Water quality is good, with less than 1,000 mg/l of total dissolved solids. Isotopic determinations of the age of the groundwater in the Wajid aquifer show that it is fossil water dated at more than 30,000 years old (Saudi Arabia Ministry of Agriculture and Water, 1979). Transmissivity varies between $5.7x10^{-4}$ and $2.1x10^{-2}$ m^2/sec, while storativity ranges from $2x10^{-4}$ to $2x10^{-1}$ in the larger confined part of the aquifer (Authman, 1983).

4. Tabuk Aquifers and Aquicludes

The Tabuk Formation (Ordovician-Silurian) consists of more than 1,700 m of rhythmically alternating marine shales and continental to marginal-marine sandstones, representing periods of transgressions and regressions in the formation (see Powers et al., 1966; Al Laboun, 1986). Within the Tabuk Formation, there is an alternation of aquifers and aquicludes. In ascending sequence, three major aquifers can be readily recognized, namely the Lower Tabuk Sandstone aquifer, the Middle Tabuk Sandstone aquifer, and the Upper Tabuk Sandstone aquifer (Figs. 4.1 and 4.2).

a) Lower Tabuk Aquiclude and Aquifer

The Hanadir Shale Member (Early Ordovician) is the lowest member of the Tabuk Formation, and ranges in thickness from 54 to 100m. It consists of greenish gray to dark gray shale, with minor greenish gray, silty shale interbedded in its upper part. The Hanadir Shale is well-developed in northwestern Saudi Arabia and is of considerable hydrogeological significance as an Aquiclude overlying the Saq Sandstone aquifer developed in that area.

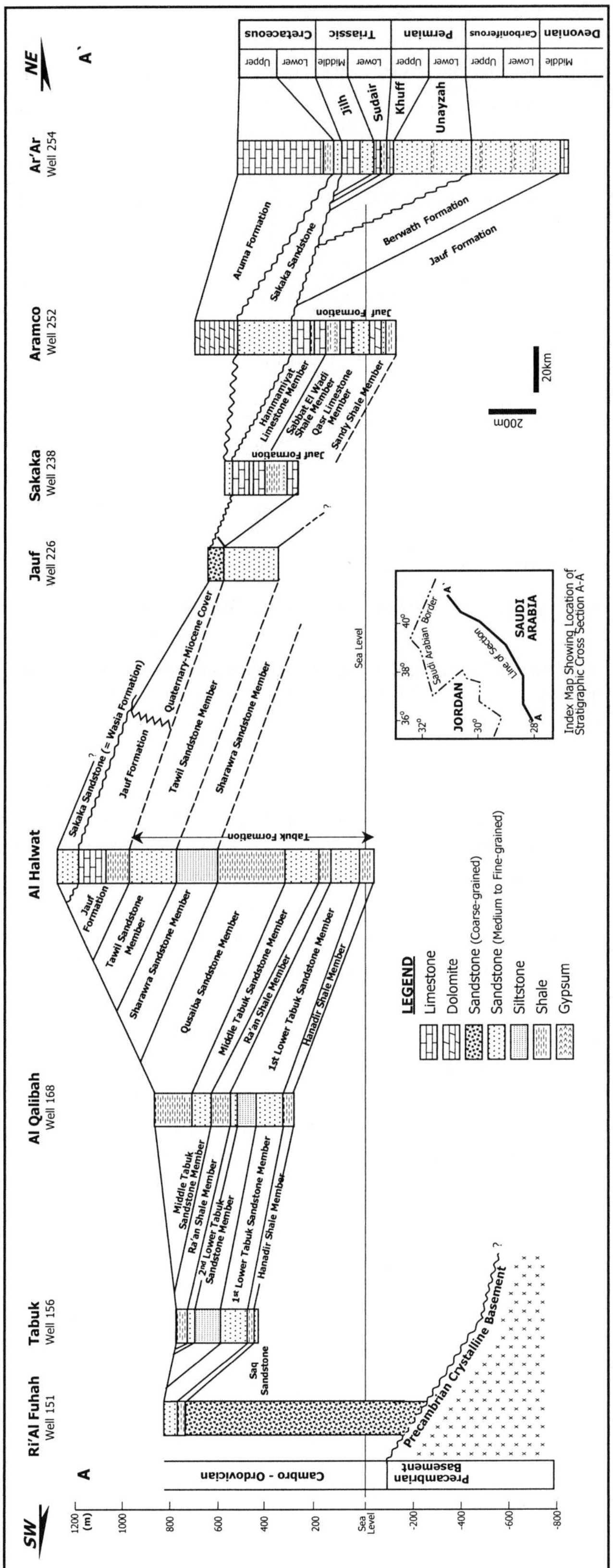

Fig. 4.5. Stratigraphic-sedimentologic cross section of Tabuk and Widyan basins in Saudi Arabia (modified from Edgell, 1997).

The Lower Tabuk Sandstone aquifer is the Lower Ordovician sandstone member of the Tabuk Formation. In northwestern Saudi Arabia it ranges in thickness from 54 to 390m, and consists of fine- to medium-grained, micaceous sandstone with thin, minor siltstone and shale intercalations in its upper part. It has suitable characteristics as a confined aquifer, with an effective porosity from 8 to 20%. Transmissivity of the Lower Tabuk aquifer averages about $2x10^{-3}$ m^2/sec, while its storativity varies from $2x10^{-2}$ (where the aquifer is unconfined) to $6x10^{-4}$ (where it is confined).

The Ra'an Shale Member is about 14 to 100m thick of purplish gray to greenish gray micaceous, graphitic shales. It rests directly overlies the Lower Tabuk Sandstone aquifer, and acts as its upper aquiclude.

b) Middle Tabuk Aquifer

The Middle Tabuk Sandstone aquifer is comprised of the Tabuk sandstone member and the Qusaiba shale member. The Tabuk sandstone member (Early Silurian) has a thickness of 83 to 242m in the Tabuk basin and consists mainly of fine- to medium-grained, micaceous, cross-bedded, quartz sandstones with a basal conglomerate. The latter has striated boulders derived from Late Ordovician glacial beds (McClure, 1978). The Tabuk sandstone member is a confined aquifer with an effective porosity of at least 10%. Transmissivity is about $1.7x10^{-3}$ m^2/sec, and its storativity is $2x10^{-4}$, where unconfined, and $1x10^{-3}$ where confined (King Fahd University Staff, 1987).

The Qusaiba Shale member (Early Silurian) consists of 50-206m thick of varicoloured, red, gray, and gray-green, graptolitic shales with some ferruginous siltstones in its upper part. It forms an important aquiclude confining groundwater in the Middle Tabuk Sandstone aquifer.

c) Upper Tabuk Aquifer

Two sandstone members of the uppermost part of the Tabuk Formation comprise the Upper Tabuk Sandstone Aquifer. They include the Lower Silurian Sharawa Sandstone Member and the disconformably overlying Lower Devonian Tawil Sandstone Member (Powers, 1968; Alsharhan and Nairn, 1997). The Sharawa Sandstone Member is quite thick, ranging from 380 to 765m of fine-grained sandstone, with much interbedded shale in the lower part. The hydrological characteristics of the Tawil Sandstone are much the better of the two (10 to 20% effective porosity), as it consists of about 190m of gray to brown, medium- to coarse-grained, sometimes pebbly, dominantly cross-bedded, quartz sandstone. Transmissivity of the Upper Tabuk Sandstone aquifer ranges from $10x10^{-3}$ to $7x10^{-4}$

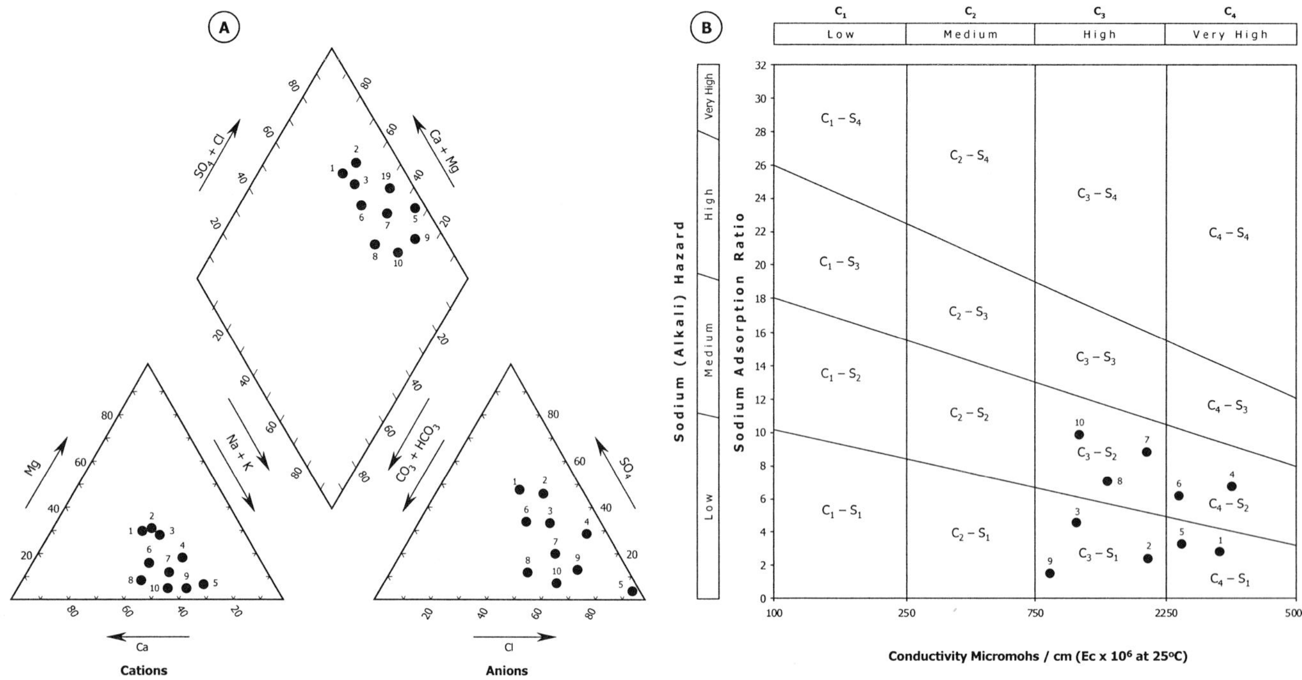

Fig. 4.6. A trilinear plot (A) and Sodium hazard (B) of some groundwater samples from Al-Mustawi area, Al-Qasim Province, northern Saudi Arabia (modified from Khalifa and Rizk, 1994).

m^2/sec, while storativity varies between $1.4x10^{-3}$ (confined) to $3.5x10^{-2}$, (unconfined) parts (King Fahd University Staff, 1987).

5. Jauf Aquifer and Aquiclude

The Devonian Jauf Formation directly overlies the Upper Tabuk Sandstone aquifer in the Tabuk basin (Figs. 4.1 and 4.5). The Jauf sandstone together with the sandy shale of the Shaibah Member constitutes a local aquiclude for the Upper Tabuk Sandstone aquifer. The aquifer yields good quality water with total dissolved solids reaching only 300 mg/l. Transmissivity of the Jauf Aquifer is generally low ($3x10^{-3}$ m^2/sec), where the aquifer is confined, and higher at $1.1x10^{-2}$ m^2/sec, where unconfined. Storativity ranges from $2x10^{-2}$ (unconfined part) to about $2x10^{-3}$ (confined part).

6. Berwath Aquifer

The Carboniferous Berwath Formation occurs mainly in the subsurface of the Midyan Basin margin, to the north of Turabah (Al Laboun, 1986) (Fig. 4.5). It consists of fine- to coarse-grained, argillaceous sandstone, which provide fair to moderate yields of good quality water in the area. An average transmissivity is $3x10^{-3}$ m^2/sec, and storativity is about $1x10^{-3}$.

7. Unayzah Aquifer

In Saudi Arabia, the Carboniferous to Lower Permian clastics of the Unayzah Formation range from 33 to 45m in thickness, and comprise cross-bedded, fine- to coarse-grained fluvial sandstones with siltstones, green-gray and purple claystones, and thin beds of impure limestones (Al Laboun, 1987) (Fig. 4.5). This formation has yielded only modest amounts of groundwater, and has a relatively low storativity.

8. Haushi Aquifer

In Oman, the Permo-Carboniferous clastic Haushi Group includes an important aquifer (Fig. 4.4). Haushi Group sediments thin towards the north and east, and their deposition has been associated with the thermal doming phase that preceded the separation of the Indian craton from the Gondwana continent.

9. Khuff Aquifer

In Saudi Arabia, the Upper Permian Khuff Formation is extensive in the subsurface, from the southern Rub al Khali northwards into southern Iraq (Figs. 4.1 and 4.5). It ranges in thickness from 250 to 600m and lithologically, it consists of limestones and dolomites, with some anhydrite aquicludes, which divide it into four permeable carbonate units. These carbonates act as rather poor aquifers (3% effective porosity) yielding mineralized water. Aquifers of the Khuff Formation in the north (Fig. 4.2) are secondary due to low storativity and low yield.

10. Ru'us Al Jibal Aquifer

Rocks of equivalent age of the Khuff Formation known as the Ru'us Al Jibal Group (Permian to the

Early Triassic), having a thickness of 1,500m well-developed in northern part of the United Arab Emirates (Fig. 4.2). This group consists of medium to thick-bedded, bioturbated dolomite occasionally intercalcated with skeletal and partly oolitic grainstone and some minor porcellaneous dolomitic limestone. The Ru'us Al Jibal Group contains the main aquifer in Ras Al Khaimah Emirate and contains considerable amounts of recent and old waters. The hydraulic conductivity of the Ru'us Al Jibal limestone aquifers varies considerably, due to their fractured and cavernous nature, as described by Todd (1980) for similar aquifers from other areas.

In Wadi Al Bih (Ras Al Khaimah Emirate) water occurs in these rocks in which hydraulic conductivity ranges between 32.65m/day and 67.30 m/day, transmissivity was 1,324m^2/day to 2,729m^2/day, and storativity was 8.26x10^{-5} to 1.53x10^{-3}. Heavy groundwater pumping during the last three decades at an average annual rate of 11.23 Mm^3 has resulted in a steady rise in groundwater salinity, and a continuous decline in hydraulic head. The hydraulic head map of Wadi Al Bih limestone aquifer shows that the water table decreases from 28m above sea level in the upstream of Wadi Al Bih dams to 5m below sea level in the Al Burayrat area at the mouth of the wadi (Fig. 4.7).

Oxygen and hydrogen isotopic data from most of the groundwater in the Ru'us Al Jibal aquifers plots close to the meteoric water line, with a deuterium excess of +20. Those data plotting to the right of the meteoric water line are consistent with evaporitic effects, mixing of different waters, or salt-water intrusion from the Arabian Gulf. The aquifers contain groundwater with a ^{3}H, δ^{18}O isotopic composition more depleted than the average values of the stable isotopic composition of the atmosphere, indicating that the possible source of moisture is the Mediterranean region which recharges groundwater at higher elevations (Akiti et al., 1992).

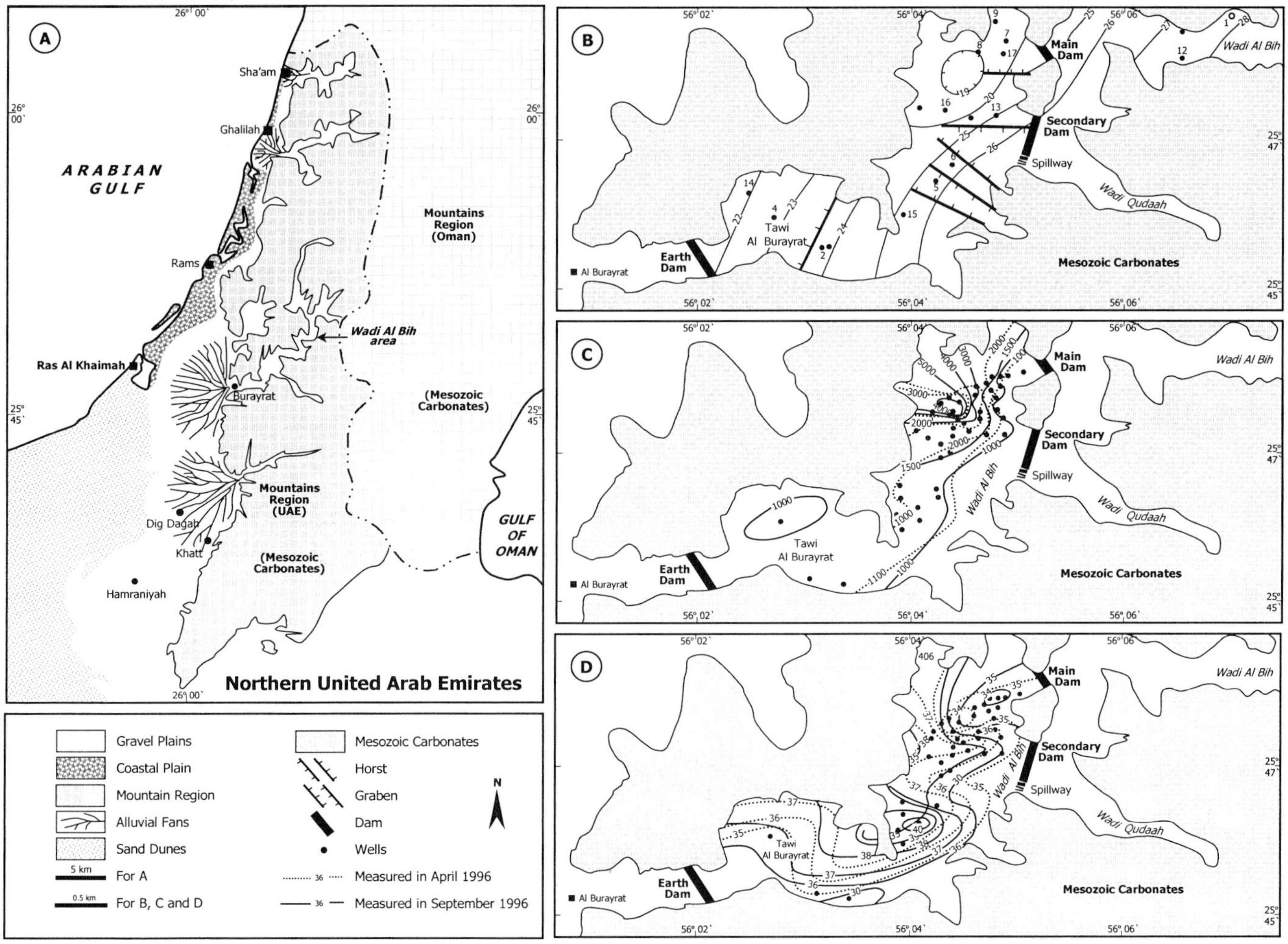

Fig. 4.7. A) Main geomorphological features of northern United Arab Emirates showing the location of Wadi Al Bih area; B) Hydraulic-head map (in meters) relative to the mean sea level of Wadi Al Bih limestone aquifer (measurement taken in September 1997); C) Isosalinity (mg/l) contour map of groundwater in Wadi Al Bih limestone aquifer (measurement taken in April and September 1996 represent by dot and solid lines respectively; D) Iso-temperature (°C) contour map of groundwater in Wadi Al Bih limestone aquifer (measurement taken in April and September 1996, represent by dot and solid lines respectively). (B,C, and D modified from Al Asam, 1997, and Al Wahedi, 1997).

The parallel increase in Electrical Conductivity and $\delta^{18}O$ from 3,200 μS/cm /l and -3.54‰ (in 1985) to 15,070 μS/cm/l and -2.89‰ (in 1986) at Wadi Sahawat shows the effect of mixing between sea water and groundwater. It was also noted that the ^3H content also decreased from 6.7 to 3.7 TU. In Dohreen, the increase of Electrical Conductivity from 2,400μS/cm, (in 1984), to 6,000μS/cm (in 1986), with the $\delta^{18}O$ remaining constant at around -3.50‰ suggests dissolution of salt in the soil during groundwater infiltration. The above two processes lead to deterioration of groundwater quality. Additionally, overpumping can induce salt-water intrusion into fresh groundwater through subsurface fractures.

B. Mesozoic Aquifers and Aquicludes

Mesozoic aquifers contain fresh water in some parts of Saudi Arabia, while in the United Arab Emirates these aquifers contain brines which can not be used for most domestic and agricultural purposes. The following is a summary of the Mesozoic aquifers in the Arabian Gulf region.

1. Sudair Shale Aquiclude

The Early Triassic Sudair Formation averages 200m in thickness and is mainly composed of brick-red to dark-red, massive shale with minor gypsum interbeds. This formation acts as a widespread aquiclude for the Khuff aquifers and for the Upper Tabuk Sandstone aquifer in northern Saudi Arabia (Figs. 4.1 and 4.5).

2. Jilh Aquifer

In Saudi Arabia, the Jilh Formation (Middle Triassic) ranges from 100 to 400m thick, and consists of thin-bedded limestone, with minor shale and gypsum. It is generally forms a secondary aquifer, owing to its poor water quality, as a result of dissolved calcium sulphate from anhydrite and gypsum layers in the formation (Fig. 4.1). East of Riyadh, a number of deep wells were drilled in 1984 to tap the Jilh Aquifer, with yields averaging 63 l/sec from sandstones (Saudi Arabia Ministry of Agriculture and Water, 1984).

The Jilh aquifer in Al-Qasim Province of northern Saudi Arabia, consists of 180 to 250m thick, well sorted, fine to medium-grained sandstone. The potentiometric surface indicates that the groundwater flow in the Jilh aquifer is from southwest to northeast in northern Saudi Arabia (Fig. 4.3). The isosalinity contour map shows that total dissolved solids range from 800 to 1,200mg/l, increasing in the direction of groundwater flow. Chemical analysis of water samples from the Jilh aquifer indicate that its water is NaCl type, possibly indicating a significant age. The quality of the Jilh aquifer within this region is generally good for almost all purposes including domestic uses (Khalifa and Rizk, 1994).

3. Minjur Aquifer

The Minjur Formation is 400m thick, and consists of cross-bedded, coarse- to very coarse-grained, quartz sandstone separated by middle Minjur shales and mudstones (Fig. 4.1). It comprises

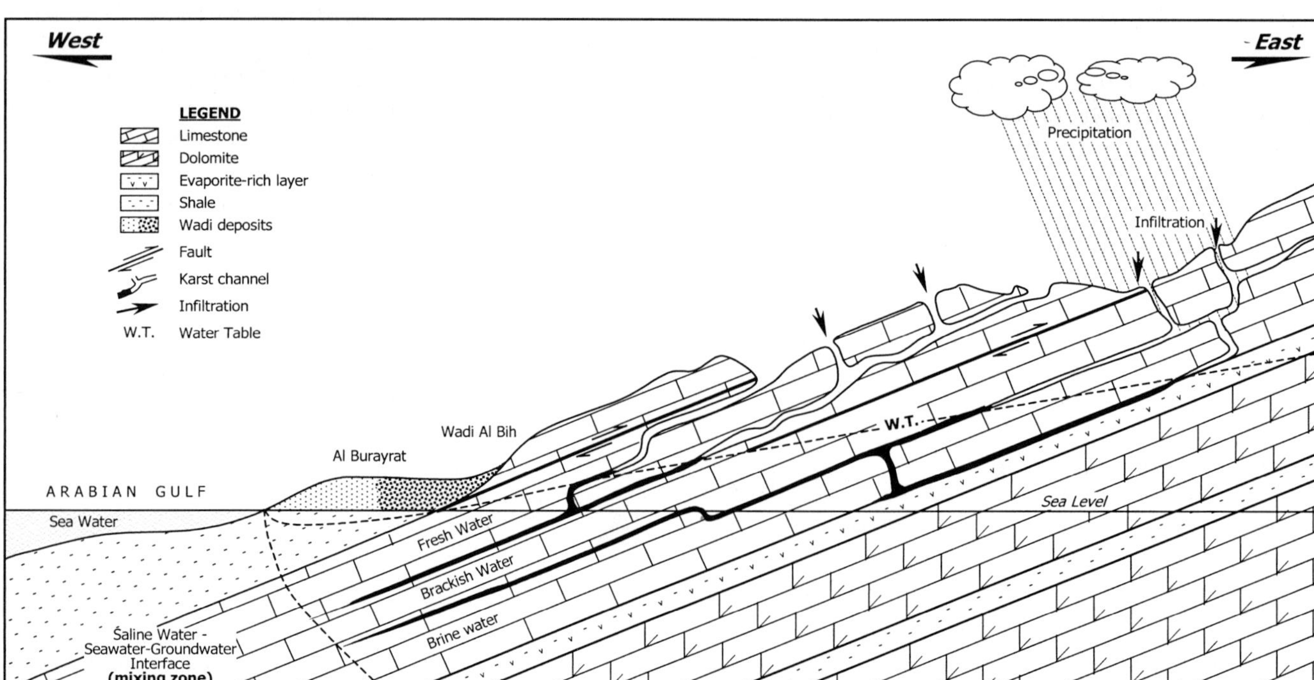

Fig. 4.8. Schematic diagram showing the hydrologic conditions of the Wadi Al Bih limestone aquifer in northern United Arab Emirates (modified from Al Asam 1997, Rizk and Alsharhan, 1999).

two sandstone aquifers: Upper Minjur aquifer and the Lower Minjur aquifer. These two aquifers are of considerable importance, since they yield 90% of the groundwater supply to the capital of Saudi Arabia (City of Riyadh). In the Sudair-Riyadh-Al Aflaj area, it is estimated that the Minjur aquifers contain 460×10^6 m^3 of sufficiently good quality water for general use (El Khatib, 1980). As an aquifer, the Minjur has good transmissivity and good water quality, 1,400 mg/l TDS and 1,600 mg/l TDS at Riyadh and Sudair areas respectively. However, the water quality becomes poorer southward and, at Al Aflaj, it has as much as 4,100 mg/l TDS. The Minjur Sandstone acts primarily as a confined aquifer, although south of 23°N, it forms a single aquifer complex with the sandy Dhruma Formation. Transmissivity of the Minjur Sandstone in the Riyadh area varies from 1.7×10^{-3} to 7.2×10^{-3} m^2/sec, while its storativity as a confined aquifer is 1.3×10^{-4}. Because of its limited outcrop area, recharge of the Minjur aquifer from direct rainfall is very small. On the other hand, the hydraulic connection between the aquifer and the underlying Jilh aquifer and overlying Middle Jurassic and Quaternary aquifers, provides recharge sources for the Minjur aquifer (Sakr, 1989).

4. Marrat Aquiclude

The Lower Jurassic Marrat Formation, measuring 120m thick, directly overlies the Minjur Sandstone north of Riyadh and consists of aphanitic and calcarenitic limestone, with interbedded shale, siltstone, and sandstone towards its base (Fig. 4.1). The lower thin sandstones generally contain small quantities of fair to good quality groundwater, and are considered as the uppermost part of the Minjur aquifer. The shales, siltstones, and aphanitic limestones of the Marrat Formation generally act as an aquiclude for the important Minjur aquifer.

5. Dhruma Aquifer

Middle Jurassic Dhruma Formation in Saudi Arabia south of Latitude 22°N, along the Interior Homocline is about 375m thick and is dominated by sandstone, with limestone interbeds (Fig. 4.1). It has moderate to good yields of high quality groundwater. In the southern area, this formation joins with the Minjur Sandstone to form one interconnected aquifer, often referred to as the Minjur/Dhruma aquifer (Saudi Arabian Ministry of Agriculture and Water, 1984), with transmissivity values from 1×10^{-2} to 1.6×10^{-2} m^2/sec.

6. Upper Jurassic Aquitard and Aquifer

In Saudi Arabia, the Upper Jurassic formations comprising the Tuwaiq Mountain Limestone, Hanifa, Jubaila, Arab and the Hith Anhydrite have very limited hydrogeological significance, since they are either dense, aphanitic and argillaceous limestones, which act as aquitards such as Tuwaiq Mountain Limestone and the Jubaila Formation, or they contain much anhydrite, as with the Arab Formation and Hith Anhydrite, which both yield only limited amounts of highly mineralized water. The Hanifa Formation has good permeability downdip, and provides limited amounts of groundwater from wells in Wadi Hanifa of Central Saudi Arabia.

In the United Arab Emirates, a massive, monotonously consistent sequence of about 1,475m thick of dark grey shallow marine carbonates of Jurassic to Lower Cretaceous age covers the bulk of the Musandam Peninsula in the northern Oman Mountains. These carbonates are known as the Musandum Group, and they represent the main recharge source for groundwater in Wadi Al Bih basin in Ras Al Khaimah Emirate (Fig. 4.8). This recharge is indicated by seasonal changes in groundwater temperature and salinity (see also Fig. 4.7).

7. Sulaiy-Yamama-Buwaib Aquifers

The carbonates of the Sulaiy, Yamama and Buwaib, representing the lower part of the Lower Cretaceous in Central Saudi Arabia, are generally unimportant as groundwater sources, with a few local exceptions (Figs. 4.1 and 4.3). Thus, in Wadi Sulaiy, about 20 km southeast of Riyadh, calcarenite in the lower part of the Sulaiy Formation forms a minor local aquifer, with water quality of 1,000 to 3,000 mg/l of total dissolved solids. Elsewhere, the Sulaiy Formation consists of about 150m thick of low permeability, aphanitic and calcarenitic limestones. The overlying Yamama Formation comprises about 50m of tightly cemented, pellet-calcarenites with aphanitic limestones in the lower part. It is also of little importance from a hydrologic viewpoint, except in the Layla lakes area ('Uyun Al Aflaj), where dissolution of the underlying Hith Anhydrite by groundwater has caused the Sulaiy, Yamama, and Buwaib formations to collapse and become brecciated, so that good secondary permeability is developed and groundwater circulates freely (Saudi Arabian Ministry of Agriculture and Water, 1984). This area was developed in ancient times by falaj, or qanats, around a lake formerly about 4 km^2 in area (Ritter, 1981), although the lake is now reduced to seventeen small lakes. The Buwaib Formation consists of 18 to 100m thick of intimately interbedded shale, siltstone, dolomite, calcarenite, and aphanitic limestone, and has no known groundwater potential other than the local occurrence at Uyun al Aflaj.

8. Biyadh- Wasia Aquifer

In Saudi Arabia, the sandstone of the Biyadh Formation is a major aquifer in the northeastern and eastern provinces. The Biyadh Formation consists of 425m of cross-bedded, quartz sandstone, with some shale and conglomeratic layers (Figs. 4.1, 4.2 and 4.3). The quality of water from the Biyadh Sandstone is very good in the Nisah Graben (500-900 mg/l TDS) and fair at Al Kharj, but is poor at depth, and in the Eastern Province (De Jong et al. 1989). It provides good yields of groundwater in Wadi Nisah.

Although, the Biyadh Sandstone is unconformably overlain by sandstones of the Wasia Formation, the two formations form one immense aquifer system in eastern and northern Saudi Arabia, with an estimated recharge in 1967 of 480 Mm^3/yr from runoff from the Tuwaiq Mountains, and from wadi underflow (Saudi Arabian Ministry of Agriculture and Water, 1984). A short distance downdip, fluid communication between the Biyadh and Wasia formations is restricted and each operates as an isolated hydrogeologic system, due to presence of shales towards the northeast at the top of the Biyadh Sandstone. In Riyadh region, near Al Kharj, water in this aquifer was found to be 8,000 years old, whereas it is 16,000 years old downdip at Al Khurais.

In northern Saudi Arabia (around Sakaka), the Wasia Formation is about 200m thick and is a good aquifer with high effective porosity ranging from 10 to 29%. It also has favorable hydraulic characteristics, with an average transmissivity of $3x10^{-4}$ to $2.8x10^{-3}$ m^2/sec, increasing northeastward to $15x10^{-3}$ m^2/sec. The average storage coefficient ranges from $3x10^{-2}$ to $5x10^{-2}$ in the unconfined part, and is about $7x10^{-3}$ in its confined part (Edgell, 1990). This united Wasia-Biyadh Aquifer is one of the largest regional potential aquifers holding very large amounts of stored water, estimated at several billion cubic meters in the Riyadh area alone (El Khatib, 1980). It supplies untreated water to Riyadh and has excellent hydrological properties. In Al Khurais area, it provides large yields of groundwater containing 550 to 1,500 mg/l of total dissolved solids. Basinward, the Wasia Formation divides into seven members, and the groundwater quality becomes progressively poorer towards the northeast, where it ultimately contains oil field brines in many areas of offshore Saudi Arabia.

9. Aruma Aquifer

In Saudi Arabia, the Aruma Formation has a thickness range from 60 to 140m, and consists primarily of fine-grained to chalky, or calcarenite limestone, with some shale and dolomite towards the top (Fig. 4.1).

The Aruma Aquifer is a secondary aquifer with relatively poor quality groundwater, though its groundwater is used by wells on the pipeline stations of Badanah and Rafha in the northeastern province. Downdip, water quality decreases, so that the eastern edge of potable water can be taken as a line from Hafr al Batin to the southwestern end of Ghawar. In the Wadi Al Sirhan Basin, the Aruma aquifer is mostly shale and shelly marl, with a few thin limestone interbeds. It acts as an aquitard, allowing water from the underlying Wasia Formation to seep upwards into Tertiary aquifers (Khouri, 1982). In the area east of Wadi Dawasir, the Aruma Formation occurs in sandy facies and

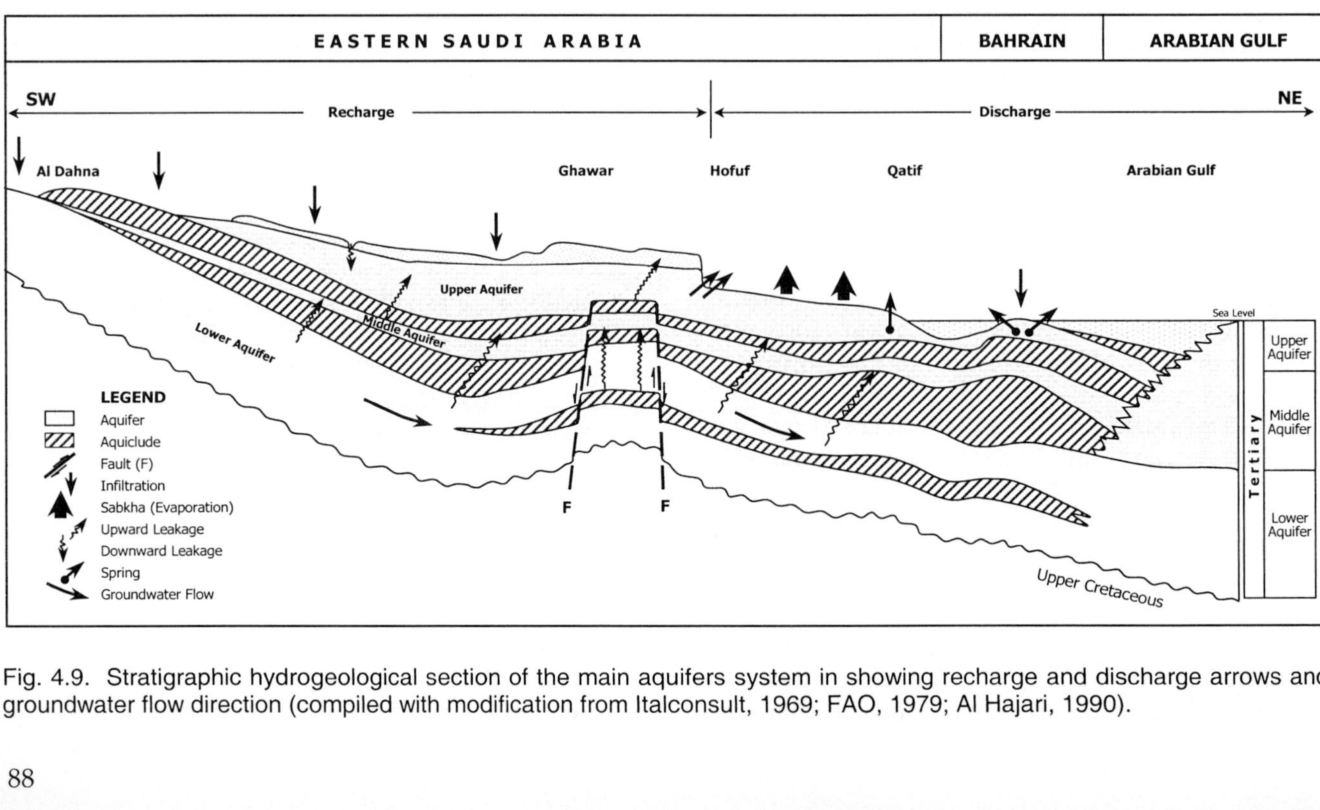

Fig. 4.9. Stratigraphic hydrogeological section of the main aquifers system in showing recharge and discharge arrows and groundwater flow direction (compiled with modification from Italconsult, 1969; FAO, 1979; Al Hajari, 1990).

combines with the Wasia and Biyadh sandstones to form the Cretaceous Sand aquifer (Saudi Arabian Ministry of Agriculture and Water, 1984).

This latter aquifer has usable groundwater with 1,800 mg/l of total dissolved solids. Transmissivity is $7x10^{-3}$ m^2/sec, and storativity, where the aquifer is unconfined, is about $2x10^{-2}$.

In the United Arab Emirates carbonate sequence of the Simsima aquifer reaches a thickness of about 120m and is composed of chalky, shaly packstones and wackestones, with the middle unit being composed of dolomitic and chalky packstones and wackestones, and dense clean limestone. The aquifer's hydraulic properties in the Abu Dhabi Emirate are characterized by porosity in the range of 6 to 35%, intrinsic permeability $8.5x10^{-8}$ to $3.4x10^{-4}$ cm/sec, transmissivity $1.6x10^{-5}$ to $2.4x10^{-4}$ m^2/sec. and storage coefficient $4.0x10^{-4}$. The aquifer represents a small pocket and is not developed into a true aquifer. The water in this aquifer may result from seepage from the Quaternary sand and gravel into the karstic cavities and voids of the Simsima Formation.

C. Cenozoic Aquifers and Aquicludes

The Cenozoic aquifers identified throughout the Arabian Gulf Region can be divided into the Tertiary aquifer systems and the Quaternary aquifer systems (see Figs. 4.2 and 4.3). Because of the wide variations in their geological setting and hydrogeological conditions (Fig. 4.9), these aquifers have widely variable thicknesses, hydraulic parameters, water chemistry and water quality.

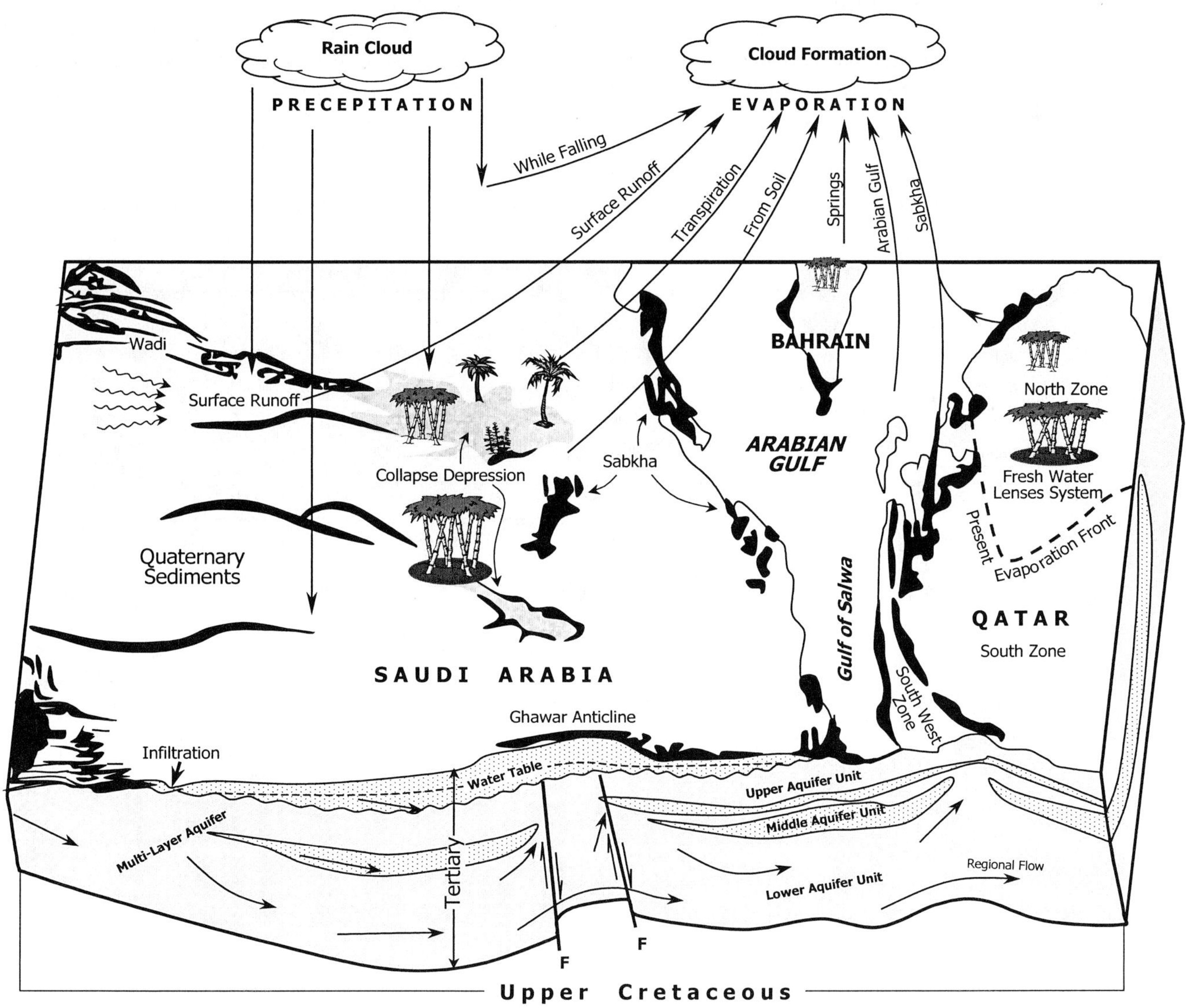

Fig. 4.10. Hydrogeological model of the Tertiary aquifer system in eastern Arabian Peninsula (modified from Al-Hajari, 1990).

The Tertiary aquifer systems are very important in the Arabian Gulf region because they are highly transmissive, highly productive, and underlie the entire Eastern Arabian basin. Because the aquifer units of the Paleogene carbonate cycle and Neogene detrital cycle of the Tertiary, are hydrogeologically connected with each other to varying degrees, these rocks form one aquifer system. Interaction across intervening confining layers has been facilitated by the absence of evaporite units of the Rus Formation, and the shale and marl units of the Dammam Formation, due either to non-deposition or dissolution, and/or by fracturing associated with slumping or tectonic activity, especially over the crests of domal and anticlinal structures where the karstification, faults and joint systems become abundant.

The Tertiary aquifers in the Arabian Gulf area can be divided into two distinct systems: a multi-layered system and a fresh water lenses system. The multi-layered system is composed of several aquifer and aquiclude or aquitard layers which are locally connected to each other. The multilayer system exhibits both confined and unconfined conditions. The fresh water lenses system is composed of isolated fresh water lenses (floating on brackish water) which occur in the upper saturated zones of the multi-layered system at some localities beneath collapse depressions. The hydrogeologic model of this system is shown in Fig. (4.10) and the conceptual model of salt-water intrusion in the northern part of Qatar is presented in Fig. (4.11). From a hydrogeological point of view, and using several cross sections, isopach and structure contour maps, the Tertiary sequence is demonstrably composed of several aquifer and discontinuous aquitard layers as shown in Fig. (4.12).

Umm er Radhuma and upper Rus Formation (Dolomite Unit) aquifer vary in their lateral extent, water quality and capacity to yield water. The pair forms the principal aquifer unit over a large area of the Arabian platform. It is present at shallow depth, extending across the Arabian Gulf States to Iraq in the north, and to Oman and Yemen in the south. The Umm er Radhuma-Rus aquifer unit has been described and exploited as a freshwater aquifer in Saudi Arabia, Oman and extends to southeast Yemen, but is a poor quality aquifer in Qatar, United Arab Emirates, Bahrain and Kuwait. In general, this aquifer unit, consists mainly of alternating layers of limestone and dolomitic limestone with intercalations of anhydrite and argillaceous shales increasing in clay content downwards. Deposition of the argillaceous fine carbonates occurred in

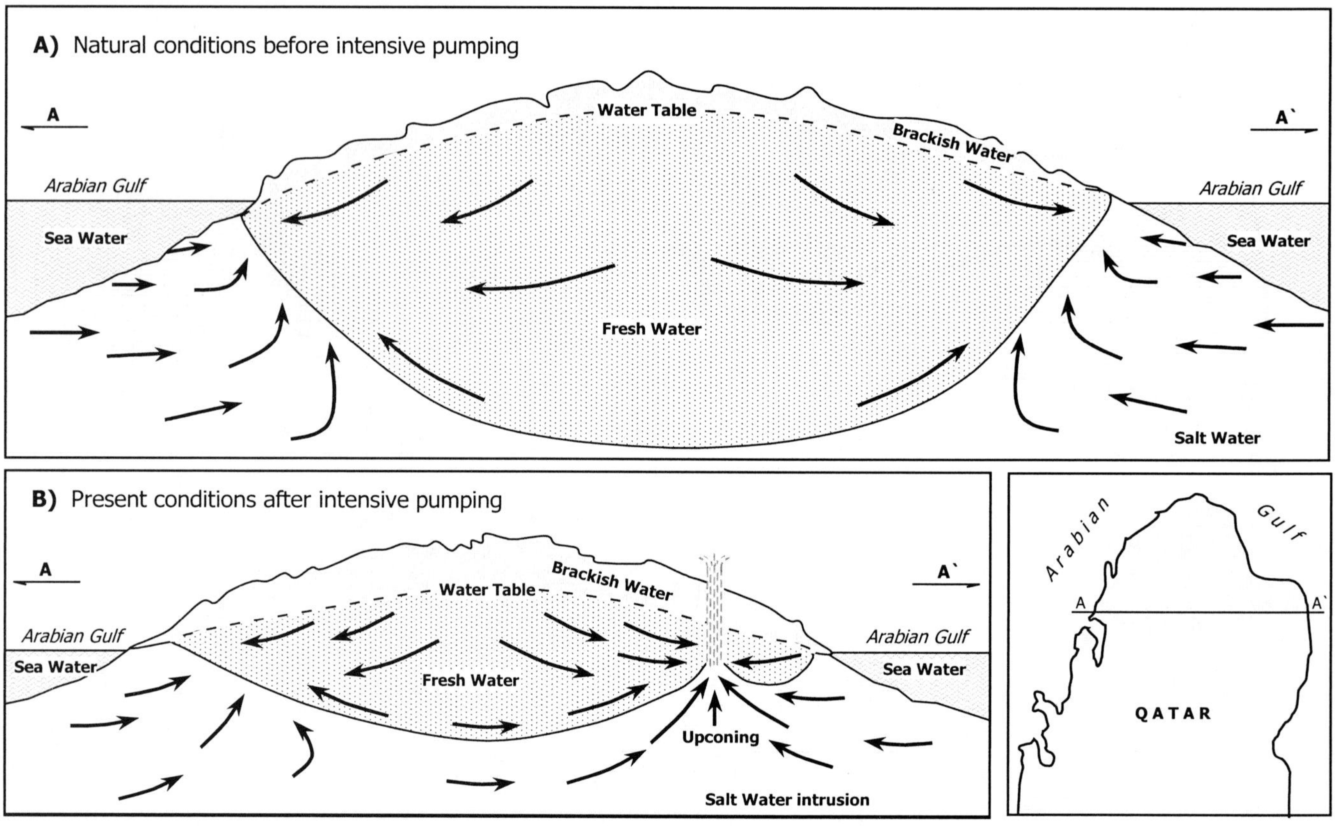

Fig. 4.11. Conceptual model of salt-water intrusion in the northern zone of Qatar as a result of intensive groundwater pumping (modified from Al-Hajari, 1990).

topographically low areas while deposition of coarser carbonate sediments was taking place on high areas. This variation in lithology has caused transmissivity in the structurally low areas to be much lower than in the positive areas. In consequence, this aquifer unit is confined below the basal shale-marl member of Umm er Radhuma, which inhibits leakage between the Tertiary aquifer system and the Cretaceous aquifer system, except around some structural high areas. The top of this aquifer unit is bounded by the most pronounced discontinuous evaporite and shale aquitard layer in the Eastern Arabia. This confining layer is formed by the evaporitic unit of the Rus Formation, and the Midra Member and the Shale Member of the lower Dammam Formation. For example, the uppermost 50-100m of the Umm er Radhuma/Rus aquifer unit exhibits varying amounts of chemical and physical alteration and is heavily fissured exhibiting karstification. This causes loss of circulation in test holes at depths varying between +20 and -175m relative to sea level. Natural dissolution has resulted in interstratal karst, and sheet-like dissolution occurs over extensive areas beneath the covering layers of rocks. Due to its intensive dissolution this aquifer unit is regionally a principal aquifer unit, commonly yielding large quantities of water to wells.

The middle aquitard unit includes the evaporitic unit of the Rus Formation and the Midra Shale of the Lower Dammam Formation, and forms the most important confining unit of the Tertiary aquifer system in the region. It is present in the shallow subsurface throughout Eastern Arabia. Where its top is close to the surface, this unit has been subjected to extensive karstification. The thickness of the Middle Aquitard Unit varies sharply over the study area, however it is thick and widespread in the southern part of Qatar, and elsewhere in Eastern Arabia, while it has been largely or totally dissolved over much of the structurally high areas. This dissolution effect causes the aquifer to change from a confined one to a permeable one since it enables a hydrogeologic connection between the lower aquifer unit and the upper aquifer unit.

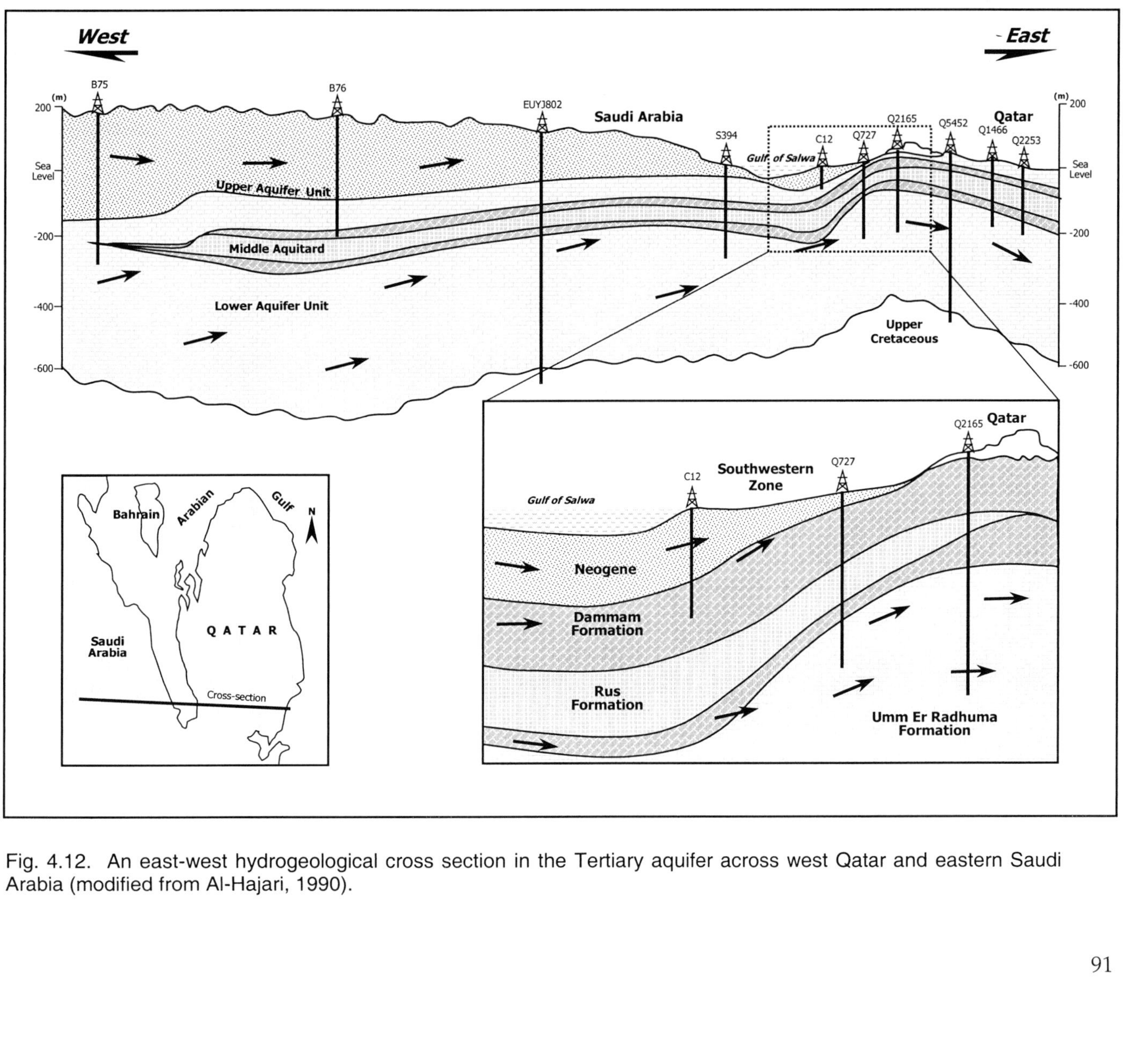

Fig. 4.12. An east-west hydrogeological cross section in the Tertiary aquifer across west Qatar and eastern Saudi Arabia (modified from Al-Hajari, 1990).

The upper aquifer unit is defined by intercalations of light gray, more or less dolomitic chalk, with a thin marl unit. This aquifer unit has limited thickness and is rarely of any importance as an aquifer, except in southwest Qatar. However, it has been identified as a major fresh water aquifer in Bahrain and Kuwait, and in part of eastern Saudi Arabia. The upper aquifer unit has been exposed at the surface for a long period of time and has been very heavily weathered and karstified, except in southwest Qatar where it underlies the Dam Formation. The Neogene detrital deposits rarely have an adequate thickness, nor do they extend over a wide enough area to act as an aquifer of any consequence in Qatar. They have been identified as an aquifer of limited potential in eastern Saudi Arabia. In the United Arab Emirates the main water-bearing formations in Abu Dhabi Emirate (for example) are the consolidated carbonate rocks of Eocene to Miocene age, and unconsolidated river alluvium and eolian deposits of Quaternary age. In some areas in the eastern province of Abu Dhabi (Al Ain region), the Tertiary carbonate rocks comprise a complex folded belt, which was disturbed by the uplift of the Oman Mountains and do not contain freshwater aquifers (Fig. 4.13).

Umm er Radhuma Aquifer

The Paleocene Umm er Radhuma Formation in eastern Saudi Arabia is a principal fresh water aquifer. However, the aquifer becomes less important in Kuwait, Qatar, Bahrain, United Arab Emirates, and Oman where it contains high-salinity water. The formation dips towards the east and also increases in thickness in that direction. The Umm er Radhuma aquifer is composed of limestone and dolomite in Saudi Arabia (Fig. 4.1), but the facies change to marly limestone in Kuwait, and extends beneath almost the entire Rub al Khali from eastern Yemen and western Oman to the Saudi/Iraqi border. Across these areas the formation varies in thickness from 245 to 700m, and consists of calcarenite limestone with some dolomitic limestone, dolomite, minor marl and shale, and aphanitic, argillaceous limestones in the lower parts.

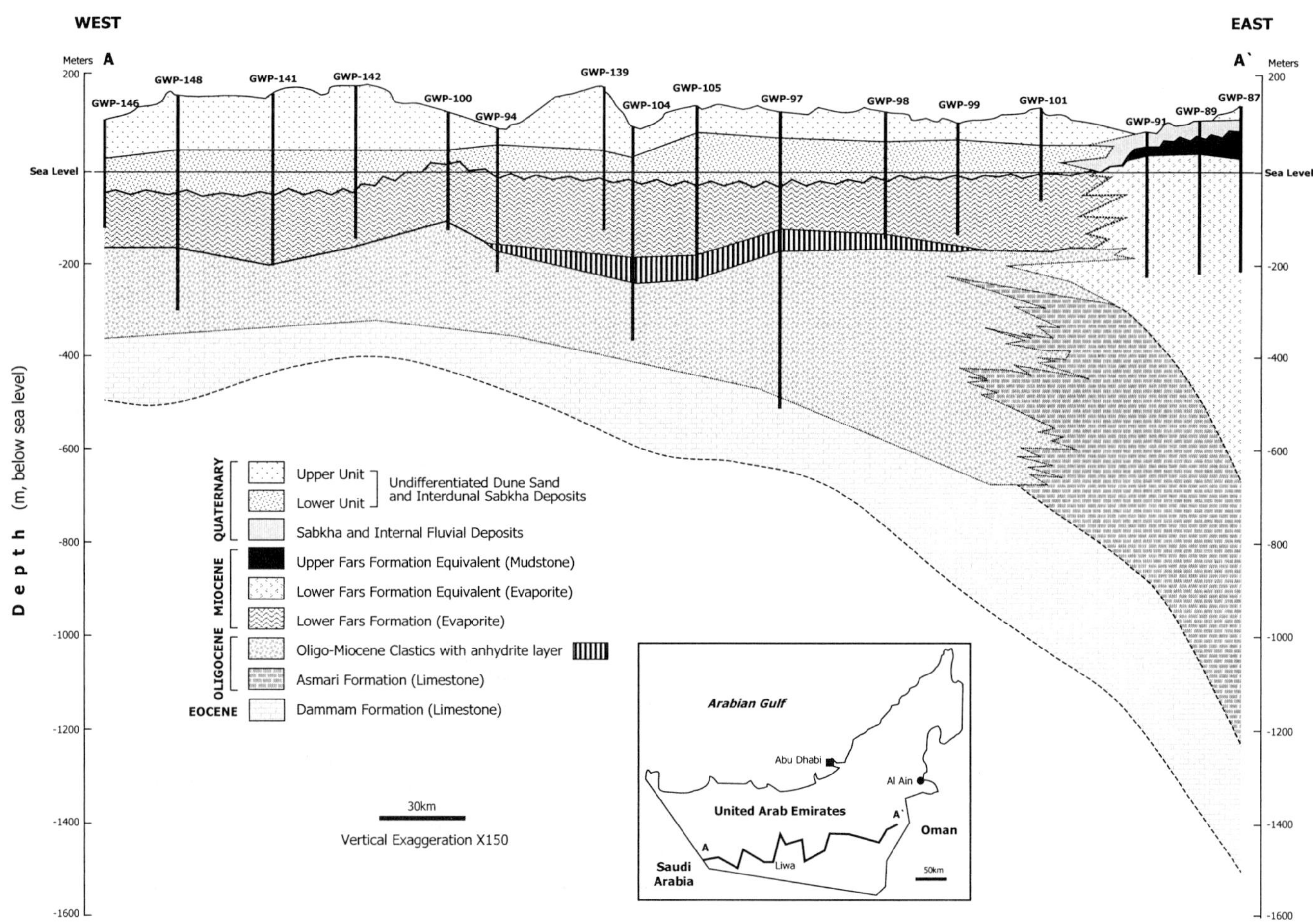

Fig. 4.13. Tertiary stratigraphic section in the Liwa region of Abu Dhabi Emirate, United Arab Emirates (modified from US Geologic Survey and United Arab Emirates National Drilling Co. Groundwater Project, 1996).

The formation does not constitute one complete interconnected aquifer, due to the presence of intercalated marls, shales, and argillaceous limestones. It may also contain thin stringers of anhydrite. It yields good to poor quality groundwater, mostly from aquifers in the upper third of the formation, where horizons of porous foraminiferal calcarenite, comprise the zones which yield up to 95 l/sec, in eastern Saudi Arabia. The latter yields are of fossil water, estimated to be 10,000 to 28,000 years old. Springs in the Al Hassa province of eastern Saudi Arabia provide large volumes of water via Neogene strata (Hötzl et al., 1978), though this originates mostly from the Umm er Radhuma aquifer. The quality of groundwater from the upper part of the Umm er Radhuma Formation is generally good, but may decline in the lower part of the lithostratigraphic unit. Water quality also deteriorates towards the northeast from less than 1,000 mg/l TDS near outcrop to 6,000 mg/l TDS along the coastal area (Naimi, 1965), though it is always of better quality over structural highs, such as the Dammam Dome, due to the absence of anhydrite in the Rus Formation. Shales and argillaceous limestones at the base of the formation form an aquitard between the Aruma and Umm er Radhuma aquifers. The vertical connection between the Wasia-Biyadh aquifer system and the Umm er Radhuma aquifer system has been calculated to be only 0.05 l/sec per km^2 (Bureau de Recherches Géologiques et Miniéres, 1977), and therefore these aquifer systems behave essentially independently. There is, however, considerable vertical leakage from the Umm er Radhuma upward into the Dammam Formation aquifers in places where the Rus Formation is thin, as at Wadi al Miyah. In the Ghawar area, transmissivity ranges from $3x10^{-3}$ to $5x10^{-4}$ m^2/sec. Storativity ranges from $5x10^{-5}$ to $5x10^{-3}$ where the Umm er Radhuma aquifer system is confined, and from $2x10^{-3}$ to $7x10^{-2}$ where it is unconfined.

In northwestern Saudi Arabia, the Paleocene-Lower Eocene Hibr Formation consists of thin-bedded, chalky and cherty limestone and marl from 150 to 485m thick. The Hibr Formation acts as an aquitard in the wadi Al Sirhan Basin, apart from containing a minor chert aquifer.

In Kuwait, the Radhuma aquifer consists mainly of anhydritic, dolomitic, and marly limestone. The thickness of the aquifer varies between 600m in the north to 420m in the south. The water salinity increases to the east and northeast. On the basis of

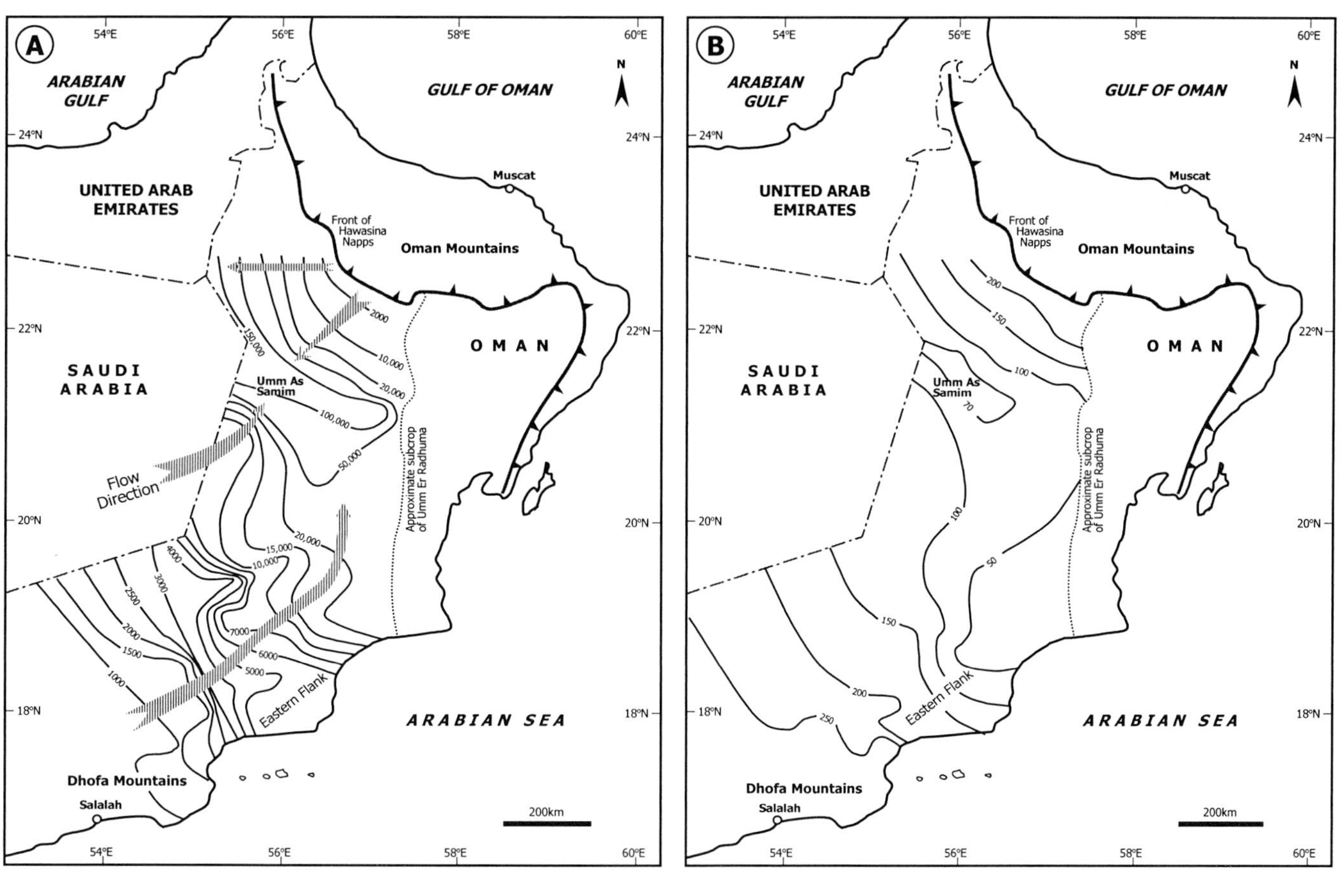

Fig. 4.14. The Umm er Radhuma aquifer in Oman: A) Groundwater salinity (mg/l); and B) Equipotential contours (m) above sea level (modified from Al Lamki and Terken, 1996).

water chemical analyses from wells located in the west and southwestern part of Kuwait, the water quality varies considerably from one area to another. The salinity of the Radhuma aquifer in southwestern Kuwait varies between 4,000 and 5,000 mg/l. Water salinity is slightly higher than for the water of the overlying aquifers. The sequence of anion and cation dominance is $SO_4^{2-} > Cl^- > HCO_3^-$ and $Ca^{2+} > Na^+ > Mg^{2+} > K^+$. The ratio of SO_4^{2-}/Cl^- is greater than 1. The water type is sulphate + chloride. The Sodium Adsorption Ratio (SAR) is in the range of 2.6 to 5.1‰. The concentration of sulphate and calcium ions in Radhuma water is greater than for waters of Dammam Formation and Kuwait Group, while chloride is less. Dissolved H_2S has been recognized in all the wells of this formation. Water salinity is greater east and northeast of Kuwait, where water salinity exceeds 35,000 mg/l in southeast Kuwait Bay. In that area the sequence of anion and cation dominance is $Cl^- > SO_4^{2-} > HCO_3^-$ and $Na^+ > Ca^{2+} > Mg^{2+} > K^+$, and the ratio of SO_4^{2-}/Cl^- is less than 0.1. The water is dominated by sulphate and chloride ions.

In Bahrain, the Umm er Radhuma-Rus aquifer or "C" aquifer zone consists of fractured chalky limestones and dolomites contains brackish to highly saline water with typical salinity of around 12,000 mg/l, with NaCl varying from 6,000 to 17,000 mg/l, but with a decrease in salinity northwest of Awali.

In Qatar, the Umm er Radhuma Formation consists of 300-500m thick of alternating sequence of limestones and dolomites. The top 30-50m is karstic dolomite, the marl content increases downwards. The aquifer contains brackish water beneath the whole country.

In the United Arab Emirates only in Abu Dhabi Emirate, the Umm er Radhuma aquifer ranges from 300 to 600m thick and is composed of four lithological units. The upper unit is of shaly lime mudstones, wackestones, and fine-grained packstones with dolomites, the second unit consists of mudstones, packstones, grainstones and wackestones, the third unit is composed of dolomitic mudstones and wackestones and the basal unit is represented by shales and marls. Hydraulic properties of the Tertiary aquifers in Abu Dhabi are summarized in Table 4.1.

In Oman, the Umm er Radhuma limestone aquifer is recharged through its exposures on the western side of the Oman Mountains. The beds forming the aquifer dip towards the west and attain a thickness of 600m in southern Oman. The map of total dissolved solids (Fig. 4.14a) shows a TDS pattern similar in appearance to the equipotential map (Fig. 4.14b). The salinity trends observed in both the Umm er Radhuma and the Paleozoic aquifers (Fig. 4.4) are in accord with the present-day direction of movement of water in the basin. The equipotential map of the Umm er Radhuma aquifer, confirms that the main recharge areas are the Dhofar Mountains and the Oman Mountains with a component from southeastern Saudi Arabia. The main discharge area is the inland Umm as Samim sabkha. Although it appears from Fig. (4.14 b) that discharge is also taking place in the Eastern Flank of southeast Oman, this is misleading, as elevation in this area is 150m above sea level, implying that the water would have to overcome a 100m head in order to appear at the surface (Al Lamki and Terken, 1996). This is an area where water stagnates and groundwater under such conditions is likely to have high total dissolved solids, as is the case for a few wells close to the Umm er Radhuma subcrop. It is also possible that Tertiary water in this area enters the Paleozoic aquifers through sinkholes in the Middle Cretaceous Natih Formation (Al Lamki and Terken, 1996).

Rus Aquiclude

In Saudi Arabia, the Lower Eocene Rus Formation, consisting of interbedded marl, dolomite, limestone and anhydrite, forms an aquiclude which generally confines groundwater to the Umm er Radhuma Formation over most of the Eastern Province. Over structural highs, the Rus Formation is mainly composed of dolomite and may yield small amounts of poor quality water by upward leakage from the Umm er Radhuma aquifer, whereas in structural troughs this formation mainly consists of impermeable anhydrite, gypsum and marl.

Table 4.1. Hydraulic properties of the Dammam, Umm er Radhuma and Simsima aquifers in onshore oil fields of Abu Dhabi (modified after Hassan and Al Aidarous, 1995).

Aquifer	Porosity (%)	Intrinsic Permeability		Transmissivity (m^2/day)	Storage Coefficient
		Darcy	cm/sec.		
Simsima	06 - 35	0.0001 – 0.4	$8.5x10^{-8} – 3.4 x 10^{-4}$	1.4 – 20.4	$4.0 x 10^{-4}$
Umm er Radhuma	15 – 25	0.01 – 0.05	$8.5x10^{-6} – 4.3x10^{-6}$	3.7 – 13.9	$9.0 x 10^{-4}$
Dammam	22 – 25	0.02 – 0.04	$1.7x10^{-5} – 3.4 x 10^{-5}$	2.8 – 5.6	----

In Qatar, the Rus Formation is characterized by about 10-100m of anhydrite with marl and some thin limestones, with overlying and underlying about 10-20m of limestone, dolomite and some marl. The anhydrite facies represents an aquiclude, while the carbonate facies is important aquifer containing large reserves of freshwater, and is in hydraulic continuity with Umm er Radhuma formation below.

In the United Arab Emirates, the Early Eocene carbonates of the Rus Formation constitute an aquifer in the eastern province of Abu Dhabi (Al Ain region). These carbonates are characterized by extensive dolomitization. Porosity is virtually nil in these rocks except for infrequent unfilled fractures, vugs and oomolds, due to cementation and chertification (Whittle and Alsharhan, 1994). Water wells tapping this formation in Al-Ain region have produced 21,000m^3/day of thermal (36-52°C), brackish water (3,900-6,900 mg/l). Geophysical well logs show that the limestone belongs to the confined-flow carbonate aquifers (White, 1977), and that the thermal water enters the aquifer from a fractured limestone interval between 93 and 102m deep in most the wells drilled (Khalifa, 1995). A conceptual model explaining the origin of thermal brackish water in Jabal Hafit area is illustrated in Fig. (4.15). Water-level measurement in June 1996 indicated a general decline in groundwater level as a result of pumping. However, the water-level rise observed in some wells is the result of downward percolation of recharge water from surface canals and ponds, constructed for recreational purposes. Chemical analyses of groundwater samples from the Rus Formation show that the water is slightly alkaline (pH 7-8) and sodium-chloride rich. Two water types were distinguished; the first has a relatively low temperature and high total dissolved solids, while the second has a high temperature and total dissolved solids. Stiff diagrams show that sodium is the dominant cation, while chloride is the dominant anion (Fig. 4.16). High radon and radium content in groundwater of the Lower Eocene limestone around Jabal Hafit area is attributed to the presence of high Uranium content (21 ppm) along joints and bedding plains (El Shami, 1990). Environmentally sensitive isotopes, such as tritium (^{3}H), deuterium (^{2}H), and oxygen-18 (^{18}O) are relatively lighter in the groundwater in other wells in the Al Ain area, suggesting possible local recharge through Jabal Hafit.

Dammam Aquifer

The Dammam aquifer is basically composed of middle Eocene limestone and dolomite, with a shale intercalation near the base and a chert crust at the top. In most of the Arabian Gulf region, the Dammam aquifer is isolated by impermeable confining layers. However, where these layers are absent, the Dammam aquifer comes into direct

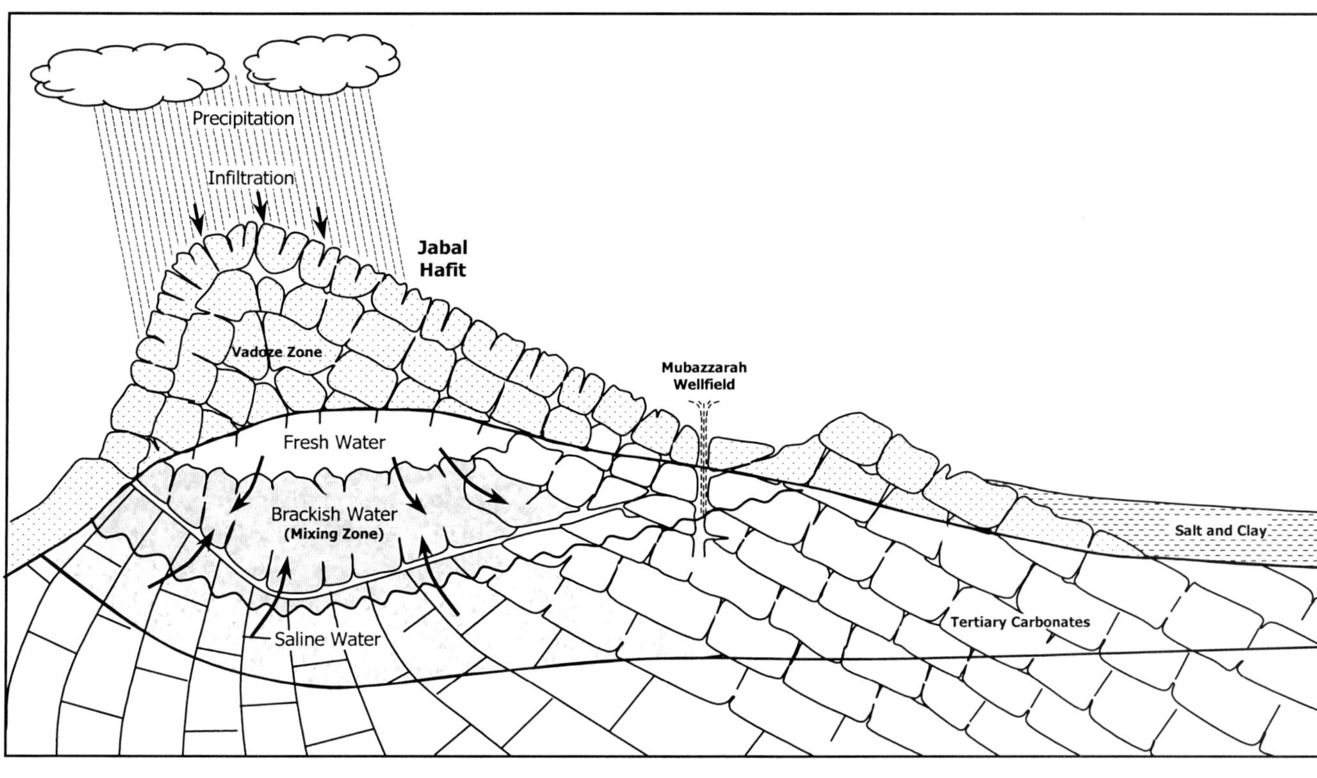

Fig. 4.15. Diagrammatic cross-section of Jabal Hafit (eastern Al Ain), United Arab Emirates, showing conceptual model of recharge and discharge of the Tertiary limestone aquifer (modified from Khalifa, 1997).

contact with the underlying Umm er Radhuma aquifer and the overlying Neogene and Quaternary aquifers. Groundwater flow direction in the Dammam aquifer in Arabia is from west to the east and northeast. The factors affecting groundwater flow in the Dammam aquifer are the geologic structure, lateral facies change, and variations in thickness of confining layers below and above it. The groundwater level above mean sea level in the Dammam aquifer varies throughout the region. It is 250 m in western Saudi Arabia, 80 to 100m in United Arab Emirates, 5 to 200m in Oman, 100 to 160m in Kuwait, 5 m in Qatar and 5m in Bahrain.

The isotopic investigations show that the groundwater of the Dammam aquifer, in the Arabian Gulf region, is old and ranges from 10,000 to 22,000 years. For example, in Kuwait the groundwater is about 22,000 years old and has not received any recent recharge. The groundwater in the aquifer was recharged during ancient pluvial periods. Groundwater in the aquifer is in direct hydraulic communication with groundwater in the Kuwait Group.

In Saudi Arabia, groundwater of the Dammam aquifer is old (16,000 to 20,000 years) and has not received recent recharge. The aquifer has hydraulic contacts with the Neogene aquifer above and Umm er Radhuma aquifer below.

In Qatar, the lower Dammam Formation consists of about 12m of compact fossiliferous chalky limestone and laminated fossiliferous shale. This shale has confined the Rus groundwater. The absence of shale from northern Qatar has a great recharge significance and controlling factor in the solution of gypsum from the underlying Rus Formation. The Upper Dammam is about 10-65m of dolomitic limestone, it is an important artesian aquifer in southwestern Qatar only, but elsewhere is unimportant. The Dammam aquifer is connected to the Umm er Radhuma aquifer, and contains old water (10,000 to 17,000 years).

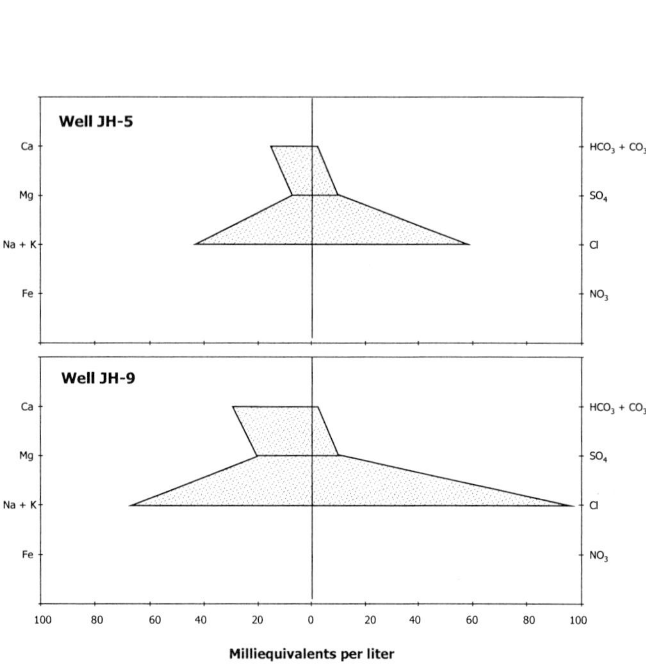

Fig. 4.16. Relative temperatures and concentrations of major ions in groundwater from two wells in Tertiary aquifer at Jabal Hafit, U.A.E. (modified from Khalifa, 1997).

In Bahrain, the groundwater in the aquifer is old (13,000 years) and also does not receive present-day recharge. The aquifer is hydraulically connected to the Umm er Radhuma aquifer in the west, but is affected by salt-water intrusion from the Arabian Gulf in the east. The Dammam aquifer system consists of two aquifer zones (Fig. 4.17). The Alat aquifer (zone A) about 15-25m thick, and Khobar aquifer (zone B) about 40m thick are the main two aquifer systems, generally having constant chemical properties, and typical salinity of 700 ppm. The "A" zone has limited hydraulic properties, with an average transmissivity of about 350 m^2/d; while "B" zone developed in highly fractured limestones and dolomites, with high transmissivity of about 10,000 m^2/d (Zubari et al., 1998). These aquifers sourced from eastern Saudi Arabia, contribute quantities of sulphate and other ions such as: total dissolved solids (1915 ppm), Na^+ (381 ppm), K^+ (16.56 meq/l), Mg^{+2} (77 ppm), Cl^- (715 ppm), SO_4^{-2} (361 ppm), HCO_3^- (174 ppm) (Doornkamp et al., 1980). The aquifers are found at very shallow depths, and are often artesian. Both seawater intrusion and saltwater intrusion from the underlying Umm er Radhuma aquifer caused by a falling level in the aquifers due to increased extraction, has introduced dissolved ions into the groundwater. Near surface groundwater in Bahrain contains large quantities of dissolved salts, rarely less than 0.3%. Due to increase pumping the groundwater level decreases with time. The hydraulic gradient in the range of 0.2 to 0.05×10^{-3}. Pumping tests across both the Alat and Al Khobar aquifes indicate transmissivities in the range of 1.4×10^{-2} to 6.7×10^{-3} m^2/sec. The average piezometric elevation on southern Bahrain is 2-3m above mean sea level, while in the northern part the water level attains an unconfined elevation of 6m above sea level (United Nations, 1982).

In Saudi Arabia, the Middle Eocene Dammam Limestone Formation extends throughout northeastern and eastern Arabia and is generally from 50 to 75 m thick. There are five members of this formation, however, only the upper two are water-bearing. These are the Alat and Khobar aquifers, poorly separated from each other by the Alat Marl aquitard. They yield moderate amounts of groundwater in the Al Hasa area and in Wadi Al

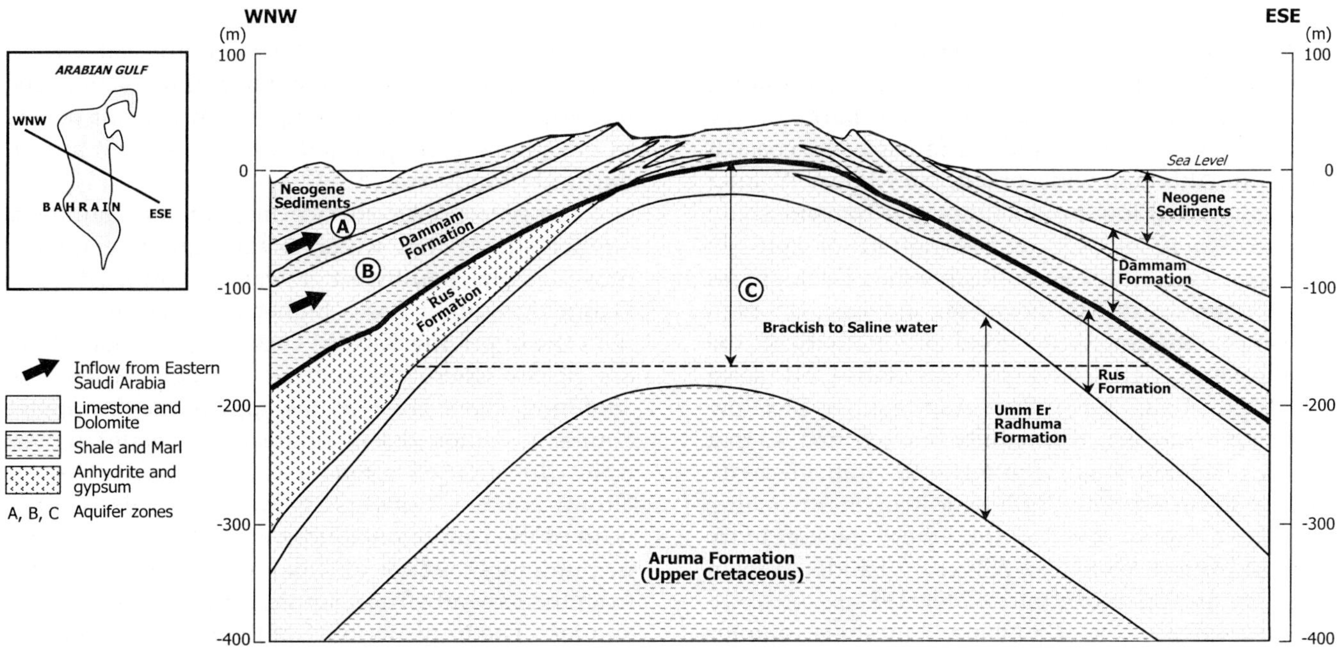

Fig. 4.17. Hydrogeological-sedimentological cross-section showing the Tertiary aquifer system in Bahrain (modified from Zubari, et al., 1998).

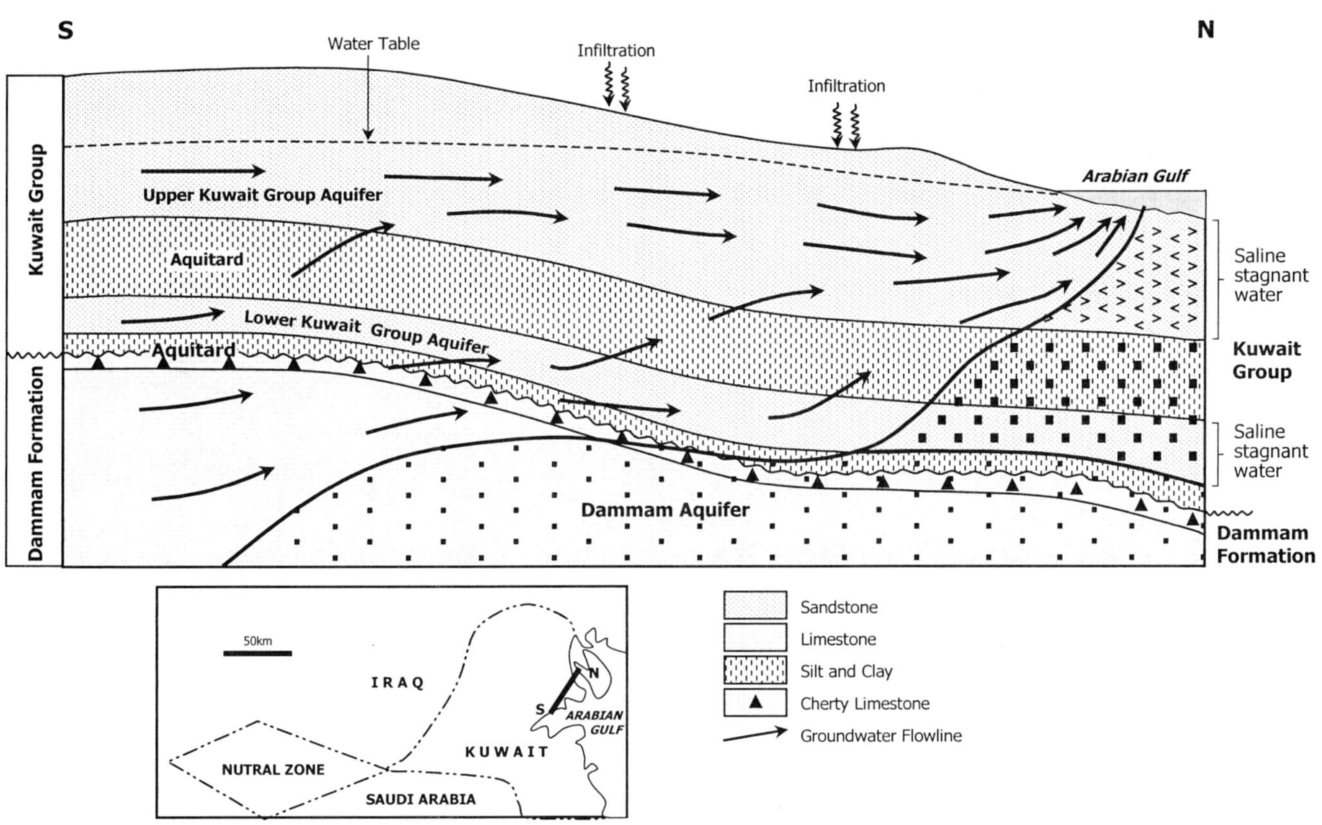

Fig. 4.18. Conceptual model of groundwater flow and aquifer system in Kuwait (modified from Mukhopadhyay et al., 1996).

Miyah. The Alat aquifer has a more limited extent, and the Khobar aquifer, which is karstified and fissured, provides most of the water for domestic and agricultural purposes in many parts of the Eastern Province from Wadi as Sahba to the coast at Al Qatif. Although widespread, the Khobar Aquifer generally has a low productivity and transmissivity of $2x10^{-3}$ to 10^{-4} m^2/sec, and storativity of $1x10^{-3}$ to less than $1x10^{-4}$ (Bureau de Recherches Géologiques et Miniéres, 1977), however it has a high transmissivity zone with good productivity in the northeastern part along the Arabian Gulf coast (Yazicigil et al., 1986). The Alat aquifer has low matrix porosity and is generally fine-grained, so that it is a relatively minor aquifer with transmissivity of $2.6x10^{-5}$ m^2/sec at Wadi Al Miyah, and reaching $2.3x10^{-3}$ m^2/sec at Ras Tanura, where its storativity ranges from $1.32x10^{-4}$ to $5.34x10^{-4}$ (Saudi Arabian Ministry of Agriculture and Water, 1984).

In Kuwait, the Dammam limestone aquifer consists of 183 to 213 m of soft and chalky, shelly and porous limestone and hard crystalline dolomitic limestone. A chert layer about 9m thick at the top or within the aquifer may act as an impermeable layer which stops upward or downward movement of groundwater (Al-Hajji, 1976). Karstification in the Dammam aquifer was recorded by Burdon (1966). The paleokarst sinkholes may control the present-day groundwater movement, the groundwater yield and its chemical composition.

The vertical hydraulic gradient is upward from the Dammam aquifer to the main Kuwait Group aquifer to the Dibdibba aquifer (Fig. 4.18). The recharge of the Dammam aquifer in Kuwait occurs in the southern Iraqi desert, west of Kuwait and in northeast Saudi Arabia.

The piezometric-surface map of the Dammam aquifer shows that water in the aquifer generally moves from southwest towards northeast. The Dammam aquifer has a higher piezometric surface than the overlying Kuwait Group aquifer, but there are places where the reverse is true. Where the confining chert layer at the top of the Dammam aquifer is either absent or fractured, water moves upward from the Dammam aquifer into the Kuwait Group aquifer. Conversely, when the piezometric head of the Dammam aquifer is sufficiently reduced by pumping, the flow of water is reversed and water from the Kuwait Group aquifer enters the Dammam aquifer. The transmissivity of the Dammam aquifer varies between 1000 and 100,000 m^2/day (Sakr, 1989) and the average storage coefficient is 2.0 x 10 (Al-Hajji, 1976).

The Dammam water varies in salinity from brackish (2,500 mg/l) in the southwest of Kuwait to brine (150,000 mg/l) in the northeast (Fig. 4.19). Local anomalies in total dissolved solid content are possibly due to variable karst development and infiltration rates (Burdon and Al-Sharhan, 1968). On the basis of water quality, the groundwater of the

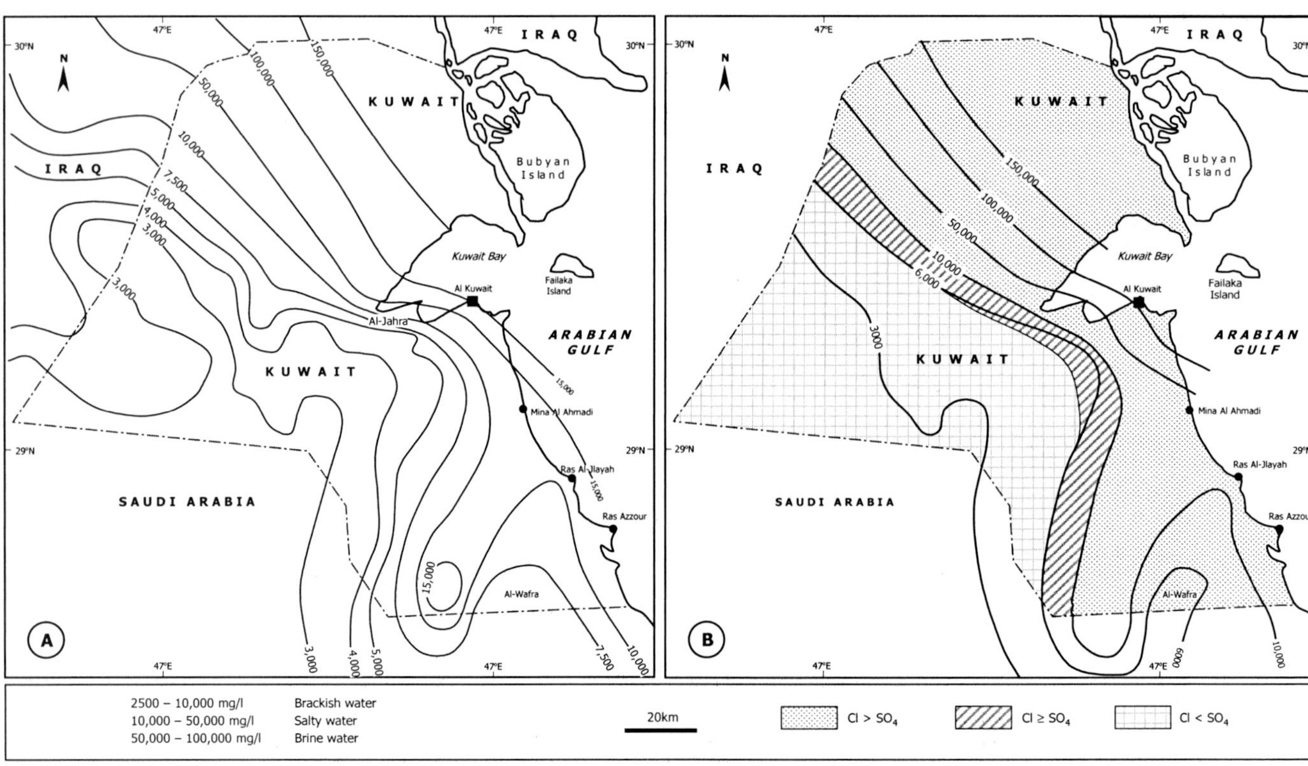

Fig. 4.19. Dammam Aquifer in Kuwait, a) isosalinity map (mg/l); and b) relative abundance of SO_4^{2-} and Cl^- in groundwater (modified from Omar, et al., 1981).

Dammam aquifer can be classified into brackish (2,500-10,000 mg/l), salty (10,000-50,000 mg/l) and brine (50,000-150,000 mg/l).

In the United Arab Emirates, The Eocene Dammam aquifer is composed of dolomite with anhydrite in the upper part, foraminiferal packstone and grainstone in the middle part, and shales and argillaceous limestone in the lower part. The thickness of the Dammam aquifer ranges from 60 to 490m, with a gradual thickening from west to east in the Abu Dhabi Emirate. Despite the high average porosity of the core samples, the intrinsic permeability of the aquifer is low. The hydraulic properties of the aquifer are summarized in Table 4.1.

In Oman, the Dammam aquifer reached its thickness of about 150m thick and composed of crystalline, chalky limestone and dolomitic limestone, with shale and marl. The aquifer has a low transmissivity, ranging from 1.3 to 8.8 m^2/day.

Chapter 5

HYDROGEOCHEMISTRY

INTRODUCTION

Over the vast extent of the Arabian Peninsula, the average rainfall ranges from, less than 50 mm to 160 mm depending upon location, exceeding that amount only in the marginal mountains of Oman, United Arab Emirates, Yemen and Saudi Arabia. Even in these more favoured localities 400 mm to 600 mm is a heavy rainfall (see Chapter 2). The rain tends to be concentrated in a few intense storms during which as much as 75% of the annual precipitation may fall in a very few hours. One result is that the surface sediments are rapidly saturated and the runoff rate is high. Water is channeled in the wadis and tends to disappear into the surface sediments forming the alluvial fans and seldom reaches the sea except where the mountains and sea are in close proximity as in eastern and northern Oman and the Red Sea area of Saudi Arabia. In western and southwestern Saudi Arabia these wadis provide important but limited sources of groundwater, and in Oman and United Arab Emirates occasional lens of potable water are found in depressions following wadi channels. Temperatures in the region are high and evaporation is correspondingly high, several times the total precipitation, under these circumstances perennial surface water cannot exist for long and surface streams are conspicuously absent. Wadi flow is restricted to a few days following a heavy storm. The dry climate, low rainfall and extremely high evaporation limit aquifer recharge, and lead to a sluggish hydrologic cycle. Aquifer lithology on the other hand, has a strong signature of groundwater chemistry. So, one can reconstruct the groundwater flow path through careful examination of the groundwater-dissolved solids relationship. The hydrogeologic conditions, represented by the nature of existing groundwater flow systems, aquifer recharge and closeness to recharge areas are, in effect, the resultant water chemistry.

The hydrochemistry of natural water in the Arabian Gulf area, as related to the prevailing arid climate, as well as, the dominant geological and hydrogeological conditions, are discussed here and interpreted, using chemical and isotope analyses of water sample from rain, springs, falajes, shallow aquifers and deep aquifers, and are represented on relevant graphs and diagrams.

Two groups of parameters are measured to characterize the chemistry of any water analysis: field-measured parameters and lab-measured parameters. The field-measured parameters include water temperature (°C), hydrogen-ion concentration (pH) and electrical conductivity (EC in micro-Siemens per centimeter or µS/cm). The values of these parameters change when they are not directly measured in the field. The number lab-measured parameters depend on the purpose of study. However, the measurement of major cations (K^+, Na^+, Mg^{2+} and Ca^{2+}) and anions (CO_3^{2-}, HCO_3^-, SO_4^{2-} and Cl^-) are determined in most chemical analyses. Analysis for stable (^2H and ^{18}O) and radioactive isotopes (^3H and ^{14}C) in rain and groundwater are also included.

1. Hydrogeochemistry of Rain Water

The rainwater is usually depleted in ions. However, the chemical analyses of rainwater samples from the Arabian Gulf arid region show relative enrichment in ion concentrations compared with rainwater in wet regions. This fact is further supported by the enrichment in stable isotopes in rainwater of the region.

The longest series of isotopic measurements (between 1963-1993), were made in Bahrain in conjunction with an International Atomic Energy Agency global survey. The stable isotope data available from this station provide the basic characteristics of the stable isotopic composition of present day meteoric water of the Arabian Gulf coastal regions (Yurtsever, 1992). A plot of oxygen-18 (^{18}O‰) versus deuterium (^2H‰) contents in rain water samples of United Arab Emirates collected during 1985-1991 period shows that the weighted average mean values of ^{18}O =-1.99‰ and ^2H = -0.4‰. According to Rizk and Alsharhan (1999), the regression line that fits all data points can be described by the equation: $\delta D = 4.26\ \delta^{18}O + 9.23$ (Fig. 5.1). However, the line best defining the ^{18}O versus ^2H, for months having more than 20 mm rain has a slope of 8, and an intercept (d or deuterium excess) of 16. This relationship is the best estimate of the stable isotope composition for groundwater of meteoric origin being replenished from precipitation under the present-day climatic conditions in United Arab Emirates. The tritium (^3H) content in rainfall events for the 1984-1987 averages about 4.7 Tritium Units (TU), which corresponds to present rainfall.

In their attempt to establish an isotopic water line for northern Oman, Macumber et al. (1997) reached the relationship: $\delta D = 5.1\ \delta^{18}O + 8$, which is very similar to that of Bahrain (Yurtsever, 1999) and United Arab Emirates (Rizk and Alsharhan, 1999) (Fig. 5.2). Macumber et al. (1997) also observed that the rainfall records for northern Oman show no significant depletion of δD or $\delta^{18}O$ with altitude, in contrast to groundwater. They attributed isotopically depleted groundwater (δD <-10; $\delta^{18}O$ <-3) to a plume originating in the high limestone areas of Jebel Akhdar and Jebel Nakhl (part of Oman Mountain chain) passing through a gap in the ophiolites at Jebel Nakhl, across the Batinah coastal plain.

To incorporate the isotopic analysis of United Arab Emirates precipitation with similar areas in the Gulf region, local data was plotted with those of Bahrain and Oman (Fig. 5.2). The isotopic composition of rainfall in Bahrain, the nearest long-term station to the United Arab Emirates, was monitored for the period 1963-1993 within the scope of the International Atomic Energy Agency/World Meteorological Organization (IAEA/WMO) global survey. The stable isotope data available from this station was used to provide basic characteristics of the stable isotopic composition of the present-day meteoric water, especially in the coastal areas of United Arab Emirates. Stable isotope data of the Al Buraimi area, Oman, was also used to characterize precipitation of the eastern mountainous area of United Arab Emirates (Fig. 5.2). This figure shows that Oman rain has the most depleted stable isotopes and falls on the Mediterranean Meteoric Water Line (MMWL) of $\delta D = 8\ \delta^{18}O+20$. Stable isotopes of Bahrain rain lies mostly between the MMWL and the Global Meteoric Water Line (GMWL) of $\delta D = 8\ \delta^{18}O+10$. In the United Arab Emirates, the winter precipitation falls on the MMWL and the summer rain, relatively enriched in stable isotopes, falls on the GMWL. Rizk and Alsharhan (1999) suggested that there are two sources of precipitation in the United Arab Emirates; the Mediterranean Sea and the Indian Ocean. The scatter of stable isotope data points in the Arabian Gulf region suggests that the raindrops are affected by evaporation during the fall of the droplets (Yurtsever, 1992). The maximum, minimum and average values of deuterium (‰) and oxygen-18 (‰) in rainwater are listed in Table 5.1.

2. Hydrogeochemistry of Spring Water

The isotopic data can also be used to determine the age of the groundwater (Table 5.2). Water drawn from the alluvial sediments, has ages measured in terms of a few years, in contrast to water pumped from the main aquifers, where ages are commonly in the range 10,000 to 30,000 years BP, the age increasing with distance from the recharge region. The figures are another measure of the over exploitation of groundwater, as pumping draws upon fossil water emplaced during the more pluvial phases of the Ice Age. They indicate the low level of recharge at the present time.

Water resources are restricted to wells and springs, and a traditional system of channels or falajes, which tap into groundwater in the

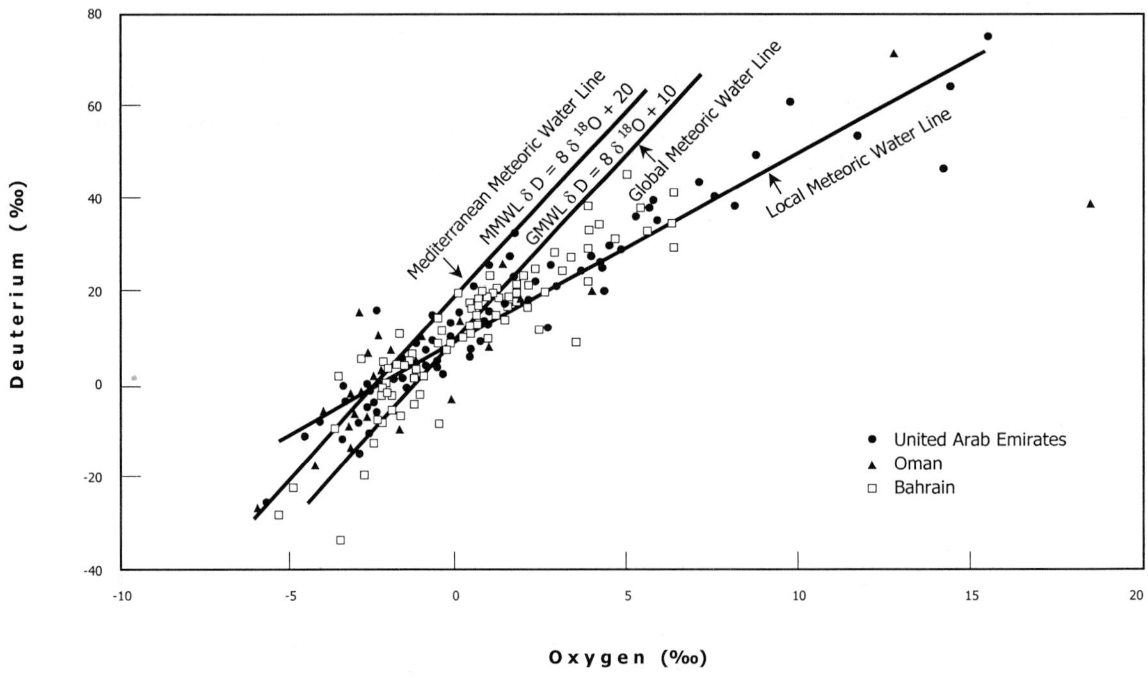

Fig. 5.1. Stable isotopes in rainwater of the United Arab Emirates, Oman and Bahrain (compiled by Rizk and Alsharhan, 1999.

mountainous areas of Oman, United Arab Emirates and Saudi Arabia. Mayboom (1966) has defined a spring as a groundwater outcrop; a definition expanded by Todd (1980) as a concentrated discharge of groundwater appearing at the surface as a flowing current and as such is an indicator of discharge areas in arid and semi-arid regions. Springs may occur in many forms, and have been classified according to cause, rock structure, temperature and discharge. The simplest breakdown is based upon discharge is into seasonal or ephemeral springs (Fetter,1988), which normally discharge only during those times of year, when there is sufficient groundwater recharge to maintain flow, as distinct from perennial springs, which drain extensive aquifers and maintain flow throughout the year. Meinzer (1923) recognized eight categories of springs, according to the rate of discharge, and consequently, a spring might change category according to the rainfall/recharge characteristics of a particular year.

A more geological classification of springs in the United Arab Emirates distinguishes gravity flow springs, which includes artesian, and thermal springs associated with caves or sinkholes in limestone rocks. The range of water temperatures of the springs in the United Arab Emirates is not regularly recorded, however recordings in different years indicate little change. The Siji spring water temperature during 1977 is recorded as 30.1 °C and as 33 °C in 1994, the figure for the Bu Sukhnah spring was 39.5 °C in 1994, the Khatt South spring for the same time was 40 °C. The highest values have been attributed to the presence of a radioactive source. However, Toth (1963) distinguished local, intermediate and regional groundwater flow patterns, although the actual system present in an area depends upon local topography and basin geometry. The water discharged from local flow systems is normally of low salinity with a temperature close to the mean annual air temperature, whereas water discharging from regional groundwater flow patterns may be highly mineralized and at elevated temperature (Fetter, 1988). On this basis, the high temperature groundwater in the United Arab Emirates, may be derived from a deeply circulating intermediate flow system, recharged far to the east in the northern Oman Mountains, with a contribution from radioactive heating, due to the decay of uranium and high radon activity, which has been reported in some of these springs.

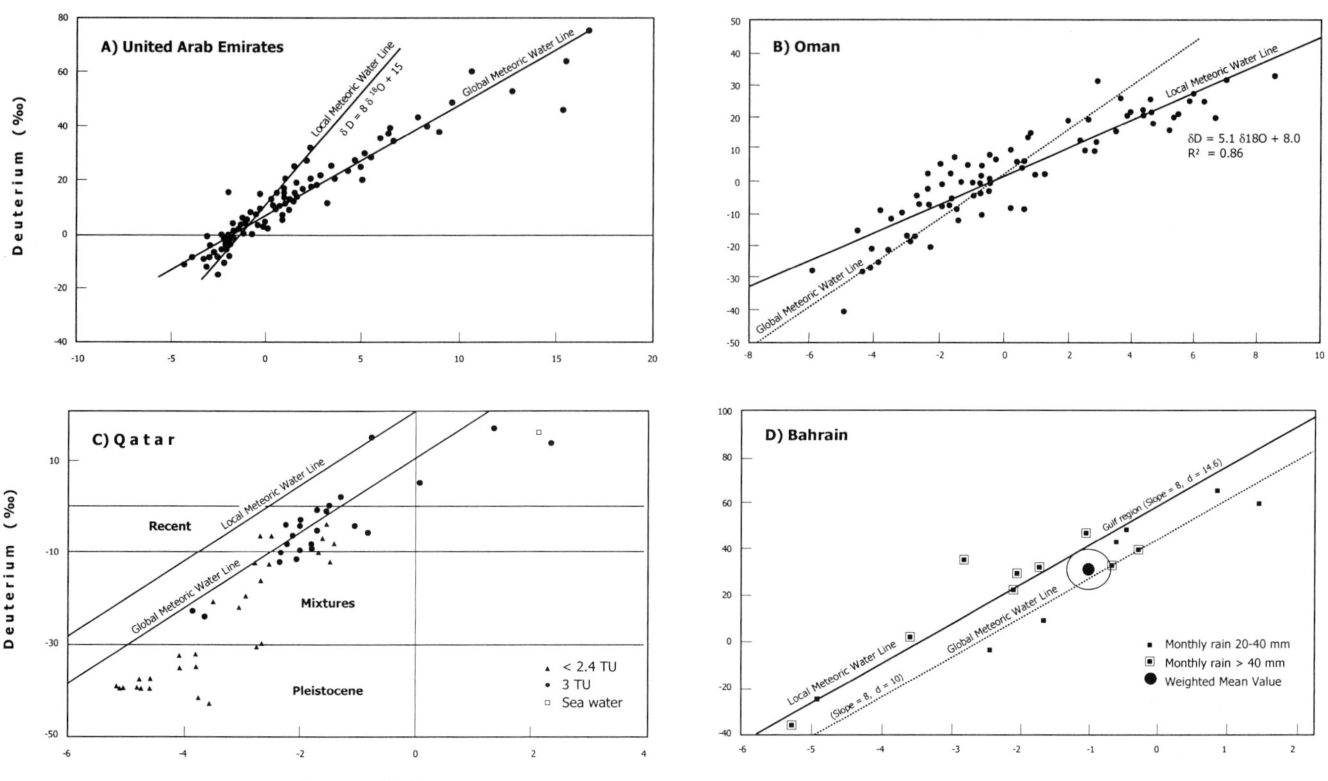

Fig. 5.2. Stable isotopes in the form of ^{18}O-^{3}H relationships from waters in the United Arab Emirates, Oman, Qatar and Bahrain compiled from different sources as follow:

a) Rainwater in the United Arab Emirates (Rizk and Alsharhan, 1999).
b) Precipitation, runoff and groundwater in northern Oman (Macumber et al., 1997).
c) Groundwater in Qatar (modified from Yurtsever, 1999).
d) Precipitation in Bahrain (modified from Yurtsever, 1999).

There is evidence of an overall secular change, with an increase in the total dissolved solids between 1976 and 1987 seen in springs both in Saudi Arabia and United Arab Emirates mainly due to excessive groundwater extraction and low recharge. In Saudi Arabia, the change is of the order of 23% from 1336 mg/l to 1567 mg/l. In United Arab Emirates, between 1991 and 1994, the mean overall increase can vary from 10% (in Khatt South Spring) to 50% in the Bu Sukhnah spring. The increase in the Bu Sukhnah spring from 1977 to 1994 is from 5,500 mg/l to 10,228 mg/l. This spectacular rise has been attributed to the solution of the Miocene Fars gypsum. The magnitude of the change can be seen in the rise of the SO_4^{2-} ion from 165 mg/l in 1991 to 560 mg/l in 1994 in Al Khatt springs contrasted with the rise from 288 mg/l to 1,896 mg/l over the same time period. In contrast the chloride ion showed only a small increase, from 4,000 mg/l to 4,040 mg/l between 1991 and 1994. In 1993 the Khatt springs had relatively low bicarbonate ion values (200 mg/l) compared with the values recorded in 1991-1992 and 1994 suggestive of younger water. Thus the change in ionic proportion with increasing total dissolved solids is in part related to the local geology and in part to the local groundwater flow regime. In a rapid circulation system as in Al Khatt springs in United Arab Emirates there is a direct relationship between rainfall and total dissolved solids, but in slowly circulating groundwater system, groundwater flow is independent of rainfall and water table fluctuation.

In United Arab Emirates, concentrations of the major ions vary from one spring to another according to the local hydrological and geological conditions (Table 5.3). Local groundwater usually has low salinity, and temperature close to the mean annual air temperature. Serial measurements over the period 1991-1994 show small increases in the concentration of all ions, within the same spring, but great variation in the concentration of the same ion in different springs, for example both Al Khatt and Bu Sukhnah springs drain limestone rocks, but the concentration of the Ca^{2+} varies from 60 mg/l at Al Khatt South to 1,100 mg/l in Bu Sukhnah (during 1991). In Al Khatt springs water circulation is rapid, and is directly related to rainfall, whereas water circulation in the Bu Sukhnah spring is slow, and independent of rainfall and water table fluctuation. The same contrast is seen in the Na^{+} from 2 mg/l in Al Siji spring to 1,600 mg/l in the Bu Sukhnah in the same year (1991). The differences in concentration reflect differences in groundwater flow pattern, a rapid, shallow flow system operating at Al Siji, and a deeper groundwater flow pattern at Bu Sukhnah. The high sulphate ion concentration, increasing from 288 mg/l in 1991 to 1,860 mg/l in 1994 suggests relatively old water, the higher bicarbonate ion concentration in Al Khatt springs is consistent with relatively young water. The Piper diagram plots of the Bu Sukhnah water composition, is distinct from that of the local groundwater, showing that, the source of water is not related to local recharge

Table 5.1. Stable isotopes values of hydrogen and oxygen of rainwater from Bahrain, Oman and United Arab Emirates (compiled by Rizk and Alsharhan, 1999).

Country	Deuterium (‰)			Oxygen-18 (‰)		
	Maximum	Minimum	Average	Maximum	Minimum	Average
Bahrain	45.3	-69.1	11.64	6.3	-10.1	0.4
United Arab Emirates	75.2	-25.4	12.4	15.5	-5.7	0.8
Oman	71.4	-26.5	3.3	18.4	-5.9	-1.0

Table 5.2. Summary of chemical and isotopic characters of groundwater flow systems in the United Arab Emirates.

Parameter	Flow system		
	Local	Intermediate	Regional
Total dissolved solids (mg/l)	500 – 1500	1500 – 10000	>10000
Water type	HCO_3^-	SO_4^{2-}	Cl^-
Dominant cation	Mg^{2+}	Ca^{2+}	Na^+
Major dissolved salt	$Mg(HCO_3)_2$	$CaSO_4$	NaCl
Tritium (^3H) (TU)	>10	>5 - <10	<5
^{14}C activity (% PMC)	>50	50 – 10	<10
^{14}C age (Year)	<5,000	5,000 – 15,000	>15,000
Rare Earth Elements (Bb)	0.73	0.51	0.20

consistent with other findings (see Figs. 5.3, 5.4 and 5.5).

Regional flow systems may be characterized by highly mineralized water and high temperature (Fetter, 1988). Radioactive heating may contribute to the higher water temperatures, El Shamy (1990) records the presence of 20.7 ppm uranium and high radon activity has been noted at the Bu Sukhnah spring.

All United Arab Emirates springs were substandard, according to drinking water criteria established by the World Health Organization, with the exception of Al Siji spring, although the water may be used for irrigation.

In Saudi Arabia, the grouping of artesian springs has been expanded to recognize springs, which arise from such different geological situations, as alluvial, sub-basaltic, interstratified, and solution openings (to

Table 5.3. Results of chemical analysis of water samples collected from the studied United Arab Emirates springs and local water wells during 1991-1994 period.

Spring Name/ Year	pH	Electrical conductivity (μS/cm)	Total dissolved Solids (mg/l)	Units	Ca^{++}	Mg^{++}	Na^{+}	K^{+}	Sum cation	HCO_3^-	SO_4^-	Cl^-	Sum anion
Khatt North Spring (1991)	7.4	2420	1490	mg/l	64.0	39.0	379.0	10.0	23.2	226.0	165.0	541.0	23.4
				epm	3.2	3.3	16.5	0.3		4.5	3.4	15.5	
				epm%	13.8	14.0	71.1	1.1		19.3	14.7	66.	
Khatt South Spring (1991)	7.4	2380	1450	mg/l	60.0	37.0	372.0	9.0	22.5	276.0	165.0	516.0	22.7
				epm	3.0	3.1	16.2	0.2		4.5	3.4	14.7	
				epm%	13.3	13.7	71.9	1.0		19.9	15.1	64.9	
Siji Spring (1991)	7.5	750	488	mg/l	26.9	54.0	39.8		7.6	248.9	113.4	40.4	7.6
				epm	1.3	4.4	1.7	0.0		4.1	2.4	1.1	
				epm%	17.7	58.6	22.8	0.9		53.8	31.1	15.0	
Bu Sukhnah Spring (1991)	7.5	8462	5500	mg/l	1100	158.0	1600	58.0	139.2	100.0	561.0	4000	127.6
				epm	55.0	13.2	69.6	1.5		1.6	11.7	114.3	
				epm%	39.5	9.5	50.0	1.1		1.3	9.2	89.6	
Khatt North Spring (1992)	7.8	2650	1605	mg/l	76.0	47.0	397.0	8.0	25.2	322.0	200.0	541.0	24.9
				epm	3.8	3.9	17.3	0.2		5.3	4.2	15.5	
				epm%	15.1	15.6	68.5	0.8		21.2	16.7	62.1	
Khatt North Spring (1992)	7.4	2500	1535	mg/l	92.0	29.0	374.0	7.0	23.5	312.0	190.0	510.0	23.7
				epm	4.6	2.4	16.3	0.2		5.1	4.0	14.6	
				epm%	19.6	10.3	69.3	0.8		21.6	16.7	61.6	
Siji Spring (1992)	8.1	1040	655	mg/l	36.0	46.0	84.0	5.0	9.4	244.0	90.0	125.0	9.5
				epm	1.8	3.8	3.7	0.1		4.0	1.9	3.6	
				epm%	19.1	40.7	38.8	1.4		42.3	19.9	37.8	
Bu Sukhnah Spring (1992)	7.2	11980	7667	mg/l	584.0	163.2	1550	90.0	112.3	140.0	1152.0	3988	138.8
				epm	29.1	13.4	67.4	2.3		2.3	24.0	112.5	
				epm%	26.0	12.0	60.0	2.0		1.7	17.3	81.1	
Khatt North Spring (1994)	7.3	2695	1630	mg/l	104.0	19.5	438.5		25.9	200.0	288.0	580.0	25.9
				epm	5.2	1.6	19.1	0.0		3.3	6.0	16.6	
				epm%	20.1	6.3	73.6	0.0		12.7	23.2	64.1	
Khatt South Spring (1994)	7.3	2618	1563	mg/l	104.0	19.5	371.5	0.0	23.0	200.0	288.0	580.0	25.9
				epm	5.2	1.6	16.2	0.0		3.3	6.0	16.6	
				epm%	22.6	7.1	70.3	0.0		12.7	23.2	64.1	
Siji Spring (1994)	7.4	1373	877	mg/l	56.0	108.0	46.0	6.0	14.0	255.0	165.0	225.0	14.1
				Epm	2.8	9.0	2.0	0.2		4.2	3.4	6.4	
				epm%	20.1	64.5	14.3	1.1		29.8	24.5	45.8	
Bu Sukhnah Spring (1994)	7.3	14905	10228	mg/l	793.0	154.5	17778	0.0	129.8	90.0	1862	5550	198.9
				epm	39.7	12.9	77.3	0.0		1.5	38.8	158.6	
				epm%	30.5	9.9	59.5	0.0		0.7	19.5	79.7	

Table 5.4. Average physical Properties of Saudi Arabian Spring water (compiled from Bazuhair and Hussien, 1990).

Spring category	Location	Water temperature (°C)	Electrical conductivity (μS/cm)	Average discharge (m^3/sec.)	Total dissolved solids (mg/l)
Alluvial	Wadi Fatimah, wadi Al Fara'a, Al Khamel, Khulais, Al Taif	23.6 – 31	300 – 1920	0.17 – 1.00	300-1920
Sub-basaltic	Al Madinah Al Munawarah, Khayber, Al Khamel	28.4 – 38.7	560 – 1920	0.008 – 0.17	560-1920
Fracture	Al Lith, Jizan	45 – 80	2200 – 5670	25-21 l/s	2000-3690
Solution openings	Al Aflaj, Al Kharj, Al Qatif	30.4 – 34.0	3620 – 6240	0.012-0.042	2200-4120
Intrastratified	Al Hasa	31.5 – 38.0	2280 – 2500	1.2-1.8	1440-1760

include both karst and sink-holes), in addition to fault and fracture related springs (Bazuhair and Hussein, 1990). The spring waters may contain dissolved minerals and gases and usually are at temperatures close to the mean annual air temperature. The highly mineralized springs, and springs with elevated water temperatures, even close to boiling, are usually associated with faulting and fracturing, have been thought to have therapeutic value and some have been developed as tourist or recreational sites. The majority of springs in the Arabian Gulf region are gravity and artesian and most are associated with sinkholes in limestone, and fractures or faults in mountainous areas. The dissolved solids present in the spring water are a reflection of the nature of the aquifer rocks. The two physical properties commonly measured are the water temperature and electrical conductivity. The latter number can be converted into a measure of the total dissolved solids present in the water (Table 5.4).

Al-Haddad et al., (1998) carried out measurement of Radon-222 (^{222}Ra) concentrations in some spring waters from Al Hasa province in eastern Saudi Arabia (Table 5.5). They found that a radon concentration ranging between 2,664 to 13,264 Bq m^{-3} largely complying with the United States Environmental Protection Agency maximum contaminant level of 11,100 Bm^{-3} limit for radon in drinking water.

The magnitude of total dissolved solids (TDS) in springs shows a seasonal change rising from about 11% to 36% during the summer months as groundwater is drawn from deeper levels. The increase is particularly marked in the major ion content. The total dissolved solids can indeed vary from month to month, and although the difference in a single spring may not be great, the differences between springs, even those in the same region, can be considerable.

In Saudi Arabia, groundwater may contain about 10% to 15% from the Paleogene Dammam and Umm er Radhuma aquifers. The geochemistry of springs results from this mixing. A significant increase in the total dissolved solids values in the Al Hasa springs illustrate the correlation between the geochemistry and the age of groundwater. The difference is also attributed to increasing dissolution along the flow path, and the decreasing impact of recharge along the hydrological gradient.

3. Hydrogeochemistry of Falaj Water

A falaj can be described as a man-made stream which intercepts the groundwater table through a single or several wells in the footslopes of the mountains bringing the water to the surface along a slope gentler than the natural hydraulic gradient. In Oman about 55% of the currently cropped land is irrigated by falaj water (Wallender, 1989), and the 4,800 active falajes deliver some 900 Mm^3 of water, representing 60% of the country's total water usage (Abdel Rahman and Omezzine, 1996). In the United Arab Emirates falajes are also part of the agricultural heritage, but recently many have run dry because of low rates of recharge and excessive groundwater pumping.

Table 5.5. Summary of the results for Al-Hasa water springs, Saudi Arabia (after Al-Haddad et al., 1998).

Spring Name	Radon-222 (Bq m^{-3})
Umm Saba'a	4,955 ±1,266
Al Mansour	13,264 ± 886
Al-Jouharyyah	9,620 ± 3,106
Al Khudood	4,125 ± 696
Al-Harra	Below the minimum detectable activity
Al-Howayrrat	Below the minimum detectable activity

No detailed study has been carried out on the falaj water sources or geochemistry. In an effort to fill that gap in natural water chemistry in the region, United Arab Emirates falajes were sampled in 1995. The samples were chemically analyzed for major and minor constituents, and the results were presented on relevant charts and graphs. A summary of the general chemical characteristics of falaj waters is given below:

The electrical conductivity values are low in water samples collected from the falajes draining ophiolite rocks, east of Al Ain and Al Fujairah areas, indicating low water salinity. In contrast, the electrical conductivity values are relatively higher in water samples of the falajes draining limestone rocks in Ras Al Khaimah and west of Al Ain areas (Table 5.6). The iso-electrical conductivity contour map shows that the electrical conductivity values of falaj waters are low near the water divide of the eastern mountains and increase further east and west, with distance from the recharge area (Fig. 5.6). The iso-electrical conductivity contours also shows that the groundwater salinity increases from the water divides towards the east and west.

The electrical conductivity of falaj samples varied between 450µS/cm and 10,940µS/cm. Generally, conductivity values are low in water samples collected from the falajes draining ophiolite rocks, whereas the electrical conductivity values are relatively higher in water samples of the falajes draining the limestone rocks. The salinity of falaj waters is low near the water divide of the United Arab Emirates mountains and increases further east and west, with distance from the recharge area. The electrical conductivity in open-channel falajes increases with increasing the falaj length as the result of the high natural evaporation rates from falaj channels. Also, the longer the channel, the greater the contact surface with the bedrock, and thus, the larger amount of total dissolved solids contributed from the bedrock. In tunnel-type falajes, the electrical conductivity does not correlate with the falaj length because of the variation in rock type and source of water in these falajes.

Trilinear plot of the chemical analysis of water samples presented in Fig. (5.7), and show that the appearance of most of the samples in the upper triangle of the diamond-shaped field, points to the dominance of Na-Mg and chloride and bicarbonate water types. The falajes water is enriched in Mg^{2+} which is dissolved from Mg-rich ophiolitic and dolomitic rocks. Because the water of the falajes is mainly used for irrigation purposes, values of electrical conductivity and sodium adsorption ratio were plotted on Fig. (5.8), and show that the water of these falajes except Khatt South and Habhab falajes are good to fair for irrigation purposes. Hydrochemical coefficients of major anions and cations were calculated for falajes water and presented in Table (5.7). According to Hounslow (1995) if the salinity is low and Mg is greater than Ca, this indicates probable dissolution of ferromagnesian minerals from mafic and ultramafic rocks. Waters of these falajes run across the Semail ophiolites in northern Oman mountains. The Ca is greater or equal to Mg in waters of the falajes draining limestone rocks such as falajes of Khatt South, Habhab, Usayli and Bu Sukhnah.

Table 5.6. Results of chemical analysis of water samples collected from United Arab Emirates falajes. Concentrations of ions are expressed in milliequivalent per liter. Hardness and alkalinity are expressed in milligrams per liter).

Falaj No.	Falaj Name	pH	Electrical Conductivity (µS/cm)	CO_3	HCO_3	SO_4	Cl	Total Anion	Ca	Mg	Na	K	Total Cation	Balance
Eastern Drainage Southern Area														
1	Rafaq	7.31	1700	0.15	3.06	4.09	9.40	16.70	2.75	2.48	11.40	0.13	16.76	-0.18
2	Munnai	8.89	780	0.68	4.25	0.76	2.14	7.83	1.24	2.29	4.20	0.10	7.83	0.00
3	Howeilat	9.08	1250	0.68	3.57	2.79	5.71	12.75	1.65	2.07	8.90	0.13	12.75	0.00
4	Warah	8.60	710	0.41	3.06	1.90	2.20	7.57	1.54	3.60	2.50	0.07	7.71	-0.92
5	Masfut	8.72	750	0.50	3.46	0.86	2.70	7.52	1.54	3.55	2.30	0.07	7.46	0.40
6	Dofdah	9.52	610	0.31	1.04	2.60	2.18	6.13	1.03	2.89	2.20	0.07	6.19	-0.49
7	Sagheer	9.53	700	1.00	3.68	0.56	1.76	7.00	1.34	4.56	0.87	0.23	7.00	0.00
8	Sahreeha	10.07	900	1.12	2.65	1.53	3.70	9.00	1.34	5.35	1.95	0.36	9.00	0.00
Western Drainage Northern Area														
9	Khatt South	8.16	2250	0.34	3.74	4.03	14.30	22.41	4.30	2.40	15.50	0.21	22.41	0.00
10	Habhab	8.12	2769	0.34	3.90	5.55	16.90	26.69	3.95	3.72	18.80	0.22	26.69	0.00
11	Usayli	8.20	2200	0.34	3.57	8.86	14.30	27.07	3.95	2.42	15.50	0.20	22.07	10.18
Western Drainage Central Area														
12	Fili	8.85	1020	0.41	2.44	3.30	5.03	11.18	1.50	2.10	8.10	0.17	11.87	-2.99
13	Asimah	7.12	450	0.08	1.53	1.19	2.00	4.80	0.30	3.03	1.13	0.08	4.54	2.78
14	Murrad	8.54	750	0.40	3.16	1.11	2.87	7.54	1.34	3.95	2.20	0.05	7.54	0.00
15	Dhaid	8.98	1180	0.60	3.37	2.12	5.62	11.71	1.00	2.30	8.26	0.15	11.71	0.00
16	Siji	8.97	750	0.61	3.47	2.36	1.14	7.58	1.34	4.44	1.73	0.07	7.58	0.00
17	Mualla	7.33	1840	0.20	4.49	4.90	9.50	19.09	1.29	4.59	13.04	0.23	19.15	-0.16
18	Manama	8.47	680	0.40	3.37	0.80	2.38	6.95	1.44	4.15	1.30	0.06	6.95	0.00
Eastern-Central Area														
19	Farfar	8.25	900	0.34	3.40	1.09	4.00	8.83	1.72	3.38	3.65	0.08	8.83	0.00
20	Mamduk	8.69	600	0.50	3.57	0.17	1.90	6.14	1.20	2.40	2.48	0.06	6.14	0.00
21	Museiriya	9.22		0.82	3.88	1.07	4.66	10.43	1.44	4.93	3.91	0.15	10.43	0.00
22	Habeesa	9.57	740	0.60	2.04	1.40	3.33	7.37	0.93	2.76	3.53	0.12	7.34	0.20
23	Bithna (R/B)	10.00	1050	0.82	2.04	2.81	4.76	10.43	1.44	4.93	3.91	0.15	10.43	0.00
24	Awaina	8.04	1030	0.20	2.14	1.82	6.04	10.20	1.65	0.70	7.80	0.06	10.21	-0.05
25	Sham	8.20	650	0.40	4.40	0.25	1.80	6.85	1.03	4.47	1.30	0.05	6.85	0.00
26	Farah	8.63	1000	0.68	5.45	0.97	4.70	11.80	2.90	4.30	4.57	0.09	11.86	-0.25
Al Ain Area														
27	Ain Sukhnah	8.86	10940	0.50	3.06	12.60	103.53	119.69	30.38	12.25	80.40	1.41	124.44	-1.95
28	Maziad	8.86	960	0.50	3.06	2.40	4.20	10.16	1.29	4.10	4.70	0.13	10.22	-0.29
29	Hili	8.44	750	0.40	3.47	1.14	2.57	7.58	1.75	2.25	3.48	0.10	7.58	0.00
30	Daudi	9.51	520	0.60	2.14	0.77	1.70	5.21	1.13	2.10	1.90	0.08	5.21	0.00
31	Al-Aini	9.87	620	0.82	2.25	0.77	1.70	5.21	1.13	2.10	1.90	0.08	5.21	0.00
32	Buraimi	9.86	380	0.60	1.63	0.38	1.19	3.80	1.03	1.81	1.20	0.06	4.10	-3.80
33	Saarrah	10.87	390	1.00	1.33	0.33	1.24	3.90	0.02	1.61	1.30	0.07	3.90	0.00

The high concentration of HCO_3^- indicates that the water is young and can be partially derived from rain. Most of falaj water, however is obtained from the Quaternary aquifer. The high SO_4^{2-} in the water is a result of dissolution of gypsum ($CaSO_4.2H_2O$) from the evaporite deposits through which the water of the falaj moves.

4. Hydrogeochemistry of Groundwater

Based on more than 10,000 chemical analyses of groundwater samples from Australia, Chebotarev (1955) observed that the groundwater tends to evolve chemically toward the composition of seawater. He also observed that this evolution is associated with regional changes in dominant anions (CO_3^{2-}, HCO_3^-, SO_4^{2-} and Cl^-). Freeze and Cherry (1979) illustrated the Chebotarev's (1955) general anion evolution in groundwater as follows:

Travel along flow path ⟶

$HCO_3^- \rightarrow HCO_3^- + SO_4^{2-} \rightarrow SO_4^{2-} + HCO_3^- \rightarrow SO_4^{2-} + Cl^- \rightarrow Cl^- + SO_4^{2-} \rightarrow Cl^-$

Increasing age ⟶

On the other hand, the cation dominance generally evolves in the order of: $Ca^{2+} \rightarrow Mg^{2+} \rightarrow Na^+ + K^+$, but this evolution is not consistent as the concentration of cations may exhibit a wide range of fluctuations in groundwater and is not as steady as the changes in anion dominance.

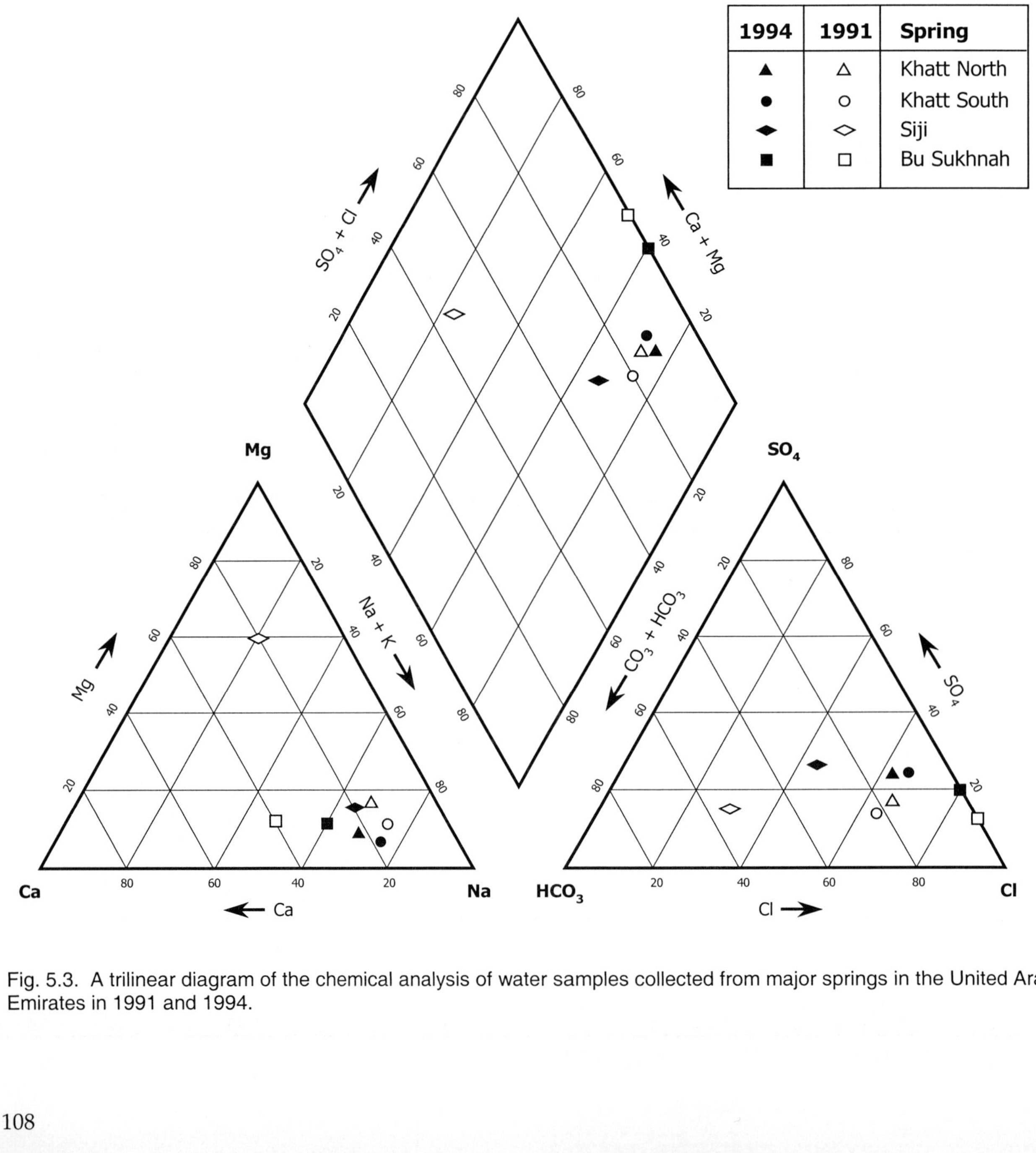

Fig. 5.3. A trilinear diagram of the chemical analysis of water samples collected from major springs in the United Arab Emirates in 1991 and 1994.

Due to the differences in groundwater chemistry, the Paleozoic-Mesozoic aquifers, Tertiary aquifers, and Quaternary aquifers in the Arabian Gulf region will be discussed separately. The Paleozoic-Mesozoic aquifers are particularly developed as fresh-water producers in the western parts of the Arabian Peninsula and the northern part of the United Arab Emirates. The Tertiary aquifer is the main fresh-water producer in the whole Gulf States, while the Quaternary aquifer is the main source of fresh water in the United Arab Emirates.

a) Paleozoic-Mesozoic Aquifer

Salinity in the Saq-Tabuk aquifer in Al-Qasim Province, Saudi Arabia shows that the total dissolved solid content of groundwater in the aquifer ranges from 1,300 to 3,000 mg/l. However, high salinity values of about 14,500 mg/l was measured in some wells rendering groundwater in the aquifer unfit for use prior to desalination and treatment (Khalifa and Rizk, 1994). A trilinear plot of groundwater samples shows that the dominant water type is sodium chloride (Fig. 5.9a). A plot of the electrical conductivity versus sodium adsorption ratio indicates that the examined water samples are unsuitable for irrigation of traditional crops before treatment (Fig. 5.9b).

A recent study by Sagaby and Moallim (1999) for the Saq aquifer estimates it may store 280 billion m^3 of water. This has been heavily exploited for agricultural purposes since 1970's, but because the water in the aquifer is fossil water 10,000-30,000 years old (Sharaf and Hussein, 1996), the aquifer underwent a noticeable decline in hydraulic head and a sharp rise in groundwater salinity.

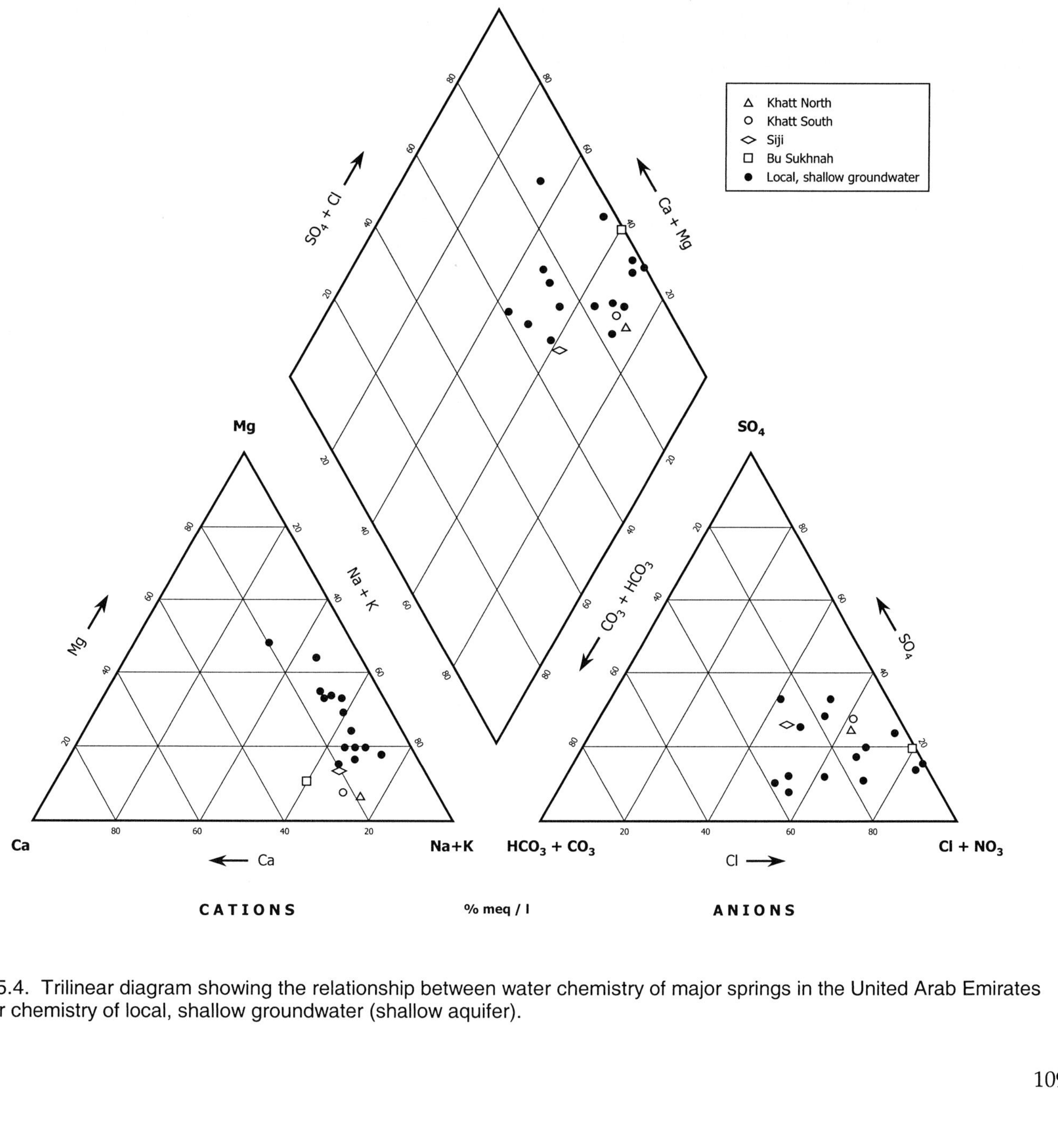

Fig. 5.4. Trilinear diagram showing the relationship between water chemistry of major springs in the United Arab Emirates water chemistry of local, shallow groundwater (shallow aquifer).

Sagaby and Moallim (1999) compared older data (collected by Bureau de Recherches Gèologiques et Minières in 1985) with their recent data for the same water wells, which indicated that the sequence of cations and anions: $Na^{+}>Ca^{2+}>Mg^{2+}$ and $Cl^{-}>SO_4^{2-}>HCO_3^{-}$, and water type (NaCl) remained the same. However, the total dissolved solid content in the aquifer between 1985 and 1996 shows a steady increase in the direction of groundwater flow, as water moves from recharge area towards the discharge area, and an additional increase of 170 mg/l (21%) in groundwater salinity occurred between 1985 and 1996. As shown on figure (5.10) all water samples are located along the right of diamond shape field, which reflect the amount of total dissolved solids, and the type of water involved in the aquifer is the chloride facies and sodium facies.

In the United Arab Emirates, the Paleozoic-Mesozoic Ru'us Al Jibal Group constitutes the main aquifer in Ras Al Khaimah Emirate and contains considerable amounts of recent and old waters. Most of groundwater in Ru'us Al Jibal aquifer plots close to the meteoric water line with a deuterium excess (d) of + 20. The groundwater plot to the right of this meteoric water line suggests evaporitic effects, the mixing of different waters or salt-water intrusion from the Arabian Gulf (Akiti et al., 1988). The aquifer contains groundwater with an isotopic composition more depleted than the average values of the stable isotopic composition of the atmosphere, indicating a possible source of moisture in the Mediterranean region, which recharges groundwater at high elevations (Akiti et al., 1988).

The parallel increase of total dissolved solids and $\delta^{18}O$ from 3,200 mg/l and -3.54‰ in 1985 to 15,070 mg/l and -2.89‰ in 1986 at Wadi Sahawat (in the vicinity of Al Bih area, Ras Al Khaimah Emirate) shows the effect of mixing between seawater and groundwater. It was also noted that the ^{3}H content also decreased from 6.7 to 3.7 TU. Also in Dohreen (in the vicinity of Wadi Al Bih, Ras Al Khaimah Emirate), the increase of total dissolved solids from 2,400 mg/l in 1984 to 6,000 mg/l in 1986 while the $\delta^{18}O$ remained constant around -3.50‰, suggests the dissolution of salt in the soil zone during groundwater infiltration (Gonfiantini, 1992). The above two processes lead to deterioration of groundwater quality. Additionally, over-pumping of freshwater can induce salt-water intrusion into fresh groundwater through subsurface fractures.

b) Tertiary Aquifer

The Tertiary aquifer systems are very important in the Arabian Gulf region because they are highly transmissive, highly productive, and underlie the entire Eastern Arabian basin. As the aquifer units of the Paleogene carbonate cycle and the Neogene detrital cycle, are hydrologically connected to each other to varying extent, these rocks are considered as a single aquifer system, although in most of the earlier studies the communication between adjacent aquifer units in the same system has been neglected. Interaction across the intervening confining layers of shale and marl units in the Middle Eocene Dammam Formation, and the absence of evaporite units in the Lower Eocene Rus Formation is further facilitated either by dissolution or non-deposition, and/or fracturing associated with slumping or tectonic activity, especially over domal and anticlinal crests where karstification, fault systems and joints become abundant.

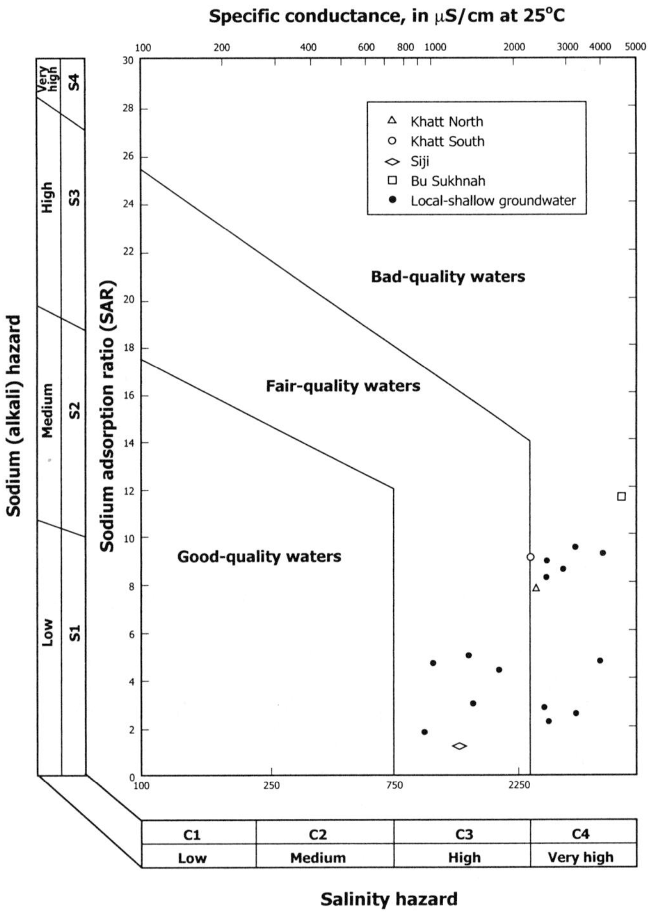

Fig. 5.5. Graph showing the classification of sodium and salinity hazards in water of major springs and local shallow groundwater (shallow aquifer) in the United Arab Emirates.

Local and regional studies of the Tertiary aquifers show a division into two distinct systems, a multi-layered system and a system of fresh water lenses. The multi-layered system, is composed of several aquifers and aquicludes or aquitard layers, locally connected to each other, sometimes confined sometimes unconfined. The freshwater lenses tend to be isolated, and occur in the upper saturated zones of the multi-layer system. At some locations, they accumulate in collapse depressions and float on brackish water. The Dammam is an important aquifer in eastern Arabia. The Dammam groundwater in Saudi Arabia is in the age range 16,000 to 20,000 years BP, and does not receive any recent recharge, but is in direct hydraulic contact with the Neogene aquifer above, and the Umm er Radhuma aquifer below. In Bahrain the groundwater is 13,000 years BP old, and is also without recent recharge, here it is in hydraulic contact with the Umm er Radhuma aquifer to the west, and to the east it is affected by salt-water intrusion. The age of the water is about the same in Qatar, from 10,000 to 17,000 years BP, but it does receive some recent recharge.

Water salinity of Paleocene Radhuma aquifer in the southwestern Kuwait varies between 4,000 and 5,000 mg/l. This salinity is slightly higher than the salinity of groundwater of the overlying Middle Eocene Dammam aquifer. SO_4^{2-} is the dominant anion and Ca^{2+} is the dominant cation. The sodium absorption ratio ranges from 2.6 to 5.1. Water salinity of Radhuma aquifer increases to the east and northeast of Kuwait, where water salinity southeast of Kuwait Bay, is more than 35,000 mg/l with Cl^- the dominant anion, and Na^+ is the dominant cation.

The Dammam Formation is the major aquifer exploited in Kuwait and extends all over the country. The groundwater is old (22,000 years BP) and does not receive any recent recharge, in fact the aquifer

Table 5.7. Calculated hydrochemical ratios in water samples collected from United Arab Emirates falajes during early 1996.

Falaj No.	Falaj Name	Sodium Adsorption Ratio	Hardness (mg/l)	Alkalinity (mg/l)	Ca/Mg	Na/K	Cl/SO_4	Na/Cl	Mg/(Ca+Mg)	$Ca/(Ca+SO_4)$	$(Ca+Mg)/SO_4$	Cl/ Total Anion	HCO_3/ Total Anion
Eastern Drainage Southern Area													
1	Rafaq	7.20	261.50	153.13	1.11	87.69	2.30	1.21	0.5	0.4	1.3	0.6	0.2
2	Munnai	3.20	176.50	212.68	0.54	42.00	2.82	1.96	0.6	0.6	4.6	0.3	0.3
3	Howeilat	6.50	186.00	178.65	0.80	68.46	2.05	1.56	0.6	0.4	1.3	0.4	0.2
4	Warah	1.56	257.00	153.13	0.43	35.71	1.16	1.14	0.7	0.4	2.7	0.3	0.2
5	Masfut	1.40	254.50	173.15	0.3	32.86	3.14	0.85	0.7	0.6	5.9	0.4	0.2
6	Dofdah	1.57	196.00	52.04	0.36	31.43	0.84	1.01	0.7	0.3	1.5	0.4	0.1
7	Sagheer	0.50	295.00	184.16	0.29	3.78	3.14	0.49	0.8	0.7	10.5	0.3	0.2
8	Sahreeha	1.00	334.50	132.61	0.25	5.42	2.42	0.53	0.8	0.5	4.4	0.4	0.2
Western Drainage Northern Area													
9	Khatt South	8.50	335.00	187.16	1.79	73.81	3.55	1.08	0.4	0.5	1.7	0.6	0.2
10	Habhab	9.60	383.50	195.17	1.06	85.45	3.05	1.11	0.5	0.4	1.4	0.6	0.2
11	Usayli	8.70	318.50	178.65	1.63	77.50	1.61	1.08	0.4	0.3	0.7	0.5	0.2
Western Drainage Central Area													
12	Fili	6.40	180.00	122.10	0.71	47.65	1.52	1.61	0.6	0.3	1.1	0.4	0.1
13	Asimah	0.90	166.50	76.57	0.10	14.13	1.68	0.57	0.9	0.2	2.8	0.4	0.1
14	Murrad	1.40	264.50	158.13	0.34	44.00	2.59	0.77	0.7	0.5	4.8	0.4	0.2
15	Dhaid	6.40	165.00	168.64	0.43	55.07	2.65	1.47	0.7	0.3	1.6	0.5	0.2
16	Siji	1.00	289.00	173.65	0.30	24.71	0.48	1.52	0.8	0.4	2.4	0.2	0.2
17	Mualla	7.60	294.00	224.69	0.28	56.70	1.94	1.37	0.8	0.2	1.2	0.5	0.3
18	Manama	0.80	279.50	168.64	0.35	21.67	2.98	0.55	0.7	0.6	7.0	0.3	
Eastern-Central Area													
19	Farfar	2.30	255.00	170.15	0.51	45.63	3.67	0.91	0.7	0.6	4.7	0.5	0.2
20	Mamduk	1.80	180.00	178.65	0.50	41.33	11.18	1.31	0.7	0.9	21.2	0.3	0.2
21	Museiriya	2.20	318.50	194.17	0.29	26.07	4.36	0.84	0.8	0.6	6.0	0.4	0.2
22	Habeesa	2.40	184.50	102.09	0.34	29.42	2.38	1.06	0.7	0.4	2.6	0.5	0.1
23	Bithna (R/B)	2.20	318.50	102.09	0.29	26.07	1.69	0.82	0.8	0.3	2.3	0.5	0.1
24	Awaina	7.20	117.50	107.09	2.36	130.00	3.32	1.29	0.3	0.5	1.3	0.6	0.1
25	Sham	0.80	275.00	220.19	0.23	26.00	7.20	0.72	0.8	0.8	22.0	0.3	0.3
26	Farah	2.40	360.00	272.73	0.67	50.78	4.85	0.97	0.6	0.7	7.4	0.4	0.3
Al-Ain Area													
27	Ain Sukhnah	17.40	2131.50	153.13	2.48	57.02	8.22	0.78	0.3	0.7	3.4	0.9	0.2
28	Maziad	2.90	269.50	153.13	0.31	36.15	1.75	1.12	0.8	0.3	2.2	0.4	0.2
29	Hili	2.50	200.00	173.65	0.78	34.80	2.25	1.35	0.6	0.6	3.5	0.3	0.2
30	Daudi	1.50	161.50	107.09	0.54	23.75	2.21	1.12	0.7	0.6	4.2	0.3	0.1
31	Al-Aini	1.90	181.50	112.60	0.45	25.00	1.85	1.22	0.7	0.5	3.3	0.3	0.1
32	Buraimi	1.10	142.00	81.57	0.57	20.00	3.13	1.01	0.6	0.7	7.5	0.3	0.1
33	Saarrah	1.20	126.50	66.56	0.57	18.57	3.76	1.05	0.6	0.7	7.7	0.3	0.1

was recharged during past pluvial periods, and the aquifer water is in direct hydraulic connection with groundwater in the Kuwait Group rocks. Groundwater varies from brackish (2,500 mg/l) in southwest Kuwait to brine (150,000 mg/l) in the northeast. Local anomalies in total dissolved solids content are possible due to variable karst development and infiltration rates. On the basis of water quality, the groundwater of Dammam aquifer can be classified into brackish, saline and brine waters. Brackish groundwater underlies a large area of Kuwait and includes the well fields of As-Sulaibiyah, Umm Al Aish, Wafra and Al-Abdaliyah. Within this area groundwater salinity varies between 2,500 and 10,000 mg/l. The dominance of SO_4^{2-} changes to Cl^- in the east and northeast, the dominant cation is Na^+ and water salinity is generally more than 6,000 mg/l. In southern Kuwait, water salinity of the Dammam aquifer ranges from 5,000 to 7,000 mg/l, the dominant anion is Cl^- and the dominant cation is Na^+. Saline water ranges in salinity from 10,000 to 50,000 mg/l and it bounds the brackish water to the north, northeast and east. The water occurs also in southern Kuwait where the water salinity reaches to more than 20,000 mg/l, where the dominant anion is Cl^- and the dominant cation is Na^+. The brine ranges in salinity from 50,000 to more than 150,000 mg/l, and extends to the northeast of salty water. The dominant anion is Cl^- and the dominant cation is Na^+. In contrast the average total dissolved solids content of the Dammam aquifer in Umm Gudair (southwestern Kuwait), is 3,480 mg/l and the dominant anion is Cl^-. The average sodium absorption ratio value is 6.98, indicating the suitability of the aquifer's water for irrigation of some crops (Al-Ruwaih, 1995).

In Saudi Arabia as groundwater in the Tertiary aquifer system moves down flow paths from outcrops in central Saudi Arabia, the total dissolved solids gradually increases as the water evolves from dominantly calcium-bicarbonate to dominantly calcium-sulphate composition. The possible reasons are upward leakage from a deep saline aquifer and an over-extraction groundwater for all purposes. The groundwater quality system in central Saudi Arabia deteriorates progressively from less than 1,000 mg/l at the outcrop to more than 5,000 mg/l at the Arabian Gulf coast. Also the chemistry of the water column in the aquifer is not homogeneous. The total dissolved solid content increases with depth due to variation in lithology and increase in temperature. In eastern Saudi Arabia the trend of increasing groundwater salinity shows a close agreement to distribution of the hydraulic head, with least concentrations occurring in the northern zone but increasing in the southern and southwestern zone. The total dissolved solid distribution clearly demonstrates a low concentration of dissolved constituents in the area of the groundwater mounds and recharge. There is an increase in total dissolved solids down gradient in all directions. The fresh groundwater body in the northern zone is mainly concentrated in the central part of the field and is surrounded by a thick body of salt water that underlies the entire northern zone. The interface between fresh water and salt water forms one of the boundaries of the fresh water lens system. However, the increases in pumping during last few decades have induced encroachment of the salt water front towards the fresh water zone.

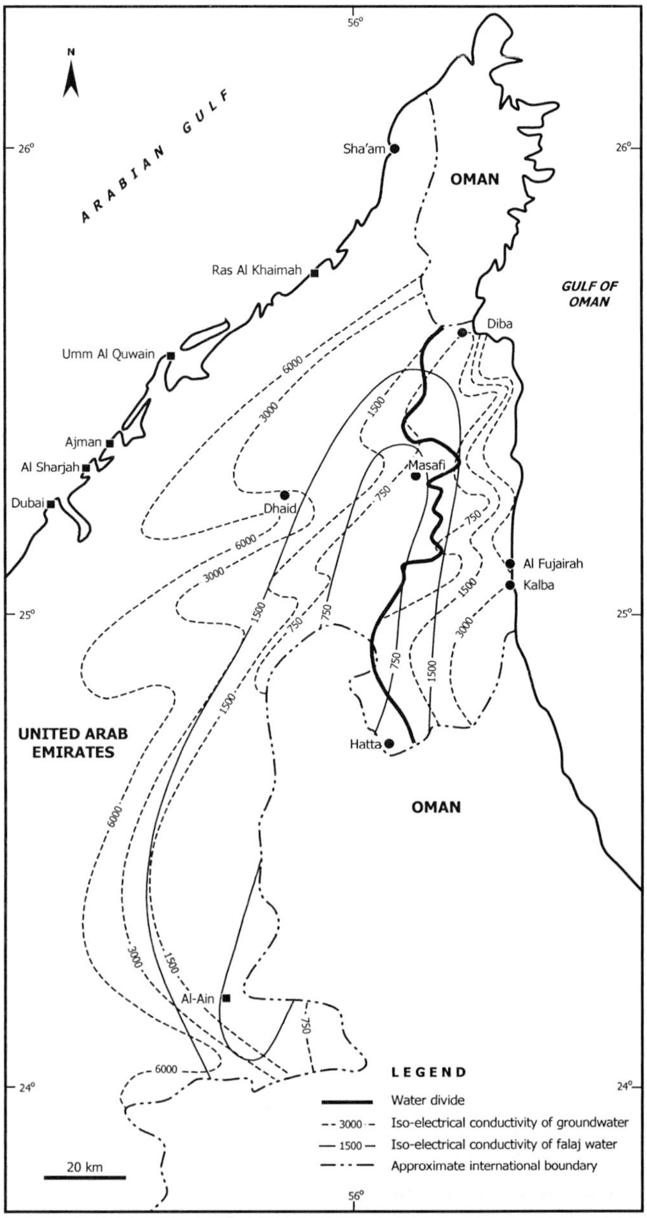

Fig. 5.6. Iso-Electrical conductivity (μS/cm) contour map of groundwater and falaj water during early 1996 in northeastern region of the United Arab Emirates.

The chemistry of the Khobar aquifer (a part of the Dammam aquifer system) in eastern Saudi Arabia has high HCO_3^- concentrations measured in groundwater of upstream areas (Al Hofuf and Wadi

Al-Miyah), evolving into SO_4^{2-} then Cl^- dominant downstream (towards the Arabian Gulf) (Hassan, 1998). This can be attributed the evolution of groundwater chemistry to mixing, cation exchange, and dissolution of sulphate minerals. The cation (Ca^{2+}, Mg^{2+}, Na^+ and K^+) concentration increases downstream towards the Arabian Gulf, in the direction of groundwater flow. The trilinear plot of the Khobar aquifer (Fig. 5.11) depicts the presence of five hydrogeochemical facies (Fig. 5.12): Ca-SO_4 facies; Na-Ca-SO_4-Cl facies; Na-Ca-Cl-SO_4 facies; Na-Ca-Cl facies; and Na-Cl facies. The total dissolved solids and concentrations of most major ions increase in the direction of groundwater flow. The aquifer is generally oversaturated with respect to carbonate minerals, and undersaturated with respect to sulphate minerals. The groundwater evolves from SO_4-Cl facies to Cl facies, in the direction of groundwater flow (Hassan, 1998).

In the United Arab Emirates, the Lower Eocene Rus Formation in Jabal Hafit (Al Ain region) produced 21,000 m^3/day of thermal (36-52°C), brackish water (3,900-6,900 mg/l). Geophysical well logs showed that the limestone belongs to the confined-flow carbonate aquifers, and that the thermal water enters the aquifer from a fractured interval between 93 and 102m deep (Khalifa, 1997). Chemical analysis of groundwater samples from the Rus Formation showed that the water is slightly alkaline (pH 7-8) and sodium-chloride rich. Two water types were distinguished; the first has a relatively low temperature and total dissolved solids, while the second has a high temperature and total dissolved solids. Stiff diagrams show that sodium is the dominant cation, while chloride is the dominant anion. High radon and radium content in groundwater of the Lower Eocene limestone in Jabal Hafit (Table 5.8) is attributed to the presence of high

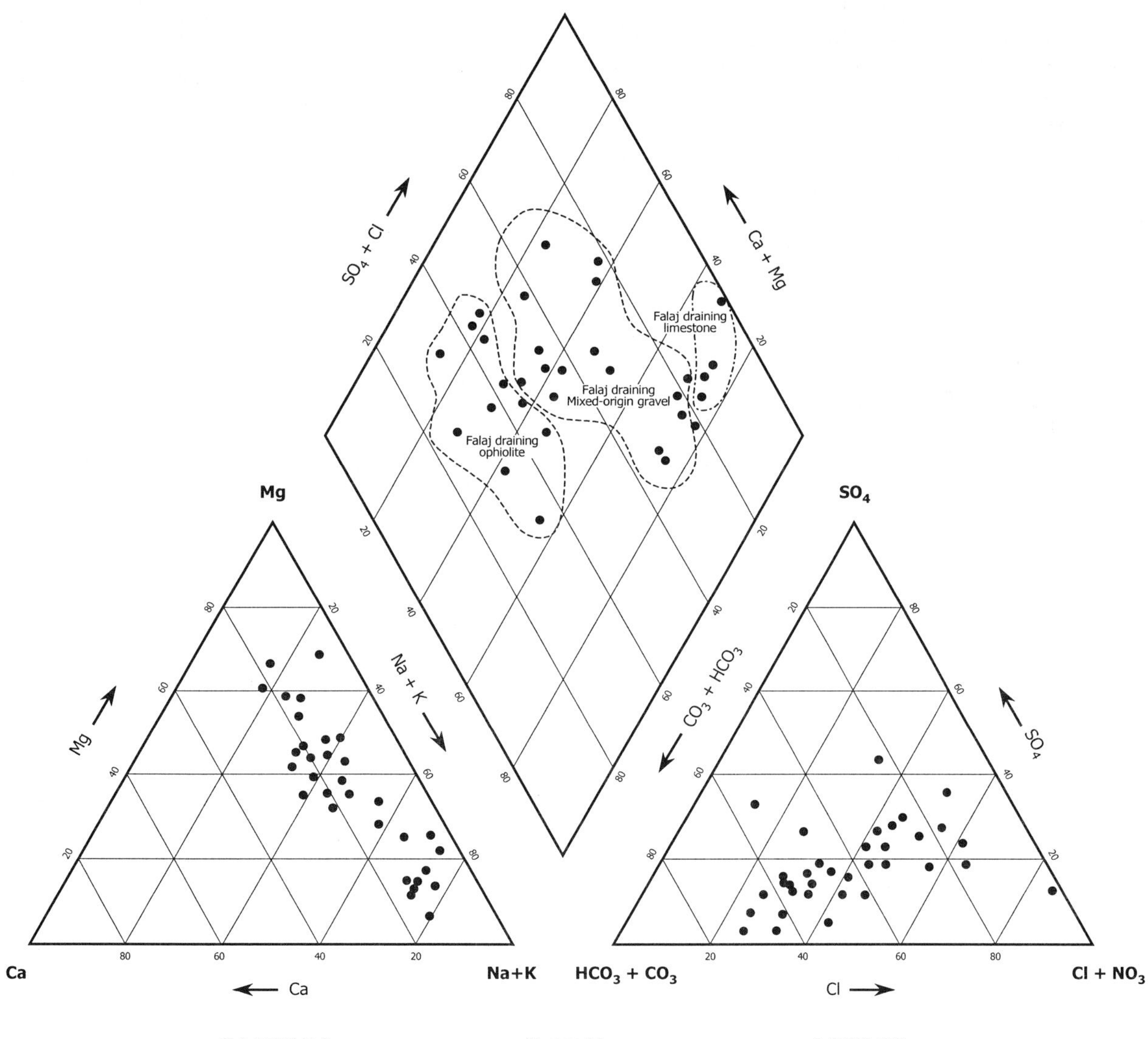

Fig. 5.7. The trilinear plot of chemical analyses of water samples collected from United Arab Emirates falajes during early 1996.

uranium content along joints and bedding plains (El Shami, 1990). Environmental isotopes; tritium (^{3}H) deuterium (^{2}H) and oxygen-18 (^{18}O) are relatively lighter than in other groundwater in the Al Ain area, suggesting possible local recharge through Jabal Hafit; negative ^{2}H and ^{18}O reflect recharge at high elevation.

The Tertiary aquifers in the western region of United Arab Emirates contain saline to brine waters that are only used for injection into reservoirs to maintain pressure. Table 5.9 shows the Tertiary hydrogeologic units in onshore Abu Dhabi oil fields. It also shows that the groundwater salinity ranges from 70,000 mg/l in the Miocene clastics to 160,000 mg/l in the Paleocene Umm er Radhuma aquifer (NDC-USGS, 1996).

In Qatar, it is clear that the recharge from rainfall is greatest in the north, and shallow wells in this area have lowest total dissolved solid concentrations. The southern and southwestern parts of the country get less recharge, and have higher total dissolved solids values. In the southern part, the conditions of low recharge and poor groundwater circulation are reflected in the general poor quality of groundwater. The transition in groundwater salinity in central Qatar coincides with the transition between carbonate facies in the north and evaporite facies in the south. Groundwater in the northern Qatar is bicarbonate rich, having lower salinity varying from 400 to 2,000 mg/l, whilst the waters in the southern and southwestern zones are sulphate rich, with salinity varying from 3,000 to 6,000 mg/l. Using tritium monitoring of wells in the Tertiary limestone aquifer in Qatar (Fig. 5.2), indicated the area where the groundwater was being effectively replenished by the direct rainfall. Based on tritium results, an average replenishment rate was in the range of 7-24 Mm^3 per annum (Yurtsever, 1999). The increased salinity in these aquifers is due to upward leakage from underlying confined Umm er Radhuma aquifer or seawater intrusion.

In Bahrain, the Dammam aquifer over the whole country has experienced a noticeable increase in groundwater salinity. A study carried out by Zubari and Madany (1992) shows salinity changes between 1980 and 1990. The salinity increase in the eastern area is caused mainly by seawater intrusion, while in the northern, western and southern areas it is caused by upward leakage of the Umm er Radhuma aquifer saline waters. The aquifer is not suitable for direct use for domestic or agricultural purposes. Half of the original Dammam aquifer in Bahrain has been polluted due to its over-exploitation during the development process (Zubari et al., 1996). A continuous increase in the rates of abstraction from the Dammam aquifer, at the present rate, could lead to its total contamination and eventually the destruction of the natural groundwater resources.

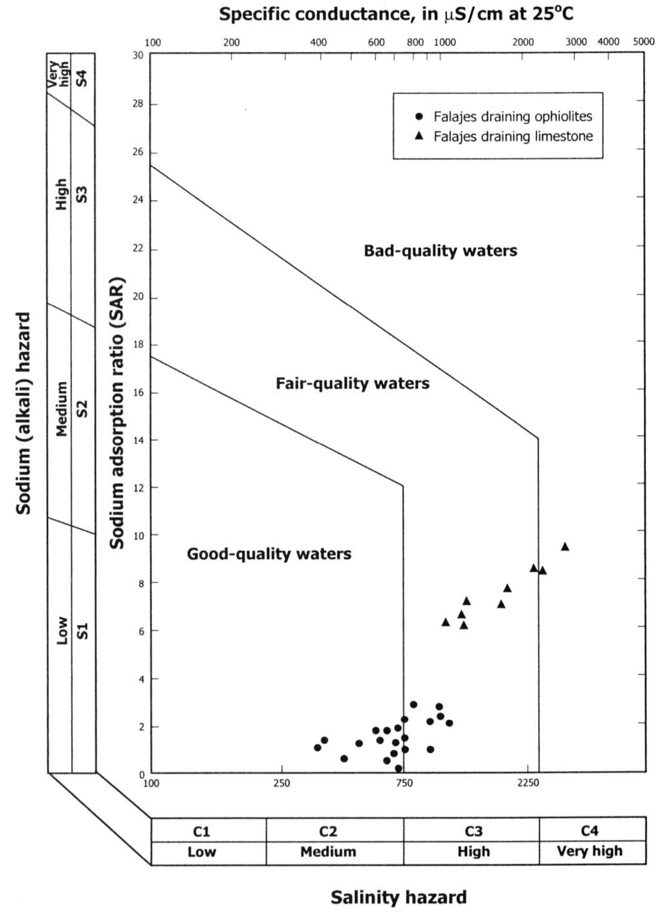

Fig. 5.8. Classification of falajes water in the United Arab Emirates.

Spatial and temporal water quality analysis carried out by Zubari et al. (1996) for 254 wells tapping the Dammam aquifer found two zones with anomalous total dissolved solid values. The first extends over the north central region (total dissolved solids =11,000 mg/l), and the second is located in the western region (total dissolved solids = 8,000 mg/l). Salinization of the aquifer in these two zones is caused by the upward flow from the underlying formations. In southwestern Bahrain, salinization of the aquifer with total dissolved solids of 7,000 mg/l is most probably caused by the flow of sabkha water into the aquifer. In western Bahrain, a salinity anomaly (total dissolved solids = 5,000 mg/l) may be attributed to aquifer contamination by irrigation drainage waters. The increase in total dissolved solid values over the period 1979-1992 is due to the steady rise of abstraction rates from the Dammam aquifer from 138 Mm^3 in 1979 to 187 Mm^3 in 1992 (Al Noaimi, 1993); with a calculated safe yield range from 90 to 112 Mm^3/year. This means that, the present abstraction rates approaches twice the recommended yield for the aquifer, which explains the continuous deterioration of the Dammam aquifer water quality.

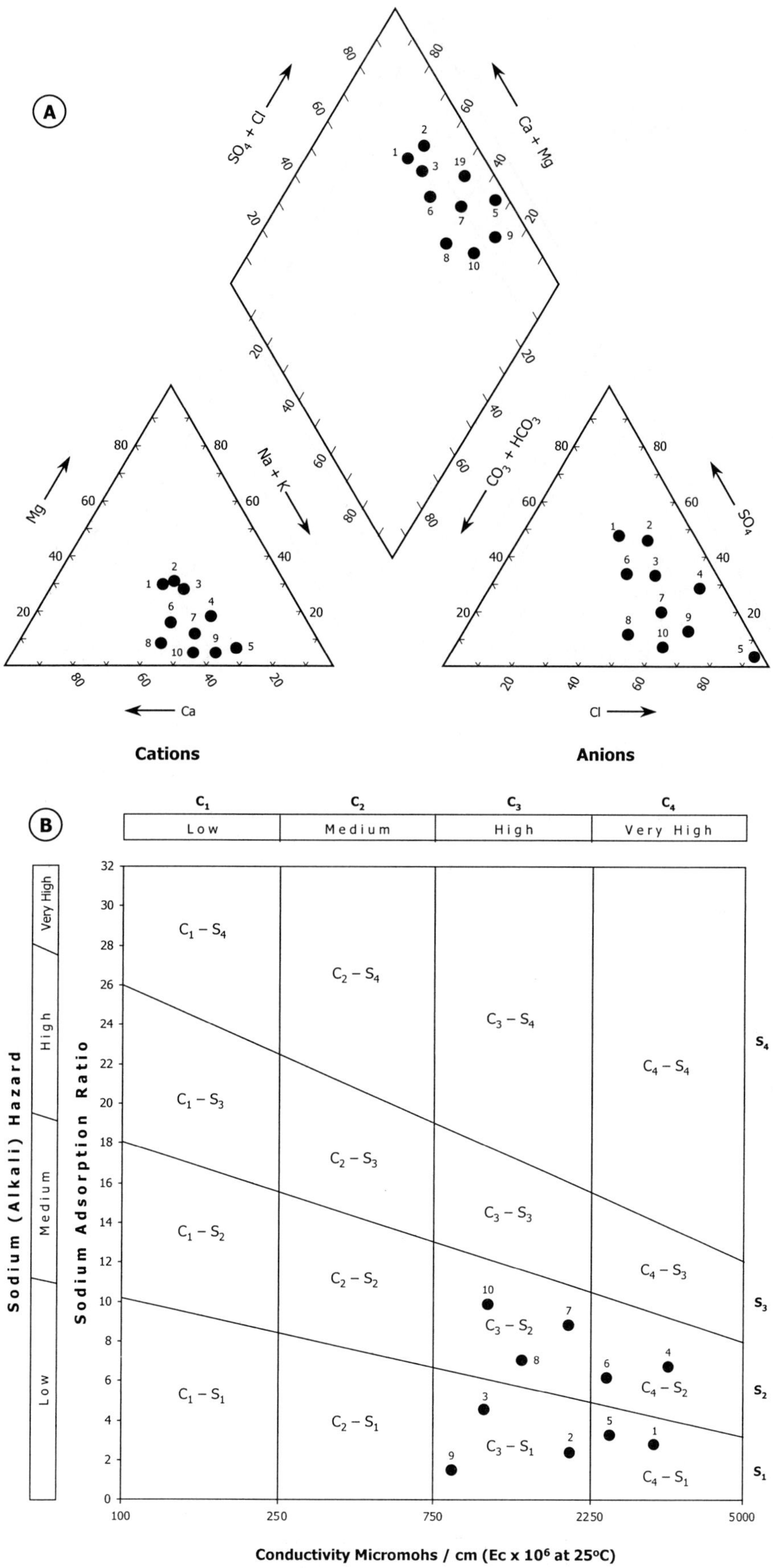

Fig. 5.9. Analyses of groundwater from Saq-Tabuk aquifer in Al-Qasim province in Saudi Arabia (modified from Khalifa and Rizk, 1994). (A) trilinear plot of groundwater samples; and (B) plot of electrical conductivity versus sodium adsorption ratio.

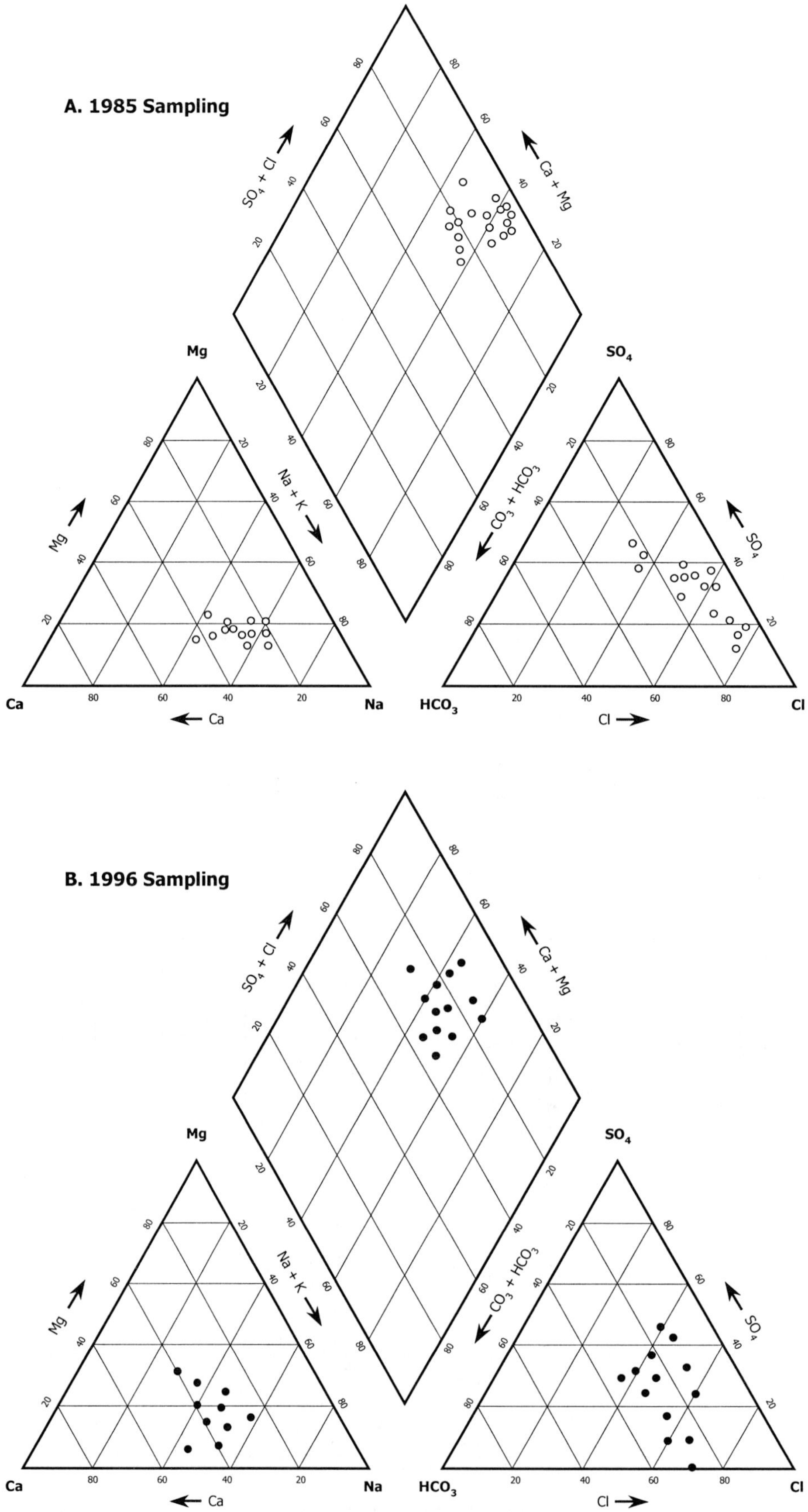

Fig. 5.10. Piper diagram showing chemical facies for the period (1985 and 1996) sampling sets of groundwater for Saq aquifer in Saudi Arabia (modified after Al-Sagaby and Moallim, 1999).

In Oman, the Tertiary marine limestone constitute the principal aquifer in central Oman (known in Arabic Al Wusta region), covering about 80,000 km^2. Because Al Wusta region is a discharge area, groundwater salinity is brackish to saline. However, fresh groundwater lenses are scattered throughout the area and are the only source of potable water (Table 5.10). The groundwater salinity varies between 500 and 1,500 mg/l, and overlies regional water with a salinity of 12,000 mg/l. Local freshwater lenses overlie the regional saline groundwater in central Oman. These lenses occur in the Tertiary marine limestone aquifers. Recharge occurs during rare high-intensity rainfall events, arising from tropical cyclones or frontal storms, that may occur once or twice a decade (Macumber, 1995). Freshwater lenses contain excellent-quality water, with a total dissolved solid content as low as 160 mg/l. The high intensity-rainfall events, are capable of generating sufficient runoff, and hence recharge to form and maintain freshwater lenses. Average rainfall cycle over a 10-years period is about 140 mm, and it is during these cycles, that groundwater recharge occurs.

Rare cyclonic rain can produce as high as 50 mm of rainwater in a single day. Runoff water then concentrates in topographic lows, recharging groundwater lenses underneath. From an average annual rainfall of 20 mm, 10% can recharge groundwater (2 mm/yr/km^2). Then, a catchment

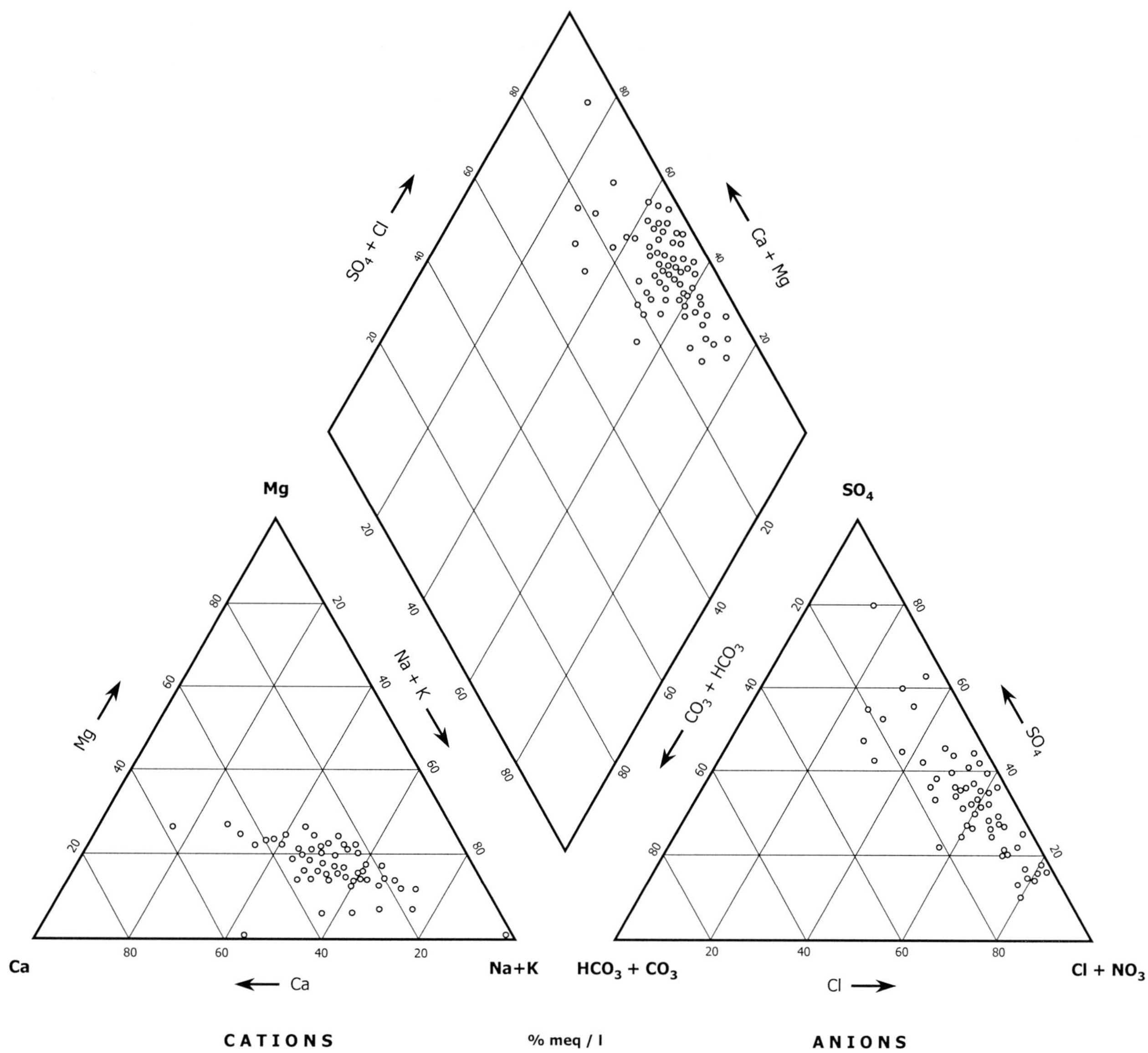

Fig. 5.11. Trilnear diagram of water samples from Khobar aquifer in eastern Saudi Arabia (modified after Hassan, 1998).

area of 1,000 km^2 can have a potential recharge volume of 2 Mm^3/year. The isotopic character of the cyclonic rainfall recorded shows that the early rains were isotopically enriched, while the latter rains were strongly depleted (Macumber et al., 1998). The isotopic character of the runoff water fell on a mixing line and matched the character of the regional and local groundwater found throughout Al Wusta Region, which is isotopically depleted and lies close to, or parallel to, the meteoric water line (Fig. 5.13) (see Macumber et al., 1995 and 1998).

c) Quaternary Aquifer

Quaternary aquifer systems are widely utilized in the United Arab Emirates, and contain most of the available good-quality groundwater. The electrical conductivity of groundwater samples collected from the Quaternary aquifer in United Arab Emirates in 1996 varied between 252 μS/cm east of the Al-Ain city and 173,000 μS/cm in a sabkha area along the Al-Ain - Abu Dhabi road (Rizk et al., 1998). Salt water intrusion as a result of heavy groundwaterpumping is noticed south of Dubai, west of Suweyhan and southwest of Al-Ain. A local fresh water lens is observed along the Madinat Zayed - Liwa road and the potential for presence of similar fresh water pockets exist within the Al-Wagan, Liwa and Umm Al Zamoul triangle (Fig. 5.14).

The main groundwater-dissolved salts in the Quaternary aquifer are $Ca(HCO_3)_2$, $Mg(HCO_3)_2$, $Na_2(SO_4)$, $CaSO_4$, $MgSO_4$, $MgCl_2$ and NaCl. These salts evolve in the direction of flow, according to the Chebotarev series (Freeze and Cherry, 1979). In eastern United Arab Emirates, the aquifer is characterized by $Mg(HCO_3)_2$ and $Ca(HCO_3)_2$ water types, the central part is characterized by $CaSO_4$ and $MgSO_4$ water types, and the western part of the aquifer is dominated by the NaCl water type.

Groundwater of the northern and eastern parts of the Quaternary aquifer in the United Arab Emirates has high ^3H and ^{14}C activities, indicating ages from modern to 7,000 years BP, while the groundwater in the western and southwestern parts has low ^3H and low ^{14}C activities, indicating ages of 15,000 years or older. Stable isotopes in groundwater of Al-Ain and Liwa are distinctly different. However, they seem to have different evaporation rates and their projection to the meteoric water line indicates recharge from higher elevations. Table 5.11 shows the δD and $\delta^{18}O$ values in the Quaternary aquifer systems in the United Arab Emirates.

The groundwater in the Quaternary aquifer of the Al-Ain area of Eastern Province of Abu Dhabi Emirates in the United Arab Emirates is recharged from the Northern Oman Mountains the east and flows towards the Arabian Gulf in the west. The results of hydrogeological and hydrogeochemical interpretation is shown in Fig. (5.14). The total dissolved solid contents of groundwater in this aquifer at the Al-Ain area increases from <1,000 mg/l in the east to >10,000 mg/l in the west; in the direction of groundwater flow. The sequence of cation dominance is: Mg^{2+} $>Na^+$ $>Ca^{2+}>K^+$ in the eastern part, $Na^+>Mg^{2+}>Ca^{2+}>K^+$ in the central part and $Na^+>Ca^{2+}$ $>Mg^{2+}>K^+$ in the western part. While the sequence of anion dominance is: $HCO_3^->Cl^->SO_4^{2+}$ $>CO_3^{2-}$ in the eastern part, $SO_4^{2+}>Cl^->HCO_3^->CO_3^{2-}$ in the central part and $Cl^->SO_4^{2+}$ $>HCO_3^-$ $>CO_3^{2-}$ in the western part. Calculated hypothetically dissolved salts are consistent with the prevailing geological and hydrogeological conditions and can be distinguished into: $Mg(HCO_3)_2$, $Ca(HCO_3)_2$, $NaHCO_3$ and $CaCO_3$ salts in the eastern part, $MgSO_4$, $CaSO_4$ and Na_2SO_4 in the central part and $MgCl_2$ and NaCl salts in the western part. Trilinear plots show that groundwater types in the Al-Ain area can be distinguished into: $Mg(HCO_3)_2$ water type in the eastern part, $CaSO_4$ and $MgSO_4$ water types in the central part and NaCl water type in the western part

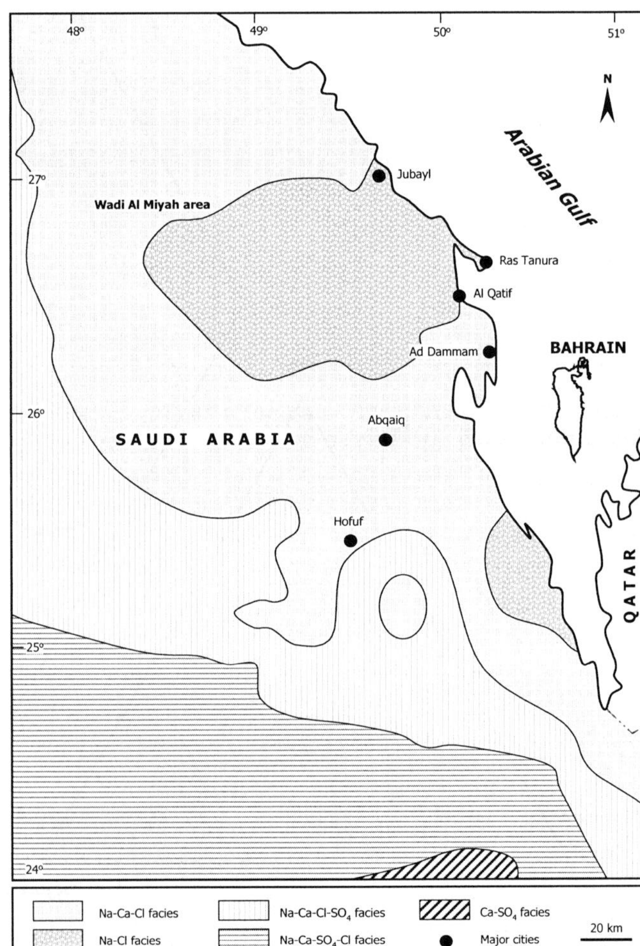

Fig. 5.12. Distribution of the hydrogeochemical facies of Khobar aquifer in eastern Saudi Arabia (modified after Hassan, 1998).

Water Salinity Variation

Waters of springs discharging from local groundwater flow systems are normally of low salinity (fresh water) and at a temperature close to the mean annual air temperature. The dominant anion is HCO_3^- and the groundwater circulation is rapid. Waters of springs discharging from intermediate groundwater flow systems are normally of moderate salinity (brackish water) and temperature. The dominant anion is SO_4^{2-} and the groundwater age is intermediate. Waters of springs discharging from regional groundwater flow systems are normally of high salinity (saline water) and temperature. The dominant anion is Cl^- and the groundwater circulation is slow. In fact, the three flow systems can exist in a single aquifer, causing variations in groundwater chemistry and quality with depths. Consequently, the depth at which a spring originates can control salinity and chemical composition of its water.

Table 5.8. Radioactive element and environmental isotopes in water from Jabal Hafit Mubazzarah well field (Wells JH-1 to JH-9), at Al Ain region located in the Eastern Province of Abu Dhabi, United Arab Emirates (after Khalifa, 1997). Unit abbreviation as follows: pCi/l (picocuries per liter), μg/l (micrograms per liter), TU (tritium units).

Well Number	Radium-226 (^{226}Ra) (pCi/l)	Radon-222 (^{222}Ra) (pCi/l)	Uranium (^{238}U) (μg/l)	Tritium (^3H) (TU)	Deuterium (^2H) (‰)	Oxygen (^{18}O) (‰)
JH-1	3.2	489	3.0	0.89	-8.57	-2.33
JH-2	457	2,836	---	0.30	-10.10	-2.11
JH-4	428	1,743	0.5	0.55	-10.70	-2.08
JH-5	1.3	1,698	2.1	1.39	-8.13	-2.20
JH-6	0.9	2,064	---	0.45	-8.34	-2.43
JH-7	7.2	1,557	---	0.85	-9.86	-2.16
JH-8	349	2,399	1.0	0.20	-10.99	-2.01
JH-9	81	6,063	---	0.78	-11.64	-2.16

Table 5.9. The Tertiary hydrogeologic units and groundwater use in onshore Abu Dhabi oil fields (modified from NDC-USGS, 1996).

Hydrogeologic unit	Thickness (m)	Nature	Lithology	Salinity (mg/l)	Importance and use
Lower Fars	160	Confining unit	Evaporite and clastic	5,000	Prevents vertical flow
Miocene	100	Aquifer	Sandstone	100,000	Receives injected brines
Dammam	265	Aquifer	Limestone	70,000	Source of water for injection into deeper reservoirs
Rus	215	Confining unit	Anhydrite/limestone	Very high	Prevents vertical flow
Umm er Radhuma	415	Aquifer	Limestone	160,000	Source of water for injection into deeper reservoirs

Table 5.10. Groundwater chemistry in the Tertiary aquifer Al Wusta Region, Central Oman (compiled from Macumber, 1995 and Macumber et al., 1998).

Area	Well	Depth (m)	Total Dissolved Solids (mg/l)	HCO_3^-	SO_4^{2-}	Cl^-	Ca^{2+}	Mg^{2+}	Na^+	K^+
Maabar	M-7	84	550	121	128	171	48	23	102	11
	M-7	108	725	128	159	237	37	52	154	13
	M-7t	134	7,773	174	875	3,872	390	244	2,210	53
	M-8	99	12,955	149	1,260	6,064	560	380	3,590	75
Wadi Rawna	WCR-1	70-175	255	203	29	31	44	20	28	---
	WCR-1s	226	10,162	133	2,275	4,276	1,277	430	1,770	---
	WCR-2	80-160	190	130	32	28	27	17	15	---
	WCR-3	40-60	123	80	15	21	20	11	12	---

Table 5.11. The maximum, minimum and average values stable isotopes of hydrogen and oxygen in groundwater of the Quaternary aquifers in the United Arab Emirates (modified from Rizk and Alsharhan, 1999).

Isotope	Deuterium (‰)			Oxygen-18 (‰)		
	Maximum	Minimum	Average	Maximum	Minimum	Average
Western gravel aquifer	19.2	-17.2	-4.2	2.3	-1.9	-0.7
Eastern gravel aquifer	-3.6	-9.5	-6.1	-2.1	-3.0	-2.5
Liwa sand aquifer at Liwa Crescent	43.1	-16.7	2.4	105	1.9	2.9
Liwa aquifer at Bu Hasa area	-14.4	8.7	-1.1	6.0	1.7	2.9
Al Ain falajes	0.0	-20.1	-7.3	1.5	-1.9	-1.3

The chemical analyses of water samples collected from the United Arab Emirates permanent springs, and the records of the Ministry of Agriculture and Fisheries, show that the total dissolved solids content has increased since 1968 to the present day by 10% to 50%. This increase can be mainly attributed to excessive groundwater abstraction in the recharge areas of these springs and low rainfall rates during the last few years. Geochemical analysis show that major ions in water samples from permanent springs belong to the chloride of sodium and potassium type. The analysis showed that the waters of Khatt North, Khatt South, and Maddab spring are similar to shallow local ground water, whereas water of Bu Sukhnah spring is distinctly different from shallow local ground water. According to Chebotarev (1955) sequence, the relatively high HCO_3^- concentration in Al Khatt springs (200-220 mg/l) indicates a relatively recent water, whereas the high SO_4^{2-} concentration in Bu Sukhnah spring (560-1590 mg/l) indicates a relatively old water. Water from the Al Khatt springs has similar compositions to local shallow groundwater. The high total dissolved solids contents in water of the springs can be attributed to dissolution of carbonate rocks that dominate their recharge area. This dissolution occurs under the effect of rainwater that is commonly depleted in dissolved ions. Calculated sodium absorption ratio for water sampled from the springs varies between 0.74 and 11.8, indicating that, except for Al Siji spring, water have a moderate to harmful effect on the soil when used for irrigation.

The salinity of some Bahrain springs has doubled or tripled during the last four decades. The salinity increase in Bahrain springs was moderate between 1950's and the end of 1970's. In 1980's and 1990's, the salinity has sharply increased reaching a maximum of 9,750 mg/l. According to Alqubaisi et al. (1999), the total spring discharge in Bahrain was 49.9 Mm^3 in 1953 decreased to 14.7 Mm^3 in 1980 (Table 5.12). The wide variation in salinity relates to the variable distances between springs and groundwater well fields, in addition to the variation in discharge rates of such springs. However, it is also clear that the changes in water salinity coincide with the changes in groundwater salinity of the Dammam aquifer feeding such springs (Fig. 5.15). In 1997, most of natural springs in Bahrain ceased to flow. The depletion of the Dammam aquifer, the source of recharge for most springs, due to excessive groundwater pumping has led to saline-water intrusion, especially in the northern and eastern parts of the Bahrain islands where most of the springs exist. Also, the uprise of high-salinity groundwater from the Umm er Radhuma aquifer, which underlies the Dammam aquifer, has contributed to the steady increase of the salinity of groundwater feeding the Bahrain springs.

Abderrahman (1990) studied the results of chemical analyses of water samples collected from 32 major springs in the Al Hasa area, in eastern Saudi Arabia. The relative ionic concentrations are Na^+> Ca^{2+}> Mg^{2+} and Cl^-> SO_4^{2-}> HCO_3, respectively. The concentrations of Na^+ and Cl^- ions comprise about 52% of the total ionic composition, and the concentrations of Ca^{2+}, Mg^{2+}, SO_4^{2-}, and HCO_3^- comprise about 48%. The high Na^+ and Cl^- contents reflect the dominance of old groundwater in the Neogene limestone aquifer feeding the springs. This was confirmed by radio isotope analysis (Bureau de Recherches Géologiques et Minières, 1977), which showed that the age of groundwater ranges from 9,000 to 14,000 years in the southwest, and 20,000 to 26,000 years in the northeastern part of the Al Hasa oasis. The average total dissolved solids in spring water ranges from 1,567 mg/l in the Khodood spring in the southwestern part of the Al Hasa Oasis, to 1,846 mg/l in the Hagege spring in the north. (Fig. 5.16, Table 5.13). The total dissolved solid values increase in the direction of hydraulic gradient of groundwater from south and southwest to the north and northwest and with it an increase in age of water. The increase of total dissolved solids in groundwater occurs due to the dissolution processes of sedimentary rock materials along the path of groundwater flow.

The value of total dissolved solids varies from month to month in the water of each spring, and varies from spring to spring in the Al Hasa Oasis. The total dissolved solids values of all springs

Table 5.12. Discharge of natural springs in Bahrain in m^3/yr (compiled from Alqubaisi et al., 1999).

Spring Type	Heim (1924)	Bahrain Petroleum Company (1953)	Groundwater Development Consultants (1980)
Land Springs	70	40.4	8.1
Subsea Springs	17	9.5	6.6
Total	**87**	**49.9**	**14.7**

Table 5.13. Annual average of total dissolved solids values for some selected major springs in Al-Hasa between 1981 and 1987 in mg/l (after Abderrahman, 1990).

Spring	Year		
	1963/1976	1981	1987
Khodoud and Hagel	1,336	1,360	1,567
Johariah	1,462	1,466	1,739
Umm Saba'ah	1,440	1,443	1,677
Hagege	1,515	1,554	1,846
Ain Naser	1,354	1,519	1,805
Harrah	1,389	1,509	1,736

increase from about 11 to 36% towards the summer months due to withdrawal of groundwater with the drop in the water levels during the same period. The seasonal increase in water consumption during the summer months (April-September) has resulted in an increase in the salinity level of spring water during summer by 11 to 36%. The overall increase in water consumption by about 58% during the 1976-1987 periods has resulted in increasing the total dissolved solids values of spring water by about 23%. The average value of total dissolved solids of all spring waters increased from 1,414 mg/l in 1976 to 1,737 mg/l in 1987. There was a total increase of 23% in total dissolved solid values, about 77% of which occurred between 1981 and 1987. This shows the negative effects of over extraction of groundwater in the Al Hasa Oasis on the quality of spring water. The continuous increase in total dissolved solid levels will have a long-term detrimental effect on the soils and the yield of crops.

In southern Oman Salalah Plain, the electrical conductivity measurements show that 0-2,000 µS/cm (freshwater); 2,000-4,000 µS/cm (transition), and 4,000-16,000 µS/cm (brackish water). Spring discharges occur at the foot of the Jebel al Qara in the Salalah Plain of South Oman and reach their peak discharge during October, after the monsoon recharge. The total dissolved solids of Oman springs are always less than a 1,000 mg/l, indicating the suitability of the water for almost all purposes.

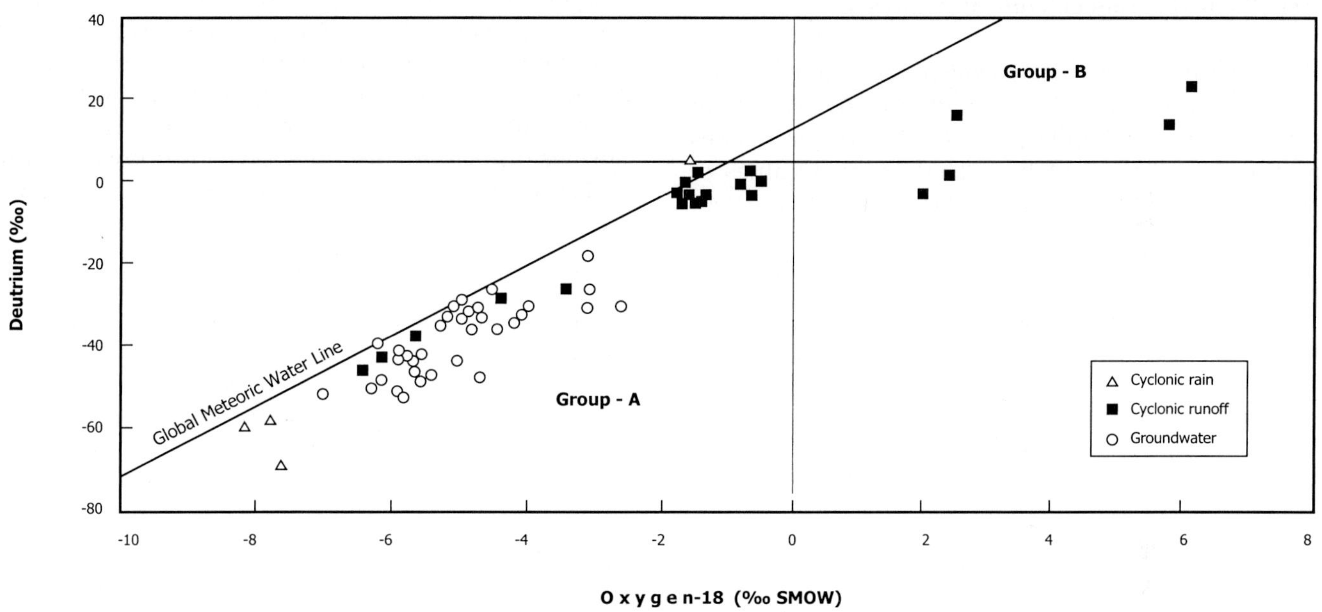

Fig. 5.13. Deuterium versus oxygen-18 for groundwater and surface water in the Al Wusta Region (Central Oman). Group A runoff samples lie between the early-enriched rainfall and runoff, and the later depleted rainfall. Group B is water resampled after 11 days at the surface, showing evaporative enrichment (modified after Macumber, 1995).

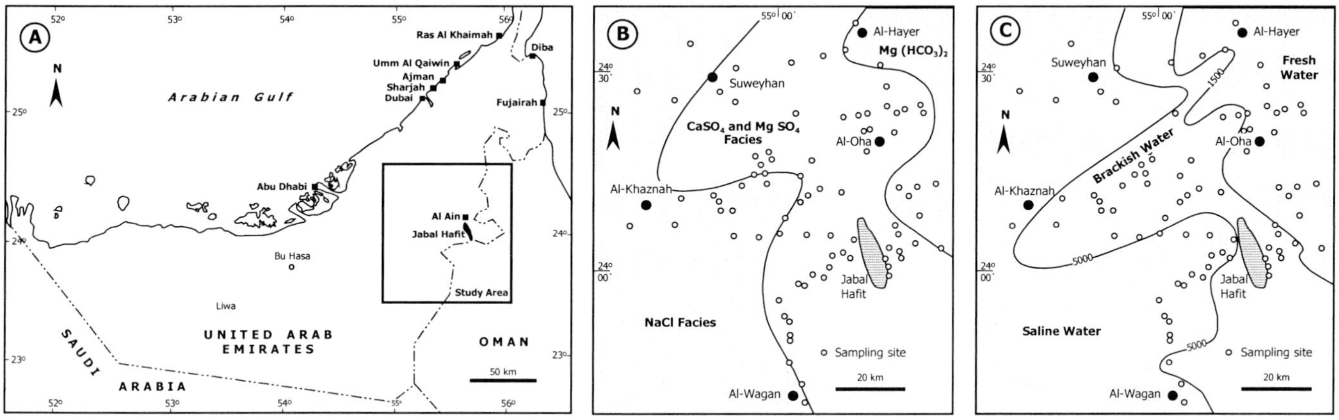

Fig. 5.14. (A) Location map of the United Arab Emirates, showing the localities mentioned in the text. (B) Interpreted groundwater types in the Quaternary aquifer in the Al Ain Region (Abu Dhabi Emirate) based on hypothetically dissolved salts and trilinear plots. (C) Groundwater types in the Quaternary aquifer in the Al Ain area, based on the total dissolved solid contents (mg/l).

Despite the salinity of spring waters in Oman, United Arab Emirates, and Saudi Arabia, water can still find use in some domestic purposes and agriculture. However, the total dissolved solid contents of almost all springs in the Arabian Gulf region have escalated with the elapse of time. Over exploitation and depletion of the aquifer feeding the springs along with the absence of recharge are the root causes of the problem. The water of Bahrain springs is mainly saline and can not be used without prior treatment. The sharp decline in hydraulic heads in the Dammam aquifer recharging springs in northeastern Bahrain, from 4 m above mean sea level to about the zero level (Alqubaisi et al., 1999), has caused the majority of springs to cease flowing.

Results of Hydrogeochemical Analysis

Concentrations of hydrogen (δD) and oxygen ($\delta^{18}O$) isotopes in rainwater of the Arabian Gulf region help determining the sources of air masses responsible of precipitation. The depleted values of such isotopes indicate that the raindrops are affected by evaporation before they reach the ground surface. These facts are important in interpreting the isotopic composition of groundwater in various aquifers. Characteristic stable isotope composition of some of the major aquifer systems in the Arabian Gulf States is shown in Fig. (5.17). Some of the regional aquifer systems contain paleowaters, confirmed by the radiometric dating indicating the water such as in Umm er Radhuma and Neogene aquifers of Saudi Arabia and Dammam aquifer in Kuwait (Yurtsever, 1999) were replenished during Pleistocene.

The climate, geologic setting, and nature of the groundwater flow system feeding springs determine their water temperature, salinity, dominant ions, water type and water quality. Springs discharging from local groundwater flow systems have low temperature and salinity, a high HCO_3^- content and young groundwater. Springs discharging from intermediate groundwater flow systems have moderate salinity and temperature, high SO_4^{2-} concentration and groundwater of intermediate age. Springs discharging from regional groundwater flow systems have high salinity and temperature, high Cl^- value, slow hydrologic cycle, and old groundwater.

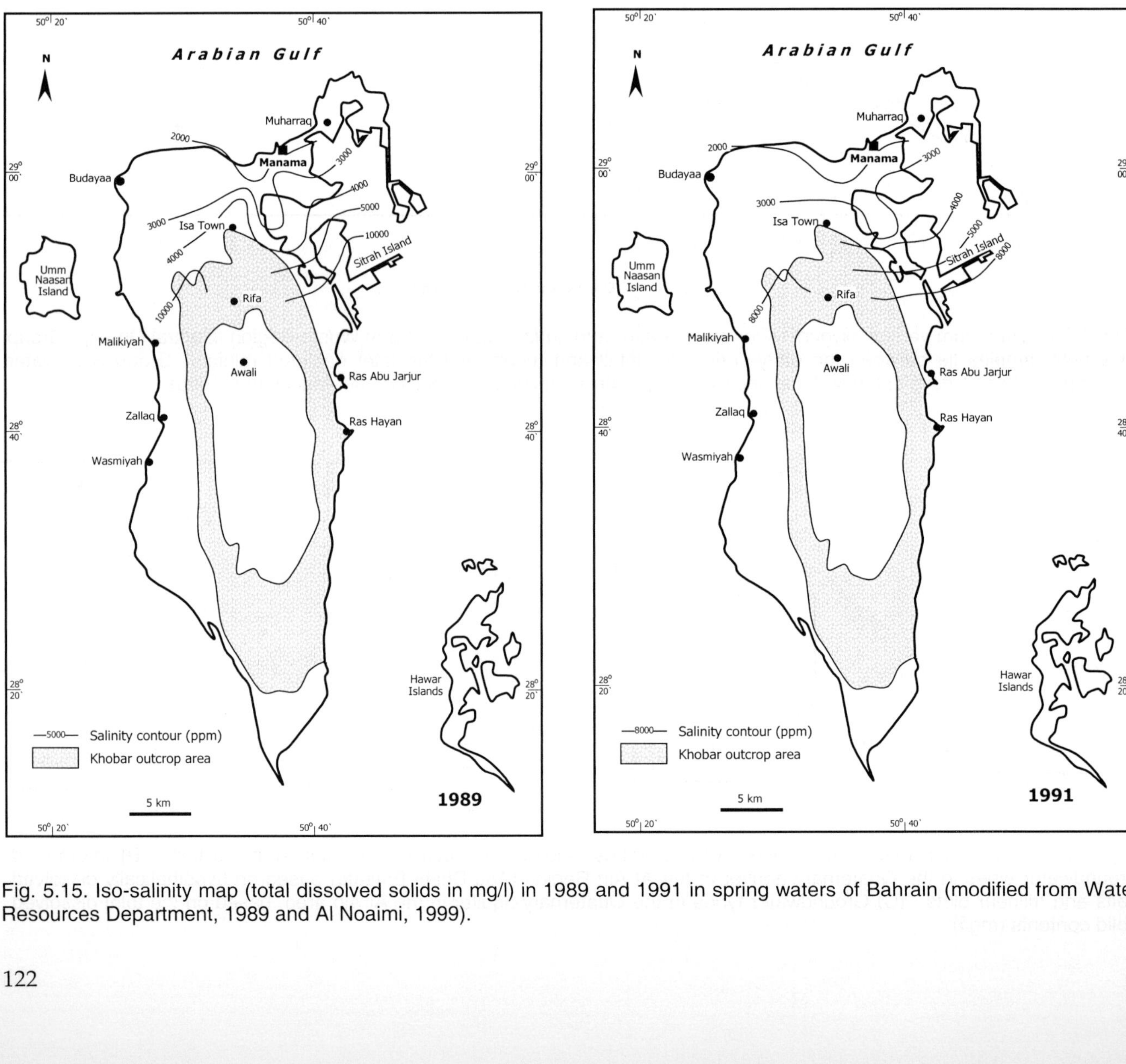

Fig. 5.15. Iso-salinity map (total dissolved solids in mg/l) in 1989 and 1991 in spring waters of Bahrain (modified from Water Resources Department, 1989 and Al Noaimi, 1999).

Springs at the footslopes of high mountains in the western parts of the Arabian Peninsula discharge good-quality water. Similarly, springs on both sides of Oman Mountains discharge fresh water. Springs located further inland from the recharge areas discharge waters of variable quality. Due to the prevailing arid climate and wide groundwater exploitation, most springs in Arabia experience a continuous increase in salinity and decline in discharge.

The salinity of falaj waters depends on the rock types traversed and the nature of the aquifer they discharge. The salinity of the Al-Gheli falajes in the United Arab Emirates increases with increasing the falaj length, because the longer the channel, the greater the contact surface with the bedrock, and thus, the larger amount of total dissolved solids contributed from the bedrock. The salinity of the Al-Daudi and Al-Hadouri falajes in the United Arab Emirates does not correlate with the falaj length because of the variation in rock type and source of water. Waters of the Al-Gheli falajes have low Na^+ content and high HCO_3^- concentration because they are generally young and partially recharged by rain. Falaj waters are enriched in Mg^{2+} content derived from dissolved Mg-rich ophiolitic and dolomitic rocks dominating in their intake areas. The high concentration of SO_4^{2-} and Cl^- in water of Al-Hadouri and Al-Daudi falajes is a result of dissolution of gypsum ($CaSO_4.2H_2O$) and halite (NaCl) from the sediments they penetrate. A plot of the electrical conductivity versus sodium adsorption ratio shows that the water of all the United Arab Emirates falajes, except Khatt South and Habhab falajes, are good to fair for irrigation purposes. For moderately salt-tolerant crops like date palms, and well-drained soils, the medium to high salinity of falaj waters is not a significant hazard in the United Arab Emirates agriculture.

Decline in hydraulic head, increasing salinity, and deterioration of quality are common problems affecting groundwater in the Arabian Gulf region. Extensive groundwater pumping has induced salt-water intrusion from the sea, underlying formations, and nearby sabkhas. Limited groundwater recharge is a result of the prevailing arid climate, and groundwater contamination arising from human-related activities adds to the problem. Renewable groundwater resources are limited to the mountains and flanking sediments. Fresh groundwater in Saudi Arabia is mainly obtained from Paleozoic and Mesozoic aquifers. Oman, Qatar, Kuwait and Bahrain pump fresh groundwater from Cenozoic aquifers. The Quaternary aquifers are the main fresh water source in the United Arab Emirates. Groundwater in the Arabian Gulf region has a widely variable salinity, but NaCl is the most common water type, indicating old age. Recent application of isotope techniques in hydrogeologic investigations has enabled the identification and determination of groundwater sources and ages. The limited groundwater modeling studies provide insights on the water balance, budget and amounts of recharge for aquifers within the modeled areas. Both techniques are powerful groundwater management tools.

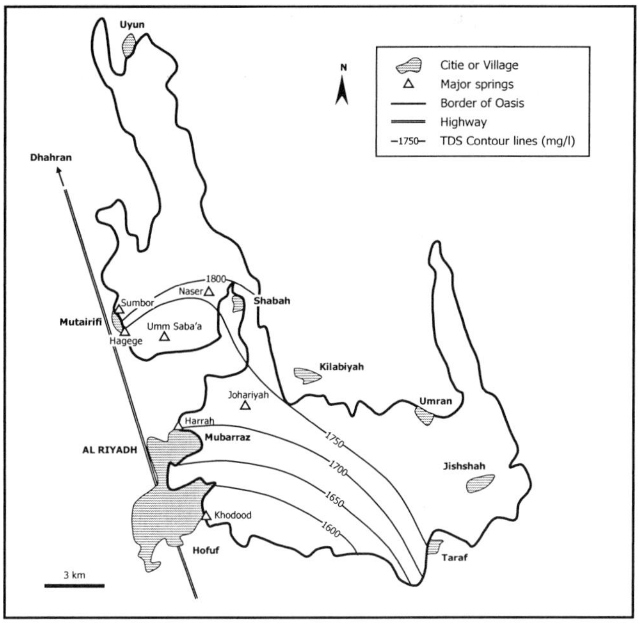

Fig. 5.16. Salinity map of spring water in Al-Hasa Oasis in eastern Saudi Arabia (compiled from Abderrahman, 1990).

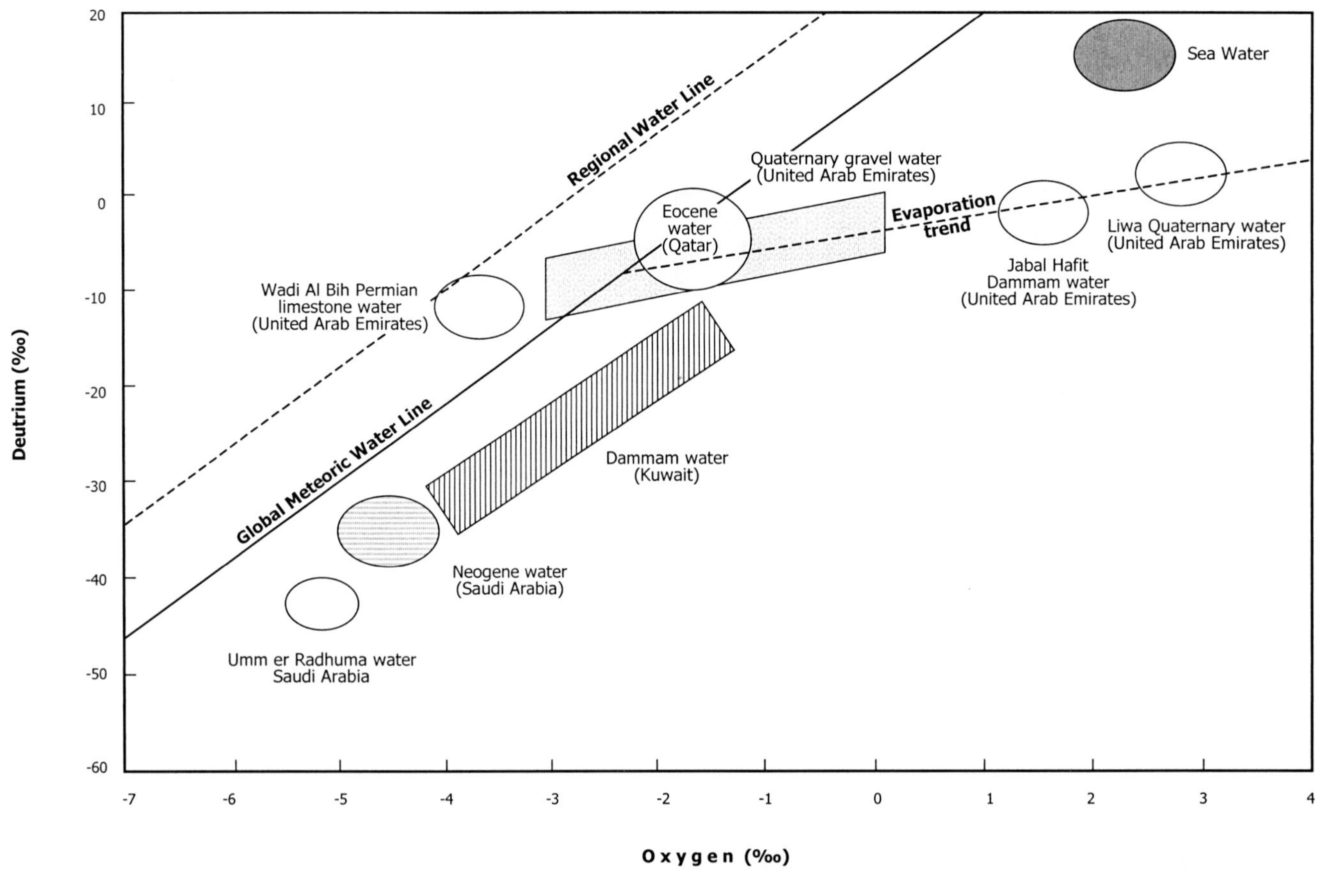

Fig. 5.17. Characteristic stable isotope concentrations of some major aquifers in the Gulf States (compiled and modified from Yurtsever, 1999; Rizk and Alsharhan, 1999).

Chapter 6

TRADITIONAL WATER RESOURCES: SPRINGS AND FALAJES

INTRODUCTION

Despite the dry arid climate and lack of permanent rivers in Arabia, springs provide reasonable amount of water which are used for various purposes. These springs are located in different rock types, at variable topographic elevations, within different drainage basins and along definite structural elements. Springs occur in many forms and have been classified on the basis of rock structure, temperature and discharge. Spring water may contain dissolved minerals and gases, with temperatures close to the mean annual air temperature or lower or higher or even boiling. Mineralized and thermal springs have been thought to have therapeutic value.

The falaj is a man-made stream which intercepts groundwater at the footslopes of mountains and brings it to the surface at a lower level for irrigation purposes. Falajes represented the main arteries of life in the eastern United Arab Emirates and Oman. At the outlets palm oases have flourished, permanent communities developed with an agricultural way of life dependent upon their water. At present time, falajes are part of agricultural heritage. Many falajes have gone dry because of the low rainfall and excessive groundwater pumping. Despite their limited amount, falaj waters are a renewable resource which originated from rainfall. The total falaj discharge varies between 3 to 10% of the total water used in United Arab Emirates to 60% in Oman. Except for a few cases, the discharge is directly related to the antecedent rainfall in the region.

SPRINGS

Springs have been historically important sources of water for Arabia since early historic times. Some of these springs were known to occur at the time of Prophet Ibrahim (peace be upon him), or even earlier. Most of the settlements along Arabian caravan routes and bedouin migration depended on springs as an important source of water. Throughout time, grass and water have been the major concerns of the bedouin in Arabia. Due to rapid development socially and economically, water demand has increased, and due to over pumping the flow of most of the springs has decreased and some of them have become dry.

Water resources in Arabia are therefore restricted to wells and springs, and a traditional system of channels or falajes, which tap into groundwater in the mountainous areas of Oman, United Arab Emirates and Saudi Arabia. Mayboom (1966) has defined a spring as a groundwater outcrop, a definition expanded by Todd (1980) as a "concentrated discharge of groundwater appearing at the surface as a flow current". Because of their importance, springs have been classified in a variety of ways. The simplest breakdown is based upon discharge is into seasonal or ephemeral springs (Fetter, 1988), which normally discharge only during those times of year, when there is sufficient groundwater recharge to maintain flow, as distinct from perennial springs, which drain extensive aquifers, and maintain flow throughout the year. Meinzer (1923) recognized eight categories of springs according to the rate of discharge, and consequently in this classification a spring may change category, according to the rainfall/recharge characteristics of a particular year.

Geologic Setting

A more geological classification of springs in the United Arab Emirates distinguishes gravity flow springs and thermal springs. In Saudi Arabia the grouping of artesian springs has been expanded to recognize springs which arise from such different geological situations, as alluvial, sub-basaltic, interstratified, and solution openings (to include both karst and sink-holes), in addition to fault and fracture related springs (Bazuhair and Hussein, 1990). The spring waters may contain dissolved minerals and gases and usually are at temperatures close to the mean annual air temperatures, even close to boiling. Mineralized springs, usually associated with faulting and fracturing, have been thought to have therapeutic value, and some have been developed as tourist or recreational sites as for example Ain Al Faydah and Ain Khatt in the United Arab Emirates and the Al Lith and Jizan group of springs near the Red Sea coast in Saudi Arabia. The dissolved solids present in the spring water is a reflection of the nature of the aquifer rocks. The location of permanent springs in Bahrain, Oman, Saudi Arabia and the United Arab Emirates are shown in Fig. 6.1. Bazuhair and Hussein (1990) inventoried most of the springs in Saudi Arabia reporting their location, geological features with field measurements of temperature, electrical conductivity and discharge together with analyses of the total dissolved solids (Table 6.1).

Table 6.1. Location, name and field-measured parameters of permanent springs in Saudi Arabia (after Bazuhair and Hussein, 1990).

Area	Name of Spring	Temperature (°C)	Electrical Conductivity (μs/cm)	Total Dissolved Solids (mg/l)	Discharge (m^3/s)
A. Alluvial Springs					
Wadi Fatimah	Mudiq	31.5	1,092	800	-
	Salah	33.5	1,023	680	-
	Al Sharaich	32	1,894	1,440	-
Wadi Al Fara'a	Al Yaseerah	29	1,563	-	1
	Umm Al-Eyal	30.2	1,360	880	0.07
	Al Mudiq	30.7	1,250	760	0.5
	Um Abu-Dhuba	31.3	1,122	760	1
Al-Khamel	Al-Khamel	31.1	593	300	0.17
Khulais	Ghurrah	23.6	2,540	1,920	Variable
	Jadha	28.5	2,860	-	0.56
Al-Taif	Al Wahat	-	790	560	-
	Al Waheet	-	987	680	-
	Shibra	-	1,083	840	-
	Al-Jar	-	925	640	-
B. Sub-Basaltic Springs					
Khayber	Raiya	30.3	1,940	1,160	0.008
	Bahaira	30.6	2,680	1,720	0.033
	Sanpoura	28.4	2,880	1,920	0.033
Al-Khamel	Al Guharya	33.7	871	560	0.17
C. Fracture Springs					
Al-Lith	Al-Harra	80	3,500	2,200	2.1 l/s
	Jum'at Bani Hilal	45	5,670	2,000	3 l/s
	Markub	58	5,360	3,690	25 l/s
Jizan	Al-Wagrah (1)	57.5	4,000	-	-
	Al-Wagrah (2)	55.1	5,900	-	-
	Khulab	50.1	3,300	2,080	-
D. Solution-Opening Related Springs-					
Al-Aflaj	Al-Rayas	-	5,240	4,120	-
	Lake 2	-	5,640	4,120	-
	Lake 14	-	4,990	3,760	-
	Lake 15	-	5,130	3,960	-
Al-Kharj	Al-Dhila	-	3,840	3,400	-
	Safwi	32	6,240	4,000	-
	Awamiyah	32.1	5,110	3,600	-
	Umm Jidiv	33.4	5,200	-	-
	Guda-Taibah	33.8	5,470	-	-
Al-Qatif	Udah	34	3,870	-	-
	Labbaniyah	33.6	4,380	2,600	0.042
	Gharra	33	4,380	2,600	0.012
	Khabbabah	30.4	3,950	2,400	-
	Gushoriyah	32.2	3,310	2,200	-
	Milishiat	30.8	3,620	2,400	-
E. Interstratification-Related Springs					
Al-Hasa	Juhariah	33.5	2,500	1,480	-
	Bahlah	32	2,430	1,640	-
	Umm Saba'a	37.4	2,430	1,680	-
	Harah	34.5	2,470	1440	-
	Khudood	31.5	2,350	1,760	1.7
	Buhairiyah	32	2,280	1,560	1.2
	Najem	38	2,470	1,720	-

Spring Discharge

The water discharged by springs may be derived from aquifers in virtually any part of the stratigraphic column, however the correlation in many instances between rainfall and discharge indicates a dependence upon rainfall recharge and in the absence of rainfall a spring may dry up (Fig. 6.2). It also makes a distinction between springs in which discharge is independent and those which are dependent on groundwater levels (Fig. 6.3). The presence of a dam by maintaining a steady and continuous groundwater recharge, can have a stabilizing effect on spring discharge, such is the case for example with the Siji spring (United Arab Emirates). The area of recharge may vary considerably, from a few hundred square meters to more than a thousand square kilometers. The relation between discharge and groundwater levels underlines the difference between springs, those which have a high rate of discharge, when the groundwater level is high, and those where discharge rates are independent of groundwater level. While in general, discharge increases in parallel with increasing recharge area, this is not always the case depending upon local hydrology and climatic conditions. There may be an appreciable lag time also between a rainfall event, and its influence on spring discharge for a spring, such as seen in the Bu Sukhnah spring in the United Arab Emirates which may only receive a fraction of its discharge from local rain.

Many of the springs which provide potable water are alluvial and associated with wadi deposits which may reach a thickness of 50m, or with the bajada deposits formed by the coalescing of alluvial fans, where the sediment thickness may be ten times that amount. These are the springs most responsive to water table elevation. The total dissolved content can vary between 300 and 1,920 mg/l. Sub-basaltic springs are restricted to the areas characterized by thick lava flows, with the water issuing from either fractured or weathered basalt or from intra-basaltic alluvial deposits.

Solution opening springs are an important group, mainly found in the Al Aflaj, Al Kharj and Al Qatif areas of Saudi Arabia, where the underlying rocks are carbonates of Late Jurassic to Early Cretaceous in age. The action of groundwater on the jointed limestone and anhydrite present has led to the formation of solution cavities, which collapsed under the weight of overlying sediment, and in the case of Al Aflaj produced springs in the middle of the desert. The same mechanism is responsible for many of the karst depressions in Qatar, Bahrain, Kuwait and other areas. In some coastal areas, water in the limestone channels discharges into the ocean, and off-shore springs are found in Bahrain, Qatar and on both the east and west coasts of the United Arab Emirates. Many of these springs have ceased to flow, as a result of the drawdown of groundwater levels, resulting from excessive pumping but the presence of channels obviously facilitates salt water intrusion, as in Bahrain springs (Table 6.2).

Most springs in Bahrain have disappeared at the present time, and the agricultural lands, which were long fed by water from these springs, are now deserted and abandoned. The sites of former springs are now dump site for different wastes. The diverse ecological environments around these springs are seriously affected. Because they have been for a long time discharge areas of the Dammam aquifer system, the disappearance of Bahrain springs is a reflection on the depletion of this aquifer in Bahrain. The excessive pumping of groundwater from the aquifer beyond its safe yield (100 Mm^3/yr) has lowered the hydraulic heads in the aquifer. Thus, in turn, has caused the Bahrain springs to stop flowing and dry up.

Fracture related springs are found in both the United Arab Emirates and Saudi Arabia. Often the springs may be aligned parallel to the fractures. As the water is derived from greater depths the temperatures and total dissolved solids tend to be higher than is found in the alluvial springs. Their composition reflects the nature of the rocks through which they pass with total dissolved solids values ranging between 2,000 and 4,000 mg/l. The waters from the ophiolites tend to be highly alkaline, as seen in the so-called blue ponds of Oman, or they may be enriched in radioactive minerals as at Al Ain. The discharge of these springs does not show the same direct correlation with precipitation.

Table 6.2. Discharge (liter/second) and salinity (mg/l) of the main natural springs in Bahrain (compiled and modified from Alqubaisi et al., 1999).

Spring Name	1953		1971		1979		1997	
	Discharge	Salinity	Discharge	Salinity	Discharge	Salinity	Discharge	Salinity
Azari	267	2,750	200	2,836	180	2,750	0	5,020
Umm Shaourn	9	2,480	3	3,100	25	2,800	0	3,830
Dabassa	18	4,060	0	4,500	0	---	0	4,824
Al Raha	73	3,300	0	7,960	2	7,000	0	8,020
Al Kubra	110	3,860	3	3,700	8	4,900	0	8,880
Al Sughra	34	3,590	2	3,300	17	3,400	0	9,750

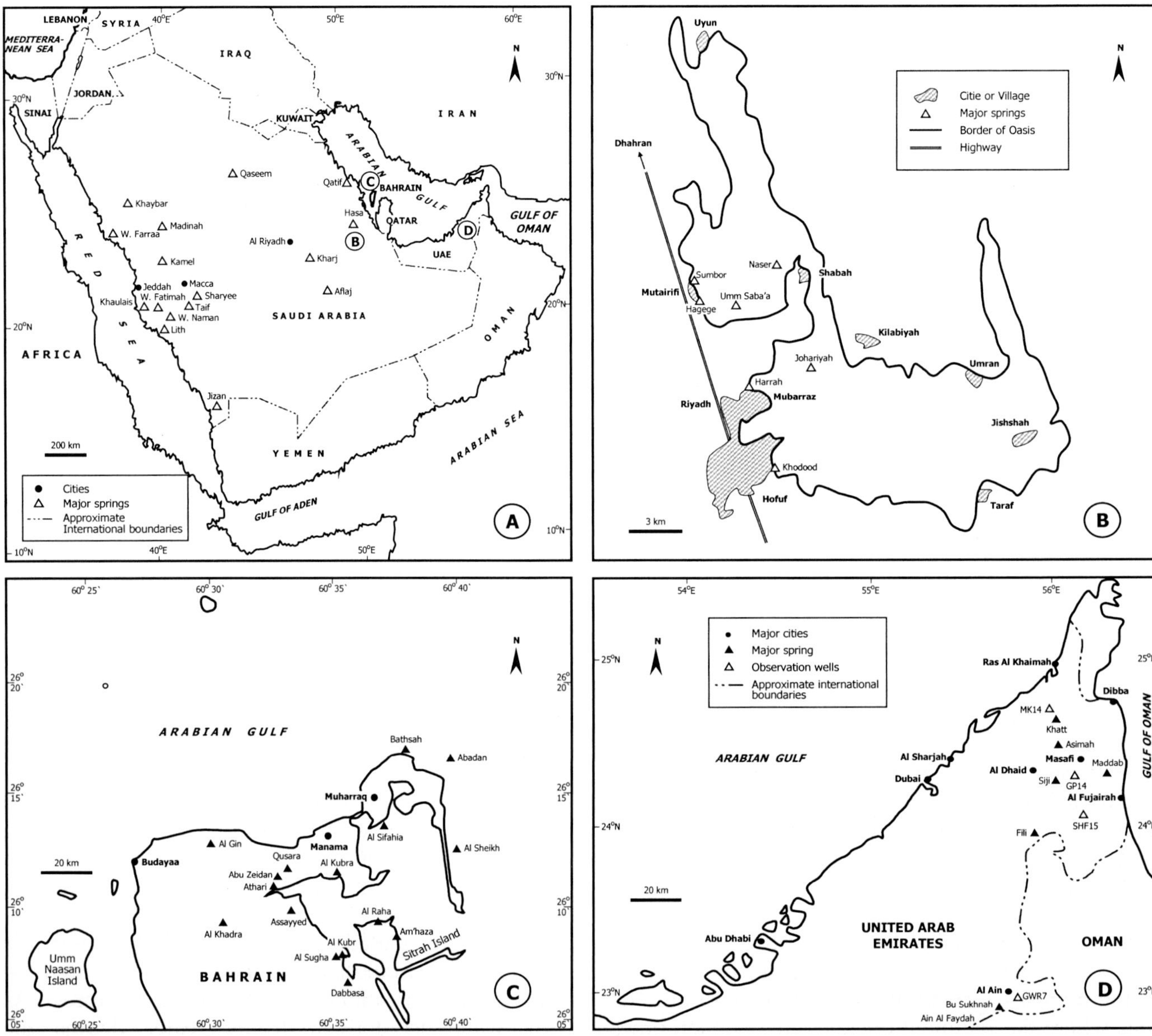

Fig. 6.1. Locations of major springs in Arabian Peninsula (A). Saudi Arabia – Al-Hasa Oasis springs (B). Bahrain springs (C). United Arab Emirates (D), (compiled from Abderrahman, 1990; Bazuhair and Hussein, 1990; Rizk and El-Etr, 1997 and Alqubaisi et al., 1999).

FALAJES

Falajes in Arabic means the division of an ownership into shares among those who have water rights. They vary in size from those supplying a small number of families with their water needs, to those providing water for several thousands of people and several hundred gardens. The falaj systems of Oman and United Arab Emirates have provided communities with water for irrigation and domestic purposes for the last 1,500 to 2,000 years. These systems raise groundwater to the surface without any mechanical device or costly expenditure of fuel.

Although falaj (qanats) are wasteful of water, they have the great advantage of deriving their water high up on the alluvial fan, where the supply is fresh and continuously replenished. The falaj is an idea of Persian origin and dates back more than 2,000 years. The palace city of Persepolis is throughout to have been supplied by qanats about 500 years B.P. (Cressey, 1958). Near the Mediterranean, qanats are erroneously attributed to Romans. The term karez is Persian but most widely used in southwest Asia is equivalent to the Arabic term qanat, and in north Africa the usual term is foggara.

Some karez are present near Al Kharj (southeast Riyadh) and Al Qatif (north of Dhahran). At Al Qatif the tunnels pass through a sand dune area, and the shafts are lined with cemented stones. In Bahrain Islands, some of the surface canals or aqueducts are roofed with slabs of stone to keep out the sand.

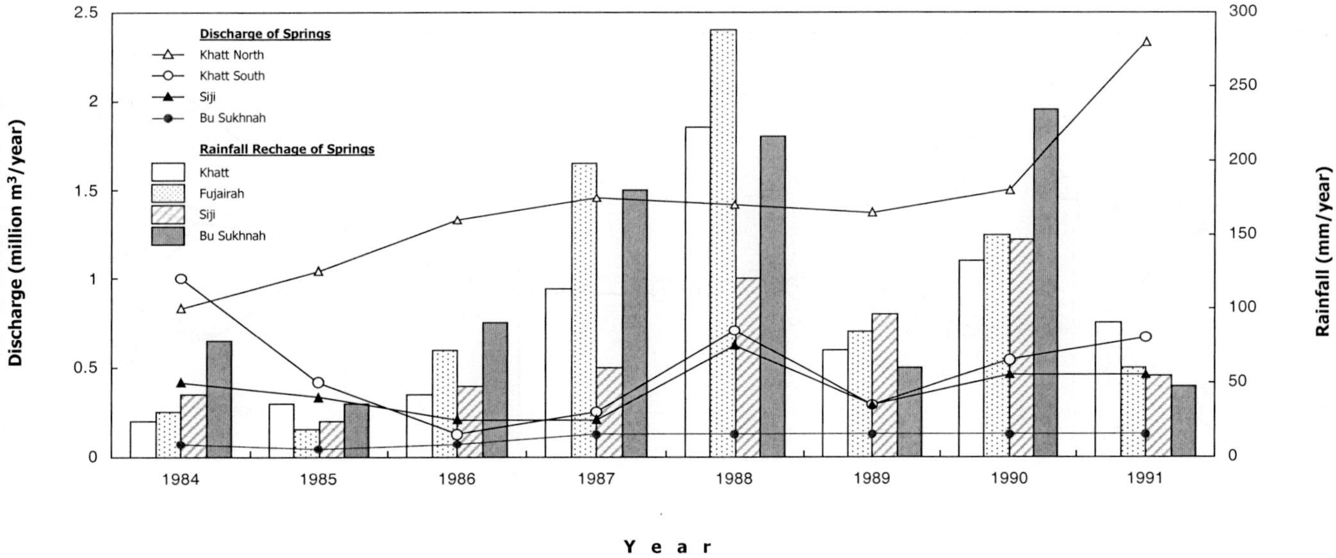

Fig. 6.2. Relationship between discharge-rainfall of the permanent springs in the United Arab Emirates (based on data from the UAE Ministry of Agriculture and Fisheries, 1993).

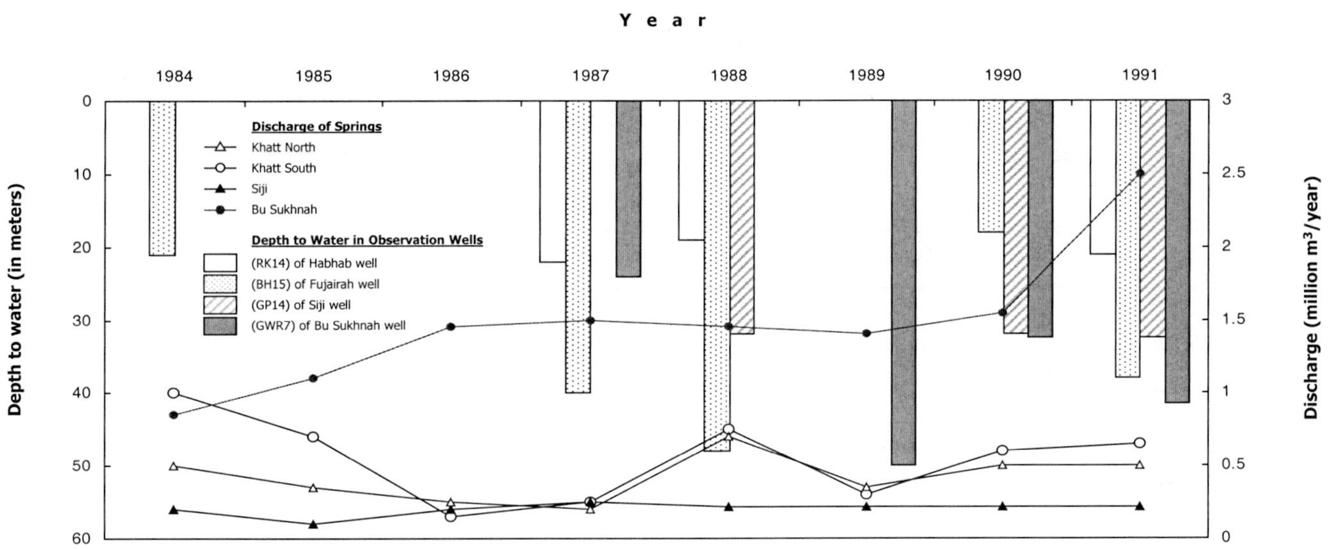

Fig. 6.3. Relationship between depth-to-water (below ground surface) in local observation wells and discharge of the permanent springs in United Arab Emirates (based on data from the UAE Ministry of Agriculture and Fisheries, 1993).

Oman and United Arab Emirates has underground infiltration galleries, locally known as aflaj. Similar structures are present in Yemen, where they are known as felledj. In Morocco, foggaras occur on both sides of the Atlas Mountains. To the north they are especially well-developed near Marrakesh, where they are also known as khottara or rhettara, about 85 systems supply city's water.

Falaj Administration

A falaj is defined as a man-made stream which intercepts the groundwater table through a single or several wells at the footslopes of high mountains. It brings water to the surface through a tunnel which has a slope gentler than the natural hydraulic gradient. As a falaj intersects the ground surface, it splits into several branches (called awamid, the columns). These are narrow, deep, open, and cement-lined small channels which deliver water from the main tunnel of the falaj to farm lands and palm oases (Fig. 6.4).

For the small falajes one man, Al Wakil, may undertake all management and operation, but for the large falaj systems a management committee is required. This committee consists of (a) al wakil, (b) al qab'th, (c) two arif and (d) a workforce of bidars (Sutton, 1984). The wakil is the overall administrator, who is responsible for the organization of falaj affairs, the ownership and rental of water rights, the

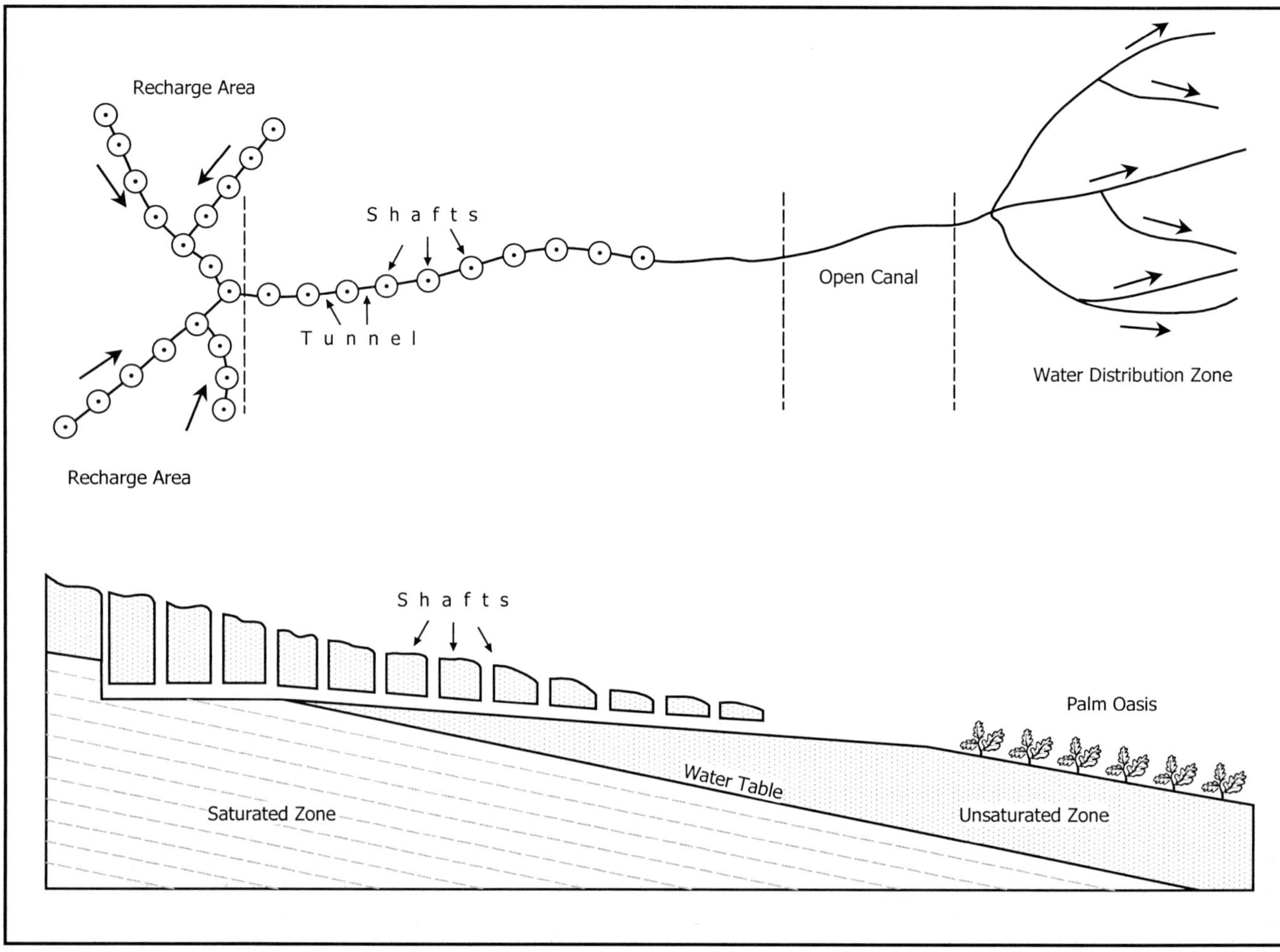

Fig. 6.4. Map view and a vertical cross section of a falaj (modified after United Arab Emirates National Atlas, 1993).

arrangement for distribution of water according to such rights, the maintenance and sale of falaj property, and policy decisions on falaj repairs. The qab'th is the treasurer dealing with receiving money for falaj funds and spending them. The arif is equivalent to the 'foreman'. He is in charge of the physical structure and knows its weaknesses, assessing the method and timing of repairs and maintenance. In the largest systems two people fulfill this role, one being responsible for the underground section and one for the above ground part. Below the management committee are the bidars, a team of labourers, who are responsible for the repair of the whole falaj and for water distribution to each garden. If a dispute over water rights arises which the wakil cannot resolve, the qadi (expert on quranic and/or civil law) or the wali (governor) will give judgement.

Historically in the United Arab Emirates and Oman falaj water was used for all purposes, however its use is currently restricted to irrigation. Its distribution is a complicated process depending on star ascensions at night and the hour of sunrise. The water is divided between participants according to determined irrigation needs on a four to eight day frequency according to a pre-determined schedule and according to the nature of the soil. The time is fixed, but its length can be varied depending upon the water supply, rate of flow and the time of year. The unit on which water distribution is founded depends on size of the falaj. In smallest falaj the ba'adas are used, while in the largest Qamas are used. In medium size falajes the rabiyas and athars are the most common (Table 6.3). The Al Arif (the knowledgeable one) is responsible for the distribution in return for a share of the falaj water. Each falaj usually belongs to a single community and its management is in the hands of community. In Oman the Ministry of Water Resources has encouraged some farmers to fill storage tanks during their period of water use. This serves a dual purpose, in water economy when modern irrigation techniques are installed and the ability to grow vegetable crops which require more frequent watering, but which can provide an additional source of income.

The Ministry of Agriculture and Fisheries in the United Arab Emirates now monitors and manages over 40 active falajes (Fig. 6.5). These falajes are confined to the Northern Oman Mountains in the United Arab Emirates and the gravels plains flanking these mountains from the east and west. The falaj lengths range from 0.5 km (e.g., Falaj Khatt at Ras Al Khaimah) and 15 km (e.g., Falaj Al Daudi at Al Ain).

Table 6.3. Time basis for distribution of falaj water in Oman (compiled from Sutton, 1984; Al-Marshudi, 1995).

Term	Time (minutes)	Time (athar)
Ba'ada	720	24
Rabiya	180	6
Athar	30	6
Qama	7.5	0.5

Water Ownership in Falaj Systems

In addition to water, falaj owns properties such as houses and palm trees which are rented by auction every 6 month. This brings income which can be used to maintain the falaj. Falaj maintenance is done by bidars (workers) supervised by arif and it involves cleaning the tunnel by removing roots and repairing minor collapses of roofing stones or walls (Al-Yahyai, 2000). Falaj water is the main source of income, but water price varies from one place to another and from year to year. For example, in Kabanat (a village near Ibri), water price rose from 2 Riyal Omani for 2 athars/six months in 1970, to 14 Riyal Omani in 1973 and 42 Riyal Omani in 1974 (Al-Marshudi, 1995).

The falaj system is an integral part of villagers life. It provides water for agricultural and domestic uses and because the community depends upon it, codes of social behaviour have become established (Sutton, 1984). One user does not pollute the system of another. For example, the uppermost access point in the settlement is set aside for the collection of drinking water, and no-one should wash in or otherwise dirty the water above this point.

Falaj Construction

Falajes are a historically important adaptation of water use in dry environments tapping groundwater and conducting it through channels, often covered, to the areas of cultivation. Falajes formerly represented the main arteries of life in eastern Arabia. Usually each falaj belonged to a single community with the management of the water in the hands of that community. Every individual in that community had rights to the water for drinking and for livestock, however its use in agriculture and irrigation remained in the hands of the traditional owners through inheritance or purchase. Such water was distributed either according to time but in many mountain villages falaj owners obtain their shares as volumes of water. The time during which each owner could draw water is fixed, the amount of time varying from place to place depending upon the water supply and time of year. Funds for the maintenance of the falajes, cleaning of channels and repairing minor wall collapse, were collected from the water rents which in turn reflect the abundance of water and the type of agriculture.

The falaj system in Oman has provided water for communities and irrigation and domestic purposes for the last two millennia, and even today about 55% of the currently cropped land is irrigated by falaj water (Wallender, 1989), and the 4,800 active falajes

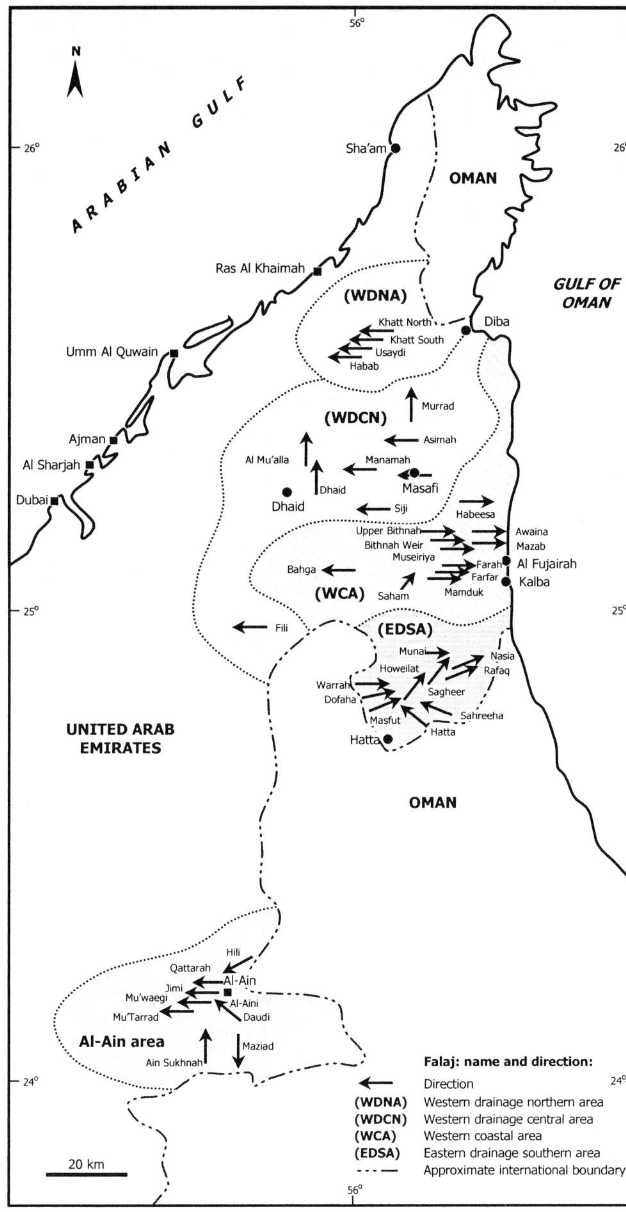

Fig. 6.5. Location and drainage areas of the United Arab Emirates falajes.

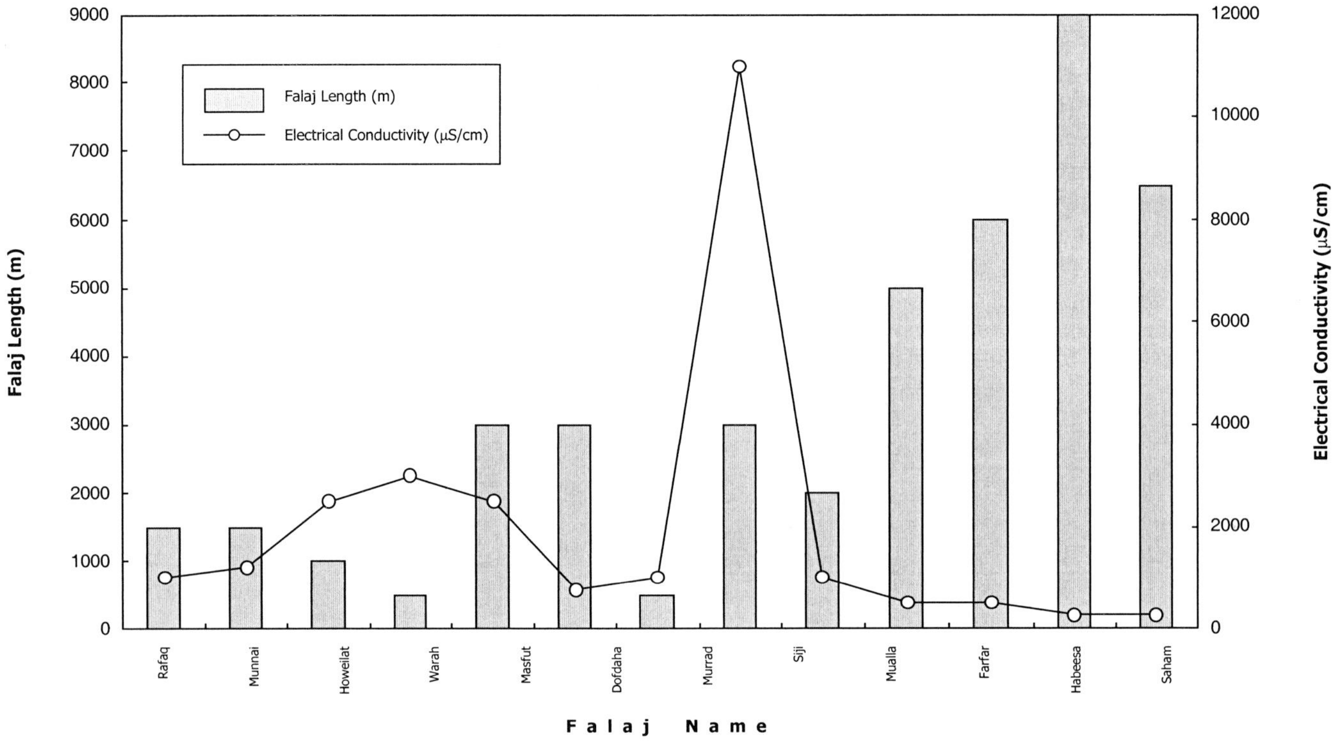

Fig. 6.6. Relationship between length (m) and electrical conductivity (EC) (µS/cm) of Al Gheli falajes in the United Arab Emirates.

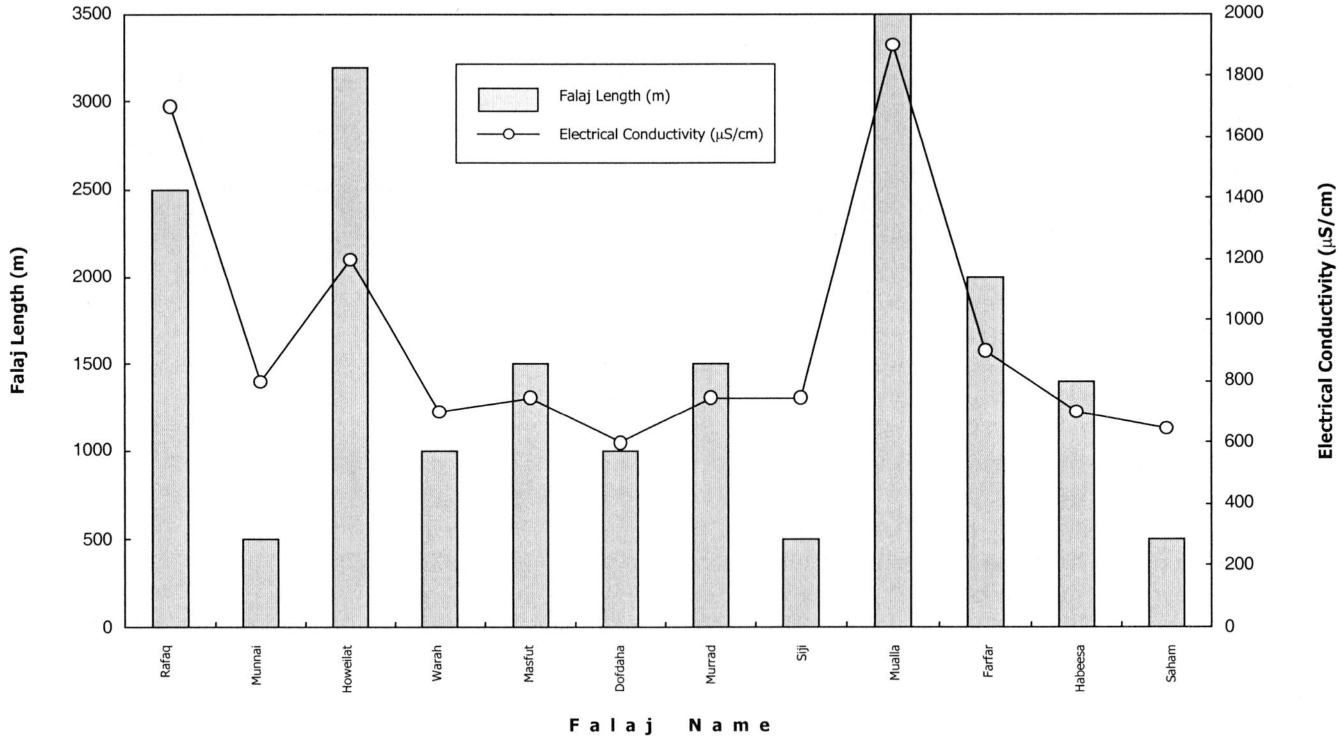

Fig. 6.7. Relationship between length (m) and electrical conductivity (EC) (µS/cm) of Al Daudi and Al Hadouri falajes in the United Arab Emirates.

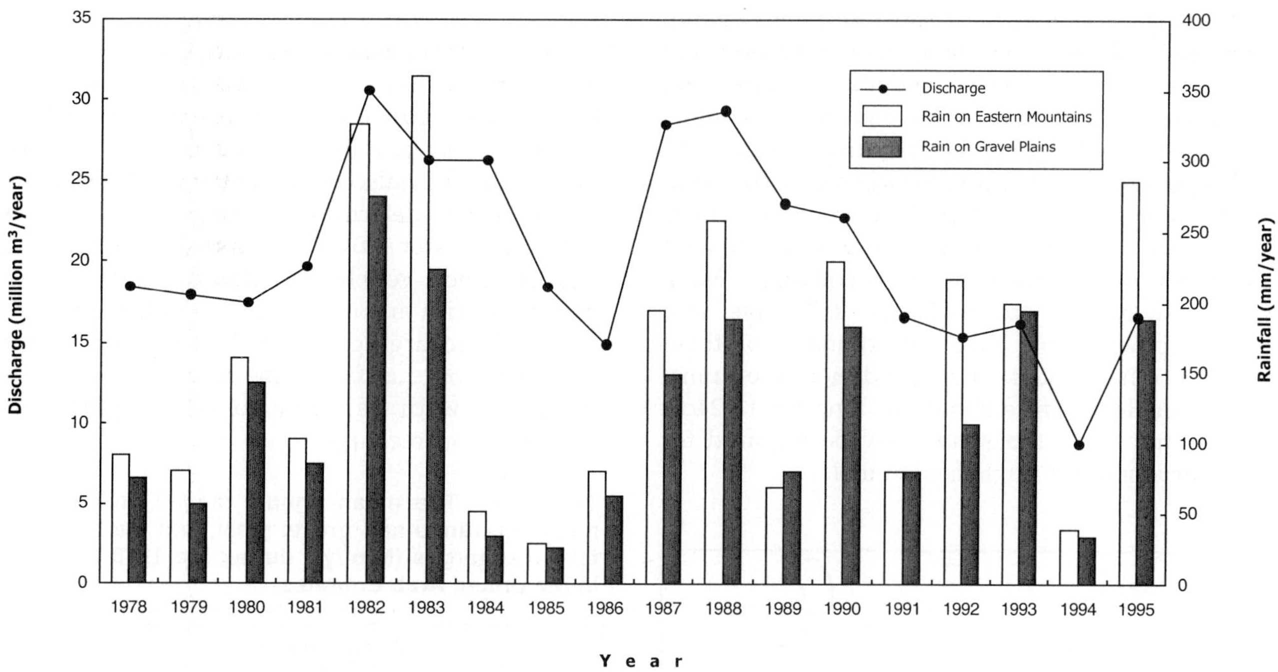

Fig. 6.8. Total annual discharge (Mm^3/y) of the United Arab Emirates falajes versus the mean annual rainfall (mm) on the eastern mountain ranges and gravel plains.

deliver some 900 Mm^3 of water representing 60% of the country's total water usage (Abdel Rahman and Omezzine, 1996). In the United Arab Emirates falajes are also part of the agricultural heritage, but recently many have run dry because of low rates of recharge and excessive groundwater pumping. A falaj can be described as a man-made stream which intercepts the groundwater table through a single or several wells in the foot slopes of the mountains bringing the water to the surface along a slope gentler than the natural hydraulic gradient. They are to be found throughout North Africa and the Middle East along the northern margin of that great desert belt, the Sahara. In the Fezzan some 700 miles south of Tripoli, the Garamnates established a system of underground channels or foggara, linked to the surface every ten years or so, and in Roman times these supported a rich agriculture until the groundwater declined through low recharge, and foggara were replaced by a few wells.

Falajes are designed to bring groundwater to the surface without any mechanical device or expenditure of fuel. They vary in length, width and in the quality and quantity of water they carry and the nature of the ground through which they pass. The upper part of the falaj tunnel, below the water-table serves as an infiltration gallery, and may have several branches to increase flow. The slope of the tunnel is gentle enough to avoid erosion. Prior to construction an expert (Al Basheer or the farsighter) based upon his experience selects a site for the mother well (Umm al Falaj). As the falaj intersects the ground surface connecting tunnels are constructed and aerated through wells spaced at about 10m intervals. The well heads are usually surrounded by a ring of baked clay and may be covered to prevent flooding and the influx of debris. At the surface the falaj is split into a number of distributary channels which deliver water from the main tunnel of the falaj to the farm lands and palm oases. In some cases they are covered to reduce losses through evaporation.

The main tunnel of the falaj may reach a depth of 30m, but decreases gradually as the falaj approaches the surface. Vertical shafts are constructed at distance from 5 to 8m. The tunnel is extended by digging horizontally between shafts. Tunnel are chosen in preference to open channels to reduce evaporation losses especially when the channel would be long (channel lengths range between 3 to 20 km). A plot of the electrical conductivity, a measure of the total dissolved solids of water in the open channels of the Al Gheli type of falaj shows an increase in total dissolved solids with increasing channel length attributed to increased evaporation and greater contact with the bed rock (Fig. 6.6). There is an absence of such a relationship in the tunnel type falajes (Al Daudi type) because of the variation in rock lithology and water source (Fig. 6.7). To build a new falaj, it is necessary to define the most suitable starting point at which there is a plentiful supply of water at relatively shallow depth (Umm al Falaj). For each falaj there may be more than one mother well resembling upstream tributaries carrying water to the main stream (tunnel of the falaj). The width of the falaj tunnel is from 0.6 to 1.2m, and the height varies between 0.9 and 2.1m.

The main construction problems concern either the presence of compact rocks hard to penetrate with hand tools, or soft rocks liable to collapse. More recently more modern tools as drill rigs and electrical power tools have solved many of these problems of hard rocks. In soft sediment the tunnel may be lined with tiles about 1m long by 12-24cm wide so that many thousands may be required for the construction of a single falaj tunnel.

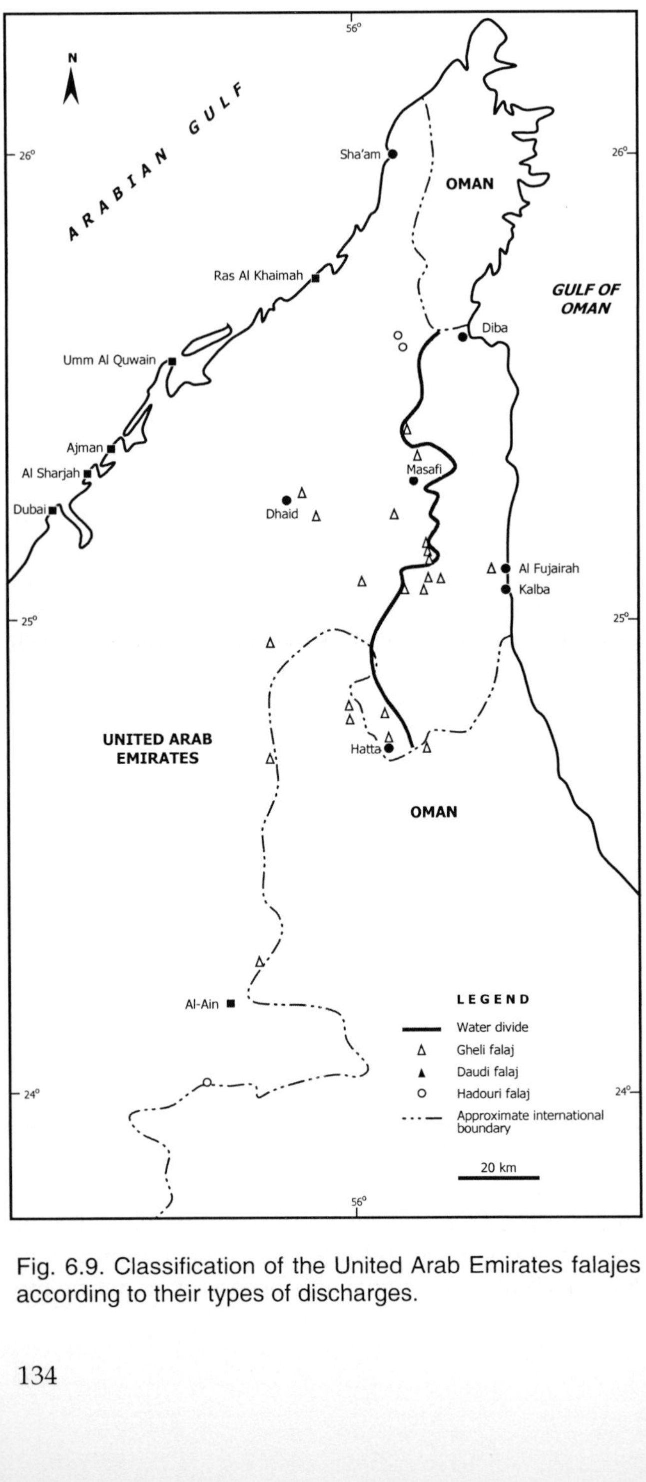

Fig. 6.9. Classification of the United Arab Emirates falajes according to their types of discharges.

Falaj Discharge

Water movement in falaj channels is controlled by many factors (Parks and Smith, 1983) and indicate that the orientation of qanats or aflaj are approximately parallel to direction of groundwater flow lines. They also modeled the water flow in falaj channels to determine and assess the factors governing water movement in relation to the source aquifer. They concluded that the falaj discharge increases with increasing falaj gradient and areal extent of recharging aquifer. The blanket recharge, resulting from rainfall in the highlands and a flood down the wadi, is likely to be the predominant mechanism of recharge.

Table 6.4. The mean annual rainfall on the eastern mountain range and gravel plain, and the total annual falaj discharges (Mm^3/yr) during the 1978-1995 period in the United Arab Emirates.

Year	Rainfall (mm/yr)		Discharge (10^6 m^3)
	Eastern Mountains Range	Gravel Plain	
1978	94.3	73.3	18.8
1979	75.6	60.4	18.2
1980	160.6	137.5	17.9
1981	110.8	100.2	21.1
1982	339.7	283	31.2
1983	353.9	225.9	26.2
1984	46.1	27.2	26.3
1985	32.0	30.5	18.4
1986	75.9	61.3	14.0
1987	191.9	153.0	28.7
1988	262.3	188.0	29.2
1989	70.3	79.4	23.2
1990	230.8	184.2	22.4
1991	81.4	80.6	16.9
1992	198.3	105.7	13.5
1993	189.9	184.1	15.0
1994	40.9	33.7	9.0
1995	284.4	185.0	16.5

Discharge from the falajes depends mainly upon mean annual rainfall, a plot of the total annual discharge of the United Arab Emirates falajes which are predominantly of the Al Gheli type, show a direct correlation with the mean annual rainfall on the Eastern Mountain Ranges and gravel plains (Fig. 6.8 and Table 6.4). Discharge can vary from falaj to falaj depending upon the location of the main well, nature of the aquifer, and the amount of seepage from the walls of the tunnel. They may be classified upon the basis of their yield in to Al Daudi, Al Gheli, and Al Hadouri types (Fig. 6.9). The Daudi falajes have a large groundwater supply and maintain a permanent discharge throughout the year with little change in their discharge rates. This latter type account for more than half of the Oman falajes. The Al Gheli falajes carry seasonal water with discharge directly related to rainfall, and may become dry when the rainfall ceases. The water flow for these falajes which are located in the mountains or close to them is through coarse wadi rubble, it is lifted either by a geological obstruction or a subterranean dam. In contrast the Al Hadouri or Al Aini falajes are usually fed directly from springs, and may produce hot water. Where they arise from limestone they provide good quality water and are liable. Where they emerge from ophiolites the water is usually strongly alkaline, and are usually connected with deep artesian aquifers, draining water which rises along fractures and fissures. The Al Hadouri (Maddah in Fujairah), Bu Sukhnah (in Al Ain) and the Ain Hammam Al Ali spring (near Muscat) belong to this category.

In the United Arab Emirates, overpumping of the Quaternary alluvial aquifer in the Al Ain area during the last three decades has lowered the aquifer's hydraulic head at an average rate of 1m/yr. As a result, the Mu'Tarrad, Maziad, and Jimi falajes went dry in 1997, 1982 and 1983, respectively (Fig. 6.10).

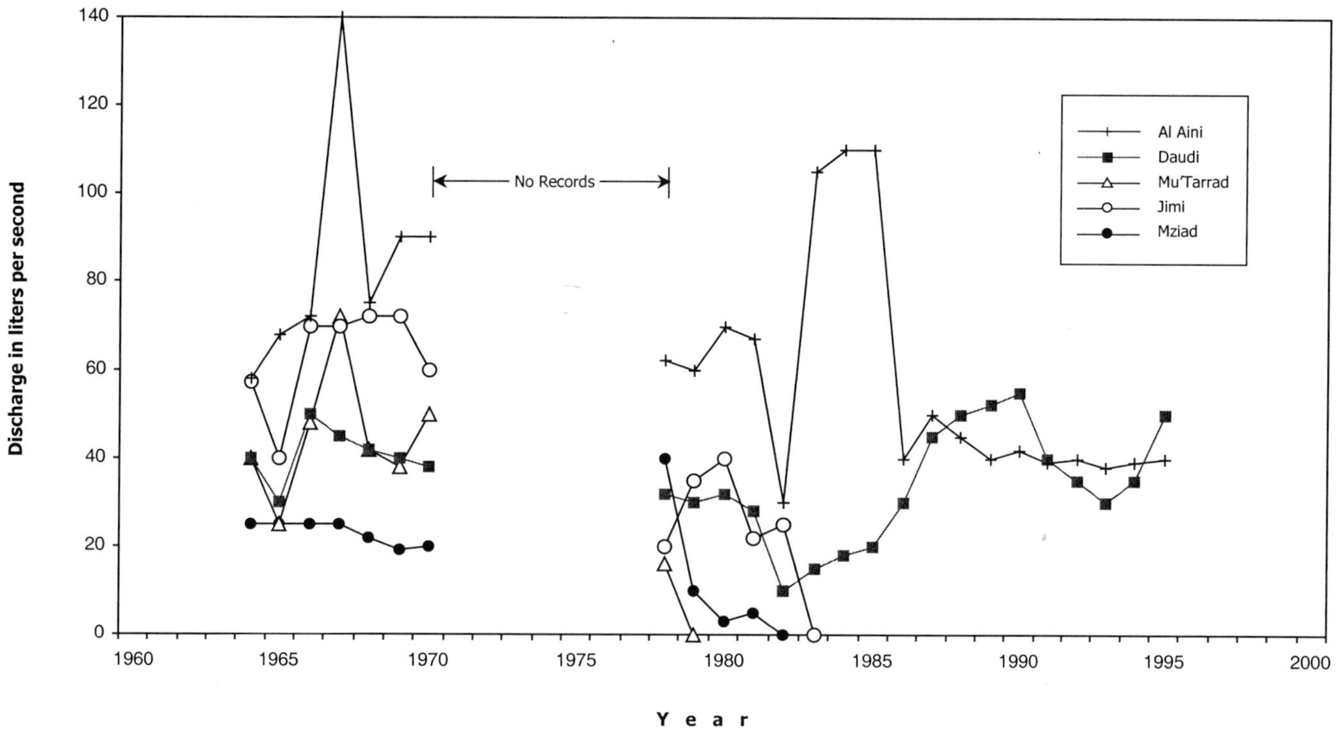

Fig. 6.10. Hydrographs of Al Ain falajes for the period 1964-1996 (compiled data based on studies of Gibb and Partners, 1970; Ministry of Agriculture and Fisheries 1993; and the authors).

Chapter 7

NON-TRADITIONAL WATER RESOURCES: DESALINATION AND TREATED WASTEWATER

INTRODUCTION

Despite being the region with the poorest per capita water availability, and the highest per capita cost of water supply in the world, and in spite of the absence of any international water agreement among its countries, there has been no documented case of violent dispute over water, among the Arabian Gulf countries. Instead, the Gulf States are working together and developing non-conventional alternative water sources, to bridge the gap between supply and demand. Desalination of seawater and brackish groundwater, has been the main alternative adopted in the region since the 1950s, and currently coastal and inland water desalination plants, are now scattered all over the Arabian Gulf countries (Fig. 7.1). Treated wastewater has recently been used, and the volumes applied are continuously rising. Importing water from water surplus countries, using supertankers or towing of large capacity water bags, has also been proposed, and interbasin transfer of water from Turkey and Iran, through pipelines has been studied. Water harvesting, weather modification and saline agriculture, are other non-conventional water supply alternatives. Per capita water availability from natural sources in the world is about 7,690m^3 compared to the Middle East which is about 1,360m^3 and in the Gulf States about 1790m^3 (Dolatyar and Gray, 2000).

Among the non-conventional water resources alternatives, is the importation of water from outside the Arabian Peninsula. The idea of exporting water by tankers, has been propagated by countries of water surplus such as France, Japan and United Kingdom, which have well established shipping infrastructure. However, the high cost involved makes this idea less favorable (Starr and Stoll, 1988). There have been also suggestions, about transfer of water from Iran to Qatar. The main part of the project consists of a pipeline, which takes water from the Karun river in the southwest of Iran, to the coast and a 200 km underwater pipe, which carries the water across the Arabian Gulf to Qatar.

Turkey also proposed to build what is called a "Peace Pipeline", to convey potable water from the Seyhan and Ceyhan rivers, in the southeast of the country through two pipelines: the western line which would deliver water to Syria, Jordan and carry water to Saudi Arabia, Kuwait, Bahrain, Saudi Arabia, and the eastern line which would supply Qatar, United Arab Emirates and Oman (Table 7.1). Besides its \$21 billion estimated coast, the main obstacle facing the peace pipeline proposal is that, the Gulf States do not want to place their water security at the mercy of Turkey. In addition, it is more secure to build a number of desalination plants to produce water at a capacity of 7.5 Mm^3/d, compared with 2.5 Mm^3/d, the proposed pipeline would deliver (Khordagui, 1996).

Desalination Processes

During the last three decades, extensive development, rapid population growth and substantial improvement in the standard of living in the Arabian Gulf Countries have all increased the imbalance between rising water demand and very limited existing water resources. Most of the Arabian Gulf Countries have experienced a 20 to 30% increase, in domestic and industrial water demand over the 1980-2000 period.

Since early 1950's Kuwait embarked on desalination as the most practical solution, to its water supply problems, and has used it successfully ever since (Dabbagh et al., 1994). Thus, in the Arabian Gulf area, desalination of seawater can no longer be regarded as a new technology, because it has been used on a commercial basis for nearly 50 years.

The capacity of desalination plants in the Arabian Gulf Countries, and the world is very high, and the growth of world quantity of desalinated water shows changes in the proportion of the total contributed by commercially available processes, distillation and membrane processes, multi-stage flash distillation (MSF), reverse osmosis (RO) and multi-effect distillation (MED). At the end of 1991, the world produced about 15.6x10^6 m^3/day of desalinated water, 7.8x10^6 m^3/day (or 50%) of which was in Arabian Gulf Countries.

The most important distillation methods are multi-stage flash distillation and multi-effect distillation. Both involve the evaporation of saline feed water, and its condensation into fresh water, leaving dissolved substances in the waste brine. In multi-stage flash, a stream of brine flows through the bottom of up to 25 stages or chambers.

Table 7.1. The proposed capacity of "Peace Pipeline" and delivery points in Turkey, Syria, Jordan and the Gulf States (after Brown and Root International, Inc., 1990 in Dolatyar and Gray, 2000).

Receiving points	Quantity (1000 m^3/day)
Western Pipeline Capital	
Turkey	300
Syria:	
Aleppo	300
Hama	100
Homs	100
Damascus	600
Damascus	300
Jordan :	
Amman	600
Saudi Arabia:	
Tabuk	100
Medina	300
Yanbu	100
Jeddah	500
Total	**3,500**
Eastern Pipeline Capital	
Saudi Arabia:	
Jubail	200
Ad Dammam	200
Al Khobar	200
Al Hofuf	200
Kuwait :	
Kuwait	600
Bahrain:	
Al Manama	200
Qatar:	
Doha	100
United Arab Emirates:	
Abu Dhabi	280
Dubai	160
Sharjah - Ajman	120
Ras Al Khaimah	120
Fujairah – Umm Al Quwain	40
Oman :	
Muscat	200
Total	**2,500**

The pressure in each chamber is maintained at a lower level than the saturation vapor pressure of the water, where a part of it flashes into steam and is then condensed.

In multi-effect distillation, evaporation takes place as a thin film of feed water moves over a heat transfer surface, which is usually the outside of horizontal tubes. The vapor formed in each effect is condensed in the next, providing a heat source for further evaporation. Energy saving is made if the vapor from the last effect is re-compressed thermally or mechanically. In multi-effect distillation method, fewer stages are involved than in multi-stage flash distillation method.

The two membrane processes are reverse osmosis and electrodialysis. In reverse osmosis, a pressure greater than the osmotic pressure of the feed-water, is applied to the feed-water side of a semi-permeable membrane, producing a flow of fresh water through the membrane. In electrodialysis feed-water flows through a stack of membranes to which an electric voltage is applied. Ions migrate to charged electrodes formed by the membranes so that an ion-depleted product and a concentrated reject brine stream are formed in alternate spaces between the membranes. While the feedstock is generally either seawater or saline groundwater, the use of sewage effluent as a feedstock, suitably treated to remove harmful bacteria, is becoming widely acceptable.

The annual capacity in the Arab world and the Arabian Gulf Countries for the period 1965-1991 shows increase in the predominance of the multi-stage flash distillation method in the area, where it accounts for 80.25% of the capacity. Reverse osmosis method provides only 16.25%, and multi-stage flash distillation and multi-effect distillation methods account for the remaining 3.5%. However, as Dabbagh et al. (1994) show these proportions differ significantly between the Arabian Gulf Countries. The total capacity in Saudi Arabia at the end of 1991 was $3.8x10^6$ m^3/day, and multi-stage flash distillation method represented $2.8x10^6$ m^3/day or 70.25%, reverse osmosis $0.96x10^6$ m^3/day (25.25%) and others the remaining 4.5%. However, in Kuwait, of a total capacity of $1.4x10^6$ m^3/day, multi-stage distillation represented $1.35x10^6$ m^3/day (97.1%), reverse osmosis 29,098 m^3/day (2.1%) and other methods the remaining 0.8%. Table 7.2 and Figure 7.2 shows the desalination capacity in the Gulf States for the year 2000. It also shows that the desalination capacity in Saudi Arabia rose in the year 2000 to $5.43x10^6$ m^3/day, increasing by about 43% over desalination capacity at the end of 1991. In Kuwait, the year 2000 desalination capacity was $1.65x10^6$ m^3/day, an increase of only about 18% over the 1991 capacity. The relatively low increase in desalination capacity in Kuwait between 1991 and 2000 can be attributed to the damage of desalination plants which occurred during the Gulf War-II. Reconstruction and rehabilitation of desalination facilities in Kuwait seem to have retarded the pace of desalination process in the country, which started in the early 1950's.

The desalination of seawater in the Gulf States is, in part, facilitated by having long coasts (about 5,395 km) that lie on the Arabian Gulf, Gulf of Oman and Red Sea. Saudi Arabia has 1,770 km on the Red Sea and 485 km on the Arabian Gulf; Oman has a total of 1,700 km on the Gulf of Oman and Arabian Gulf; United Arab Emirates has a total of 600 km on the Arabian Gulf and Gulf of Oman; Qatar has about 420 km on the Arabian Gulf; Bahrain has about 120 km on the Arabian Gulf; and Kuwait has about 300 km on the Arabian Gulf.

Water metering and charging for water has been widely adopted throughout the Arabian Peninsula, but has not lowered water consumption, because the water price is low so that costs to the

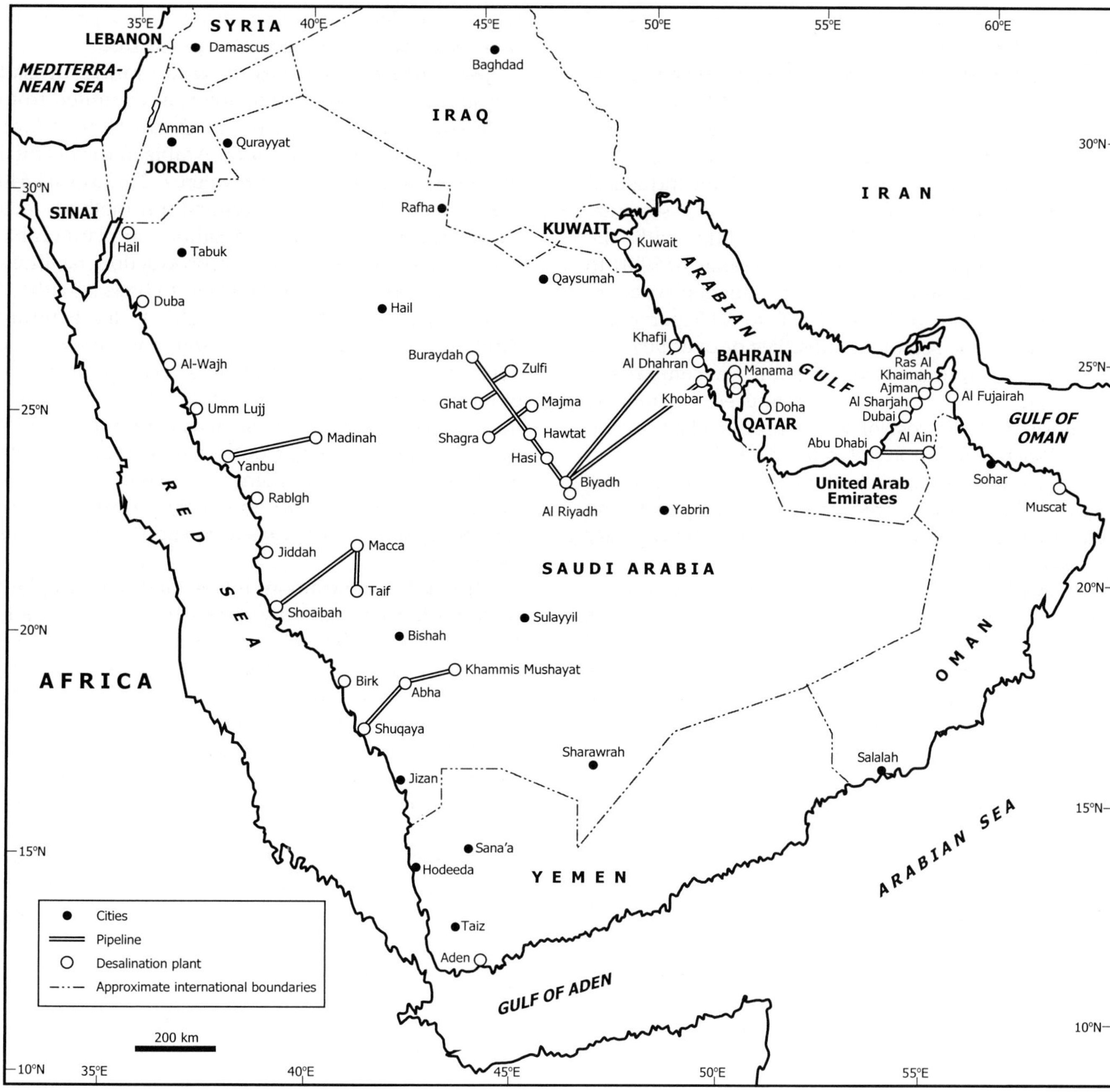

Fig. 7.1. Location of major desalination plants in the Arabian Peninsula.

Table 7.2. Desalination capacity (in 1000 m^3 per day) in the Gulf States (extracted from the 2000 desalination inventory, Middle East Desalination Research Center, Oman).

Type / Country	Desalination process					
	Multistage flash distillation	**Multi-effect distillation**	**Vapor compression distillation**	**Reverse Osmosis**	**Electro-dialysis**	**Total**
Bahrain	581	1	47	141	14	784
Kuwait	1,469	12	0.2	166	5	1,652
Oman	330	4	14	29	0.9	378
Qatar	783	4	21	14	-	822
Saudi Arabia	3,487	18	76	1,751	98	5,429
United Arab Emirates	4,469	9	474	175	5	5,132

consumers do not effectively discourage excessive water use or waste.

Tariffs per cubic meter of potable water in the lowest charge band range from $0.04 in Saudi Arabia to $1.21 in Qatar (although water is free for Qatari citizens). The highest charge rate per cubic meter in any of the countries is $1.71 for industrial usage in Oman. Some countries have a separate charging structure for brackish water with the monthly cost per cubic meter in the range of $0.01 to $0.23. Since desalinated water production costs are currently on the order of $0.50 to $3.5 per cubic meter, it can be seen that subsidies of over 90% are common. Figures (7.3a, b) shows a comparison between the cost of water produced by reverse osmosis and the cost of water produced by multi-stage flash distillation in United Arab Emirates

Desalination plants in the United Arab Emirates operate based on shared production of generating electricity and drinking water. Other low-capacity plants apply the reverse osmosis methodology. Advances in water desalination techniques have reduced the production costs of water from 8 Dirhams (about 2.45 $) for 1 m^3 in the 1980 to 4 Dirhams (about US $ 1) in the 1995. Despite this price reduction, its usage in irrigation is still uneconomical. Due to the large investment required in water desalination projects, the water price for the consumer has to be re-evaluated. The use of a solar energy as an alternative source of energy in water desalination is an approach to be seriously considered in the Arabian Gulf region.

Economic Constraints

From the economic viewpoint, the major disadvantage of water desalination is its high coast. Desalination is still too expensive to use except in the Gulf States where oil revenues provide the initial investment in addition to the energy source that is readily available. Khordagui (1997) argues that under the present circumstances desalination is one of the most effective solutions to water deficit in the Gulf States.

Environmental Impact

The brine discharge from desalination plants into the marine environment may have physical, chemical and biological impacts (Khordagui, 1997). The physical impact arises from the temperature difference between intake water and the reject brine. The chemical impact results from chemicals remaining in the brine which were added during the pre-treatment chlorination and antiscalant stages. The biological impact is represented by the high Biochemical Oxygen Demand (BOD) in the reject brine, leading to low dissolved oxygen in the seawater adjacent to the brine release area.

The reject brine has a wide range of salt concentrations. Reject brine analyses compiled from different desalination plants in Qatar, United Arab Emirates, Bahrain and Saudi Arabia indicate total dissolved solid contents between 8,276 and 63,640 mg/l, depending on the desalination method, feed water and efficiency of the desalination plant (Table 7.3). In the case of seawater production facilities, this brine can usually be returned to the sea without much adverse effect, although a temperature differential between intake water and reject brine may affect marine organisms. To protect the environment against harmful effects, various measures can be adopted, such as careful placement of intake and discharge outlets with regard to the local marine environment as well as to the turbulence, temperature and major constituents of the reject brine, which vary according to the process used.

If reject brine from an inland desalination plant is deposited haphazardly it can have considerable adverse effects and pollute groundwater. To avoid this the brine can be deposited in lined ponds, spread out over the desert, or injected into the ground at great depths where it is not expected to harm the environment. Disposing of reject brine safely can cost between 5% and 30% of the total cost of plant installation, depending on its type and location (Mandeel, 1992). Alsajwani and Lawrence (1995) indicated that water desalination has provided drinking water in areas which would otherwise have depopulated, but they stressed the importance taking alleviative measures to rescue the ecosystems into which the reject brine will be discharged.

Environment protection measures necessary when using desalination are attainable, although there is still room for improvement in this field. There are even prospects for utilizing beneficially the reject brine that is the main pollutant of desalination.

Security Problems

Desalination plants are extremely vulnerable to both accidental and deliberate damage. During the Gulf War I (1980-1988) and the Gulf War II (1990-1991), it was demonstrated that desalination plants can easily be destroyed or seriously threatened by oil pollution. However, precautions against oil spill can bring about unexpected environmental problems. For example, in an effort to protect desalination plant seawater intake area from oil pollution, DUBAL (Dubai Aluminium Company) installed a breakwater in 1991. Unfortunately, this barrier has lead to an increase in chemical consumption, reject circulation and accumulation of silt in the breakwater basin.

Since desalinated water has become an integral part of water resources in the Gulf States, and almost half of the world's production takes place in the Gulf region, issues such capacity building and indigenous training are essential aspects. This can be achieved through the application of the practical experience in installation, operation and maintenance of desalination plants during the last 50 years. Local experience in desalination industry has to be promoted and national manufacturing of certain parts and spare parts for desalination plants must be seriously considered. The different desalination technologies, types of feed-water and environmental conditions necessitate the exchange of knowledge and experts between the Gulf States. Manufacturing of pumps, measurement and control devices and reverse osmosis membranes reduces the costs and leads to establishment of national industries supporting the desalination industry.

Table 7.3. Characteristics of reject brine water from desalination plants in the Gulf States (compiled from Khordagui, 1997; Al Mutaz and Al Sultan, 1997; Al Noaimi, 1999; Hassan, 1998).

Station / Parameter	Qatar	Bahrain	United Arab Emirates				Saudi Arabia	
	Abu Finate	Sitrah	Ajman	Umm Al-Quwain	Quidfa-1	Quidfa-2	Manfouha	Al Jubail
Temperature (ºC)	42	42	30.6	32.4	32.2	29.1	41	42
pH	8.20	8.90	7.46	6.70	6.97	7.99	7.5	7.46
Electrical Conductivity (μS/cm)		87,000	16,490	11,325	77,000	79,600		68,600
Ca^{2+}	1,350	1,700	312	173	631	631	160	701
Mg^{2+}	7,650	10,400	413	282	2,025	2,096	468	2,200
Na^+		22,800	2756	2315	17,294	18,293	3,511	
HCO_3^-	3,900	210	561	570	159	150	73	133
SO_4^{2-}	3,900	4,708	1,500	2,175	4,200	4,800	5,368	4,950
Cl^-	29,000	38,750	4,572	2,762	30,487	31,905		29,350
Total Dissolved Solids (mg/l)	52,000	12,100	10,114	8,276	54,795	57,935	12,721	63,640
Total Hardness (mgt/l)				32	198	207		10,800
SiO_2		0.13	23.7	145	1.02	17.6		

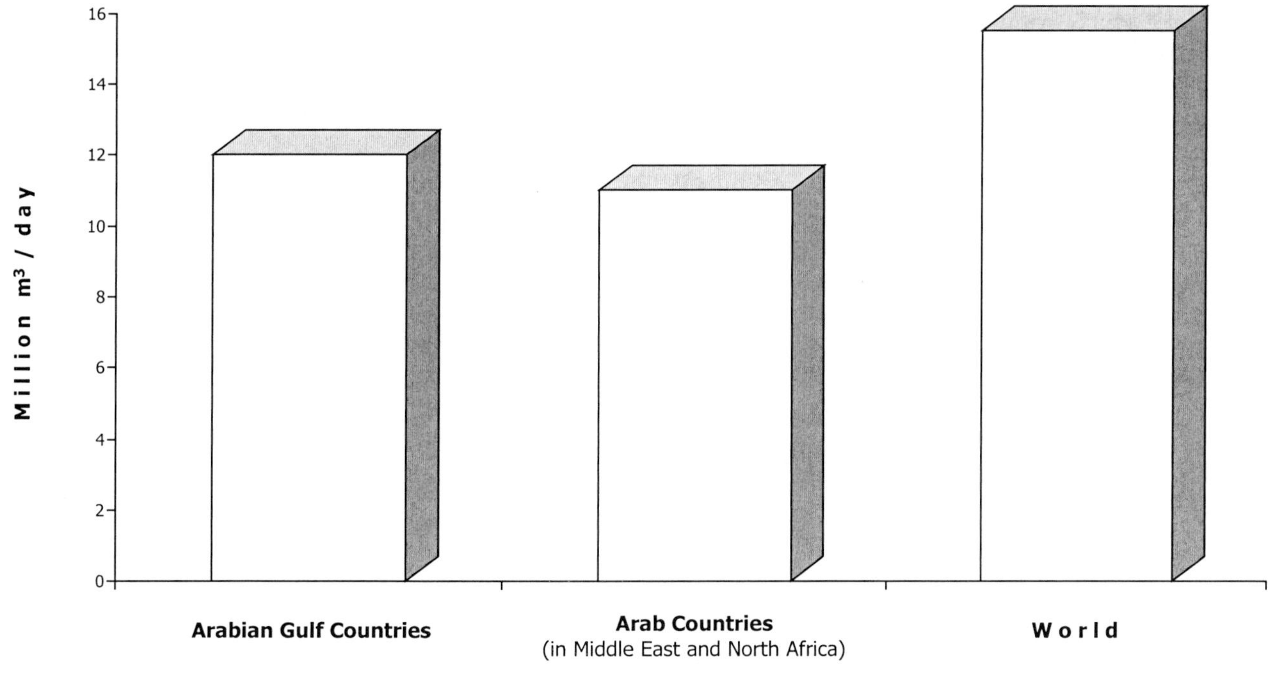

Fig. 7.2. Capacity of desalination plants (in Mm^3/d) in the Gulf States, Arab countries and the World.

Treated Wastewater

Water is scarce in the Gulf States; therefore, every drop of water must be carefully used in an economically feasible manner so that no higher quality water should be used for a purpose that can tolerate a lower quality. In these countries, treated wastewater is now a widely accepted non-conventional source of water that can supplement the traditional water resources and desalinated water. Since early 1980's, there has been increasing use of treated wastewater in irrigation and landscaping as a result of the availability of relatively large volumes of this water due to the completion of urban wastewater treatment facilities and the expansion of sewage networks in most of the Gulf States major cities. Because of their environmental and health hazards, wastewater has been treated, completely or partially, regardless of their ultimate utilization. As a substitute to the limited fresh water in the Gulf States, treated wastewater can play an important role in water resources management and lessen the present and long-term demand versus supply imbalance.

The objectives of this section are to present the current recycled volumes of treated wastewater in the Gulf States, predicted volumes of wastewater production, advantages and limitations of wastewater use and criteria of treated wastewater.

As a substitute for groundwater used in agriculture and industry, treated wastewater has an important role to play in the Gulf States water resources management. In 1997, the Gulf States recycled no more than 35% of their total treated wastewater, which contributed 2.2% of their total water supply. Treated wastewater is now used in landscaping, fodder crop irrigation and has some industrial uses. However, the main problems facing the wide application of treated wastewater are both psychological and technical (microbiological pollution and possible accumulation of heavy metals in irrigated soil). If only 50% of domestic water supplies were treated and recycled in agriculture, such recycled water has the potential to meet more than 11% of the Gulf States total water demands, and could satisfy more than 14% of their agricultural sector demands (Zubari, 1997). It could reduce withdrawal of non-renewable groundwater by more than 15% by 2020.

The estimated treated sewerage water annually produced in United Arab Emirates is 80 Mm^3. This water is now to used irrigate public parks and beautify streets and roundabouts. It is predicted that the production of treated sewage water in the country will reach 175 Mm^3 in the year 2000. Tertiary treated sewerage water could be used in irrigation, especially when poisons and heavy minerals have been removed. The sewage water can be used for industrial purposes, to recharge the aquifer and as an additional source of agricultural irrigation after assurance of its good quality for such usage.

The economic feasibility of treated-sewage water and its usage depend on the cost of treatment, the degree of required treatment and the cost of producing from an alternative source for the same usage. At the present time, the cost of producing 1 m^3 of desalinated water is 5 Dirhams, while the cost of production of 1 m^3 of treated sewage water is 2 Dirhams. There still remains a question about the degree of treatment of sewage water for irrigation and the possibility of chemical and biological pollution to the plants, soil and groundwater. Additionally, work and health hazards related to the use of sewage treated water in the irrigation are possible. However, such consequences can be avoided by safe treatment processes which insure getting rid of pollution and purifying the system for ideal water treatment to be used for plant irrigation. Periodic analysis and field studies must be made to follow the safe use of treated sewage water in irrigation and to educate the community in the same field to avoid adverse effect of working and health environment related to it.

Application of treated wastewater in most of the Gulf States was introduced during early 1980s. The treated wastewater is becoming more important as the sewage water treatment facilities expand and the volumes of produced water become progressively larger. This water is completely or partially treated to meet environmental requirements, regardless of their utilization.

Table 7.4 displays the current volume of treated wastewater, reused volumes, available facilities, and type of utilization in the Gulf States. At present all the six countries are operating modern treatment facilities with advanced tertiary treatment capability. The total designed treatment capacity of the major facilities is more than 728 Mm^3/yr, with a present total treated wastewater volume of about 720 Mm^3/yr. However, the recycled volume of these waters is about 252 Mm^3/yr, which represents about 35% of the total treated wastewater produced. In most of the countries, the remaining unused waters are discharged in the sea. Recycling is used mainly in urban areas (irrigation of gardens and roads ornaments), fodder crop irrigation and highway landscaping.

Contribution of Treated Wastewater to Total Water Demands

The groundwater represents the main source of water supply in most of the countries (89.3%), used mainly in agriculture and is supplemented by desalinated water (8.5%) for domestic and drinking water supply. With respect to total water supply, recycled water contribution ranges between 1.4% (Saudi Arabia) and 21% (Kuwait), with an overall average of 2.2% of the total water utilization in the Gulf States.

These figures indicate that water recycling in the Gulf States is still in its early stages. However, all the Gulf countries have ambitious future plans for the expansion in utilization of the treated wastewater, as an alternative source to complement their future demands. For example, Bahrain is planning to utilize 42 Mm^3/yr of its tertiary treated wastewater in crop irrigation, landscaping, industry and groundwater artificial recharge by the year 2005 (Al-Aradi, 1994). Kuwait also is planning to utilize about 140 Mm^3/yr of its tertiary treated water in irrigation and landscaping by the year 2010 (Al-Muzaini and Ismail, 1994). The Saudi Arabia is planning to utilize about 254 Mm^3/yr of reclaimed wastewater in the eastern region for crop irrigation and landscaping (Al-Saati, 1995).

Advantages of Wastewater Reuse

In addition to its role as an additional water resource, treated wastewater is also cheap compared to desalination water. Al Noaimi (1993) estimated the cost of wastewater in Bahrain at 0.164 US$/$m^3$, and its tertiary treatment costs 0.317 US$/$m^3$. On the other hand, the cost of desalinated water in Bahrain ranges from 0.661 US$/$m^3$ (MSF) to 1.164 US$/$m^3$ (RO), with an average cost of 0.794 US$/$m^3$.

Agriculture in the Gulf States consumes between 80 and 90% of the total water used. This water is mainly drawn from aquifers which are suffering from serious depletion. Reclaimed wastewater can be considered as a source that can ease the pressure on groundwater and significantly improve the situation. In many areas of the Gulf States, tertiary treated wastewater has a better quality than the groundwater actually used for irrigation. Because most soils in the Gulf States are have sandy texture, traces of organic matter and are deficient in major nutrients, treated wastewater use can provide nutrients to soil and plants, especially nitrogen and phosphorus, and reduce the applied amounts of fertilizers.

Gur (1995) described three levels of water treatment for industrial uses: 1) utility water for general cleaning, highway landscaping, greenery irrigation and fire-fighting: 2) process water for crude oil desalting and cooling tower use; and 3) high quality feed water for boilers. Because most of the water supplied to industry in the Gulf States is used principally for cooling, industries can recycle their own wastewater or rely on urban treated wastewater. One of the best examples of treated wastewater utilization by industry in the Gulf States is the Riyadh Refinery. About 7.3 Mm^3/yr (20,000 m^3/d) of Riyadh wastewater treatment plant effluent is pumped to the Refinery's water reclamation plant.

Recharging groundwater with treated wastewater can control salt-water intrusion in coastal areas. All Gulf States aquifers are experiencing quality deterioration due to seawater intrusion or upconing of saline water from deeper formations caused by over-pumping. Moreover,

Table 7.4. Treated water, Reused water, Treatment facilities, and Type of utilization in the Gulf States (compiled by Zubari (1999) from Al-Muzaini and Ismail, 1994; Al-Shaqsi, 1994; Al-Assam and Abdul-Rahim, 1994; Al-Saati, 1995; and Al-Hajj, 1995).

Country	Treated Water (m^3/d)	Reused water (m^3/d)	%	Treatment Facilities			Type of Utilization
				Number of plants	Total capacity (Mm^3/y)	Treatment level	
Bahrain	154,000	25,000-30,000	16-20	1 (large)	158,000	Tertiary	Irrigating fodder, crops, garden and highway landscaping
Kuwait	208,000	129,400	62	4 (large)	208,000 (354,000)	Tertiary	Irrigating crops, highways, coastal zones and Kuwait zoo
Oman	20,300	10,850-17,350	54-86	2 (large) 53 (small)	24,000 50-5,000	Tertiary	Irrigating landscape areas and parks, recreational activities and fountains
Qatar	75,000-80,000	69,000	92-86	2 (large) 9 (small)	80,000 120-3,000	Secondary & Tertiary	Fodder crops, gardens and landscaping
Saudi Arabia	1,230,000	275,000	22	30	>1,230,000	Secondary & Tertiary	Crop irrigation, highway irrigation, landscaping and artificial recharge
United Arab Emirates	280,000	170,000	61	4 (large)	295,000	Tertiary	Irrigating parks, golf courses, highways and urban ornamentals
Total	1,972,30 (720 Mm/^{3}y)	690,750 (252Mm^3/y)	35	12 (large) 52 (small)	1,995,000 (728.2 Mm^3/y)		

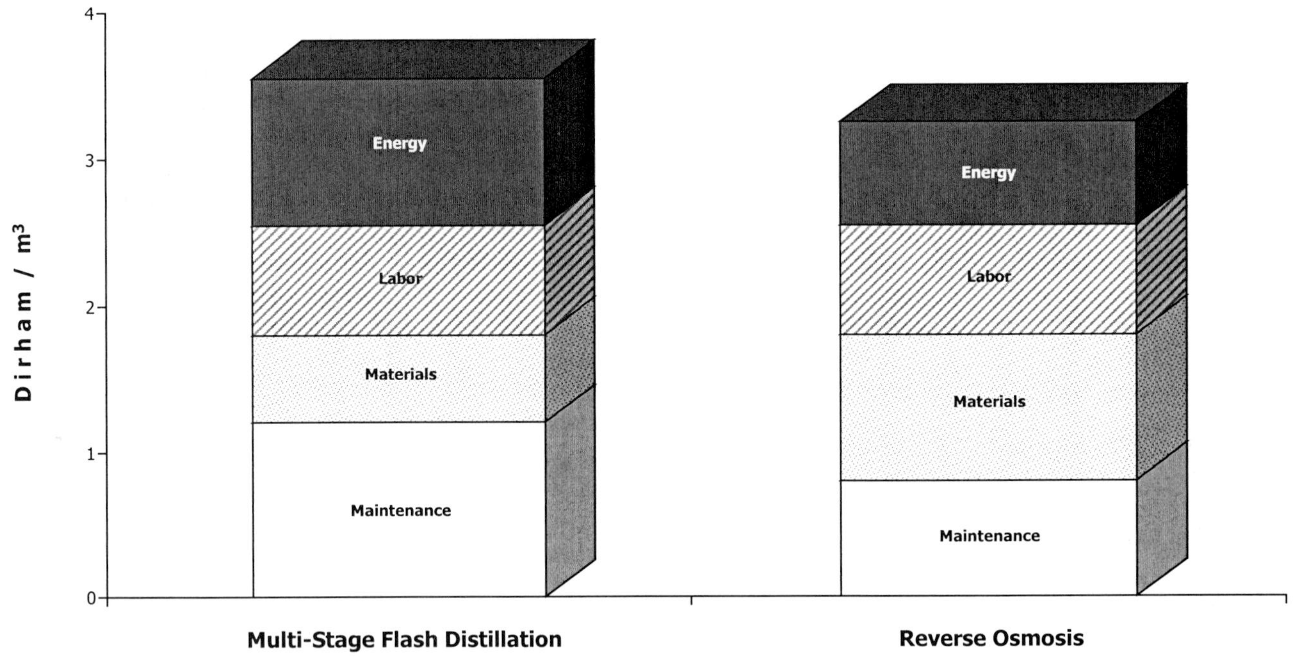

Fig. 7.3a. Comparison between the cost of water production by multi-stage flash distillation and reverse osmosis in the United Arab Emirates.

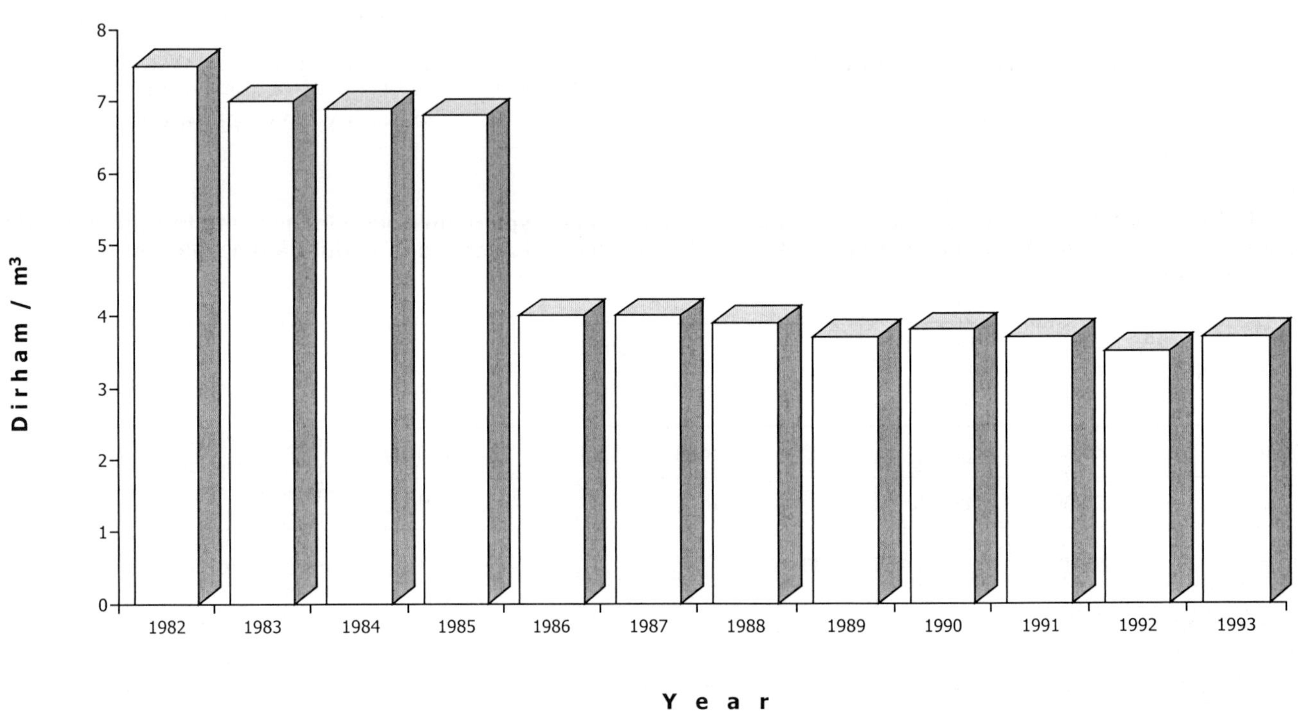

Fig. 7.3b. Evolution of production cost of desalination water in the United Arab Emirates for the 1982-1993 period.

presently large portions of unused treated wastewater in Gulf States are discharged into the sea. Instead of discharging large volumes of treated waster water to the sea in Bahrain (38 Mm^3/yr) and Saudi Arabia (350 Mm^3/yr), this water can be used to recharge groundwater and combat saline water intrusion. However, recharged treated wastewater should not be extracted before the elapse of >400 days in the ground, the time which is needed for natural self-purification (Shuval, 1969).

Constraints on Wastewater Reuse

The main constraints in reusing wastewater in the Gulf States can be divided into public attitude toward this water reuse and technical problems that affect the quality of produced treated wastewater.

1. *Public Attitude*

Public attitude towards reuse of treated wastewater is generally negative. Madani et al. (1992) assessed the public awareness in Bahrain towards application of reclaimed water in specific uses. Most questioned individuals strongly opposed the use reclaimed water regardless its degree of purity. Further, they were willing to pay more money to avoid using such water. The main reasons for opposition were health risks, psychological repugnance and religious beliefs. However, the lack of public awareness about various aspects of wastewater treatment needs promotion of these aspects among general population. Where treated wastewater has a better quality than available groundwater, farmers in Bahrain were willing apply treated wastewater in irrigation. Conveyance of reclaimed wastewater by government distribution system represented a further incentive.

2. *Technical Problems*

Microbiological pollution is the main concern over the reuse of treated wastewater effluents in the Gulf States. Shuval et al. (1985) have expressed significant concern about diseases associated with the use of raw wastewater in irrigation, especially

Table 7.5. Maximum contaminant levels (mg/l) in treated wastewater of Saudi Arabia, Oman and Kuwait (compiled from Al Dhowaila and Al Mutlak, 1999; Al Sabahi, 1997; and Al-Awadhi et al., 1994).

Parameters \ Country/ Organization	Saudi Arabia	Oman	Kuwait	Organizations: Food and Agricultural Organization (FAO)	Organizations: Environmental Protection Agency (EPA) – Agriculture	Organizations: Environmental Protection Agency (EPA) – Landscaping
Biochemical Oxygen Demand	10	15	30	-	-	20
Total suspended solids	10	15	15	-	-	15
Total dissolved solids	-	1500	4500	-	-	-
Sodium adsorption ratio	-	10	10	-	-	8 - 18
pH	6.0 – 8.4	6 - 9	6 - 9	-	-	6 – 9
Aluminium	5	5	5	5	5	5
Arsenic	0.1	0.1	0.1	0.1	0.1	0.1
Berylium	0.7	0.1	-	0.75	-	-
Boron	-	0.5	0.75	-	-	-
Cadmium	0.01	0.01	0.01	0.01	0.01	0.01
Chloride	280	650	-	-	-	100 - 200
Chromium	0.1	0.05	0.1	0.1	0.1	0.1
Cobalt	0.05	0.05	0.05	0.05	0.05	0.05
Copper	0.4	0.5	0.2	0.2	0.2	0.2
Cyanide	0.05	0.05	0.1	-	-	-
Fluoride	2	1	-	1	1	2
Iron	5	1	5	5	5	5
Lead	0.1	0.1	5	5	5	5
Lithium	2.5	0.07	2.5	2.5	2.5	2.5
Manganese	0.2	0.1	0.2	0.2	0.2	0.2
Mercury	0.001	0.001	0.1	-	-	-
Molybdenum	0.01	0.01	0.01	0.01	0.01	0.01
Nickel	0.2	0.1	0.2	0.2	0.2	0.2
Oil and grease	Absent	0.5	5	-	-	Nil
Phenols	0.002	0.001	0.1	-	-	50
Selenium	0.02	0.02	0.02	0.02	0.02	0.02
Sulphate	-	400	-	-	-	200 - 400
Vanadium	0.1	0.1	0.1	0.2	0.1	0.1
Zinc	4	5	2.0	2	2	2
Fecal coliform (MPN) in 100 ml	2.2	200	23	-	1000	2.2

where the infecting agents were not removed completely by conventional wastewater treatment processes. The infection poses a great risk to farmers and to consumers of farm products. The maximum contaminant levels in treated wastewater in Gulf States are listed in Table (7.4). Unfortunately intestinal parasitic diseases are widespread in the Gulf States (Gur, 1995) where they represent a real challenge to engineers to design conventional treatment plant which complies with the requirement of the Environmental Protection Agency health guidelines.

Table 7.5 shows that the observed heavy minerals concentrations in treated wastewater in Saudi Arabia, Oman and Kuwait are generally low and below the maximum allowable limits in irrigation water, the concern remains about the accumulation of these minerals with time in the irrigated soil. In alkaline desert soils, such as those dominant in Gulf States, certain heavy minerals accumulate and their concentrations start to increase progressively with the use treated wastewater in irrigation. In many cases metal concentrations exceed the maximum permissible limits and become hazardous to human and environment. Preventive measures are more effective in minimizing accumulation of heavy minerals in soils irrigated by treated wastewater. Discharge of industrial waste into domestic waste network introduces toxic organic substances or increases the concentration of some heavy metals to a level that they would be harmful to workers and toxic for plants and soils. Therefore, the discharge of industrial wastewater must be handled separately away from the domestic waste networks.

3. *Environmental Concerns*

Although the treated sewage from Al Ardheyah treatment plant in Kuwait appeared to be less saline than the brackish groundwater in the same area, both waters were considered by Al Ruwaih (1985) to be of very poor quality. Continuous irrigation with such waters is predicted to increase sodium hazard, especially in fine-textured soils. However, the concentrations on boron and trace elements are within the permissible limits for irrigation water.

Potential of Treated Wastewater

Ismail (1985) indicated that the water projected water needs of the Gulf States would reach about 28,000 Mm^3 by the year 2020, taking into account several water conservation measures. Projection also indicates that almost all the agricultural water needs will be met by withdrawal of groundwater, whereas the domestic water demand will be met via water desalination. The treated wastewater reused in the Gulf States is predicted to increase from 250 Mm^3/yr (1.15%) in 1990 to 650 Mm^3/yr (2.33%) by the year 2020. Zubari (1997) claims that if only 50% of domestic water supplies are treated and recycled (about 3,000 Mm^3/yr) by the year 2020, the treated wastewater would increase its contribution to 11% of the total water demands in Gulf States. Furthermore, if reclaimed wastewater were used in crop irrigation, it could satisfy about 14% of the agricultural sector demand, and could reduce groundwater withdrawal by about 15%.

Guidelines for Wastewater Reuse

It is essential for the Gulf States to establish strategies for usage of treated wastewater taking into account the suitability of different crops to different levels of wastewater treatment, potential of effluent reuse for agriculture, industry and urban landscaping, reclaimed water reuse for recharging depleting aquifers and conducting research to specify limits of sustainable use of treated wastewater in agriculture for long periods under different conditions.

The successful reuse of treated wastewater in agriculture depends on the reliability of reclaimed wastewater as an alternative source for groundwater in irrigation through high level of treatment (tertiary treatment), quality assurance and monitoring, promotion of public awareness and setting national standards such as those of Saudi Arabia, Oman and Kuwait (Table 7.5).

Chapter 8

CASE STUDIES ON THE HYDROGEOLOGY OF THE CENOZOIC AQUIFER SYSTEMS IN THE ARABIAN PENINSULA

The Arabian Peninsula hydrogeological basin has been divided into depleting or recharging aquifers, and non-depleting or charging aquifers (Italconsult, 1969). This classification was modified using the piezometric data for the entire aquifer system. FAO (1979) established two distinct sets of piezometry a) Wasia-Biyadh (Cretaceous) aquifer which maintains an extremely low gradient and high hydraulic head from its recharge area to the coastline, b) the Umm er Radhuma, Alat and Khobar Members of Dammam Formation and Neogene complex (Cenozoic) which show regionally similar gradients, and nearly identical artesian heads. Anomalies in one aquifer are reflected in the others.

The present study focuses on the Cenozoic system, since this highly transmissive aquifer is the most productive in the area and underlies the entire Eastern Arabian basin.

Since the aquifer units of Paleogene carbonate cycle, and Neogene detrital cycle of the Tertiary are hydrogeologically connected with each other to varying degrees, these rocks are considered as one aquifer system. Interaction across any intervening confining layers has been facilitated by the absence of evaporite units of the Rus Formation, shale and marl units of the Dammam Formation, due either to non-deposition or dissolution, and/or by fracturing associated with slumping or tectonic activity, where the karstification, faults and joint systems become abundant.

Cenozoic Hydrogeological System

The Cenozoic hydrogeological system can be divided into two distinct systems:
a) Multi-layer system: This is comprised of several aquifer and aquiclude or aquitard layers which are locally hydrogeologically connected to each other. It exhibits both confined and unconfined conditions; and b) Fresh water lenses system: A system of isolated fresh water lenses which occur in the upper saturated zones of Multi-layer System at some localities beneath collapse depressions, floating on brackish water.

The Tertiary sequence is composed of several aquifer and discontinuous aquitard layers (from oldest to youngest) is as follows:

1. *Lower Aquifer Unit:* it includes Umm er Radhuma Formation and Dolomite Unit of Rus Formation. Despite varying lateral extent, water quality variations and capacity to yield water, this Lower Aquifer Unit can be classified as a single aquifer unit over a large area of the Arabian platform. It is present at shallow depth, stretching throughout the Gulf States. It has been exploited as a fresh aquifer is Saudi Arabia and Oman, but is a poor aquifer in Qatar, United Arab Emirates, Bahrain and Kuwait.

This Lower Aquifer Unit, consists mainly of alternating layers of limestone and dolomitic limestone, with intercalations of anhydrite and argillaceous shales, increasing in clay content downwards. Deposition of the argillaceous fine carbonates occurred in the topograpically low areas, while coarser carbonate sediments were forming over high areas. This variation in lithology caused transmissivity in the structurally low areas to be much lower that in the positive areas. In consequence, this aquifer unit is confined below by the basal shale-marl member of Umm er Radhuma, inhibiting leakage between the Tertiary aquifer system and Cretaceous aquifer system, except around some structural high areas. Top of the Lower Aquifer Unit is bounded by the most pronounced discontinuous evaporite and shale aquitard layer in the Eastern Arabia. This confining layer is formed by the evaporite unit of the Rus Formation, and the Midra and Saila Shale Members of the Dammam Formation. The uppermost part of the Lower Aquifer Unit exhibits varying amounts of chemical and physical alterations. This aquifer unit is heavily fissured and exhibits karstification, which causes loss of circulation.

Natural dissolution has resulted in interstratal karst, and sheet-like dissolution occurs over extensive areas beneath the covering layers of rocks. Due to its intensive dissolution this aquifer unit is regionally a principal aquifer unit. It commonly yields large quantities of water to wells.

2. *Middle Aquitard Unit:* it includes evaporite unit of the Rus Formation and Midra and Saila Shale of the Dammam Formation. This unit forms the most important confining unit of the Tertiary aquifer system in the region. It is present in the shallow subsurface throughout the Eastern Arabia, and is exposed at the surface. Where its top is close to the surface this unit has been subjected to extensive karstification. The thickness of the Middle Aquitard Unit varies sharply over the study area, but it is thick and widespread in the southern part of Qatar and elsewhere in the Eastern Arabia, while it has

been largely of totally dissolved over much of the structurally high areas. This latter changes the unit a confining one to a permeable one, since it enables a hydrogeologic connection between the Lower Aquifer Unit and the Upper Aquifer Unit.

3. The Upper Aquifer Unit: This unit formed by intercalations of light gray dolomitic chalk, and thin marl unit. This aquifer unit has limited thickness and is rarely of any importance as an aquifer except in the southwest Qatar. However, it has been identified as a main fresh water aquifer in Bahrain and Kuwait and part of eastern Saudi Arabia. This Neogene detrital deposits rarely have an adequate thickness or extend over wide enough area to act as an aquifer of any consequence in Qatar but it has been identified as an aquifer of limited potential in neighbouring regions, such as in the United Arab Emirates, when the greatest supplies are carried in the Quaternary alluvial sands and gravels.

1. CENOZOIC AQUIFER SYSTEM OF KUWAIT

INTRODUCTION

Kuwait, in the northwestern part of the Arabian Gulf, is an arid region which underlies the importance of groundwater supply both in quantity and quality. The main sources of fresh water and brackish water are in the sediments of the Kuwait Group and the Dammam Formation.

The annual average rainfall in Kuwait is 100 mm, most rainfall occurring between October and April, but with an annual evaporation in excess of 2,500 mm, rainfall events which can produce runoff or groundwater recharge are very rare. However, in depressions such as Ar-Raudhatain and Umm Al Aish conditions are favorable for some recharge to occur, creating isolated fresh groundwater lenses floating on more saline water (Al Sulaimi et al., 1988).

Kuwait has an area of 9840 km^2 consisting mainly of desert and low offshore islands on the western side of the Arabian Gulf. Topographically it is an area of gently undulating sandy and gravelly plains sloping towards the sea (Fig. 8.1) broken only by low hills, escarpments and depressions. The maximum elevation in the southwestern corner of the country is about 275m dropping to sea level at a rate of 3m/km. The major features are the Jal-as-Zor escarpment bordering the northern side of Kuwait Bay and the Al Ahmadi Ridge. The major depression, the broad, shallow Wadi Al Batin which extends southwest-northeast across the country from its western boundary, is a relatively old structure. Small playas occur throughout the northern, western and central parts of Kuwait become filled with water following the rare rains. In the coastal regions large areas of salt marsh develop, and in the northeastern

Table 8.1 . Hydrogeological column of the Hasa and Kuwait Groups in Kuwait (compiled from various sources).

Age	Group	Formation	Description	Thickness (m)	Environment	Water Quality
Recent			Beach sands and limestones, wind blown sand, playa silts and clays, wadi alluvium	30	Continental and evaporitic with shallow marine	Above saturation or locally has brackish or freshwater, depending on drainage and topographic conditions
Pleistocene	Kuwait	Dibdibba	Coarse upland gravels, gravel and sand, mainly conglomeratic sandstone, siltstone, shale	110	Continental and shallow marine	Most permeable of shallow aquifers, water locally fresh beneath wadis and depressions, brackish at depth
Miocene	Kuwait	Lower Fars	Fine to conglomeratic calcareous sandstone, variegated shales, fossiliferous limestone, gypsiferous	100	Continental and evaporitic with shallow marine	Low permeability, water generally brackish, where formations are near surface the upper water is locally fresh beneath wadis and depressions
Miocene	Kuwait	Ghar	Quartzose sandstone and conglomerate, some shale in lower part	275	Continental and shallow marine	Groundwater deep, generally brackish
Early Eocene (Lutetian)	Hasa	Dammam	Limestone, chalky limestone, dolomitised limestone with chert bands towards the top and shelly chnes at bay	180-210	Shallow marine with continental at boundaries between units 2 and 1	Moderately brackish change gradually to saline water from southwest to the northeast ranges from 2,500 to 150,000 mg/l
Early Eocene (Ypresian)	Hasa	Rus	Anhydrite with limy and shaly intercalation	75-120	Evaporitic	Brackish water
Paleocene-Early Eocene	Hasa	Radhuma	Limestone, calcarenite dolomitized in places with intercalating anhydrite and shaly partings	180-425	Shallow marine with recurrent evaporitic	Brackish to saline with H_2S
Maastrichtian	Aruma	Tayarat	Limestone		Shallow marine	Saline water

part of the country a few barchan dunes up to 25m high are found. Al Sulaimi et al. (1988) identified three drainage systems in Kuwait; the northern, western and southern systems (Fig. 8.2). The northern drainage system is well developed with long, closely spaced and dense streams. It has a dendritic pattern the main streams converging into large depressions. The western drainage system is less developed with long, shallow streams flowing towards the northeast. The southern drainage system is poor with no prominent drainage lines. The general stratigraphy of the Tertiary succession in Kuwait and its hydrogeologic units shows in Table (8.1).

There is a tri-partite division of the Kuwait Group (Owen and Nasr, 1958) into the Dibdibba, Lower Fars and Ghar formations which can be recognized in northern Kuwait due to the presence of the Lower Fars beds, but where these beds are absent or questionable the sequence is referred to the undifferentiated Kuwait Group as in southeastern Kuwait.

Lithologically the Dibdibba Formation is made up of fluviatile, ungraded and often cross-bedded sand and gravel with subordinate, lenticular sandy clay, sandstone, conglomerate and siltstone locally cemented by lime or gypsum. It is little different

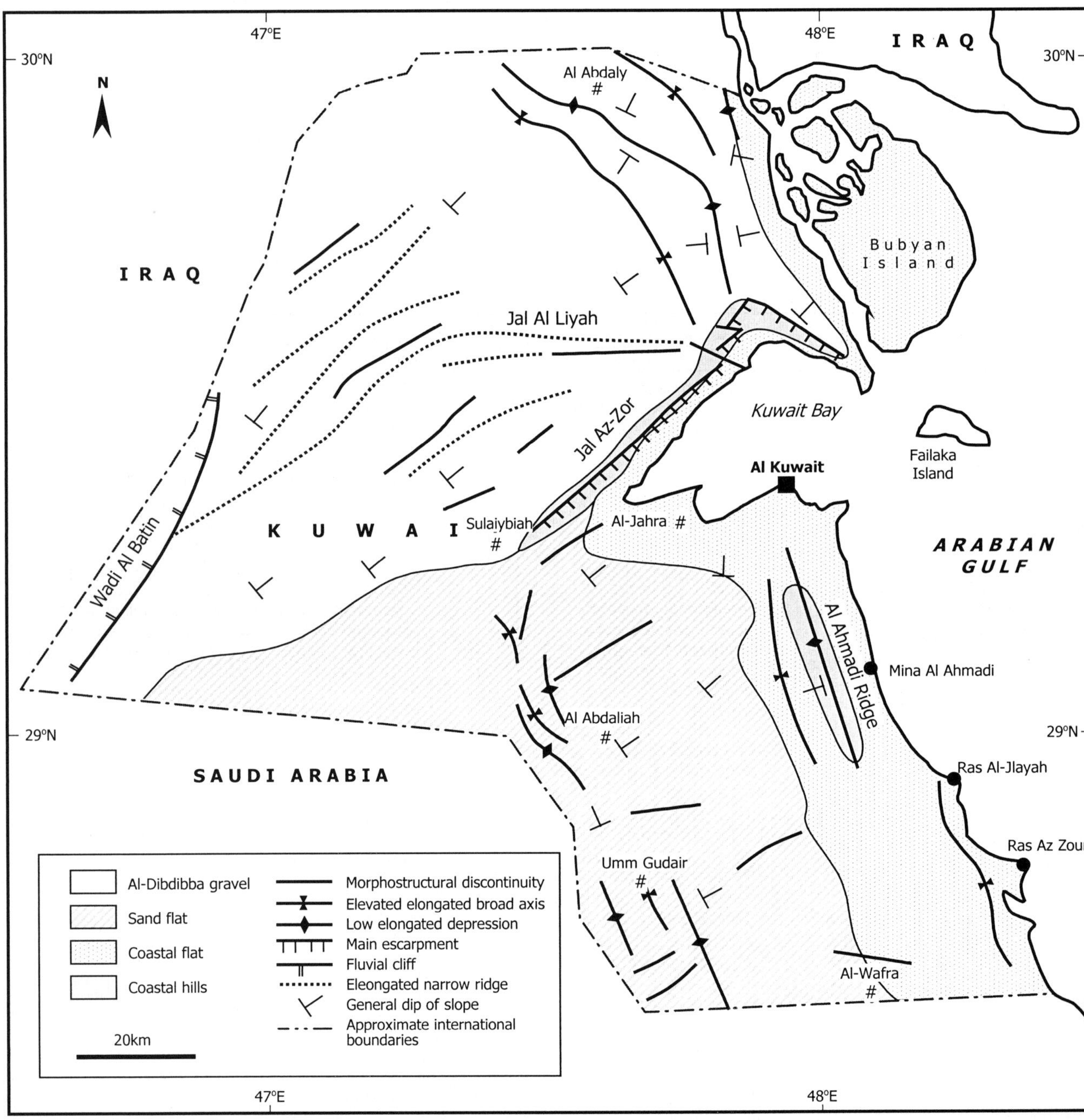

Fig. 8. 1. Physiographic provinces and morphostructural zonation map of Kuwait (modified from Al Sulaimi and El Rabaa, 1994; Al Sulaimi et al., 2000).

from the coarse marine to terrestrial sands of the Ghar Formation although the latter has some limestone streaks and may be better cemented. The two cannot be readily distinguished except where separated by the finer grained clastic sediment and thin bands of fossiliferous limestone, especially in the southern part of the country.

The Hasa Group is composed of Paleocene formations, the Dammam Formation limestones, the Rus Formation anhydrites and the Radhuma Formation limestones (Owen and Nasr, 1958). It consists mainly of carbonate rocks, partly dolomitized in the lower part with middle, evaporite (anhydrite) division. The succeeding Oligocene was a period of erosion and at the top of the Hasa Group is a pronounced unconformity surface. The Dammam Formation was subject to erosion during the late Eocene and during almost the whole Oligocene. The Rus Formation is composed of low porosity chalky limestone with marl and gypsum intercalations and subordinate amounts of sand and anhydrite.

Hydrogeology and Groundwater Occurrence

The groundwater resources in Kuwait are limited to the Kuwait Group and the Dammam aquifers. Water in both aquifers is brackish (TDS

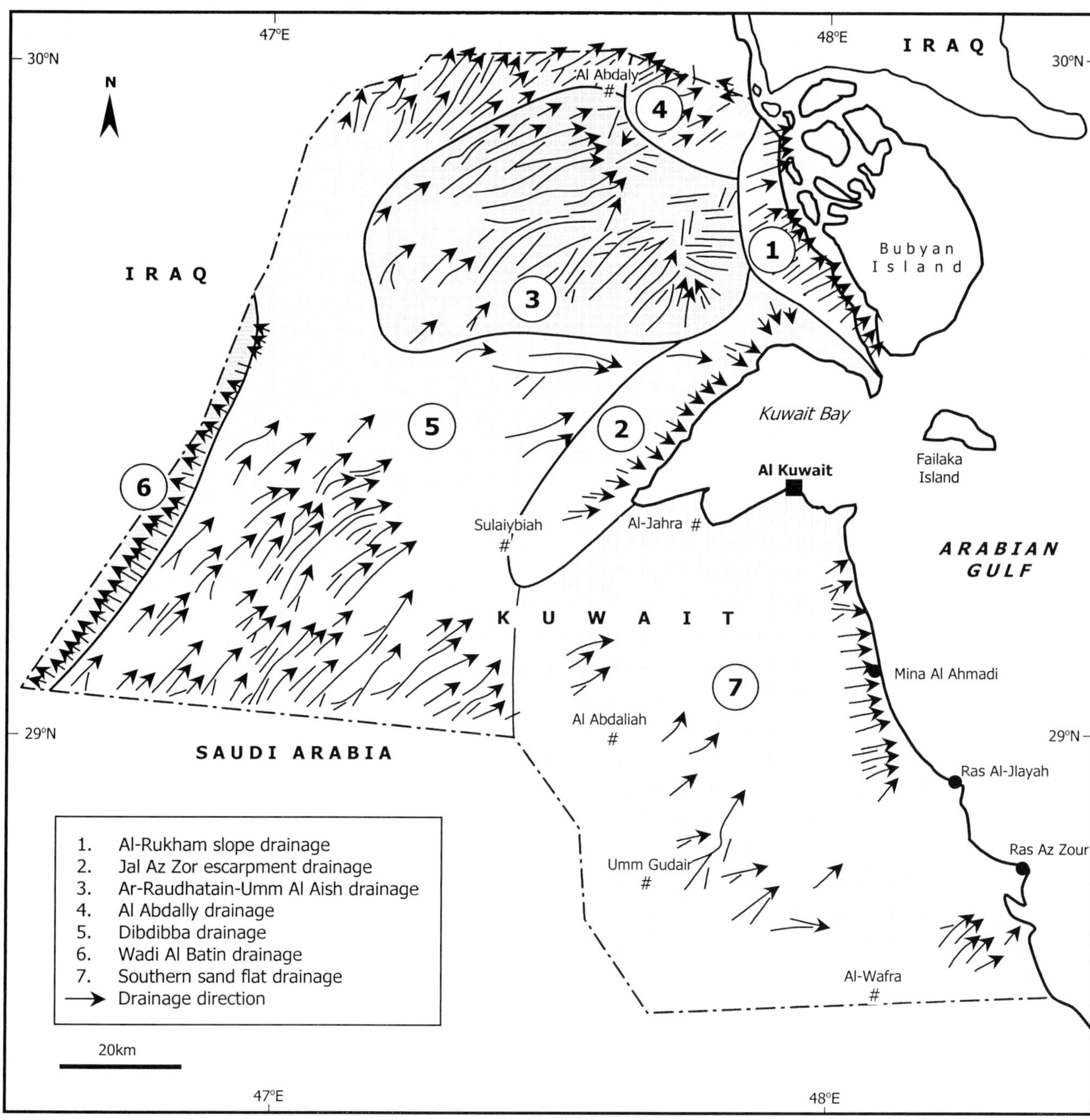

Fig. 8.2. Generalized map of drainage zones and directions in Kuwait (modified from Al Sulaimi and Akber, 1999).

contents 2,500-7,500 mg/l) and is used as raw water for some desalination plants. The recharge for groundwater in Kuwait occurs in Saudi Arabia and southern Iraq. Then groundwater moves from reacharge areas towards the north and east and becomes more saline as it reaches the discharge zone along the coast of the Arabian Gulf and the groundwater table varies from about 90m above mean sea level in the southwest, to near zero at the Arabian Gulf coast (Fig. 8.3).

Kuwait Group Aquifer

The clastic units of the Kuwait Group serve as aquifers separated by an aquitard. The lower aquifer is believed to extend under most of Kuwait. The upper Kuwait Group aquifer ranges in thickness from 25 to 45m and consists mainly of alternating sandstones, clay, and minor interbeds of gypsum and calcareous limestone. The lower Kuwait Group aquifer ranges in thickness from 15 to 40m and consists of coarse-grained sandstone with a few clay

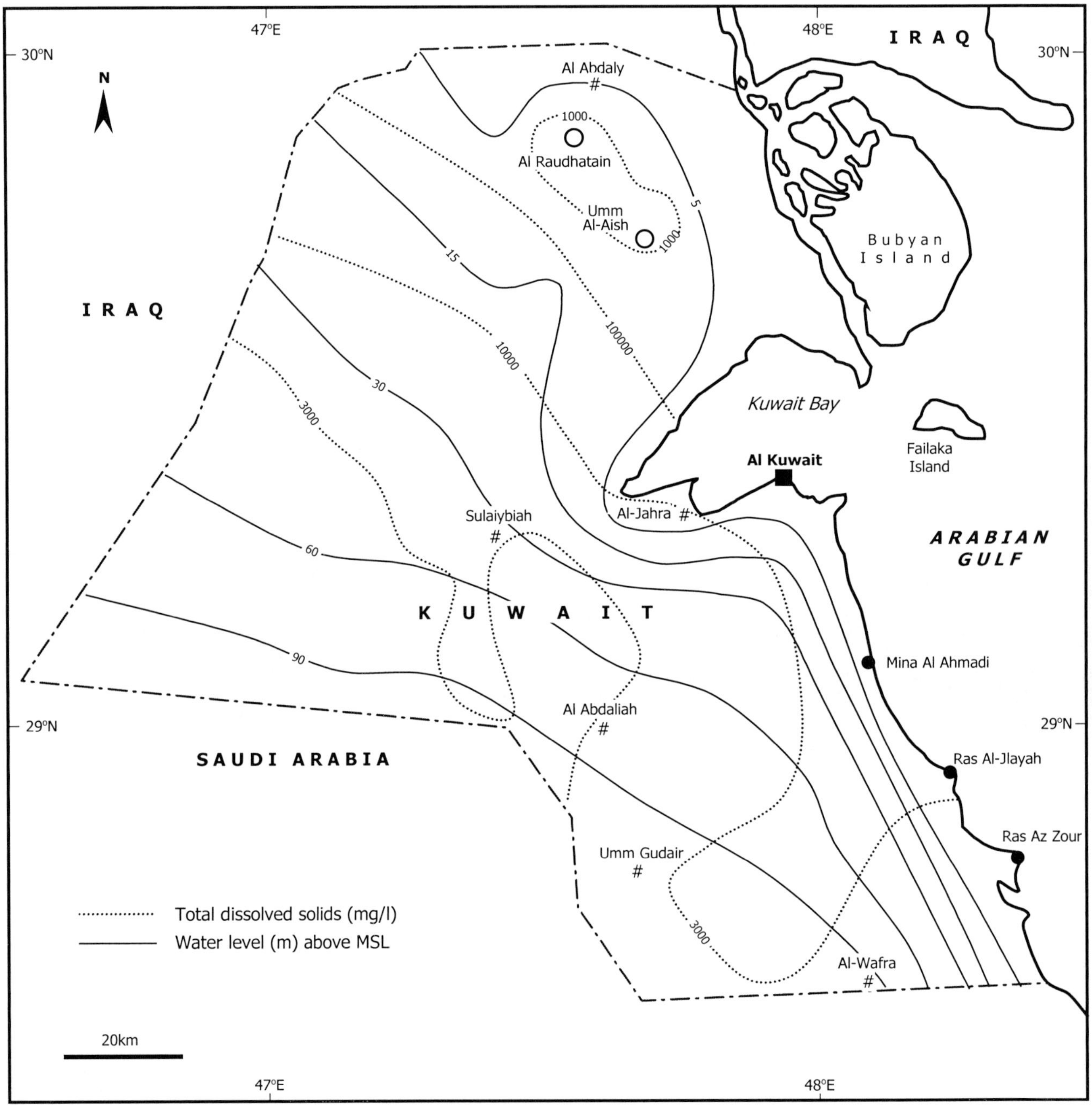

Fig. 8.3. General distribution of total dissolved solids (mg/l) and groundwater level (m) in Kuwait (compiled from various sources cited in the text).

beds. The lower Kuwait Group aquifer is semi-confined under most of Kuwait and disappears in the extreme southwestern part of the country. The piezometric surface of the Kuwait Group aquifers varies between more than 60m above sea level in the west and to than 5m in the northeast (Fig. 8.4).

In the Al-Raudhatain and Umm Al Aish areas occasional rainstorms create fresh water lenses overlying, but separated from, the upper Kuwait Group aquifer. Therefore, Al Sulaimi and Akbar (1999) suggested further classification of the Kuwait Group into three aquifer units: the upper Dibdibba aquifer, the lower Dibdibba aquifer and Lower Far/Ghar aquifer. The saturated thickness of the upper Dibdibba aquifer in the Al Abdaly area decreases from 30m in the southwest to less than 1m in the northwest (Fig. 8.5). The downward evolution of groundwater salinity and quality from the Kuwait Group fresh water aquifer is illustrated in figure 8.6.

The silicified topmost part of the Dammam Formation with basal shale layers of the Kuwait Group, form an aquitard that separates the Dammam aquifer from the overlying Kuwait Group aquifer. Both the Lower Fars, the Ghar and the

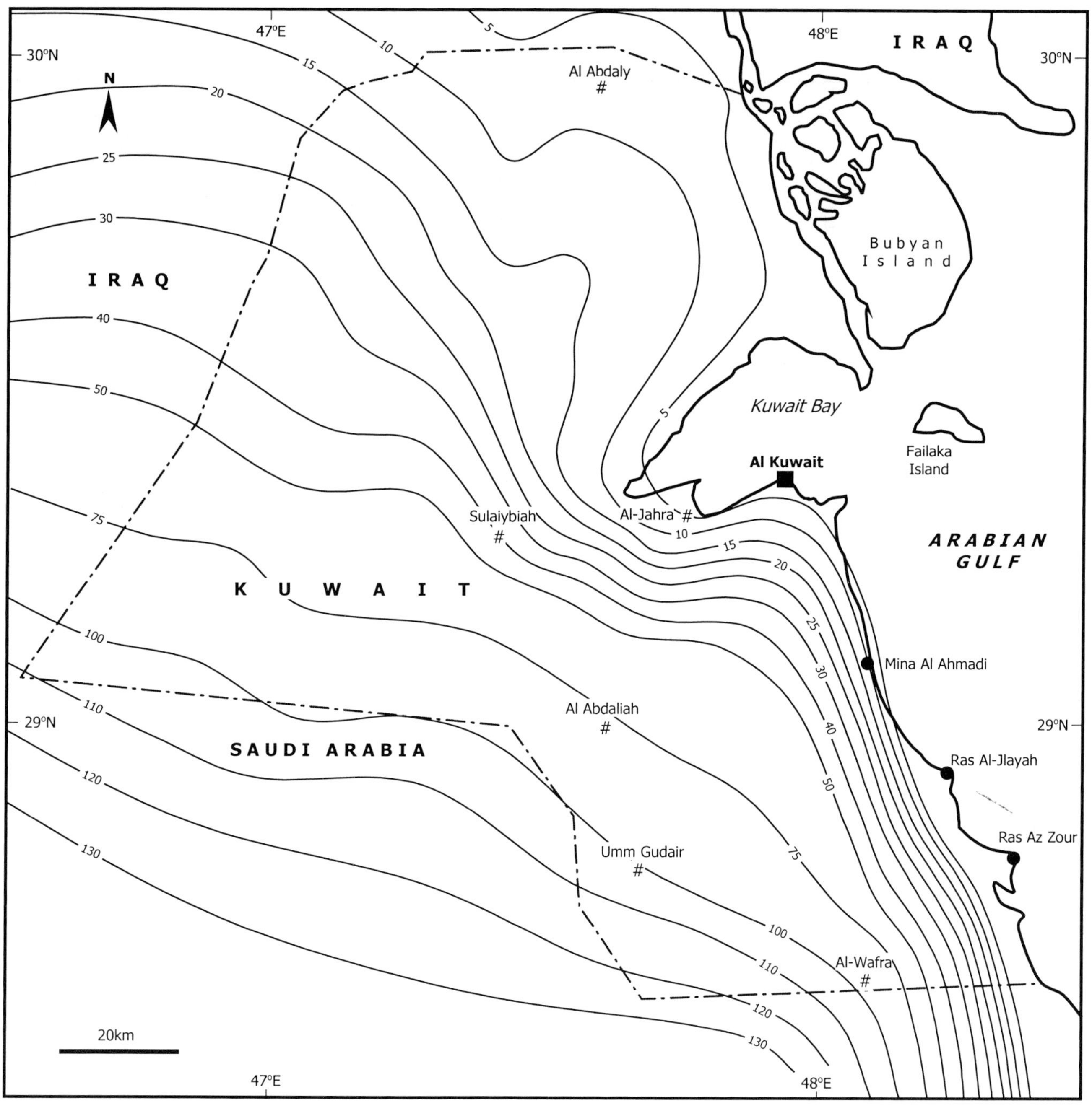

Fig. 8.4. Initial (pre-1960) potentiometric heads (m) above main sea level in the Kuwait Group, before the start of the large scale exploration of the aquifers (modified from Mukhopadhyay et al., 1996).

Dibdibba formations constitute the Kuwait Group aquifer. A sandy shale unit acts as an aquitard dividing the Kuwait Group into a semi-confined lower aquifer and an unconfined upper aquifer in the central and southern part of the country. The average transmissivity of the Kuwait Group aquifer increases from 10 m^2/day in the southwest to 1,500 m^2/day in the northeast. The saturated thickness of the aquifer increases from less than 1m to 400m in the same direction. The storage coefficient ranges from 4.9 x 10^{-2} in unconfined southern part to 1 x 10^{-4} in the confined northern part. The Dibdibba Formation overlies the Ghar and the Lower Fars formations in northeastern Kuwait, forming the uppermost aquifer in the Kuwait Group aquifers (FAO, 1979). The Dibdibba aquifer is unconfined (storage coefficient = 8 x 10^{-2}), its transmissivity increasing from 10 m^2/day near the Kuwait Bay to 175 m^2/day in the northeastern corner of the country. The aquifer thickness increases in the same direction.

Dammam Aquifer

In Kuwait, the Dammam aquifer is generally composed of Upper Eocene limestone and dolomite, with shale intercalation near the base and chert crust at the top. The Dammam limestone aquifer consists of soft and chalky, shelly and porous limestone and hard crystalline dolomitic limestone. A hard 6 to 9m thick chert layer at the top or within the aquifer

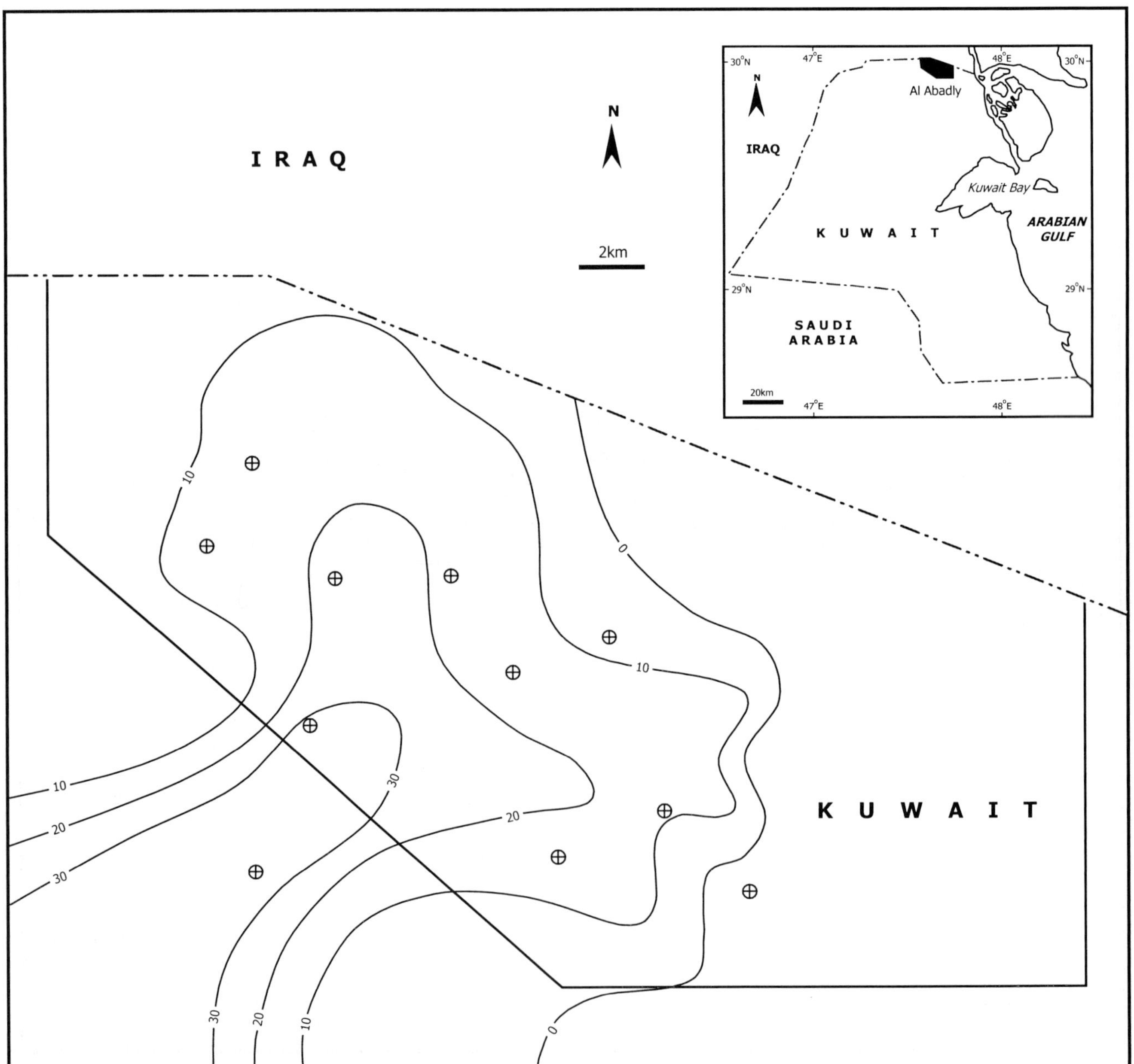

Fig. 8.5. Thickness of groundwater salinity (m) with total dissolved solids of less than 5000 ppm in Kuwait Group at Al Abdaly field wells in northern Kuwait (modified from Al Sulaimi and Akber, 1999).

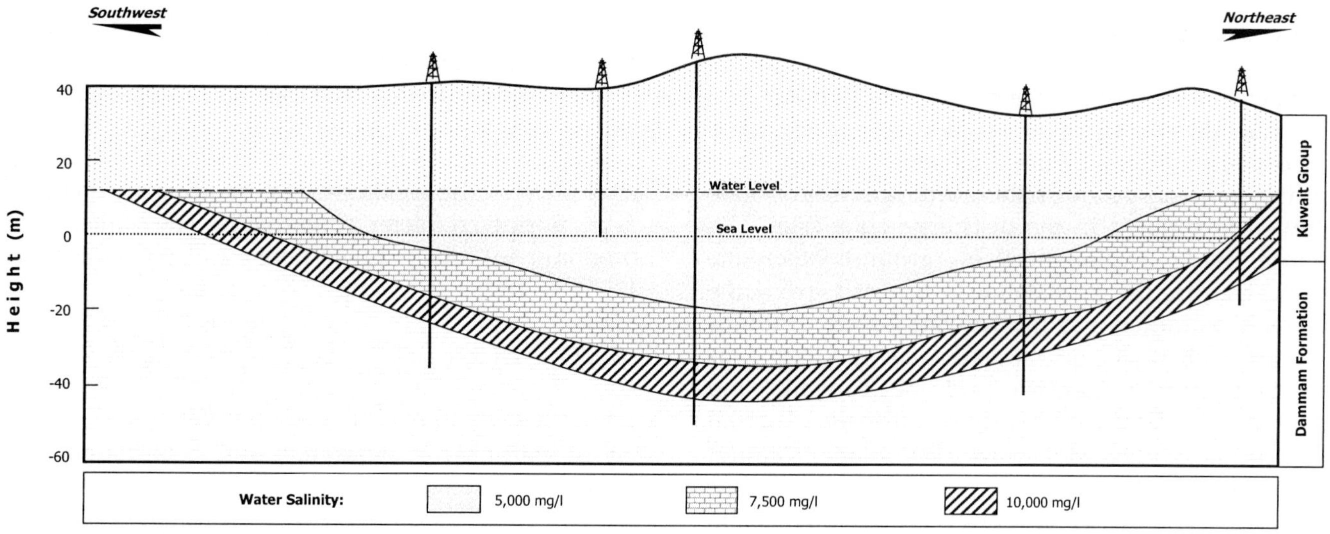

Fig. 8.6. Cross-section showing the distribution of groundwater in the Kuwait Group salinity (ppm) in Al Raudhatain and Al Abdaly field wells in northern Kuwait (modified from Al Sulaimi and Akber, 1999).

acts as an impermeable layer which stops any upward or downward movement of groundwater (Al-Hajji, 1976).

Anhydrites of the Rus Formation together with the basal shale of the Dammam Formation act as an aquitard, separating the underlying Radhuma aquifer from the Dammam aquifer. Based on lithology, hydraulic properties and karstification, the middle part of the Dammam Formation represents the main aquifer in Kuwait (Al Awadi and Mukhopadhyay, 1995). The thickness of the aquifer varies from 120m in the southwestern corner of the country, to 280m in the Al Mutla area in the north. The Dammam aquifer has a thickness of 225m in the synclinal area west of the Ahmadi ridge, and more than 350m to the east of it. Along the ridge itself, the aquifer has a constant thickness of 200m.

In most of the area, the Dammam aquifer is isolated by confining layers. However, where these layers are absent, the Dammam aquifer comes into direct contact with the underlying Radhuma aquifer, and the overlying Neogene and Quaternary aquifers.

The groundwater level in the Dammam aquifer is at 100 to 160m. Flow in the aquifer occurs from west to the east and northeast. The factors affecting groundwater flow in the Dammam aquifer are the geologic structure, lateral facies change and variation in thickness of confining layers below and above it (Sakr, 1989).

Karstification in the Dammam aquifer was recorded by Burdon (1966). The paleokarst sinkholes may control the present-day groundwater movement, the yield in wells and the chemical composition of the groundwater.

The recharge of the Dammam aquifer in Kuwait occurs in the southern Iraqi desert, west of Kuwait and in Saudi Arabia (400 km south of Kuwait), where the exposures of the Dammam aquifer cover 1200 km^2. The piezometric-surface map of the Dammam aquifer (Fig. 8.7) shows that, water in the aquifer generally moves from southwest towards northeast. The Dammam aquifer has a higher piezometric surface than that of the overlying Kuwait Group, but the reverse is true, where the confining chert layer at the top of the Dammam aquifer is either absent or fractured, and water moves upward from the Dammam aquifer, into the Kuwait Group aquifer. In contrast, when the piezometric head is sufficiently reduced by pumping, the flow of water is reversed and water moves from the Kuwait Group aquifer into the Dammam aquifer.

The transmissivity of the Dammam aquifer in Kuwait varies between 1,000 and 100,000 m^2/day (Sakr, 1989) and the average storage coefficient is 2.0 x 10^{-4} (Al-Hajji, 1976). The transmissivity of the Dammam aquifer decreases towards the north and east, and increase in the southwestern part of the country. The Dammam aquifer storage coefficient averages $1x10^{-4}$ (Mukhopadhyay et al., 1994).

Radhuma Aquifer

In Kuwait, the Radhuma aquifer varies between 600m in thickness in the north to 420m in the south. It consists mainly of anhydritic, dolomitic, and marly limestone, with minor amounts of crystalline sulphur of secondary origin, in the aquifer south of Kuwait. The Radhuma aquifer is not exploited due to the higher salinity, low transmissivity (low productivity) and H_2S content. The anhydrite of the Rus Formation and the shales in the lower part of the Dammam Formation act as an aquitard, separating the Radhuma aquifer from the Dammam aquifer (Mukhopadhayay et al., 1996; Omer et al., 1981).

Groundwater Flow

Two hydraulic gradients exist; horizontal and vertical. With respect to the horizontal gradient, almost all aquifers in the Arabian Peninsula have initial gradients towards the north and east, where Kuwait is situated in the discharge zone (Fig. 8.8). Before wide exploitation of groundwater, the vertical hydraulic gradient was upward, from the Dammam aquifer into the Kuwait Group aquifers, but by 1988, this vertical hydraulic gradient was reversed in the central part of Kuwait, and groundwater started flowing from the Kuwait Group aquifers into the Dammam aquifer (Mukhopadhyay et al., 1994).

The direction of groundwater movement in Kuwait can be determined from the hydrochemistry of water. On the basis of quality and type of groundwater in Kuwait, it is clear that groundwater flows from areas with a lower ion concentration to areas of higher ion concentration. The chemical analyses of water from different aquifers in Kuwait show that, the water salinity as well as cation and anion concentrations increase generally from southwest to northeast, the direction of groundwater flow.

Hydrogeochemistry

Water salinity in Kuwait Group and Dammam Formation increases from southwest to northeast with a change from a sulphate to chloride water type. The only exceptions are the Al Raudhatain and Umm Al Aish fresh water basins in the northern part of Kuwait.

During the natural flow the groundwater, as the depth increases and as the groundwater moves further away from recharge, a change in the predominant ions takes place. During recharge groundwater has a bicarbonate character, but during its flow from the recharge area, it ultimately changes to a chloride character at discharge. The sequence of predominant ions is given here (se also Chebotarev, 1955):

$$HCO_3^- \rightarrow HCO_3^- + SO_4^{2-} \rightarrow SO_4^{2-} + Cl^- \rightarrow Cl^- + SO_4^{2-} \rightarrow Cl^-$$

In Kuwait the HCO_3^- and $HCO_3^- + SO_4^{2-}$ water types occur only in some wells of Al Raudhatain and Umm Al Aish water fields, where during the rainy seasons water accumulated in the near surface recharges groundwater. SO_4^{2-} and Cl^- water type is the dominant type of groundwater in the different aquifers in southwest Kuwait. This type changes gradually to $Cl^- + SO_4^{2-}$ and Cl^- -type, respectively to the east and northeast, where groundwater having reached its point of discharge, is almost stagnant or has a minimum rate of flow. The chemical characteristics of the groundwater, of the Kuwait Group and Dammam Formation are presented in Table (8.2).

For the Dammam aquifer, natural isotopes were used to examine aquifer recharge, the hydraulic connection between the Dammam and other aquifers, determination of groundwater age, identification of recharge and discharge areas, and estimation of groundwater flow velocity. The isotopes used were oxygen-18 (^{18}O), deuterium (^{2}H), tritium (^{3}H), and carbon-14 (^{14}C). The results of these studies are summarized below:

Table 8.2. Chemical characteristics of groundwater of Tertiary aquifers in Kuwait (after Mukhopadhyay et al., 1996).

Aquifers and Location	Total Dissolved Solids (mg/l)	Major Cations	Major Anions	Sodium Adsorption Ratio	SO_4/Cl	Water Type
Dibdibba Aquifer (top part)	< 500	HCO_3> SO_4> Cl	Ca> Na> Mg	6-12	>1	$HCO_3 + SO_4$
Dibdibba Aquifer (lower part)	500 – 2,000	SO_4> Cl> HCO_3	Na> Ca> Mg	6-12	>1	SO_4 + Cl
Kuwait Group Aquifer (south-west and central Kuwait)	3,000 – 5,000	SO_4> Cl> HCO_3	Ca> Na> Mg	4-10	>1 to<1	SO_4 + Cl & Cl + SO_4
		Cl> SO_4> HCO_3	Na> Ca> Mg			
Kuwait Group Aquifer (southern Kuwait)	4,000-10,000	Cl> SO_4> HCO_3	Na> Ca> Mg	7-19	<1	Cl + SO_4
Kuwait Group Aquifer upper part (north and northeastern Kuwwait)	10,000-20,000	Cl> SO_4> HCO_3	Na> Ca> Mg	--	<0.5	Cl + SO_4
Kuwait Group Aquifer lower part (north and northeastern Kuwait)	≥ 100,000	Cl> SO_4> HCO_3	Na> Ca> Mg	--	<0.1	Cl + SO_4 & Cl
Dammam Aquifer (southwest and central Kuwait)	3,000 – 5,000	SO_4> Cl> HCO_3	Na> Ca> Mg	--	>1	SO_4 + Cl
Dammam Aquifer (southern and eastern Kuwait)	5,000 – 7,000	Cl> SO_4> HCO_3	Na> Ca> Mg	--	<1	Cl + SO_4
Dammam Aquifer (north and northeastern Kuwait)	20,000-100,000	Cl> SO_4> HCO_3	Na> Ca> Mg	--	<0.1	Cl + SO_4 & Cl
Radhuma Aquifer (southwest Kuwait)	4,000 – 5,000	SO_4> Cl> HCO_3	Ca> Na> Mg	--	>1	SO_4+Cl
Radhuma Aquifer (north and east Kuwait)	5,000- 150,000	Cl> SO_4> HCO_3	Na> Ca> Mg	--	--	Cl + SO_4 & Cl

The isotope investigation in Kuwait shows that the groundwater in Dammam Formation is old (22,000 years). It has not received recent recharge, and the groundwater in the aquifer was recharged during last pluvial periods. Groundwater in the aquifer is in direct hydraulic connection with groundwater in the Kuwait Group. The brackish water salinities at Al Sulibiyah and AshShigaya ranges from 2,500-10,000 mg/l. Brackish water at Wafra is of the SO_4^{2-} and Na^+-type becoming Cl^- and Mg^{2+} type and eventually Cl^- and Ca^{2+}-type in the northeast where salinities reach 150,000 mg/l (Robinson and Al Ruwaih, 1985).

Water Quality in the Kuwait Group Aquifers

The overall salinity of the Kuwait Group aquifer ranges from 4,000 mg/l in the southwest to 18,000 mg/l and higher in the northeast. The hydrochemical facies of groundwater are sodium-calcium and chloride-sulphate, and hypothetically water-dissolved salts are NaCl and $MgCl_2$ (Al-Ruwaih, 1994). Carbon-14 (^{14}C) age dating of groundwater in the Kuwait Group aquifers south of Kuwait indicated ages of 7,000 to 10,000 years. Groundwater in the Kuwait Group aquifers is supersaturated with respect to $CaCO_3$ and CaMg

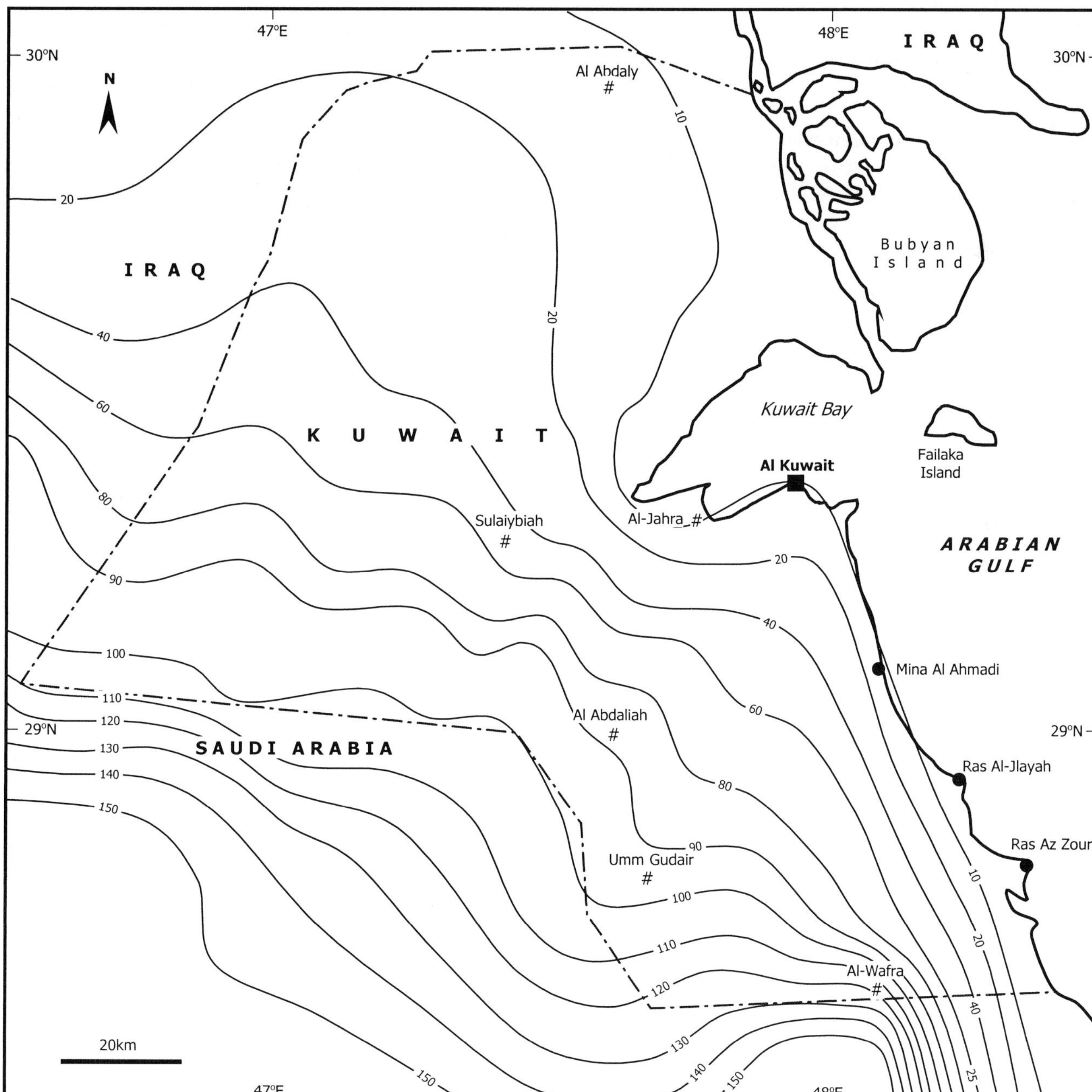

Fig. 8.7. Initial (pre-1960) potentiometric heads (m) above main sea level in the Dammam Formation aquifer in and around Kuwait, before the start of the large exploration of the aquifers (modified from Mukhopadhyay et al., 1996).

$(CO_3)_2$ (Mukhopadhyay et al., 1996). An isosalinity map of the Kuwait Group aquifer shows water-quality deteriorating in the northeast in a pattern similar to that of the Dammam limestone aquifer (Fig. 8.9). The groundwater salinity of Kuwait Group increases generally from southwest (brackish water) to northeast (salty and brine water), except in some areas in the northeast where fresh water is being produced from the upper part of Kuwait Group (Fig. 8.10). The chemical analyses carried out by Al Ruwaih (1984, 1985) have been plotted on a Piper (1944) diagram (Fig. 8.11), which shows that alkali earths exceed alkalis and strong acids exceed weak acids, suggesting the Dibdibba aquifers has been recharged and diluted by infrequent rainstorms and freshwater of meteoric origin. Groundwater of Kuwait Group is classified on the basis of its quality to the following:

a. *Fresh Water.* Fresh water exists in the upper part of Dibdibba Formation of Kuwait Group in Ar-Raudhatain and Umm Al-Aish basins north of Kuwait. Water salinity in the Al Raudhatain and Umm Al-Aish fields ranges from 400 and 2,000 mg/l, surrounded and underlain by brackish and salty water. The sequence of anion dominance of this water is mainly $SO_4^{2-} > Cl^- > HCO_3^-$ and sometimes $SO_4^{2-} > HCO_3^- > Cl^-$ and $HCO_3^- > SO_4^{2-} > Cl^-$, while the sequences of cation dominance is $Na^+ > Ca^{2+} > Mg^{2+} > K^+$ and $Ca^{2+} > Na^+ > Mg^{2+} > K^+$. The ratio SO_4^{2-}/Cl^- is more than 1.Water types according to Schoeller (1962) are sulphate + chloride, sulphate + bicarbonate, and bicarbonate + sulphate. The sodium adsorption ratio is in the range of 6 to 12, indicating that water will not cause a harmful effect when used for irrigation of traditional crops.

b. *Brackish Water.* South of Kuwait, where the Al-Wafra farms are located, the water salinity of Kuwait Group (55m in thickness) ranges from 4,000 to 9,000 mg/l, increasing generally to the north and northwest. The sequence of anion and cation dominance is $Cl^- > SO_4^{2-} > HCO_3^-$ and $Na^+ > Ca^{2+} > Mg^{2+} > K^+$, respectively. The ratio of SO_4^{2-}/Cl^- is less than 1. Water type is sulphate + chloride. Sodium adsorption ratio is in the range of 4 to 10. East of Kuwait, where some private hand dug wells are located near Abu Halifa-Al Managish coastal road, the water salinity of Kuwait Group (9-30m) ranges from 5,000 to 12,000 mg/l. The sequence of anion and cation dominance is $Cl^- > SO_4^{2-} > HCO_3^-$ and $Na^+ > Ca^{2+} > Mg^{2+} > K^+$, respectively. The ratio of SO_4^{2-}/Cl^- is less than 1. Water type is chloride + sulphate. Sodium adsorption ratio is in the range of 9 to 17. In the northwest of Kuwait, where wells no. NW-1 and NW-2 are located, the water salinity of the upper Kuwait Group (122-152m) ranges from 12,000 to 15,000 mg/l. The sequence of

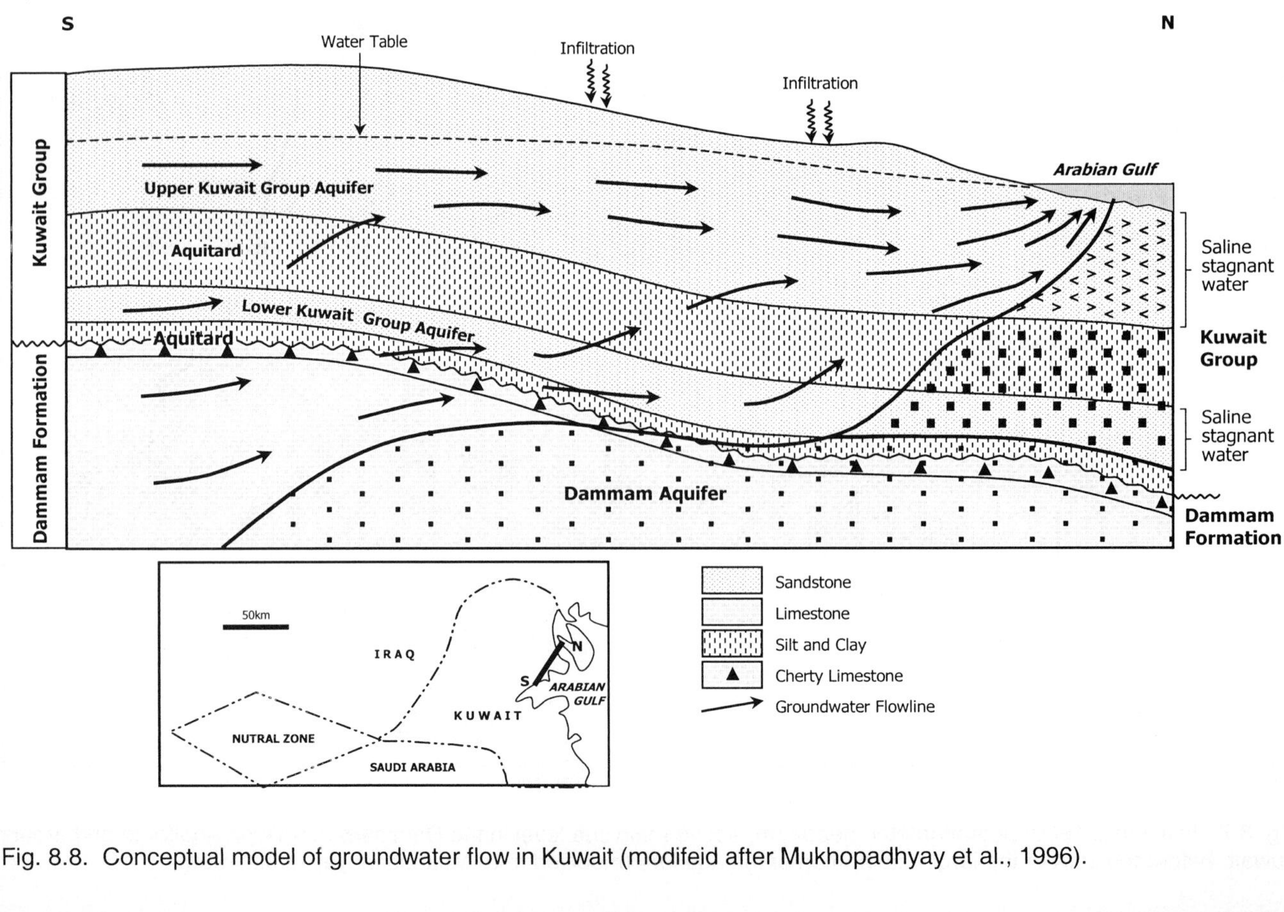

Fig. 8.8. Conceptual model of groundwater flow in Kuwait (modifeid after Mukhopadhyay et al., 1996).

anion and cation dominance is $Cl^- > SO_4^{2-} > HCO_3^-$ and $Na^+ > Ca^{2+} > Mg^{2+} > K^+$, respectively. The ratio of SO_4^{2-}/Cl^- is less than 1. Water type is chloride + sulphate. Sodium adsorption ratio is in the range of 11.2 to 20.7. On the basis of water chemical analysis of Parsons wells (N and M wells) which are located in the northeast of Kuwait, represent the upper part of Kuwait Group. Water salinity in the area ranges from 4,000 to 14,000 mg/l, increasing in the northeastern direction as well as with increasing depth. The water chemical types of these wells is given in the following:

i) In Al-Raudhatain field, the sequence of anion dominance is $SO_4^{2-} > Cl^- > HCO_3^-$. Ratio of SO_4^{2-}/ Cl^- is more than 1. The water type is sulphate + chloride. With increasing depth the sequence changes to $Cl^- > SO_4^{2-} > HCO_3^-$, while the ratio of SO_4^{2-}/Cl^- becomes less than 1, and water type is chloride + sulphate.

ii) In Al-Mutlah wells, the sequences of anion dominance is $Cl^- > SO_4^{2-} > HCO_3^-$, while the ratio of SO_4^{2-}/Cl^- is less than 1 and water type is chloride + sulphate. Unfortunately no records for cation concentration are available.

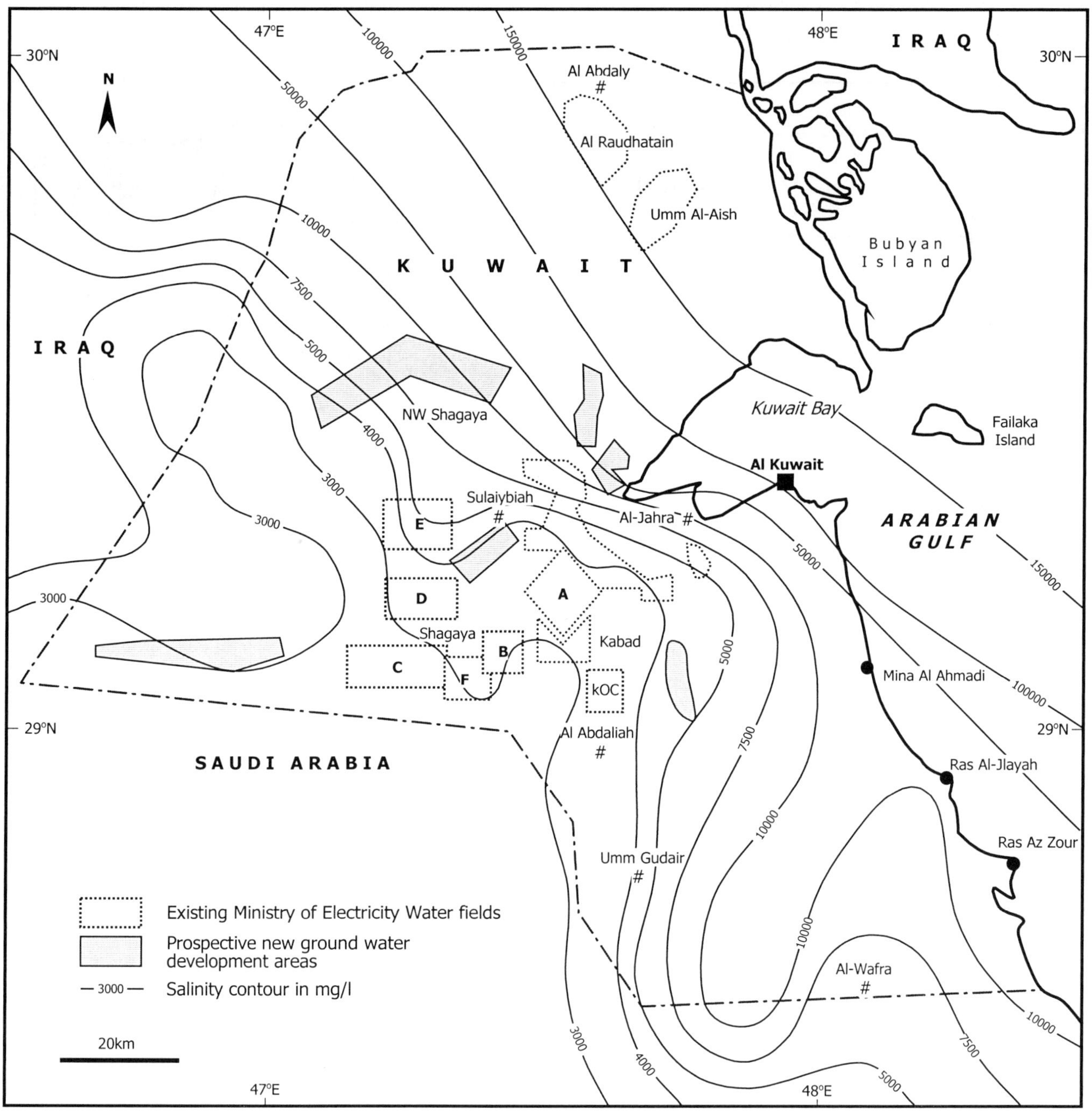

Fig. 8.9. Variation in the quality of the groundwater in the Dammam Formation and Kuwait Group aquifers and major water fields in Kuwait (modified from Mukhopadhyay et al., 1996).

c. *Salty and Brine Water*. Water salinity of Kuwait Group increases generally in northeast of Kuwait as well as with the increase of depth where it reaches more than 100,000 mg/l. The dominant sequences of anions and cations are $Cl^- > SO_4^{2-} > HCO_3^-$ and $Na^+ > Ca^{2+} > Mg^{2+} > K^+$, respectively. The ratio of SO_4^{2-}/Cl^- is very low (less than 0.1), where chloride concentration reaches more than 50,000 mg/l. The water types are chloride + sulphate and chloride.

Water Quality in the Dammam Aquifer

The Dammam Formation is the major aquifer which is being exploited in Kuwait. It underlies Kuwait Group and extends all over the country. Its water varies in salinity from brackish (2,500 mg/l) in the southwest of Kuwait to brine (150,000 mg/l) in the northeast (Fig. 8.9). The hypothetically water-dissolved salts include $CaSO_4$, $CaCO_3$ and NaCl. The aquifer is supersaturated with respect to $CaCO_3$ and is undersaturated with respect to $CaSO_4$. Plumer et al. (1991) attributed these conditions to dolomitization. Computation with the help of the WATEQ4F software (Ball and Nordstrom, 1992) suggests dissolution of anhydrite and precipitation of calcite in the Dammam aquifer.

Local anomalies in total dissolved solids content are possible due to variable karst development and infiltration rates (Burdon and Al-Sharhan, 1968). On the basis of water quality, the groundwater of Dammam aquifer can be classified into the following:

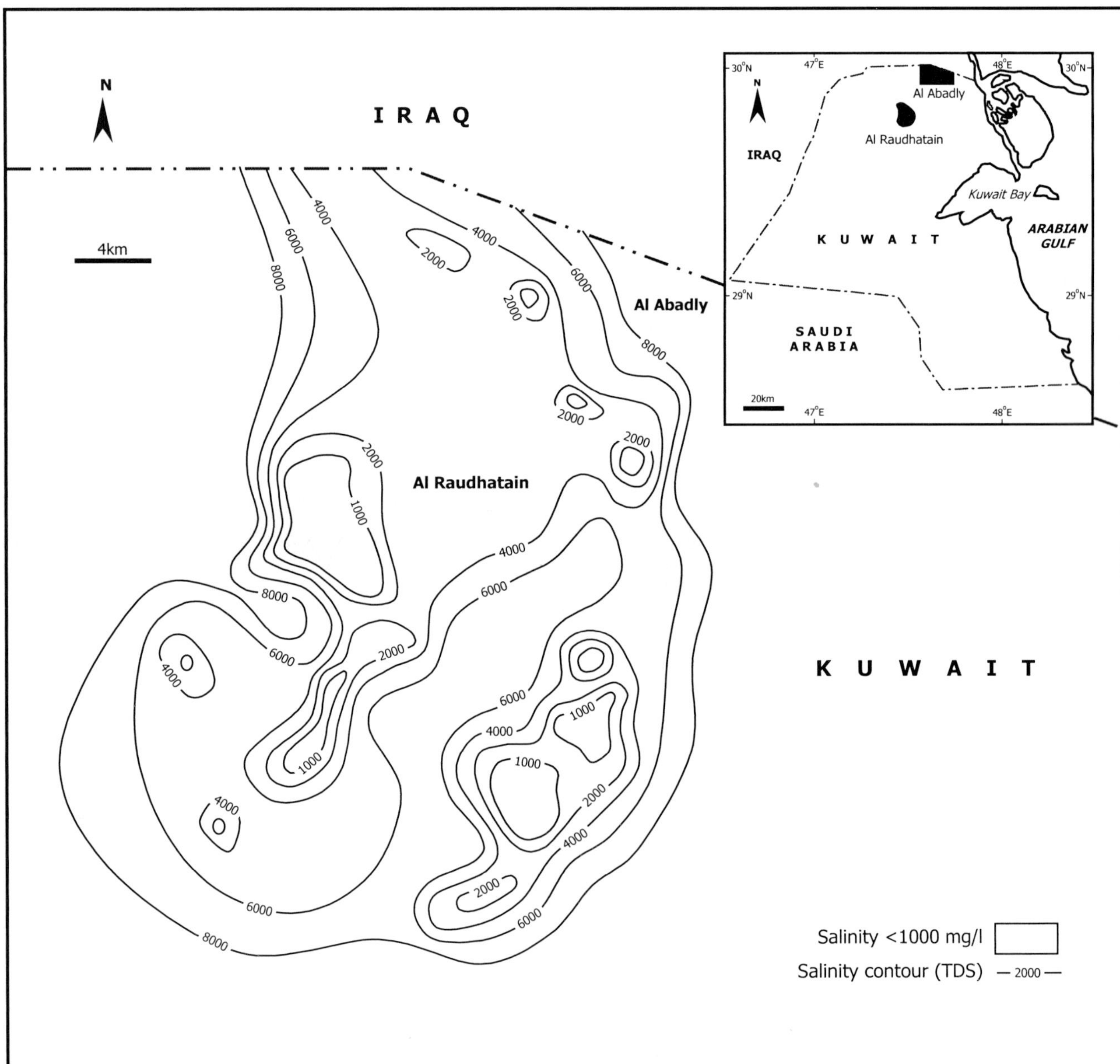

Fig. 8.10. Contour map of salinity (total dissolved solids in mg/l) of the groundwater in Al Raudhatain and Al Abdaly areas, Kuwait (modified from Al Sulaimi and Akber, 1999).

Brackish Water. This water covers a large area of Kuwait and includes most groundwater well fields in Kuwait (i.e., As-Sulaibiyah, Al-Wafra and Al-Abdliyah fields). Within this area groundwater salinity varies between 2,500 and 10,000 mg/l. The sequence of anion dominance is SO_4^{2-}> Cl^-> HCO_3^-, but changes to Cl^-> SO_4^{2-}> HCO_3^- to the east and northeastward where water salinity is generally more than 6,000 mg/l (Fig. 8.12). The sequence of cation dominance is Na^+> Ca^{2+}> Mg^{2+}> K^+. Ratio of SO_4^{2-}/Cl^- is more than 1 and decreases gradually towards east and northeast directions where the ratio is less than 1. Water type is sulphate + chloride and changes to chloride + sulphate to the east and northeast. In southern Kuwait, where Al-Wafra wells (Ministry of Electricity and Water wells) are located, water salinity of Dammam aquifer ranges from 5,000 to 7,000 mg/l. Water salinity increases slightly with increasing depth. The sequence of anion and cation dominance is Cl^-> SO_4^{2-}> HCO_3^- and Na^+> Ca^{2+}> Mg^{2+}> K^+, respectively. The ratio of SO_4^{2-}/Cl^- is less than 1. Water type is chloride + sulphate. Sodium adsorption ratio is in the range of 8 to 17.

Salty Water. This water ranges in salinity from 10,000 to 50,000 mg/l and it bounds the brackish water from the north, northeast and east. The water occurs also in southern Kuwait (west and northwest of Al-Wafra wells) where the water salinity reaches to more than 20,000 mg/l. The sequence of anion and

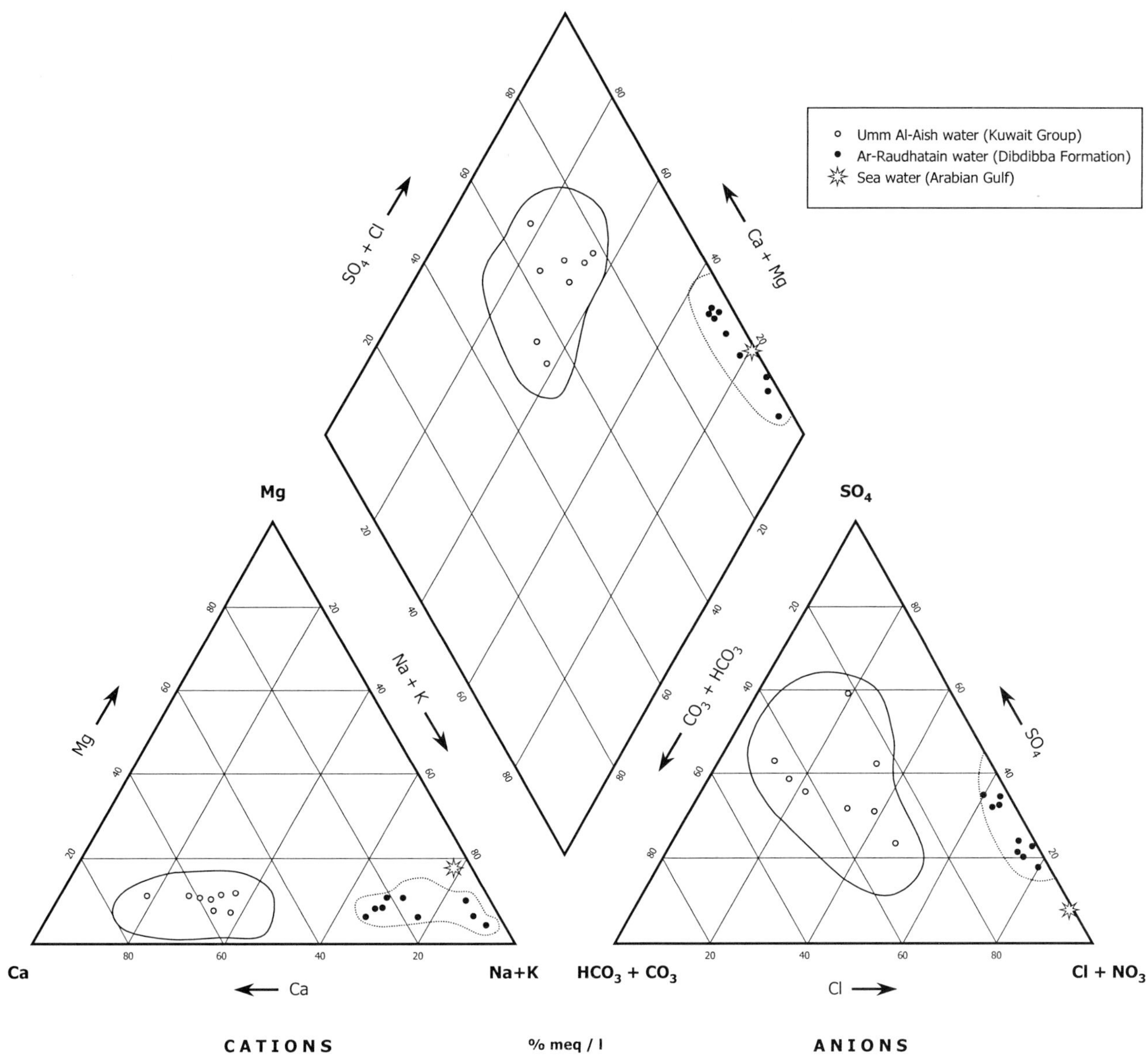

Fig. 8.11. Groundwater analysis of Kuwait Group from Umm Al-Aish and Al Raudhatain field wells, Kuwait (compiled from Al Ruwaih, 1984, 1985).

cation dominance is $Cl^- > SO_4^{2-} > HCO_3^-$ and $Na^+ > Ca^{2+} > Mg^{2+} > K^+$, respectively. The ratio of SO_4^{2-}/Cl^- is less than 1, and the water type is chloride + sulphate. No records for cations are available.

Brine Water. This water ranges in salinity from 50,000 to more than 150,000 mg/l, it extends to the northeast of salty water. The sequence of anion and cation dominance is $Cl^- > SO_4^{2-} > HCO_3^-$ and $Na^+ > Ca^{2+} > Mg^{2+} > K^+$, respectively. The ratio of SO_4^{2-}/Cl^- is less than 0.1, and the water type is chloride + sulphate (Fig. 8.13).

Water Quality in the Radhuma Aquifer

The water salinity of Radhuma Formation increases generally to the east and northeast. On the basis of water chemical analysis of Radhuma wells, in the west and southwestern part of Kuwait, the following are considered:

Water salinity of Radhuma Formation in the southwestern Kuwait (depth drilled ranges from 510 to 795m), varies between 4,000 and 5,000 mg/l. Water salinity is slightly higher than the water of the overlying aquifers. The sequence of anion and cation dominance is $SO_4^{2-} > Cl^- > HCO_3^-$ and $Ca^{2+} > Na^+ > Mg^{2+} > K^+$, respectively. The ratio of SO_4^{2-}/Cl^- is more than 1. The water type is sulphate + chloride. Sodium adsorption ratios in the range of 2.6 and 5.1. Concentration of sulphate and calcium ions of Radhuma water, is higher than in the water of Dammam Formation and Kuwait Group, while chloride content is less. Dissolved H_2S was indicated in all wells of this formation. The water in the Radhuma Formation becomes salty in the east and northeast of Kuwait, where the water salinity of wells southeast of Kuwait Bay, is more than 35,000 mg/l, and the sequence of anion and cation dominance is $Cl^- > SO_4^{2-} > HCO_3^-$ and $Na^+ > Ca^{2+} > Mg^{2+} > K^+$, respectively. The ratio of SO_4^{2-}/Cl^- is less than 0.1. The water is dominated by sulphate + chloride.

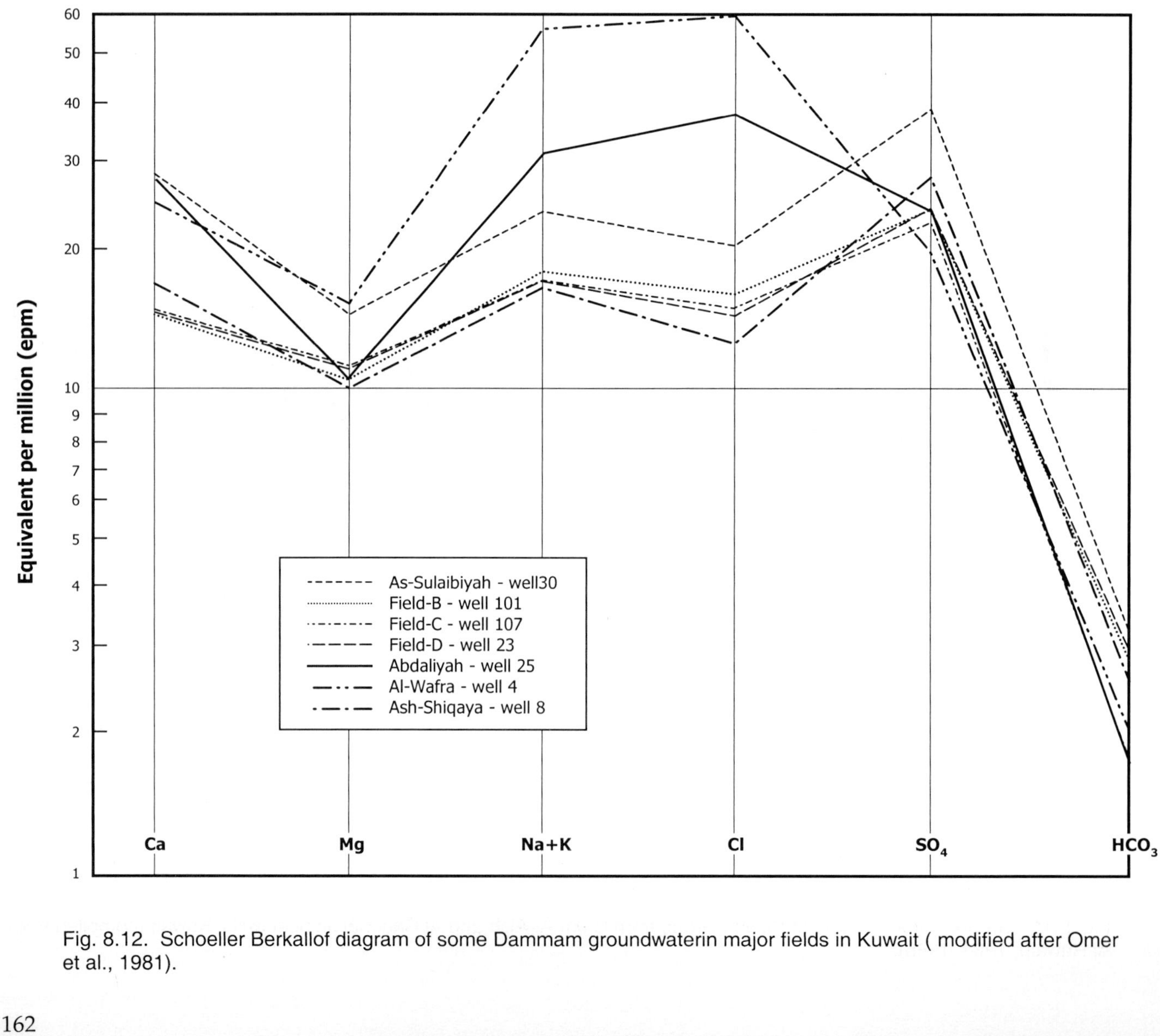

Fig. 8.12. Schoeller Berkallof diagram of some Dammam groundwaterin major fields in Kuwait (modified after Omer et al., 1981).

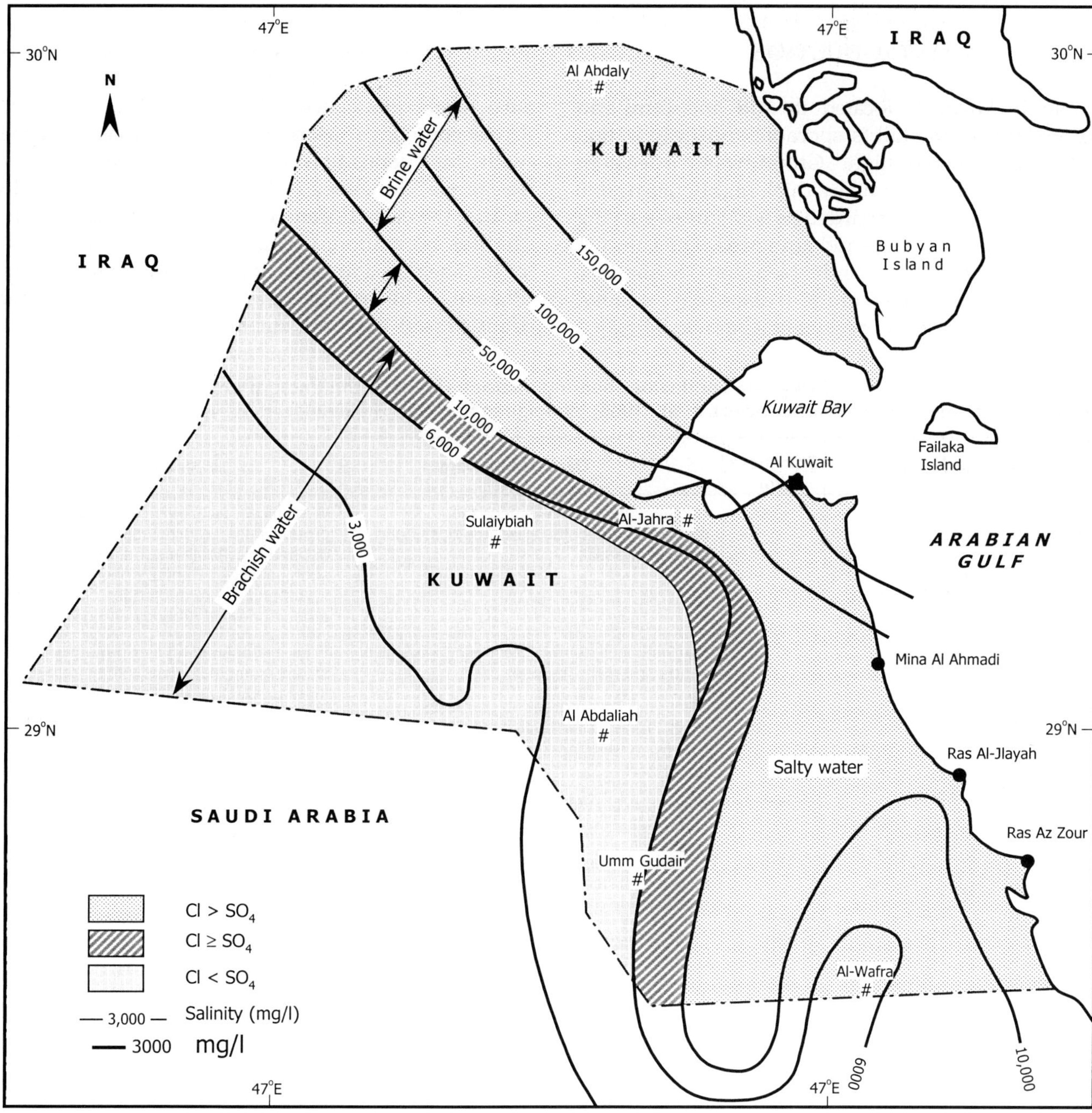

Fig. 8.13. Relative abundance of SO_4^- and Cl^- and iso-salinity (mg/l) in Dammam groundwater in Kuwait (modified from Omer et al., 1981).

2. CENOZOIC AQUIFER SYSTEM IN SAUDI ARABIA

INTRODUCTION

Saudi Arabia is located in arid and semi-arid regions, where rainfall is sporadic and evaporation losses are extremely high. Groundwater in eastern Saudi Arabia found in many thick highly permeable Tertiary sediments is regarded as a plexus of varying compositions of aquifers and aquitards. The Umm er Radhuma, Dammam and Neogene formations contain groundwater of a reasonable quality, transmission, storage capacity and characteristics.

In central and eastern Saudi Arabia the early Tertiary was marked by a continuation of the structural quiescence which prevailed since late Cretaceous time. Marine shelf conditions continued with the deposition of the Paleocene Umm er Radhuma limestone. Arid conditions prevailed during the early Eocene when a thick evaporitic unit, the Rus Formation, was deposited over virtually all of the shelf area to the east and north of the Summan Plateau. In middle Eocene time the sea again transgressed over the stable platform area and the

Table 8.3. Tertiary geological sequence and water-bearing characteristics in the Eastern province of Saudi Arabia (compiled with modification from Powers et al., 1966; Yazicigil et al., 1986; Bakiewicz et al., 1982).

Age	Formation	Member	Thickness (m)	General Lithology	Hydrogeology
Quaternary			± 30	Aeolian sands, wadi-fill deposits, sheetwash deposits, alluvial deposits and sabkha deposits	Wadi-fill deposits may contain localized groundwater, but availability is seasonally dependent. Aeolian sand dune belts, such as the Ad Dahna, pond-up surface runoff and induce recharge. Sabkhas are areas of natural groundwater discharge
Neogene	Hofuf		10-30	Marl with limestone interactions of fluviatile sands and marls in upper parts	Poor, unconfined aquifer-generally but locally along major wadis may form a more productive aquifer
	Dam		60-110	Hard, compact chalky to marly limestone. Extensive fissuring and karstification in the upper part.	Excellent aquifer
	Hadrukh		25-90	Clean sands at the base followed by marly sands, siltstone and sandy limestone	Excellent aquifer
Paleogene	Dammam	Alat limestone	15-50	Limestone often fissured with cavities infilled with Neogene sands common. Chert bands in top part common.	Moderate aquifer
		Alat marl	10-20	Light reddish brown colorations	Aquitard where present
		Khobar limestone	20-45	Calcarenitic and dolomitic limestone, locally fissured	Aquifer
		Khobar marl	5-15	Mainly marl, with subordinate shales and thin limestone layers	Aquitard where present
		Alveolina limestone	± 15	Thin limestone interbedded with marls or shales	Complete section forms an aquitard. Effectiveness as aquitard reduces over Ghawar anticline
		Saila-Midra	5-10	Dark-grey shale	Aquitard where present
	Rus		20-200	Two main facies exist: Anhydritic facies consist of relatively thick layers of anhydrite with subordinate gypsum intercalated with relatively thin layers of marl and limestone. Non-anhydritic faceis consist of limestone, in places dolomitic and marl; locally fissured	Anhydritic facies constitute an aquiclude. Non-anhydritic facies constitute an aquifer in hydraulic continuity with the Umm er Radhuma Formation
	Umm er Radhuma		300-600	Monotonous limestone and dolomite in varying proportions with anhydrite facies. Dolomitic limestone locally karstified but subsequently infilled with argillaceous sediments. Calcarenitic limestones, frequently fissured, and this grades downwards into dolomitic faceis and more argillaceous limestone with shales/marls at the base	Calcarenite facies constitute an excellent aquifer, particularly if fissuring is well developed. Fine-grained and anhydritic facies constitute a very poor aquifer. Basal shales form aquitard between Umm er Radhuma and Aruma. Dolomitic zones are only moderate aquifer if fissured

limestone sequence of the Dammam Formation was deposited. The Oligocene has not been found here, when the region had very stable relief. The Miocene and Pliocene sediments of the Hadrukh, Dam and Hofuf formations are erratic and their lithologies include sandy limestone, lacustrine limestone, marl and sandstone (Table 8.3).

Hydrogeology and Groundwater Occurrence

The Tertiary of eastern Saudi Arabia contains good aquifers but there are wide variations in their geological setting, hydrogeological conditions, thicknesses, hydraulic parameters and water chemistry (Tables 8.3 and 8.4). The main aquifers of the sedimentary provinces of Saudi Arabia can be classified, by origin, into two broad groups, namely aquifers of primary and secondary origin. Aquifers of primary origin include the Quaternary sands of the wadi systems which are quartzose sandstones, and conglomerates with primary porosity; and calcarenites, coquinites and oolitic limestones with primary porosity. Quaternary sand aquifers are found in Wadi ar Rimah and Wadi Al Batin drainage systems, where shallow supplies of poor quality water (specific conductivity 2,000 to 5,000 μS/cm) are used locally for irrigation. Quartzose sandstones of Hadrukh Formation all have high primary intergranular porosities and form the most important aquifer.

Aquifers of secondary origin consist primarily of limestones, which have undergone secondary solution or dolomitization, and karstified limestones found in the Umm er Radhuma, Dammam and Dam formations. Calcarenites of the Dammam and Umm er Radhuma formations also form extensive and important aquifers, with much of the primary intergranular porosity still preserved. Coquinite horizons, composed of larger foraminiferids, constitute aquiferous zones of high primary porosity in the top third of the Umm er Radhuma Formation.

Secondary solution of the limestones and evaporites is a major cause of permeability of the Khobar and Alat members of the Dammam Formation and the Rus Formation, as well as in the Umm er Radhuma Formation, where lost circulation cavities are commonly encountered in drilling. Karstified and fissured limestones act as aquifers, (Fig. 8.14) as in the Dam Formation and the Alat Member. The northern outcrop area of the Alat limestone is extensively karstified, with many sinkholes and enlarged fissures and joints which trap local surface runoff. The Dam Formation is an aquifer with good permeability in Al Hasa area.

The very low rainfall conditions that prevail for most of the Saudi Arabia do not allow substantial recharge of most aquifers in their exposed and unconfined parts. This is borne out by isotopic evidence showing that most aquifers contain fossil groundwater, which is tens of thousands of years old, and was evidently recharged during previous pluvial intervals during the Quaternary. The fragility of most aquifers in Saudi Arabia cannot be overemphasized, and their rapid exploitation has led in some places to dramatic falls in the groundwater table.

Mean annual recharge from rainfall, for the Paleogene Umm er Radhuma Formation has been calculated at 1,048 Mm^3 (Bakiewicz et al., 1982), but is probably supplemented by considerable upward flow, from the Aruma Aquifer, and downward flow from the Dammam and Neogene, which plus lateral flow, makes a total recharge of 2,256 Mm^3. Recharge by rainfall on the Middle Eocene Dammam aquifers, is also low and much of their groundwater comes from the underlying Umm er Radhuma aquifer,

Table 8.4. Tertiary aquifer characteristics in Al Hasa region, Eastern Province of Saudi Arabia (compiled with modification from Edgell, 1990; 1997; Dabbagh and Abderrahman, 1997).

Aquifer	Age	Lithology	Thickness (m)	Porosity (%)	Transmissivity (m^2/sec)	Storativity	Water Quality	Recharge (Mm^3)	Salt Content (mg/l)	Reserve (Mm^3)
Hadrukh	Lower Miocene	Sandstone with marl	20-120	>3	$7x10^{-4}$ to $4x10^{-2}$	$1x10^{-2}$ unconfined $2x10^{-4}$ confined	Good	360	2,400-4,000	130,000
Dam	Middle Miocene	Karstified limestone	30-100	3-10	$1x10^{-2}$ to $70x10^{-3}$	-	Good			
Alat	Middle Eocene	Karstified limestone	10-50	>4	$2.6x10^{-5}$ to $5.1x10^{-3}$	$1.3x10^{-4}$ to $2.6X10^{-5}$	Moderate	200	2,600-6,000	45,000
Khobar	Middle Eocene	Karstified limestone and dolomite	80	>3	$1x10^{-2}$ to $3x10^{-6}$	$1x10^{-3}$ to $1x10^{-4}$ confined	Good			
Umm er Radhuma	Paleocene - Lower Eocene	Limestone and Dolomite	300-700	>5	$1x10^{-2}$ to $1x10^{-3}$	$5x10^{-5}$ to $5x10^{-3}$ confined	Good	406	2,000-5,000	190,000

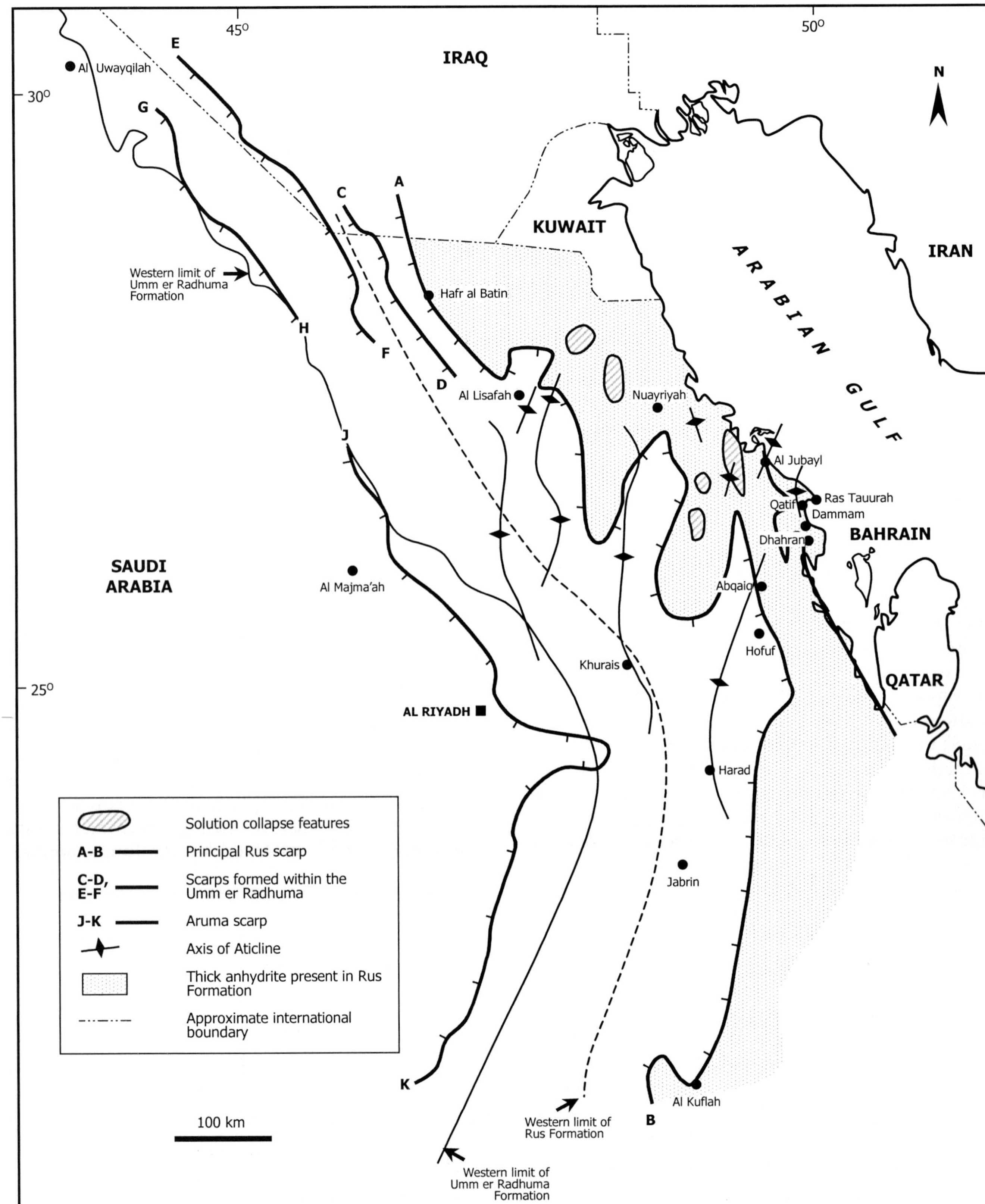

Fig.8.14. Evaporite solution collapse, principal scarps and major structural elements in eastern Saudi Arabia (modified from Bakiewicz et al., 1982).

where the Rus Formation is thin and act as a leaky aquitard. Most of the water in the Dammam and Umm er Radhuma aquifers infiltrated into these aquifers, between 30,000 and 50,000 years ago, during pluvial Quaternary climatic conditions. The Hadrukh and Dammam aquifers have a low annual recharge by rainfall, but are supplemented by water, through an erosional window in the Rus Formation, in south Ghawar anticline.

Infiltration and recharge through sand dunes has been studied by Dincer et al. (1974) and Dincer (1978) as a mechanism for aquifer recharge, and movement of water through coarse-grained sand dunes, which has been traced by tritium measurements.

Table 8.5. Estimates of annual recharge by rainfall for the Tertiary aquifers in eastern Saudi Arabia (Mm^3/year) (after Bakiewicz et al., 1982).

Aquifer / Year	Umm er Radhuma	Dammam	Neogene	Total (Mm^3)
1952	220	0	0	220
1953	3,665	82	740	4,487
1954	456	0	0	456
1955	4,079	134	1,210	5,423
1956	86	0	0	86
1957	2,133	28	254	2,415
1958	649	0	0	649
1959	1,128	13	122	1,263
1960	100	0	0	100
1961	792	0	0	792
1962	218	0	0	218
1963	387	0	0	387
1964	1,553	40	356	1,949
1965	169	0	0	169
1966	180	0	0	180
1967	264	0	0	264
1968	712	4	39	755
1969	3,131	100	905	4,136
1970	8	0	0	8
1971	1,173	8	74	1,255
1972	1,047	19	170	1,236
1973	800	0	0	800
1974	836	16	141	993
1975	1,095	6	54	1,155
1976	2,884	149	1,346	4,379
1977	477	3	31	511
1978	64	0	0	64

However, the movement is very slow, and rarely complete, and the little water seeping through dunes often encounters less permeable substrata, and does not necessarily contribute to major aquifer recharge. The use of tritium has been demonstrated by Hötzl et al. (1980), who showed that at Al Qatif and Al Hasa, the water is almost tritium free due to their age, while Wadi Hanifah wells are an exception and show relatively high tritium concentrations, proving recent (<50 years B.P.) local rainfall recharge to groundwater.

Umm er Radhuma Aquifer

The Umm er Radhuma Formation crops out as an arc parallel to the middle part of the Arabian Shield adjacent to the Al Dahna Sand Sea. The aquifer is composed of coralline and fine-grained limestone. Karst features were recognized in outcrops at different levels (Fig. 8.14). In subsurface these features can influence the hydraulic properties of the aquifer, and represent conduits for water to flow from one aquifer system to another. The aquifer dips from west towards east, its depth increases in the same direction. At Al Hofuf, the aquifer thickness is 500-700m, but decreases toward Hafr Al Batin, where its thickness is 240m, and to 110m in Al Sahba'a.

Paleo-rivers at Wadi Al Batin, Wadi Al Sahba'a, Wadi Al Dawasir and Recent sabkhas and dune sands are the most important geomorphic features in eastern Saudi Arabia. Sabkha bodies are closed evaporation pans for shallow groundwater, and for deep groundwater, migrating upward through karst pathways in overlying confining layers. The presence of these pathways is indicated by loss of circulation when drilling. The sand dunes (e.g., Great Nefud) swing in an arc along the western front of Umm er Radhuma outcrops. The dunes act as a dam retaining floodwater of major wadis and enhance groundwater recharge through high infiltration rates.

Hydrogeology

The Umm er Radhuma is the only aquifer in which groundwater flows under both lateral and vertical gradients. Despite its high porosity (≥30%), the hydraulic conductivity of the aquifer is low because it is fine-grained. Locally, however, the aquifer's transmissivity is high, due to the presence of large secondary cavities and leaching, of the argillaceous and anhydritic limestone facies, especially in the dolomitized parts of the aquifer. The average hydraulic conductivity of the Umm er Radhuma aquifer is 0.32 m/day and 32 m/day, in the unfissured and fissured portions of the aquifer, respectively. The piezometric contours of the Umm er Radhuma aquifer follow the outcrop trend in the west and the coast of the Arabian Gulf in the east, with an easterly regional hydraulic gradient. Local variation in magnitude and trend of this gradient, exist as a result of changes in aquifer's recharge, discharge, hydraulic conductivity and transmissivity. Groundwater recharge for the Tertiary aquifers shown on (Table 8.5), is based on many factors such as rainfall amount, intensity and duration, evaporation and transpiration, infiltration rates, soil capacity and runoff.

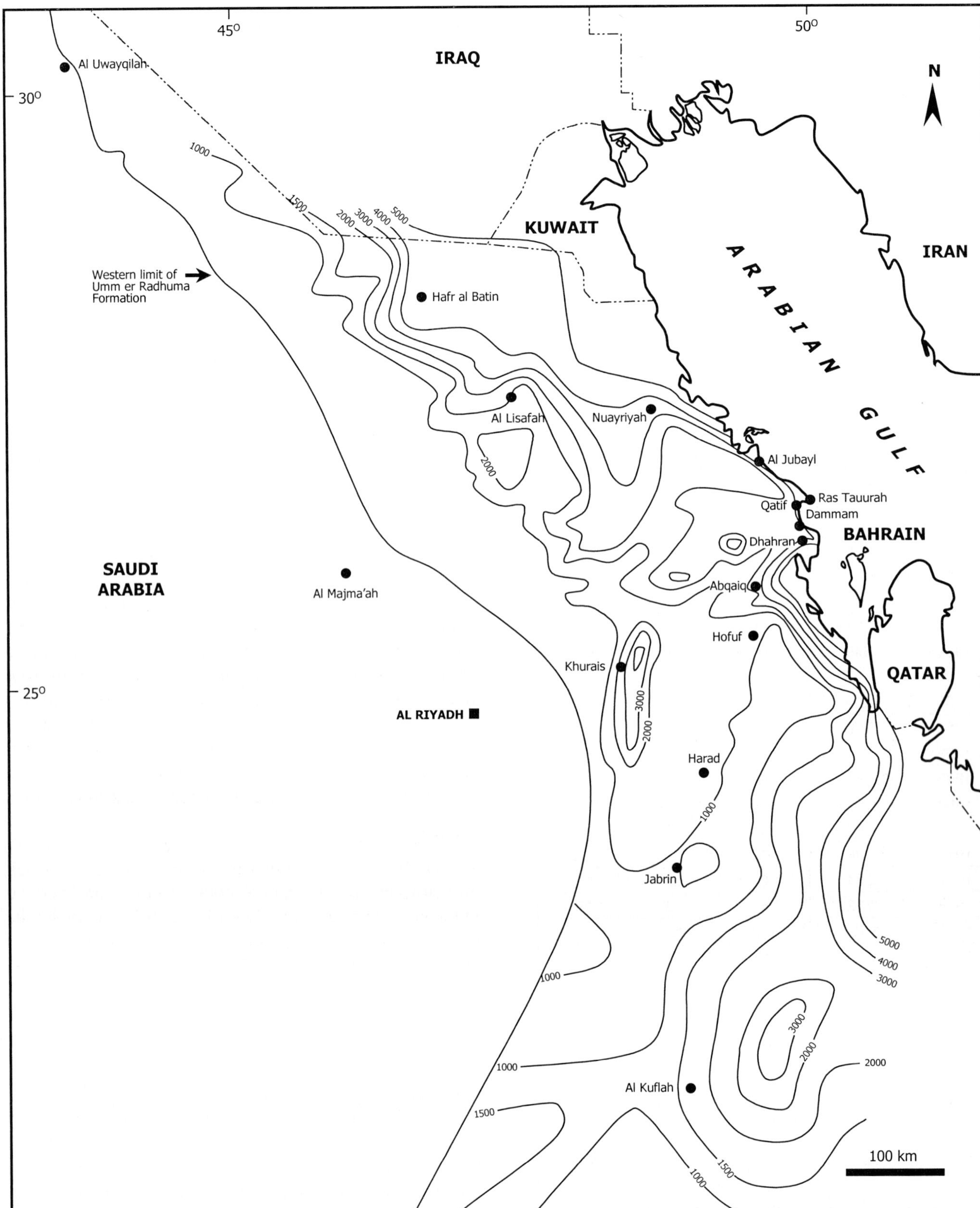

Fig. 8.15. Water quality showing total dissolved solids (in mg/l) in the Umm er Radhuma aquifer in eastern Saudi Arabia (modified from Bakiewicz et al., 1982).

Natural discharge from the Umm er Radhuma aquifer occurs from artificial groundwater abstraction, evaporation and transpiration from shallow water table, and spring discharges. These discharges occur at Al Hasa Oasis, Al Qatif coastal strip and at the northern part of Bahrain. The estimated quantities of all the natural discharges are summarized in Table 8.6.

The estimated values of recharge (1,272 Mm^3/year), and discharge (1,311 Mm^3/year) from the Umm er Radhuma aquifer in eastern Saudi Arabia seem to balance, before the present excessive artificial groundwater exploitation from the aquifer. This balance ignores the amounts of water exchanged between the Umm er Radhuma aquifer and the underlying Aruma aquifer, and the overlying Dammam and Neogene aquifers.

Table 8.6. Estimates of annual discharges from the Umm er Radhuma aquifer in eastern Saudi Arabia (Mm^3/year) (after Bakiewicz et al., 1982).

Discharge Mechanism	Discharge amount (Mm^3/year).
Sabkha discharge	855
Transpiration from water table	158
Land spring discharge	285
Offshore spring discharge	13
Total	1,311

As water moves from recharge area towards the discharge area, through the Umm er Radhuma aquifer, the salinity increases and the water-dissolved chemical species change, from calcium biocarbonate through calcium sulphate to sodium chloride. The anomalously low salinity distribution reflects preferential paths of groundwater flow. In contrast, areas of anomalously high salinity represent regions of high hydraulic resistance and very low groundwater movement. High tritium (^3H) content at or near the Umm er Radhuma outcrops indicate recent recharge. The ^{14}C age of groundwater samples generally increases from west to east, in the direction of groundwater flow toward the Arabian Gulf.

Water Quality

The water quality in the Umm er Radhuma aquifer varies widely with a variation in total dissolved solids from 600 to 900 mg/l. In Al Harad area, the aquifer salinity ranges from 600-1,300 mg/l, while in Al Qatif the salinity ranges from 1,300-2,200 mg/l. The wide variation in aquifer salinity is attributed to the dissolution of easily soluble thick evaporite of the Rus Formation, which overlies the Umm er Radhuma. Leaching of salts existing in the aquifer by groundwater movement from west to east causes a gradual salinity increase towards the Arabian Gulf. Distribution of total dissolved solids concentration of the groundwater (Fig. 8.15) shows pattern of anomalously low salinity due to preferential paths of groundwater flow, while high salinity are due to high hydraulic resistance and very slow groundwater flow (Bakiewicz et al., 1982). Tritium and ^{14}C (Fig. 8.16) shows that groundwater containing significant tritium occurs below the unsaturated Dammam and Neogene Formations and at or near Umm er Radhuma outcrops, which proves conclusively recent recharge. The ^{14}C age of groundwater generally increases from west to east in the general direction of natural flow (Bakiewicz et al., 1982).

Hydrogeologic Properties

The piezometric contours (Fig. 8.17) generally follow the trend of the outcrop of Umm er Radhuma in central Arabia toward the east, showing an easterly general hydraulic gradient. The formation is characterized by high porosity and permeability, which increases aquifer storage, however, water quality is highly dependent on the nature of aquifer facies and lateral and vertical changes in their mineralogical and chemical composition. The hydraulic properties of the Umm er Radhuma are affected by several processes such as: The dolomitization of limestone, which leads to the replacement of Ca^{2+} with Mg^{2+}, and formation of dolomite crystals. This process increases porosity and improves the aquifer properties of the formation.

The fissures, fractures and joints which affect several areas in the Umm er Radhuma Formation, also increase porosity, permeability and aquifer's ability, to store and transmit large amounts of water. The karst phenomena resulting from partial dissolution of limestone, also increase the porosity, permeability and storativity of the aquifer. Karstification is mainly found in northern Hafr Al Batin, the Ghawar anticline, and in the Rub Al Khali. In outcrop areas karstification can lead rainwater to move directly into the formation causing aquifer recharge, and contributing to its storage.

Dammam Aquifer

The Eocene Dammam aquifer is generally composed of limestone and dolomitic limestone with shale intercalations near its base. The area of the aquifer is about 20,000 km^2 (Fig. 8.18). The Dammam aquifer dips from west to east, extending under the Neogene and Quaternary sediments. The Dammam Formation is usually subdivided into five members which are from base to top: the Midra, Saila, Alveolina, Khobar and Alat Members (Table 8.3).

Hydrogeologically, the Dammam Formation can be subdivided into three units from base to top are: The lower unit composed of shale and shaly limestone. It includes the Midra, Sila and Alveolina

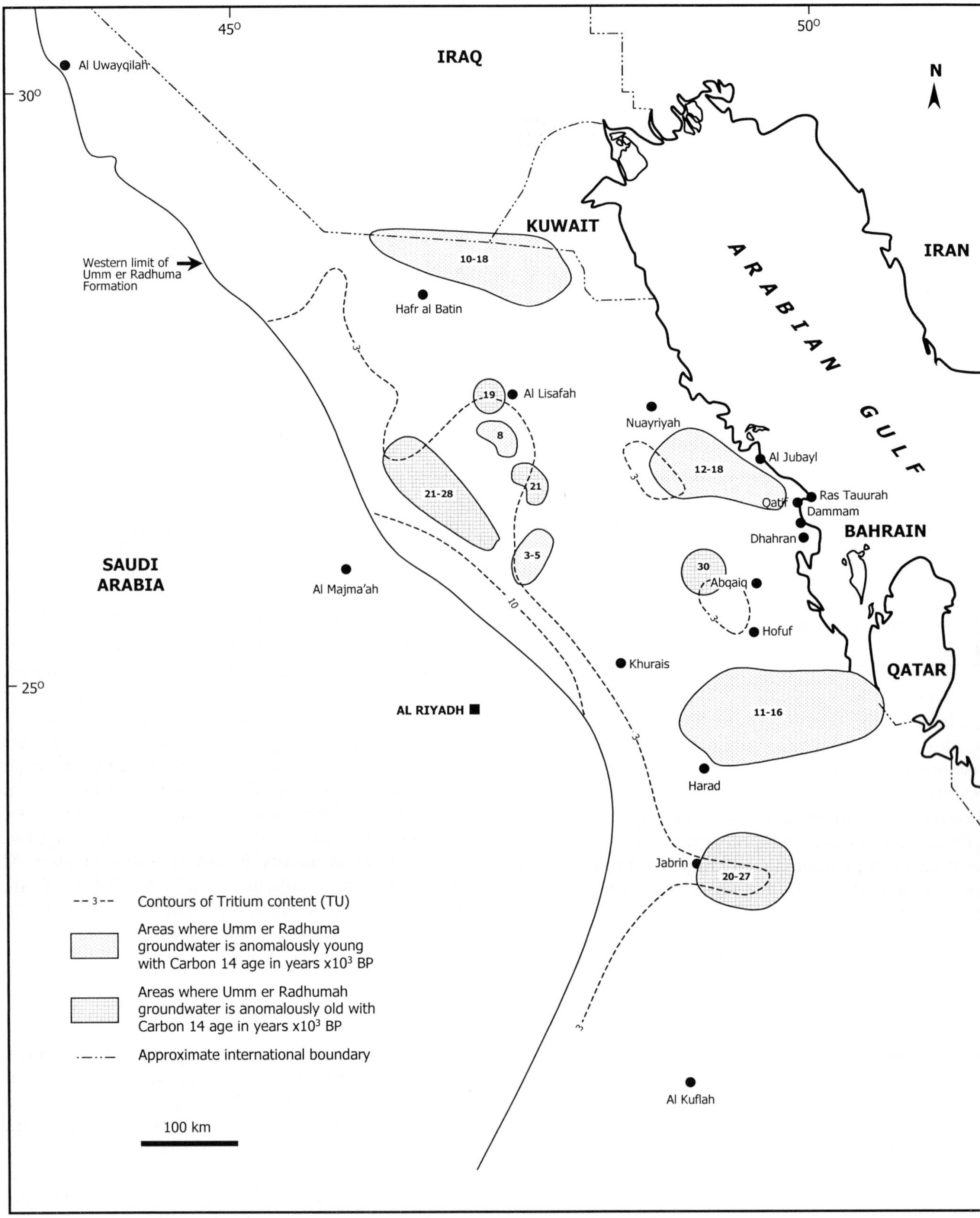

Fig. 8.16. Stable isotopes in the Umm er Radhuma groundwater aquifer in eastern Saudi Arabia (modified from Bakiewicz et al., 1982).

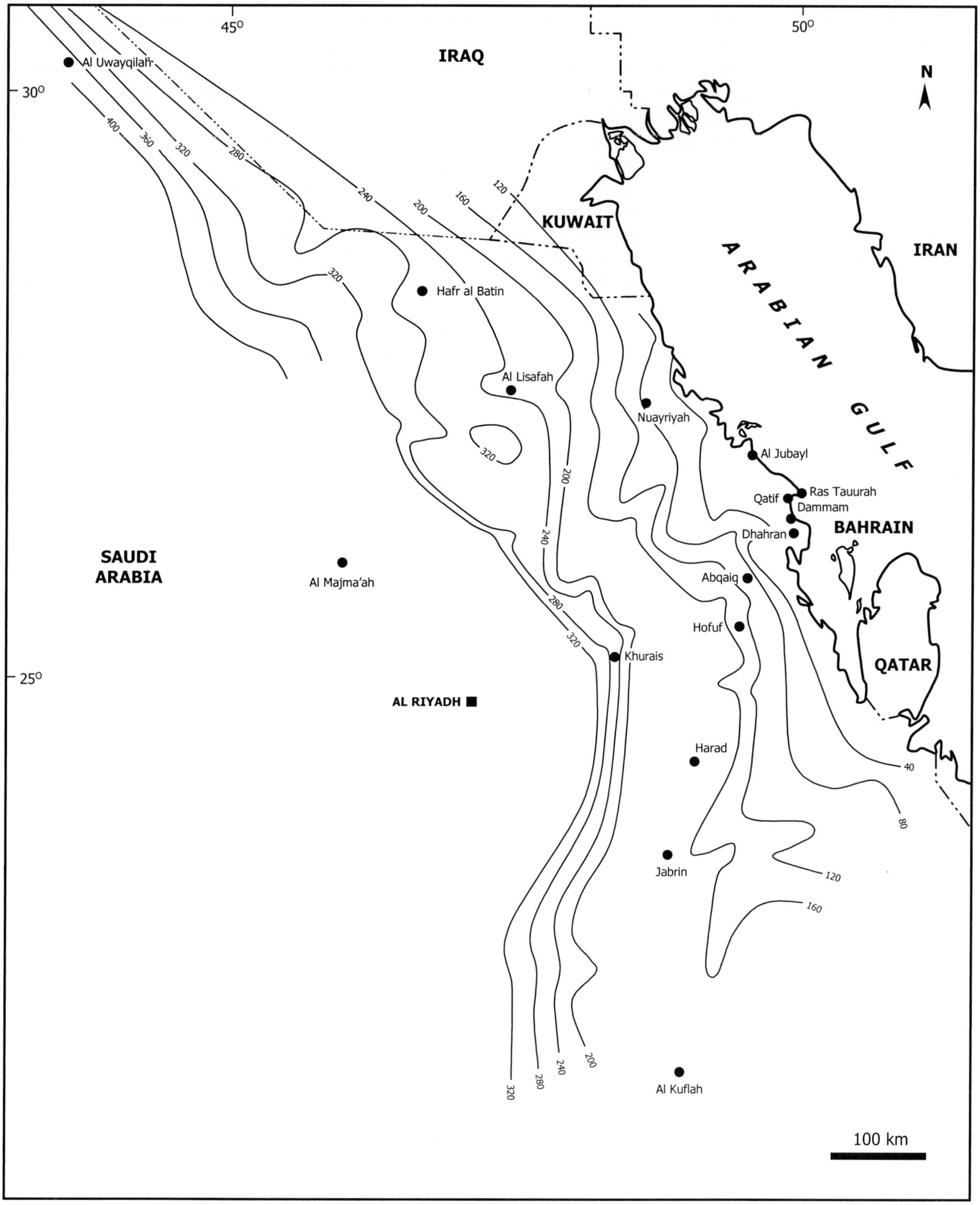

Fig. 8.17. Piezometric contours (in meters) in the Umm er Radhuma aquifer in eastern Saudi Arabia (modified from Bakiewicz et al., 1982).

members and represents the base of the Dammam aquifer.

The middle unit represents the lower part of the Dammam aquifer (Khobar Member). The thickness of the Khobar unit decreases in anticlines and increases in synclinal structures. This unit may be absent in some areas. The Khobar Member overlies the lower unit from which it is separated by the thin layer of limestone and marl of low-porosity of the Alveolina Member.

The upper unit, the Alat Member forms the upper part of the Dammam aquifer. It overlies the Khobar Member and is separated from it in many areas by marl. The upper surface of the Alat is erosional and affected by karst features in places.

Hydraulic Properties

The Dammam aquifer is recharged from rains falling directly on the outcrops or through the overlying Neogene and Quaternary sediments. The dominance of fissures, fractures and joints, and the absence of confining layers enhances aquifer recharge.

The Khobar Member is most utilized part of the Dammam aquifer (Tables 8.3 and 8.4). A marl layer separating the Khobar and Alat members plays an important hydraulic role in the Dammam aquifer, where it is present the Khobar and Alat are treated as separate aquifer units. In areas where the marl layer is absent, the Khobar and Alat members are treated as one aquifer.

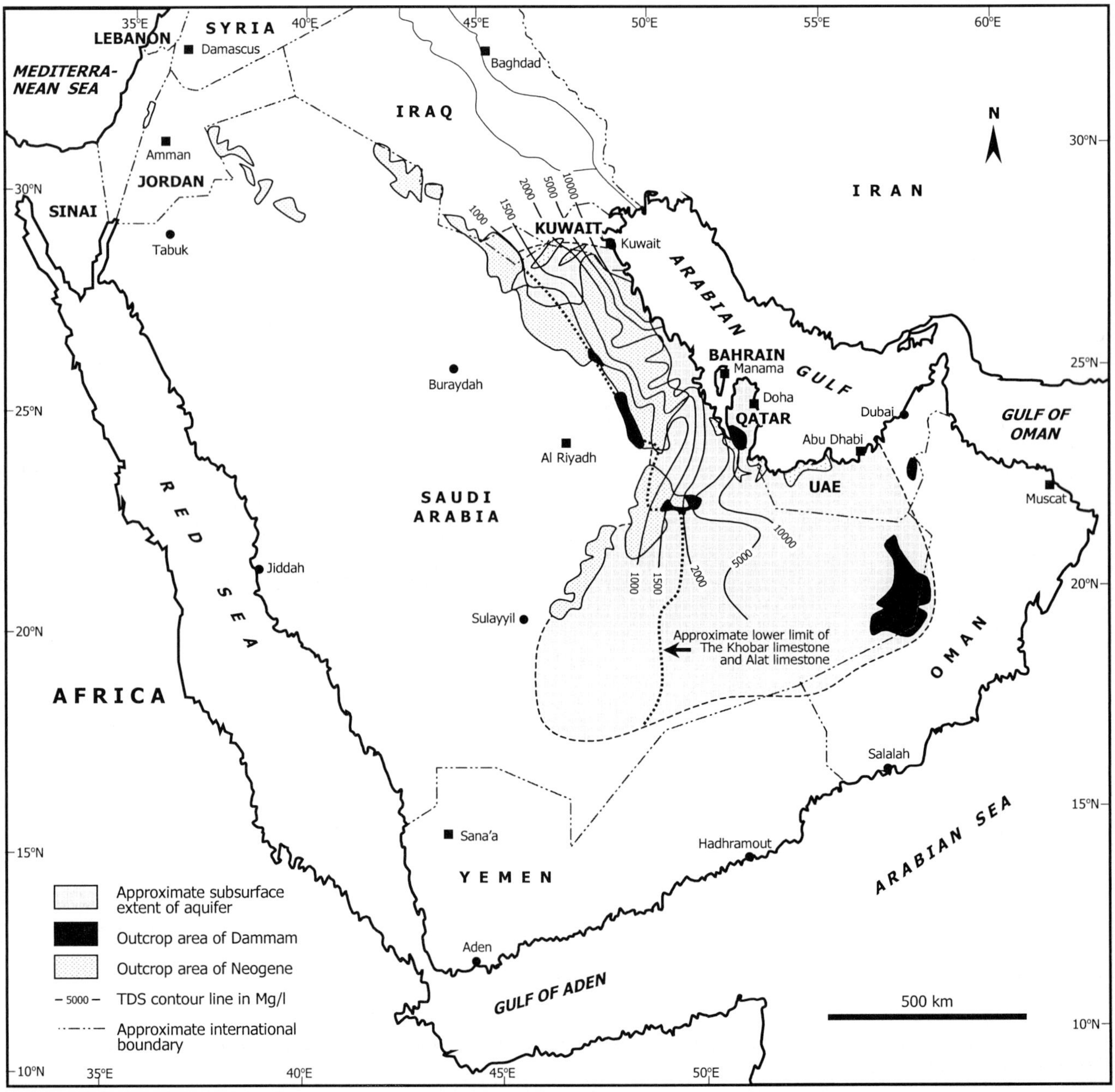

Fig. 8.18. Outcrop and subsurface extent of Dammam and Neogene aquifers and total dissolved solids in mg/l in eastern Saudi Arabia (modified from Al-Bassam et al., 2000).

The main factors affecting the hydraulic properties of the Dammam aquifer are: the impermeable units (confining layers) that separate the Khobar from the Alat member. The thickness of confining layers, their continuity and lack of fractures determines their effectiveness as sealing units between the different aquifers. The confining units separating the Dammam aquifer from underlying and overlying formations also affect aquifer recharge, water quality, groundwater flow and its relation with the underlying Umm er Radhuma aquifer.

Dolomitization increases the aquifer porosity and permeability, while, silicification leads to the decrease of porosity. Deposition of silica has led to the formation of an impervious siliceous limestone horizon in the upper Dammam aquifer. The epeirogenic movements led to the formation of several anticlinal and domal structures, which uplifted the formation to a level where it was subjected to erosion and development of karstification. The tectonics have also led to intensive faulting, fracturing and jointing, which contributed to improvement of hydraulic properties of the aquifer.

Two-dimensional groundwater flow code of Trescott et al. (1976), along with field and calculated parameters were used to construct a numerical model for the Dammam aquifer over an area of 176,000 km^2 in eastern Saudi Arabia in order to obtain calibrated values of the Dammam hydraulic parameters of the aquifer and its boundary conditions, and to evaluate aquifer's response to projected pumping plans (Yazicigil et al., 1986). The model assumes that the Khobar and Alat members be considered as a single aquifer and that the Dammam aquifer is hydraulically linked to the Umm er Radhuma aquifer. The hydraulic head map of the Dammam aquifer by Italconsult (1969) and Bureau de Recherches Gèologiques et Minières (1977) were used as a pre-development condition. The aquifer trnasmissivity applied varied between 0.1 m^2/day and 6.0x10^{-1} m^2/day. The abstraction rates for steady calibration were estimated at 6.60 m^3/sec (207x106 m^3/year). The model results indicate that 13.4 m^3/sec of water enter the western boundary aquifer modeled area. The Dammam aquifer also receives 0.24 m^3/sec of water through the upward leakage from the Umm er Radhuma aquifer, and loses 0.26 m^3/sec to Umm er Radhuma aquifer elsewhere.

Hydrogeologic Properties

Saudi Arabia exploits 360 Mm^3 of water from the Dammam aquifer every year. In eastern Saudi Arabia, the Dammam aquifer is largely utilized in agricultural activities. The aquifer salinity varies between 900 and 6,000 mg/l. The main hydrogeologic properties of the Dammam aquifer are:

1. The Dammam aquifer at Wadi Al Miyah is about 70m thick, and its depth from the ground surface ranges from 25 to 80m. The aquifer is artesian and its hydraulic head is at 110m above sea level. At Hafr Al Batin, the aquifer is artesian and its thickness varies between 35 and 180m. In Al Hasa area, the aquifer thickness ranges between 35 and 65m, the aquifer is artesian, and the hydraulic head is 120 to 150m above sea level. In the Rub Al Khali, the aquifer is about 90m thick, and the water wells penetrating the aquifer in this area are either artesian or flowing artesian wells.

2. The groundwater movement in the Dammam aquifer is controlled by the geologic structures, the lateral changes in geologic and hydrogeologic properties, the thickness of the aquifer and the nature of overlying and underlying confining layers. In this aquifer, the groundwater generally moves from the central Arabian Peninsula in the west towards the Arabian Gulf in the east, and the aquifer flow direction resembles that in the underlying Umm er Radhuma and the overlying Neogene aquifers. The Dammam aquifer transmissivity decreases eastwards, leading to a higher hydraulic gradient, and a longer residence time. The groundwater in the Dammam aquifer is generally under artesian conditions, except for areas where the overlying siliceous layer, and other younger formations disappear. The aquifer naturally discharges along the sabkha line, at the coast of the Arabian Gulf, and via sub-sea and onland springs. The artificial pumping for different purposes and local pumping, may affect the directions and velocity of groundwater flow. Vertical groundwater movement from the Umm er Radhuma, into the Dammam and Neogene aquifers, can occur under favorable structural and hydrogeologic conditions. This condition occurs in the Al Hasa area, where the hydraulic head of the Umm er Radhuma aquifer is higher than that of the Dammam and Neogene aquifers consequently, groundwater moves from the Umm er Radhuma aquifer upward into the Dammam and Neogene aquifers. Hydraulic connection of aquifers is likely to occur in the anticlinal areas, where the confining layers are thin or missing.

3. The lateral groundwater movement from west to east in the Dammam aquifer is controlled by the hydraulic head distribution. The average hydraulic head of the Dammam aquifer changes from 250m in western Saudi Arabia, to less than

1m in the eastern coastal area. Usually, the hydraulic head of the Khobar aquifer is higher than that the overlying Alat aquifer.

4. The Dammam aquifer is non-homogeneous both lithologically and hydrogeologically. The rocks of the aquifer change laterally from dense limestone to dolomitic, fractured and jointed limestone, and this has a great influence on the aquifer's transmissivity and storativity. The transmissivity of the Alat aquifer in Al Hasa and Ras Tannorah areas is 25 m^2/d. Maximum transmissivity of the Alat aquifer is 25,000 m^2/d along the eastern coast near the Arabian Gulf. The transmissivity of the Khobar aquifer is highest in Al Hofuf area as 8,000 m^2/d. The storage coefficient of the Alat aquifer varies between 1.3×10^{-4} and 5.7×10^{-4}, while the storage coefficient of the Khobar aquifer ranges from 2.9×10^{-4} to 1.0×10^{-3}.

Water Quality

The water quality in the Dammam aquifer is affected by its lithology, porosity, transmissivity and hydraulic connection, with other underlying or overlying aquifers. The groundwater moves from outcrop areas in the north and northeast, where the

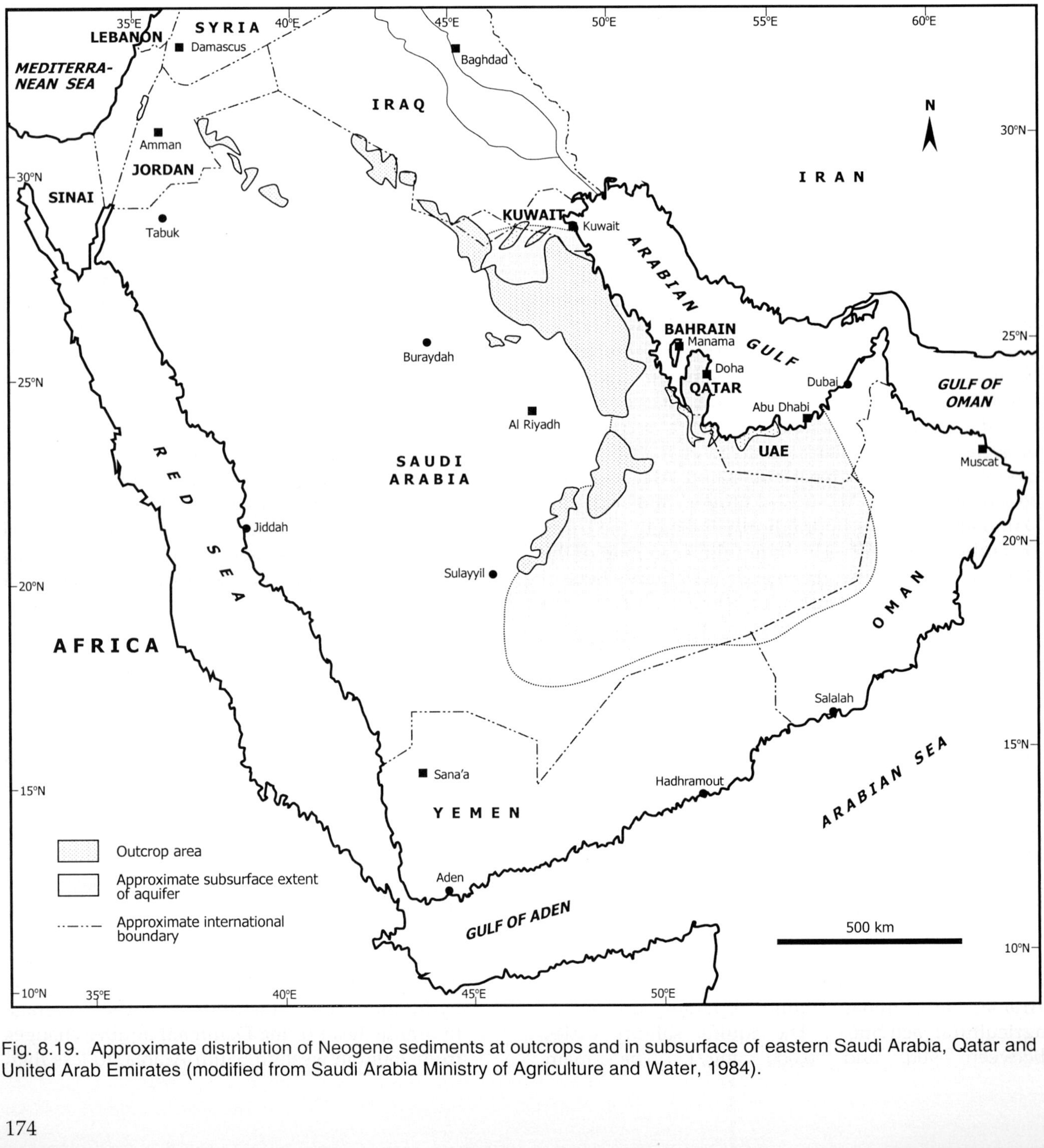

Fig. 8.19. Approximate distribution of Neogene sediments at outcrops and in subsurface of eastern Saudi Arabia, Qatar and United Arab Emirates (modified from Saudi Arabia Ministry of Agriculture and Water, 1984).

aquifer is recharged from rain and flood waters towards the Arabian Gulf. The groundwater salinity is generally less than 1,000 mg/l in recharge area, increasing to 7,000 mg/l near the Arabian Gulf. In Hafr Al Batin, the Khobar salinity ranges from 3,700 to 4,500 mg/l, and in Wadi Al Miyah the Khobar's salinity is 1,550-5,000 mg/l. In the Alat aquifer, the salinity varies between 1,000 and 7,000 mg/l. In Al Hasa area, the groundwater salinity in Khobar and Alat aquifer units is less than 2,500 mg/l. The hydrochemical facies vary from a chloride facies west of the Ghawar anticline to sulphate north of it, as a result of dissolution of gypsum and anhydrite within the Rus Formation. Along the Arabian Gulf the facies of the Dammam aquifer is dominantly sodium chloride.

Isotope Hydrology

In order to assess different hydraulic parameters such as estimation of the aquifer recharge; hydraulic connection between the Dammam and other aquifers; groundwater age dating; and determination of groundwater flow directions and velocity, the radioisotopes naturally occurring in groundwater were applied to the hydrogeologic investigations of the Dammam aquifer.

Stable isotopes ^{18}O and ^{2}H and radioisotopes ^{3}H and ^{14}C, naturally occurring in groundwater of the Dammam aquifer, were used and the results of investigations indicate that, the absence of ^{3}H means that the groundwater in the aquifer is old, and does not receive significant recent recharge.

The groundwater in this aquifer is 16,000 to 20,000 years old based on ^{14}C activities. The ^{18}O and ^{2}H content reveals a hydraulic connection between groundwater in the Dammam aquifer, and the groundwater in the Neogene aquifers, and with groundwater in Umm er Radhuma aquifer.

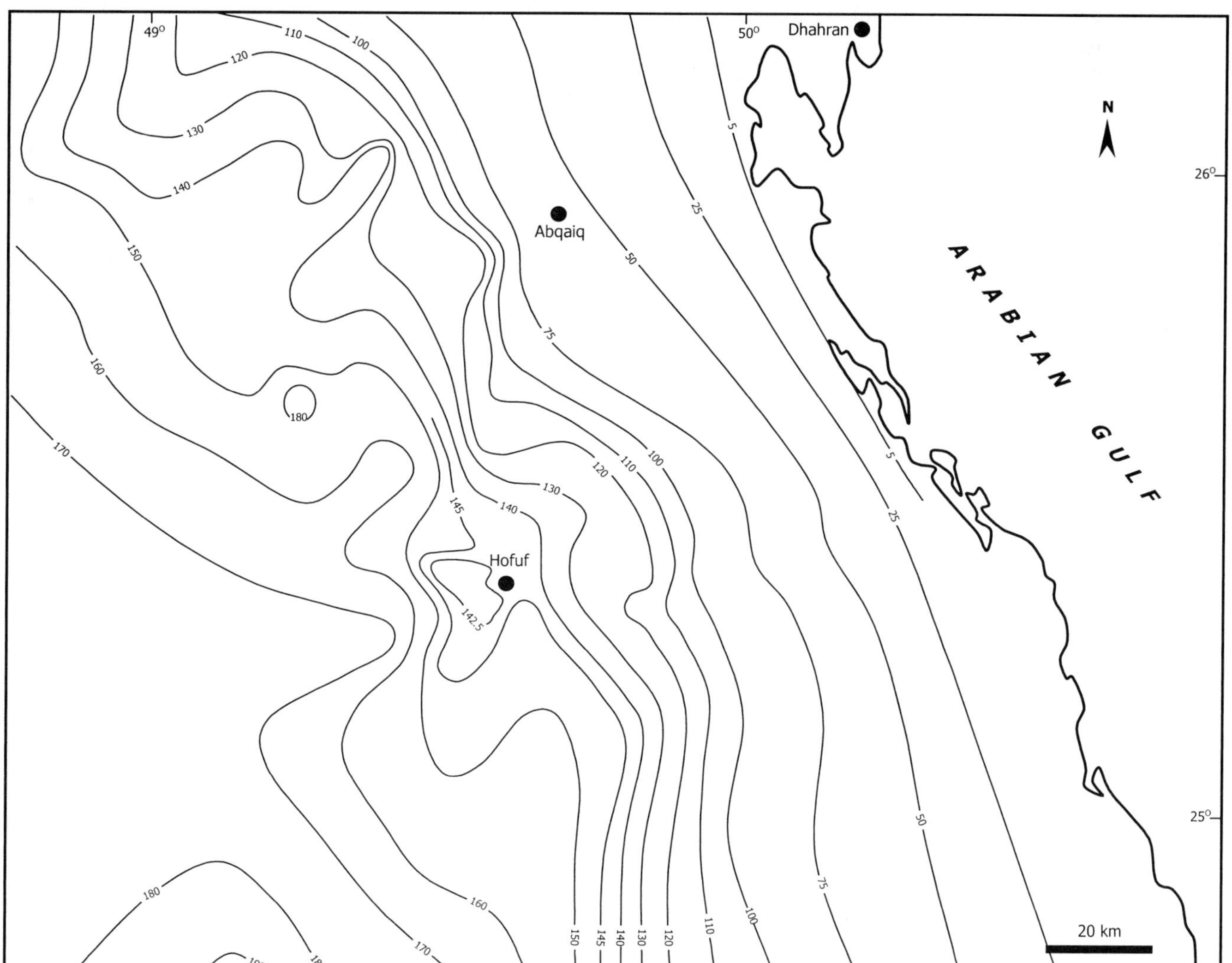

Fig. 8.20. Contour intervals of Neogene water level (in meters) in subsurface of eastern Saudi Arabia (modified from Bureau de Recherches Gèologiques et Minières, 1976).

Neogene and Quaternary Aquifers

The Neogene and Quaternary aquifers comprise the sediments unconformably overlying the Dammam Formation and range in thickness from more than 190 m in central Arabia to less than 5m toward the Arabian Gulf (Figs. 8.19 and 8.20). These sediments are composed of siliceous, chalky and argillaceous limestone, alternating with silty, argillaceous and gypsiferous evaporitic deposits, gravel, sand and conglomerate. The Neogene sediments originated under terrestrial and transitional environments and witnessed several marine transgressions and regressions. This resulted in wide lateral and vertical variations in lithology, and in turn hydrogeology and groundwater quality. The Neogene sediments were subdivided into three formations from base to top the Hadrukh, Dam and Hofuf (Tables 8.3 and 8.4).

Despite the fact that groundwater of the Al Hasa Oasis is from the Neogene aquifer, the deuterium values (^{2}H = -39.6) suggest an origin in an old reservoir. This hypothesis is substantiated by the lack of tritium and low ^{14}C (1.4% of modern carbon). This favors the hypothesis that a part of groundwater in the Neogene aquifer comes from the upward leakage from the Umm er Radhuma aquifer.

In Wadi Al Miyah, the water-bearing units include sandy marls and clays of the Miocene Dam and Hadrukh formations. The Quaternary wadi fill consists of aeolian sand, silt and fluviatile sand with very shallow groundwater or sabkhas may cover wadi floor. The hydrogeologic condition of the different aquifers in the Wadi Al Miyah area, eastern Saudi Arabia is presented in Table 8.7.

Stable and radioisotope data from the Al Qatif and Al Hasa oasis, eastern Saudi Arabia (Fig. 8.21), suggest groundwater in both areas is more than 20,000 years old, and therefore does not contain tritium. In contrast, water of springs in Wadi Hanifah has a high tritium content indicating the contribution of present rainfall to the shallow groundwater feeding such springs.

Table 8.7. Some hydrogeologic conditions of different aquifers in Wadi Al Miyah area, eastern Saudi Arabia (data from Italconsult reports in Jöb, 1978).

Aquifer	Depth (m)	Temperature (°C)	Total Dissolved Solids (mg/l)	Static level (m) (above sea level)
Neogene	0 - 83	27	3,520	98.6
Alat	83 -109	27	2,220	109.8
Khobar	129 -171	29.5	2,190	111.5
Umm er Radhuma	241	34.5	1,820	117.3

Groundwater samples from Al Qatif and Al Hasa Oases were recharged under climatic conditions different than those prevailing at present (Table 8.8). The ^{14}C content supports the stable isotopes (^{2}H and ^{18}O) and ^{3}H contents in indicating that the springs in eastern Saudi Arabia discharge old reservoirs, most probably of the Umm er Radhuma aquifer system.

Table 8.8. Results of ^{14}C and ^{3}H measurements of groundwater samples from Al Qatif and Al Hasa Oases (after Möser et al., 1978).

Well Place	^{14}C Content (% modern)	^{14}C – Age (years B.P.)	^{13}C (‰)	Tritium Content (TU)
Ain Labaniyah-1 (Al Qatif)	< 5.9	> 22,000	-9.0	< 0.8
Ain Labaniyah-1 (Al Qatif)	< 1.2	> 34,500	-8.8	< 2.7
Well-32 (Al Hasa)	< 1.4	> 33,000	-10.3	< 0.9
Ain Mansur (Al Hasa)	< 1.4	< 33,000	-10.0	< 0.9

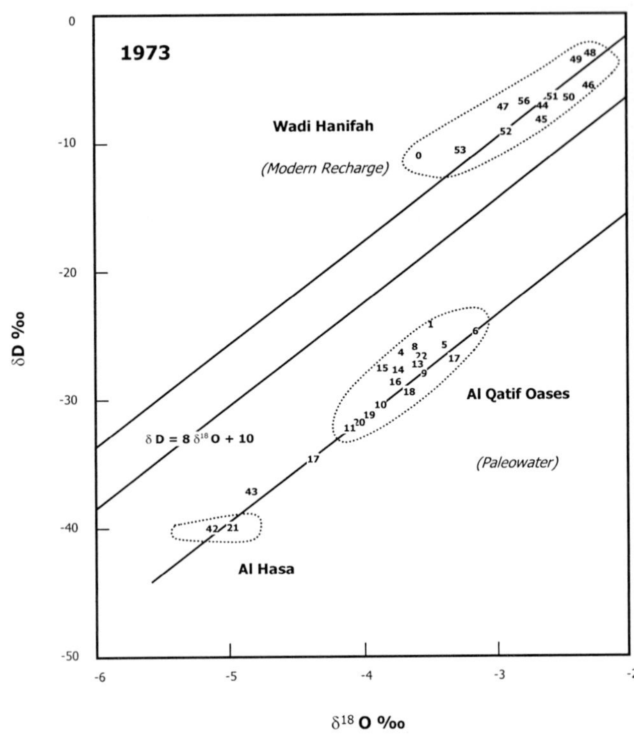

Fig. 8.21. Relationship between δD and $\delta^{18}O$ content in water samples from Wadi Hanifa, Al Qatif and Al Hasa area of Saudi Arabia (modified from Möser et al., 1978).

Where the Neogene and Dammam aquifers are hydraulically connected, water moves from the Neogene aquifer downwards recharging the Dammam aquifer. Fractures and the karstic nature of the Neogene formations improve their hydraulic properties, especially within the Dam and Hadrukh formations. The area of the confined Neogene aquifer in eastern Saudi Arabia is 30,000 km^2 while the unconfined Neogene area is about 18,000 km^2. The groundwater flow in the Neogene aquifers occurs from the southwest towards the north and northeast. The aquifer hydraulic head less than 1m near the Arabian Gulf rising to 25m above sea level north of Al Hasa. Due to excessive groundwater pumping from the Neogene aquifer at Al Hofuf, the discharge of springs in Al Hasa Oasis is noticeably reduced.

The Neogene aquifer transmissivity at Al Hofuf ranges from $7x10^{-4}$ to $4x10^{-2}$ m^2/sec and its storativity varies between $2x10^{-4}$ in the confined part and $1x10^{-2}$ in the unconfined part. The Neogene aquifer receives annual recharge from rainwater. It is the main source of water feeding the Al Hasa springs. More than 119 Mm^3 of fresh potable water estimated by Bureau de Recherches Gèologiques et Minières (1985) can be pumped annually from the Neogene aquifer in eastern Saudi Arabia without introducing an harmful effect on spring discharge in Al Hasa Oasis.

The average saturated thickness of the Neogene aquifer is 240m, and its average effective porosity is 3%. The groundwater stored in the unconfined part of the aquifer is about 130,000 Mm^3, while the aquifer storage in its confined part is 270,000 Mm^3.

Water Quality

Groundwater salinity in Neogene aquifer ranges from 1,200 to 1,450 mg/l. The aquifer salinity decreases towards the unconfined part, where recharge takes place. Because of its proximity to recharge area, southwest of Al Hofuf, water in the Neogene aquifer has total dissolved solids of 500 mg/l (Table 8.9). Groundwater is this area is suitable for all purposes, including domestic uses.

Table 8.9. Total dissolved solids content (mg/l) in groundwater samples collected from different wells in eastern Saudi Arabia (data compiled from Al-Sayari and Zötl, 1978).

Area	Depth (m)	Total Dissolved Solid (mg/l)
Al Hasa region	50-100	1,170-4,300
Southeastern region	150	1,500
Northeastern region	80	500-800
Hafr Al Batin	100	750

3. PALEOGENE AQUIFER SYSTEM IN BAHRAIN

INTRODUCTION

Bahrain consists of an archipelago of 33 islands located in the Arabian Gulf about midway between Saudi Arabia to the west and Qatar to the east (Fig. 8.22). It has a total area of 695 km^2. Bahrain like most of the Gulf States, has experienced accelerated development. This occurred as a direct result of the increase in the country's oil revenues, which led to a rapid increase in its economic base and an improvement in the standard of living. Accordingly, a rapid increase in country's population occurred. The fast growth in population and the associated development processes, represented by urbanization, expansion of irrigated agriculture and industrialization, has placed major stresses on a country of limited natural water resources.

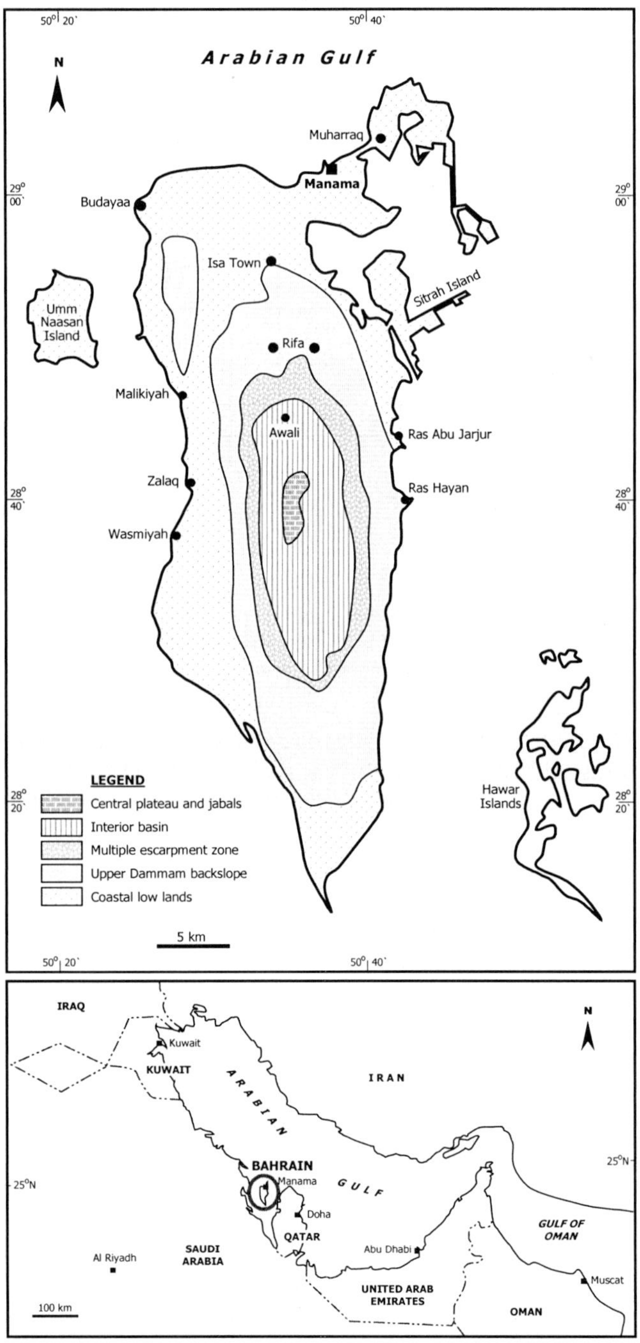

Fig. 8.22. Location map of Bahrain showing main physiographic features (modified from Doornkamp et al., 1980).

Due to the quantitative and qualitative deterioration of the groundwater resource caused by over-exploitation, more than half of the original aquifer has been polluted by saline water invasion, and the rest is threatened. One of the greatest problems and challenges Bahrain is facing is the provision of fresh water to supply demands for domestic, agriculture and industry. The present situation with respect to the groundwater resources is both dangerous and alarming and this study evaluates and reviews the status of the groundwater, the only natural fresh water resource in Bahrain, and monitors the deterioration which has occurred in terms of hydraulic head and salinity changes during the economic development process.

Bahrain like most of the countries in the Arabian Peninsula, has an arid to extremely arid environment and is characterized by irregular, scanty rainfall, and high evapotranspiration rates. The average annual rainfall is less than 80 mm (Raveendran and Madany, 1991), while the potential evapotranspiration averages about 1,850 mm/yr (Zubari, 1987), leading to a high negative deficit in the water budget, and absence of perennial surface water.

Prior to 1925, the population of Bahrain depended entirely on the naturally flowing fresh water from onshore and offshore springs to meet its domestic and agricultural needs. The estimated natural spring discharge, from the aquifer was 90 Mm^3/yr (Ferguson and Hill, 1953; Hill, 1953) from about 15 onshore and 20 offshore springs. Mechanized well drilling, and water extraction was introduced to the Bahrain Islands in 1925, along with oil exploration activities (Hamilton, 1965). The oil discovery in 1932 was accompanied by a rapid growth in population and urban development, as well as a dramatic increase in water demands and consumption (Fig. 8.23). This demand had been met mainly by abstraction from the Dammam aquifer replacing the natural springs, which experienced a significant reduction in their discharge with most of them ceasing to flow. The total aquifer withdrawal for domestic purposes increased from about 5 Mm^3/yr in the 1950's (Porritt, 1953) to about 16 Mm^3/yr in the 1960's (Sutcliff, 1966) and to about 21 Mm^3/yr in the early 1970's (Italoconsult, 1971).

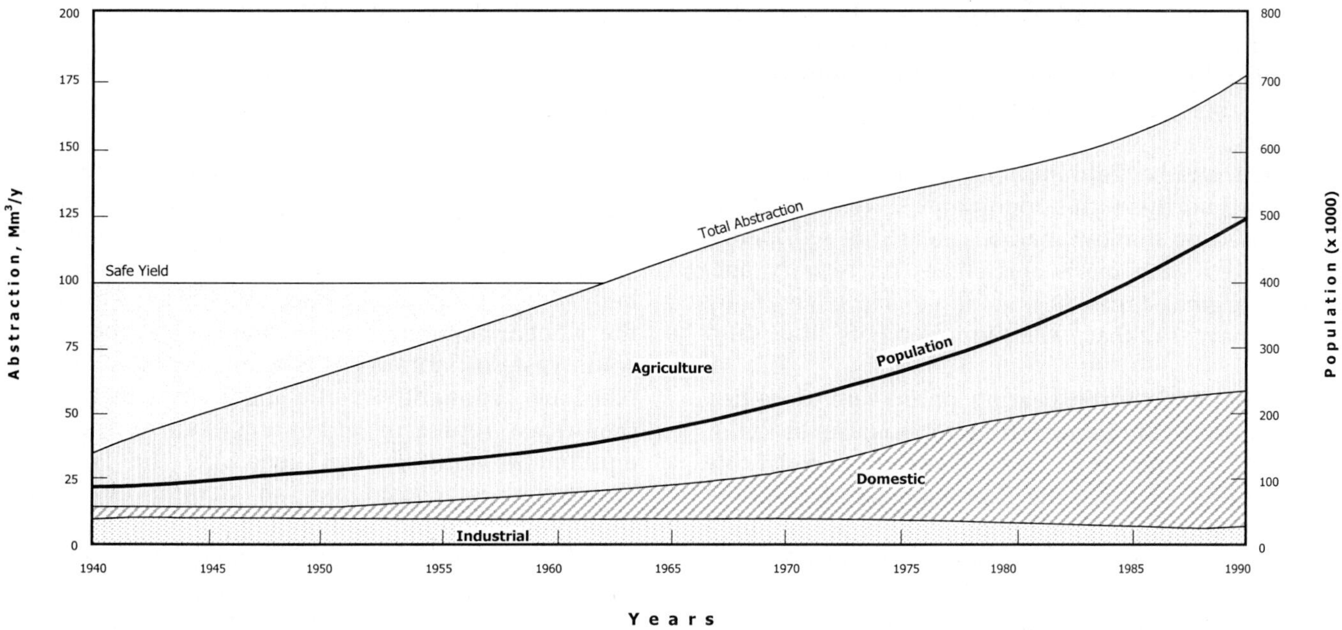

Fig. 8.23. Groundwater abstraction (Mm^3/yr), domestic consumption and population growth in Bahrain, 1940-1990 (compiled and modified from Zubari and Junaid, 1992; Al Noaimi, 1999).

Fresh water demands in Bahrain have therefore been met by groundwater abstraction from the Dammam aquifer system developed in the Tertiary limestone and dolomite members of the Dammam Formation. Abstraction from this aquifer accounts for more than 75% of the total consumption while the remainder is provided by desalination plants.

The government of Bahrain constructed its first desalination plant in 1975 on Sitrah Island on the east coast, to complement the domestic water needs. The plant has been expanded periodically with the increase in domestic demand to produce 41 Mm^3/yr in 1992. In 1984 the government established another desalination plant to provide an additional water source and to stop deterioration of the groundwater quality. The plant is located in the area of Abu Jarjur on the east coast and has a production capacity of 17 Mm^3/yr. The feed water for the plant is the brackish water from the Umm er Radhuma aquifer. A third desalination plant, located in the Al Dur area on the east coast, was begun in 1993 using seawater. It has the same production capacity as the Abu Jarjur plant, increasing the production capacity of desalination plants in Bahrain to 75 Mm^3/yr, which is still insufficient to meet the domestic water needs in Bahrain. The deficit in domestic water demand is still being supplied by abstraction of groundwater from the Dammam aquifer. Because the groundwater salinity is higher than the World Health Organization standards for drinking water, blending with the desalinated water provides the water needed of reasonable quality. The total domestic consumption in Bahrain in 1991 was about 110 Mm^3/yr, with about 53 Mm^3/yr obtained from groundwater.

Hydrogeology

The Bahrain Islands are characterized by a low topography, with the maximum elevation in Jabal al Dukhan (122m above sea level) and an average elevation of 20m. Bahrain comprises five physiographic units (Fig. 8.22): the central plateau, internal basin, rocky slopes, plains and coastal belts. The central plateau, higher in the north, occupies the core of the Bahrain main island, with isolated flat-topped hills with elevations between 20 and 60m. The internal basin is composed of a lowland 2 to 8 km wide, surrounding the central plateau. The surface of the plain is irregular, covered by aeolian sand and playas, and transversed by radial dry wadis. The rocky slopes form an oval belt around the internal plain. The slopes vary from simple to complex, according to lateral changes in lithology, erosion and, possibly, geologic structure. The elevation ranges from 40m in the north and west to 67m in the south and southwest. The plains are wide with gentle slopes in the north and south and narrow with steeper slopes in the east and west. The plains are crossed by shallow dry wadis that become active during rare rainstorms. The coastal belt is composed of unconsolidated argillaceous sands, covered by isolated dune sands in the north. The north and northwestern parts of the coastal belt have low salinity, well-drained soils suitable for agriculture. The northern and eastern parts are occupied by sand dunes, while the southwestern coastal belt is occupied by low-lying sabkha deposits.

Structurally, Bahrain is an asymmetrical dome, with a north-south orientation. The dome is composed of shallow marine calcareous sediments, intermixed with marls and shales of Tertiary age. The marginal areas of the dome are covered by Quaternary sabkhas, alluvium deposits and sand dunes.

Three main water-bearing units were identified in Bahrain locally known (from top to base) as the A, B and C aquifers (Fig. 8.24). These units constitute two aquifer systems; the Dammam aquifer system and Umm er Radhuma aquifer system, which are a part of the regional hydrogeology of eastern Arabian Peninsula. To the west of Bahrain, the Eocene and Paleocene carbonate aquifers are exposed in eastern Saudi Arabia, where they are recharged by rainwater. In eastern Bahrain, the upper Damman aquifer is exposed. The hydraulic heads and hydrogeochemistry of the Dammam and Umm er Radhuma aquifers in Bahrain show that they represent discharge area of the regional hydrogeologic systems in eastern Arabian Peninsula.

Aquifer Systems

1. Dammam Aquifer System

The Dammam aquifer system is the only natural fresh water source in Bahrain. Abstraction from this aquifer accounts for more than 75% of total water use, while desalination plants provide the rest. The Dammam aquifer system consists of an upper A aquifer (Alat aquifer) and a lower B aquifer (Khobar aquifer). Both aquifers disappear over the crest of the Bahrain dome, but their thicknesses increase to the east and west, averaging 15-25m for aquifer A and 20-45m for aquifer B (Fig. 8.24). The Dammam aquifer system in Bahrain is mostly confined by Neogene claystone (10-60m) at its top, and shale beds (8-20m) with anhydrites at the base. However, the Khobar aquifer is unconfined in northern and southwestern Bahrain (Al Noaimi, 1999). Where the Neogene sediments overlie the Dammam aquifer, they may represent a continuation to the Alat aquifer (A aquifer), but generally, the Neogene Formation is considered an aquifer of minor importance, and most commonly acts as an aquitard. Hand-dug wells in the Neogene are used for irrigation in northeastern Bahrain. The Alat and Khobar aquifers are separated by the Orange Marl Member (9-15m) which acts as a confining layer. The hydraulic properties of Alat aquifer are as follow: transmissivity 350 m^2/d, hydraulic conductivity 14 m/d and storativity $6.5x10^{-2}$; while for Khobar aquifer the hydraulic conductivity is 458m/d, transmissivity 21,000 m^2/d and storativity $5x10^{-5}$.

The relationship between groundwater withdrawal and hydraulic head in three observation wells is illustrated in figure 8.25. Despite the overall decline observed in the three hydrographs, reduction of groundwater pumping from the Dammam aquifer during the last quarter of 1984 associated with the commencement of the Abu Jarjur desalination plant, resulted in noticeable groundwater recovery.

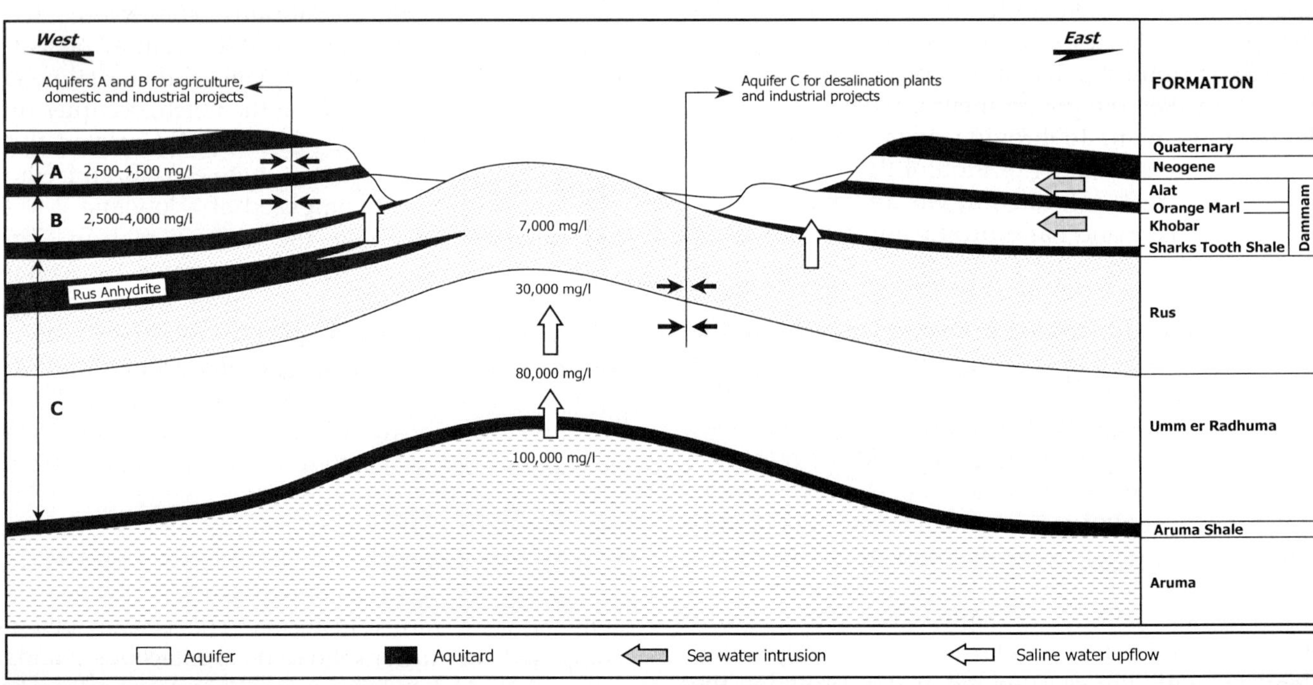

Fig. 8.24. Simplified regional hydrogeological cross section showing recharge and discharge areas of the Cenozoic aquifer in Bahrain (modified from Zubari and Junaid, 1992; Zubari et al., 1997).

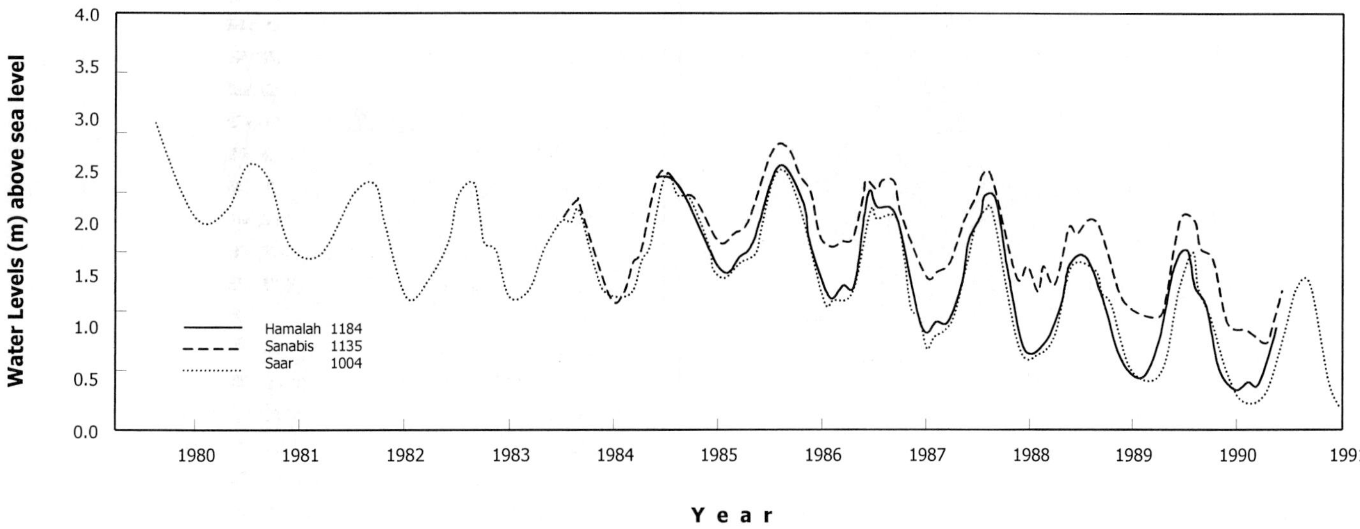

Fig. 8.25. Hydrographs of three observation wells showing the decline in hydraulic heads of the Dammam aquifer as a result of groundwater abstraction (modified from Al Noaimi, 1999).

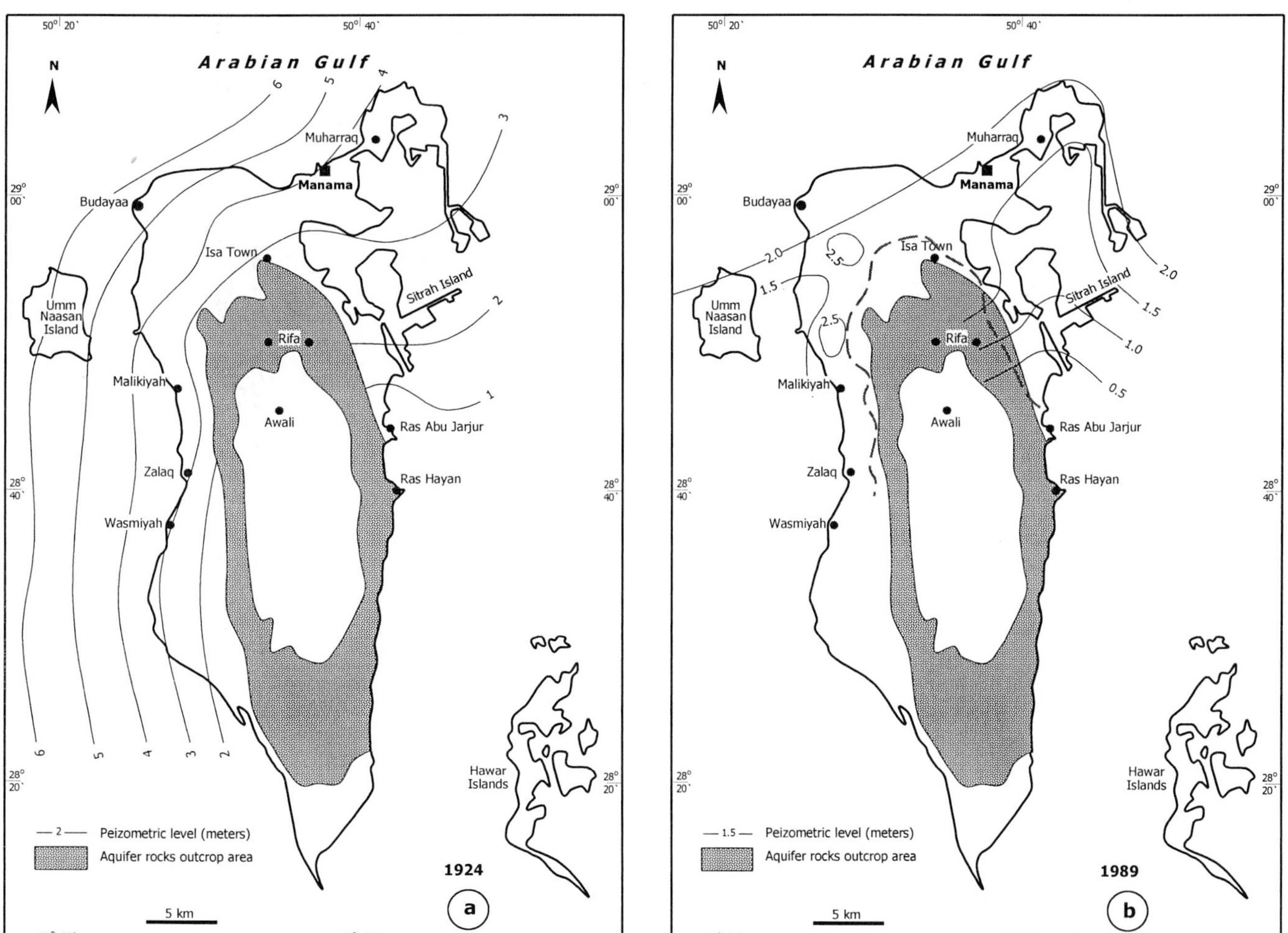

Fig. 8.26. Comparison of the hydraulic heads of the Alat aquifer (upper Dammam aquifer system) in 1924 and in 1989 (modified from Groundwater Development Consultants, 1979; Al Noaimi, 1999).

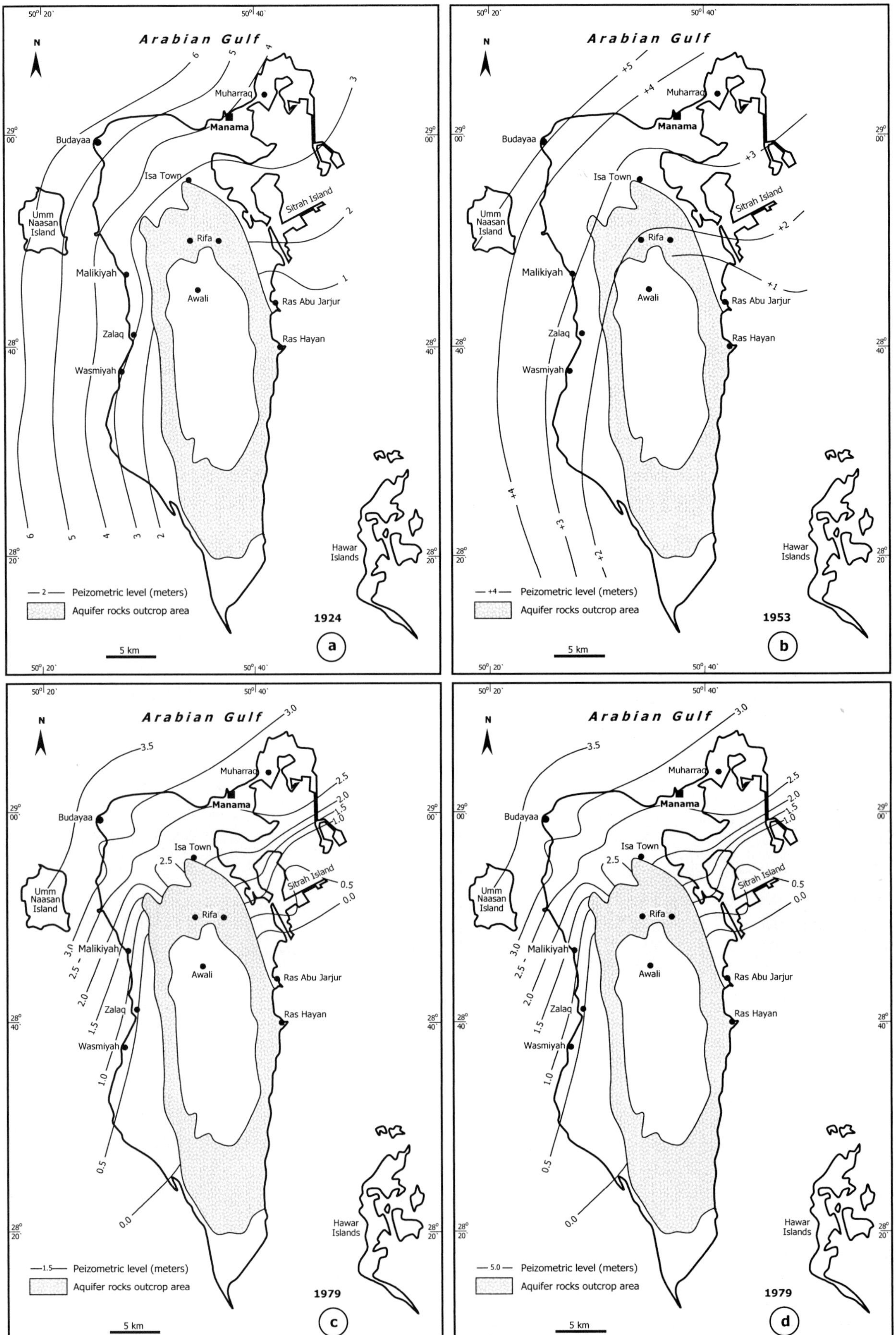

Fig. 8.27. Comparison of the hydraulic heads of the Khobar aquifer (lower Dammam aquifer system) in 1924, 1953, 1979 and in 1990 (modified from Groundwater Development Consultants, 1979; Bahrain Petroleum Company, 1952; Al Noaimi, 1999).

The regional groundwater flow direction is from northeast (from the recharge area in Saudi Arabia) towards southeast (discharge area in Bahrain). This is the case in the Damman aquifer but, in the Umm er Radhuma aquifer, the vertical hydraulic gradient becomes more pronounced, and groundwater moves from it upward, into the overlying Dammam and Neogene aquifers. The hydraulic head maps show that the hydraulic head in the upper Dammam Alat aquifer decreased by 1 to 3m between 1924 (Fig. 8.26a) and 1989 (Fig. 8.26b), respectively. Hydraulic head maps for the lower Dammam Khobar aquifer in 1924 and 1953 (Groundwater Development Consultants, 1979) are illustrated in figure 8.27a,b. These maps indicate that the hydraulic head in the Khober aquifer decreased by 4 to 5m between 1924 and 1953. Al Noaimi (1999) reported a further decline on 1 m in the hydraulic head of the Khobar aquifer occurred between 1979 and 1990 (Fig. 8.27 c, d). The maximum drop in the hydraulic head of the Khobar aquifer of about 5m was observed in western Bahrain, while the minimum drop (2m) was observed on the east coast, where the aquifer water is in direct contact with sea water.

Development activities in Bahrain have significantly increased abstraction rates from the Dammam aquifer. The total abstraction from the aquifer, which was approximately 65 Mm^3/yr in the early 1950's, increased to about 112 Mm^3/yr in the mid-1960's and reached about 145 Mm^3/yr in the mid-1980's. In the early 1990's the total abstraction from the aquifer has reached about 179 Mm^3/yr (Fig. 8.23). The safe yield of the Dammam aquifer is about 100 Mm^3/yr, estimated from numerical models (Groundwater Development Consultants, 1980; Zubari, 1987), which is equal to the recharge rate of the aquifer, by underflow from Saudi aquifers under steady-state conditions. The abstraction rate has exceeded the suggested safe yield of the aquifer since the early 1960's (Fig. 8.23), and presently is about twice that rate, indicating that a large proportion of the water abstracted is being taken from the aquifer storage.

2. *Umm er Radhuma Aquifer System*

The Rus Formation is composed of early Eocene chalky and siliceous limestone with anhydrite intercalations, of variable thickness, but which disappear in the center, east and northeast due to solution. Where the anhydrite exists, the Rus Formation acts as a confining layer, separating the Dammam aquifer above, from the Umm er Radhuma aquifer below. In contrast, when anhydrite is not present the Rus and Umm er Radhuma form a single aquifer known as the C aquifer. The average thickness of the Rus Formation is 105m and that of Umm er Radhuma averages 350m. The hydraulic properties of C-Aquifer are transmissivity 24,600 m^2/d, hydraulic conductivity 25 m/d and storativity $1.8x10^{-1}$ to $2.3x10^{-4}$.

In 1979, the Groundwater Development Consultants constructed a map showing the distribution of hydraulic heads in the Umm er Radhuma aquifer (Fig. 8.28a), which were markedly higher than that of the lower Dammam Khobar aquifer (Fig. 8.28b). This situation has led to an upward movement of water from the Umm er Radhuma aquifer into the Khobar aquifer. Records from the three observation wells tapping the Umm er Radhuma aquifer reveal that the hydraulic head has declined at an average of 2m between 1980 and 1991 (Fig. 8.29), a decline attributed to groundwater extraction and its use as feed water for the Abu Jarjur desalination plant.

Hydrogeochemistry

Groundwater salinity in Bahrain is generally high (minimum salinity is 2,200 mg/l TDS) compared with the World Health Organization (WHO) standards for drinking and industrial waters. The main reason for high groundwater salinity returns to the fact that the Bahrain Islands lie in the discharge area of the regional aquifer systems in eastern Arabian Peninsula.

Groundwater types in Bahrain are dominated by the sodium chloride, while bicarbonates, which mostly prevail in recent and renewal groundwater, occur in low concentration. Groundwater Development Consultants (1979 cited in Al Noaimi, 1999) distinguished four groundwater types in Bahrain: (1) highly saline confined water, (2) water of old pluvial periods, (3) recent rain water, and (4) sea water invasion of coastal aquifers.

Dammam Aquifer Salinity

Groundwater salinity in the Dammam aquifers increases from the west (2,200-2,500 mg/l TDS) to the east (5,000-20,000 mg/l TDS), in the direction of groundwater flow. In addition to sea-water intrusion, excessive groundwater pumping from the Dammam aquifer has increased upward movement of more saline water from the Umm er Radhuma aquifer (Fig. 8.24). The groundwater salinity in the Khobar aquifer increased by 2,000 to 6,000 mg/l in the north, and by 500 to 2,500 mg/l in the west between the years 1965 and 1992, respectively (Fig. 8.30). The continuous rise in groundwater salinity in the Dammam aquifer has led to abandonment of farms in north central areas.

Umm er Radhuma Aquifer Salinity

The Groundwater Development Consultants (1979) isosalinity map of the Umm er Radhuma aquifer, measured in 1979 shows an aquifer salinity

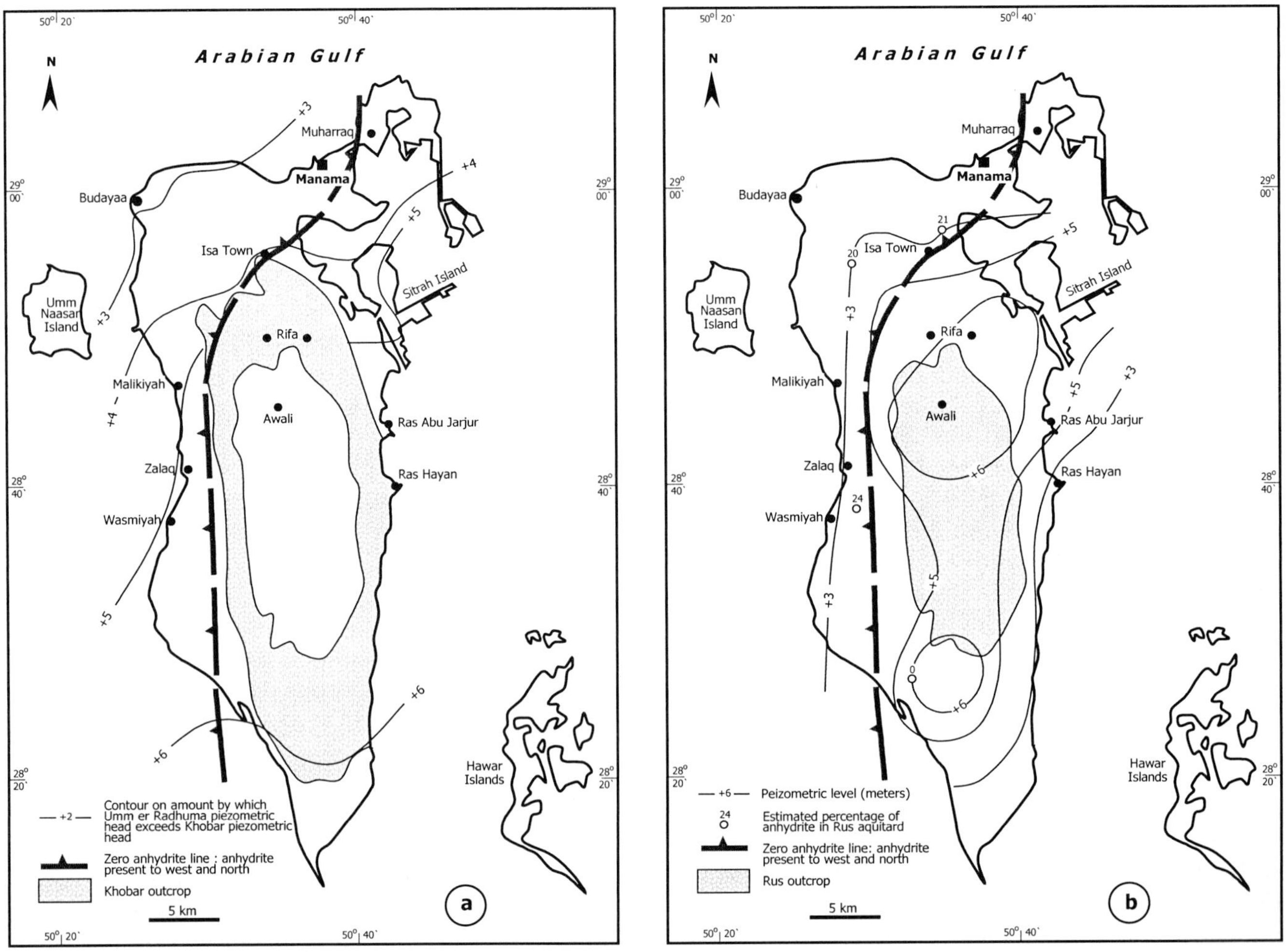

Fig. 8.28. Piezometric level measured in 1979 between the Umm er Radhuma-Rus and Khobar aquifer (a) and the Umm er Radhuma-Rus aquifers (b) (modified from Groundwater Development Consultants, 1979; Al Noaimi, 1999).

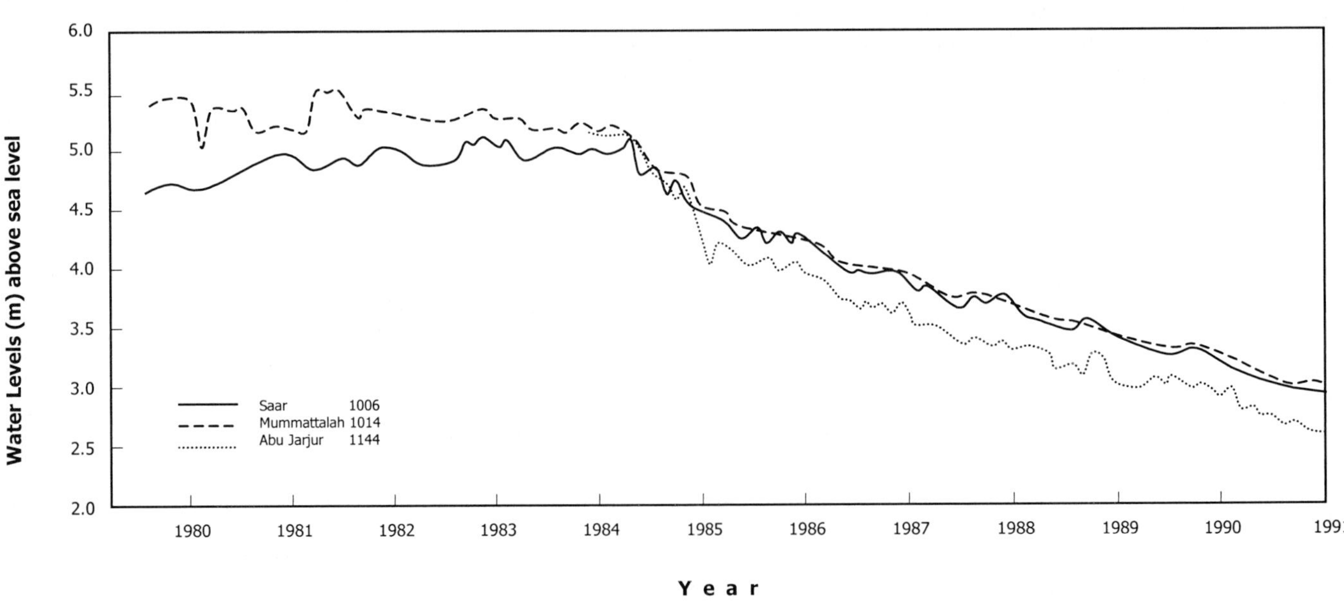

Fig. 8.29. Hydrographs of three observation wells showing the decline in hydraulic heads of the Umm er Radhuma-Rus aquifer during the period 1980-1991 (modified from Al Noaimi, 1999).

range from 2,000 to 12,000 mg/l (Fig. 8.31), but most commonly the aquifer contains brackish to saline water, with salinity of around 10,000 mg/l. The aquifer salinity increases downward, and reaches a total dissolved solids value of about 40,000 mg/l. Due to the high salinity of the aquifer water, aquifer utilization is restricted to industrial purposes, and feeding desalination plants. It is this regional aquifer that provides Bahrain with its water by lateral underflow (Groundwater Development Consultants, 1980). Here, the two are considered together to represent the Dammam aquifer. The Rus Formation is composed of fractured chalky dolomitic limestone with subsidiary shale and anhydrite intercalations in its upper section, and in the central and eastern areas of Bahrain, extensive solution of the anhydrite has led to the collapse of overlying rocks. More importantly, it has reduced the effectiveness of the upper confining layer, permitting relatively easier migration of water into the upper Dammam aquifer.

Interpretation of Groundwater Chemistry

Based on records of groundwater salinity between 1925 and 1979, Groundwater Development Consultants (1979) concluded that, the landward side of the seawater front along eastern coast of Bahrain has moved inland at a rate of 100 m/yr. This seawater intrusion has increased groundwater salinity, in the Dammam aquifer system. The saline water up-flow from the Umm er Radhuma aquifer into the Khobar aquifer in the north central areas adds to the salinity problem. The rate of upward migration of the Umm er Radhuma saline water increases with increasing the difference in hydraulic heads, between the Dammam aquifer and the Umm er Radhuma aquifer.

The degree of increasing aquifer salinity, can be illustrated by comparing isosalinity maps in figure 8.30, which shows the evolution of salinity levels in the Dammam aquifer during the 1965-1992 period. The comparison indicates that almost all areas of

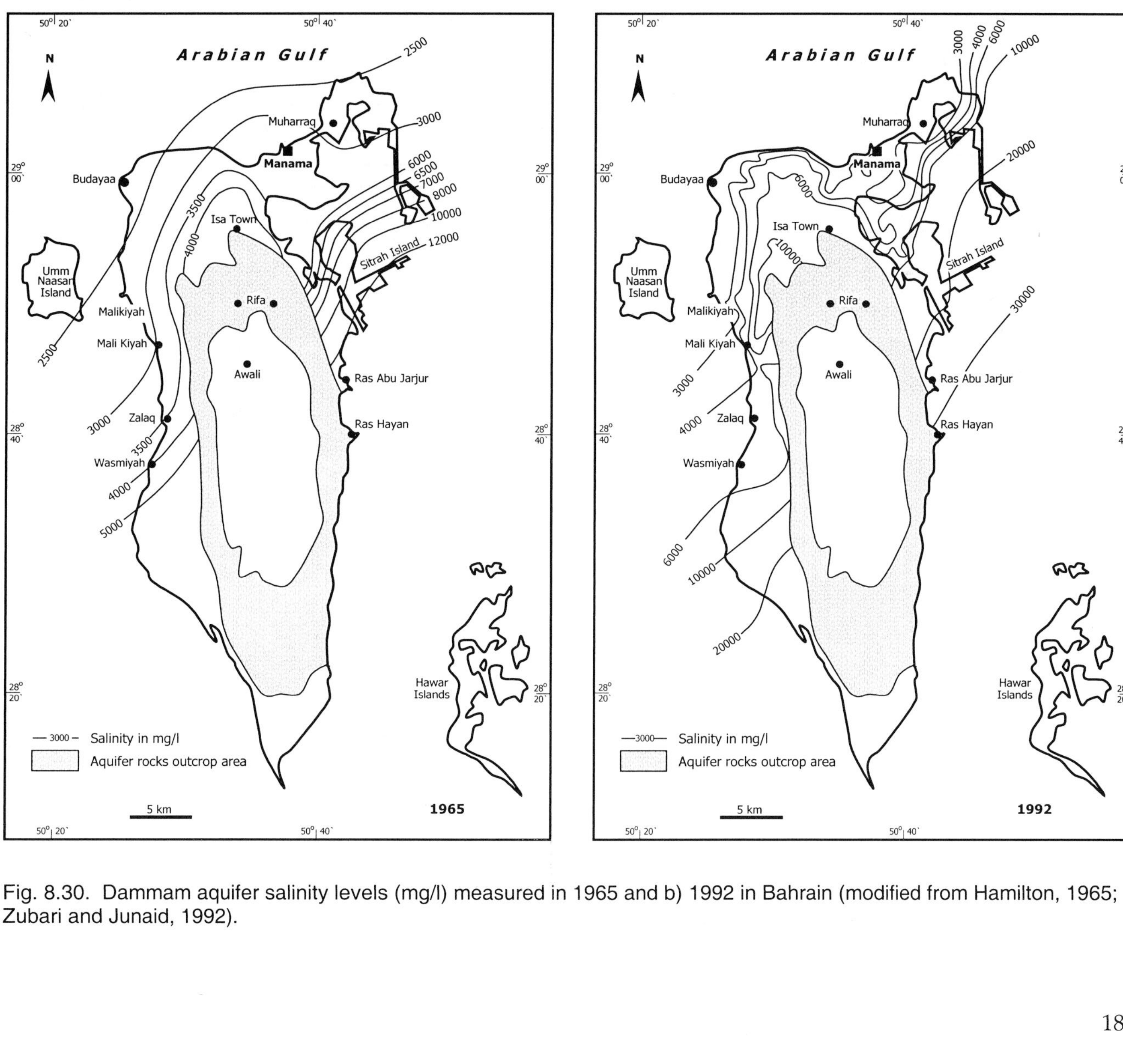

Fig. 8.30. Dammam aquifer salinity levels (mg/l) measured in 1965 and b) 1992 in Bahrain (modified from Hamilton, 1965; Zubari and Junaid, 1992).

Bahrain have experienced an increase in aquifer salinity. The salinity increase in the eastern area is caused mainly by sea-water intrusion, while in the north central area it is caused by upward leakage of saline water from the Umm er Radhuma aquifer. In the western area, the southern part is affected by seawater intrusion, and to the north by upward migration from the Umm er Radhuma aquifer.

Areas of the Dammam aquifer that are unsuitable for direct water supply for use for domestic or agricultural purposes are shown in figure 8.32. More than half of the original Dammam aquifer in Bahrain has been polluted, due to over-exploitation. This leaves a very small aquifer area available for competing agriculture and domestic sector demands for fresh water, exacerbating the depletion of the groundwater reservoir. Any increase in the rates of abstraction from the Dammam aquifer from the present rate, might lead to total contamination and eventually its destruction as the natural groundwater resources in Bahrain.

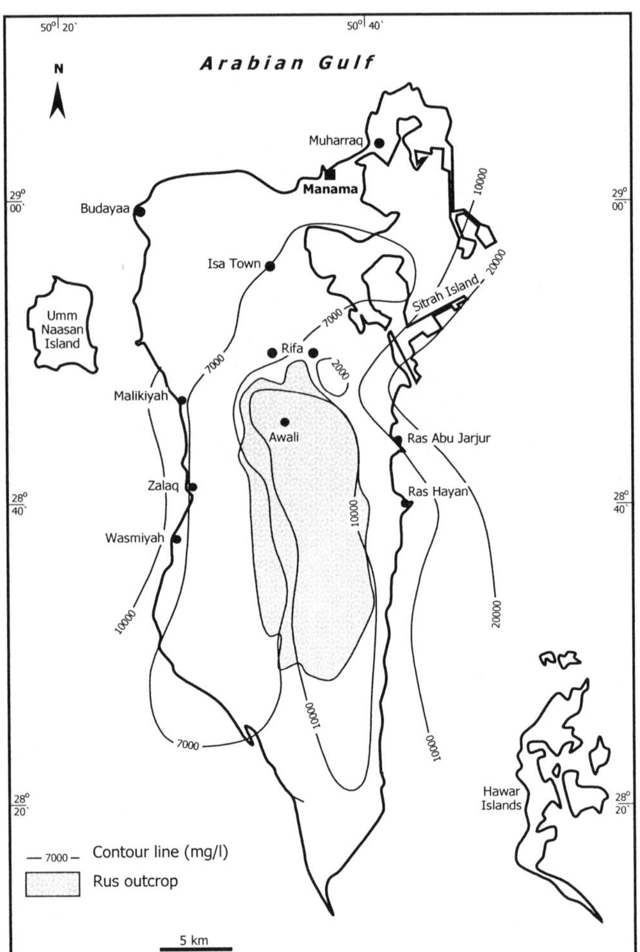

Fig. 8.31. Isosalinity map (total dissolved solids in mg/l) in Rus and Umm er Radhuma aquifers in 1979 (modified from Groundwater Development Consultants, 1979; Al Noaimi, 1999).

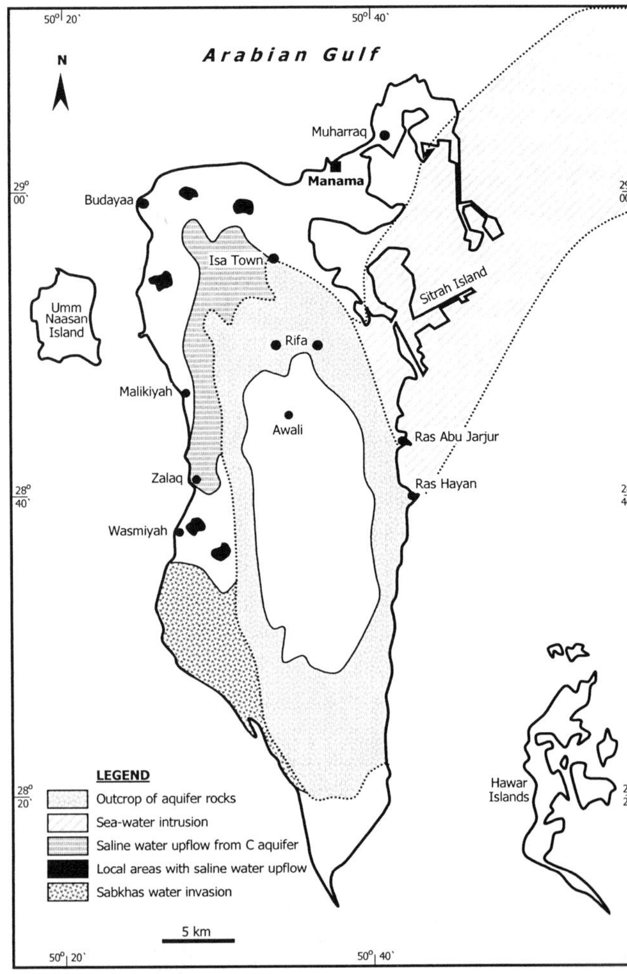

Fig. 8.32. Extent of pollution in Dammam aquifer, Bahrain (modified from Zubari and Madany, 1993).

Spatial and Temporal Changes in Groundwater Salinity

A detailed survey of water quality in the Dammam aquifer from 254 wells across Bahrain was carried out by Zubari et al. (1997) and summarized below. Out of the total 254 water samples, 110 samples were analyzed for major cations (Na^+, Ca^{2+}, Mg^{2+}, K^+) and major anions (Cl^-, SO_4^{2-}, HCO_3^-), total alkalinity ($CaCO_3$), total hardness ($CaCO_3$), pH, electrical conductivity and total dissolved solids. The rest of water samples (144) were only analyzed for electrical conductivity and their total dissolved solid content was calculated from an established relationship between the total dissolved solids and electrical conductivity measured in first 110 wells (Fig. 8.33). The results of this survey are illustrated by the frequency plots shown in figure 8.34.

Spatial Trend Analysis

The groundwater flow in Bahrain from Saudi Arabia generally has background quality limits, characterized by moderate salinity (total dissolved solids <3,000 mg/l). The sequence of anion dominance is $Cl^- > SO_4^{2-} > HCO_3^- >$, while the cation dominance is $Na^+ > Ca^{2+} > Mg^{2+} > K^+$ (Groundwater Development Consultants, 1980; Hassan and Cagatay, 1994; Sen and Al-Dakheel, 1986). A comparison between background concentrations and the mean concentrations of the 1997 survey carried out by Zubari et al., shows an overall increase in the concentration of major ions in the groundwater of Bahrain. Deviation of the sampled concentrations from the background values is shown in figure 8.35. The results indicate that the major part of the groundwater sampled in Bahrain suggest a widespread inland contamination by higher concentration waters. The spatial distribution of the groundwater major ion chemistry can be represented by the contour maps. The maps indicate that the aquifer recharge comes from eastern Saudi Arabia and approaches the Bahrain Islands from the northwest direction. The isosalinity contours indicate a rise in groundwater salinity in areas marked A, B, C, D and E. Because the groundwater in southeastern Bahrain is hydraulically connected with the sea (Wright, 1967), the salinization process in zone A is essentially attributed to seawater encroachment. Two major salinity anomalies are also displayed this figure. Zone B extends over most of the north central region where the total dissolved solids has reached about 11,000 mg/l, and zone C is located in the western region where total dissolved solids has reached about 8,000 mg/l. The reasons behind the high salinity in both zones are municipal pumping and associated upward movement of saline water from the Umm er Radhuma aquifer which has an average salinity of about 12,000 mg/l (Groundwater Development Consultants, 1983). The high groundwater salinity in zone D, in southwestern Bahrain, is attributed to dissolution of the nearby sabkha deposits containing saline water with total dissolved solids of over 10,000 mg/l (Groundwater Development Consultants, 1980). Therefore, salinization of the aquifer with total dissolved solids of 7,000 mg/l is most probably caused by the flow of sabkha water into the aquifer.

In western Bahrain, the zone E has a high groundwater salinity (5,000 mg/l) resulting from aquifer contamination by irrigation drainage water. Unlike other parts of Bahrain, the aquifer in this area is under water table conditions, due to the erosion of its upper confining layer. This layer is assumed to protect the Dammam aquifer from surface pollution sources in most of Bahrain (Groundwater Development Consultants, 1980). In addition, the area contains massive agricultural activities manifested by heavy flood irrigation, frequent washing for soil salinity, heavy application of fertilizers and lack of an efficient irrigation drainage. The irrigation drainage water in this area contains elevated concentrations of Ca^{2+} and SO_4^{2-}, with averages of 730 and 2,140 mg/l, respectively (Raveedran and Madany, 1991). The high

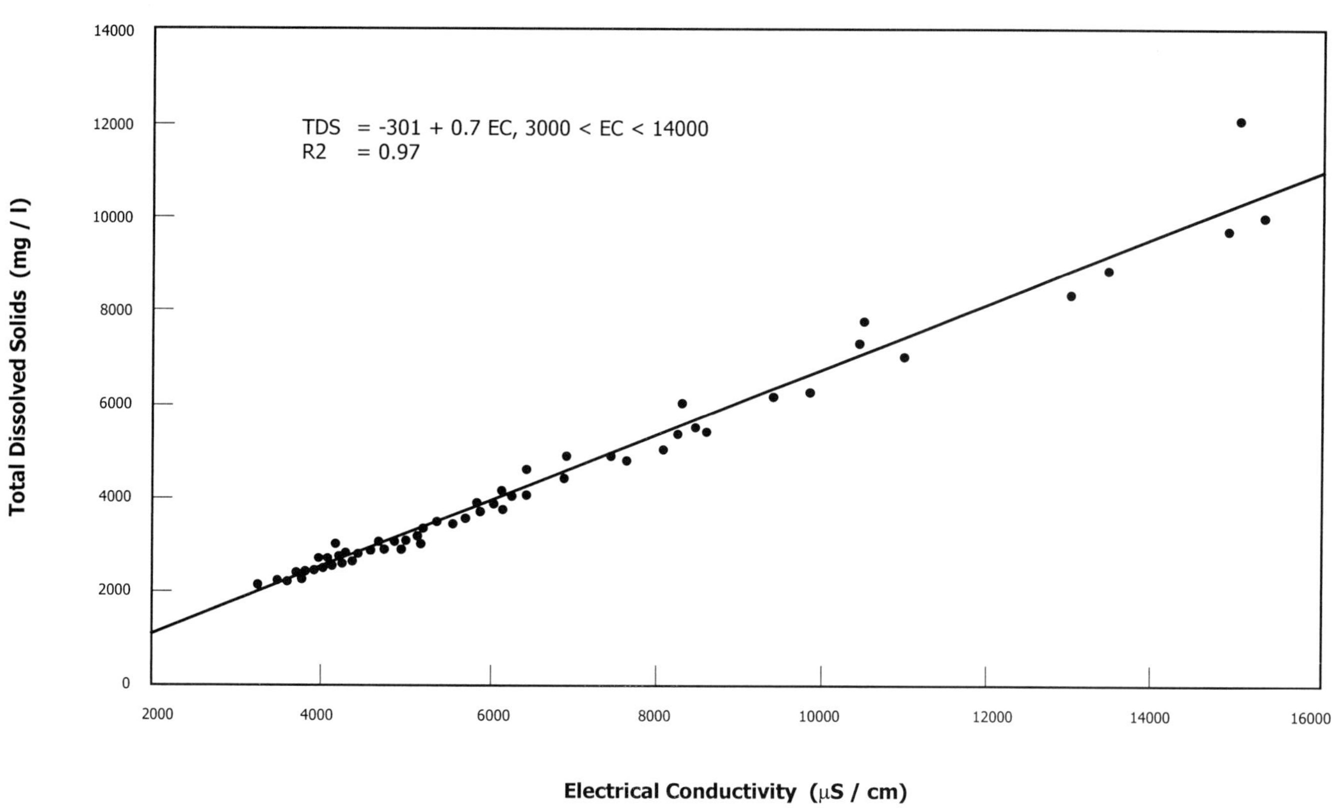

Fig. 8.33. The salinity (total dissolved solids in mg/l) versus Electrical Conductivity (μS/cm) regression line for Dammam groundwater in Bahrain (after Zubari et al., 1997).

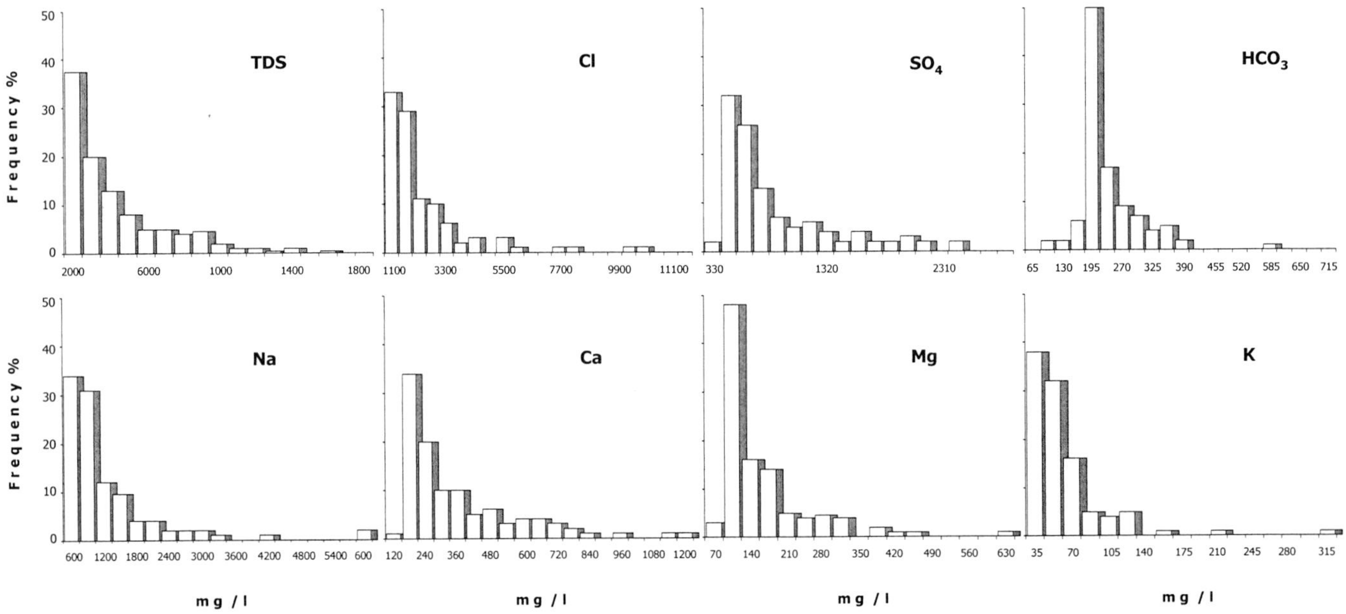

Fig. 8.34. Frequency distribution of groundwater total dissolved solids and major ions chemistry in the Dammam aquifer, Bahrain, 1992 (after Zubari et al., 1997).

concentration of these two ions in the irrigation drainage water was interpreted by Zubari et al. (1997) as due to soil composition, mainly limestone ($CaCO_3$) and gypsum ($CaSO_4.2H_2O$), and to the use of sulphate fertilizers.

Temporal Trend Analysis

Table 8.10 indicates that the mean total dissolved solids value for the Dammam aquifer in Bahrain has increased by about 25% over the period 1979-1992. Figure 8.36 shows that the total dissolved solids have increased in 79% of the 187 compared wells, reaching a maximum increase of 9,140 mg/l. The remaining 21% have shown a decrease in total dissolved solids reaching a minimum of 5,980 mg/l. On the other hand, a review of the abstraction rates from the Dammam aquifer reveals that the 138 Mm^3 pumped from the aquifer in 1979 (Groundwater Development Consultants, 1980) has increased by about 27% to reach 187 Mm^3 in 1992 (Al-Noaimi, 1993). The increased abstraction has taken place mainly in the northwest region towards the natural recharge front, where better water quality exists. The suggested safe yield from the Dammam aquifer in Bahrain ranges from 90 to 112 Mm^3/year (Al-Noaimi, 1993; Groundwater Development Consultants, 1980; Wright, 1967; Zubari, 1987), which means that the present abstraction approaches twice the recommended safe yield from the aquifer, and explains the continuous deterioration of the Dammam aquifer water quality.

The deterioration of groundwater quality in most of the aquifer areas in Bahrain shows that, the 16 major areas of increase in total dissolved solids, coincide with the main pumping areas for municipal and agricultural purposes in north-central, western

Table 8.10. Statistical summary of the total dissolved solids and major ion concentrations in the Dammam aquifer in 1992 (after Zubari et al., 1997).

Constituent (mg/l)	Number of samples	Mean	Standard deviation	Mode	Median	Minimum	Maximum
Total dissolved solids	254	4,679	2,723	2,240	3,455	2,120	1,6640
Cl^-	110	2,017	1,761	990	1,323	788	10,615
SO_4^{2-}	110	807	542	441	581	225	3,293
HCO_3^-	109	229	59	212	214	104	610
CO_3^{2-}	93	0	0	0	0	0	0
Na^+	110	1,037	907	525	688	421	5,819
Ca^{2+}	110	366	207	220	273	177	1,172
Mg^{2+}	110	140	92	81	107	57	667
K^+	98	53	40	28	40	24	321

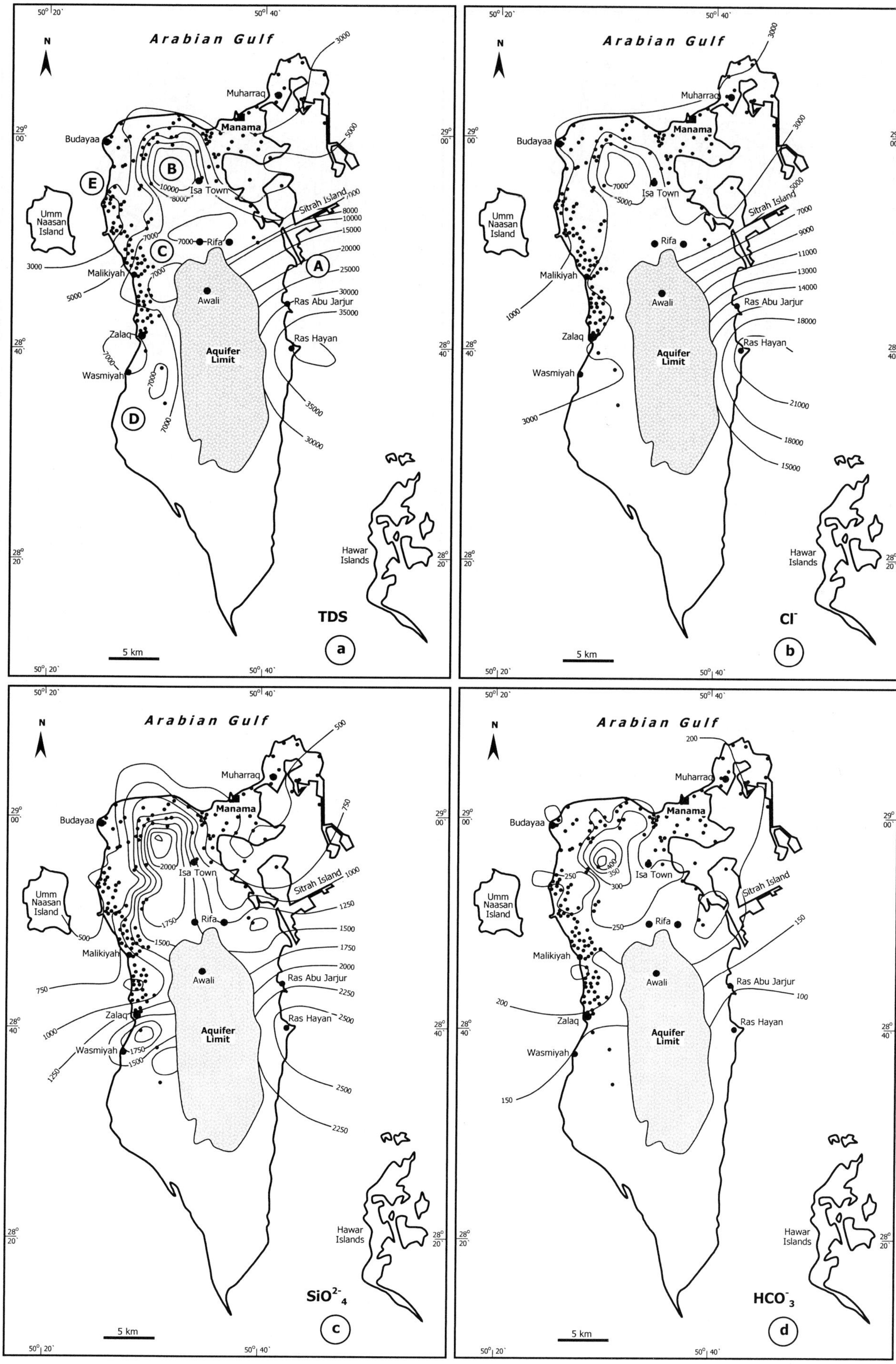

Fig. 8.35. Spatial distribution of groundwater major ions chemistry measured in 1992 in the Dammam aquifer, Bahrain (after Zubari et al., 1997). (a) Salinity expressed in terms of total dissolved solids in ppm; (b) Cl^- ion; (c) SO_4^{2-} ion; (d) HCO_3^- ion; (e) Na^+ ion; (e) Ca^{2+} ion; (g) Mg^{2+} ion; and (h) K^+ ion. Contour interval in mg/l.

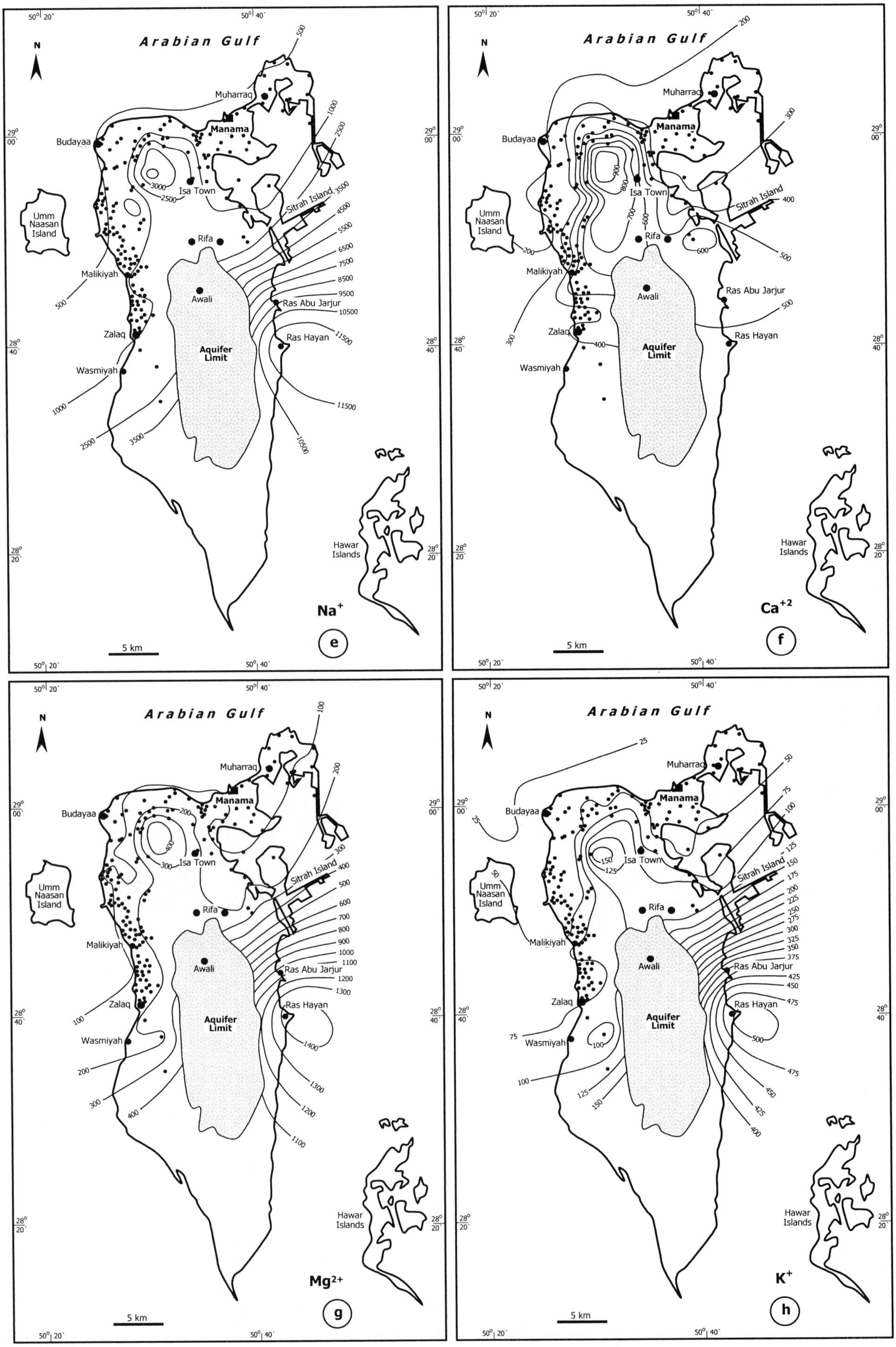

Arabian Gulf
N
Muharraq
Manama
Budayaa
Isa Town
Sitrah Island
Umm Naasan Island
Rifa
Malikiyah
Awali
Ras Abu Jarjur
Zalaq
Ras Hayan
Aquifer Limit
Wasmiyah
Hawar Islands
5 km
50° 20'
50° 40'
29° 00'
28° 40'
28° 20'
Na+
e
Ca+2
f
Mg2+
g
K+
h

Fig. 8.35. (Cont.)

and southwestern areas of Bahrain. The map indicates that the upward migration of brackish groundwater from the underlying formations, has extended to become the main source of quality deterioration in the Dammam aquifer. This implies that the management scheme began in 1980's to reduce the upward migration of brackish water, has not been effective. The control mechanism for this management scheme was based on reducing the difference in hydraulic heads between the Dammam aquifer and the underlying Umm er Radhuma aquifer.

In contrast, at two eastern coast localities, north Manama and Sitrah Island, a decrease in the groundwater salinity can be clearly observed. The north Manama decrease in the total dissolved solids levels of about 500 mg/l is attributed to the reduction of municipal groundwater abstraction by about 20 Mm^3/year (from 56 Mm^3/y to 37 Mm^3/y), and its replacement by desalinated water in 1985 (Statistical Data, Bahrain, 1991). In Sitrah Island, the abandonment of agricultural lands and cessation of aquifer abstraction in the 1980's, resulted in the stabilization of the hydraulic heads of Dammam aquifer (Zubari et al., 1993), and a consequent reduction in sea-water intrusion. This indicates that the reduction in municipal abstraction from the Dammam aquifer in the east coast in 1984 has been effective in decreasing the aquifer salinity.

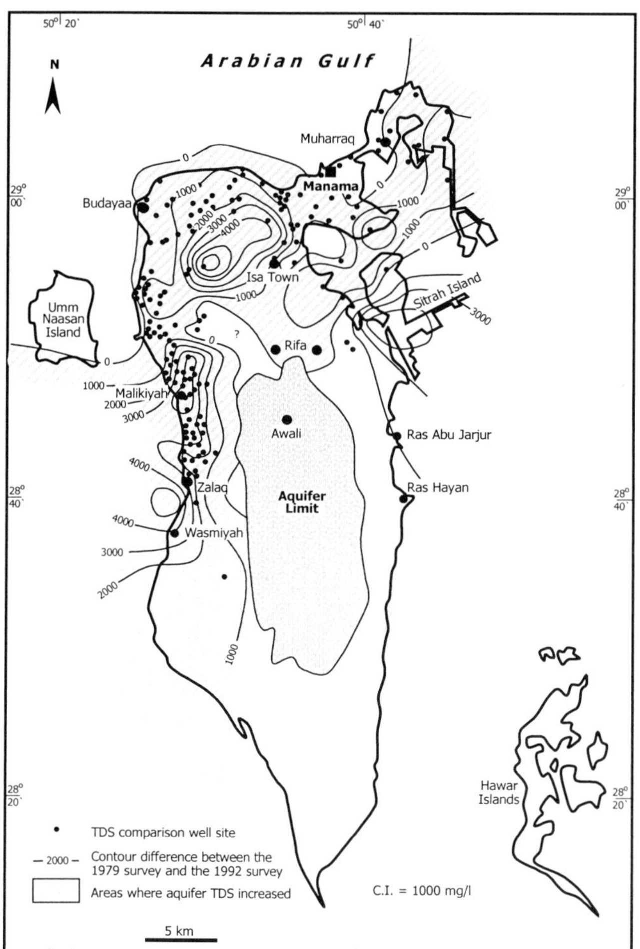

Fig. 8.36. Contour map of total dissolved solids (in mg/l) differences measured in 1979 and 1992. Negative contour values indicate decrease in total dissolved solids. Positive contour indicate increase in total dissolved solids (modified from Zubari et al., 1997).

Water Quality

Several hydrogeologists in Bahrain have identified four sources of contamination contributing to groundwater degradation of the Dammam aquifer in Bahrain. These are: sea-water intrusion in eastern Bahrain; brackish to saline upward flow from the underlying Umm er Radhuma aquifer in north-central and western Bahrain; migration of sabkha water in the southwest; and agriculture drainage water in local areas in western Bahrain (see Fig. 8.32).

Comparison between the measured total dissolved solids in a 1992 survey and the previous 1978 survey shows deterioration of groundwater quality in about 80% in 187 of the well sites (see Al Noaimi, 1999; Zubari et al., 1997). The quality deterioration identified over the comparison period, reveals that, upward flow of more saline water from the Umm er Radhuma aquifer, into the Dammam aquifer has expanded to become the dominant source of contamination. Meanwhile, agricultural drainage water has become an additional source of aquifer contamination, due to the prevailing hydraulic conditions, that favor the infiltration of surface water into the aquifer. The results obtained from this investigation suggest that more attention must be given to the vulnerability of the Dammam aquifer, to pollution from surface sources.

Temporal changes in groundwater quality, are attributed to the continuous increase of abstraction rates from the Dammam aquifer. Accordingly, the aquifer heads have fallen, permitting brackish and saline water from surface and subsurface contamination sources, to migrate into the aquifer. The hydrochemical characteristics of the recharge flow received from Saudi Arabia at the northwestern parts of Bahrain main Island, has remained unchanged. Moreover, the slight improvements in groundwater quality achieved in certain areas in the east and northeast coasts of Bahrain (Manama, west Muharraq Island and Sitrah Island), are the result of reducing abstraction rates in those areas.

Investigation of groundwater quality of the Dammam aquifer in Bahrain has shown the sources of increasing groundwater salinity. To control this rise in groundwater salinity, and overcome quality deterioration, the industrial sector in Bahrain must make more use of brackish water. The groundwater abstraction from the Dammam aquifer has to be reduced, especially in areas affected by sharp salinity rise. Artificial recharge of the Dammam aquifer by rainwater or treated sewage water can be assessed. Construction of additional desalination plants is needed to satisfy the ever-increasing domestic water demands.

4. TERTIARY AQUIFER SYSTEM IN QATAR

INTRODUCTION

The State of Qatar is peninsula without natural running water, extending into the Arabian Gulf. It runs 600 km north-south and is 65 km wide at its broadest point. To the south its border with Saudi Arabia lies in a zone of sabkha and sand dunes. The annual rainfall lies between 10 and 200 mm/yr (Fig. 8.37) and the annual surface runoff has been estimated at 1.35 Mm^3. Two thirds of the land surface is made up of some 850 contiguous depressions of interior drainage, with catchment areas varying from 0.25 to 4.5 km^2. Direct recharge may occur during some particularly heavy storms, but most is indirect through runoff, from surrounding catchment areas. The most important source of fresh and potable water, is obtained from freshwater lens, floating on brackish and saline

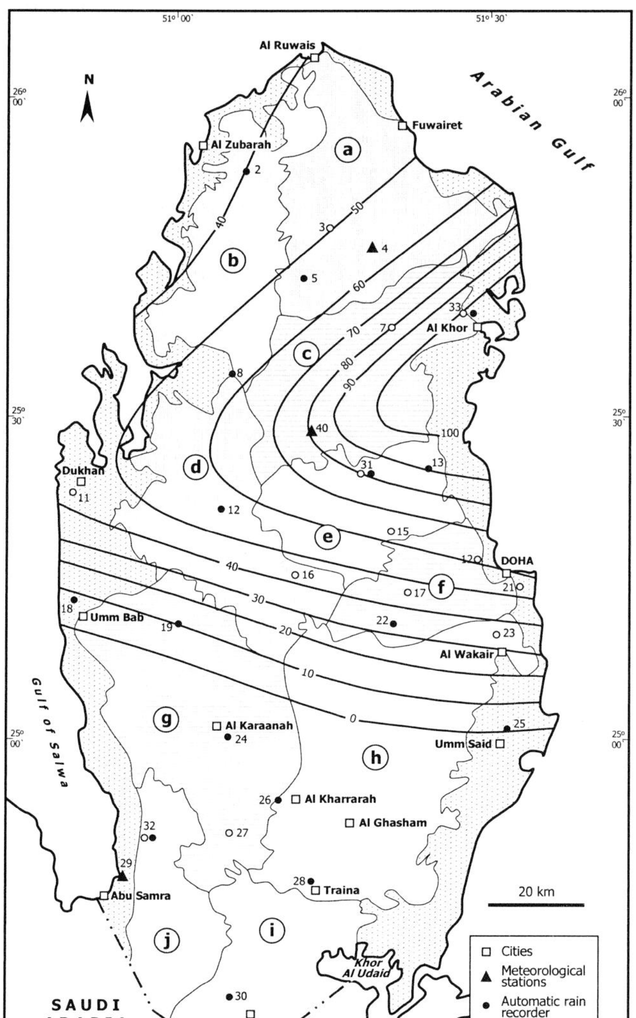

Fig. 8.37. Isohyet map of Qatar (rainfall in mm).

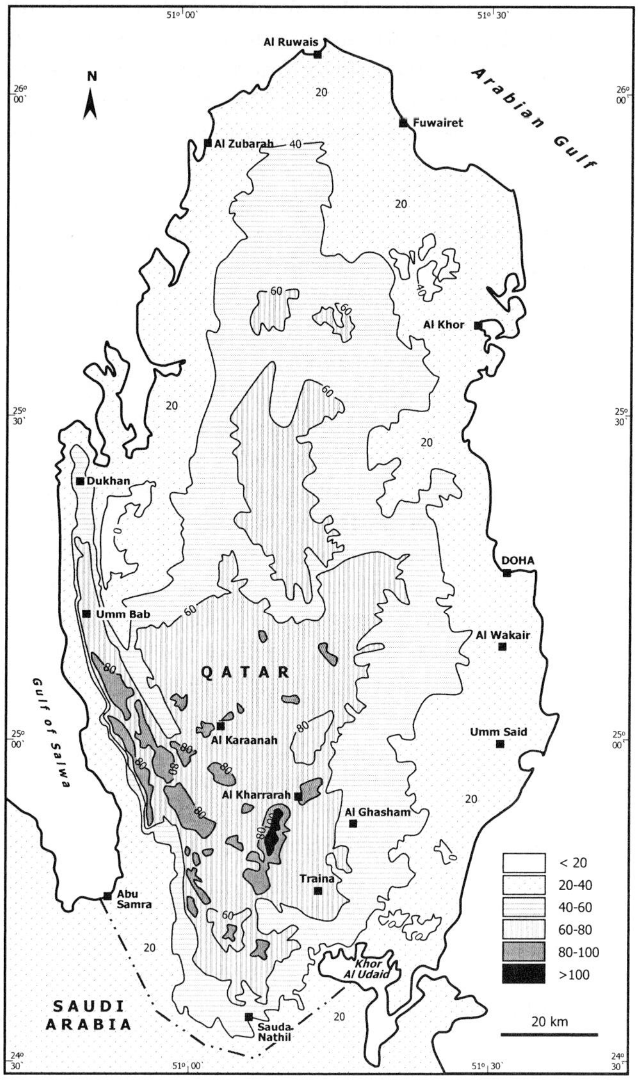

Fig. 8.38. Topographic map of Qatar (elevation in meters)

water. Some recharge is possible from storm water flowing into collapse depressions. Twenty offshore springs were listed by Walton (1962), but few are still flowing due over-pumping, especially during the last few decades. In an otherwise featureless landscape (Fig. 8.38), the most significant topographical features are the large number of shallow depressions, which are surface expressions of shallow collapse structures, a karst topography through which some recharge of the shallow aquifer, through the drainage of winter storm water may occur. The stony desert surface is composed mainly of alluvium in the depressions, calcareous sands, continental gravels, silts, muds, aeolian sand and sabkha deposits.

The two main aquifers underlying Qatar are recharged in Saudi Arabia. Over most of Qatar, the Dammam Formation contain only minor quantity of water because of its altitude. It dips in the southwest, and contains water in its lower part (the Alat Member). The underlying Umm er Radhuma Formation has an estimated safe yield of 10 Mm^3/yr, based on the annual flow from Saudi Arabia. In northern and central part of the Rus Formation is a partly unconfined aquifer, recharged by rainfall and return flows of agricultural water.

The Tertiary carbonates (dolomites, limestones and evaporates) and clastics (shales and sandstones) are interbedded with thin layers of marl and calcareous claystone underlie Qatar and crop out at the surface. The limestones, dolomites and sandstones act as aquifers, and the evaporites, shale and marls form aquicludes and aquitards.

The first comprehensive study of the hydrogeology of northern Qatar was carried out by the Qatar Petroleum Company and le Grand Adsco in 1957-1959, which included core drilling and resistivity survey, of some of the depressions. With the rapid growth of Doha (capital of Qatar) fresh groundwater became limited and the government commissioned a new survey of groundwater resources (1960-1961), the Parsons Corporation recommended exploratory drilling to locate higher quality groundwater (Al-Mojil, 1963). Subsequently three wells drilled indicated that the deeper aquifers yield saline water unsuitable for most purposes. Naimi (1965) presented clear evidence that the salinity of all Mesozoic and Cenozoic aquifers increased towards the east in the Arabian Peninsula consistent with the hydraulic gradient and the distance from the source of recharge. Further research by Italconsult (1967-1969) confirmed the existence of the salinity of the deeper aquifers, indicating that no potable water could be obtained from these aquifers. Songreah (1966) proposed that, the brackish groundwater in the middle Eocene sediments of Abu Samrah in southwest Qatar, could be piped to Doha and blended with desalinated water. This proposal was shelved and additional well fields in northern Qatar were drilled which, coupled with an increase in capacity of the desalinated plants, could meet the supply requirements.

During the 1970's a series of studies were undertaken by the government of Qatar with the aid of the UN Development Program and the UN Food and Agricultural Organization, to provide a quantitative assessment of the hydrogeological balance in Qatar, as well as a complete reconnaissance of the soils. One result of these surveys was to modify previous concepts of a floating freshwater lens to a more complex, two layers aquifer system. Ecclestone and Harhash (1982) have divided the aerial extent of the two layers aquifer model, into two broad hydrologic provinces, a northern and southern. To this two province model a small southwesterly zone is added (Fig. 8.39). Later in 1988, two deep wells drilled by the Ministry of Industry and Agriculture in the Sinneha area and in Wadi Lakhouane have shown that, the water tapped in the Dammam and Rus formations has the best quality, even though the salinity level is greater than that desired for agriculture. One of the wells which penetrated the Aruma Formation showed water with good quality potential.

The aquifer system of Qatar is an integral part of the Eastern Arabian aquifer system. The hydrogeological system of Qatar is heterogeneous. The varied distribution of depositional systems and their component facies imparts heterogeneity to hydrogeological conductivity, transmissivity, and lithology within the aquifer. Variations of climate, topography and artificial discharge within the area, also contribute to hydrogeologic heterogeneity. To detect relationships between geology and groundwater flow systems, as well as to delineate aquifer response to other controls, the hydrogeology of Qatar can be divided into three hydrogeologic zones (Fig. 8.39) (see Ecclestone et al., 1981), each zone with a distinct set of hydrogeologic properties (Table 8.11).

Table 8.11. Hydrogeological summary of Qatar aquifer system (modified from Al Hajari, 1990).

Zones	Hydrogeologic lithology	System	Water-produce characteristics	Transmissivity (m^2/day)	Storage coefficient
Northern Zone	Limestone, dolomitic limestone and chalk limestone	Freshwater lenses	Principal freshwater aquifer supplies small moderate fresh slightly saline water for agriculture moderate saline to poor a depth	Varies between 2-58	1.26×10^{-8}
Southern Zone	Dolomitic limestone, shale and thick evaporite	Multi-layered aquifer. It exhibits both confined and unconfined conditions	Brackish to saline water with a thin lense restricted beneath the collapse depressions	37	0.2×10^{-8}
Southwestern Zone	Dolomitic limestone, marl and evaporite	Artesian aquifer system	Brackish to saline water	200	10×10^{-4}

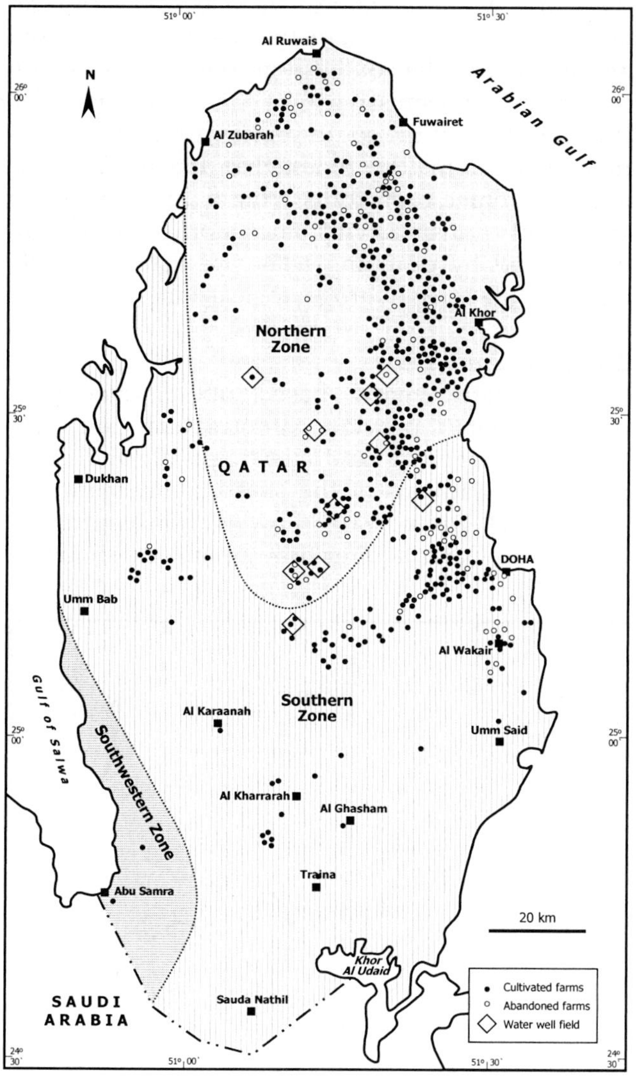

Fig. 8.39. The hydrogeologic zones, farms, and water wells in Qatar (modified from Al-Hajari, 1990).

Northern Hydrologic Zone

The northern groundwater zone or province has an area of 2180 km^2 and is the most important source of fresh and potable groundwater in Qatar. It occurs as a fresh water lens floating on brackish and salt water beneath collapse depressions. The lithology of this zone is characteristically a carbonate facies composed of gray to buff, compact, crystalline dolomitic limestone overlain by light-colored, soft, porous, chalky limestone intercalated with thin layers of marls, chert bands and calcareous claystone. The northern zone is limited by an evaporite front to the south (Fig. 8.39) and by the Arabian Gulf in the other directions. The hydraulic behaviour of the water lens follows the Ghyben-Herzberg principle (Pike, 1978) which states the relationship of a floating lens type aquifer to the lowering of the water table, will cause a rise of the fresh water/saline water interface at the base of the lens, by a factor ranging from 25 to 40, depending upon the salinity concentration of the underlying saline water. This leads to a situation where over-extraction will cause a concomitant rise of the interface accompanied by upwelling of saline water (Pike, 1978).

The dissolution of the Lower Eocene Rus evaporite unit in the northern zone has led to the creation of a complex lens beneath collapse depressions. The increasing porosity, permeability, transmissivity and storage coefficient has had a fundamental effect upon the present groundwater regime in northern zone.

Southern Hydrologic Zone

The southern groundwater zone or province occurs beneath more than half of Qatar, and forms an aquifer of somewhat less importance than the one to the north. This zone is mainly dominated by evaporite facies. This evaporite is characterized by thick, impermeable, compact beds of gypsum, overlain by a thin layer of microporous dolomitic limestone of the upper aquifer unit. It is underlain by the thick carbonate of the lower aquifer unit. The presence of the evaporite unit acts as an aquitard, except where occasional collapse depressions have allowed groundwater movement between the lower and upper aquifer units.

The groundwater distribution in this multi-aquifer system is controlled by facies distributions, related to tectonically controlled sedimentation and subsequent dissolution. The aquitard mainly contains saline water, with thin lenses of fresh water, restricted beneath isolated collapse depressions, within the upper part of saturated aquifer zone. This tends to give low yields of poor to brackish water.

Southwestern Hydrologic Zone

The southwestern groundwater zone occurs at the margin of the southwest of Qatar, and forms an artesian aquifer, in beds equivalent to the Alat and Khobar members of the upper Dammam aquifer unit of Saudi Arabia. The dominant structures of the southwestern groundwater zone are the Salwa syncline, which has a gently dipping western limb, and is isolated from Qatar by the Dukhan and Sauda Nathil domes in the Abu Samrah and Wadi al Araig areas.

Lithologically this aquifer unit is contained within predominantly dolomitic limestones, interbedded with marl totalling about 30m. It rests on top of the confining shale of the lower Dammam Formation. The unconformably overlying varieties of clay, marl, limestone and shale, of the lower Dam Formation form an aquiclude.

The Relationship of Geology and Groundwater

The lithofacies distribution, thickness variations, structure and post-depositional dissolution of the

Tertiary carbonates and evaporite rocks, have had a significant influence generally on the hydrogeology in all of Eastern Arabia. Variations in these characteristics have affected:

1. The aquifer parameters
2. Groundwater flow
3. Groundwater quality
4. Groundwater recharge and discharge

1. Aquifer Parameters

Tansmissivity and storage coefficient vary considerably with the maximum transmissivity and storage coefficient occurring at the margins of collapse depressions. Porosity and permeability decrease towards the southern groundwater zone (Eccleston and Harhsh, 1982; Harhash and Nasser, 1982). Transmissivity in the northern groundwater zone varies between 2 m^2/day and 5800 m^2/day with the highest values of 3600-4500 m^2/day occurring in zones where the aquifer is fractured and jointed. In the southern groundwater province the mean transmissivity is 37.2 m^2/day. The transmissivity of the southwestern zone varies from more than 312 m^2/day to less than 156 m^2 /day. The average storage capacity is around $10x10^{-4}$ m^2. The values of transmissivity and storage coefficient determined by pumping tests, are listed in Table (8.11) for the three hydrogeological zones.

2. Groundwater Flow

Piezometric maps, based upon water level measurements provide information on the hydraulic gradient and its regional and local variation. Gradients of the pressure head and the direction of groundwater flow within the Eastern Arabia and Qatar aquifer flow system are shown in

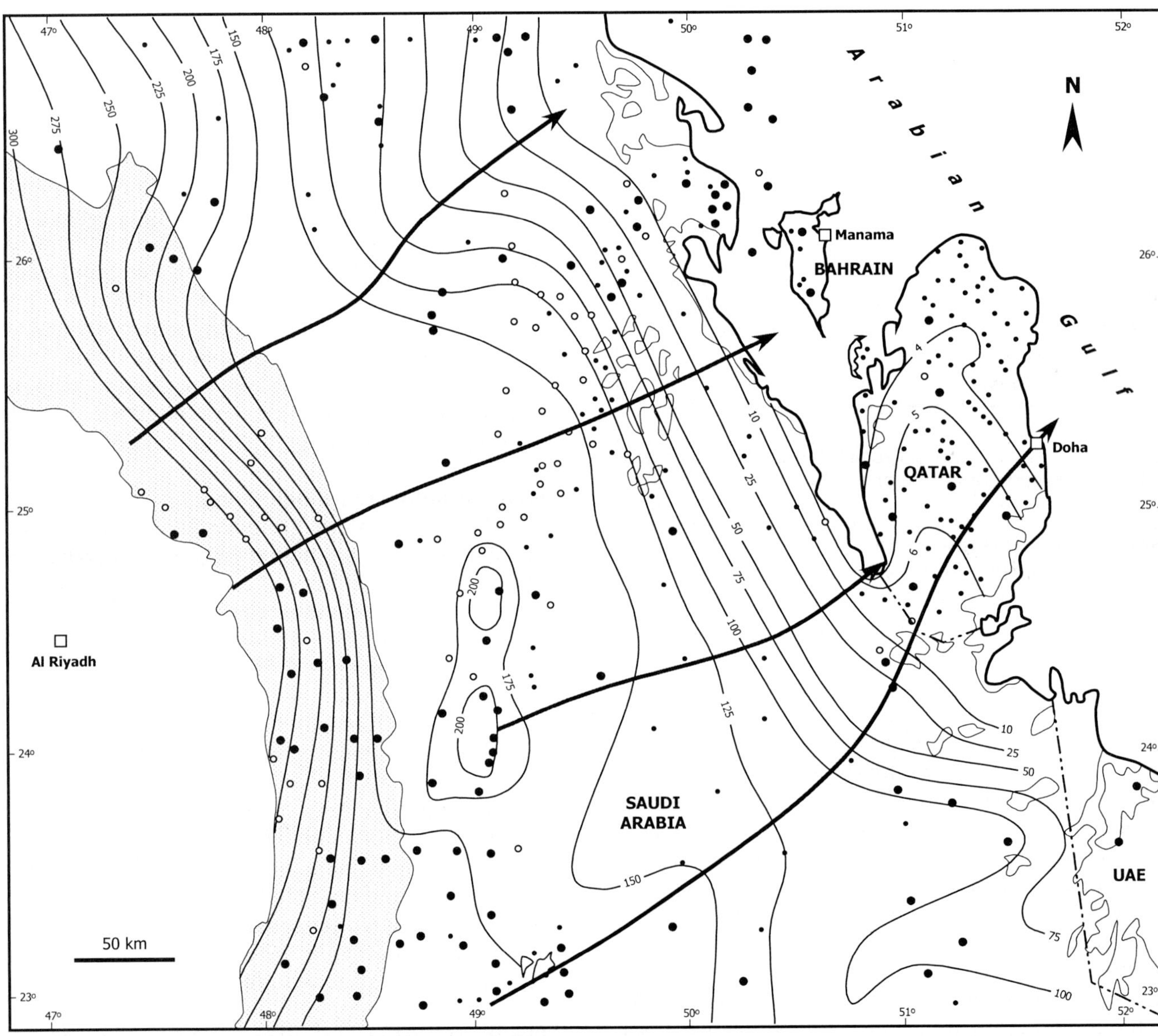

Fig. 8.40. Potentiometric surface map (in meters relative to sea-level) and flow direction of the Tertiary (lower aquifer unit), Eastern Arabia (modified from UN Food and Agricultural Organization, 1979).

Figure 8.40. This figure illustrates the regional direction of groundwater movement, within the eastern Arabia Tertiary aquifer system, at right angles to the lines of equipotential head. This movement extends from the center of Saudi Arabia and radiates towards the Arabian Gulf, a flow pattern that persists despite regional topographic features.

Groundwater circulation is essentially controlled by climate, topography, geology, and human activity. In Qatar, the geology (karst springs, evaporation through sabkhas and leakage from deep saline aquifer) and human activity, are the dominant controls on flow in the Tertiary aquifer system. The water level in Qatar varies with respect to the mean sea-level. It is about 9m in the southern zone and about 4m in the northern zone (Fig. 8.41), controlled by the hydrostatic head in Saudi Arabia. The groundwater flows radially outwards from recharge areas, centered over higher land-surfaces in the northern and southern zones and discharges into the adjacent low lying sabkhas and the Arabian Gulf, reflecting changes in land-surface topography and the elevation of Qatar (Fig. 8.41).

In one interpretation, the pattern of groundwater flow can be inferred from the distribution of hydraulic head in the northern aquifer zone. Figures (8.42 and 8.43) show the potentiometric surfaces, where high and low flow occur reflect water level measurements made during 1958 and 1988. This natural flow pattern suggests that, the northern aquifer zone has been changed, by heavy agricultural pumping over the last 30 years. Intensive groundwater production from the fresh water lens system, has resulted in brackish water intrusion, into the northern groundwater zone (Fig. 8.43). At the present time, the water levels have declined, to a new stable or near steady state condition. The extracted fresh water from the system is replaced laterally and vertically by saline water, without important changes in hydraulic head (Ministry of Electricity and Water - Qatar, 1987).

Sabkhat Dukhan as shown in Figure (8.41) has a significant impact on the aquifer system of Qatar. Discharge by evaporation through this sabkha, has created a regional cone of depression in the potentiometric surface and water flows from all directions toward the center of the cone. Groundwater flow from northern and southern zones flows in a curved path toward the sabkha.

The groundwater flow regime within the southern zone is dominated by groundwater mounds shown in Figure (8.41), which extend to 9m above sea level. Existing data does not show any water level decline in this area, although the shape of the mounds are slightly disturbed, because of variations in the distribution of middle aquitard, transmissivity, and vertical leakage. The higher potentiometric surfaces of these water mounds observed in Figure (8.41) are a reflection of the upward leakage, from the lower aquifer unit to the upper aquifer unit, through the middle aquitard bed.

3. *Groundwater Quality*

Regional trends in chemical composition of groundwater are mappable and very predictable, as shown by the isosalinity contour map (Fig. 8.44). Water quality in the Eastern Arabia varies geographically and vertically, and does always coincide with depositional distribution. As groundwater moves down flow paths from outcrop in central Saudi Arabia, a systematic hydrochemical evolution occurs; the total dissolved solids (TDS) gradually increases, and water evolves from dominantly calcium-bicarbonate, to dominantly

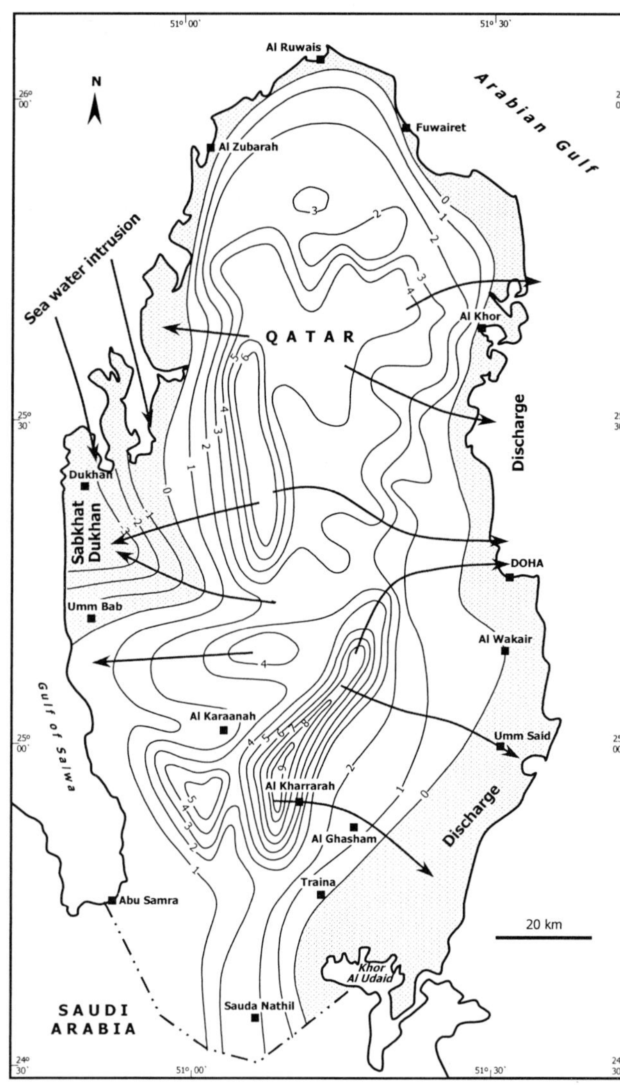

Fig. 8.41. Potentiometric surface map (in meters relative to sea-level) and flow direction of the Dammam aquifer in 1980, Qatar (modified from Al-Hajari, 1990).

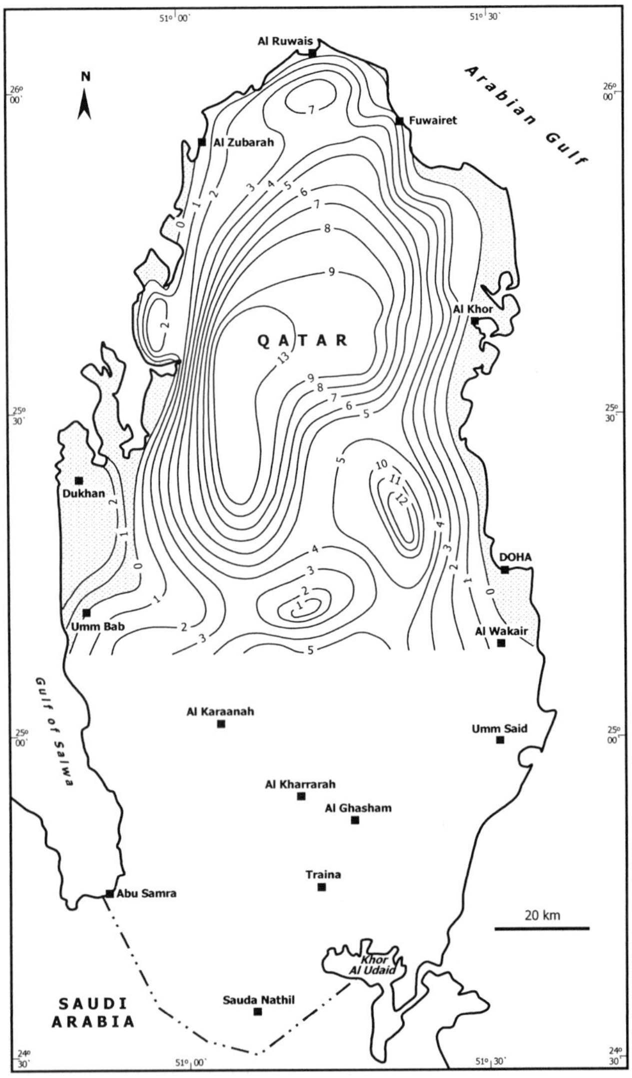

Fig. 8.42. Potentiometric surface map (in meters relative to sea-level) of the Dammam aquifer in 1958, Qatar.

calcium-sulphate composition (Fig. 8.44). This evolution and increase in the total dissolved solids values in the direction of groundwater flow is an expected result of the increase of the dissolution process, with distance and time from the contact between the groundwater and the rock matrix. The other possible reasons are upward leakage of deep saline aquifer, and over-extraction of water in the direction of flow.

Groundwater quality deteriorates progressively from less than 1,000 mg/l at the outcrop in Saudi Arabia, to more than 5,000 mg/l at the Arabian Gulf coast. The chemistry of the water column in the aquifer is not homogeneous, for the total dissolved solids content, increases with depth, due to variation in lithology and increase in temperature.

In Qatar and elsewhere in the Gulf states, water quality is a particularly important consideration in evaluating groundwater. As was indicated earlier, sea-water intrusion and deep saline water contamination is a significant problem. The salinity increase has been most marked in the coastal areas, but even the major onshore springs at Adhari in Bahrain, have more than doubled their salinity during the past 30 years, with a now undrinkable concentration of 3,000 mg/l NaCl (Walton, 1962).

The distribution of total dissolved solids in Qatar is shown in Figure (8.45). Local hydrochemical anomalies in this figure can be related to variations in recharge characteristics, groundwater mixing, and aquifer lithology. The most important processes controlling hydrochemical evolution within the aquifer are calcium- sulphate dissolution, and saline water contamination. The isosalinity trends show a close agreement with the equipotential with lowest concentrations occurring in the northern zone, increasing in the southern and southwestern zones. The total dissolved solids distribution shown on

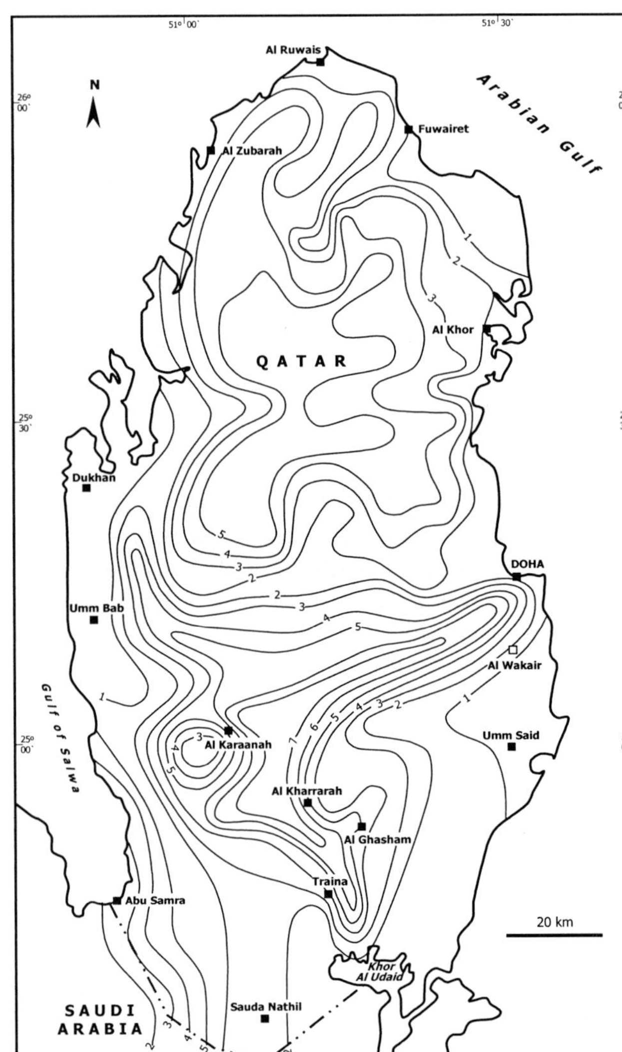

Fig. 8.43. Potentiometric surface map (in meters relative to sea-level) of the Dammam aquifer in 1988, Qatar.

Figure (8.45) clearly demonstrates a low concentration of dissolved constituents in the area of the groundwater mounds and recharge. There is an increase in total dissolved solids down-gradient in all directions; the sole exception to this being the mound underlying the Sauda Nathil dome, where the concentration of total dissolved solids is relatively high and amounts to 5,000 mg/l. In contrast to the conclusions of the piezometry and geology studies, which suggest that there is potential for groundwater movement beneath Sauda Nathil dome, chemical studies provide conclusive evidence that upward movement through and between aquifer units is taking place (Fig. 8.45).

The fresh groundwater body in the northern zone is mainly concentrated in the central part of the field, and is surrounded by a thick body of salt water which underlies the entire northern zone. The interface between fresh water and salt water is one of the boundaries of the fresh water lens system (Fig. 8.41). However, increases in pumping and production rates in more recent times has induced the movement of the salt water front towards the periphery of the fresh water zone. The saline invasion front is clearly seen in observation wells which indicate that it has moved about 4 km since artificial abstraction began (Fig. 8.43).

4. *Recharge and Discharge*

According to the precipitation data, it is clear that the rainfall recharge is greatest in the northern zone. Shallow wells in this zone have lowest total dissolved solids concentrations (Fig. 8.41). The southern and southwestern zones get less recharge and are lithologically more heterogeneous and have

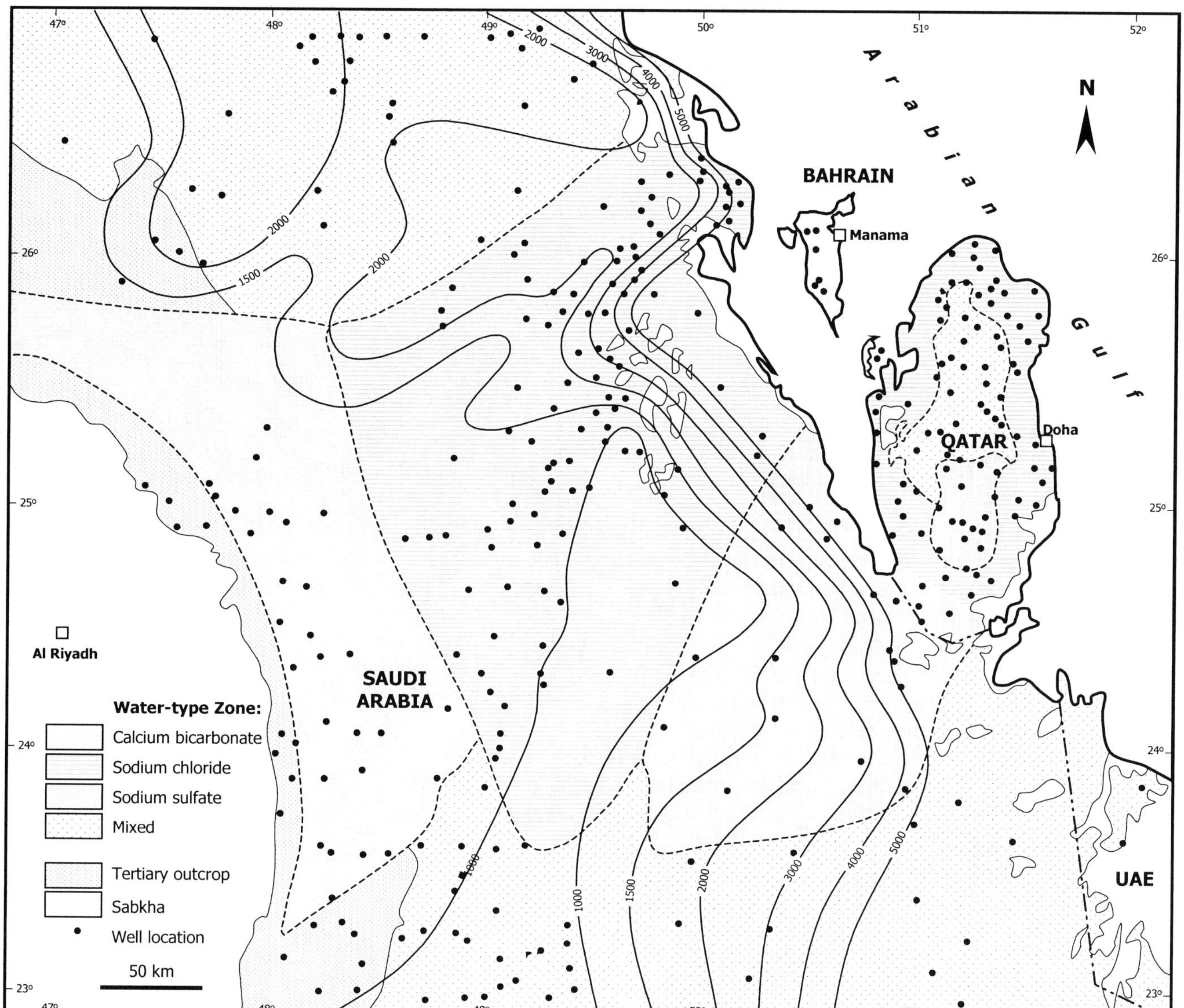

Fig. 8.44. Distribution of salinity and water-type zones in the Tertiary aquifer system, eastern Arabia (modified from UN Food and Agricultural Organization, 1979; Zekai and Al-Dakheel, 1986).

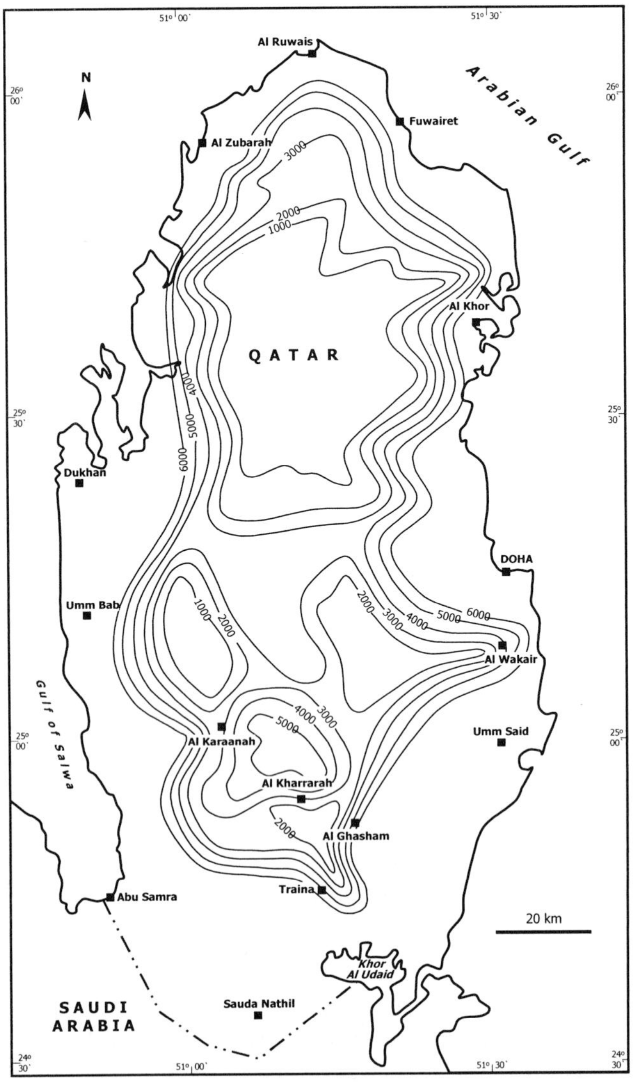

Fig. 8.45. Isosalinity contour map (mg/l) of groundwater in the Tertiary aquifer system in 1982, Qatar (modified from Al-Hajari, 1990).

higher total dissolved solids values. In the southern zone, the conditions of low recharge and poor groundwater circulation, are reflected in the general poor quality of groundwater.

The high salinity of the waters of central Qatar in the zone intermediate between the northern and southern zones, approximately coincides with the transition between carbonate and evaporite facies. The high salinity of the waters in the southern and southwestern zones coincides with the north-south Dukhan anticline axis (Fig. 8.46).

Hydrochemical analysis (Fig. 8.47) and facies maps (Figures 8.43 and 8.34) illustrate the compositional evolution that occurs as the groundwater moves through the aquifer system. These maps indicate that the waters in the northern zone are bicarbonate rich, with a lower salinity varying from 400 to 2,000 mg/l, whilst the waters in the southern and southwestern zones are sulphate-rich with salinity varying from 3,000 to 6,000 mg/l. This variation is probably due to the lithological differences between the northern and southern zones. For example, the high calcium concentration in the waters reflects the influence of carbonate lithofacies in the northern zone, while the sulphate waters also have higher calcium and sulphate levels, which indicate that the major source is the gypsum of the southern zone.

Under arid climatological conditions of Eastern Arabia, where potential evaporation greatly exceeds rainfall, infiltration to deep aquifers is one of the most controversially discussed issues. There is a continuous debate over the issue of fossil gradients, and whether the deep aquifer systems of North Africa and Arabian Peninsula are in receipt of any component of modern recharge (Burdon, 1977; Lloyd and Farag, 1978; Burdon, 1982 and Bakiewicz et al., 1982). Studies in this field have shown that

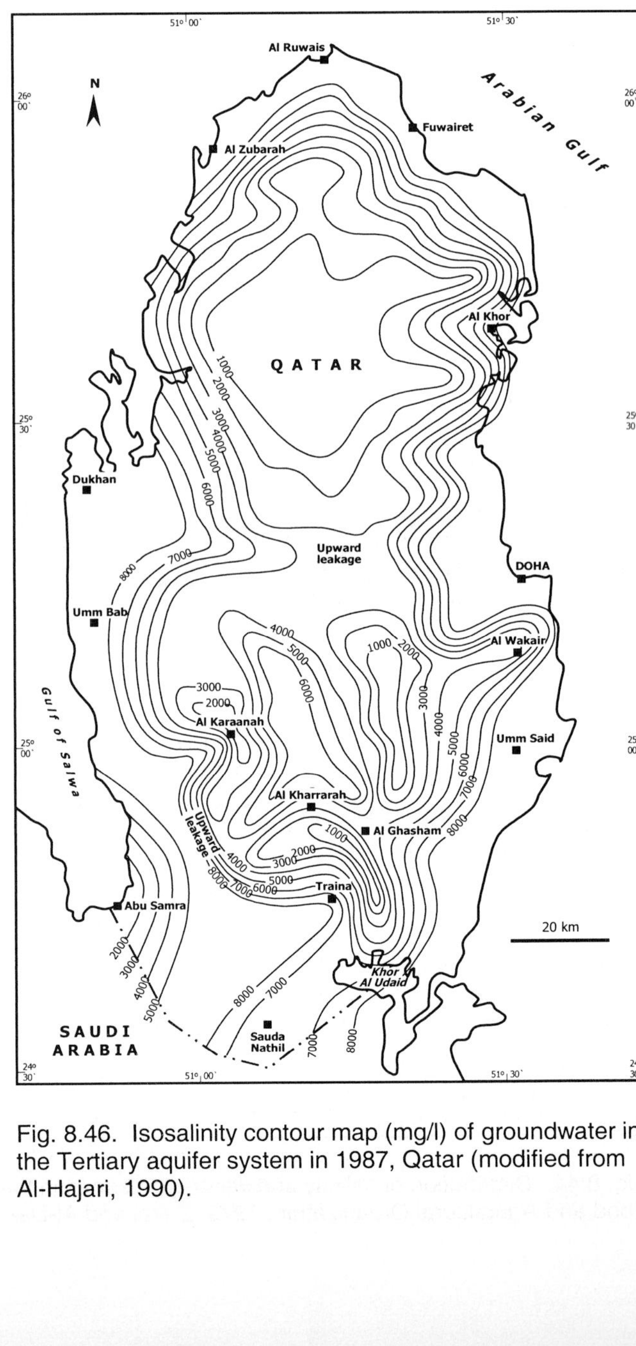

Fig. 8.46. Isosalinity contour map (mg/l) of groundwater in the Tertiary aquifer system in 1987, Qatar (modified from Al-Hajari, 1990).

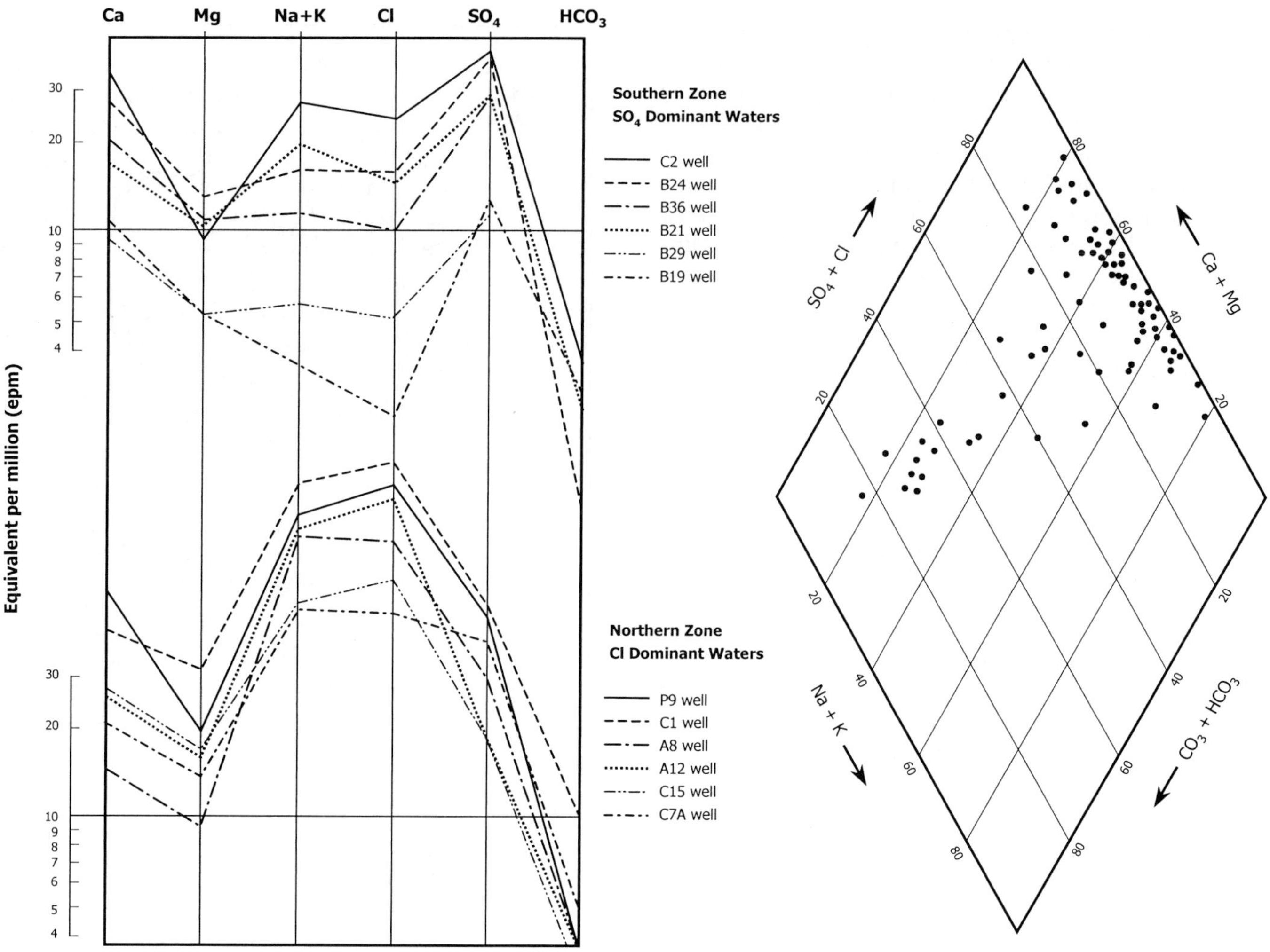

Fig. 8.47. Average chemical characters of the groundwater groups and piper diagram, Qatar (compiled and modified from UN Food and Agricultural Organization, 1979; Al-Hajari, 1990).

most of recharge of the Eastern Arabia aquifer systems was received during the past pluvial periods, and that present recharge is considered as meager (Cavelier et al., 1970).

These waters may have been modified somewhat by such processes as mixing with brines or surface waters, evaporation, hyperfiltration, and oxygen isotope exchange with rocks (Robinson and Al Ruwaih, 1985). Age determination based on ^{14}C analyses indicates that water being produced at Sabsab (near Marmul in Oman) from the lower aquifer unit (Umm er Radhuma Formation), some 130 km from the only recharge area in the Jabal Qar (South Oman), has an age between 9,000 and 13,000 years BP (Parker, 1985). Edgell (1990, 1997) indicates that 75-80% of the total spring water originates as fossil water from the upper and lower aquifer units through by-pass connections between the three aquifers in the truncation area on top of the Ghawar anticline southwest of Al Hofuf City in eastern Saudi Arabia. Evidence from isotopes showed groundwater in Arabia was originally recharged as rainfall on outcrops many thousands of years ago, when a more humid climate prevailed in the region. However, owing to the widespread karst conditions in Qatar, a great deal of natural recharge to the shallow water lens aquifer system can be expected. This occurs mainly where seasonal streams flow across the open karst areas. After intense storms, water can be seen flowing into many of the collapse depressions (Fig. 8.48) and the water collected in the temporary ponds partly infiltrates into the subsurface. The amount of infiltration in these areas depends on degree of karstification, near-surface geology conditions, topography, permeability and specific retention of the soil as well as rainfall distribution.

The results of Tritium (^3H) monitoring of wells in the upper shallow limestone aquifer in Qatar is shown in Figure (8.49). It also shows areas where groundwater is effectively being replenished at the present time (Yurtsever, 1992).

Most of the investigations on groundwater recharge show that infiltration to the aquifer system can occur under the present arid climate in the open karst areas. Studies of spring water using the ^{14}C method have shown that water from springs located in Al Hasa oasis in eastern Saudi Arabia has a young age (Abderrahman, 1979). Meteoric origin of groundwater in the shallow aquifer (Dibdibba and Dammam formations) in southwest Kuwait were reported by Hamida and Yaqubi, (1979); Sulin, (1946) and Collins, (1975). A study in an arid karst area in Saudi Arabia by Al-Saafin et al. (1989) proved that a considerable amount of recharge can be expected in open karst areas.

The geology, chemistry, and hydraulic head data show that regional groundwater circulation in Qatar is controlled primarily by geology, topography, karst features and rainfall distribution. Most recharge coincides with topographic highs in open karst, while major discharge areas coincide with major springs and inland and coastal sabkhas.

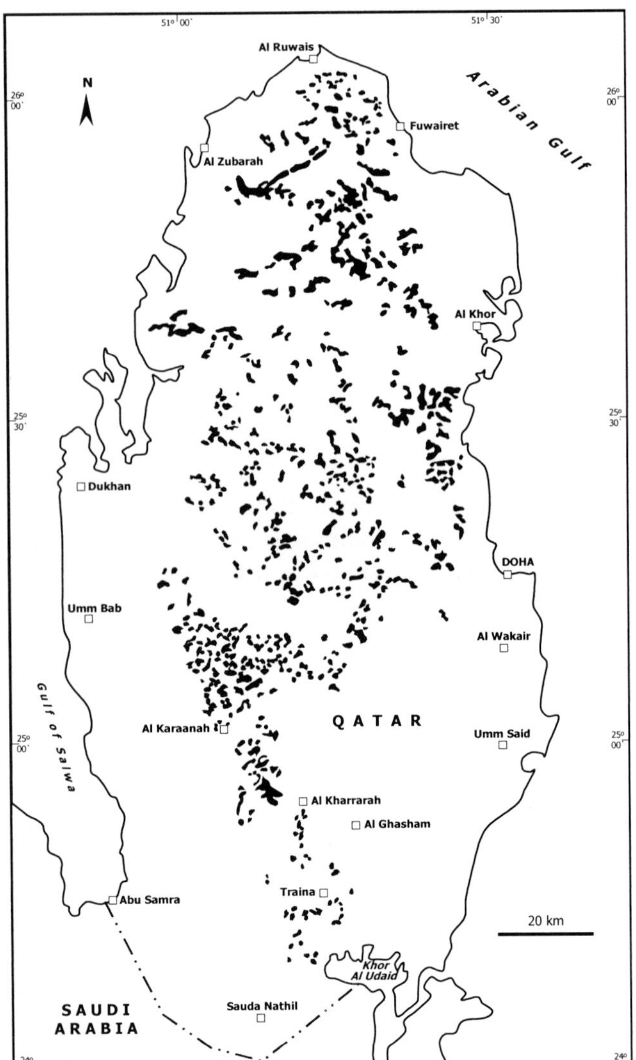

Fig. 8.48. Collapse depression, which serve as recharge areas for the Tertiary aquifer system in Qatar (modified from Al-Hajari, 1990).

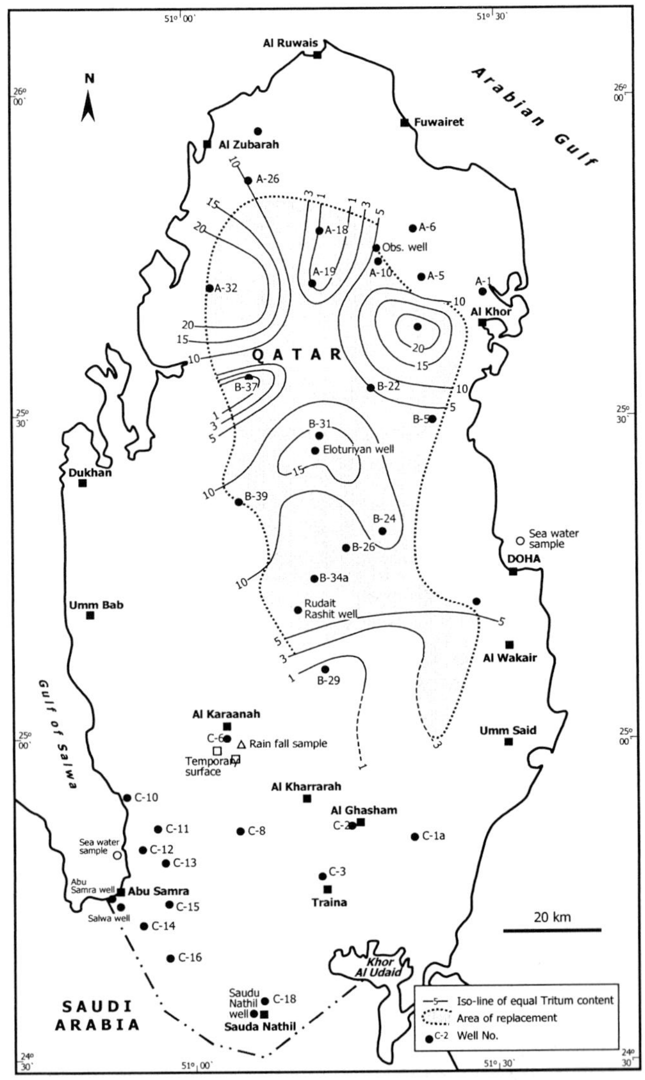

Fig. 8.49. Isolines of equal tritium content for shallow limestone aquifer in Qatar (modified from Yurtsever, 1992).

Groundwater recharge of the shallow water lens aquifer system in Eastern Arabia occurs in several forms: infiltration through open karst and collapse depressions, deep upward leakage through fractures and joints present in the rocks, particularly where underlying structure and/or confining beds, have been partly or totally dissolved, and laterally effected by sea water invasion. On the other hand, discharge occurs at the shoreline, in inland and offshore springs, and through areas where the water table intersects the land surface.

Experiments by Ball et al. (1981), using energy balance equipment for direct evaporation measurement, estimate the annual losses by evaporation from sabkhas and springs to be 1,050 Mm^3/yr.

Recharge and discharge of the aquifer system in Qatar can be classified into three sources for groundwater input and two zones of groundwater discharge. In the northern groundwater zone,

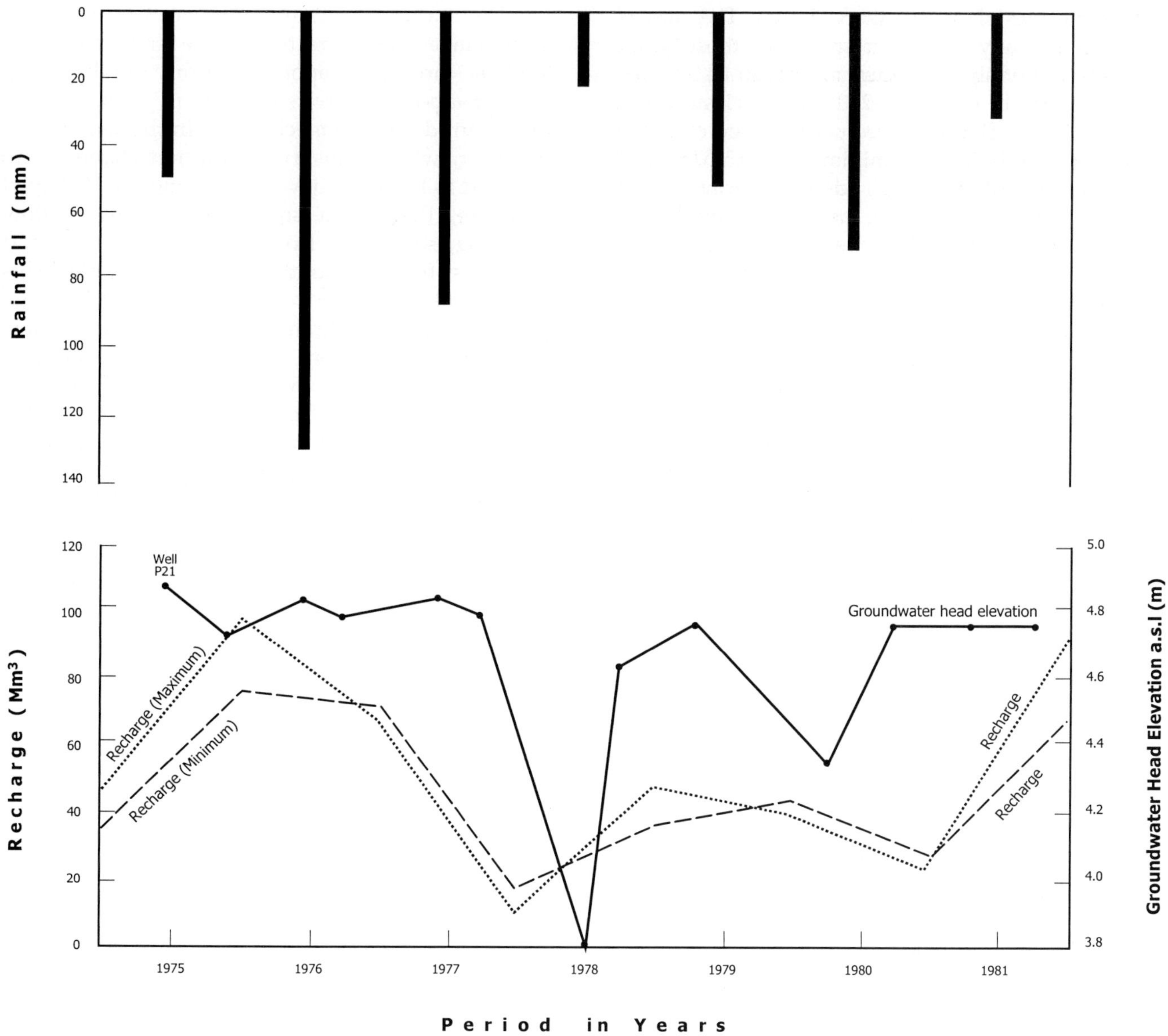

Figure 8.50. Diagram showing the relationship between rainfall, recharge, and groundwater head in Qatar (modified from Al-Hajari, 1990).

groundwater recharge occurs from rainfall, through hundreds of collapse depressions in the open karst, which serve as a connection between the surface drainage system and subsurface water lenses aquifer complex (Fig. 8.46). Most farms in Qatar are in the northern zone, and the aquifer system in that area receives recharge from the direct infiltration of excess irrigation water. The third source of groundwater recharge is marine water intrusion and underflow of groundwater from the intake areas beyond the borders of Qatar, under a natural gradient, through lower aquifer units in the southern zone, and both lower and upper aquifer units in the southwestern zone (Figs. 8.39 and 8.47).

The southern and southwestern zones get less recharge from rainfall because the dissolution of evaporite unit at shallow depth has not gone to completion. The occurrence of about 6 m of "Midra Shale Member" and the thick evaporite unit prevents the vertical infiltration of water to the aquifer system. However, there are a few depressions along the main anticlinal axis and around Sauda Nathil dome in the southern zone, formed in response to fractures and dissolution, which break the surface layers and permit infiltration, and enhance water circulation.

Atkinson and Eccleston (1986) state that, recharge is unlikely to occur from storms during which the rainfall is less than 10 mm. The rate of recharge is likely to vary from 1% for those years with rainfall of about 30 mm, to as high as 30% for rainfall years in excess of 200 mm (Eccleston and Harhas, 1982). The mean annual recharge over the northern zone is 27 Mm^3, minimum of 0.5 Mm^3 and a maximum of 86 Mm^3, derived from direct recharge which equals 2% of the annual rainfall, and 10% of annual rainfall by indirect recharge. In contrast in the southern zone, the mean annual recharge equals 6% of annual rainfall, and is estimated to have been an average of 14 Mm^3 with a minimum of 0.2 Mm^3 and a maximum of 40 Mm^3 (Eccleston and Harhash, 1982). Al Hajri (1990) reported that data from a previous storm which occurred in December, 1989 shows evidence of recharge to the shallow aquifer system (Figures 8.50 and 8.51).

The natural discharge of the Qatar aquifer system takes place directly into the Arabian Gulf, as well as through evaporation from sabkhas (Fig. 8.41). Evaporation from sabkhas is considered an important discharge mechanism. In the central part of Qatar, where subsurface solution channels are believed to be better developed along the V-shaped structure, there is a strong component of flow radiating, southwest from the northern zone and northwest from the southern zone (Fig. 8.41). This component of flow moves toward Sabkhat Dukhan, and subsequently discharges by upward leakage and high evaporation. In the extreme southeast there is an obvious component of flow throughout the area originating from the central part of the southern zone. In central east Qatar the groundwater probably flows laterally, and subsequently discharges sub-sea along the coastline between Al Doha and Al-Khor cities.

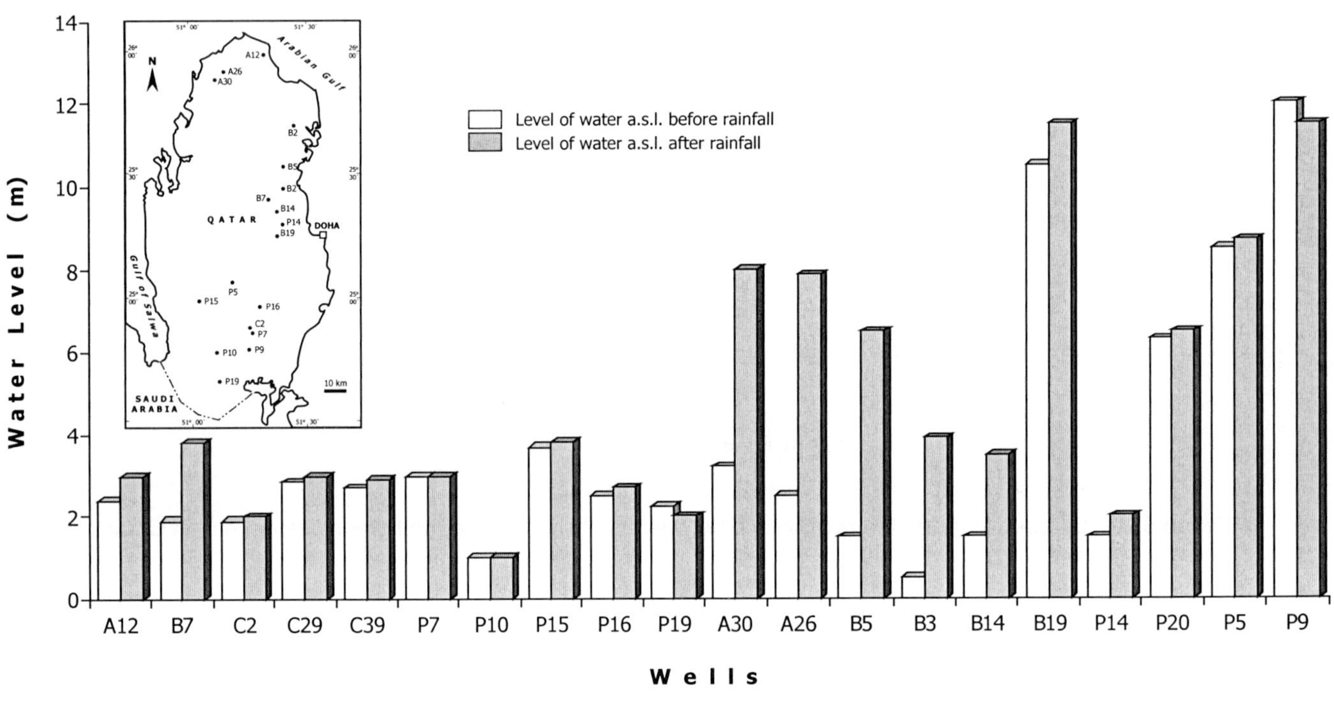

Figure 8.51. Evidence of recharge to the lower aquifer system in Qatar after heavy rains in December 1989 (modified from Al-Hajari, 1990).

5. QUATERNARY AQUIFER SYSTEM IN UNITED ARAB EMIRATES

INTRODUCTION

The Quaternary aquifer system contains the most important aquifers in the United Arab Emirates. The aquifers consist of alluvial gravels on both sides of the northern Oman Mountains in the eastern region, and the sand dunes in the western region (Fig. 8.52). These aquifers contain the largest reserve of fresh groundwater in the country.

Field measurements show that the depths to groundwater are 5m in the Liwa, Dibba, Khor Fakkan, Kalba, Shaam and Khatt areas, 10-25m in the Al-Shuayb, Madinat Zayed and Al-Madam areas, 25-50m in Al-Wagan, Al-Hayer, Jabal Hafit, Al-Faiyah, Al-Jaww plain, Hatta and Masafi areas, 50-100m in Wadi Al Bih and Al-Ain areas, and >100m in Al-Dhaid area (Fig. 8.53). Hydraulic head measurements reveal the presence of four major cones of depressions centered at Al-Dhaid, Hatta, Al-Ain and north of Liwa. Water depths in the first three cones is greater than 100m, and water depth in the center of the fourth cone is about 50m.

The presence of the cones of depression is related to excessive groundwater pumping, and the limited annual replenishment of the exploited aquifers. These cones reflect declines in groundwater level, and result in wells going dry (Al-Dhaid area), an increase of groundwater salinity and the beginning of salt-water intrusion. Two west-east progressing salt water tongues south of Dubai and north of Al-Ain, have been observed in the sand and gravel aquifers. Salt-water intrusion also occurs west of Kalba and north of Khor Fakkan along the eastern coast, and at Wadi Al Bih on the northwestern coast. Salt-water intrusion is not limited to coastal areas, because salt water can move upward upconing from deeper horizons of the aquifers (Al-Dhaid and Al-Ain areas). Saline groundwater under sabkha areas (such as Sabkhat Al-Thuwaymah, west of Al-Ain city), can move laterally under the effect of heavy pumping, to intrude into fresh groundwater in the Al-Ain area.

Field measurement of depth to water and ground elevations from topographic maps are used in the construction a rough hydraulic head map for the sand and gravel aquifers (Rizk et al., 1997), and Figure 8.54 shows a hydraulic head map for the Quaternary aquifer system in the United Arab

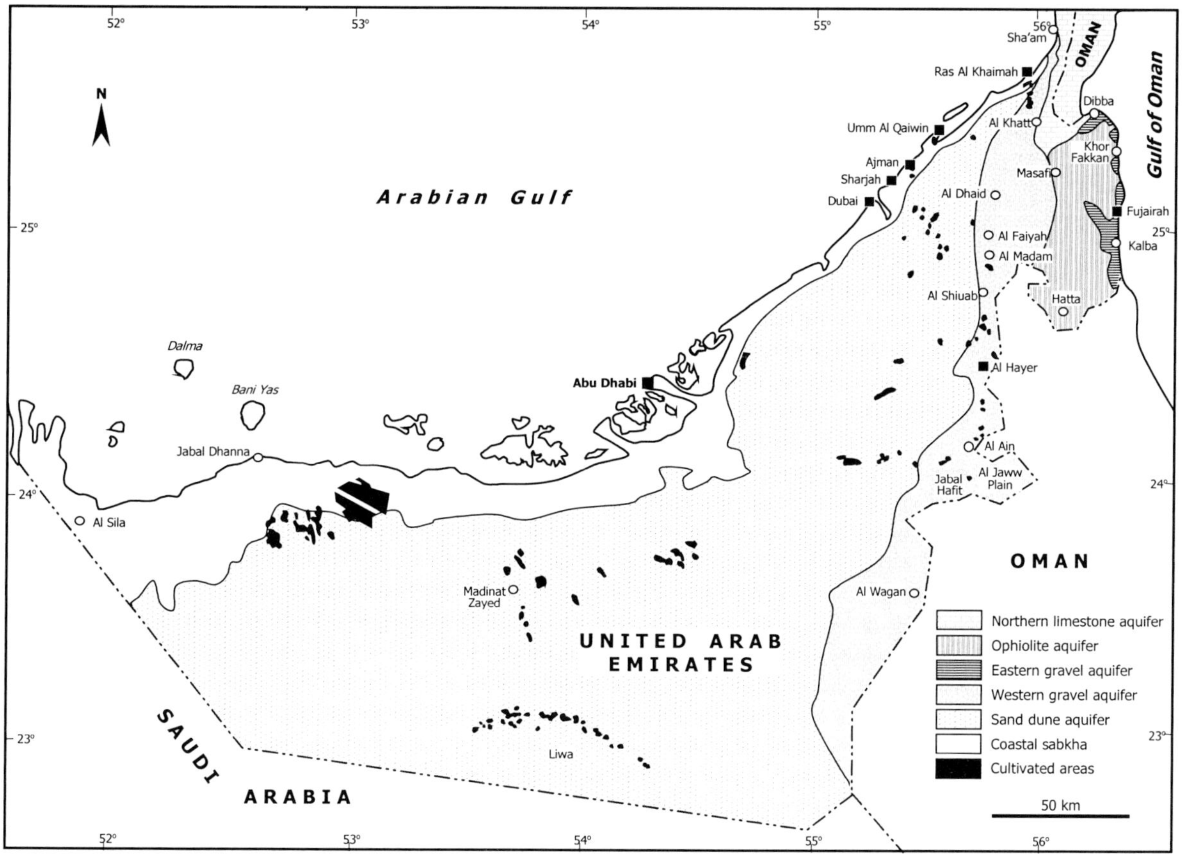

Fig. 8.52. The main water-bearing units (aquifers) in the United Arab Emirates.

Emirates during 1996. This map shows that the eastern mountains are the main recharge area for groundwater in the United Arab Emirates, whereas the Arabian Gulf and the Gulf of Oman are the main discharge areas. Local discharge areas are encountered west of Al-Ain, south and east of Liwa and in the western Abu Dhabi coastal sabkhas close the Arabian Gulf.

The groundwater flow in the northern limestone aquifer is mainly controlled by fractures, with a net flow towards the Arabian Gulf. The Khatt springs (Ras Al-Khaimah) originate where a fault structure interrupts the continuity of these fractures. Artesian conditions in the United Arab Emirates was observed in farms located southwest of the Khatt springs. Groundwater flow in the ophiolite sequence is also controlled by fractures, and the Maddab spring (Al-Fujairah) is one which originates along an east-west fault dissecting these rocks. The groundwater flow in the sand and gravel aquifers on the western side of the mountains, is generally from east to west and northwest between Latitudes 24°00` and 26°00`N, and from southeast to northwest, between Latitudes 22°00` and 24°00`N.

Flow Systems

Toth (1963) suggested that most flow nets could be separated into local, intermediate and regional systems of groundwater flow, although, the flow system actually present in an area depends on local topography and basin-shape geometry. The detailed study of groundwater flow in the United Arab Emirates is consistent with the presence of local, intermediate and regional groundwater flow systems (Fig. 8.55). Water wells and springs discharging from local groundwater flow systems, are of low salinity and water temperature are close to the mean annual air temperature. In contrast, the water of the springs discharging from regional groundwater flow systems, is highly mineralized and at a higher temperature as discovered by Fetter (1988). The local groundwater flow system is limited to the eastern mountains, where the hydrologic cycle is relatively rapid, and groundwater has a short residence time. The low salinity water of this system belongs to the HCO_3^- water type. The groundwater of local flow systems has a good quality, such as those of Masafi and AI-Jaww plain areas. It seems that the Khatt (Ras Al-Khaimah) and Maddab (Al-Fujairah) springs, discharge a local groundwater flow system. Inland sabkhas are the main discharge areas for groundwater of the intermediate flow system. In the inland discharge areas, groundwater is generally brackish, has a moderate residence time and belongs to the SO_4^{2-} water type. Because of the discharge area, groundwater has relatively high salinity,

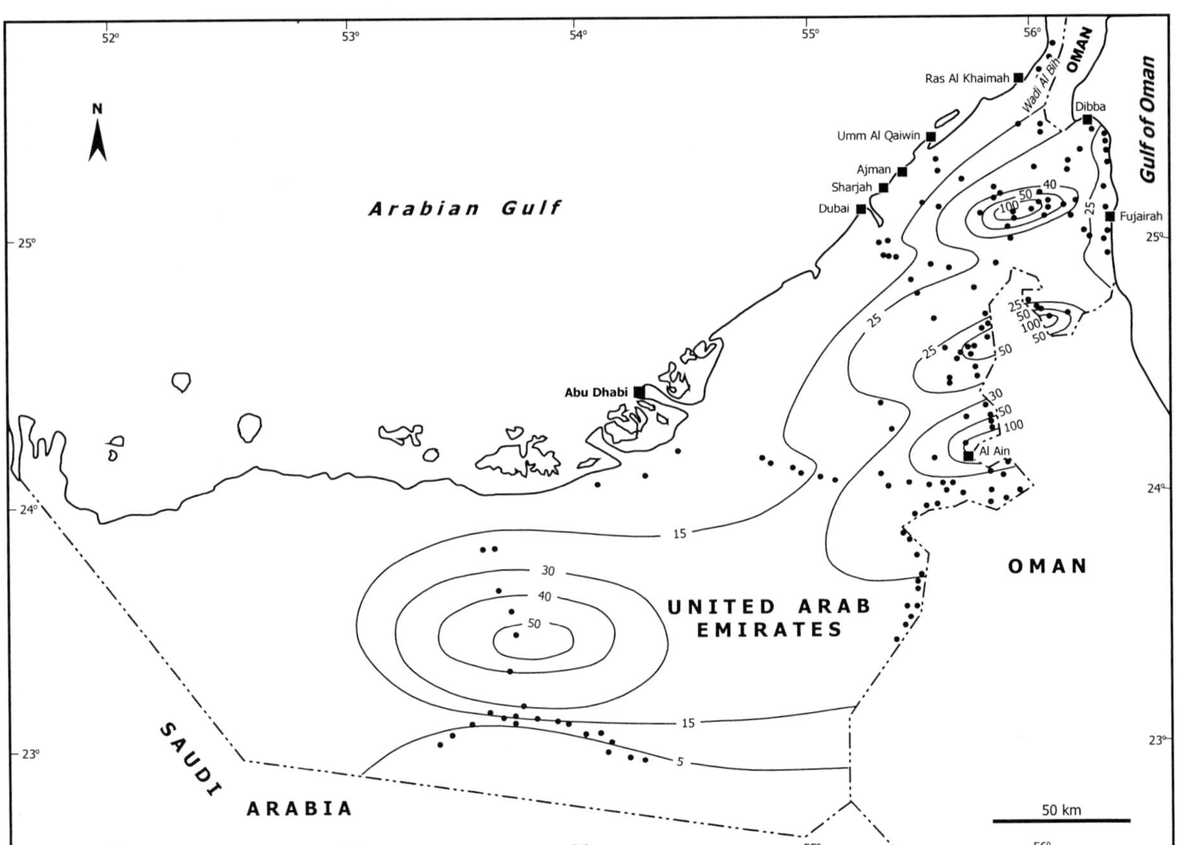

Fig. 8.53. Depth to groundwater (m) in shallow wells tapping different aquifers in the United Arab Emirates, measured in March (1996).

groundwater of this system has limited uses. Ain Al-Faydah (Al-Ain) seems to discharge from such a system. Coastal sabkhas represent the essential discharge areas, for regional groundwater flow systems. The water is highly saline, has a long residence time and belongs to the Cl^- water type.

Quaternary Aquifers

The Quaternary aquifers of the United Arab Emirates are the gravel aquifers flanking the eastern mountain ranges and the sand dune aquifers in the west and southwest (Fig. 8.52).

A. Gravel Aquifers

The largest reserve of fresh groundwater in the United Arab Emirates occurs in the alluvial deposits, of the piedmont plains bounding the eastern mountains, on the east and west. These aquifers can be distinguished into the eastern gravel aquifer, the northwestern gravel aquifer, and the western gravel aquifer. The eastern gravel aquifer is composed of a series of alluvial flats filling the embayments between promontories of rock spurs into the Gulf of Oman. South of Khor Fakkan, the flats and wadi fans coalesce, to form an almost continuous coastal plain, between the mountains and the sea. The fans near the mountains comprise rock and coarse gravel, which become progressively finer in size, as the distance from the mountains increases. The aquifer contains fresh groundwater, which drains from the apex of the wadi fans toward the sea.

The northwestern gravel aquifer occurs as thin lenses under the sands, and in places, as buried alluvial channels entrenched in the bedrock. The presence of these channels is associated with low-salinity groundwater. An example of these channels was pointed out by Rizk et al. (1998) based on their work in Al-Ain area. Deposits of gravel have been penetrated within the Quaternary sediments, in nearly all boreholes drilled in the gravel plain at depths of 60m below the ground surface, and at distances up to 70 km from the eastern mountains (United Nations, 1982).

Quaternary alluvium of the western gravel aquifer, is composed of a sequence of about 60m of sand and gravel, with thin interbeds of silt and clay. The alluvium was derived from the ophiolitic Oman Mountains. In the north and west of Al-Ain, present-day wadis are located between NE-EW trending sand dunes.

Recharge of the gravel aquifer in the Al-Ain area, comes from rain that falls on the western flank of the Oman Mountains, and runs through wadis, where it infiltrates to recharge the aquifer. Water-table maps

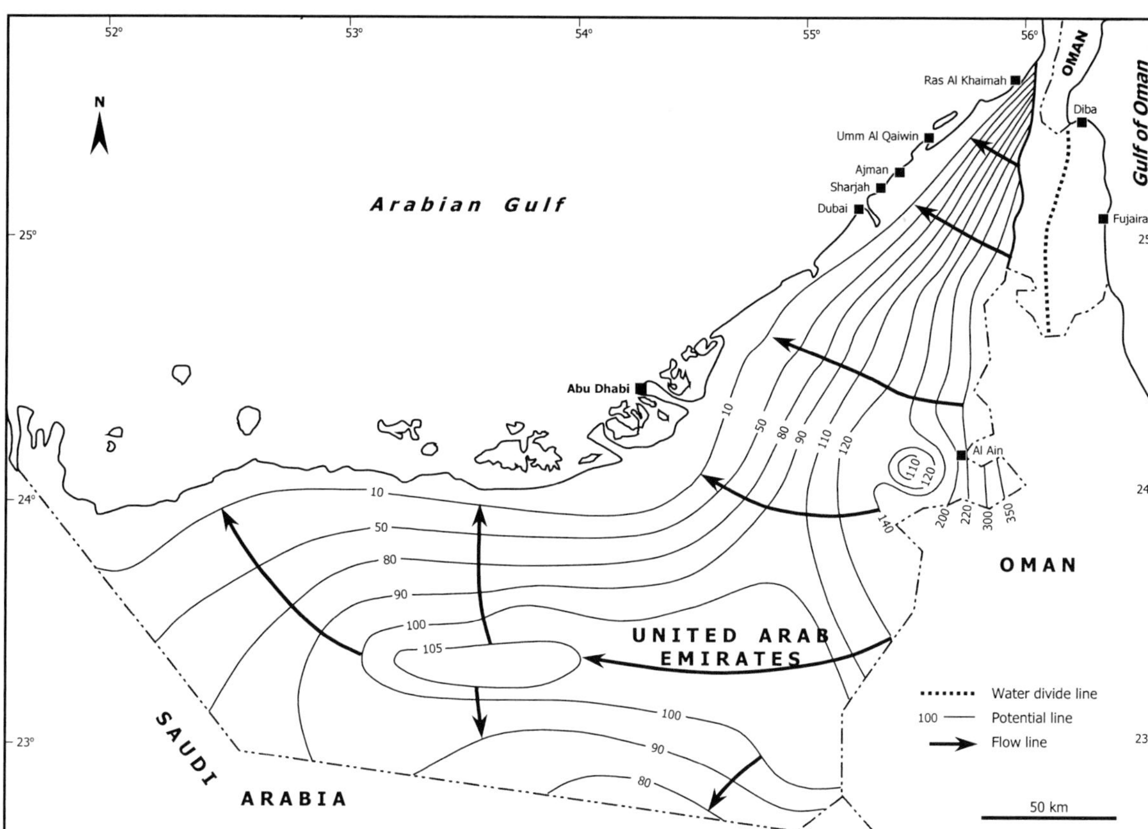

Fig. 8.54. Hydraulic head map, in meters above sea level, for the sand and gravel aquifer, 1995, United Arab Emirates (modified from US Geological Survey – Groundwater Research Project, 1996).

for the gravel aquifer in the Al-Ain area, were previously prepared by Gibb and Partners (1970), Hydroconsult (1978), and German Water Engineering (1982). Woodward et al. (1992) used oil exploration uphole seismic data to assist in constructing a water-table map of the gravel aquifer in the Al-Ain area. This map suggests a buried paleodrainage network which contains saturated alluvial fill, and may constitute major fresh water aquifers in the Al-Ain area.

B. Sand Dune Aquifer

Sand dunes cover about 74% of the total area of the United Arab Emirates, and gradually increase in elevation from sea level at the western coast, till they reach 250m above sea level at the Liwa-Al Batin basin in the south central part of the United Arab Emirates. Sand dunes are the least studied aquifer in the United Arab Emirates. The Groundwater Research Project (the National Drilling Company and the United States Geological Survey) described a fresh water aquifer in the Quaternary sand dunes between Liwa and Madinat Zayed. Based on preliminary hydrogeological investigations, we can claim the presence of a similar fresh water mound in the sand dunes of the Bu Hasa oil field. Exploration of the sand dunes between Al-Wagan and Liwa, may lead to the discovery of similar fresh water lenses. The Liwa area is covered by undifferentiated eolian sands and sabkha deposits. Eolian deposits consist of medium to very fine-grained sand with silt composed of quartz, carbonates and heavy minerals. The sabkha deposits occur between dunes, mainly south of the Liwa Crescent area, and are composed of thin sand and silt deposits. The thickness of this unit is highly variable, varying between 50 and 150m.

Physical Properties and Water Chemistry

The study of dissolved chemical constituents in groundwater of the United Arab Emirates reveals the effect of rainfall, geology and hydrogeological conditions on water quality, and suitability for different uses. The electrical conductance (EC) (μS/cm), water temperature (°C) and hydrogen-ion concentration (pH) were directly measured in the field. The samples were then analyzed for major cations (Ca^{2+}, Mg^{2+}, Na^{+} and K^{+}) and anions (CO_3^{2-}, HCO_3^{-}, SO_4^{2-} and Cl^{-}) in the Chiba University laboratory, Japan.

A. Water Temperature

Measured groundwater temperatures vary between 22°C along the eastern coast, to 51°C in water wells penetrating the Eocene limestone aquifer at Jabal Hafit (Fig.8.56). Centers of high groundwater temperatures are observed in springs and wells of Khatt (Ras Al Khaimah), Al Dhaid (Al Sharjah), Hatta (Dubai) and south Al Wagan (Abu Dhabi) areas. These centers lie along a NNE-SSW striking thrust fault, which represents the western boundary of the northern Oman Mountains. A separate high groundwater temperature center is located between Madinat Zayed and Liwa, from which groundwater temperature decreases in all directions. It is believed that, high-temperature groundwater is related to: (a) the groundwater flow systems, or (b) radioactivity. It is likely that Ain Al-Faydah spring (Al-Ain) obtains most of its water, through an intermediate groundwater flow system, that starts at the northern Oman Mountains in the east. This explains the spring's high water temperature and total dissolved solid contents. The joints and bedding planes of some calcareous rocks of Jabal Hafit

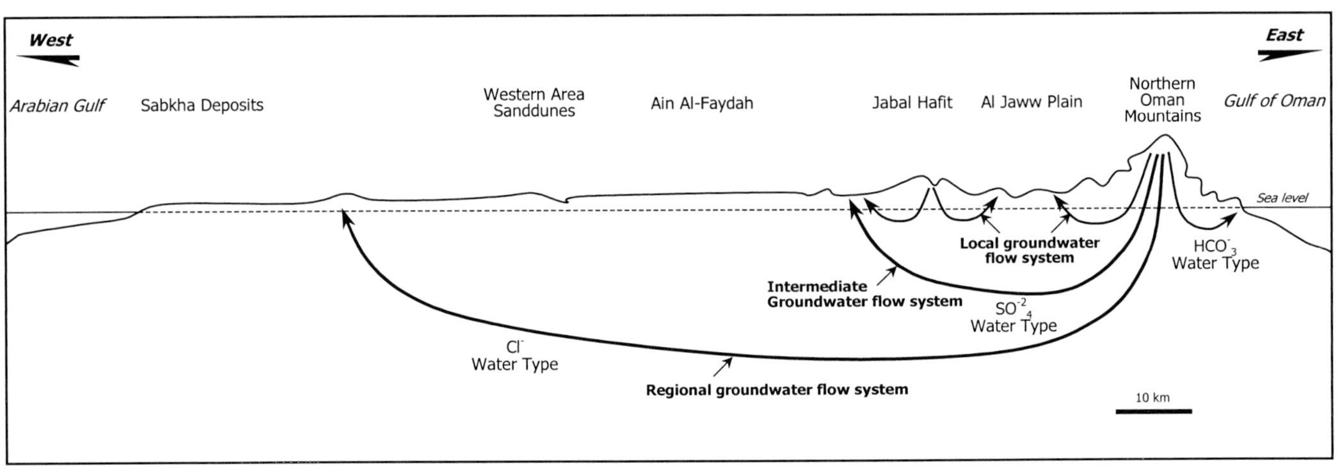

Fig. 8.55. Schematic diagram illustrating different flow systems in the United Arab Emirates, based on hydrological, hydrochemical and isotopic characteristics of groundwater (modified from Rizk and El-Etr, 1994).

contain black organic deposits that contain 20.7ppm uranium (Terratest, 1975). The lack of these black deposits in some layers is due to its leaching by water discharging into Ain Al-Faydah spring. The heat associated with radioactive decay of uranium raises the water temperature in the spring (El-Shamy, 1990). Recently, the United Arab Emirates National Drilling Company -US Geological Survey (joint project) measured high radon counts, suggesting a radioactive source.

B. Electrical Conductivity

The electrical conductance of groundwater samples collected from the United Arab Emirates during early 1996, varied between 252 μS/cm on the Al-Jaww plain, (east of the Al-Ain city), and 173,000 μS/cm in a sabkha area along the Al-Ain-Abu Dhabi road (Fig.8.57). Salt-water intrusion, as a result of heavy groundwater pumping, is noticed south of Dubai, west of Suweyhan and southwest of Al-Ain. It is also observed north of Khor Fakkan and west of Kalba, along the eastern coast. The Dibba-Hatta line represents a water divide along which the electrical conductivity is low (500 μS/cm), increasing in the west and east directions, along the groundwater flow paths. A local fresh water lens is observed along the Madinat Zayed - Liwa road. Potential for presence of similar fresh water pockets, is possible within the triangular area between Al Wagan, Liwa and Umm Alzamoul. The electrical conductance contour map suggests the presence of hydraulic connection, between Al-Ain and Liwa areas. However, this assumption needs further investigations. The conductivity values obtained from falaj water samples vary from 450 μS/cm in falaj Asimah (Al Fujairah) to as much as 10,940 μS/cm in falaj Bu Sukhnah. The latter figure is probably due to the falaj running across gypsum deposits of the Fars Formation and is consistent with similar increase found in well and spring waters.

Values are generally low in falajes draining from ophiolitic rocks east of Al Ain and Al Fujairah, but higher in water samples from falajes draining limestone rocks. The isoelectrical map shows that like the groundwater samples from wells and springs, the conductivity values are low near the water divide, increasing both east and west of this line with increasing distance from the recharge area and increasing falaj length, e.g., Falaj Asimah (3km) has a conductivity of 450 μS/cm, Falaj Fili (5 km) has a conductivity of 1,020 μS/cm, whereas the 10 km Al Dhaid Falaj has a conductivity value of 1,180 μS/cm.

The plot of electrical conductivity (μS/cm) against falaj length shows that the conductivity of the water in open channel falajes (Al Gheli type), increases with increasing falaj length, due to a

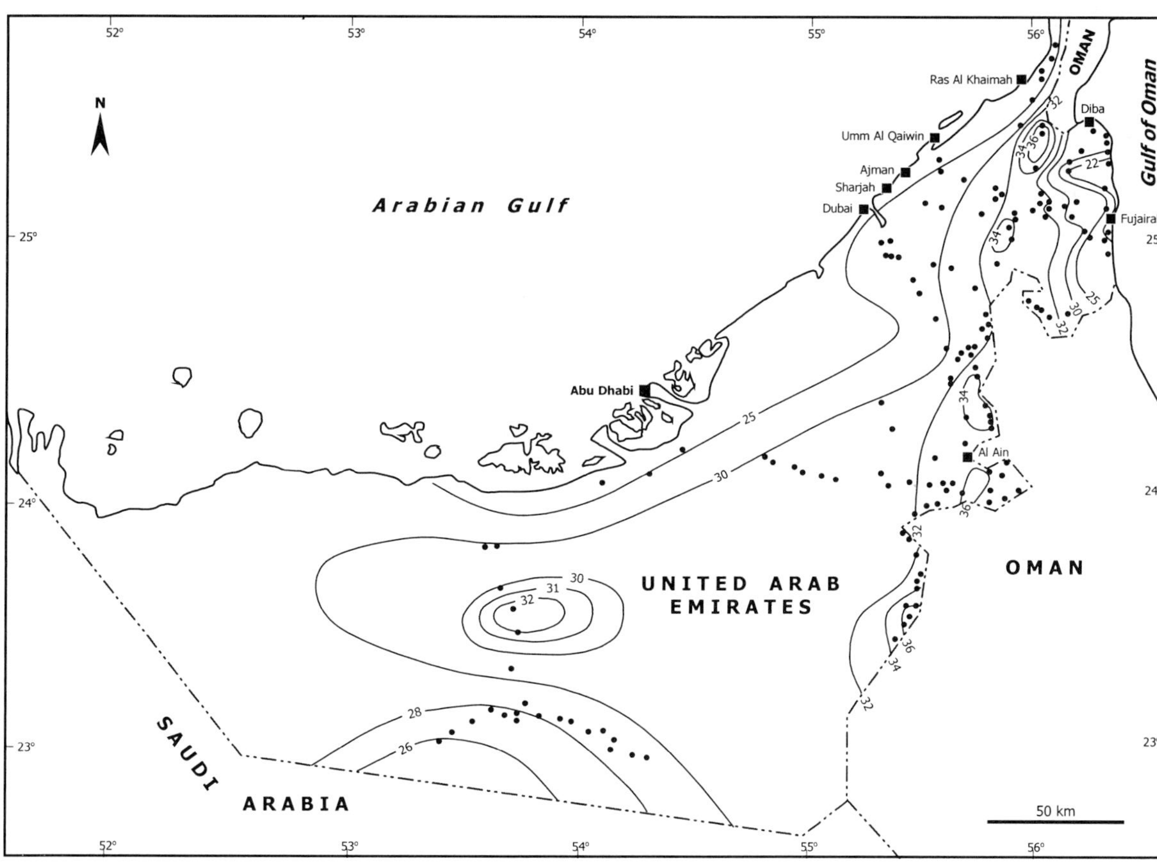

Fig. 8.56. Groundwater temperatures (°C) in the United Arab Emirates, based on field measurements in 1996.

combination of high evaporation rates, and the increased dissolution from the increased area of the falaj channel exposed, (and thus increased total dissolved solids). The same correlation does not hold for tunnel falajes (Al Daudi type) because of the variation in rock type, and the source of water. Falaj Habhab water has a high electrical conductivity, despite its short length (0.5 km), whereas Falaj Al Aini with a length of 6 km has a conductivity of only 620 μS/cm

A plot of electrical conductivity of falaj water against the sodium adsorption ratio (US Salinity Laboratory diagram) shows that, for irrigation purposes the falaj water ranges from good to poor, and for moderately salt tolerant crops as date palms the medium to high salinity water of the falajes is not a significant agricultural hazard. Despite their limited discharge falaj waters are a renewable resource related, in most cases, to rainfall on the eastern mountains and the eastern part of the plains. During the period 1978-1995, the total discharge ranged between 9 Mm^3/yr in 1994 and 31.2 Mm^3/yr in 1982 representing 9.7 to 2.8 of the total water used in the country.

C. Hydrogen-Ion Concentration

The hydrogen-ion concentration of water is related to its quality and affects, to a great extent, its suitability for different uses. The hydrogen-ion concentration of shallow groundwater in the United Arab Emirates varies between 5.37 near the western coast, southwest Abu Dhabi, and 8.18 west of Al-Ain city (Fig. 8.58). Generally, the hydrogen-ion concentration of shallow groundwater in United Arab Emirates decreases, from the recharge areas near the mountains, to the discharge areas near the coastal zones. High values are encountered in areas dominated by limestone aquifers, such as Wadi Al Bih basin (Ras Al Khaimah), and Jabal Hafit (Al-Ain).

D. Major Cations

The sequence of cation dominance in ground water of the United Arab Emirates has the order: $Mg^2 > Ca^2 > Na^+ > K^+$ in the eastern part; $Ca^2 > Mg^2 > Na^+ > K^+$ in the central part, and $Na^+ > Ca^2 > Mg^{2+} > K^+$ in the western part. Iso-concentration contour maps of Ca^2, Mg^2, Na^+ and K^+ ions show the same general pattern (Figs. 8.59-8.62). Differences in this pattern are related to changes in lithology, hydrogeology, and groundwater extraction rates. The Ca^2 concentrations increase towards west and northwest, as the percolation of rainwater causes dissolution of limestones dominating these areas, enriching groundwater with this ion (Fig. 8.59). In the central and southern parts, Ca^2 content increases along the

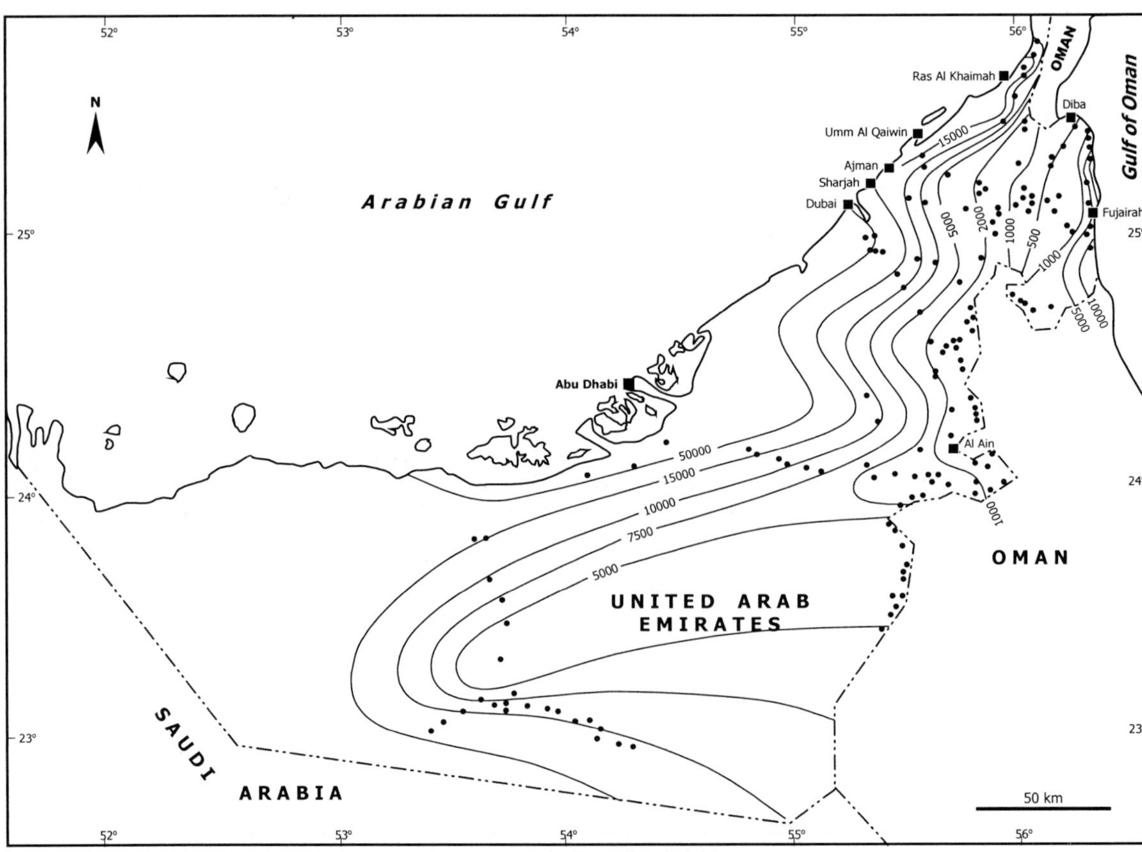

Fig. 8.57. Electrical conductivity (μS/cm) of groundwater in the United Arab Emirates, measured in 1996.

direction of groundwater flow. In the eastern gravel aquifer, however, the Ca^{2+} amounts are low, because of the lack of carbonate rocks, relatively fast groundwater flow and slow dissolution of Ca-rich ophiolitic rocks. The Mg^{2+} iso-concentration map (Fig. 8.60) shows its general increase in the direction of groundwater flow. The main source of Mg^{2+} in gravel aquifers is the dissolution of Mg-rich ophiolitic rocks, from the northern Oman Mountains. High Mg^{2+} content is also observed in groundwater, close to the eastern and western coasts. With difference in magnitude, Na^+ and K^+ contents show a similar pattern (Figs. 8.61 and 8.62). Both ions exhibit low concentrations near the water divide, increasing in the east, northwest, west and southwest directions.

The sodium ion concentration is important in classifying irrigation water, because high sodium concentrations in groundwater reduce oil permeability, and a sodium adsorption ratio has been defined to evaluate the suitability of water for irrigation:

$$\text{Sodium Absorption Ratio} = \sqrt{\frac{\text{Na}}{(\text{Ca} + \text{Mg})/2}}$$

where concentrations are expressed in meq/l.

According to the sodium absorption ratio calculated in May 1995 groundwater in the eastern part of the United Arab Emirates has no harmful effect on plants when used for irrigation, however in the western area groundwater can cause limited to moderate harmful effects.

E. Major Anions

The sequence of anion dominance in groundwater of the United Arab Emirates has the order: HCO_3- > Cl^- > SO_4^{2} > CO_3^{2-} in the eastern part; SO_4^{2} > Cl^- > HCO_3^- > CO_3^{2-} in the central part, and Cl^- > SO_4^{2} > HCO_3^- > CO_3^{2-} in the western part. High HCO_3^- concentrations are observed in groundwater of the northern and eastern parts of the United Arab Emirates, which are the areas receiving the highest rainfall in the country (Fig. 8.63).

The HCO_3^- content decreases in the directions of groundwater flow. The fresh groundwater found north of Liwa is also characterized by high HCO_3^- contents. The SO_4^{2-} concentrations are high in the eastern and western coastal plains. High SO_4^{2-} content is also observed in Al-Ain and Al Wagan groundwater (Fig. 8.62). The high-sulphate groundwater may mark discharge areas of intermediate groundwater flow systems. The Cl^- iso-

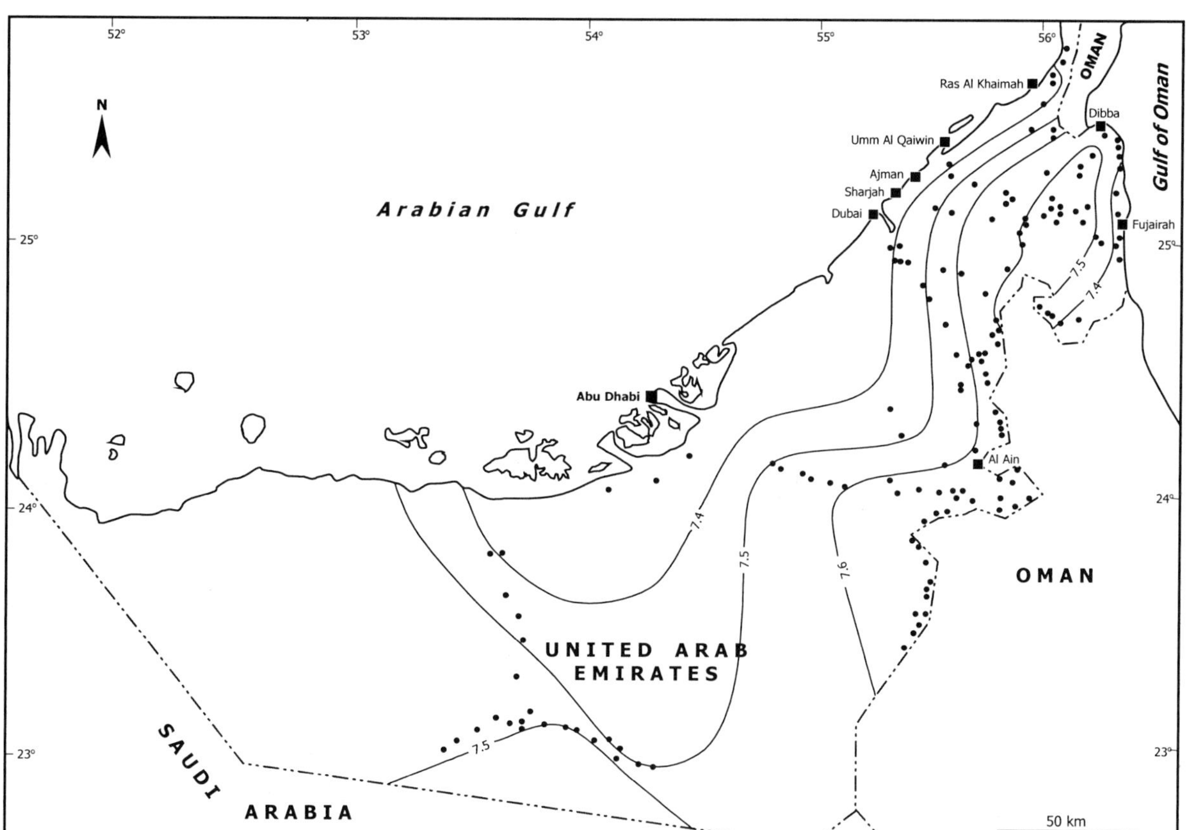

Fig. 8.58. The hydrogen-ion concentration in groundwater of the United Arab Emirates, based on field measurements in 1996.

concentration contour map (Fig. 8.65) shows a pattern similar to that of the Mg^{2+}, Na^+ and K^+. The Cl^- content is low along the Dibba-Hatta line and increases in the directions of groundwater flow. According to Freeze and Cherry (1979), nitrate ion (NO_3^-) is the most common identified contaminant in water. The World Health Organization (1971) recommended limits for nitrate in drinking water are 10 mg/l as nitrate nitrogen, and 45 mg/l as nitrate (NO_3^-). Centers of high nitrate ions are encountered in Wadi Al Bih, south of Dubai, Al-Ain, Al-Khaznah, Madinat Zayed and Liwa. Nitrate ion (NO_3^-) concentration as high as 1,000 mg/l in shallow groundwater of the United Arab Emirates were measured west of Al-Khaznah and in the Liwa areas (Fig. 8.66). Because of the close correlation between high nitrate ion contents and the presence of intensive farming, it seems that the agriculture is the main source of nitrates in shallow groundwater in the United Arab Emirates. Because of the persistent of nitrate ions in oxygenated systems, the availability of abundant oxygen, in the shallow horizons of the Quaternary aquifers, add to the nitrate contamination problem in the country.

F. Water-Dissolved Salts

The main groundwater-dissolved salts in United Arab Emirates are: $Ca(HCO_3)_2$, $Mg(HCO_3)_2$, $Na_2(SO_4)$, $CaSO_4$ $MgSO_4$ MgCl and NaCl. The relative abundance of these salts is consistent with the prevailing hydrogeological conditions. These salts evolve in the direction of flow according to the Chebotarev series (Freeze and Cherry, 1979), and confirm the presence of different groundwater flow systems. Groundwater in the northern limestone aquifer, the northwestern gravel aquifer, the eastern gravel aquifer and the ophiolite aquifer, which receive a relatively high rainfall, are enriched in Ca $(HCO_3)_2$ and $Mg(HCO_3)_2$ salts. The salts characterize groundwater of a local flow system. This water has a low salinity, a short residence time and a good quality (Figs. 8.67-8.69). Groundwater in the western gravel aquifer, are dominated by $CaSO_4$ and $MgSO_4$ salts, which mark an intermediate groundwater flow system. The groundwater of this system, is mainly brackish and of intermediate residence time (Figs. 8.70 and 8.71). In the sand dune aquifer, which occupies the western and southern parts of the United Arab Emirates, groundwater contains $MgCl_2$ and NaCl salts, indicating a regional groundwater flow system. The groundwater in this system is mainly saline, and has a long residence time (Figs. 8.72 and 8.73).

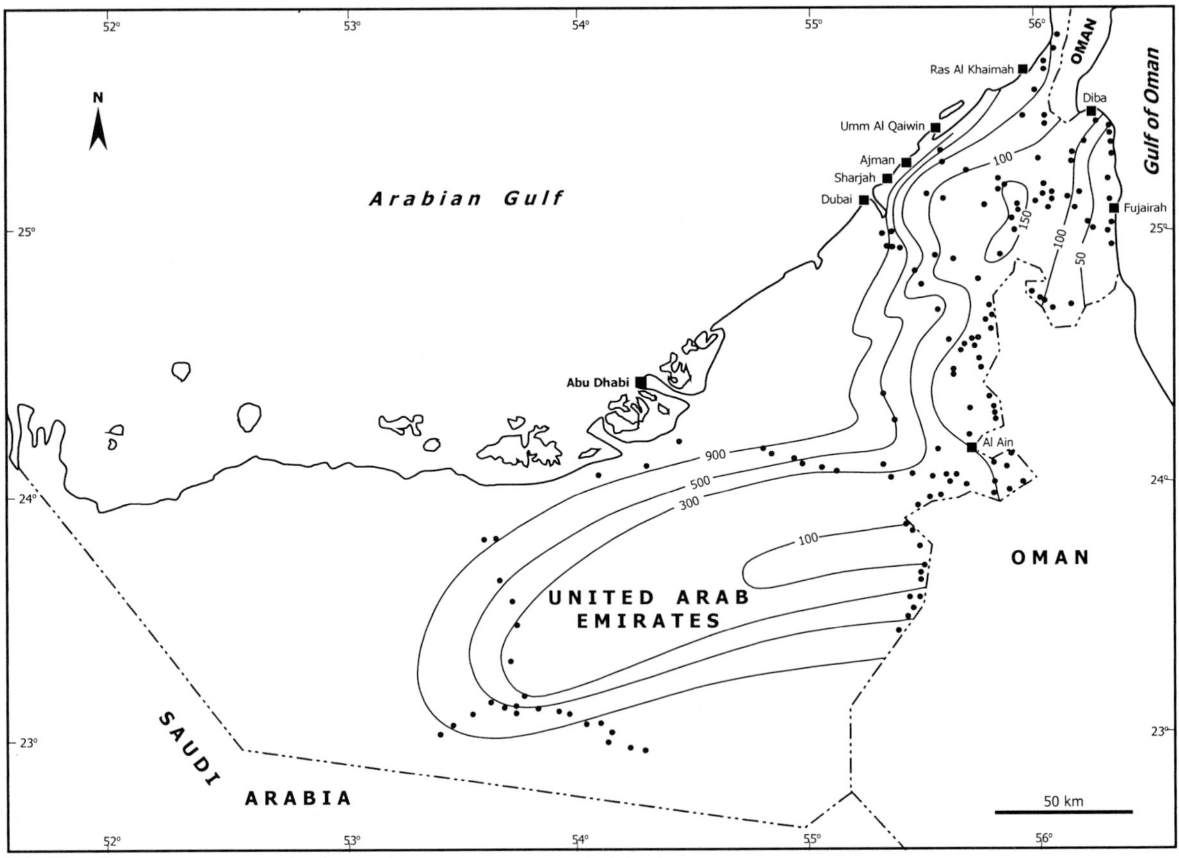

Fig. 8.59. Iso-concentration (mg/l) contour map of the calcium ion (Ca^{2+}) in groundwater of the United Arab Emirates, measured in 1996.

G. Groundwater types

Trilinear plots of the chemical analyses of water samples collected from the United Arab Emirates groundwater are shown in Figures (8.74-8.76) and presented on maps in Figures (8.77-8.82). These plots show the following:

1. Groundwater in the eastern gravel aquifer has an $MgCl_2$ type, whereas the groundwater in the northwestern gravel aquifer is a NaCl water type. This again reflects the effect of dissolution of Mg-rich ophiolitic rocks. The high chloride content in the northwestern gravel aquifer, indicates salt-water intrusion as a result of excessive groundwater pumping.
2. The western gravel aquifer shows variable water types, depending on the relative proximity to the northern Oman Mountains. On its eastern side, this aquifer is characterized by $Mg(HCO_3)_2$ and $Ca(HCO_3)_2$, in its central part, the aquifer is characterized by $CaSO_4$ and $MgSO_4$ water types, and the western side of the aquifer is dominated by the NaCl water type.
3. The sand dune aquifer in the Liwa area is characterized by the NaCl water type. Despite its old age, the low salinity of this groundwater is related to the nature of the aquifer which is composed of sand.

H. Water Quality

The iso-electrical conductivity contour map (Fig. 8.57) shows that, the groundwater in the eastern mountains, and the flanking gravels, is mainly fresh and can be used for all purposes. However, because of excessive pumping, groundwater in several areas, is now suffering from salt-water intrusion, not only from the sea, but from deeper horizons of the same aquifer, and possibly from nearby sabkha deposits. The iso-hardness contour map shows, that the groundwater is very hard in the northeastern, Al Dhaid, Kalba, Al-Khaznah and along the western coast (Fig. 8.78). Groundwater in the eastern mountains, and most of the flanking gravels, does not have hardness problem, and can be used for domestic purposes. The calculated Sodium Adsorption Ratios show that, the groundwater in the northern and eastern parts of the country, has little harmful effect on plants and soils. Groundwater along the western coast, west Al-Ain and east Liwa, has high sodium adsorption ratio values, and can be very harmful to plants and soils when used for irrigation.

I. Hydrochemical Coefficients

Hydrochemical coefficients show the relative

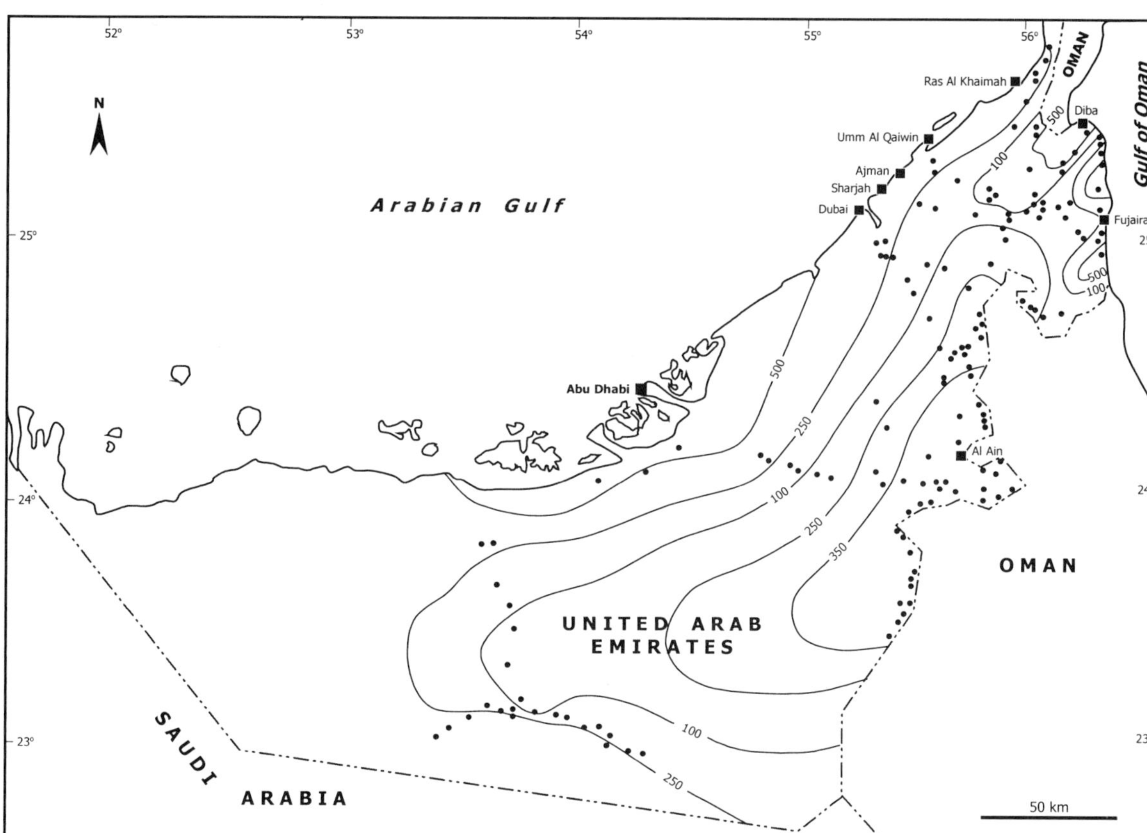

Fig. 8.60. Iso-concentration (mg/l) contour map of the magnesium ion (Mg^{2+}) in groundwater of the United Arab Emirates, measured in 1996.

concentrations of various ions, and are used to indicate the predominance of a particular ion, and to define locations of salt-water intrusion. The Ca/Mg ratio in groundwater of the United Arab Emirates shows that, Ca^{2+} is dominant over Mg^{2+} in the northern limestone aquifer, along Khatt - Al-Khaznah line, around Jabal Hafit, and in the sand dune aquifer (Fig. 8.79). The SO_4/Cl ratio in groundwater of the United Arab Emirates indicates that, the SO_4^{2-} is dominant over Cl^- at Suweyhan, between Dubai and Abu Dhabi and south of Liwa (Fig. 8.80). The $Cl/(CO_3 + HCO_3)$ ratio is used to evaluate salt-water intrusion, either from neighboring areas, or from underlying formations. The chloride-ion (Cl^-) is a dominant anion in salt water, and normally occurs in small amounts in groundwater. The bicarbonate-ion (HCO_3^-) is the most abundant anion in groundwater. Figure (8.81) shows that groundwater in most of the country is suffering from serious salt-water intrusion problems, except for the central part of the ophiolite aquifer. The Na/Cl ratio is also used to indicate areas suffering from salt-water intrusion (Figure 8.82). Salt-water intrusion problems reported in cultivated areas in Ras Al Khaimah, Al Dhaid, Dibba, Kalba, Dubai - Jabal Al-Dhanah, Madinat Zayed, Liwa and Al-Ain.

J. Isotope Techniques

Variations of stable isotopes (^2H and ^{18}O) and radioisotopes (^3H and ^{14}C) were measured in large numbers of water samples, were collected during the 1984-1990 period, by the International Atomic Energy Agency (IAEA) for the Ministry of Electricity and Water, United Arab Emirates; at laboratories in Jordan and Austria. Complete chemical analysis of these samples conducted in the Hydrochemical Laboratories of the Ministry.

1) Isotope Composition of the Atmosphere

The nearest long-term isotope monitoring station to the United Arab Emirates is Bahrain, where the isotopic composition of rainfall was monitored from the 1963-1993, within the scope of the IAEA/WMO global survey (Figs. 8.83; 8.84). The stable isotope data available from this station can be used to provide basic characteristics of the stable isotopic composition, of the present-day meteoric water in the area (Yurtsever, 1992). The plot of the data shows a scatter of the points, which suggests that raindrops are affected by evaporation during the fall of the droplets (IAEA, 1984).

A plot of oxygen-18 (^{18}O) versus deuterium (^2H) contents in 52 samples of United Arab Emirates

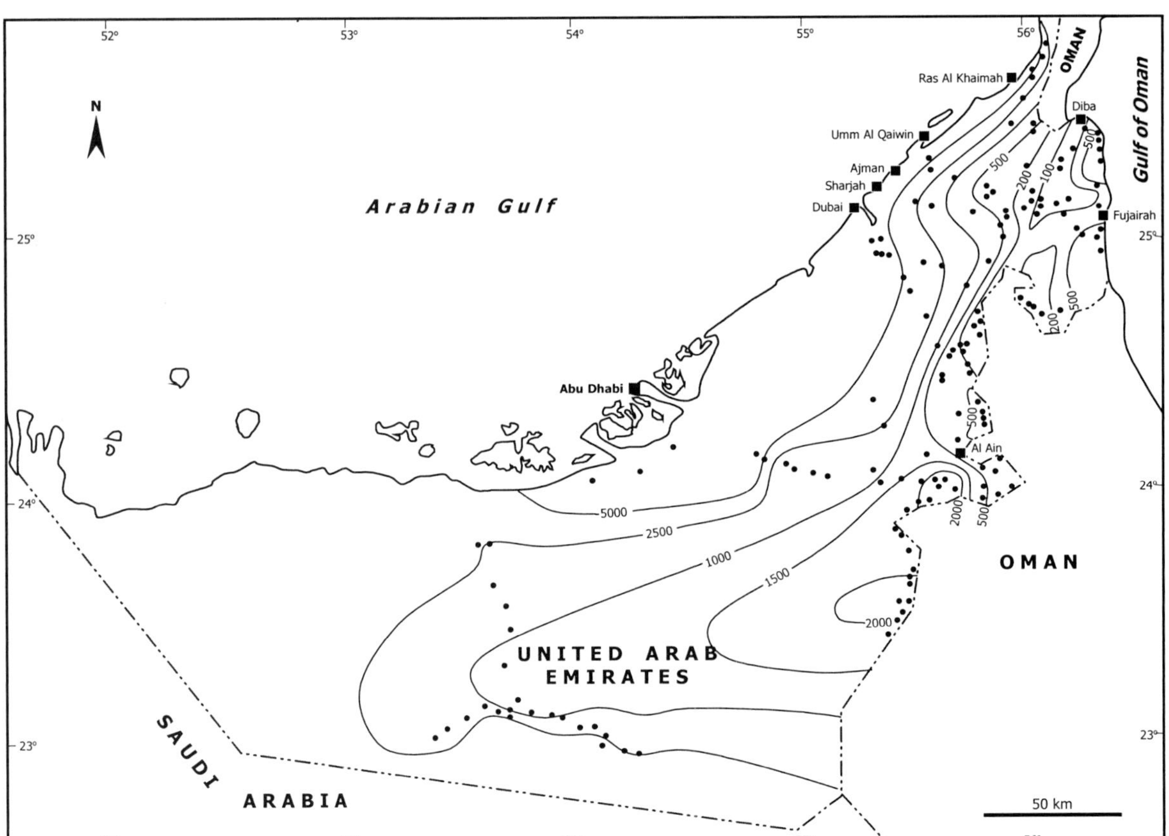

Fig. 8.61. Iso-concentration (mg/l) contour map of the sodium ion (Na^+) in groundwater of the United Arab Emirates, measured in 1996.

rainwater collected by the Ministry of Electricity and Water during 1985-1991 period is shown in Figure (8.85). The weighted average values for this data is:

Mean ^{18}O = -1.99 ‰ and Mean ^{2}H = -0.4 ‰

The line best defining the ^{18}O versus ^{2}H, for months having more than 20 mm rain, has a slope of 8 as shown in Figure (8.85), which has an intercept (deuterium excess = δ ‰) of 16. This relationship is the best estimate of the stable isotope composition for groundwater of meteoric origin, being replenished from precipitation under the present-day climatic conditions in the United Arab Emirates. The tritium (^{3}H) content in rainfall events for the 1984-1987 period averages about 4.7 ± 1.1 Tritium Units (TU).

2) Isotope Characteristics of Groundwater

The large differences in the $\delta^{18}O$ values, observed in the groundwater of United Arab Emirates, is the result of various processes and mechanisms, occurring before and during groundwater recharge, such as evaporation, before infiltration or mixing between different waters in the aquifers. Because of the distinct geomorphology and hydrogeology, of different aquifers in United Arab Emirates, striking differences were also observed in isotopic content, of groundwater in various aquifers. Consequently, it was necessary to consider each aquifer as a separate hydrogeological regime.

i) Gravel aquifer

The groundwater of the northern gravel aquifer is enriched in stable isotopes, indicating different groundwater origin, or the effect of evaporation. Figures (8.86 and 8.87) show that the isotopes undergo enrichment as the groundwater moves towards the coastline. Electrical conductivity also increases as groundwater moves downgradient. The infiltration rates of the sand dunes around Al-Ain area are three to six times those of the gravel aquifer on the Al-Jaww plain (Rizk et al., 1998).

Groundwater in the eastern gravel aquifer plot on the meteoric water line. However, few wells show the effect of evaporative enrichment. The low chloride concentrations, suggest younger water in hydrogeological terms. This would mean, the wells obtain water from a local groundwater flow system. The stable isotope contents, are relatively depleted, compared with the northern sand and gravel aquifer. The deuterium excess of 13.6 suggests that this region is, in part receiving recharge from two air masses, the winter precipitation from the Mediterranean, and the Monsoon rains of the Indian Ocean. The tritium content in groundwater, of the eastern gravel aquifer, is higher than in present-day rainfall (Fig. 8.88). It seems that this water was recharged after 1972 (which was an exceptionally wet year), and decayed in time during groundwater circulation. Groundwater in this aquifer, contains the

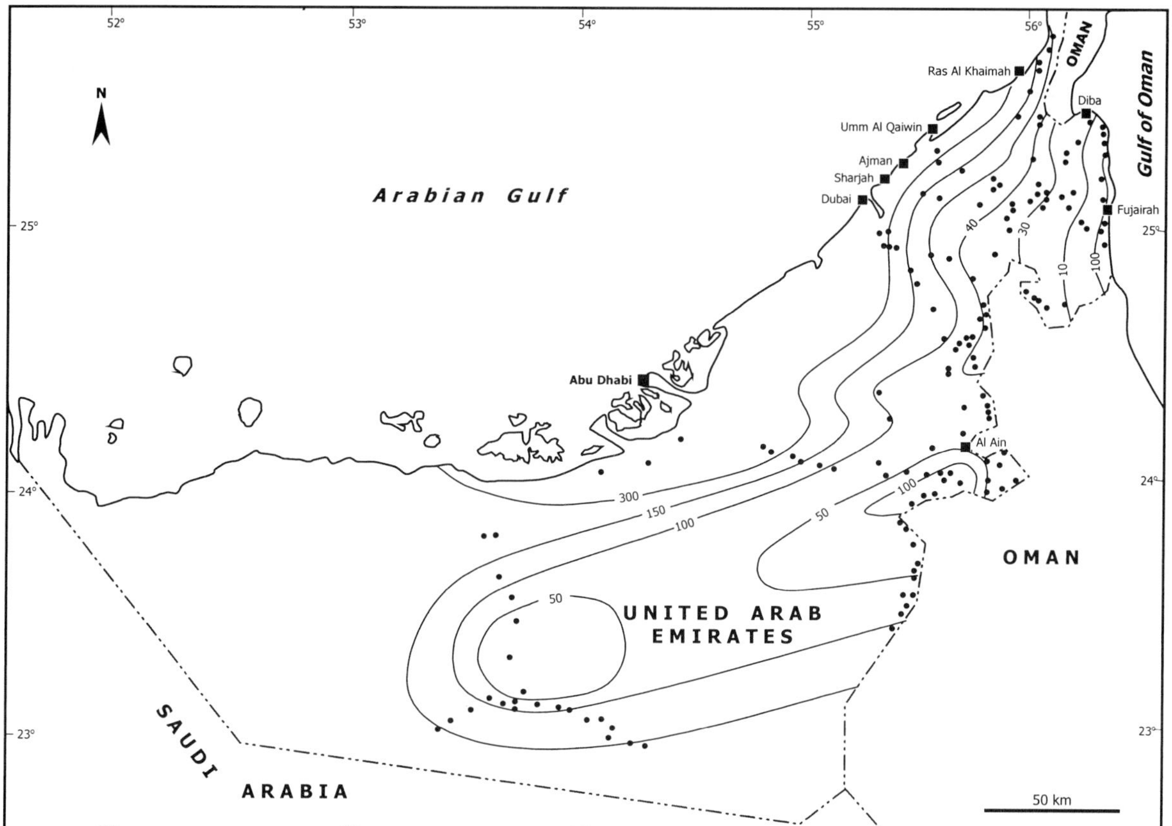

Fig. 8.62. Iso-concentration (mg/l) contour map of the potassium ion (K^{+}) in groundwater of the United Arab Emirates, measured in 1996.

highest activity level in ^{14}C, of Total Dissolved Inorganic Carbon in United Arab Emirates (Fig. 8.89). The ^{14}C ages of groundwater range from modern to 7,000 years B.P. This agrees with the high ^{3}H content in the aquifer, and confirms that this aquifer is receiving modern recharge.

The majority of groundwater samples, collected from the western gravel aquifer, plot to the right of the meteoric water line, indicating enrichment during infiltration. This enrichment could come about by the residence of water, in surface depressions before recharge. The clay in alluvium will not permit rapid infiltration, and therefore causes enrichment before the water is recharged. The high chloride content and enriched stable isotopes, confirms the effect of groundwater flow, and its dissolution of salts as it moves. Evaporation from groundwater increases the value of the deuterium excess. It is also possible that the western gravel aquifer receives old water, which mixes with infiltrating rain water, coming through fractures. Present-day recharge is restricted to the mountainous areas, and the areas adjacent to the mountains. A general increase in groundwater age is observed in the western gravel aquifer, suggesting the reduction of hydraulic head, as water moves towards the sand dunes. At the gravel-sand dune boundary at Idhn, the well United Arab Emirates 175 contains 9 tritium unit, indicating recent recharge. The well United Arab Emirates 178 (down gradient) contains no ^{3}H, suggesting that there has been no recharge since 1952. This shows that, the communication between the gravel and sand dune aquifer, can be slow or rapid depending on the prevailing routes (Akiti et al., 1992). It is possible that there is flow from the alluvium to the sand dunes or that the recharge events occurred by way of ancient wadis. This point must be considered for water-resource planning purposes. Wells in the immediate vicinity of the mountains such as those of Idhn and Manama contain high tritium. The wells at the western edge of the western gravel aquifer contain little or no tritium. The activities of ^{14}C in Total Dissolved Inorganic Carbon are very low suggesting the great age of groundwater. The ^{14}C age of groundwater in Abu Dhabi area ranges from modern to 15,000 years B. P. (Fig. 8.90).

ii) Sand dune aquifer

The groundwater in sand dune aquifers plots very far from the global meteoric line, indicating enrichment before and/or during groundwater recharge. The low salinity of water in the sand dunes of the Liwa area suggests the possible recharge by way of an ancient wadi. However, this possibility needs further investigations. The tritium values

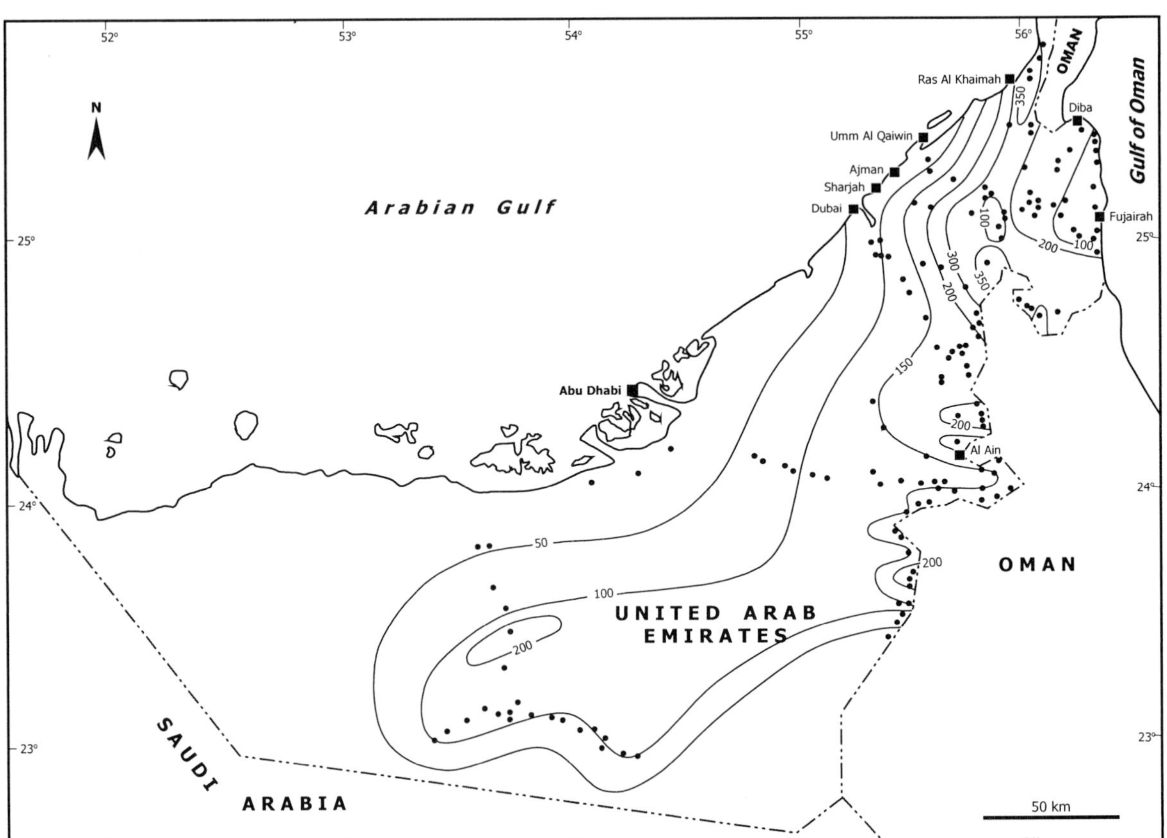

Fig. 8.63. Iso-concentration (mg/l) contour map of the bicarbonate ion (HCO_3^-) in groundwater of the United Arab Emirates, measured in 1996.

obtained range from 2.89 to 20.30. One exceptionally high value was measured in August 1996 as 47.46 TU from Madinat Zayed. This value may be a laboratory error or could indicate groundwater recharge in the year 1996. The wells tapping groundwater in sand dunes contain practically no detectable tritium. However, the samples analyzed in August 1996 contain about 4 Tritium Units (TU), indicating that these wells contain old water recharge before 1952. The lowest ^{14}C activities are found in the sand dune groundwater. These low activities are accompanied by low Cl^- and total dissolved solids contents, because the aquifer is mainly composed of sands, which usually has low salt content. The groundwater with ^{14}C of total dissolved inorganic carbon content higher than 80% Pre-Modern Carbon contain significant ^3H content (10 TU), confirming the modern ages of these waters. The values > 5 and < 10 TU, represent a mixture of young water with old water (Fig. 8.89).

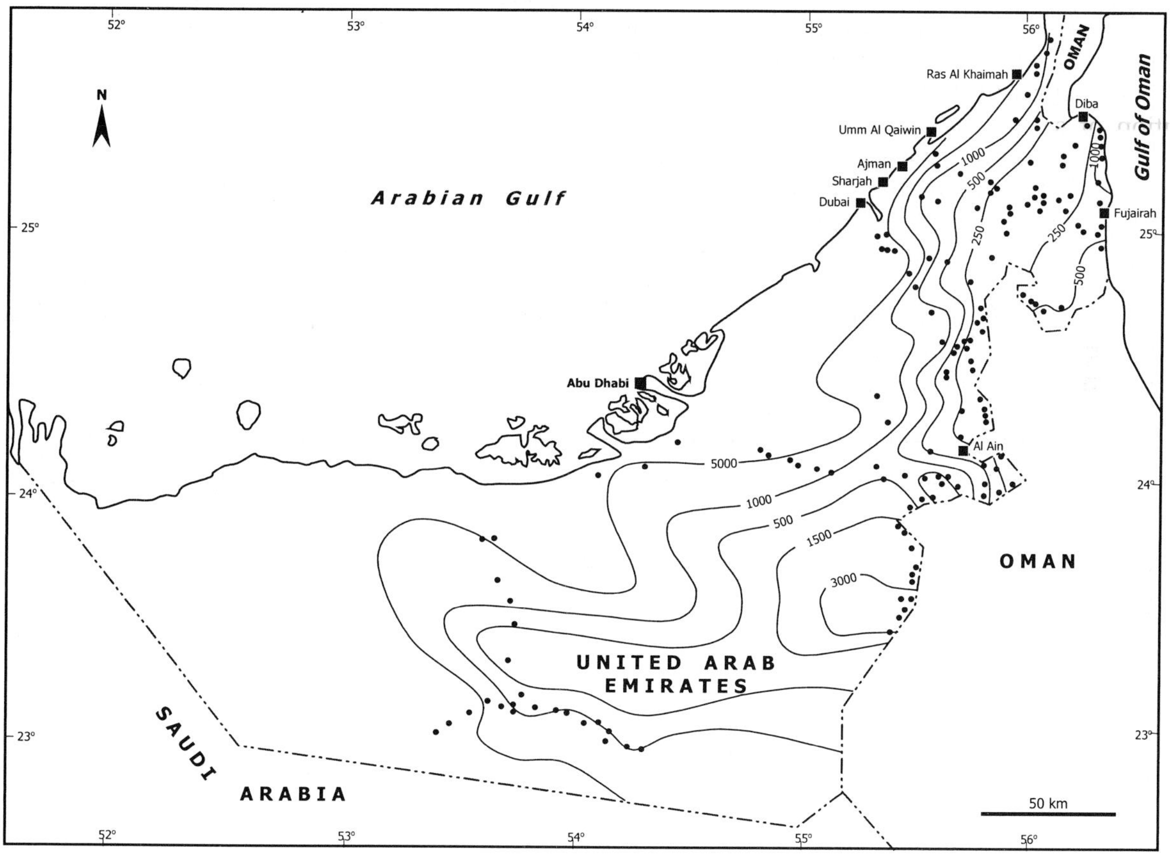

Fig. 8.64. Iso-concentration (mg/l) contour map of the sulphate ion (SO_4^{2-}) in groundwater of the United Arab Emirates, measured in 1996.

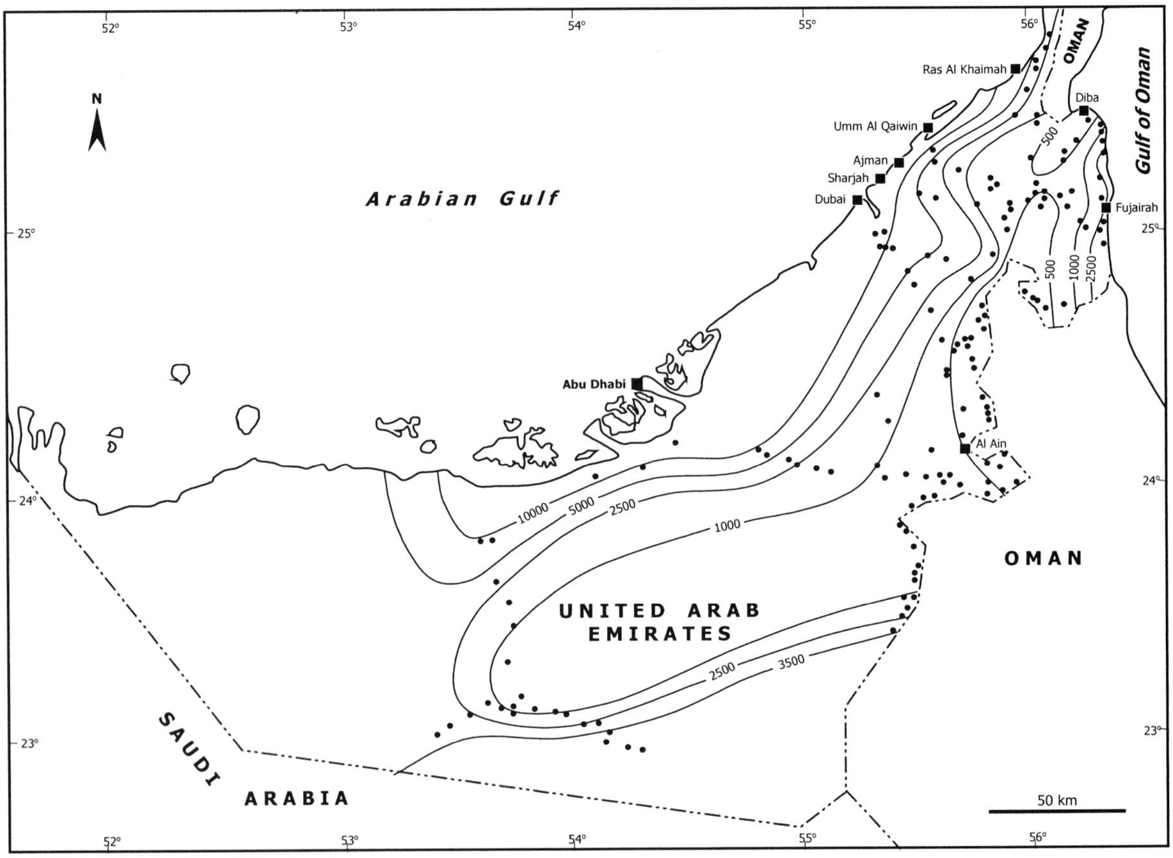

Fig. 8.65. Iso-concentration (mg/l) contour map of the chloride ion (Cl^-) in groundwater of the United Arab Emirates, measured in 1996.

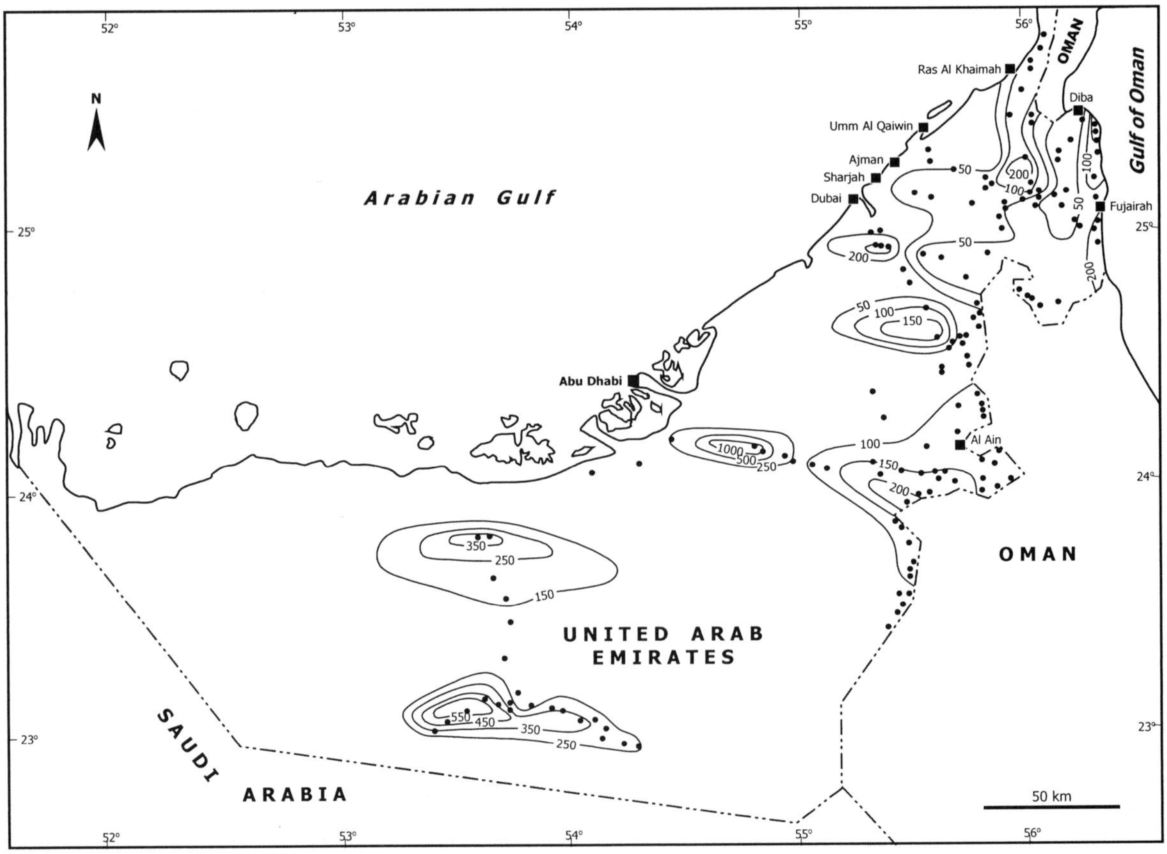

Fig. 8.66. Iso-concentration contour map of the nitrate ion (mg/l) in groundwater from the sand and gravel aquifer during 1996 in the United Arab Emirates.

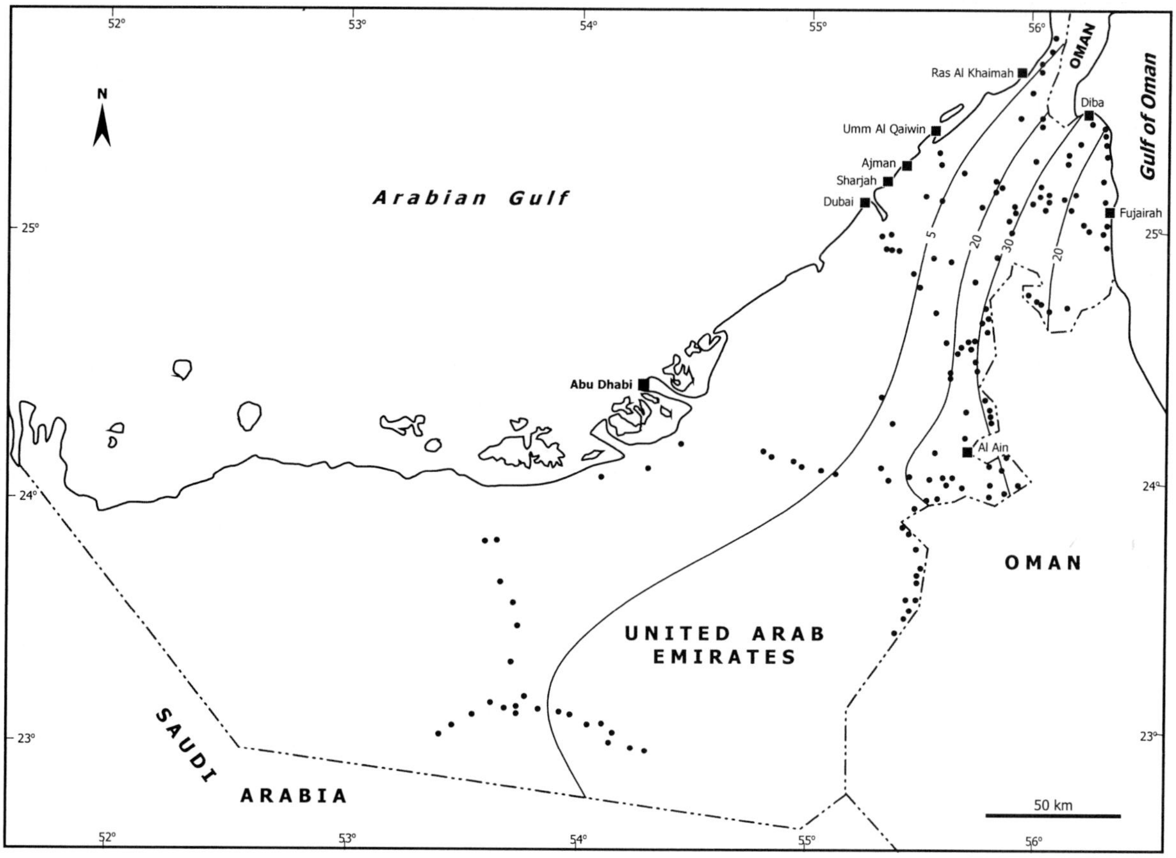

Fig. 8.67. Dominance of the calculated $Mg(HCO_3)_2$ salt (%) dissolved in groundwater of the United Arab Emirates in 1996.

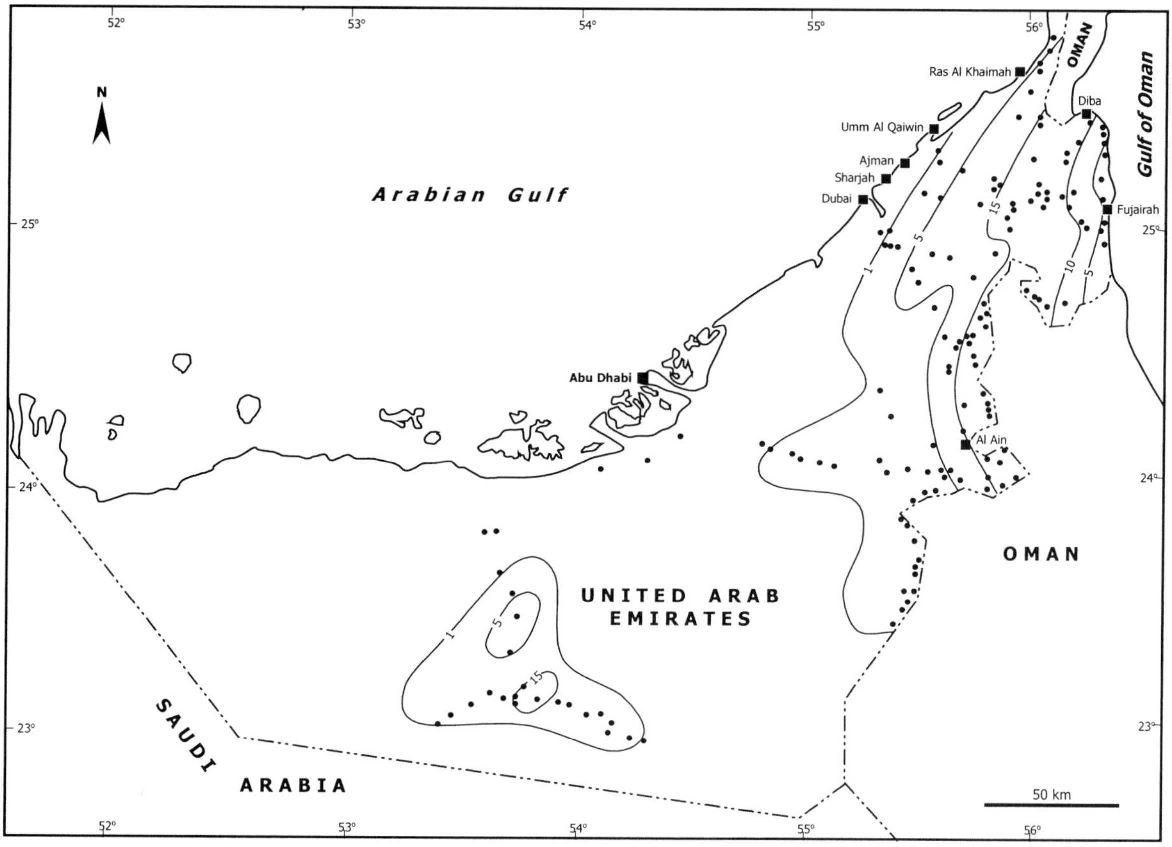

Fig. 8.68. Dominance of the calculated $Ca(HCO_3)_2$ salt (%) dissolved in groundwater of the United Arab Emirates in 1996.

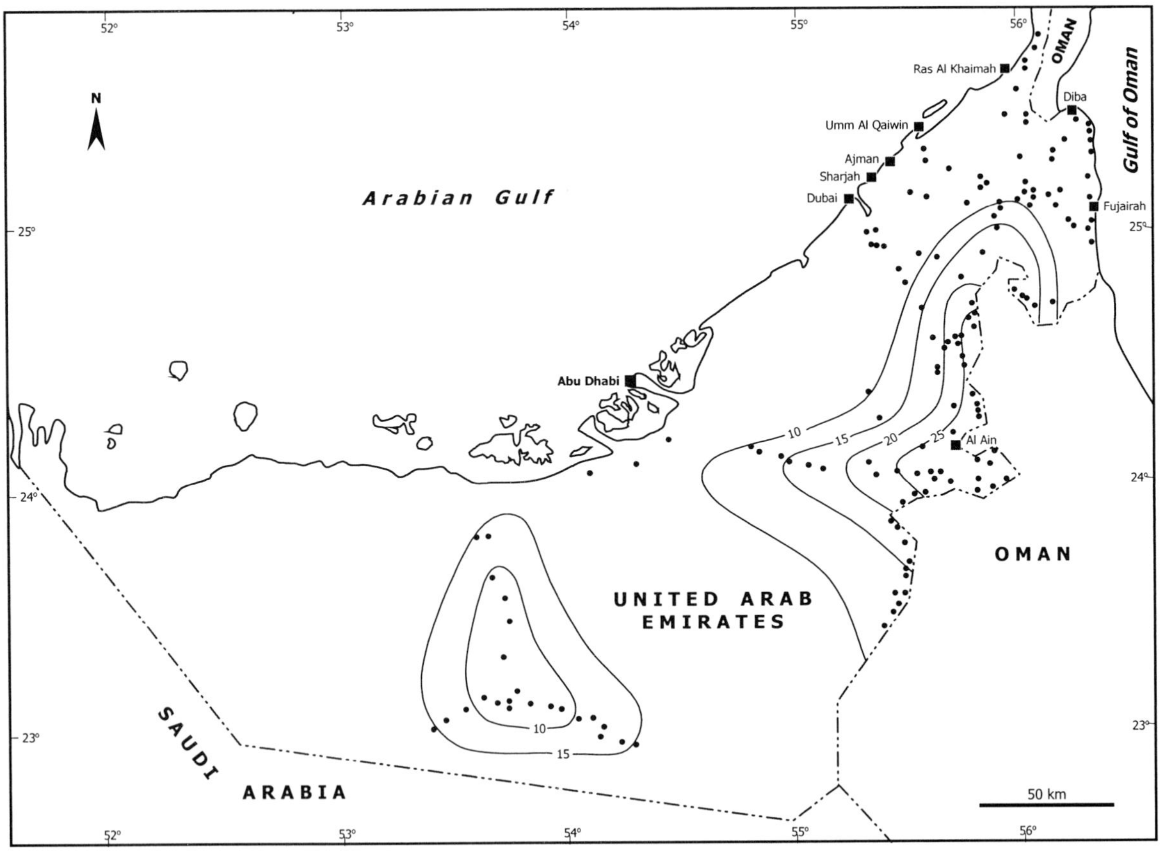

Fig. 8.69. Dominance of the calculated $CaSO_4$ salt (%) dissolved in groundwater of the United Arab Emirates in 1996.

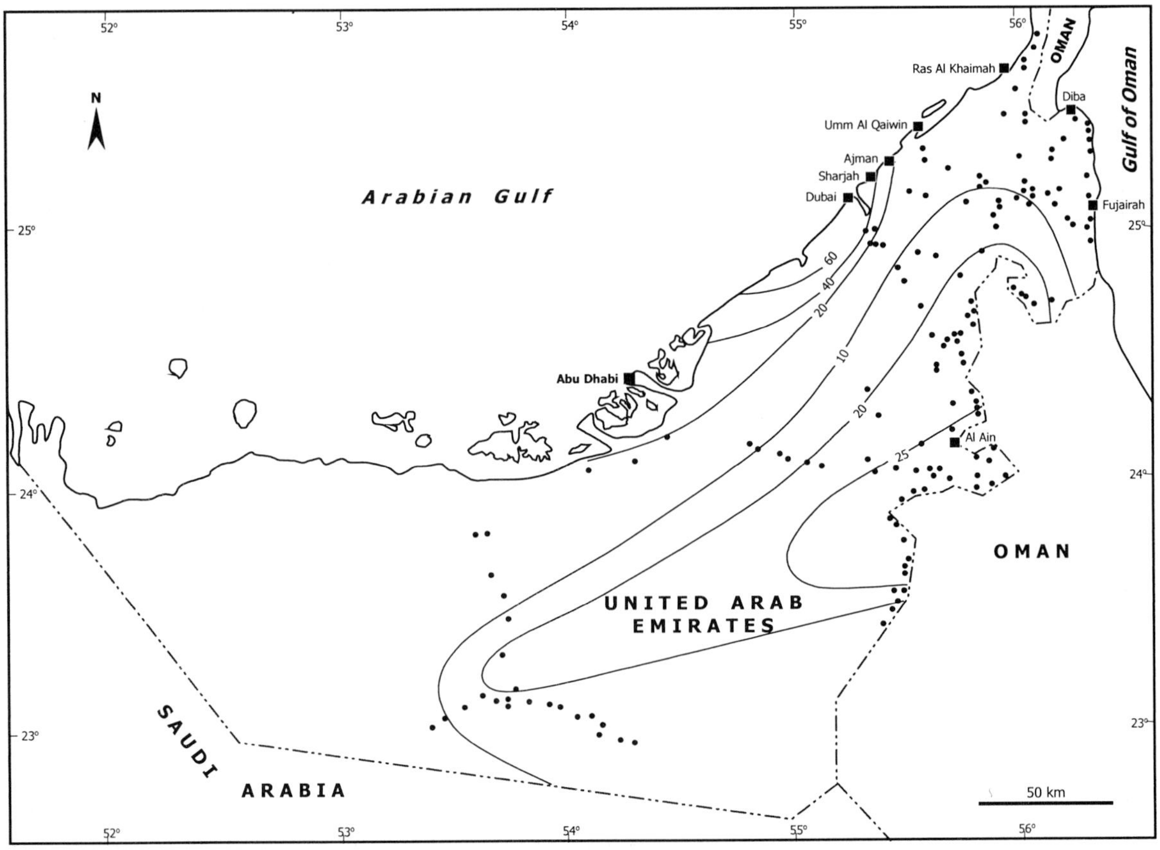

Fig. 8.70. Dominance of the calculated Na_2SO_4 salt (%) dissolved in groundwater of the United Arab Emirates in 1996.

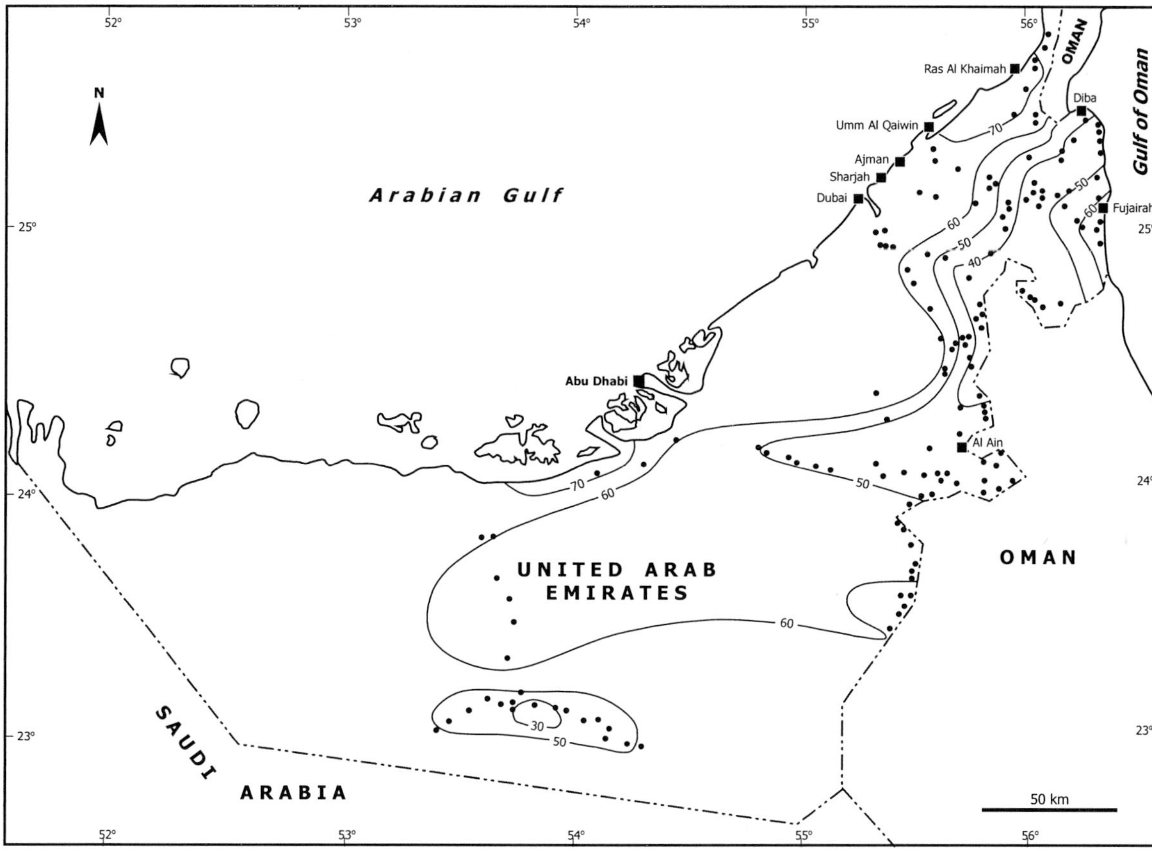

Fig. 8.71. Dominance of the calculated NaCl salt (%) dissolved in groundwater of the United Arab Emirates in March 1996.

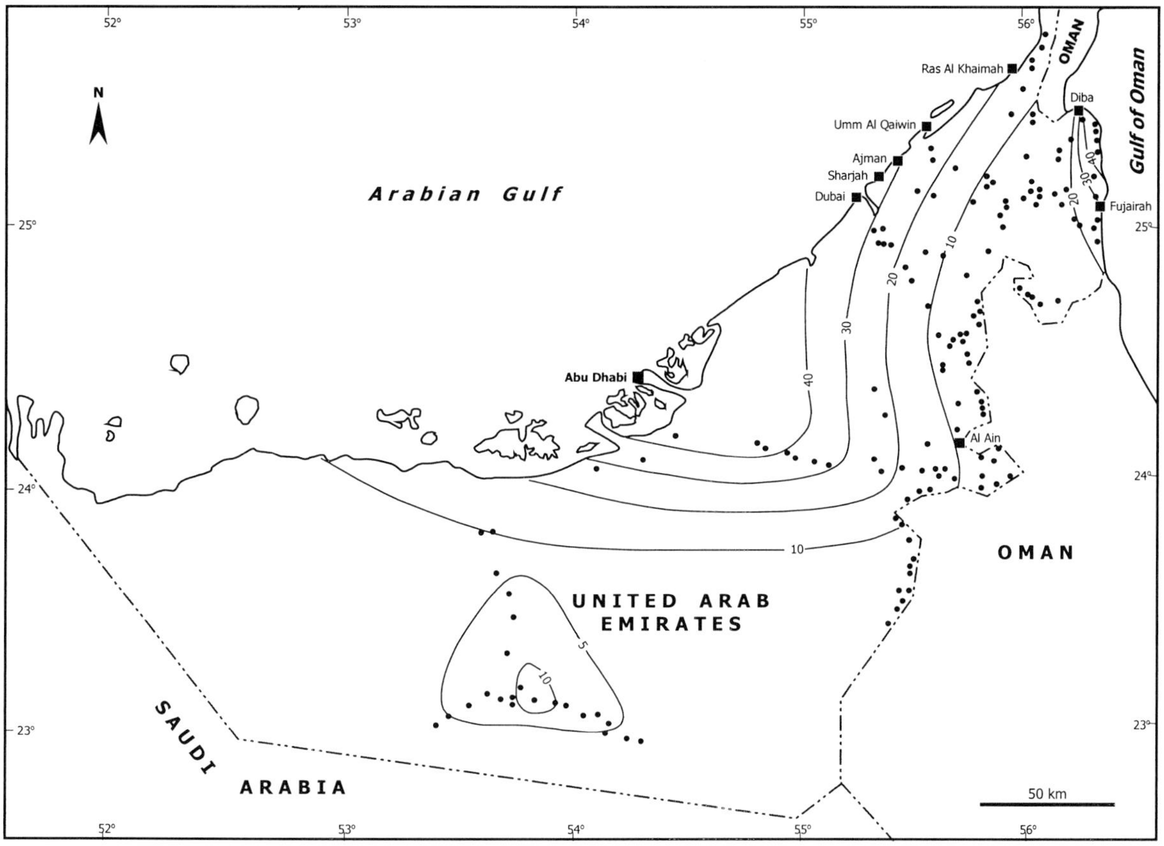

Fig. 8.72. Dominance of the calculated $MgCl_2$ salt (%) dissolved in groundwater of the United Arab Emirates in March 1996.

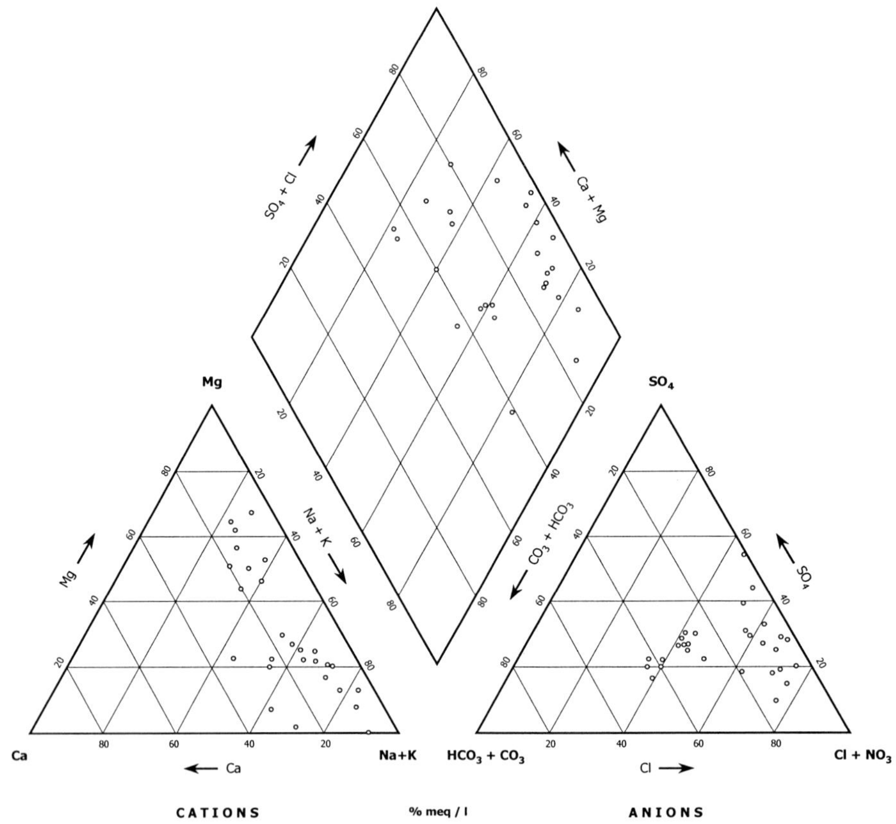

Fig. 8.73. A trilinear plot of the chemical analysis of groundwater samples collected from the ophiolite aquifer in 1996 in the United Arab Emirates.

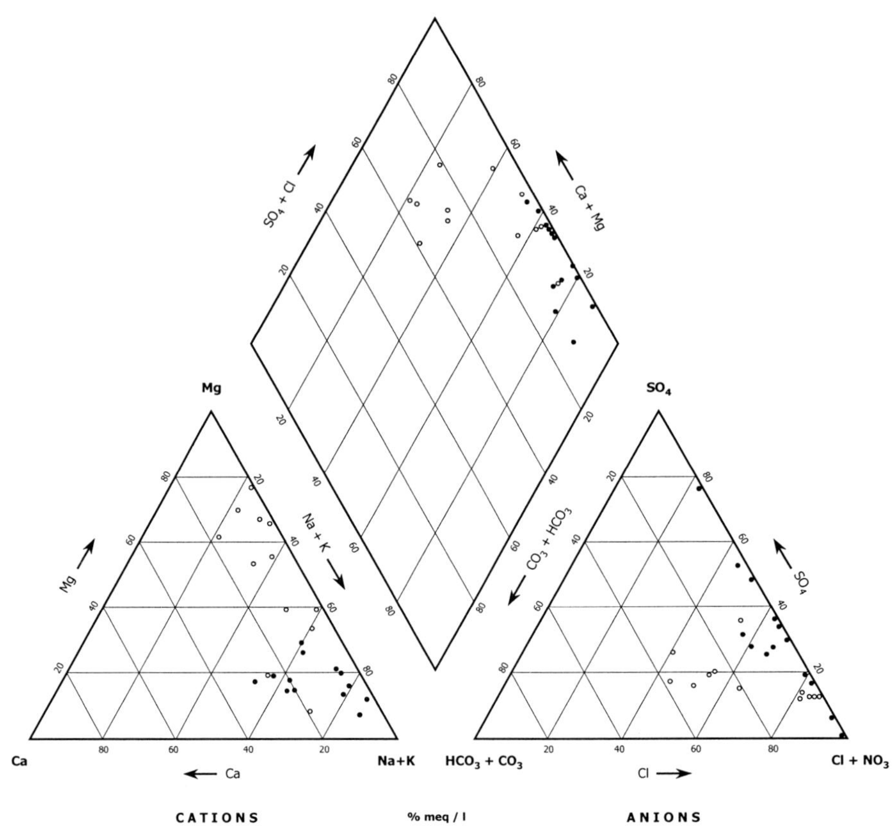

Fig. 8.74. A trilinear plot of the chemical analysis of groundwater samples collected from the eastern and northwestern gravel aquifers in 1996.

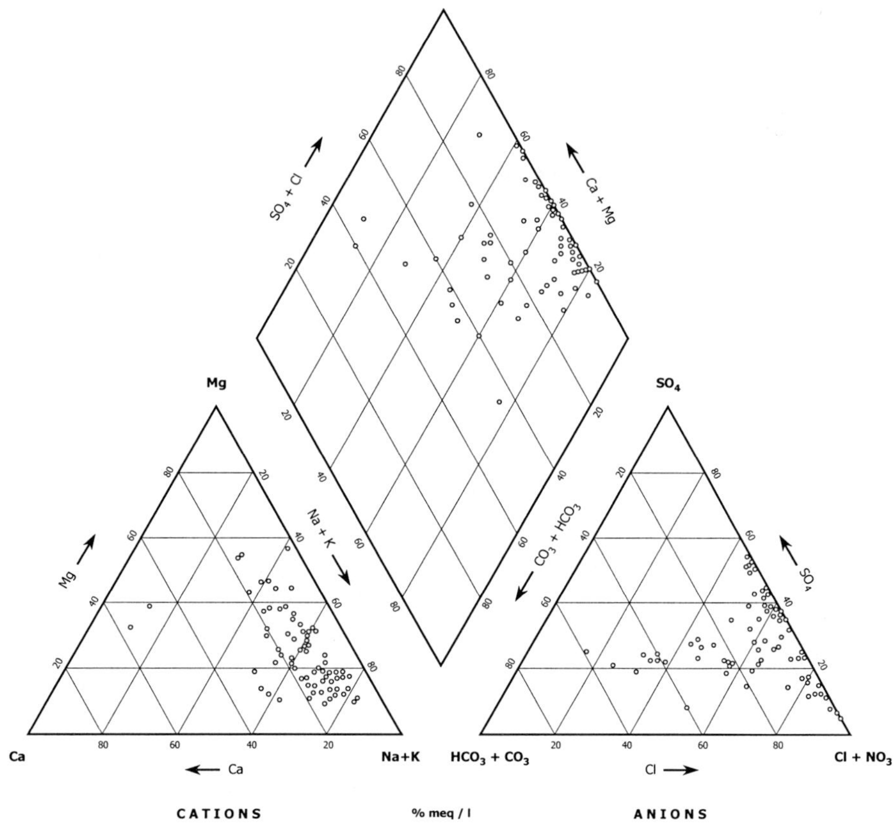

Fig. 8.75. A trilinear plot of the chemical analysis of groundwater samples collected from the western gravel aquifer in 1996.

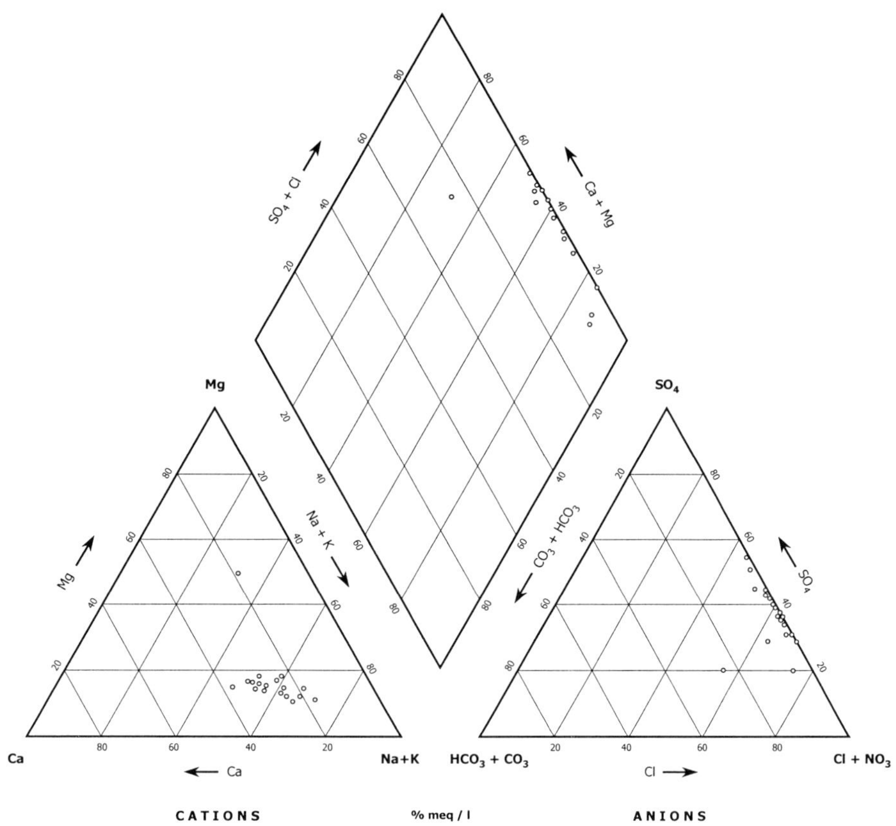

Fig. 8.76. A trilinear plot of the chemical analysis of groundwater samples collected from the sand dune aquifer in 1996.

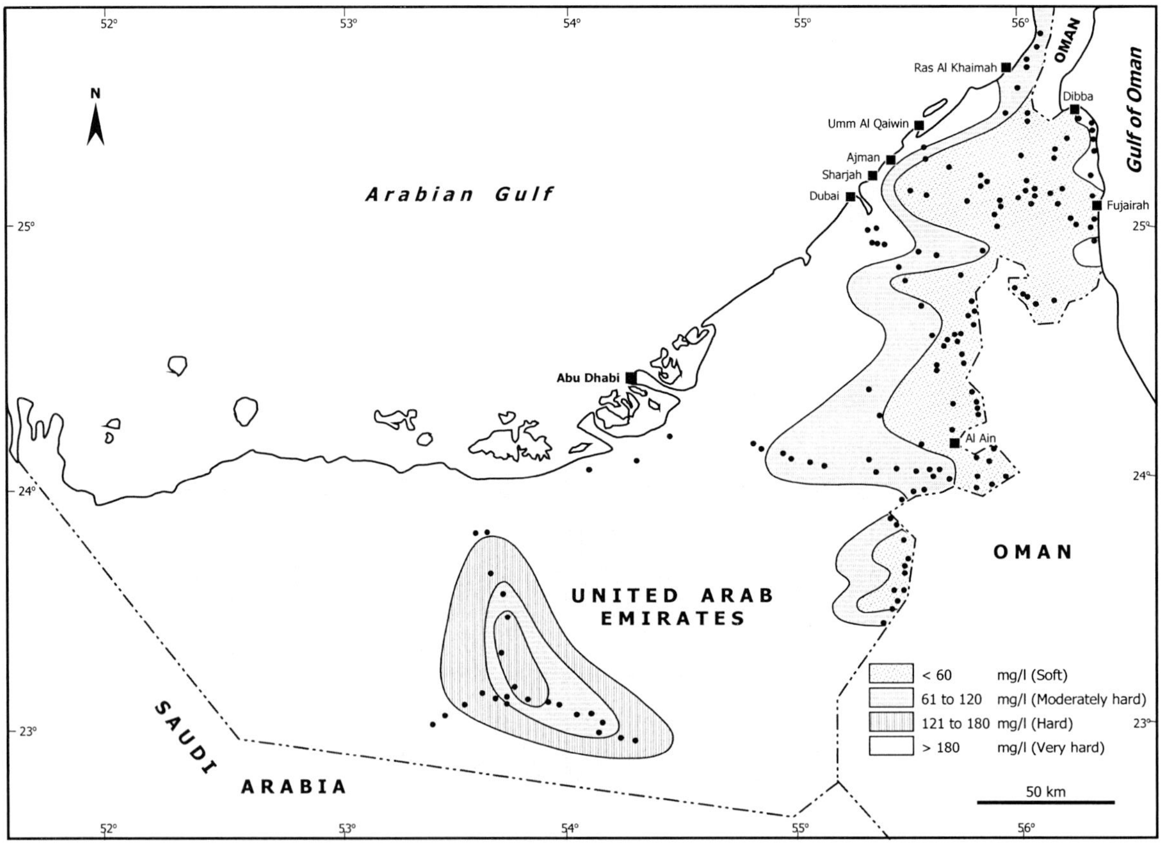

Fig. 8.77. Calculated total hardness (mg/l) in groundwater of the United Arab Emirates in 1996.

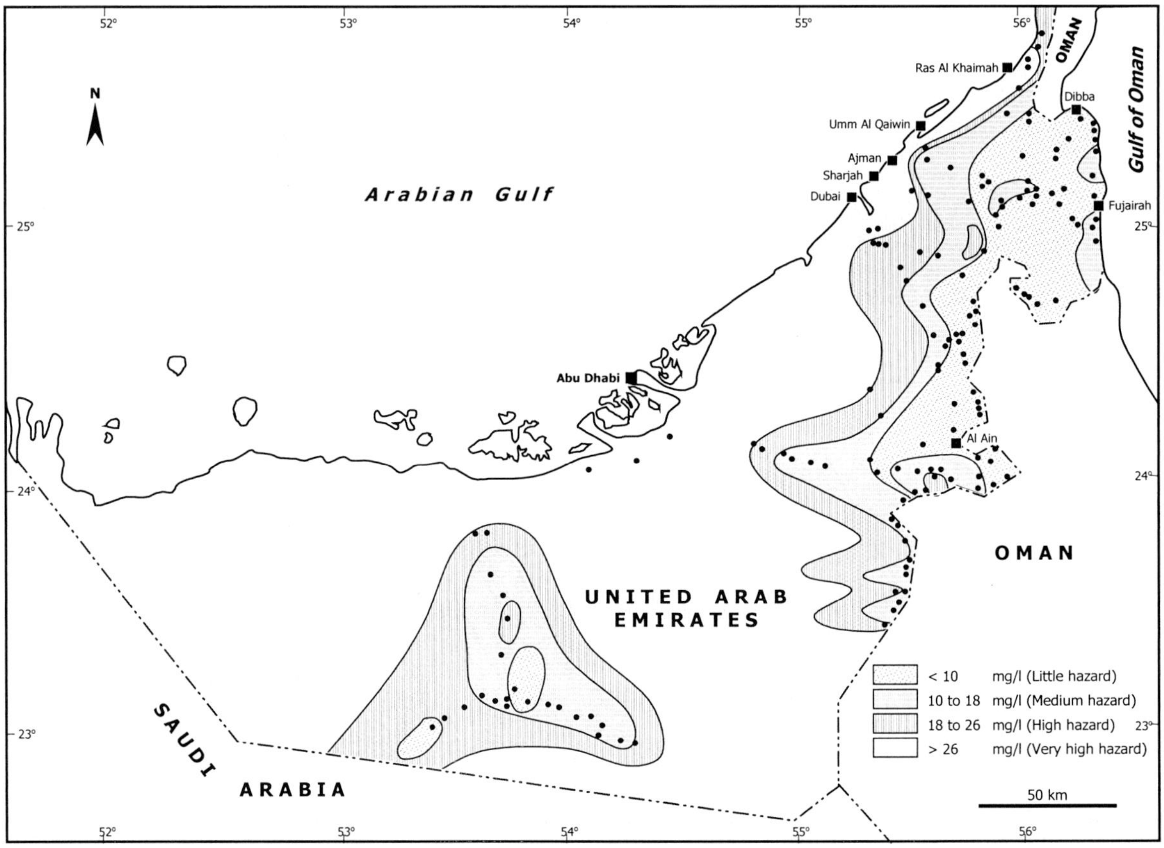

Fig. 8.78. Calculated sodium adsorption ratio (SAR) of groundwater in the United Arab Emirates in 1996.

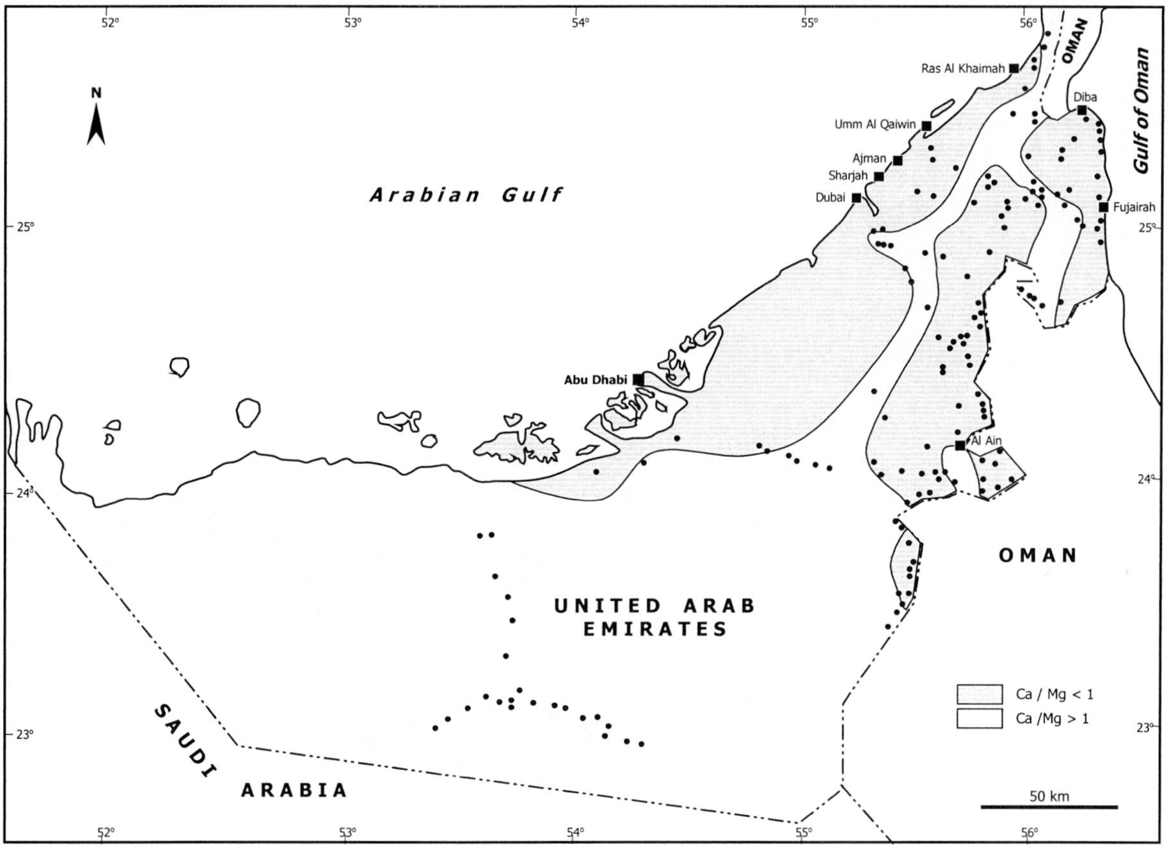

Fig. 8.79. Hydrochemical coefficients Ca/Mg in groundwater of the United Arab Emirates in 1996.

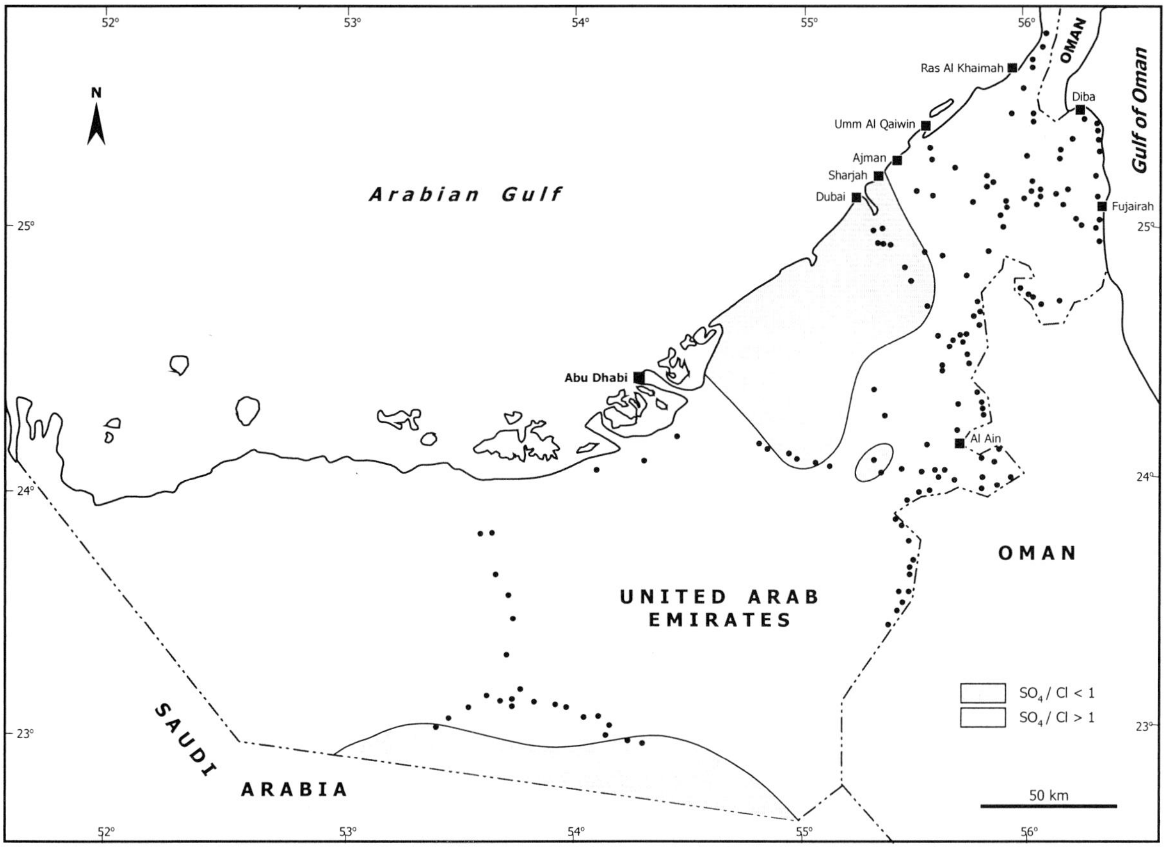

Fig. 8.80. Hydrochemical coefficients SO_4/Cl in groundwater of the United Arab Emirates in 1996.

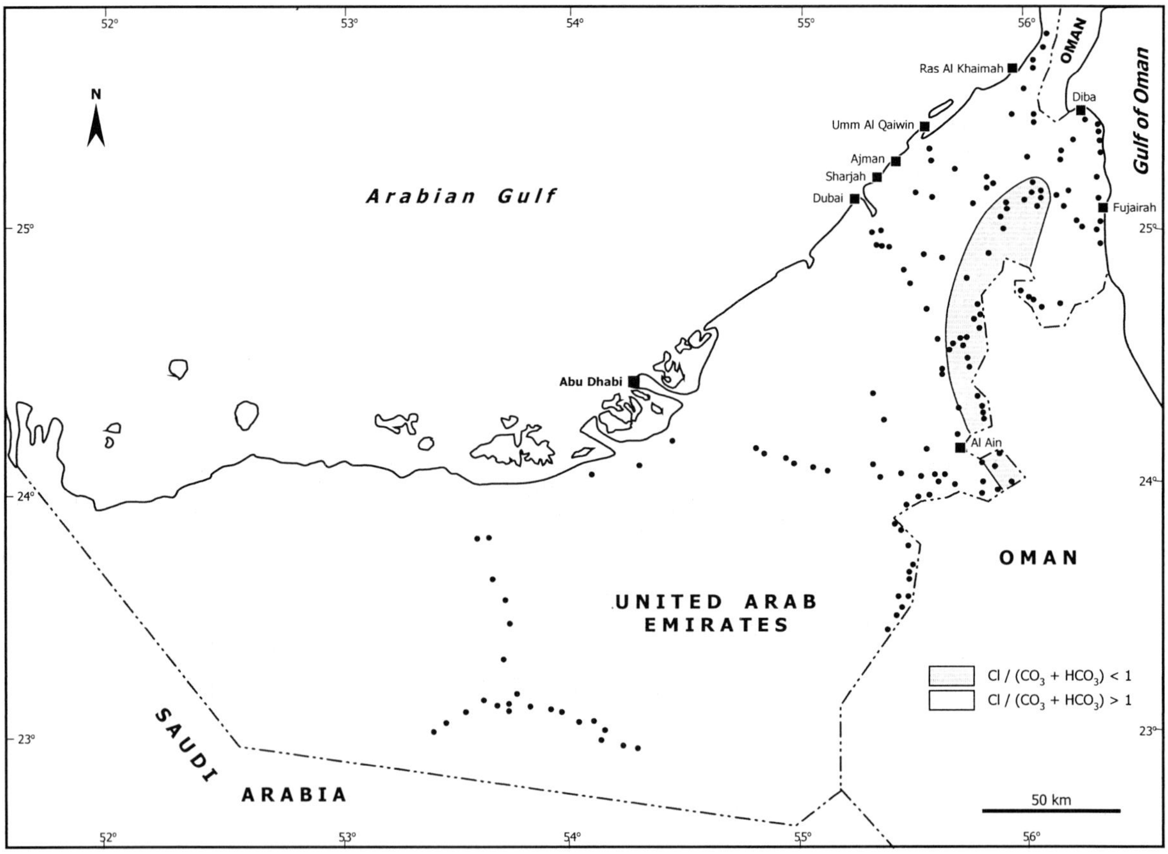

Fig. 8.81. Hydrochemical coefficients Cl/(CO_3+HCO_3) in groundwater of the United Arab Emirates in 1996.

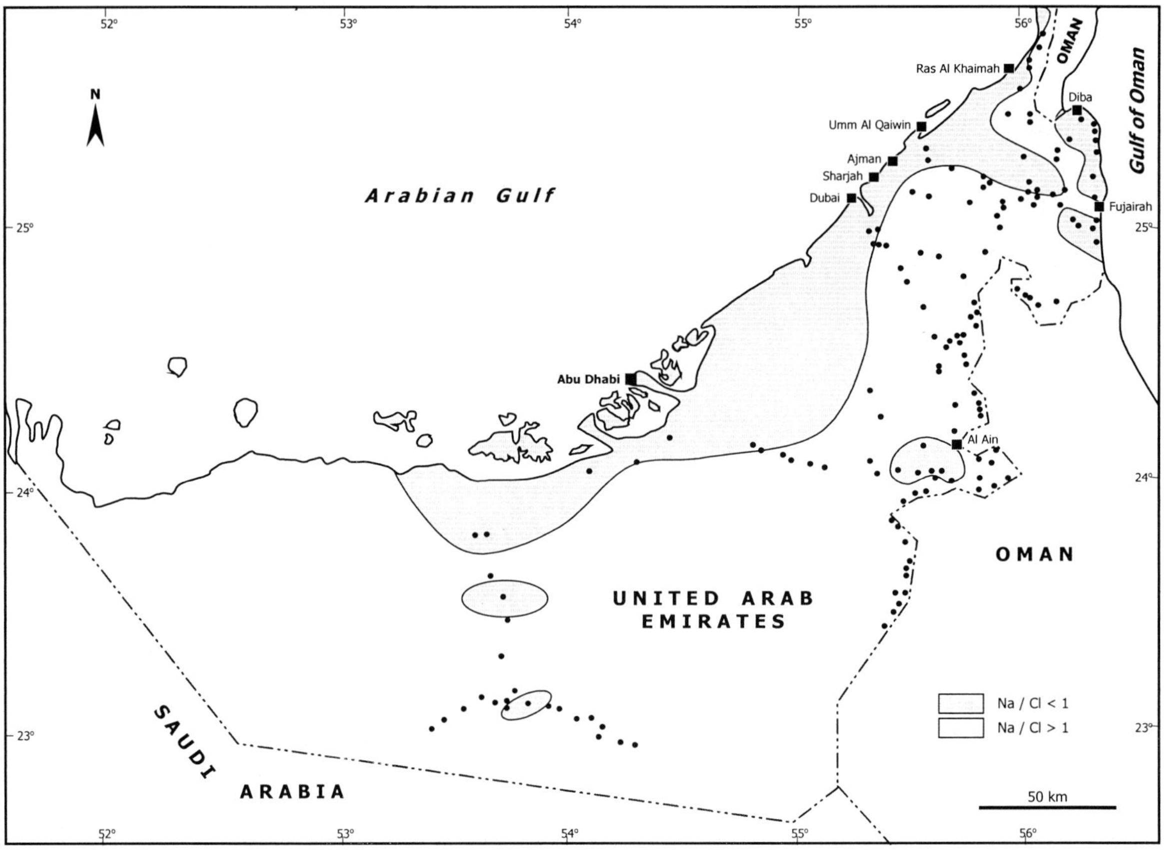

Fig. 8.82. Hydrochemical coefficient Na/Cl in groundwater from the sand and gravel aquifer during 1996, United Arab Emirates.

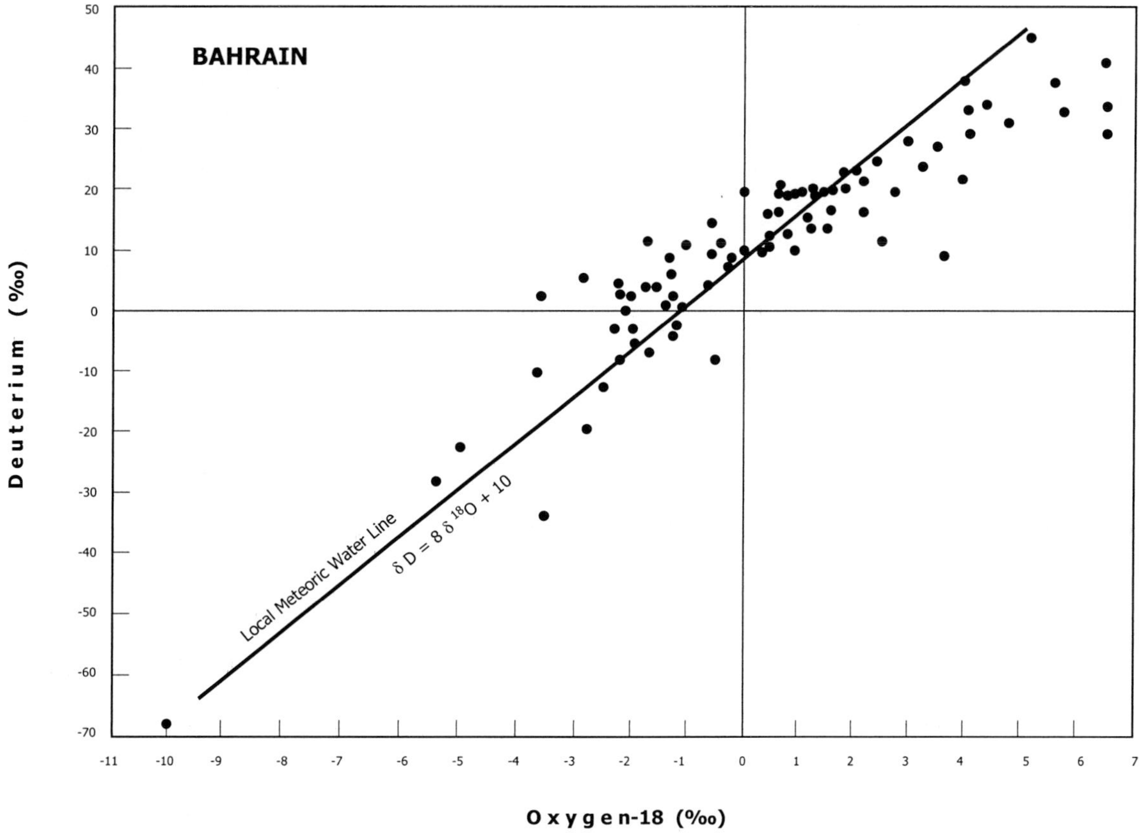

Fig. 8.83. The ^{2}H (‰) versus ^{18}O (‰) in rainfall of Bahrain IAEA/WMO station for the 1963-1993 period (after International Atomic Energy Agency Yurtsever, 1999).

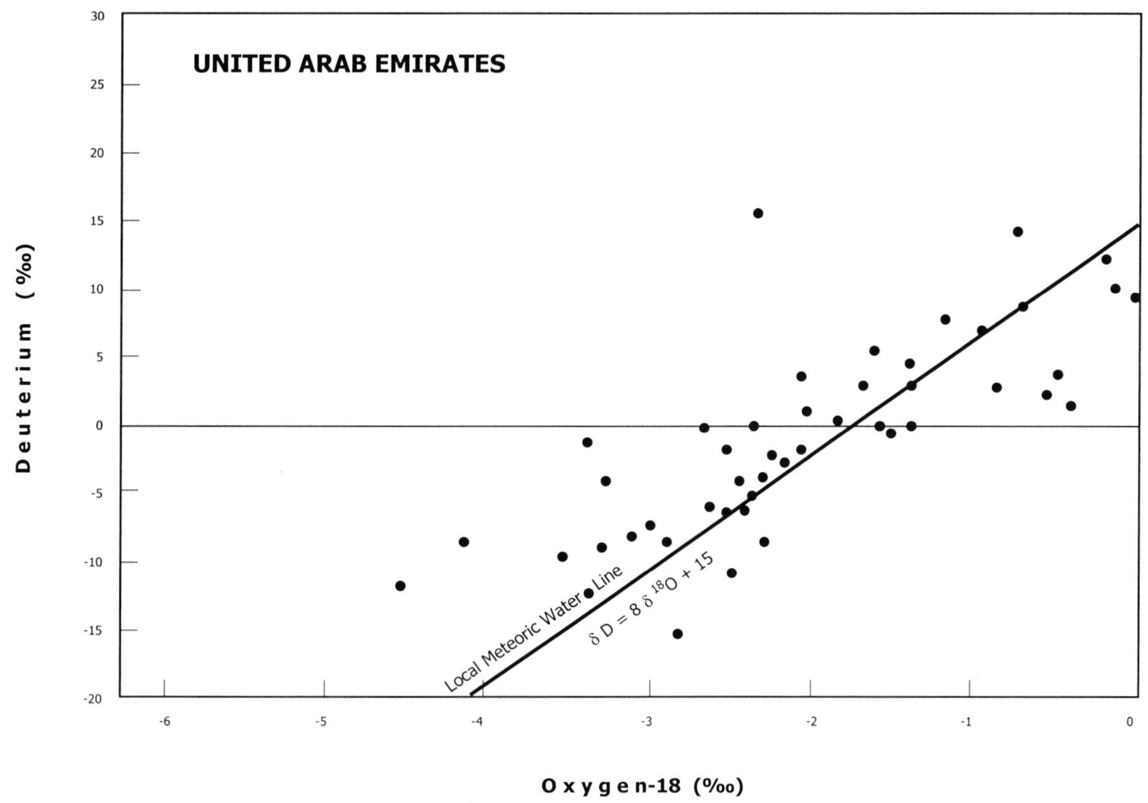

Fig. 8.84. The ^{2}H (‰) versus ^{18}O (‰) in rainfall of the United Arab Emirates for the 1984-1990 period.

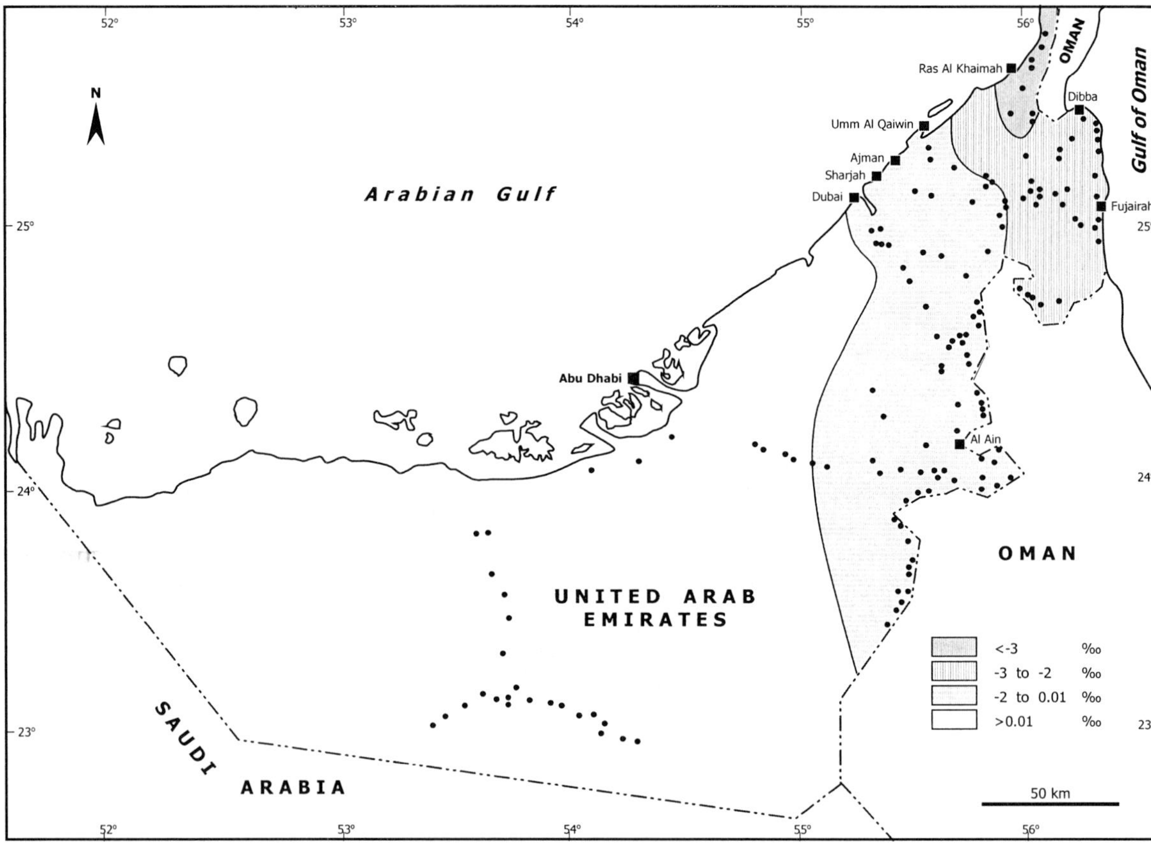

Fig. 8.85. Distribution of ?^{18}O (‰) in groundwater of the United Arab Emirates, based on data collected by the Ministry of Electricity and Water (1984-1990) and the authors (1996).

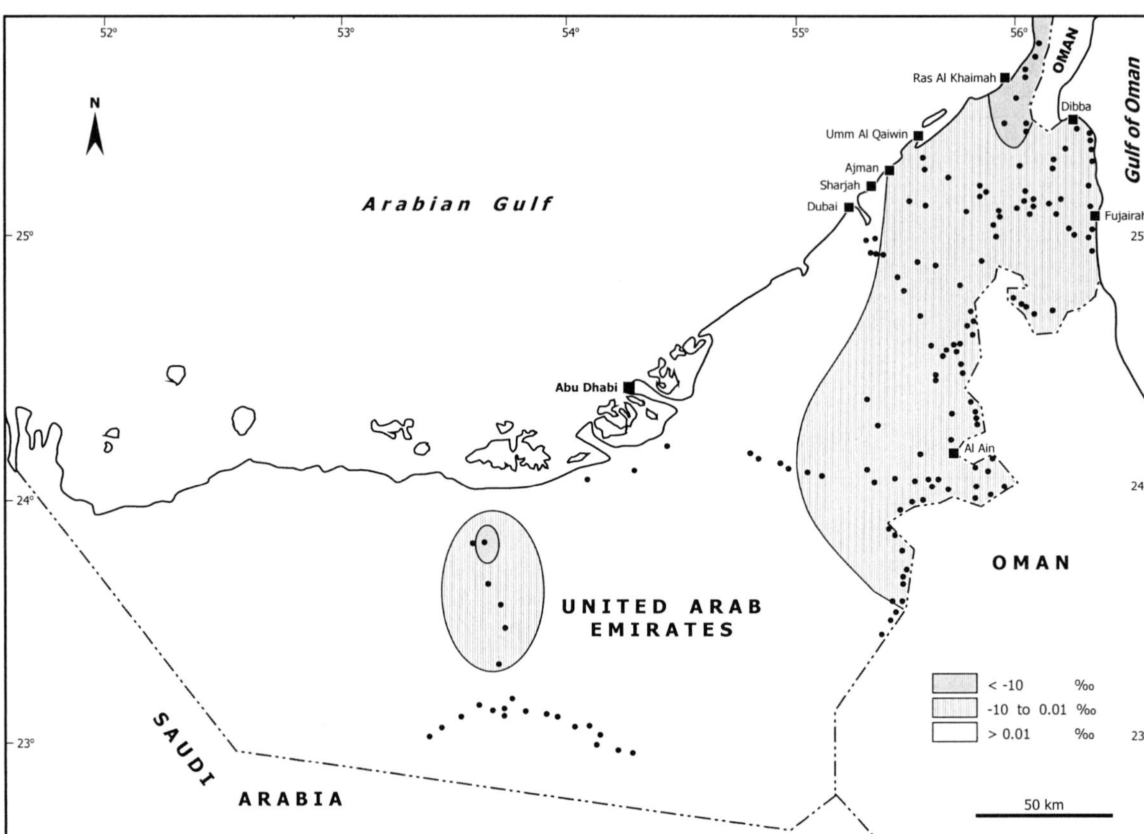

Fig. 8.86. Distribution of ?^{2}H (‰) in groundwater of the United Arab Emirates based on data collected by the Ministry of Electricity and Water (1984-1990) and the authors.

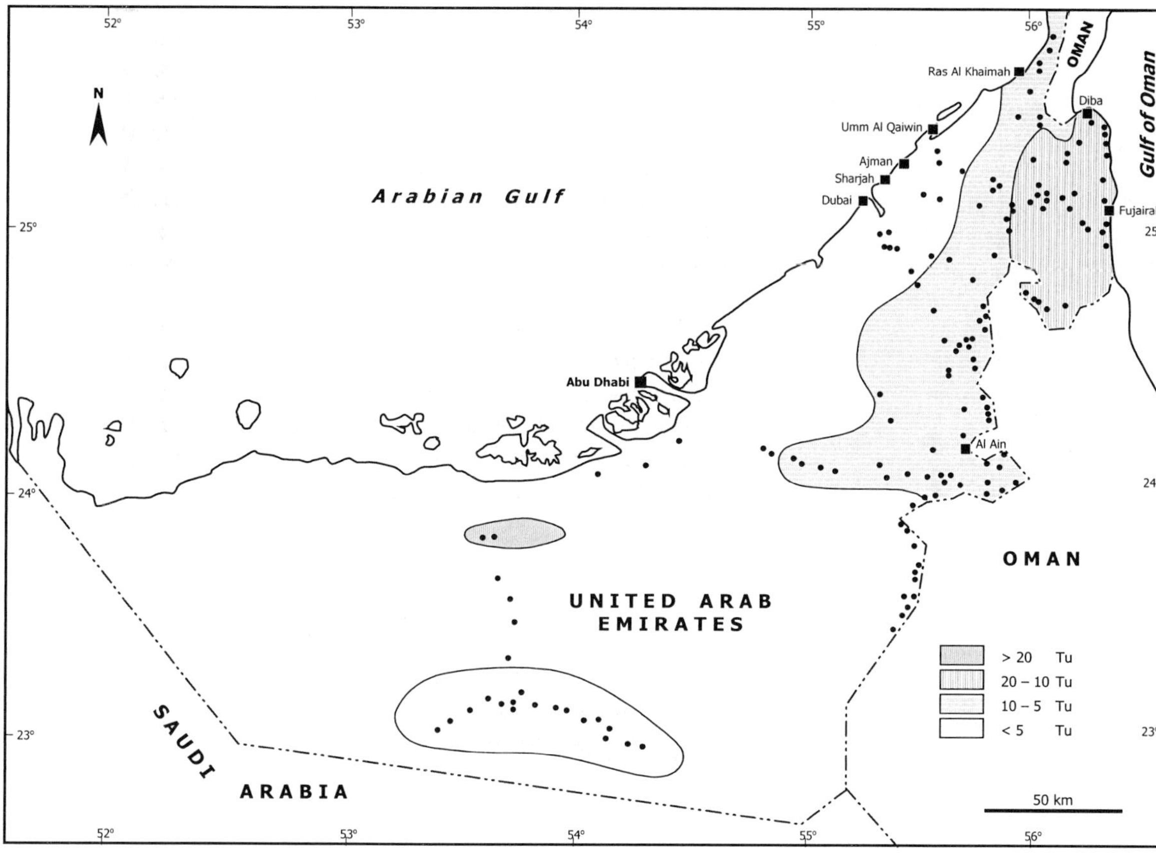

Fig. 8.87. Distribution of ^3H (TU) in groundwater of the United Arab Emirates, based on data collected by Ministry of Electricity and Water (1984-1990) and the authors.

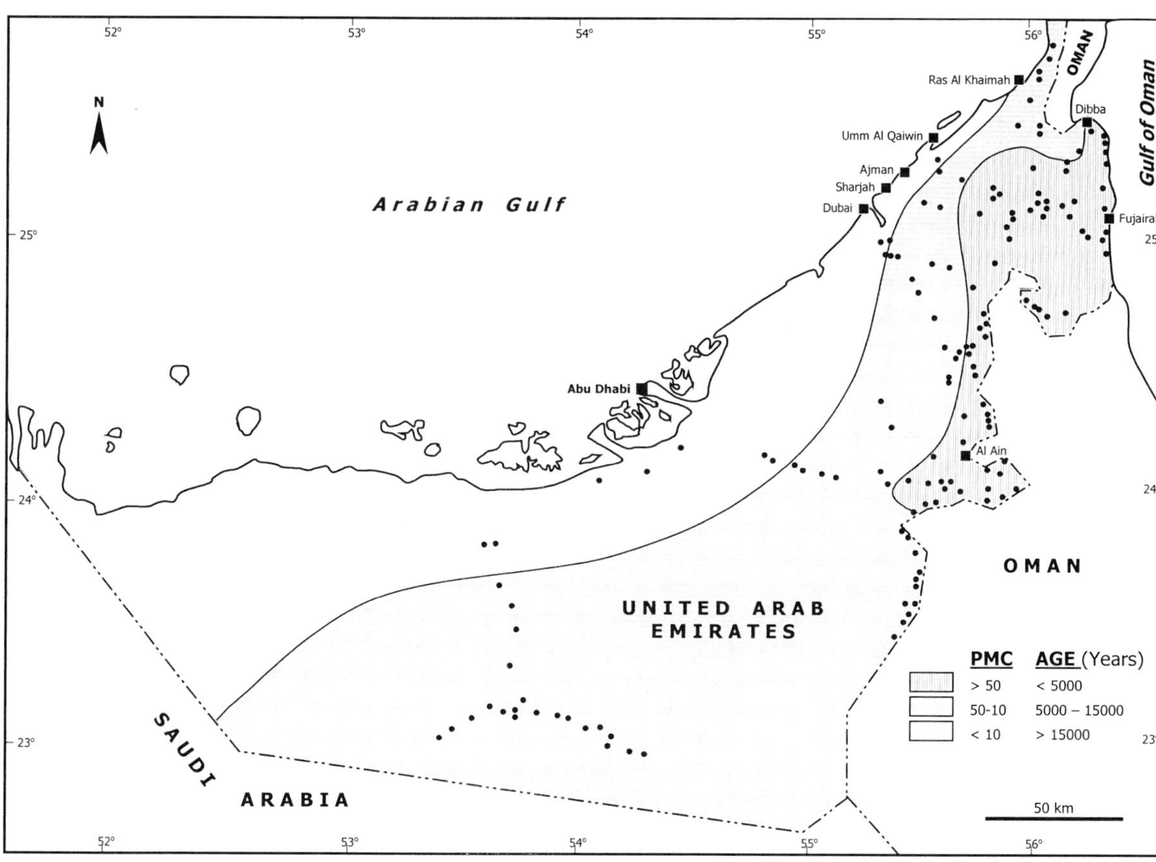

Fig. 8.88. Distribution of ^{14}C (PMC) in groundwater of the United Arab Emirates (based on data collected by Ministry of Electricity and Water during the period 1984-1990).

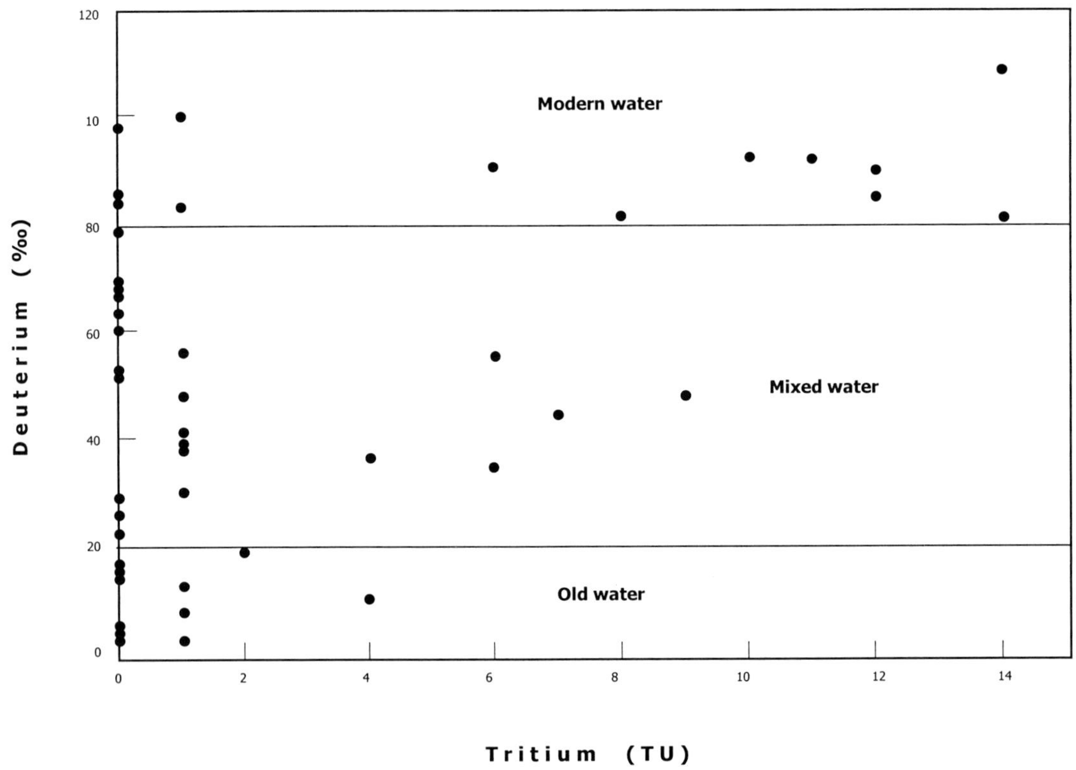

Fig. 8.89. Classification of groundwater of the United Arab Emirates, based on their ^{3}H (TU) and ^{14}C (PMC) contents.

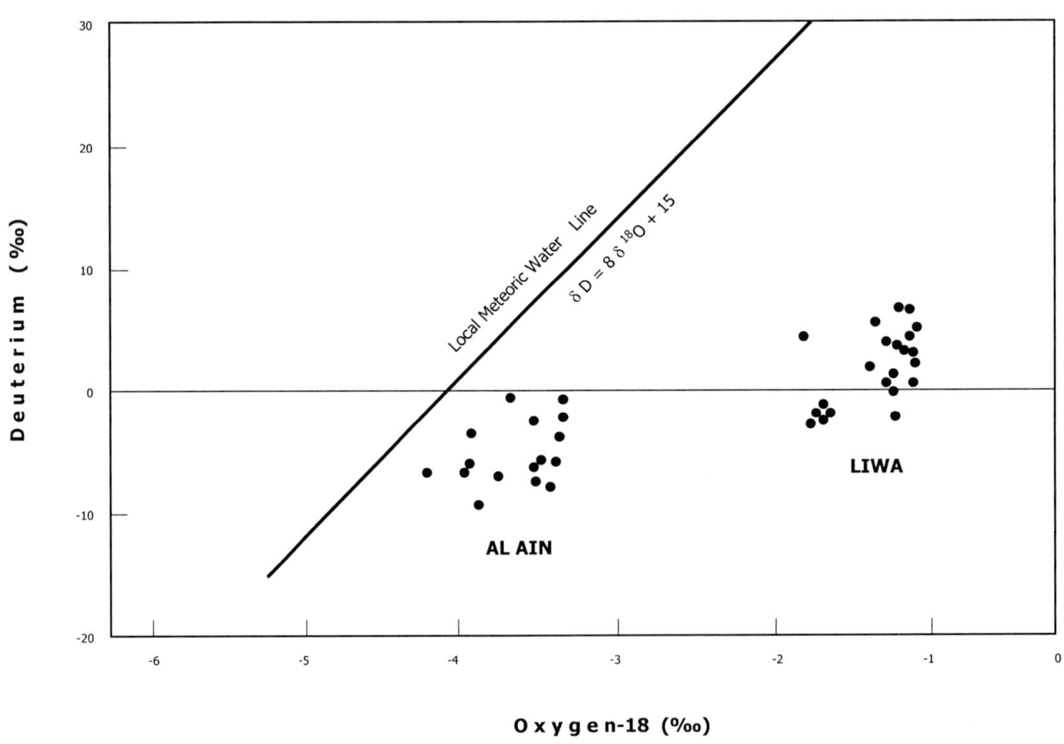

Fig. 8.90. Distribution ^{2}H (‰) and ^{18}O (‰) in groundwater samples collected from the Al-Ain and Liwa areas in March 1996.

6. CENOZOIC AQUIFER SYSTEM OF OMAN

INTRODUCTION

Oman forms the eastern margin of the Arabian Peninsula, with coastlines on the Arabian Gulf, the Gulf of Oman and the Arabian Sea. The Oman hydrogeological basin is bounded to the north by the Oman Mountains which reach an elevation of more than 3000m, to the south by the Dhofar Mountains with elevations of about 1,800m, to the east by the low-lying Huqf outcrops and to the west by Rub al Khali sand dunes (Fig. 8.91). The population of Oman is concentrated in a number of small coastal towns, which draw their water either from groundwater or from desalination plants. The groundwater supply is taken from the thick, relatively unconsolidated gravels and sand of Neogene or younger age of the coastal plains. The exploitation of the available water resources, is through a system of falajes which still provide about 55% of the water used in irrigation (Wallender, 1989), and the 4,800 active falajes provide more than 60% of the total water usage (Abdel Rahman and Omezzine, 1996). In recent years, the balance achieved over the centuries has been upset by modern methods of extraction, and in some areas the increased agricultural demand for water has placed a strain on the traditional water supplies and salt-water intrusion has been the result. For the larger volumes of water, such as are required by the oil industry, Tertiary aquifers are tapped, in particular the water from the Umm er Radhuma Formation.

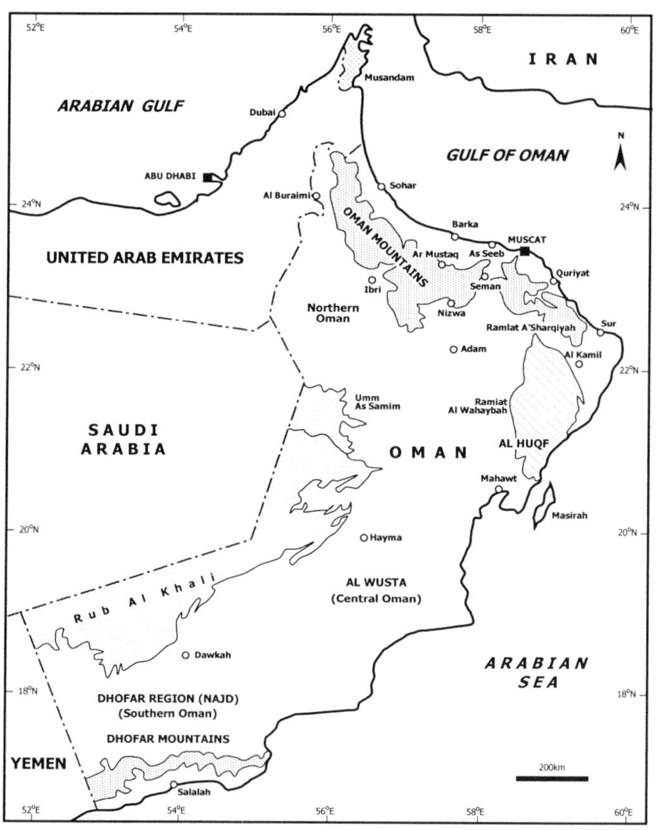

Fig. 8.91. Sultanate of Oman showing major localities mentioned in the text.

The Ministry of Water Resources, created by royal decree in January 1994 evolved out of the Water Resources Council which was established in 1975 and the Public Authority for Water Resources. The two were merged in 1986 and in 1988 the country's water resources were declared part of the natural wealth of the country and in 1994 the development, maintenance and jurisdiction, with the records of wells and falajes became the responsibility of the new ministry. The ministry met the challenge by requiring the registration of all wells, and by 1990 about 167,000 wells had been registered. Permission for the sinking of new wells, deepening of existing wells and repairs to falajes and wells have to be approved by the Ministry which also handles violations and appeals. Conservation plans call for the construction of recharge dams and so far about 20 had been completed by the end of year 2000.

Hydrostratigraphy

The stratigraphy and hydrostratigrpahic record comprises strata ranging from Infracambrian to Quaternary in age. The sequence marked by many unconformities and hiatuses and lithological variations recording extreme climatic changes through time due to the occurrence of Oman in varied latitudinal positions. The hydrogeological cross-section of Oman is shown in figure (8.92) which shows also water flow regime and the aquifers (Huqf, Haima, Al Khlata and Umm er Radhuma) and aquitards (Nahr Umr and Shammar shale of the lower Umm er Radhuma). The hydrogeological history of Oman differs from that of the other Gulf countries on the Arabian plate for two reasons, because of its position, which brings it under the influence of monsoonal circulation, and because of orographic rainfall associated with the mountains, which formed along the eastern margin of the Arabian plate. Consequently, although water is still not abundant, the shortages are not quite as critical as in the rest of the Gulf States.

The earliest hydrogeological studies were restricted to shallow wells (500m) in Tertiary rocks from which most of the water used by the oil industry was extracted. When the biodegradational effects on meteoric water were recognized, more wells were drilled mainly in the Umm er Radhuma Formation. The result has been a coherent picture of

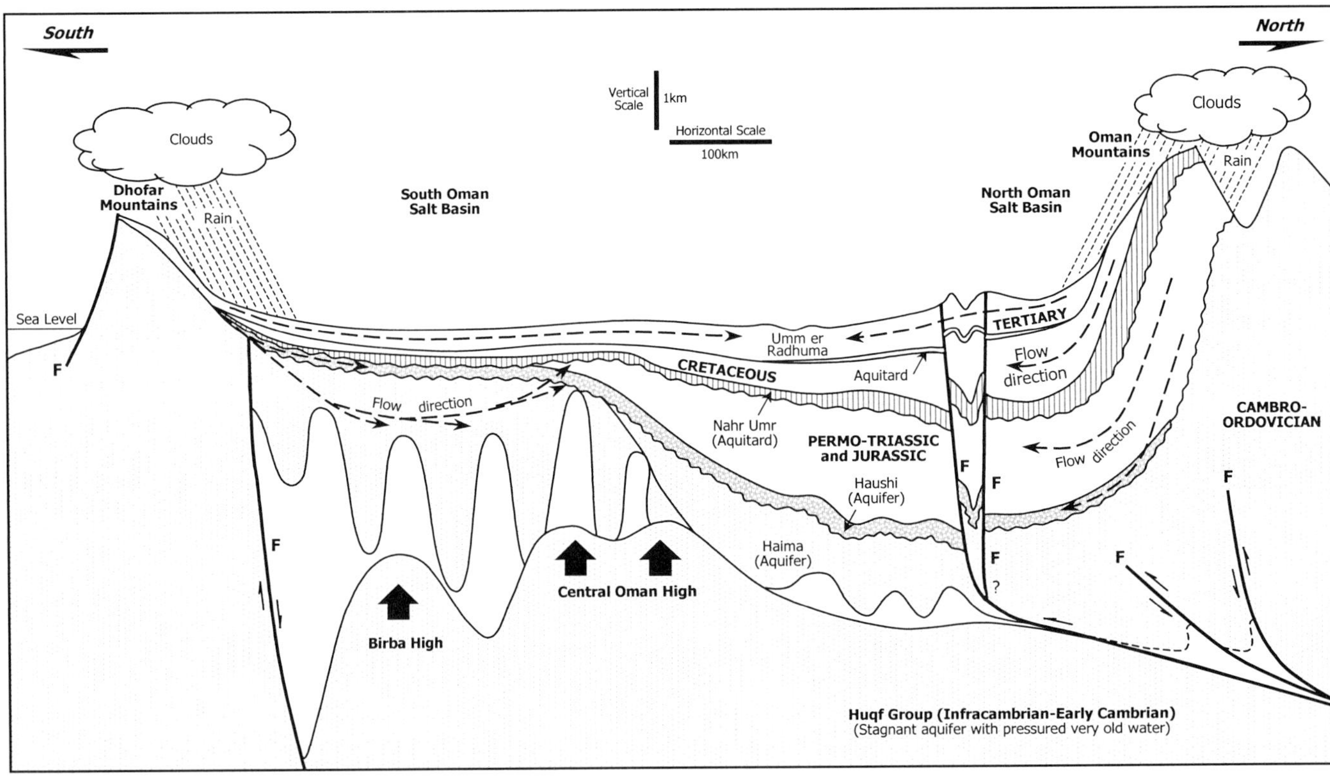

Fig. 8.92. A hydrogeological cross-section of Oman showing water flow regime (modified from Al Lamki and Terken, 1996).

the aquifer system in the Oman subsurface. Static water levels and salinity variations, have been monitored in over 500 Tertiary wells, which reflect subtle spatial variations, in the regional water flow pattern.

Aquifer temperatures have been recorded and bottom hole temperatures, from 250 exploration and appraisal wells, by means of which, it is possible to create temperature "slices" at different depths, and study the thermal structure, as a non-linear function of depth. The temperature slice at 500m show lower temperatures to the south, where recharge occurs along the Dhofar Mountains (Fig. 8.93), while the Oman Mountains thrust front, has clearly had little effect on the temperature pattern.

There are four principal aquifers in central and southern Oman, in the Tertiary marine limestones, marls and minor evaporates, which make up the Umm er Radhuma, Rus and Dammam formations of which three are confined and support flowing artesian wells. Groundwater recharge comes from the Northern Oman Mountains and Dhofar Mountains, and discharge zones are found in the Sabkhat Umm as Samim (Rub Al Khali) in the west, where gypsum flats develop, and halite caps the shallow groundwater. Four aquifers have been identified by Clark et al. (1987) in south Oman (the Najd area) in the Tertiary formations: the Dammam aquifer, upper Umm er Radhuma aquifer, upper part of lower Umm er Radhuma aquifer, and lower part of lower Umm er Radhuma aquifer. The Dammam aquifer generally has a good quality water, with electrical conductivity of <2,000 μS/cm. Degraded groundwater quality was recorded, in the lower part, due to the solution of evaporite minerals also found in the underlying Rus Formation. In the southern Najd, the upper Umm er Radhuma aquifer has groundwater of good quality, with electrical conductivity of <2,000 μS/cm. The aquifer is confined and under artesian conditions throughout the area and the aquifer has a high yield and high transmissivity (> 10,000 m^2/day). The main aquifer in the Najd area is the upper part of lower Umm er Radhuma aquifer. This part of the aquifer is highly karstified and groundwater flows under artesian pressure. The groundwater quality is good, with electrical conductivity of <1,500 μS/cm, deteriorating in the direction of groundwater flow i.e. towards the north. Groundwater is under reducing conditions with a high H_2S content. The lower part of lower Umm er Radhuma aquifer is also fractured, but to a lesser degree than in the upper part. The groundwater quality is still good, with electrical conductivity (about 2,000 μS/cm), but the aquifer transmissivity is also lower than that of the upper part. The regional groundwater is of fossil origin, verified by isotopic dating, however, scattered freshwater lenses, recharged during the rare storm events exist in wadi sediments, where favored by local topography.

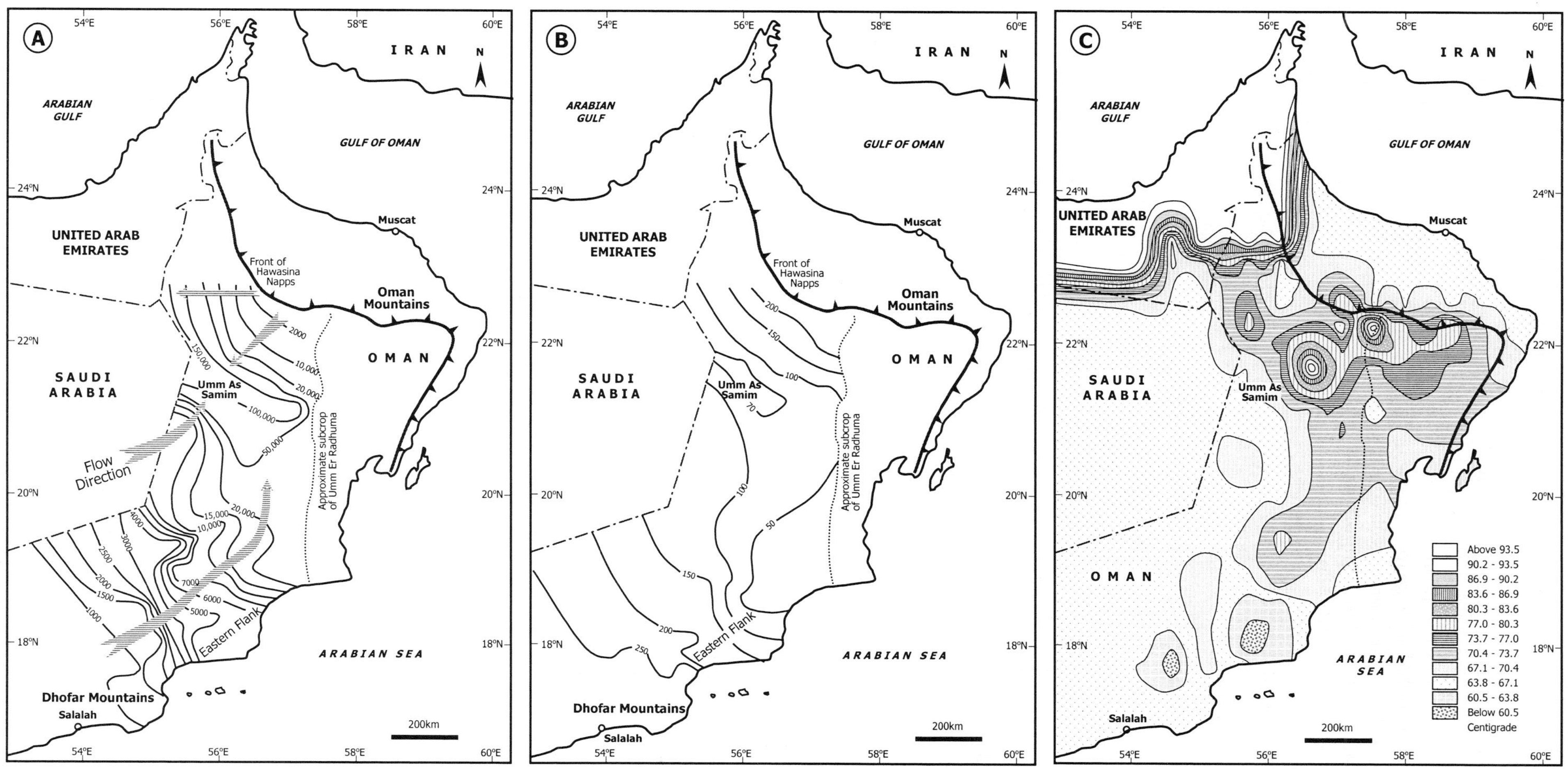

Fig. 8.93. Slice maps of Umm er Radhuma Formation in Oman (modified from Al Lamki and Terken, 1996). A) Water salinity (ppm) indicates flow from the Oman and Dhofar mountains to the Umm as Samim Sabkha. B) Equipotential map in m above mean sea level showing the main recharge areas in mountains and discharge area in the Umm as Samim Sabkha. C) Temperature (°C) slice at 500 m (approximate depth of Umm er Radhuma Formation) indicating lower temperature to the south where recharge occurs along the Dhofar Mountains.

Groundwater Flow

Groundwater flow in the Oman hydrogeological basin is gravity induced and topographically driven. The flow of meteoric water into the formations is through highland recharge with discharge in the low relief areas (Fig. 8.92). At the present time, recharge over the Dhofar Mountains is the most important, for it is estimated that as much as 300 mm of misty precipitation takes place during the monsoon period from July to September. Precipitation over the Oman Mountains is estimated at 150 mm, however, towards the end of the last Pluvial Period, between 2,000 and 10,000 years BP, the region was much wetter than it is today (Beydoun, 1980), and it is believed that the greater part of the aquifer charge dates from this time period. The equipotential map of the Umm er Radhuma aquifer indicates that, the main recharge area is in the Oman and Dhofar Mountains with discharge in the Umm as Samim sabkha (Fig. 8.93). The age of the groundwater in the Umm er Radhuma aquifer was determined from isotopic studies (^{14}C and tritium). The ^{14}C studies establish the age of the groundwater ranges from 5,000 to 17,000 years, the age varying with distance from the recharge area, but generally indicating a flow rate of about 10 m/yr (Al Lamki and Terken, 1992). Water flows through the karstified limestones of the Umm er Radhuma which with the development of sink holes, caverns and gorges features accounts for most of the fluid flow within Tertiary rocks. Local cross-formational flow from the Umm er Radhuma into the underlying mid-Cretaceous Natih carbonates is believed to occur in the fold and thrust belt of the Oman Mountains, however, the Albian Nahr Umr below the Albian-Cenomanian Natih Formation is a regional aquitard, and separates the Tertiary carbonate, aquifer from the Permo-Carboniferous platform carbonates and the Haushi and Haima Groups clastic aquifers.

A second group of aquifers comprises the fluvio-glacial and the Haushi and Haima clastic aquifers which may be in hydraulic continuity. The distinction between the two aquifer groups is based mainly upon salinity. Water flow in the deeper Palezoic aquifers, is determined solely by salinity and the trend of the Gharif Formation (Fig. 8.92), which suggests a flow direction, similar to that in the Tertiary aquifers. There are no signs of intervening large scale salt dissolution, nor of areas of local recharge-discharge. In fact, the salinity of the Gharif Formation water, close to that of the recharge area in South Oman, suggests the strata have been flushed by meteoric water. This is supported by the dating of the Gharif water at 30,000 years in the Marmul area (south Oman), 150 km from the recharge area. Although hydraulic communication can occur across the aquitards, on a time scale of 30,000 years, the young age of the Gharif water suggests the aquitard is probably discontinuous, in the recharge area and there my be direct communication between the Tertiary and Paleozoic aquifers.

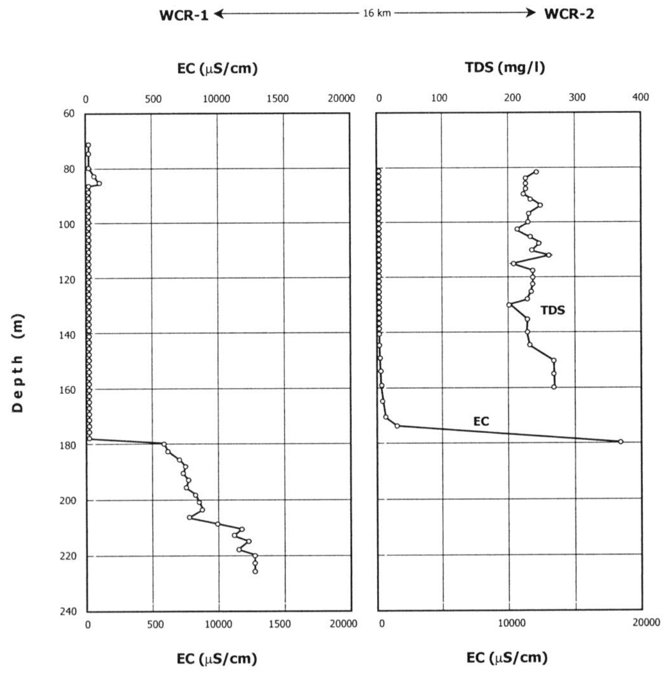

Fig. 8.94a. Electrical conductivity (EC) and salinity variation (TDS) versus depth through the wadi Rawnab (South Oman) freshwater lens into the underlying trnasition and saline zone (after Macumber et al., 1998).

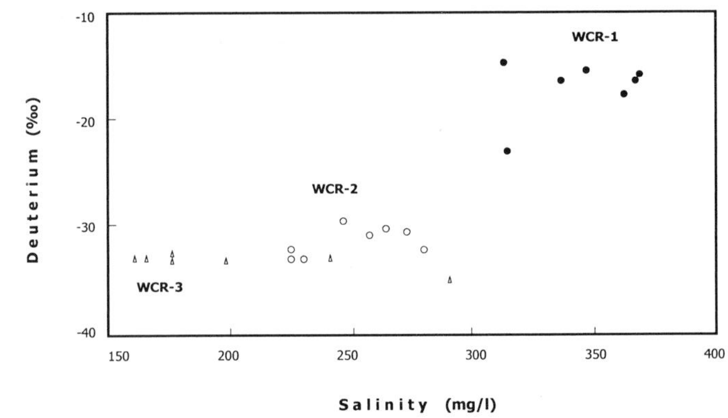

Fig. 8.94b. Deuterium (‰) versus salinity (mg/l) for the Wadi Rawnab freshwater lens (after Macumber et al., 1998).

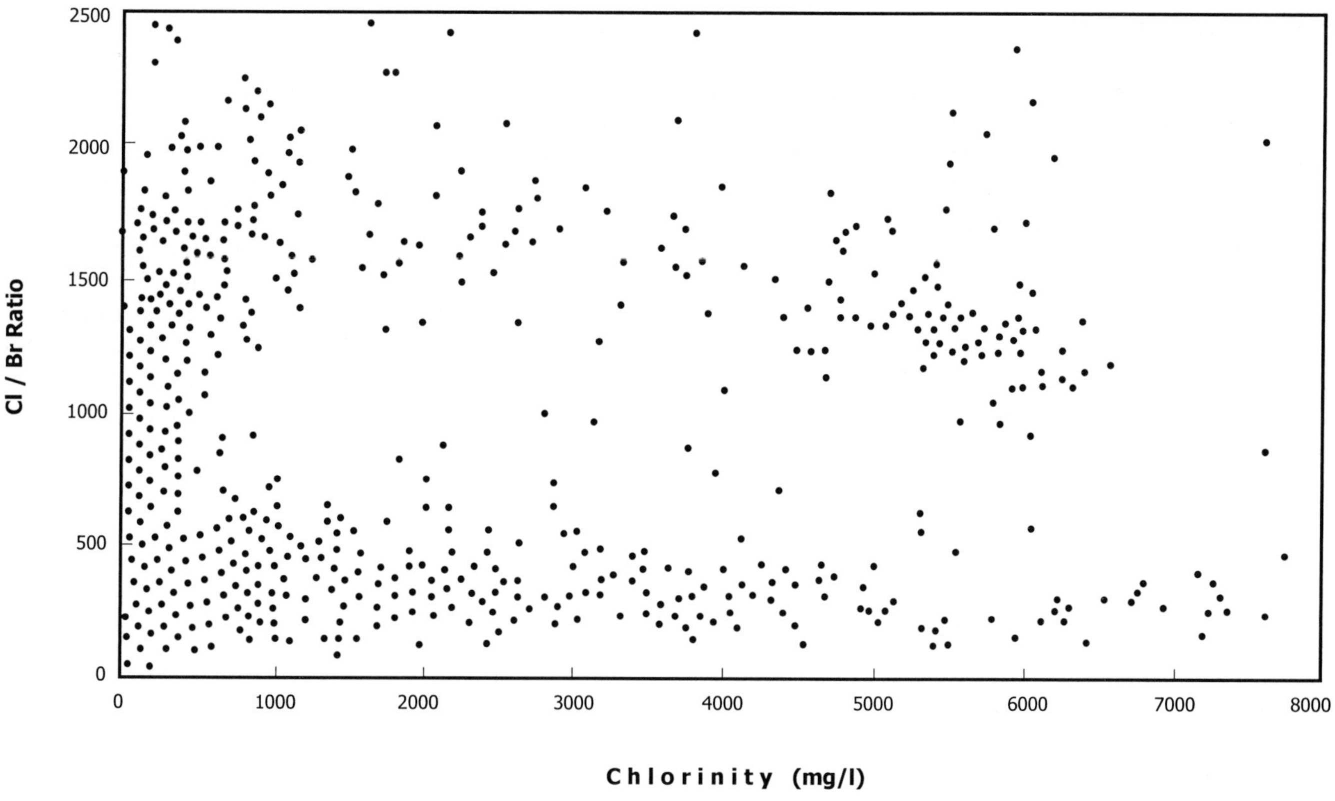

Fig. 8.95a. Chloride-Bromide ratio versus chlorinity (mg/l) for all Oman groundwaters (after Macumber et al., 1998).

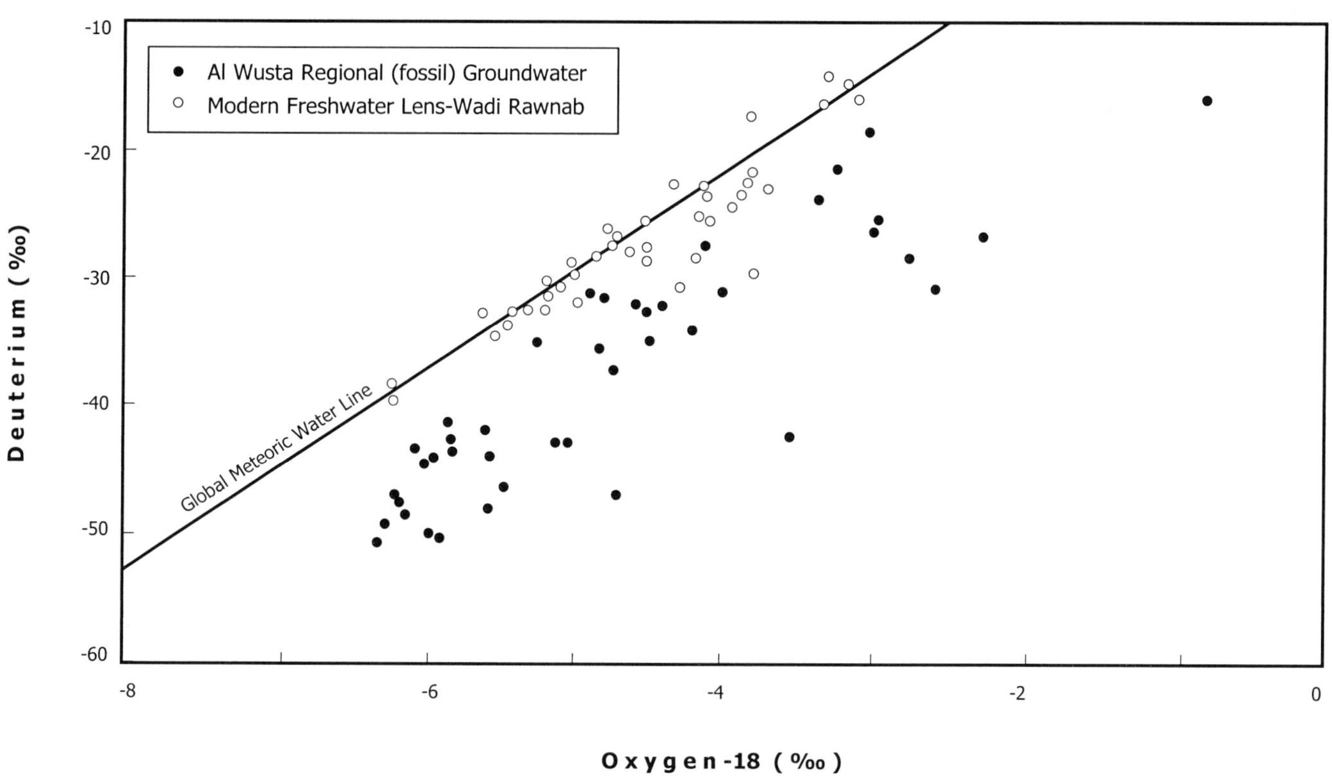

Fig. 8.95b. Deuterium (‰) versus Oxygen-18 (‰) for regional (fossil) and local (modern) groundwater. The modern groundwater lying furthest from the GMWL is transitional (after Macumber et al., 1998).

Hydrochemical Facies

Groundwater flows from the northern and southern Oman Mountains towards the Rub al Khali. As groundwater moves from central Oman (the Al Wusta area) it passes from a calcium-magnesium bicarbonate type to a sodium chloride-sulphate type, and becomes saline. In contrast, fresh groundwater of the best quality occurs in Tertiary limestones, in lenses up to 100 m thick, and extending over a wide area of several kilometers wide, and tens of kilometers long (Fig. 8.94a). These freshwater lenses overlie the saline regional groundwater.

The chloride/bromide ratio (Cl/Br) is commonly used to detect seawater intrusion. Seawater has fixed ratio of 290 to 300. In central and southern Oman, groundwater occurs in three categories (Fig. 8.95a) groundwater with high Cl/Br ratio (bromide depleted) >700 (most commonly 1,000 to 2,500) (Al Wusta area), low Cl/Br ratio (bromide enriched) 100 to 250 (Najd area), with the Cl/Br ratio near the oceanic value of about 300 (Dhofar area and Salalah Plain) (Fig. 8.95b).

Based on the Cl/Br ratio and isotopic data (^{2}H and ^{18}O) Macumber et al. (1998) suddivided groundwater in central and southern Oman into four hydrofacies. These facies reflect the monsoonal rainfall, seawater influence, cyclonic storms and frontal rain events. The salinity of freshwater lens decreases from 350 mg/l (in well-1) to 250 mg/l (in well-2) and less than 200 mg/l (in well-3). Associated with decreases in salinity is a parallel depletion in stable isotopes, with deuterium values falling from -20‰ to -40‰ (Macumber et al., 1998) (Fig. 8.94b). The groundwater in the Al Wusta and Najd has high Cl/Br ratio and depleted stable isotopes (Fig. 8.96), while fresh groundwater in the Dhofar Mountains and Salalah plain has low Cl/Br ratio (100-250) and is isotopically enriched. An oceanic Cl/Br ratio characterizes the saline and brackish groundwater underlying the freshwater zone in eastern and western Salalah plain.

A high bromide (low Cl/Br) ratio marks a meteoric origin, and arises from monsoonal rainfall, where the Cl/Br ratio decreases linearly with altitude (Macumber et al., 1998). Macumber (1990) attributed the high Cl/Br ratio in central Oman

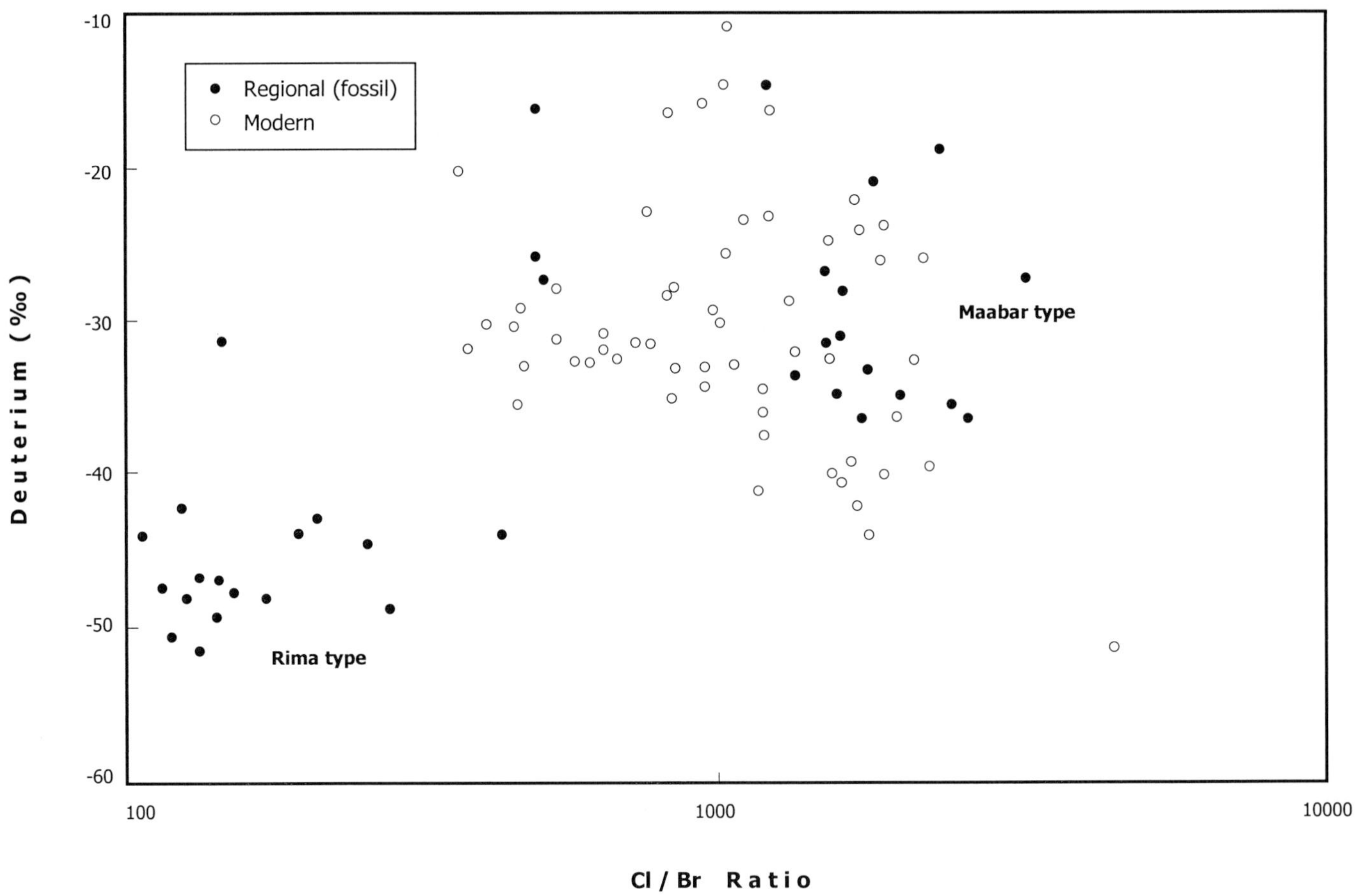

Fig. 8.96. Deuterium versus Cl/Br ratio for region (fossil) and local (modern) groundwater in central (Al Wusta) region. Note the highly depleted-low Cl/Br ratio of the Rima type fossil groundwater which differs from all other groundwater (fossil or modern) (after Macumber et al., 1998).

groundwater to re-solution of halite, while Duce et al. (1963) indicated that photochemical oxidation with the aerosols permits the escape of Br giving a higher Cl/Br ratio at higher elevations. However, high Cl/Br (700-2,500) also occur in the very low salinity groundwater lenses at Maabar and Wadi Rawnab in central Oman (TDS = 130 to 350 mg/l and Cl = 20 to 30 mg/l). Since this fresh groundwater exists in limestones and dolomites with marls and evaporite intercalations, dissolution of evaporites is most likely the source of high Cl.

Isotope Hydrology

The hydrogeology and isotope hydrology of Oman is based mainly on a comprehensive investigation of groundwater, evaluating water quality, saline water intrusion and artificial recharge (for more detailed see Clark et al., 1987; Clark, 1988; Macumber et al. 1994) and summarized below.

Precipitation in northern Oman is dominated by orographic and convective processes, which cause local intense storms of limited duration during summer, and depression systems originating in the Red Sea area, generally in winter months. Both systems are infrequent, and can be absent for months to years, causing negative impacts on groundwater resources and agriculture. In the Al Wusta area (central Oman), rainfall may originate from the Mediterranean, the horn of Africa, or from India, through tropical cyclones moving westwards across the Arabian Sea. The monsoonal rains are enriched in the stable isotopes ^{18}O (-1.66 to –0.36‰), and ^{2}H (-4.4 to 5.7‰), while the rains associated with frontal storms are depleted in ^{18}O (-8.15 to –7.67‰), and ^{2}H (-68.1 to -57.5‰) (Macumber et al., 1994) (Figs. 8.97; 8.98).

The summer and winter rains in Oman reflect a similar origin with initial $\delta^{18}O$ values of -4 to -2 ‰. However, high temperatures during summer precipitation causes a strong evaporative enrichment of the stable isotopes ^{2}H and ^{18}O, compared with winter precipitation originating in the western depression systems (Fig. 8.99).

The northeastern and southwestern sides of the Oman Mountains, have strikingly different tritium (^{3}H) levels. The rainfall on northeastern side contains between 10 and 15 Tritium Units (TU), while the southwestern side rain contains up to 22 TU. This is an effect of the higher elevation, to which these clouds reach before rainout, resulting in increased addition of ^{3}H from storage in the atmosphere.

The summer monsoon is the only reliable source of precipitation in the Dhofar of southern Oman. Stable isotopes are unusually enriched, and reflect rapid evaporation-condensation cycle. The $\delta^{18}O$ and

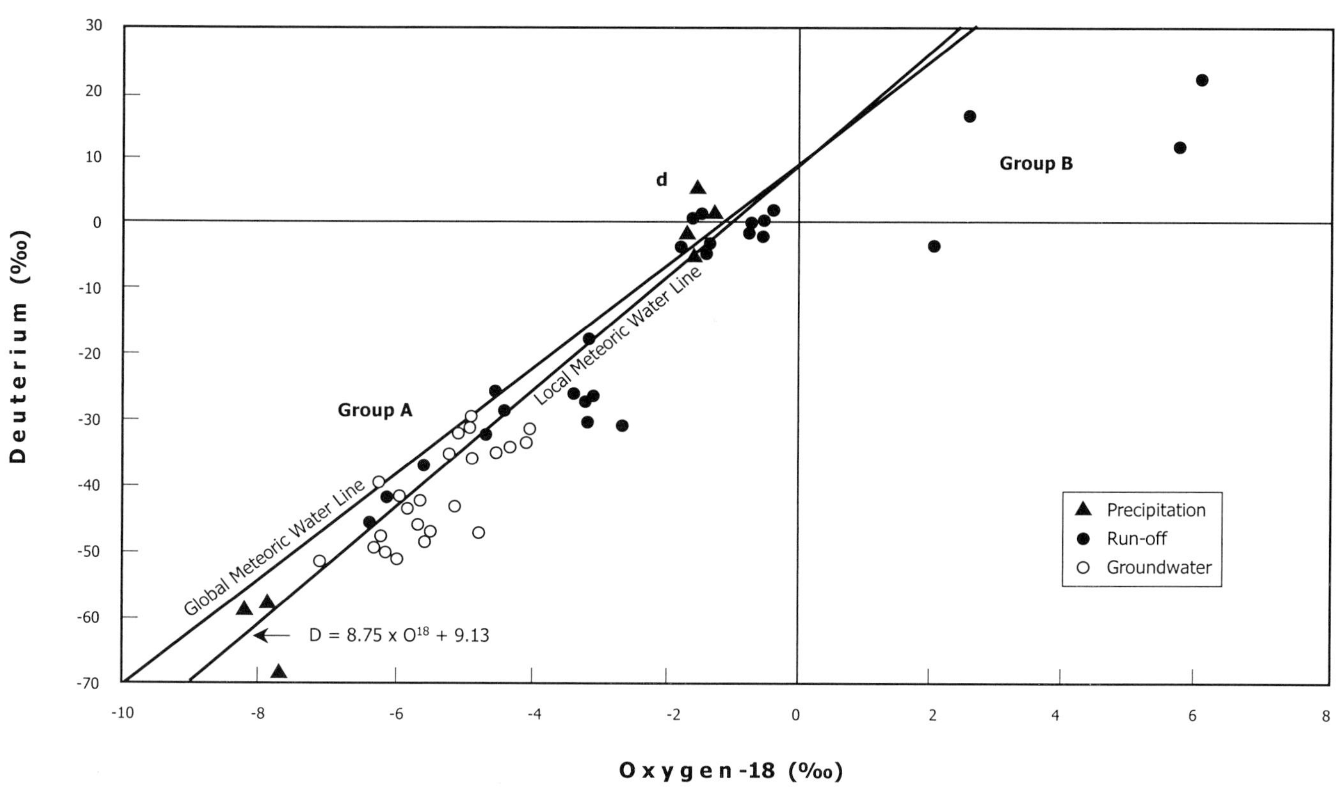

Fig. 8.97. Oxygen-18 versus Deuterium data from Al Wusta. All run-off and rainfall data are from the cyclonic event. The precipitation samples was collected several days prior to the cyclone (after Macumber, 1994).

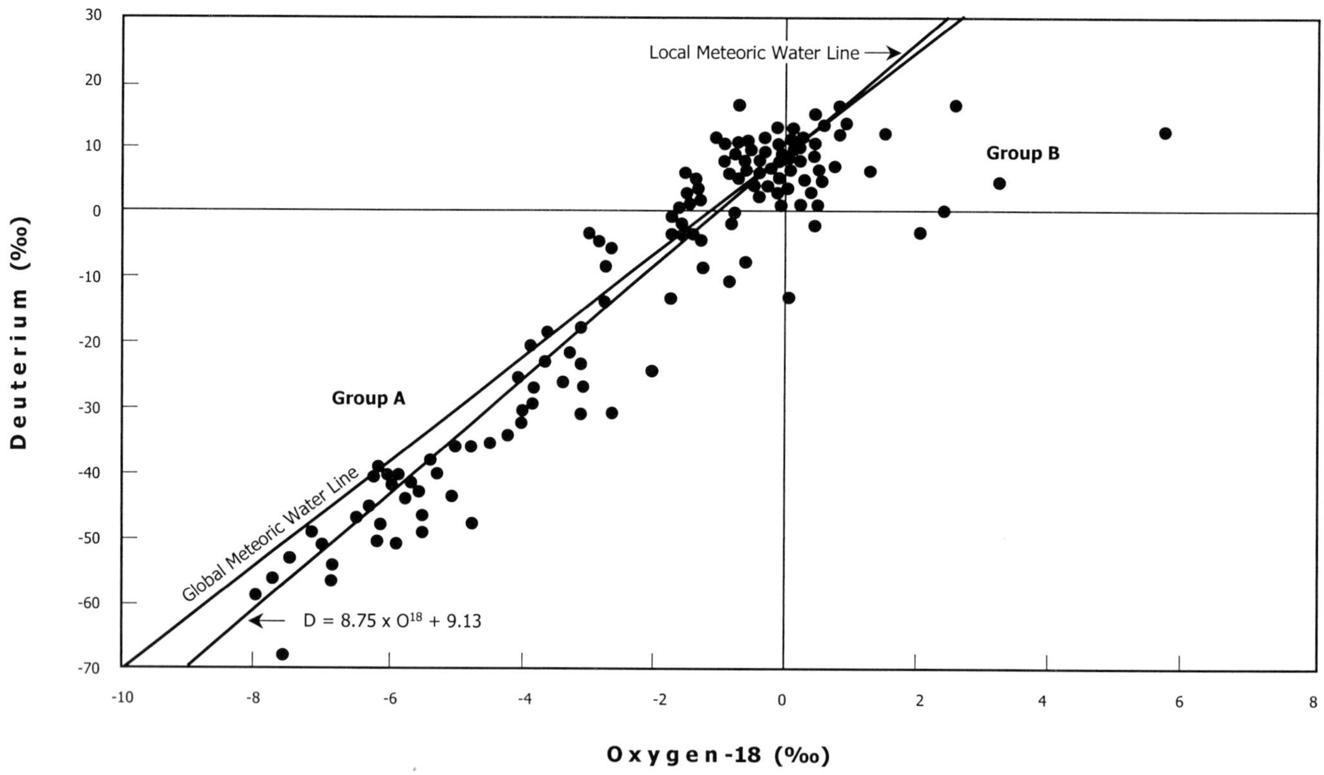

Fig. 8.98. Oxygen-18 versus Deuterium for precipitation, run-off and groundwater from southern and central Oman (Al Wusta and Dhofar) (after Macumber, 1994).

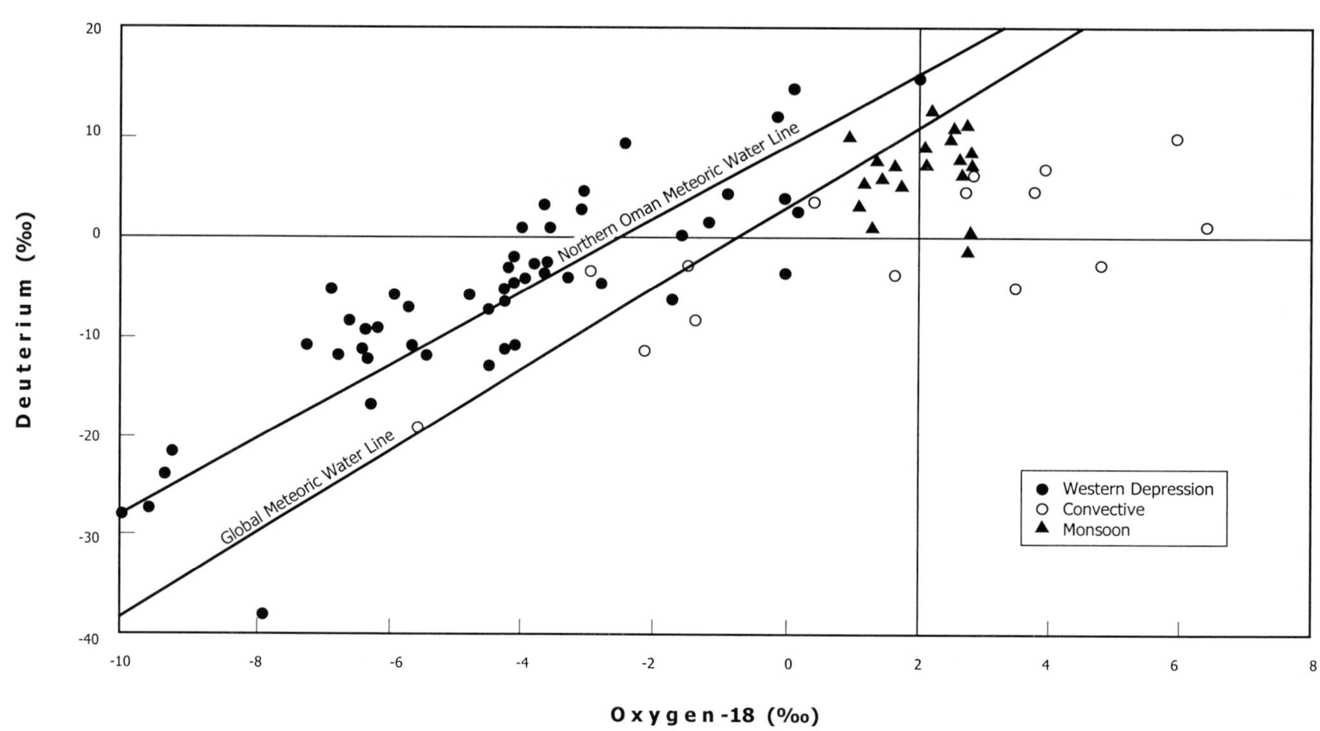

Fig. 8.99. The isotope relationship for rainfall in Oman, showing the effect of evaporative enrichment which takes place during run-off and recharge during summer (after Clark, 1988).

δ^2H values are almost identical to local sea water. The ^3H levels are also affected by this short, low altitude cycle, averaging 4 TU. Interior rainfall in southern Oman is depleted in stable iostopes ($\delta^{18}O$ -4 to -8‰) and with ^3H similar to that of northern Oman rain.

Aquifers

1. Quaternary Aquifer of Northern Oman Mountains

Stable isotope values in Northern Oman Mountains show an altitude effect, indicating two levels of recharge: at elevations in excess of 2,000m, corresponding to the upper plateau region, and on the north side of the mountains at elevations between 600 and 1,200m above sea level. Tritium shows rapid circulation with subsurface mean residence times of <5 to 10 years. Deeper groundwaters, including most thermal springs, appear to be recharged at altitude in the mountains, and have mean circulation times up to several thousand years.

The bedrock aquifers show no effect of evaporative enrichment, and are recharged from winter precipitation, associated with inferquent western depression storms (Fig. 8.100). The lack of recharge from the more frequent, localized, convective, summer rainfall, is a function of temperature/relief effect, but in all cases, little evaporation is evident, and direct recharge to bedrock aquifers is apparent.

Wadi alluvial aquifers in the mountainous Musandam Peninsula of northern Oman, show isotopic similarities to groundwater, in fractured bedrock aquifers, suggesting recharge at high elevation. However, at lower elevation, recharge is also apparent; probably from the Arabian Gulf. This is consistent with data from the northern United Arab Emirates (Gonfiantini and Akiti, 1985). Bedrock fractured aquifers along the northeastern front of the Northern Oman Mountains, and in the Musandam, represents a strong potential source of groundwater. Isotopic contents indicate aquifers are currently recharged, and that the subsurface mean residence times are less than 5 to 10 years.

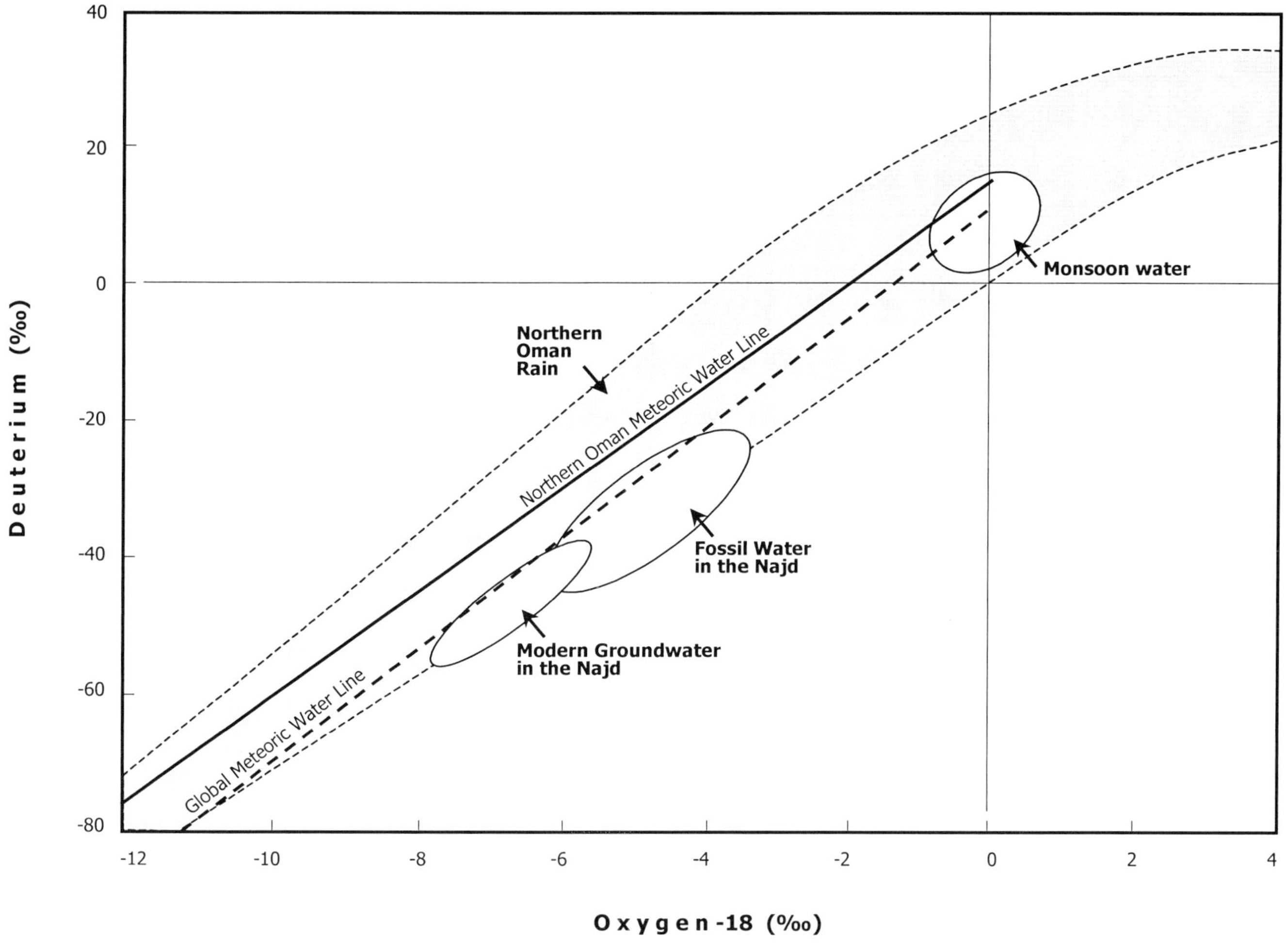

Fig. 8.100. Range in ^2H and ^{18}O contents in rainfall in Oman. Also indicated are values for deep (fossil) and modern groundwater in the Najd. The northern Oman rains includes all precipitation samples, not only those used for the calculation of the Northern Oman Meteoric Water Line (modified after Clark, 1988).

2. Quaternary Coastal Aquifer

Groundwater in coastal alluvial aquifers, and associated alluvial aquifers, and interfluvial aquifers, contain tritium and receive regular recharge. However, tritium-free groundwater does exist, indicating complicated recharge mechanisms. Because of low annual rainfall, high evaporation rate, and strong anisotropy of the alluvial aquifers, recharge mechanisms are wadi related. The modern tritium levels in groundwater of coastal alluvial aquifers suggest a mean residence times of 5 years. However, fresh groundwater at depths of 200 to 300m along the coast can be several thousands of

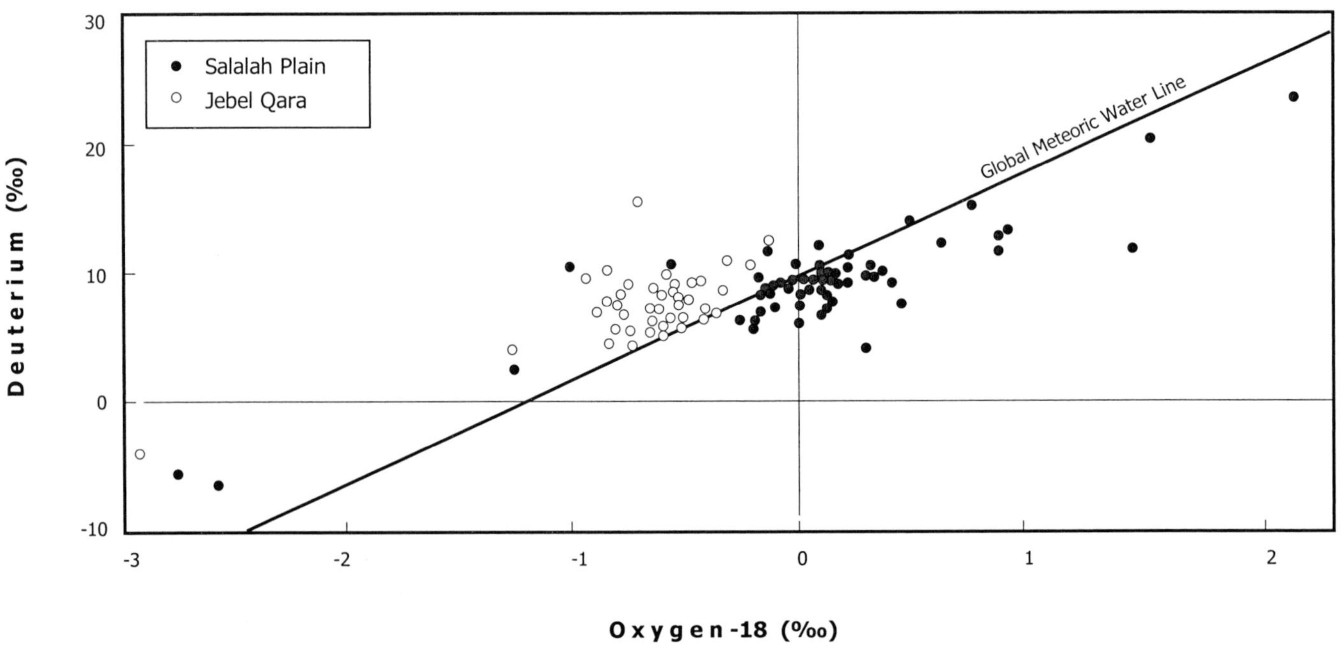

Fig. 8.101. Deuterium (‰) versus Oxygen-18 (‰) for rainfall on the Salalah plain and Jebel Qara (South Oman) (data from Wushiki, 1991 in Mcumber et al., 1998).

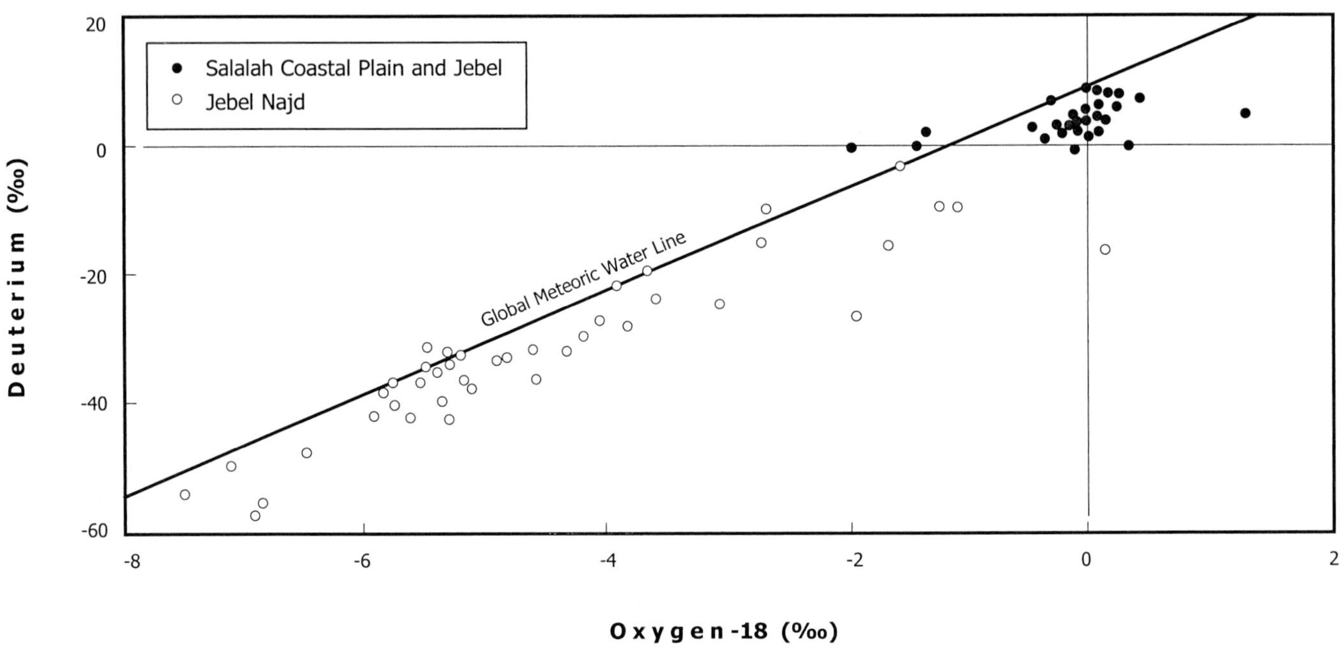

Fig. 8.102. Deuterium (‰) versus Oxygen-18 (‰) for Dhofar groundwater (data from Public Authority for Water Resources, 1986 in Macumber et al., 1998).

years old. Groundwater in shallow interfluvial aquifers along the coast are sub-modern in age, (tritium-free but with modern ^{14}C activities). Although containing a stable isotope content, similar to alluvial aquifers, it does not receive the present-day recharge, as wadi alluvium groundwater. Piezometric heads show that, these aquifers follow long term cycles, unrelated to the recent active wadi circulation.

Geochemical and stable isotope studies suggest that, the increase in groundwater salinity of coastal aquifers is attributed to dissolution of evaporite minerals in aquifer matrix, or as a result of upconing of high-salinity water, from deeper aquifers. ^{14}C activities indicate that this is a recent process associated with increasing groundwater exploitation for different purposes.

3. Quaternary Interior Aquifer

Groundwater in the interior alluvial aquifers originates principally from localized, summer orographic/convective storms. Stable isotope values show a consistent enrichment (from –3 to –3.5‰ $\delta^{18}O$ to 1 to 3‰ $\delta^{18}O$) through evaporation which characterizes runoff originating from this type of precipitation. Despite their irregularity, such storms seem to be much more effective, in recharging interior alluvial aquifers, than the rains associated with winter low pressure systems.

Although the coastal alluvial aquifers, and interior alluvial aquifers, show similar evaporative shifts from the local meteoric water line, the locus of the interior aquifers, is displaced slightly lower than that for coastal aquifers. Both aquifer systems are recharged by clouds from the same source. However, these clouds have to travel further, to recharge the interior aquifers, where they are subjected to further isotopic depletion. Tritium contents in groundwater of most interior alluvial aquifers, indicate present-day recharge, approaching maxiumum levels of 20 TU. This is higher than the ^{3}H concentrations in coastal alluvial aquifers as a result of altitude effect for clouds recharging the interior alluvial aquifers, have to rise higher while crossing mountains, before producing rainfall on interior aquifers. Tritium contents of interior alluvial aquifers vary both with distance from the recharge areas, and with distance from the principal wadi channels.

Interfluvial interior aquifers in Oman are almost exclusively tritium-free, indicating lack of recent recharge. Deeper aquifers are also affected by minimal recharge being generally tritium-free below a depth of 150 m. This becomes an important problem in areas such as the Buraimi-Al Ain border area, where substantial groundwater exploitation is not balanced by replenishment.

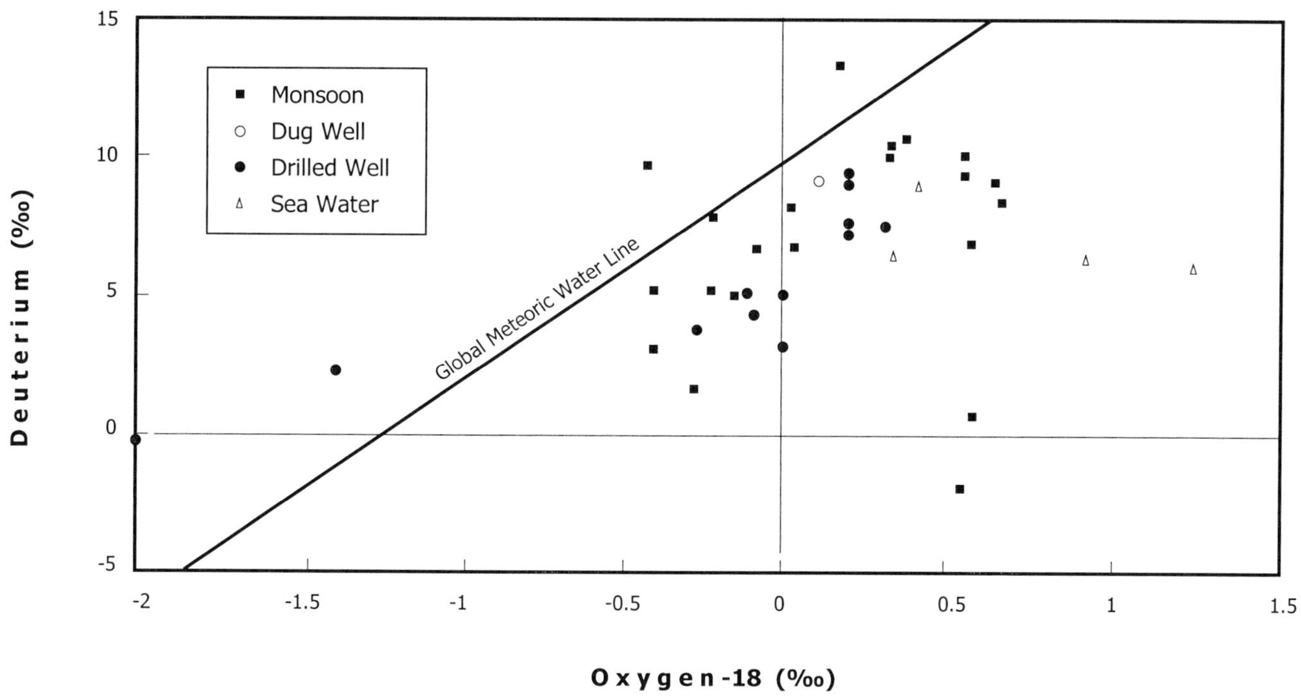

Fig. 8.103. Oxygen-18 and Deuterim in groundwater from the Salalah plain and Dhofar Mountains. Also shown are data from the monsoon and for seawater in the Arabian Sea along the Oman coast (after Clark et al., 1987).

The sand dunes such as Ramlat A'Sharqiyah of northern Oman, have no drainage systems, and recharge occurs only from direct precipitation. The ^{36}Cl studies suggest that, the rate of infiltration is less than 1% of rainfall. The groundwater from the Ramlat A'Sharqiyah aquifer, has experienced strong enrichment by evaporation ($\delta^{18}O = +2‰$), supporting the ^{36}Cl calculations. The groundwater in the Ramlat A'Sharqiyah aquifer is mostly sub-modern, (tritium-free and modern ^{14}C activities). However, one sample with 5.5 TU from southern central part of the aquifer demonstrates that modern recharge is still active at the present, but is rather limited. Groundwater flow from 160m depth was determined to have a residence time >40,000 years, representing the oldest groundwater in Oman.

The aquifers of the Salalah Plain and Dhofar Mountains, are mainly recharged by the monsoon, but storm type rainfall may contribute a limited input. The tritium contents in these aquifers are generally less than 6 TU, very close to current levels in the monsoon, indicating a residence time of about 5 years. The stable isotope data suggests that, infiltration of rain water to these aquifers is very high, with minimal evaporation losses (Figs. 8.101, 8.102, 8.103).

Despite its apparent extension beyond the topographic divide in the Dhofar Mountains, the influence of the monsoon is limited to the coastal aquifers. Figure 8.104 shows a considerable spread and depleted stable iostopes, a typical character of unconfined aquifers. The upstream aquifers in the Najd area contain modern groundwater with a wide range of tritium, ranging from 0 to 9 TU. Clark et al. (1987) attributed the tritium in the water in the Dammam aquifer to recharge by infrequent rainstorms. Deeper aquifers contain old groundwater (4,000–30,000 years) that flows under artesian pressure. Stable isotopes indicate that the recharge source for this old groundwater was also from rainstorms but during past pluvial periods.

4. Paleogene Aquifer

The Umm er Radhuma groundwater consistantly lacks tritium. Radiocarbon dating shows

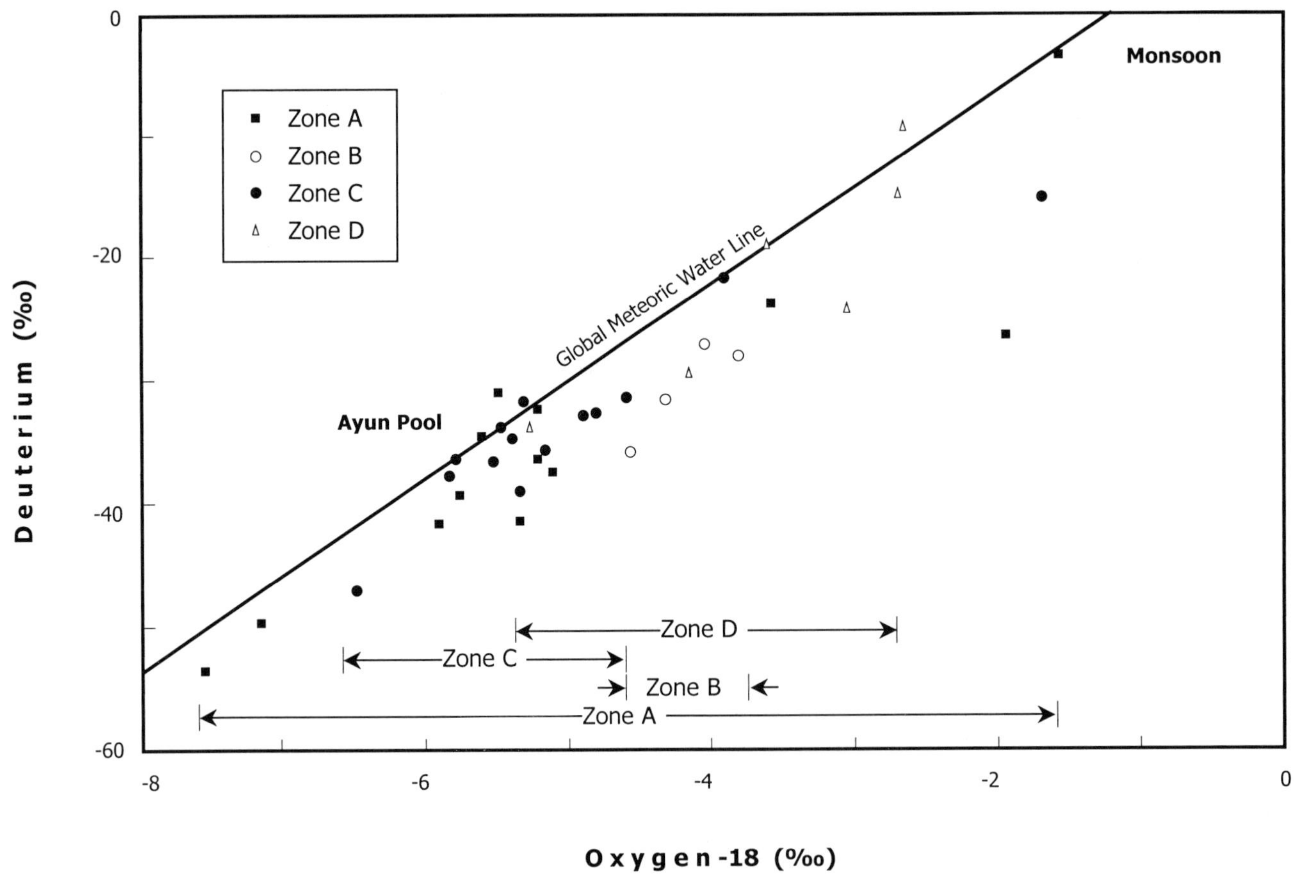

Fig. 8.104. Oxygen-18 and Deuterium diagram for groundwater from South Oman Najd aquifers. Note the difference in values from the monsoon, which occurs in the Dhofar Mountains where Zone B, C and D aquifers outcrops; this suggests a non-monsson recharge source to these aquifers (after Clark et al., 1987).

groundwater residence time ranging from 4,000 to 30,000 years BP. Stable isotope data shows that, the groundwater in the confined part of Umm er Radhuma aquifer originated as storm-type rainfall. However, mixing along the flow paths is responsible for the wide range of stable isotope values (Figs. 8.100 and 8.104). These figures show that all Najd groundwater, both modern and fossil, plot on a meteoric water line with a much lower deuterium intercept than that in northern Oman. A similarly lower line was found by Möser et al. (1978) for old groundwater in Saudi Arabia, and was attributed to recharge under a cooler climate.

The age of groundwater increases away from the Dhofar Mountains, and towards the center of the Arabian Peninsula. This gradient is consistent with the hydraulic gradient (Quinn, 1986), however, a gradation of groundwater ages from old ages in the Najd, to a sub-modern range within the mountains, needs further investigation for it, would imply that these groundwaters originated during a pluvial, epoch, and that such recharge does not exist at the present time.

The distribution of tritium in groundwater of northern and southern Oman is shown in figure (8.105). In northern Oman, the distribution is strongly bimodal, with values below the detection limit or within the range of modern recharge (last 10 years).

The lack of high ^{3}H levels (>25-50 TU) indicates that all groundwater containing tritium has been recharged. It was also possible to distingush three aquifer types in northern Oman:

Shallow aquifers of high hydraulic conductivity and transmissivity, allowing renewal of groundwater within <10 years. These aquifers are represented by the alluvial deposits in wadi channels, and a fractured bedrock aquifer along the flanks of the mountains.

Deep aquifers of low hydraulic conductivity, and transmissivity and long groundwater residence times. These aquifers result from deep groundwater circulation, as in case of some thermal springs, and interfluvial aquifers.

Leaky Aquifers which witness mixing of modern recharge with old groundwater, result in low tritium contents of 1 to 3 TU. Such tritium levels can result only from recharge prior of thermonuclear bomb testing, or by mixing. Mixing is the more plausible mechanisms, considering that groundwater circulation of more than a few decades would undoubtably have incorporated recharge of more than 2 to 3 years.

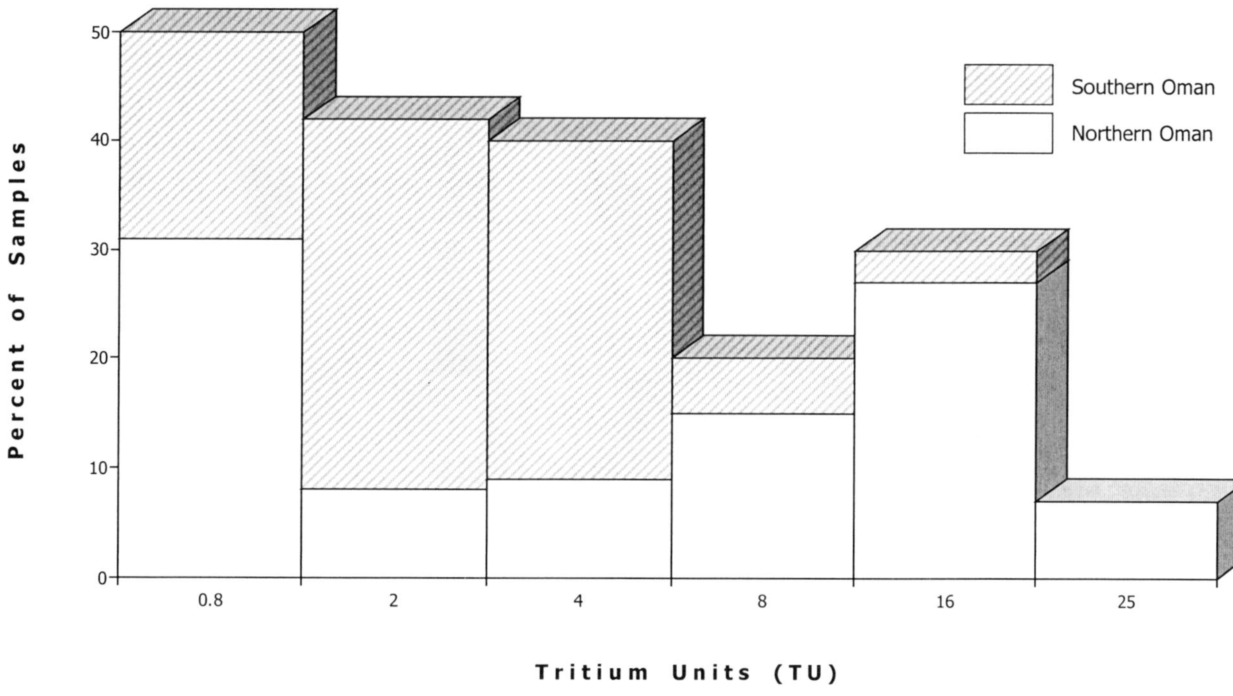

Fig. 8.105. Frequency histogram of tritium data for northern and southern Oman (modified from Clark, 1988).

Most heavily utilized groundwater resources in Oman are renewable, but problems of high exploitation will always exist, consequently shallow aquifers are characterized by low storativity and excessive pumping of water from these aquifers, creates an unbalanced situation between recharge and discharge. The wide scattering of stable isotope data is a further evidence of the low storativity of these aquifers (Fig. 8.100).

The low tritium levels in groundwater of northern Oman indicates absence of recharge, during the period of nuclear bomb testing. Mean groundwater residence time is either about 5 years or several thousands of years. The shallow Najd aquifers are exception to the bimodal distribution, and have modern ^{14}C ages and low tritium. Groundwater in these aquifers must represent a mixture of old groundwater, with components of modern reharge. The low tritium levels in coastal aquifers of southern Oman, correspond to those in the monsoon, while the low or absence of tritium in the interior region corresponds to old groundwater.

The confined part of Umm er Radhuma aquifer in northern Oman contains old, tritium-free groundwater, with a residence time of some 1,000 years, because of its low transmissivity, slow groundwater flow, and steep hydraulic gradient. In southern Oman, although the aquifers have high transmissivity, they were mainly recharged during pluvial climates.

Chapter 9

THE LEGAL BASIS FOR GROUNDWATER PROTECTION IN THE GULF STATES

PART ONE: AN INTRODUCTION TO ISLAMIC LAW APPLIED TO WATER

1. Introduction

The modernization of the Gulf area based upon oil revenues has resulted in the development of such modern western style cities as Riyadh, Jeddah, Abu Dhabi, Dubai, Doha, Manama, Kuwait, Muscat and others. The growth has been accompanied by a tenfold increase in population, considerable industrial and agricultural development and all the attendant social and administrative complexities of a modern society. Modernization brought with it the need to provide the usual and necessary services as a communication network, health services, education and the utilities, water and electricity. Of these, it is the environmental issues associated with the supply and protection of an adequate water supply which is of concern here. The latter is only one aspect of what has come to be recognized as a major issue, that of environmental control and protection of public health and the natural environment. At the present time the Gulf countries are still in the early phase of development, so although a complex problem, it is less of a problem than in areas with dense populations and a long history of industrialization and urbanization. One consequence of the early recognition of the problems which can arise, is the establishment of a regulatory authority to co-ordinate efforts in environmental protection help avoid excessive degradation of the environment. To achieve this requires the authority to develop an enforceable set of guidelines consistent with the legal and social structure of the Gulf Countries.

Prior to modernization and the influx from overseas of a population exceeding the total indigent population, the measures which society had evolved to deal with water problems, a commodity never in abundant supply, were adequate to cover the needs of a small population and subsistence agriculture. Under such conditions, the disposal of both liquid and solid waste never taxed the carrying power of the local environment. These measures however, incapable of dealing with the burden of increased water use both by the larger population and their increased "per capita" requirements brought about by the rising standard of living, augmented by the demands associated with agricultural development and industrialization in countries where the average rainfall is that of a desert zone.

Under such circumstances with demand greatly exceeding the natural groundwater supply, the deficit was met by a massive program of desalination and the mixing of such desalinated water with groundwater to keep it within the limits set by the World Health Organization (WHO). However, no comprehensive water management plan exists, nor is there an adequate body of law to resolve litigation if and when disputes arise for the existing legislation is ill equipped to meet the new circumstances.

To protect the natural resource there are technical and legal issues to be faced. Under Islamic law water is a human right to which no one can be denied access provided it occasions no harm to others. Under the constitution of the United Arab Emirates and other Gulf states no legal measure can be promulgated which contravenes the directives of Islamic Law, but as new situations arise which did not exist in former times all legislation requires careful consideration of the Sharia's and the Medhab commentaries.

There are two aspects of the water problem; the first is the technical problem concerned with the supply of water of the necessary quality, a problem that, while new in terms of magnitude, may be adequately handled through modifications of existing law. What is new and results from the magnitude of groundwater extraction, introduces the second problem, the environmental effects brought about by the falling groundwater level, the deterioration in water quality, potential contamination, as well as the problem of disposal of wastes. The complexity of the disposal of wastes resides not only in terms of the disposal of the wastewater after use, (whether domestic or industrial and/or agricultural), but of the solid and liquid industrial waste contaminated by chemical compounds resulting from modern industrial processes. Although the provision and disposal of water and waste is a technical issue and can be handled, because of the potential effects on the quality of life of uncontrolled discharge of wastewater from both domestic, agricultural and industrial processes, legislation is required to provide for, and protect the needs and rights of all members of society.

2. Principles of Islamic Law applied to water

The legal issues involved in water and waste treatment are dealt with at three levels. At the highest level the fundamental tenets upon which all law is based and with which all future legislation must conform is Islamic Law, derived from the Qur'an and the statements of the prophet Mohamed (the hadiths). Its authority is specifically recognized in the constitution of the Emirates and the other Gulf states. For example, the following paragraphs extracted from the constitution of the United Arab Emirates:

"The Supreme Court shall apply the provisions of the Islamic Shari'ah, Union laws and other laws in force in the member Emirates of the Union conforming to the Islamic Shari'ah. Likewise it shall apply those rules of custom and principles of natural and comparative law which do not conflict with the principles of the Shari'ah " (Ballantyne, 1990).

"The new code of Civil Transactions provides that the legislative provisions shall apply to all matters contained therein either expressly or by implication; that there shall be no place for construction or interpretation where a provision is clear; and if the judge does not find any provision in the law, then he must adjudicate in accordance with the Islamic Shari'ah. In so doing, the judge is directed to the Maliki and Hambali schools; if he finds no answer there, then he may revert to the Shafii or Hanafi schools as the matter may require. If the judge still does not find the guidance he requires, then he gives judgment in accordance with custom, however not to be contrary to public policy or morals, and if it is a custom peculiar to a particular emirate then his judgment applies to that emirate"

Article 27 states *"It shall not be lawful to apply principals of law designated by the foregoing provisions if such principals are contrary to Islamic Shari'ah or public policy or morals of the United Arab Emirates."* (Ballantyne, 1990).

While the Shari'ah provides the broad basis on which all decisions must be based, recourse is made to the commentaries for guidance in more specific cases. There are four main commentaries (medhabs), the Al Hambali, Al Maliki, Al Shafii and Al Hanafi, each associated with an eminent Islamic scholar. The sequence in which they should be employed is specified. They operate much like English Common Law and their precepts have been followed in case precedents upon which future decisions can be based. The precedents differ from common law only in that they are not written in formal language but rather form the basis of an oral tradition. They are nonetheless the valid basis upon which more formal modern ordinances must be formulated.

The current stage has been the development in legal format of codes, ordinances and instructions consistent with the preceding. This does not pose the problem which might be thought if Makdisi (1990) is to be believed and English Common Law is based upon or related to, Islamic Law for dependence on precedent makes the law structure seem very familiar. It may be suppose that when the code is promulgated, it will replace existing laws. Within the United Arab Emirates, of the laws, ordinances and instructions which already exist some are valid for all emirates, and thus are equivalent to federal rules and regulations, whereas, some promulgated in individual emirates, are equivalent to state laws and valid only in that particular emirate, Abu Dhabi, Dubai, Sharjah and Ras Al Khaimah for example have environmental regulations which do not exist in the other emirates.

3. Water as a public right

Everyone has the right of access and may benefit from it provided this does not infringe on the rights of others. The Prophet Mohamed stated *"People share three things, water, grass and fire"*.

This means that people have the right to share public water (but cannot own it), they have an equal right to grass grown on open range (which is not irrigated) and to share the warmth and light of a fire (but not to the fuel). Islamic Law also encourages people to conserve water, God said *"Eat, drink but do not waste"* and enshrined in one of the hadiths, Prophet Mohamed said *"Do not wastewater even if your source of water is a running stream"*.

Traditionally there are many water rights, such as the Shirb and Shurb, spring or well water, the private streams, channel rights (Hag al Magara), and the drainage rights (Hag al Maseel).

a) Shirb and Shurb water rights

In so far as water is concerned, in a region where it is a scarce commodity, it is not surprising that distinctions are made in the 'types' of water and their use. Under the Shurb (or Shifa) and Shirb rights many categories have been established according to their mode of occurrence. Water stored in private containers which is broadly interpreted to cover storage tanks and pipes, is the property of whoever was responsible for the storage and may be sold, for it cannot be used without the consent of the owner. Such a usage is consistent with Islamic Law.

b) Spring or well water rights

Such water may be owned, or it may be common or public property. When the well is owned, the Shafii medhab considers the well as much the property of the landowner, as the fruit on the trees surrounding it, the owner has the right to extract whatever he needs for his own use, but he does not have the right to deny its use to others for drinking, whether for humans or stock, when their need is urgent, even when the amount barely serves for both. If a landowner refuses under such circumstances, they have the right to use force. The advice of Al Khalifa Omer Ebn Alkhatab to people refused access was to "take your weapons to fight him." However there are differences in the Medhabs in how this should be handled. Certainly the user is not absolved from the right to pay for what was used. The digging of a well is regarded as a meritorious act, in the words of the Prophet Mohamed *"Whoever digs a well at Ruma, for him is Paradise"*. When the Prophet Mohamed came to Madinah the only sweet water was privately owned and Moslims had to purchase from the owner until Uthman (the third Khalifa) bought the well and made it a public gift (waqf).

c) Private stream rights

Even when water is on private land, it follows from the Shurb right that everyone has the right to use this water for his own needs, and that of his livestock even if it may cause minor damage. This right does not include irrigation. However the owners' permission is required, and the water may have to be paid for. According to the Al Hanafi medhab, the owner of this type of water, cannot sell irrigation rights separate from the land, although the two may be sold jointly. In the other medhabs (Hambali, Shafii and Maliki), the owner may sell the water separate from the land, even though it is more meritorious to provide it free of charge. The Shafii medhab requires the weight of water be used as a basis for charge, the others consider volume for example, the water of a major rivers as the Tigres and Euphrates is not owned, and everyone has the right to benefit from its use, for communication, domestic purposes, livestock and irrigation. The construction of canals and the use of machinery for irrigation purposes is permitted provided no damage is done, and if this rule is violated every Moslim has the right to protect the source of water.

d) Stream (or channel) rights (Hag al Magra)

Channels are referred to in Arabic as "aflaj" a word with a dual meaning referring both to a system of irrigation, and to ownership rights to the water distributed through that system.

The channel right is a property right, which permits the importing of water from one place to another across the property of another. The channels may be private or shared by many so each may irrigate separate plots. A use schedule is established so that, each has an established time for exclusive use, drawing water through a separate opening according to the size of his lot. The landowner does not have the right to impede the flow of water, or change the direction of flow without the agreement of the others users. If the channel is blocked, or its banks damaged, the users are required to rectify the damage and restore flow. No channel can be blocked, except by common agreement. The users do not have the right to plant the channel banks. Where the land to be irrigated is above stream level, mechanical devices may lift water by agreement with other users.

e) Drainage Rights (Hag al Maseel)

The drainage right covers the removal of excess water, from individual property into a public drainage system, through a surface channel or an underground drain across private property. The right is permanent, even in the case where the use of the land crossed has changed. It is clear from the proceeding that the use of water carries both rights and obligations. The established rights, based on the Shari'ah must be respected in any formal legislation. The principal points are summarized below:

- The user has to protect the source of water, whether well, spring or river, as it is supply for the common good. If the user fails to comply, then the owner may deny him access on the principal that the use should create neither "harm to himself, nor cause harm to others". The owner of a source of water cannot irrigate to the extent that a neighbor's land be flooded (harm to others).
- A water user may transfer water through a public watercourse, and where this is not possible, the landowner must permit transfer consistent with channel rights.
- Water rights can be inherited, and may include, a well from which others benefit. However the water cannot be sold independently of the land for its quantity cannot be unequivocally determined, and hence cannot be fairly evaluated.
- If water is owned by an individual, (which means the source is on his land), he has the right to use it as he wishes, but not to the detriment of others, nor should he deny access to it to people less fortunate.
- If a group owns water, it must be shared equably amongst them, using as the criterion the size of each holding.
- In a populated area, with water resulting from rainfall, from a stream or by flooding, land is irrigated from the highest to the lowest based on a statement attributed to Prophet Mohamed *"till the lands [be] irrigated or the water (supply) diminished"*.

PART TWO: SUMMARY OF THE LEGAL SITUATION IN THE GULF STATES

Water Conservation in the Gulf States

In 1998, the General Council of the Gulf Countries issued, in Arabic, the "System for conservation of water resources in the Gulf Cooperation Council (GCC) Countries", under publication number 0188-091/H/K/98, General Secretary – Riyadh, Saudi Arabia. This publication includes water conservation rules and the executive procedures of these rules. The following is a summary of this system:

A. System for Conservation of Water Resources

Chapter I: General Definitions

It includes definition of a few water-related terms, such as water resources, groundwater, and aquifer, drilling contractor, drilling criteria and abandoned wells.

Chapter II: Detailed Regulations (Items 2 to 14)

Item 1: Without violation of the Sharia'h rules, water is a general commodity, everyone has the right to use, according to the rules of this system, in addition to others systems.

Item 2: Water resource conservation and use, is the speciality of a responsible authority, which in order to achieve that has to:

a. Establish the necessary rules and procedures, for conservation of water resources and their protection against pollution.
b. Organize uses of water resources, in a way that fulfills their availability, and fair distribution among consumers.
c. Set the regulations required for drilling of water wells, and construction of dams or any other water establishment. Define areas in which drilling of water wells is allowed, and others in which drilling is prohibited, whether these wells were for agriculture, industry or any other purpose.
d. Define of the capabilities and rules that have to be met by drilling contractors and their classification into orders according to their technical, administrative and financial status. After classification, contractors can obtain drilling licenses.
e. Issue licenses for water well drilling, deepening, development, abandonment or maintenance. The license should include the criteria of each specific task such as drilling, deepening or development. These activities have to be under the supervision of a geologist staff of the licensing authority.
f. Take the necessary and appropriate measures to limit depletion of water resources or their contamination.
g. Supervise and inspect to ensure the application of the system, and its rules, and enforce necessary fines in cases of violation of the system or any of its rules.

Item 3: The priority of water use is according to the following order:

a) Human needs.
b) Animal needs.
c) Agricultural needs.
d) Industrial, urban and other needs. The priority of the other needs is defined by a law from the director of authority responsible for water resources or his deputy.

Item 4: In emergency or water shortage situations, the director of the authority responsible for water or his deputy, can decide and take the necessary technical and administrative procedures, to ensure a just distribution of water to consumers, taking into account the priorities assigned in Item 3. The water authority has the right to organize the rules of water consumption, and use in all cases to conserve water.

Item 5: Guarantee protection of water resources against pollution, waste or causing hazards to public health or soil, the water authority determines who is responsible for costs of fixing or abandonment of old wells or construction of alternative wells, or compensation for wells to achieve the same purpose, whether these wells were drilled with a license, or were prior to the issuing of licenses, for water wells drilling. In wells drilled without licenses, the water authority may decide on the abandonment of maintenance, and such works are achieved by the owner of these wells, at his cost, during a definite period of time. Where the owner refuses to do so, fines assigned in Item 9 of this system are strictly applied.

Item 6: It is absolutely forbidden to drill water well, construct a dam or any other water installation or conducting any activity affecting water resources without obtaining the required license to do so.

Item 7: The drilling contractors must refuse to drill any water well when the owner does not has the appropriate permission designated by Item 6. In case of possession of permission or license, the contractor has to strictly comply with the instructions and criteria designated in the permission. The contractor will be required to abandon the wells he drilled in violation to these rules.

Item 8: The owners of existing water wells or wells under construction report these wells according to the executive rules obtained from water authority within a month time from the date of issuing this system.

Item 9: Anyone violating this system or its rules will be fined according the executive regulations stated in Chapter 4 of this system.

Item 10: The director of the water authority or his deputy, can withdraw the license from any contractor who repeatedly violates the rules of this system. The contractor can submit a complaint within a month time since issuing this system.

Item 11: The director of the water authority issues a decision with the fines applied to the violators of the system and its rules. Inspection, investigation of fine imposition is decided by the system and takes place according to the justice system.

Item 12: The responsible minister can issue the executive rules of this system and the regulations, rules, and fines become effective a month after publication.

Item 13: This system is a basic system for water resources conservation in the Gulf Cooperation Council Countries. The member countries can reform some of its regulations to conform with local conditions. The General Council has to be informed of such modifications.

B. Executive Rules of Water Resources Conservation System

Chapter I. Registration and Licensing

Item 1: Drilling contractors have to obtain a license from water authorities by submitting an application indicating the availability of required workers, technicians and drilling equipment indicated by certificates issued from responsible authorities.

Item 2: The drilling contractor can register with the water authority after submission of certificates and complete the procedures defined in this item within a three month in period from the publication of this system.

Item 3: Drilling contractors are classified according to the following considerations:
1. Technical efficiency.
2. Administrative efficiency.
3. Financial efficiency.

Item 4: The responsible minister or his deputy can enter any ground or water establishment to conduct studies, or surveys investigation, of information gathered about water. He can also conduct any activities required by the system, to conserve water resources after informing the owner or the resident, before the start of these activities.

Item 5: The contractors obtaining the license to drill water well, must deposit an endorsed copy of their licenses with the workers at the drilling sites. The representative of the water authority has the right to see this license, at any time to make sure of its validity. The drilling contractor has also to show his name and the number of his drilling license, and the location of his drilling machine at the drilling site.

Item 6: Anyone obtained a permission to drill water well has to keep an endorsed copy of this permission at the drilling site, and give another copy to the drilling contractor. The water authority representative has the right to see this permission, and ensure its validity during or after drilling.

Item 7: It is not permitted to offer permission to drill water wells for agricultural purposes, unless the area which will utilize produced water, is within a certain limits. The distance between the proposed well and other water wells in the area or nearby farms has also to taken into account.

Item 8: If the designated agricultural area or the distance between two neighboring water wells are not satisfied, the responsible minister can allow more than one farmer to share a single well with one permit. The farmers will have a single permit, and must share the expenses of well construction according to the area owned by each of them. This ownership can be registered.

Item 9: The supervising geologist has the right to stop the drilling contractor, whenever he discovers any violation of the drilling permit, issued to the land owner or user, any violation of the regulations and criteria applied in water well construction. The drilling contractors carry the responsibility of his misconduct and violations.

Item 10: The drilling contractor has to collect samples from the drilled formation every three meters, or whenever the color or type of rock changes. Samples have to be kept in plastic sacks or sampling containers. On each sample the depth of drilling, number of well and name should be clearly registered. The samples must be kept until asked for by the responsible geologist.

Item 11: The drilling contractor has to provide the water authority, (or one of its branches) in his work area, with a well completion report according to the designated design to the water authority within 15 days of well completion. He has also to inform the water authority in advance, with the position of his drilling equipment.

Chapter II. Drilling Contracts

Item 1: The drilling contractor has to obtain a written contract, from the well owner or his deputy, signed by both parties before drilling. The contract must include the technical rules, according the well design, provided by the water authority.

Item 2: The technical rules stated in the drilling contract endorsed by the water authority, are a part of any agreement between the water well owner or user, and the drilling contractor, even if these rules were not explicitly stated in this agreement.

Chapter III. Water Uses

Item 1: The contracts for drilling and uses of water wells must be according to the following order of priority:

1. Drinking water
2. Agriculture
3. Industry

Item 2: Every farmer has to install a flow meter on all water wells used in his farm. The water authority will instruct farmers on the amounts of water allocated to their farms. Accordingly, the future needs of water wells; their design and locations can be determined. The farmers have to strictly comply with installation of water meters in all wells, and conserve and maintain of these meters.

Item 3: The owners of farms have to do the following:

1. Inform the water authority with the type of crops, the areas they desire to cultivate, and the working hours in management of their farms.
2. Construction of high efficiency irrigation canals in case of application of traditional irrigation methods.
3. Use of recent, high-performance irrigation tools.

Item 4: The water authority can shut down water wells, or abandon them, if these wells are not constructed, according to the designated regulations, or when their existence represents a threat to the public health. The water authority should determine those responsible who will pay the costs.

Item 5: Any water well or establishment constructed on government land, without permission from the water authority, will be considered a public facility, used by the water authority, for the general public, without any compensation to the constructor.

Chapter IV. Violations and Fines

Item 1: The drilling contractor who does not have a drilling license, pays a fine, which does not exceed an amount left open to be determined by water authorities in each country. Drilling activities are put on hold, until the necessary permit is obtained. The fine increases to an amount left open, to be determined by water authorities in each country, in case of repeated violations.

Item 2: The drilling contractor who carries an expired drilling license, is fined by an amount left open to be determined by water authorities in each country, and stopped from drilling until he pays the fine and renews his drilling license. The fine is doubled in case of repeated violation. The drilling contractor also pays a fine if he does not provide the water authority with a well completion report within the designated time allowed, or in providing false information, or not reporting the position and movement of his drilling equipment.

Item 3: The drilling contractor and the well owner pay a fine, of an amount left open to be determined by water authorities, in each country for each of them, in case of drilling water well without issuing a permit. If they fail that, well is abandoned at their expense.

Item 4: The drilling contractor and the well owner each pay a fine, of an amount left open to be determined by water authorities in each country, in case of drilling a water well using an expired drilling license. Drilling is stopped until the fine is paid, and the license is renewed or the well will be abandoned at their expense.

Item 5: The farm owner pays a fine of (an amount left open to be determined by water authorities in each country) in case of drilling water well, without a drilling license on his farm, or in case of drilling the well in a site different than the one determined in drilling permit.

Item 6: In case of repeated the violation in Item 4, the drilling license is withdrawn from the contractor for a period not more than a year.

Item 7: In case of discovery of violations mentioned in Chapter IV, the water authority makes a report in cooperation with a representative from the Ministry of Interior, in the same area or a nearby area. The report is submitted to the responsible authority to determine the fine. In case of absence of one of the representatives, the other can still file the violation report.

Item 8: The responsible minister or his deputy issue the fines for violators, via decisions prepared by the water authority. The decision must include the name of the violator, kind of violation, location, name of the person filing the violation and date of violation.

Item 9: The water authority collects the fines designated in the executive rules from the violators, according to the regulation followed in that country.

1. THE UNITED ARAB EMIRATES

The Shari'ah and the hadiths form a solid basis upon which to develop a modern legislative system to cope with the variety of problems, which could not have been foreseen when Islamic Law was established. These problems can be broken down into those arising from the over drafting of groundwater supplies, largely to meet the needs of agriculture, which has resulted in a falling water table, increasing total dissolved solids, and decreasing water quality, with the constant problems of sea water intrusion. To these, must be added the dangers of pollution of those groundwater supplies, by the waste products of modern industry, (from the oil industry for example) and human activity both domestic (landfills) and agricultural (pesticides and fertilizers). These are problems common to the Gulf area as a whole, but there is no coordinated policy for dealing with them, and each state has proceeded with the development of its own plan of campaign in greater or lesser detail. The most comprehensive appears to be that developed in the United Arab Emirates that will serve as an example of the complexity of the problems involved.

The water supply problem in the United Arab Emirates was, and is handled by two ministries, the Ministry of Electricity and Water and the Ministry of Agriculture and Fisheries. The Federal Water Authority of the Ministry of Water and Electricity is responsible for providing and distributing an adequate supply of water for domestic and industrial use to the northern Emirates (Al Sharjah, Ajman, Umm Al Quwain, Ras Al Khaimah and Al Fujairah) although the administrative aspects are handled by the appropriate departments of the local authority. Abu Dhabi and Dubai are the two larger Emirates, where there existed larger functional departments prior to the formation of the United Arab Emirates. The provision of water for agricultural use remains in the hands of the Ministry of Agriculture and Fisheries, and is organized in the same way; the only slight variation is that the local authority in Al Sharjah maintains a water treatment plant.

Unfortunately, the success of the Ministry of Agriculture and Fisheries in enabling the enormous expansion in agricultural development, was made at the cost of drastic over-use of groundwater supplies. In some areas, this has led to excessive salt-water intrusion, and the general increase in the total dissolved solids, present in well water has rendered some well water unpotable. These changes could perhaps have been obviated, or at least minimized, had there been a coordinated management plan between the Ministry of Electricity and Water, charged with the supply of potable water, and the Ministry of Agriculture and Fisheries responsible for the supply of water for irrigation.

For many years, the handling of water-born waste products was ignored, and no coherent policy for handling of wastes existed, nor was the true magnitude of the problems involved clearly understood. Even today the only Emirates with active waste disposal legislation are Dubai and Abu Dhabi.

Historically, the first federal measures to attempt to deal with environmental concerns was the establishment of the Higher Environmental Agency within the Ministry of Health in 1992 with a mandate to safeguard health through control of food processing and its protection from radiation contamination. Subsequently the mandate was expanded by the creation in 1993 of the United Arab Emirates Federal Environmental Authority by Federal Law No. 7/1993 (see Bakhit, 1998) with its own budget, and the express purpose to maintain the quality of life. The task of the Authority was to develop a comprehensive environmental policy through the establishment of uniform federal requirements, which, as in the USA are only superceded, where more rigorous regulations are established in the individual states/emirates. The functions of the Federal Environmental Agency therefore overlap, and to some extent will eventually supplant some of the functions of the Ministries of Electricity and Water and of Agriculture and Fisheries. Initially an organization to administer and ensure compliance with its directives was established consisting of a nine-member board of directors under the chairmanship of the Minister of Health was created under guidelines set out in Article 6 of Federal Law 7/1993 (see Bakhit, 1998). The board appointed a Director General, directly responsible to the Chairman of the

Board of Directors, assisted by a technical coordinator and an administrator of financial affairs. Within this structure three major departments were set up each reporting to the Director General, namely Assessment and Monitoring, Research and Development, and Coordination.

The Federal Environmental Agency has already developed draft regulations, which are comprehensive concerning the protection of the marine environment and of the soil but are much less explicit concerning groundwater. In fact only the broadest principles concerning the protection of drinking and subterranean water are contained in two articles (63 and 64), which are given below:

Article 63. Authorities charged with matters related to drinking water and subterranean water shall pursue the development of sources for water resources: the establishment of controls which serve this purpose: the prevention of misuse and wastage or pollution thereof, and making use of storage of such waters subject to the conditions specified by such authorities in co-ordination with the Agency.

Article 64. Authorities charged with licensing installations related to the storage of water in buildings and structures shall carry out periodic annual inspections of such installations to verify their safety, and shall notify the landlord on such procedures as should be taken to ensure the delivery of safe water to the inhabitants. In the event of his failure to implement such directives, such authorities may carry out the necessary corrections at his expense, and collect the cost thereof by way of administrative sequestration. The result of such periodical inspection shall be recorded in special registers maintained by the said authorities.

A series of articles (Articles 65-75) dealing with the protection of soil and the use and storage of fertilizers, pesticides, and wastes (generated by the oil industry), fall under the general control of the Ministry of Agriculture and Fisheries. The regulations are designed to protect the soil and soil fertility, but make no provision for water percolating from the soil into groundwater, where its impact on water quality may be considerable.

Although the legislation referenced above covers potential pollution, and provides the broadly defined objectives of groundwater protection, it lacks the necessary degree of specificity such as is found in the regulations concerning the marine environment.

Review of Current Dubai Legislation

i. Water and waste regulations

As indicated earlier the principal features of the Federal Regulations, which deal with groundwater, are covered in Articles 63 and 64. Articles 65 to 75 are concerned with the protection of soil and the use and storage of fertilizers, pesticides and wastes generated by the oil industry but make no provision for percolation of contaminated water from the soil into groundwater. Although more detailed federal regulations are being developed none have been promulgated up to this time, however, in the Dubai Emirate, Environmental Regulations have been published in Local Order 61/1991 and Administrative Orders 211/1991 and 73/1993, with explicit directions recorded in the Environmental Protection and Safety Technical Guidelines established by the Health Department of the Dubai Municipality as amended 30 June 1995. Subsequently a series of draft articles (Articles 63-67) was developed under which the Director General of the Dubai Municipality Drainage and Irrigation Department could declare any area a groundwater protection area. Under such an order point source discharges to groundwater could be controlled or stopped, the installation of further wells prevented, the amount of water extracted limited, and any activity that could adversely impact groundwater quality controlled. Further articles banned the discharge of industrial wastes into a soak away, and established the conditions under which sewage might be disposed. Finally the discharge of wastes to groundwater through a recharge bore required a permit from the competent authority. Article 67 concerned the non-point source contamination controlling the application of pesticides, herbicides and fertilizers and the use of water in agriculture to ensure that neither runoff nor infiltration of contaminated water should enter surface or groundwater.

ii. Dubai ordinances

According to Article 1 of Local Order 61/1991, H.H. Sheikh Hamdan Bin Rashid Al Maktoum, as chairman of the Dubai Municipality promulgated Administrative Order 211/1991 on the "The issue of Executive Regulations for Local Order 61/1991 concerning the Environment Protection Regulations in the Emirate of Dubai", pursuant to the decree issued by the H.H. ruler of Dubai which established the Municipality. That Local Order established the structure of the agency within the Health Department and the chain of command under a director who reported directly to the Chairman of the Municipality. The regulations cover eight specific fields as follows:

1. Regulations on the reuse and land disposal of wastewater and sewage sludge.
2. Regulations concerning the disposal of wastewater into the marine waters.
3. Regulations for air pollution control from stationary sources.
4. Occupational health and safety regulations.

5. Swimming pool regulations.
6. Regulations on the safety of toys.
7. Noise control regulations.
8. Protected areas (Wildlife Sanctuaries).

In the context of the present review, only items 1 and 2 are of concern.

The Director of the Municipality, as authorized under Local Order 61/1991 Article 3, issued Administrative Order 211/1991. This consisted of 46 articles which set forth the mode of operations to affect the controls as required by the Local Order 61/1991 through the issuance of permits, establishment of emission and effluent standards, and the limits of permissible pollution levels. Penalties were established for non-compliance, which range from simple fines to cancellation of permits. Subsequently, through Executive Order 73/1993 the coverage and limits were expanded to make for a more comprehensive coverage. The municipality also established a special Department of Environmental Protection and Safety Section, which serve to ensure compliance with local ordinances.

iii. Implementation of regulations

As stated above, Environmental Protection and Safety Section is responsible for the implementation, control, monitoring, and enforcement of the regulations for environmental protection according to Local Order 61/1991 and the Administrative Orders of 211/1991 and 73/1993.

The Municipality began implementing the regulations from the day the local order was gazetted. The main features of Administrative Order 211/1991 establish:

a) Permits are required from Environmental Protection and Safety Section for reuse and land disposal of water and sludge or for disposal of wastewater into marine water.
b) The owner of any premises must comply with all the regulations stated in the local order and the administrative orders.

For administrative purposes, Environmental Protection and Safety Section divided Dubai into districts simplifying monitoring procedures. Detailed large-scale maps are used to identify each district, each map showing the location of all industrial and residential areas, the sewerage system, the treatment plants, the landfills, and the outfalls where wastewater is discharged into Dubai creek.

A Geographic Information System is used to store all the data on Dubai. All the industrial premises and all establishments, which impact the environment, are listed in the computer and Environmental Protection and Safety Section receives weekly a list of the premises that must be inspected. All the premises are subjected to periodic inspection. The duration and frequency of the inspections depends upon the type of company, the treatment and chemicals used.

The Environmental Protection and Safety Section Administrator may enter any company premises at any time for inspection and to sample for analysis. Samples are also taken from the outfalls to determine whether the pollutants are within the regulatory limits. In case of violation, the Administrator issues a notice of violation and non-compliance with Environmental Protection and Safety Section regulations. Failure to comply with the regulations after the notice leads to fine up to 2000 Dirham (about US$ 545), but no civil action or criminal action is possible under the present regulation enforcement system. The municipality cannot take cases to court, and despite the damage to the environment, it has only authority to stop discharge and cancel the permit.

iv. Regulations on the reuse and land disposal of wastewater and sludge

In general the treatment of wastewater, sewage, ends up with two products, treated water, which may be used either for irrigation, or is discharged into the sea, and solid waste which can be used as fertilizer.

Under Article 5 of Administrative Order (211/1991) wastewater has to be recycled by the municipality wherever possible, in accordance with the conditions and specifications stipulated in the executive regulations.

Under Article 6 of Administrative Order (211/1991) the discharge of wastewater to land in any reuse scheme, the discharge of sludge or its use as a fertilizer is prohibited without a Health Department certificate.

The surface discharge of wastewater on to land in any reuse scheme, or discharge of sludge or its use as a fertilizer is prohibited without permit from Environmental Protection and Safety Section (Administrative Order 211/1991, Article 1) as indicated above.

Article 3 of the Amiri Decree 73/1993 states that *"Discharge of any substance into the sewerage system by any means which alone, or in combination with other substances, is hazardous to the normal functioning of the sewerage system is prohibited."* Treated wastewater must meet with the effluent standards (Local Order 61/1991 - Article 8).

The effluent standards are listed in the following:

- Limits for disposal of sludge:
  - Annex III (Administrative Order 211/1991).
- Effluent quality standards for trade wastewater discharge:
  - Annex IV (Administrative Order 73/1993).

Applications for disposal or reuse of wastewater or for wastewater or for disposal of sludge or its use as a fertilizer are required on a prescribed form. The regulations require that any person, who was currently disposing of, or re-using wastewater or sludge at the time the regulations were instituted must apply for a permit "within six months from the date of validity of these regulations (Local Order 61/1991, Article 11)". The application form is shown in Annex I of the Amiri Decree 211/1991.

Article 7, Administrative Order 73/1993 requires that "Any application for (the disposal of) trade wastewater submitted to Environmental Protection and Safety Section should include

(a) the design of pre-treatment facilities.

(b) further waste quality testing.

(c) a waste minimization plan.

If sludge has been stockpiled prior to use for a period of not less than five years, or if sludge is sterilized to Environmental Protection and Safety Section satisfaction, then no permit application is required (Administrative Order 211/1991, Article 6).

The municipality retains the right to approve, disapprove or cancel a permit previously issued. This is covered by Administrative Order 211, Article 2, the Director of the Health Department or his deputy may issue a decision to grant the permit, disapprove granting of a permit, or cancel any permit previously issued, as applicable, giving the grounds for the decision. The applicants who must demonstrate that the grounds for the initial rejection are covered in the new application may resubmit rejected applications.

To ensure compliance with regulations the Health Department of the Municipality is required to monitor, and control the discharge of wastewater by periodically collecting samples of wastewater for analysis. The analytical results have to be recorded in a special register, in which the compliance or non-compliance with the standards of wastewater is stated. In case of violation, the municipality will advise the concerned person of the result, and he can object to result within one month. (Local Order 61/1991, Article 10). The same article states that "Executive regulations shall specify the insurance to be paid as well as measures to be taken to settle litigation costs, fees or re-analysis required by the objector, and other fees accrued by him."

If analysis shows that the wastewater characteristics do not conform with the effluent standards specified in the executive regulations, the concerned shall, within a limited period specified by Environmental Protection and Safety Section from the date of his being advised of results, implement a method to treat the wastewater to necessary quality, conforming with the prescribed effluent standards, otherwise the license shall be canceled by a decision, from the director of municipality stating the reasons. Time limits may be extended, if the violator can present acceptable reasons. Failure to remediate will result in the municipality taking remedial action at the expense of the violator. The person concerned is required to take samples periodically, and analyze them at this expense, over a period of time as specified in the executive regulations. A permanent record of the analytical results must be kept, and submitted monthly to Environmental Protection and Safety Section (Local Order 61/1991, Article 10).

To monitor trade waste, the concerned person or business discharging trade wastewater into the sewerage system "must measure and keep records of the volume of waste discharged" under the conditions of the permit and submit an annual report on the result (Amiri Decree 73/1993, Article 9). Also Article 14 provides that: "The Director of the Health Department may require anyone seeking to discharge trade waste to the sewerage system to install batch storage tanks and provide an analysis of the waste to Environmental Protection and Safety Section before discharge of the waste from these tanks into the sewerage system".

To ensure that wastewater cannot be discharged in the system without approval, the Environmental Protection and Safety Section has the authority to lock the release valve from such storage tanks (Article 15 of the Administrative Order 73/1993).

A scale of penalties for non-compliance is listed under the general provisions of Local Order 61/1991 in Articles 88-90. The fines range from \$136 to \$545 and \$1,500 to \$4,000. One of the penalties permitted in Article 90 of Local Order 61/1991 is the "Closure of premises for a period not exceeding one month". This does not take into account what will happen if remediation requires more than a month. Also the actions in the penalties do not show what will happen if the violator does not respond to Environmental Protection and Safety Section, although by extension of the pattern established by the fines the authority could take over remedial action at the cost of the violator.

v. Regulations concerning the disposal of wastewater into marine waters

The regulations specify that the "Discharge or disposal of any wastewater from residential premises, shops, trading, industrial and tourist establishments, sewerage and other sources, into marine waters, either directly or indirectly, shall be prohibited without permit approval from the municipality" (Local Order 61/1991, Article 14). This prohibition parallels that for onshore disposal of liquid waste. Wastewater intended for disposal in the marine environment must meet, or be pre-treated to comply with, the standards for liquid effluent discharge into the marine waters as stipulated in executive regulation (Amiri Decree

211/1991, Article 10). These regulations strictly prohibit discharge of effluent if it contains:

a) Pesticides, insecticides;
b) Radioactive elements;
c) Polychlorinated biphenyl or chlorinated organic material; and
d) All materials produced for biological or chemical warfare (Local Order 61/1991, Article 18).

Environmental Protection and Safety Section requires that, the holder of a permit must have a periodic monitoring system to make the necessary measurements, and analysis of the physical, chemical, and biological elements of the wastewater (Local Order 61/1991, Article 15). Further the permit holder must monitor and control discharge wastewater into marine water. "Monthly averages, necessary measurements of the chemical and biological conditions to be observed when discharging wastewater into marine environment, and a record must be kept to register results of these periodic analyses". These are specified under the conditions of the provisions of Administrative Order 211/1991 Article 9. The provisions are listed in the same article. The records must include all parameters listed in Article 10 of Administrative Order 21/1991, and Article 23, paragraph 6 of Local Order 61/1991 states that:

"The owner of an establishment which discharge wastewater to marine water shall take and analyze specimens on his account periodically as specified by executive regulations. He shall keep a permanent record of analytical results which shall be sent monthly to Environmental Protection and Safety Section."

To monitor and control wastewater disposal into marine waters and to ensure compliance with Environmental Protection and Safety Section standards, and enforce effluent limitations, the Health Department will conduct regular analyses of wastewater from each regulated site both periodically as well as at random.

To ensure compliance a series of penalties was enacted:

Non-compliance with Environmental Protection and Safety Section provisions could result in a fine of not less than \$136 (500 Dirham) and not more than \$545 (2000 Dirham) with an immediate halt to discharge. The violator is given a grace period to comply Article (23) of Local Order 61/1991 states:

- "If the treatment does not take place by the end of the period, or if it was not appropriate, the municipality will cancel the license and discharge must cease".
- "If result of analysis shows the discharged wastewater does not conform with effluent standards, and to the requirements pursuant to the provisions of the order, and if the effluent poses an immediate risk of polluting the marine water, the concerned shall advised to remove causes of the damage immediately, otherwise the municipality will carry out remediation on his account or cancel the license and immediately stop discharge".

The technical basis for groundwater protection regulations

It is clear from the review of the current Dubai legislation, the only existing legislation, that it concentrates on wastewater disposal through sewerage and effluent control, but lacks any consideration of other sources of groundwater pollution, an essential component in maintaining an adequate supply of potable water, since the water supplied is a mix of groundwater and desalinated water, to keep it with the required limits, set by the World Health Organization. Groundwater is endangered by the growth in population, and the increasing use of irrigation in maintaining agricultural production. The 1995 groundwater production in the United Arab Emirates for both domestic and agriculture use was estimated by Ministry of Agriculture and Fisheries to be 1,000 Mm^3, over the same period the public water groundwater supply extracted was 130 Mm^3/yr. Agriculture was the major groundwater user with an estimated 700 Mm^3/yr. In the same year, an additional 170 Mm^3 of groundwater was used for landscaping and forestry. Al Hawsani (1996) estimated that the total annual extraction of groundwater, about 800 Mm^3/yr, some 200 Mm^3/yr less than the Ministry of Agriculture and Fisheries estimate, while the recharge, about 129 Mm^3 /yr, leaves a deficit of 671 Mm^3/yr.

The desalinated and treated water production does not make up the deficit (the 1995 desalinated water production was 418.5 Mm^3/yr, and treated water production was 109.5 Mm^3/yr), although these figures are about double the 1985 production.

Dubai and Abu Dhabi are the major producers of desalinated and treated water. In Dubai there are two treatment plants, the largest at Al Awir, built at a cost of Dirham 400 million (\$109 million) receives both domestic and commercial sewage which may contain organic wastes. At the present time it has a daily output of about 137,000 m^3/day (30 million gallons/day) of treated water, but has a capacity to roughly double that amount. The treatment of sewage ends up with two products, treated water and solid waste, which can be used as a fertilizer. The municipality encourages the use of this water in agriculture and horticulture. The solid wastes are dried, powdered with the inorganic part disposed, of in the form of sand grain sized granules, bagged in plastic and disposed of in landfills. Under the current regulations the use of the treated water and the discharge of sludge or its use as a fertilizer is prohibited without a permit from Environmental Protection and Safety Section. In the following section the other principal sources of pollution, both point source and nonpoint source, are reviewed as a basis for suggestions as to how they may be controlled.

There are additional potential sources of pollution which more directly affect groundwater, and which are not covered by the preceding legislation, and which can be broken down into two categories, point source and non-point source pollutants. These are considered below to provide a basis on which control measures may be based.

a) Point source pollutants:

i. Landfill
ii. Underground Storage Tanks
iii. Injection Wells

i) Landfill: Conditions in the United Arab Emirates are such that no real difficulty should be met in the establishment and control of landfills, the climate is arid so there is minimal danger from rain leaching through the landfill during most of the year. In Dubai there are two main landfills, one in Jebel Ali for both hazardous and municipal waste, the other located in Al Khawaneej. A third site for hazardous waste is under construction in the Jebel Ali Free zone.

ii) Underground Storage Tanks: These are used in United Arab Emirates mostly for hydrocarbon storage, a large percentage of which were installed prior to any environmental regulations, consequently many tanks are single walled and few, if any, arrangements for monitoring leakage were installed. As indicated in a later section the definition of underground storage tanks also includes connection tanks and the underground linking pipes.

iii) Underground injection wells: These provide the best means for fluid waste disposal if situated, operated and constructed properly. They have been used for disposal of wastes, from gas and oil extraction, to provide a barrier against saltwater intrusion or for the recharging of groundwater. In the United Arab Emirates there are currently no regulations controlling water injection, and underground injection is restricted to the disposal of high salinity water associated with the production of hydrocarbons.

The Bu Hasa Field provides a case in point. Here highly saline water is pumped from the Late Cretaceous Simsima and the Tertiary Umm er Radhuma and Dammam aquifers, and injected into the Early Cretaceous Shuaiba reservoir to maintain production. To date some 5.5 million bbls/day has been injected into this reservoir. The water produced with the oil is largely re-injected into the Miocene clastics, which underlie the Lower Fars evaporates, however there is a highly saline water cut with a total dissolved solids of 100,000 to 150,000 ppm, contaminated with some oil, which is released at the surface, into unlined ponds at a rate of 2,000 bbl/d. Thus, in addition to the danger of contaminating the Liwa aquifer from below, the downward percolating water from the unlined ponds contaminates from above. The result is that the total dissolved solids in the southern part of the Bu Hasa field area increased from 500 ppm in to 1000 ppm in 1985 but in the last decade has rapidly increased to 4,500 ppm in 1996. Thus the use of the southern part of the Liwa aquifer has been destroyed, restricting pumping of water for domestic and agricultural use to the northern part of the aquifer in the area of the Bu Hasa Field.

This type of well is the only one of the five classes of injection well defined by the Environmental Protection Agency (USA) according to the type of waste, and where it is injected relevant to the situation existing in the United Arab Emirates.

b) Non-point source pollutants:

Infiltration of irrigation water carrying dissolved fertilizers, insecticides and with it other pest control chemicals, which can pollute groundwater.

In general the regulations which currently exist concern the buying, selling and storage of fertilizers and pesticides; there are few which specify application levels for agricultural pesticides or fertilizers, which can be applied, and certainly no data exists concerning the levels at which these may infiltrate through, and contaminate groundwater. There are clear indications however, at several locations of nitrate concentrations well in excess of World Health Organization standards (of 10mg/1) having penetrated, and polluted Quaternary groundwater at the five locations as below:

West of Al Khaznah area (Abu Dhabi) (100 mg/l); Wadi Al Bih area (Ras Al Khaimah) (200 mg/l); south of Dubai (200 mg/l); between Madinat Zayed and Bu Hasa (Abu Dhabi) (350 mg/l); Liwa area (Abu Dhabi) (350-550 mg/l); and Al-Ain area (Abu Dhabi) (200 mg/l).

c) Deterioration of groundwater quality due to over-pumping:

Before comprehensive regulations can be adopted it will be necessary to undertake an extensive monitoring program to establish empirical values. Until this is done reliance will have to be placed on empirical knowledge.

In the coastal area of the United Arab Emirates over-pumping of groundwater has led to seawater intrusion in Ras Al Khaimah, Al Fujairah and Dubai. In Al Fujairah seawater intrusion has been particularly rapid extending 4km within 16 years. South of Dubai and north of Al Ain two west-east tongues of seawater have also been detected.

In the Dhaid area near (Al Sharjah Emirate) over pumping led to the exhaustion of several wells drawing from the shallow gravel aquifer in the year 1992. Since that time heavy rains have led to some recharging, but the level of total dissolved solids (at 10,000 ppm) means that the water is no longer usable. Upconing of salt water with a salinity of 10,000 ppm from deeper saline aquifers has also occurred in the Dhaid and Al Ain areas.

An inevitable secondary effect of over-pumping is the lowering of the groundwater level, and increasing the diameter of the cone of depression, with the upconing of salt water from deeper saline aquifers.

In different areas such as Dhaid, Al Ain, Hamranyah (Ras Al Khaimah), and Al Fujairah, water level was lowered 1-2 meter per year (Ministry of Agriculture and Fisheries report, 1986). The results are shown in Table 9.1. Cones of depression can range from 60 -100 km in diameter as for example at Al Dhaid, Hatta, Al Ain and Liwa.

Realistically, because desert or near desert conditions exist, over most of the available land area and because of the concentration of population in the urban areas, it is only in the urban areas that the regulations require the stringency of the US-Environmental Protection Agency regulations. Nonetheless the protection of groundwater must be regarded as a national priority. The fresh water in the United Arab Emirates is largely confined to the gravels flanking the eastern mountains where total dissolved solids values are generally <1500 ppm. However in different locations such as Al Ain and Liwa, there has been saltwater movement from the deeper saline aquifers.

Table 9.1. Depth to groundwater level and associated with salinity in selected areas in the United Arab Emirates.

Depth to groundwater (m)	Total dissolved solids (mg/l)
< 5	Dibba (1,000-3,500), Khor Fakkan (5,500), Kalba (10,000), Al Khatt (1,000-3,500)
10-25	Al Shuayb (230-1,000), Madinat Zayed (3,500-6,500), Al Madam (3,500-6,500)
25-50	Al Wagan, Al Hayer, Jabal Hafit, Al Faiyah (1,000-3,500), Al Jaww plain (230-1,000)
50-100	Wadi Al Bih, Al-Ain area (west and south 10,000), North Liwa (3,500-6,500)
>100	Al Dhaid (10,000)

Groundwater Protection

a) Regulations for point source pollutants and landfill sites

Specific regulations need to be promulgated to control all three major sources of point source pollution, although as underground injection in the United Arab Emirates is confined to the disposal of oilfield brines, and underground disposal of radioactive wastes is forbidden, regulations can be far less complex than those developed by the Environmental Protection Agency. However two special programs, the Wellhead Protection Areas and the Sole Source Aquifer program respond to the particular needs of the United Arab Emirates.

Hazardous and municipal solid waste must be regulated, and sites located to avoid ecologically vulnerable areas, although there are few sites prone to seismic sites activity or other natural disasters, unstable terrain in the proximity to major habitations or sites of industrial activity, or airports, must be avoided. The requirements of a landfill site are relatively simple and can be summarized under the following headings:

- Site selection and characterization.
- Landfill design, nature and thickness of the lining, cover of landfill deposits.
- Leacheate control and treatment, monitoring fluids within the landfill, and external monitoring to test for potential leakage due to breaches in the lining.
- Gas movement and control gas production within the landfill material.
- Permitting procedures, which establish guidelines for, landfill construction and operation.
- Operation, training and maintenance.
- Closure and post-closure procedures.

To achieve these ends a monitoring program is necessary to ensure that any water percolating out of a landfill does not exceed groundwater standards. Operating the landfill requires keeping accurate records, including an inventory of any waste deemed hazardous and its location in the site. These records are open to regular, biannual inspection. Leakages must be remedied by the operators of the landfill and usually a bond has to be posted as an assurance that the remedial resources are available. When eventually the landfill is closed monitoring the integrity of the cover, the groundwater and methane emissions as well as the leachate must continue for a sufficient length of time (25 to 30 years) to ensure public safety (Davis, 1996).

As in the USA, in the United Arab Emirates, the Department of the Environment must establish standards to be met if a permit is to be issued to establish a landfill, to serve and to ensure that the necessary safety standards will be maintained. It is not uncommon for a buffer zone of at least two miles to be established around a landfill site, although the size of the buffer zone has been the subject of considerable debate, and will depend to a large extent upon land use.

It is fundamental that a geological survey be carried out of any proposed site to establish its suitability, i.e. a site that will not readily permit potential polluted water reaching and polluting groundwater. The construction of the landfill with an impermeable lining of clay and/or plastic lined, is designed to contain drainage within the landfill, which can be regularly tested, remediated when necessary and then safely disposed.

b) Underground Storage Tank Program

We may adopt the US definition of underground storage tanks as *"any one or combination of tanks (including underground pipes attached thereto) used to contain an accumulation of regulated substances, the volume of which is 10% or more below the surface of the ground"* (South Carolina Department of Health and Environmental Control, 1990). Regulated substances *"include but (are) not limited to petroleum, petroleum based substances comprised of a complex blend of hydrocarbons derived from crude oil through processes of separation, conversion, upgrading, and finishing, such as motor fuels, lubricants, petroleum solvents, and used oil"* (South Carolina Department of Health and Environmental Control, 1990). While these substances, when they leak form the principal sources of groundwater contamination, they are not the only sources, which need to be covered by protective legislation (Nardi, 1993).

Preventive measures must not only cover design standards, performance characteristics and installation of new tanks, but also record keeping and monitoring and rules to be followed in reporting cases of leakages or accidental spills. There must also be regulations concerning remedial actions to be taken in case of accidental spills or leakages from both newly installed tanks and tanks already in place. In the latter case the situation is more complex since most of these were installed before any regulations existed. It is not sufficient for remedial action to seal leaks and removal of leaking material; it must also cover the remediation and restoration of the affected areas, taking into account the future potential use of the contaminated sites. No such regulations can expect to be effectively followed until non-compliance penalties are sufficiently rigorous and regulations effectively monitored to make it economically sound to comply.

A necessary preliminary step must be the establishment of an inventory of existing tanks. Notification requirements require all owners and/or operators of existing storage tanks to inform the Federal Environmental Agency of the existence of tanks, their location, size, type, date of installation, volume and type of stored material. All tanks should require United Arab Emirates Federal Environmental Agency approval as a means of ensuring all new installations follow the minimum requirements according to the volume, toxicity and/or corrosive nature of the stored materials. In most cases it is impossible to establish requirements to retrospectively upgrade existing tanks, reporting and testing may be made more stringent for older tanks, until, when leakages are detected, replacement of older tanks can be required as part of the remediation process.

Typically the United Arab Emirates Federal Environmental Agency minimum requirements for tanks include such features as double walled design, and installation criteria, which facilitate regular testing. The Agency approval of the completed installation will also be required.

During operation owners and operators must maintain records to show amounts and nature of toxic materials stored and used. All repairs must be reported and daily logs kept to record potential spills, overflow, and the status of corrosion equipment, test wells, and clean-up equipment. Periodic testing must be made for leakages, and this recorded for transmission to the authority. In the event of any spill or leakage, remedial action has to be immediate and all such leaks and spills must be reported within 36 hours although remediation must begin with hours of leakage detection. A modified version of the Oil Spill report can be developed for use onshore (Environmental Protection and Safety Technical Guideline Number 5). Costs of clean up are born by the owners or users of the tanks. The authority upon being informed of a leakage or spill will work in concert with the owner/operator, and should the latter be late in beginning remedial action the operation may be taken over and completed by the authority with the owners liable not only for reimbursement of the expenses incurred, but with the addition of a surcharge on the owners for delayed compliance. US experience shows that when the punitive damages are assessed, owners comply more readily and rapidly. Out-of-Service or temporary shutdowns do not relieve the owner/operator of the obligation to maintain continuous monitoring and all other statutory requirements. Closure for a time in excess of three months requires the emptying and cleaning of the tank. In the case of permanent closure the site must be assessed for pollution damage and remediated before Federal Authority approval for such action can be issued. Owners can be required to comply with a requirement for mandatory shut down for maintenance, although such a requirement may be waived if the owner can demonstrate continued compliance and an absence of reportable spills.

Violations or failure to file report mandatory reports will result in the issuance of an order from the Federal Authority Administrator to comply, or make good any deficiencies within a specified time period after which a penalty will be levied (of the order of 30,000 Dirham to 100,000 Dirham per day). Continued non-compliance may result in punitive damages being imposed.

Methods of leakage detection, and clean-up must meet the Federal Authority specified criteria.

c) Underground Injection Control Program

The goal of this program is to protect underground source of drinking water from pollution that might result from emplacement of fluid (by means of wells) over, into or below it.

Not only can underground injection wells be used for disposal of wastes but they have also been used for mineral extraction, gas and oil production and storage, as salt water intrusion barriers, and even for the recharge of groundwater. Consequently the Environmental Protection Agency regulations have to be broad to cover all the potential uses and in the USA the Environmental Protection Agency defined five classes of underground injection wells according to the type of waste and where it was to be injected (Jensen, 1993). However, the principal use of underground injection wells in the United Arab Emirates is confined to a single type of well, the Environmental Protection Agency class 11 well, for the disposal oil and gas fluids either to maintain production or dispose of a water cut.

The Underground Injection Control regulations deal with the release of the fluid underground and the mechanical integrity of the all injection equipment, it does not regulate the quality of the fluid injected. In general, injection of any fluid into or above an Underground Source of Drinking Water is prohibited. The United Arab Emirates Federal Environmental Agency must approve all injection wells. The approval depends upon meeting the minimum constructional and operational requirements established to protect Underground Source of Drinking Water. The reports to the Federal Environmental Agency must include location of injection wells, name and address of underground injection well owners' operators, a drawing of well construction, injection system, depth, composition of construction, injection system material, analysis of injected fluid, date of injection, injection rate and injectivity pressure. A monitoring system and ground water quality analysis data are required (see also South Carolina Department of Health and Environmental Control, 1993) The fluid injection pressure must meet the pressure specified by the United Arab Emirates Federal Environmental Agency to prevent any migration of the injected fluid through fractures, which could result from excessive pressure. Underground injection wells require regular testing, and the operator of such wells must maintain a record of the injection pressure. The area around an injection well must also be tested, and monitored, according to a Federal Environmental Agency schedule. If a well is found to be leaking, remedial action must be taken. Any injection well that no longer is operational, must be plugged in manner approved by the Federal Environmental Agency in presence of an inspector.

Enforcement for Underground Injection Control program requirements may require the United Arab Emirates Federal Environmental Agency to take action against any violations. The violators may be subject to penalties ranging from an administrative order to criminal or civil action.

The three programs described above are intended to control point sources of pollutants. There are also different programs to protect specific areas that are sensitive and activities that might create risk to public health which could result from the contamination of public water supply, therefore need to be regulated and controlled. These programs are listed below:

i) Wellhead Protection Areas Program: The Wellhead Protection Areas are zones of pollution prevention established to protect water wells or well fields supplying a public water system from contaminants. They are surface and subsurface areas surrounding the water well or well field and very sensitive to contaminants. The sources of contamina-tion are microbial, chemical or anthropogenic (caused or performed by human action) and direct contact in well area with pumps, pipes or casing.

The regulations cover different measures to control activities which create risk affecting the public water supply system within Wellhead Protection Areas, such as design standards (design and construction of structures, buildings and their components) and well standards for specifications, construction, operation and abandonment, operating standards for existing activities in developed area, permitting, and prohibition the usage or storing hazardous material or certain chemicals. Intentional discharge of harmful substances or interfering in the operation of a public water system is prohibited. The assigned authority is required to enforce the provisions of Wellhead Protection Areas' exacting penalties for non-compliance in addition to ensuring mandatory clean up (South Carolina Department of Health and Environmental Control, 1994).

The assigned authority is required to include non-regulatory measures such as:

- Locating all the public supply wells on map;
- Identifying and mapping all the Wellhead Protection Areas;
- Inventory the potential and the existing contaminant sources within the Wellhead Protection Areas;
- Mapping all data such as Wellhead Protection Areas, wells, potential sources and existing sources;
- Handling the potential sources, which are located within Wellhead Protection Areas, separately from others, located outside the Wellhead Protection Areas and developing contingency plans to provide alternative water supply must be included.

Education and awareness are tools to minimize the degradation of groundwater quality consequently people and business in the Wellhead Protection Areas, must be aware of the best management practices to protect the public water supply (South Carolina Department of Health and Environmental Control, 1994). Groundwater monitoring can help in detecting the contaminants, and thus help in the prevention of contaminants, from reaching the well supply, by allowing the authorities to take immediate action. It also provides an incentive to people and business to comply with the regulations.

ii) Sole Source Aquifer Program: A program to protect groundwater from contamination is described under the "Sole or Principal source aquifer" (Section 1424 (e) Source of Drinking Water Aquifer). The program identifies a critical area with a Sole Source Aquifer as a "Critical Aquifer Area" based on the criteria listed below:

- The aquifer must supply at least 50% (or any percentage recommended by the authority) of drinking water to persons in the area over the aquifer;
- It is the principal supply, with no alternative should the sole source become contaminated; and
- Aquifer boundaries must be clearly defined.

Developers of projects (such as highway improvements and new road construction, wastewater treatment facilities, agricultural projects), which are located within the critical aquifer area, and could prove to be a potential threat to the aquifer, must submit their plans to the United Arab Emirates Federal Environmental Agency for review and approval. Project activities, which would result in contamination and the creation of significant hazards, will automatically be rejected, or must be redesigned to meet the Federal Environmental Agency requirements and resubmitted. The review will include the surface and subsurface recharge area of the aquifer, the water shed area that contributes to surface flow across the aquifer, consequently the developer's plans should incorporate all measures in planning, design, construction, maintenance, monitoring and corrective action to prevent any contamination to the sole aquifer.

After the completion of the project, the Federal Environmental Agency administrator will periodically inspect the project, and the developers/ operators or owners records to comply with the Federal Environmental Agency requirements for operating, maintaining and monitoring the project for submission to Federal Environmental Agency. They are required to report to the Federal Environmental Agency any accident within their premises, which could lead to the contamination of other aquifers and must take the corrective action approved by the Federal Environ-mental Agency. Failure to comply with Federal Environmental Agency regulations will result in various actions ranging from a simple administrative order, to criminal or civil action with penalties commensurate with the magnitude of the violation and its impact on the aquifer.

d) Regulations for non-point source pollutants

This program is to establish procedures to protect groundwater quality and quantity from uncontrolled extraction that could affect the water potability. The regulations cover the procedures to obtain a permit to extract and utilize groundwater, and the submission of reports concerning the amount of water extracted, its use or its intended use, and the proposed aquifer or aquifers from which the water will be withdrawn, to the Federal Environmental Agency. It will include provisions to protect the aquifer against or reduce seawater intrusion, sinkhole formation, land compaction or subsidence, unreasonable adverse effects on other users within the capacity use area, and will establish well standards, well location and spacing, well depth, maximum pumping rate for any given well and aquifer.

The regulations are intended to outline specific procedures for the Federal Environmental Agency in the granting, denial, revocation, modification or the conditional granting, of permits to withdraw groundwater. These regulations also provide for the gathering of information on the geologic and hydrologic character of the rock at and below the surface, well data, groundwater levels and related material. The regulations include instructions concerning the implementation of water conservation policy in conjunction with the withdrawal of groundwater.

The program will deal only with the "Capacity Use Areas" which is defined under the Groundwater Use Act of 1969 as *"one where the departments find that the aggregate use of groundwater in or affecting the area have developed, or threatened to develop, to a degree which requires coordination and regulation, or exceeds or threatens to exceed or otherwise threaten or impair, the renewal or replenishment of waters, or any part of them."*

The Federal Environmental Agency has the right to delineate, declare, or modify any Capacity Use Area where protection is needed.

The Federal Environmental Agency will base its findings on various reasons which may result in or might result in falling groundwater table, sea water intrusion, interference between wells, overdrawing water from an aquifer, sinkhole formation and subsidence of the aquifer due to over pumping, capturing and diversion of natural or anthropogenic contaminants.

To declare an area as a capacity use area the Federal Environmental Agency must investigate the area in question and consult with all interested parties considering all the reasons, and prepare a report outlining their finding and delineating the boundary of the area along with their recommendations. Once an area is declared a Capacity Use Area, the Federal Environmental Agency will determine a withdrawal rate (daily, monthly and annually) and will prohibit any groundwater extraction exceeding the specified limit. To determine the amount of water withdrawn from the aquifer, the Federal Environmental Agency will consult all the relevant agencies to develop a formula, which covers all the different water uses such as agriculture, processing facilities, domestic and industrial use.

An example of the operation of this type of policy can be found in the Michigan Natural Resources and Environmental Protection Act, Section 32708 where it is stated, *"The Department of Environmental Quality and the Department of Agriculture in consultation with the Cooperative Extension Services and the Soil Conservation District shall develop a formula or model to determine the amount of water withdrawn for agricultural purposes. A person using water for an agricultural purpose who withdraws over 100,000 gallons of water per day on average in any consecutive 30-day period for irrigation shall provide information regarding the location and source of water, type of crop irrigated, and the acreage of each irrigated crop to the Soil Conservation District. The Soil Conservation District and the Department of Agriculture shall use the model or formula developed to aggregate these data and provide county and watershed totals by water source to the department by April 1st of each year for irrigation water use that occurred in the previous year."*

The United Arab Emirates Federal Environmental Agency will consider many factors while reviewing the application such as the number of users withdrawing from the aquifer, extent and necessity of their respective withdrawal and its uses; the chemical and physical nature of any impairment of the aquifer, any adverse effects on its availability or fitness for other use such as public or farm use; the size and the nature of the aquifer, the probable severity and duration of such impairment under foreseeable condition, activities and business to which the various uses are related: risk to public health, the diversion from or reduction of flow in other aquifers, the hydrologic and geologic characteristics of the aquifer system, together with any documentation of effective water conservation.

All groundwater users must file a report with the Federal Environmental Agency stating well locations, construction, well specifications, operation and maintenance which must comply with the Federal Environmental Agency well standards in construction, operation and maintenance. The report will also include groundwater withdrawal, the amount and its duration and usage. Any person wanting to withdraw more than the specified amount must apply for a groundwater use permit and submit this application to the Federal Environmental Agency justifying, the amount of water to be withdrawn, location of well, and the general nature of the use proposed for the groundwater, etc.

The Federal Environmental Agency may establish two classes of groundwater users: Class E (existing users) and Class N (new groundwater users). A Class E user pumping groundwater at greater than the Federal Environmental Agency rate and wishing to continue extracting water at the same rate would have to apply to Federal Environmental Agency for permit but could continue pumping while the application was under review. A Class N user would have to apply for a permit to drill and construct a well and withdraw groundwater. The application would have to include well location, construction, design, and any information that the Federal Environmental Agency required. The Federal Environmental Agency might grant permit conditionally or deny the application. It can also revoke a given permit. The Federal Environmental Agency must consider while reviewing the application such factors as the number of users and their needs, and the impact on public health and welfare, the condition of the aquifer and the potential contamination. The permit if granted must specify its duration, and cannot be transferred without formal Federal Environmental Agency approval. When the permit expires, it may be renewed by the Federal Environmental Agency.

Violation of the regulations may result in a simple administrative order for civil or criminal action depending on the nature of the violation. The Federal Environmental Agency reserves the right to enter any property, public or private, to investigate installation and operation of any wells and determine the withdrawal rate in the protection of the aquifer.

Fertilizers and pesticides control program: The current regulations concerning pesticides and fertilizers mostly deal with commercial issues such as the buying, selling and storage and are relatively restricted in their application. The USA-Environmental Protection Agency together with other United Nations organizations e.g. World Health Organization, concerned with the environmental impact of pollutants on drinking water standards, have specified limits for certain pollutants in drinking water such as 10mg/l for nitrate as nitrogen. The same environmental concerns exist in the United Arab Emirates and in this section of the paper are outlined the basic parameters which need to be incorporated in any legislation which seeks to prevent contamination of groundwater resulting from the use of fertilizers, insecticides and pesticides in general. Overall authority should reside with the Ministry of Agriculture and Fisheries, it should have the power to prohibit the use of any fertilizer, and pesticide or insecticide deemed an environmental hazard, while still leaving choices open to the individual farmer. The regulations may also cover setting standards covering the optimal amount, which can be applied per acre to maximize crop yield and minimize loss to groundwater through leaching or downward percolation.

To ensure groundwater is protected it will be necessary to establish a monitoring system along with regular record keeping and reporting. An adequate monitoring system may well involve the drilling of new wells to provide an adequate grid. Such a system is probably best managed through an agency of the Ministry of Agriculture and Fisheries on a national basis to ensure uniformity. A national laboratory would have to be established capable of handling the volume of analyses. The actual sampling could be left in the hands of the individual farmer subject to periodic control by ministry inspectors. Maintenance of the monitoring wells should also be left in the hands of the individual farmer subject to periodic Ministry inspection of sufficient frequency to avoid a tendency to neglect maintenance. Each farmer must of course be provided with the records of the analyses, such that there is a record of performance and if problems arise there is a working basis from which to determine the best remedial action and result in agreed JOINT ministry and individual action. Such action could range from reduced application, to complete banning of use, or corrective actions such as flushing. Examples of corrective actions may be taken from regions where they have been successfully applied.

No such system will work without some system of penalties for non-compliance, and a system of appeals. Based upon experience in other countries penalties to be successful have to be of such a magnitude that compliance is the preferred course of action.

Discussion and conclusions

a) Potable water supply

Although the principal water supply is maintained by the Department of Electricity and Water and accounts for more than 90% of the water used in Dubai, there are many private wells. The city is supplied with water drawn from well fields mixed with desalinated water to reduce the total dissolved solids to the permissible limit 500 ppm.

The water supply from private agencies is predominantly through companies which store and bottle water delivering it to customers as a commercial commodity. Typical of such companies, for example the mineral water companies which bottle drinking water for domestic use.

In the absence of permanent running water the only source of natural supply is groundwater extracted from aquifers. With the growth of population and the increase in the use of irrigation to enhance the farming potential, groundwater quality and quantity has deteriorated through uncontrolled use and severe over-pumping. Additionally, there is the danger of pollution by fertilizers penetrating into the groundwater. The only means of maintaining an adequate supply of good quality drinking water is through the construction and operation of desalination plants. Such plants exist in most Emirates, Abu Dhabi for example is totally dependent on desalinated water, and in Dubai the water from a large multi-stage flash plant (in the Jabal Ali Free Trade Zone) is mixed with groundwater from the Al Aweer well field. Incredible as it may seem the desalination plant produces more water than Dubai needs and the excess is allowed to flow into the sea. There is however no laws controlling groundwater extraction or pumping from private wells.

In the other United Arab Emirates and in most of the Gulf States there is no attempt to recover the full costs of providing water, a subsidized, fixed rate is charged. Only in Bahrain is a block rate used in charging. The Ministry of Electricity and Water provides the water service to all the northern Emirates. The cost of producing the water varies according to its source and ranges from 4-5 Dirham per 1000 gallons for groundwater to 14-15 Dirham per 1000 gallons for reverse osmosis to as much as

28 Dirham per 1000 gallons for water produced by the multi-flash process.

The principal behind adopting block rate charging is that anyone may use as much water as he wishes, but it should be paid for, and by using a graduated scale, may be partially underwriting the costs of the whole operation. The municipality is thereby following the Shari'ah, providing each according to his needs and ensuring the protection of a communal benefit, at the same time providing an accurate base for water management planning, with the flexibility to adapt to changing circumstances.

b) Waste disposal

"Enforcement is the ultimate action to seek compliance with laws and regulations and to assess damages for past non-compliance...". Obtaining compliance and deterring the regulated party from future violations are the major purposes behind penalties and related enforcement actions that is, the policy is to protect the environment from harm or damage. The weapons available to achieve these ends are both economic and legal, which can include both civil and criminal action, and ultimately the closing down of operations.

In establishing penalties, which should be both fair and effective, the actual costs, or the future impact cost resulting from the violations, should be based upon the extent of the pollution, its toxicity, the duration of the violation, the cost of remediation, and the potential impact on the health of both humans and the environment. Currently in Dubai the fines, which may be levied, are listed in Table 1 (of Article 88-90 of Local Order 61/1991), from which it can be seen that they range from $136 (500 Dirham) to about $200 (2000 Dirham). The violator is allowed a month to comply. The penalties which may be imposed, in the range 500 Dirham to 1500 Dirham are also set out in Local Order 61/1991.

These fines and penalties are insufficient to act as an incentive to comply with regulations since they are substantially below the cost of remediation, in fact it is economically more practical to pay the fine when a transgression is notified and take chances on subsequent fines. Even closure is not a real incentive to comply as a closure is for a period "not exceeding one month" during which period it is to be supposed that remedial action has to be taken. The worst action that can be taken from a company standpoint is that in the case of continued non-compliance, the local authority could take over remediation and charge the cost to the company. In essence the fines are too low to be a deterrent, and in no way compare with the punitive fines exacted in the USA where they may amount to thousands of dollars per day. To be effective fines and potential closure of plants should be sufficiently punitive that it makes more economic sense to comply with the regulations. This requires a monitoring system, which is rigorous and frequent enough to gain the employers respect, and legal action, which is prompt and enforced. Appeals must be handled with dispatch to avoid the delays of protracted court action. All of this must be set out in Local Order's, that is a scale of fines for the first offense, which increases with each subsequent offense and culminates with plant closure, and a scheme of monitoring this is rigorously adhered to, with the addition of random spot checks. To maintain fairness in the system, it will be necessary to establish pollution limits, which are reasonable on the basis of scientific study, and there must be allowance for accidents which can result from failure of equipment (when adherence to regular maintenance schedules can be demonstrated).

c) The consequence of legislation

Clearly some legislation is necessary to ensure compliance with environmental regulations, and this must be based on the Shari'ah concept, that water is right for all people. Legislation based on this premise, must be fair in accordance with the Shari'ah. Perhaps the most telling element in any legislation is the Shari'ah mandate that no damage should be caused to others, and that everyone has the right (or duty) to preserve the vital resource.

Some form or inventory is needed since no controls are possible when there is no quantitative data. The Shari'ah does not permit water rights to be sold independent of land since there is no control on quantity. Therefore legislation requires an assessment and assessment is impossible if the number of wells is unknown and if the extraction from these wells is uncontrolled.

Since the best form of control, both in terms of efficiency and in terms of plain economics is through the pocket, this ought to form one of the basic features of the law. As it is necessary to have a record of all wells, both in terms of location, rate of water extraction and salinity, depth to the water table and its variation and bacterial content of the water, type of bacteria or noxious chemicals, this may require the municipality to cap all wells, and put in place a submersible pump with a monitoring package which can provide a central computer with location, conductivity (as a measure of solutes present), depth to the water table, and measure pumping and total extraction on a daily and monthly basis. This is not only a practical health measure, but if done at community expense, raises fewer potential objections. In this way the Department of Water in the Emirates would have a record of water used, fluctuations in depth to the water table, and salinity on which to develop a water policy. There ought to be a means of obtaining water samples for periodic analysis.

A maximum salinity and a maximum extraction rate may be established by law, and wells closed down, where salinity or bacterial content are too high. Water would then be replaced by the public water supply.

Conservation of water, which is an obvious necessity, can be controlled through economic action, with a gradual increase charge based on amount used. In order to conform with the Shari'ah, it would be essential for the cheapest rate to be based on a generous estimate of average daily use. For agricultural irrigation and industrial use a clear case can be made for the use of treated water, which if inadequate could be mixed with potable water and charged at the economic rate plus some service surcharge.

The principle behind all the regulations is that everyone is free to use as much water as desired, with the proviso that cost should reflect usage with a bias towards conservation. The municipality is thereby following the Shari'ah providing each according to his needs and ensuring the protection of a communal resource. At the same time the regulations provide an accurate base for water planning with the flexibility needed to deal with changing needs.

d) Policy co-ordination

As a result of the gradual growth of ideas on water use and protection of water quality, authority over different aspects of water use has been assigned to different authorities, for example at the federal level the Ministry of Agriculture and Fisheries has had authority over the provision of water for a agricultural purposes whereas the Ministry of Electricity and Water has been responsible for the provision of water for domestic use, and the Ministry of Health for the purity of that water. A complicating factor is that each emirate has the authority to establish regulations within its own jurisdiction. This functions in a manner similar to the US Federal and State regulations where the regulations for each state take precedence over the federal, where the conditions are more stringent, in the absence of local regulations the federal regulations apply.

With the founding of the United Arab Emirates Federal Environmental Agency comes the need to group all the regulatory functions under one authority at the national level and assign to specific departments specific issues, thus there will be departments controlling the provision of potable water, which will also cover health questions, such as bacterial content of the water, a department monitoring water for agricultural use, which will also cover the use of treated water. A planning department will need to monitor well construction, and water extraction through the issuance of permits, to authorize the drilling of new wells, and establish extraction limits designed to protect the aquifers from excessive depletion and the dangers of increasing salinity. This will involve a consideration of well spacing, the amount that may be extracted and the use to which the water will be put. Since this is liable to be a sensitive issue, the Federal Environmental Agency may assign primary responsibility to individual emirates, however the Federal Environmental Agency will have overall authority in the development of a national water plan, which should also cover potential needs for increased desalinated water production, the location of such plants and their output. Within each individual emirate local issues are best handled by local authorities who have the ability to define regulations appropriate to local conditions, and as indicated above such regulations have priority where they are more stringent than the federal regulations. Where no regulation exists then the federal regulations apply. Another department will be needed to handle such matters as waste disposal establishing minimum criteria to be adopted in each emirate.

In theory, this can or should be able to avoid certain major bureaucratic problems such as over-pumping groundwater for agricultural use but destroying the aquifer in the process. However, it will require a major effort by the Federal Environmental Agency to avoid departmental rivalries, such as by holding regular meetings of department heads under the minister, as well as good communications with the appropriate agencies in the individual emirate.

2. SAUDI ARABIA

Several authorities are responsible for water in Saudi Arabia including: the Ministry of Planning, Ministry of Agriculture and Water, Ministry of Municipal and Water Affairs and the General Establishment of Water Desalination. The following is a summary of the duties of each of these authorities:

1. Ministry of Planning

The ministry prepares the five-years development plans of the country, taking into account the technical balance between economic, social and organizational sectors. Between 1967 and 2000, six five-year development strategic plans were prepared by the Ministry of Planning, including the regulations that governs the water sector.

2. Ministry of Agriculture and Water

The Ministry of Agriculture and Water is responsible for conducting groundwater investigations, surveying and exploration. The Ministry of Agriculture and Water is also in charge of the following:

a) Issuing licenses for water-well drilling, preparing well designs, supervising water-well drilling and providing criteria for water distribution networks, whether groundwater or waste-treated water.
b) Conveying desalinated water to consumers, supervision of water-connection networks, installation of water collection and distribution tanks that provide towns and villages with water, construction of waste-water treatment plants.
c) Construction of all kinds of dams and utilization of runoff water for drinking or for recharging groundwater.
d) Preparation of a national water plan which includes an inventory of all present and possible water sources, present and future demands for water, and design of policies and disciplines controlling previous activities.
e) Working on issuing the Royal Decree No. M/34 in 1979 for water resources conservation in Saudi Arabia. The Decree arranges the use, conservation, and protection of water resources.

3. Ministry of Municipal and Water Affairs

The Ministry of Municipal and Water Affairs is responsible for administration and maintenance of water connection networks, water desalination plants and wastewater treatment plants in towns and villages after their completion under the supervision of the Ministry of Agriculture and Water. The Ministry of Municipal and Water Affairs is also responsible for the following activities:

a) Construction of water projects in new suburbs of large towns in some remote villages and residential areas.
b) Supervision of extensions in water networks and operation and maintenance of water networks in towns and villages.
c) Connections of water and sewage networks to houses in towns and villages, and collection of the annual charges for water use.
d) Design, installation, operation and maintenance of wastewater treatment plants in towns and villages.
e) Preparation of adequate construction for conservation of rainwater and protection against floodwater.

4. General Establishment of Water Desalination

The General Establishment of Water Desalination is an independent organization headed by the Ministry of Agriculture and Water. The responsibilities of the establishment include the following:

a) Design, supervision, and construction of all kinds of water desalination plants, in addition to the administration, operation and maintenance of these plants.
b) Construction of pumping stations and pipe networks connecting desalinated water to consumers in towns and villages.
c) The establishment's headquarter is based in the Al Riyadh town, while branches exist in Jiddah, Al Khobar and Al Gebeel.

3. KUWAIT

The Ministry of Electricity and Water, General Authority of Agriculture and Fisheries, Ministry of General Works and Kuwait Institute of Scientific Research have all water responsibilities. The following is a brief summary of duties of each authority:

1. Ministry of Electricity and Water

The Ameri Decree established the Ministry of Electricity and Water, responsible for production of electricity and water and their distribution and connection for all purposes. The Ministry of Electricity and Water of Kuwait is also responsible for network expansion and development to meet the needs of the country. The water-related Ministry's duties include:

a) Construction, management and operation of water production facilities, including water desalination plants and their operation and maintenance.
b) Exploration, drilling and production of ground-water.
c) Conducting water and electricity-related, applied engineering research.
d) Construction, administration, operation and maintenance of drinking water mixing, pumping and storage facilities.
e) Collection of costs of water connections and uses.
f) Supervision of water-well drilling, including the criteria and specification of drilling operations.

2. General Authority of Agriculture and Fisheries

The Ameri Decree No. 49 for the year 1983 states that "the General Authority of Agriculture and Fisheries is responsible for supervision of land and water uses in agriculture and fisheries, insuring their proper use and conservation".

3. The Ministry of Public Works

The Ministry of Public Works is responsible for rainwater and floods and establishment of adequate structures for protection against floodwater. The

Ministry is also in charge of the sewage network projects.

4. Kuwait Institute of Scientific Research

The Ameri Decree of the establishment of the Kuwait Institute of Scientific Research in 1981 state that "The Kuwait Institute of Scientific Research will study sources of natural resources and their exploration and utilization of water and energy resources, improvement of agricultural uses and development of water wealth and cooperation with other authorities responsible for studies and development of water resources".

4. BAHRAIN

Bahrain consists of an archipelago of 33 islands in the Arabian Gulf about midway between Saudi Arabia to the west and the State of Qatar to the east. It has an area of 695 km^2. As most Gulf States it experienced rapid growth since the early 1960's as a result of the increase in oil revenues. The population increased from about 162,000 in 1965 to more than 700,000 in 2000. This growth in urban expansion, industrialization and the growth of irrigation have placed major stresses on the country's limited water resources, and over exploitation of the groundwater resource has led to pollution of more than half the original aquifer. The need to provide fresh water to meet all the needs of agriculture, industry and the urban population is the greatest challenge facing Bahrain.

Until the 1950's the agricultural sector depended mainly on natural springs supplemented by some extraction wells, expansion until the 1960's was gradual and accompanied by reasonable rates of groundwater extraction, however after that date there was a 24% increase in groundwater extraction matched by only a 6% increase in agricultural land. The increase resulted from the intensification of agricultural activities to meet the growing demand for agricultural products and the general deterioration of aquifer water. The use of primitive irrigation techniques and the lack of an efficient drainage system exacerbated the situation even further. There was even a drop in agricultural activities due to desertification of agricultural land, the cultivable area decrease from 38 km^2 to about 17.5 km^2 in 1979, a decrease of 54%.

This state of affairs prompted the Bahrain government to initiate a major agricultural development program of improving the irrigation methods by more water efficient methods subsidizing more than 50% of the cost and by the construction of a major drainage system to alleviate water logging and salt accumulation. They introduced the use of sewage treated water and provided an advisory service for farmers, and undertook the reclamation of new agricultural land resulting in an increase to 30 km^2 by 1990, a 71% increase with only an increase in extracted water of 2 Mm^3/yr. The use of treated water, although small in amount (7 Mm^3/yr) had a noticeable effect on lowering abstraction levels (Zubari and Lori, 1991). Although the industrial sector expanded significantly during this time and its demand for fresh water remained high, its consumption from the Dammam aquifer did not exceed 10 Mm^3/yr, only 3% of the water taken from the aquifer, because of a government decree prohibiting use of aquifer water. Most industrial plants rely on the desalination of water from the Umm er Radhuma aquifer for their water needs.

The development activities in Bahrain have raised the total abstraction from the Dammam aquifer from approximately 65 Mm^3/yr in the early 1950's to about 112 Mm^3/yr in the mid 1960's to 145 Mm^3/yr in the mid 1980's and 179 Mm^3/yr in the early 1990's. The safe yield from the Dammam aquifer is about 100 Mm^3/yr, equal to the recharge by underflow from the upgradient Saudi aquifers under steady state conditions. Thus abstraction has exceeded the safe yield since the 1960's, and is now about double that rate, indicating withdrawal from the aquifer storage. A comparison of the potentiometric levels maps of 1991 with that of 1925 which represents the steady state condition of the aquifer shows that the potentiometric surface has dropped an average of 4m, the maximum fall of 5m, is observed in the up-gradient western areas of Bahrain to a minimum of 2m in the down-gradient area on the east coast where the aquifer water is in direct contact with sea water. The reduction of aquifer storage, and the drop in the potentiometric surface, is marked by a deterioration in the aquifer water quality, due to increased upward migration of saline water, from the Umm er Radhuma aquifer and by sea water intrusion, in the eastern and southeastern areas of Bahrain. The interface between the fresh water in the aquifer and the saline sea water has migrated at a rate of 100 m/yr. More than half of the original Dammam aquifer in Bahrain, has been polluted because of over-exploitation, leaving only a very small area to satisfy the competing needs, of the agricultural and domestic sectors.

Water Policy

Water policy in Bahrain is neither well-defined, in terms of time frame nor of action plans. During the last 15 years several steps and courses of action have been considered to provide solutions to the water crisis, the most important of which have concerned the supply side and these are documented below. The other aspect of water policy concerns conservation. Apart from the more technical questions which concern water losses through leakages, and these can be substantial, the main control is economic which can be enforced through block-incremental water rates.

a. Non-traditional sources

Non-traditional sources have been introduced to alleviate stresses on the groundwater system. They all into two categories, the development of treated water and the provision of desalinated water.

In 1990 some 7 Mm^3/yr of treated sewage effluent was being used for irrigation purposes (Zubari and Lori, 1991), and it is projected that by the year 1995 that total rose to 34.7 Mm^3/yr and it is projected to reach 60.6 Mm^3/yr in 2010 about 75% of which will be earmarked for irrigation of both private and government farms.

Some 57 Mm^3/yr of desalinated water is used to meet domestic demands. The re-use of agricultural drainage water on salt tolerant crops is under consideration (Musayab, 1988). The government policy is to reduce groundwater abstraction for domestic use, and replace it by desalinated water, which involves by 1999-2000 construction of two additional desalination plants with a total capacity of 50 Mm^3/yr.

b. Water conservation

The government of Bahrain recognized that, the wastage and unwise use of waters by consumers, is a reflection of not having to pay for water, which removes all incentives for economical use. In order to keep the relatively cheap groundwater source functioning, as the basis of development, the amount abstracted must be limited to the aquifer's safe yield rate. Water demand therefore must be balanced with the natural hydraulic system, to establish the aquifer's natural conditions and modify its quality. With such a complex dynamic system, of water needs and constraints multi-objective planning is of paramount importance, and should involve integrating short- and long-term water uses, and requirements with the available traditional sources, and non-traditional resources as desalinated water, treated sewage, irrigation drainage and the appropriate cost constraints. Different ministries have developed different policies, which have been only partially implemented, and applied without regard to other sectors. This lack of cooperation is a fundamental cause for the lack of success, of what may have otherwise been good policies.

In the past Bahrain's small population shared community rights for irrigation from the larger water sources such as springs and a system of underground channels (qanats). Prior to 1925 Bahrain drew its water supply from naturally flowing wells and springs both onshore (15 wells) and offshore (20) wells, which yielded up to 90 Mcm^3/y with, water drawn from the Dammam aquifer. For domestic and municipal use, community rights were common with community wells made available by the government, or from hand dug wells made by the community. At present, privately owned water rights, are the only water rights applied in Bahrain. The governing principal is that, the groundwater is the property of the landowner, who has the exclusive right to extract, and use as much water as desired, for any purpose, without being liable for damage to the aquifer or neighbors.

Because of changes in such a system, the Bahrain government tried to establish a code of water laws, in the 1930's (Proclamation No. 48, 1933), and since then there have been various decrees, ordinances and regulations, enacted to regulate and control water use, and conserve the groundwater resource from salination and depletion. The major drawback was that, these acts concentrated on controlling the construction of new wells, without regard to the broader aspects of water resource planning.

In the 1980's as a result of comprehensive hydrological and agricultural investigations. Amiri Decree No. 12 governing the use of groundwater was issued with the aim of protecting the available resources from further salinization and deterioration. In 1982 a High Council for Water Resources was established under the leadership of the Prime Minister (Legislative Decree, 7/1982) with three objectives:

- To develop water policies in view of the results of resource studies and surveys, and coordinate plans of the different ministries.
- To protect and develop resources to increase the efficiency and prolong the availability of water from the aquifer
- To solve problems arising from the implementation of the policies.

The Ministry of Commerce and Agriculture acted as the advisory body for the council. The initial measures concentrated upon the control of new wells. Amiri Decree No. 12/1980 governed groundwater use, to protect the available water source from further deterioration. Extraction rates were metered, and charges made for excessive withdrawals. A well participation system was introduced to reduce the number of wells, and a licensing system gave the Ministry of Commerce and Agriculture, the right to determine the water quota for each plot of land, and to charge for water used in excess of that amount.

Based upon the articles and clauses set out in the decree, two Ministerial orders were issued banning the drilling of new wells into the Dammam aquifer between 1980 and 1984 to allow for aquifer recovery

(Ministerial Orders 23/1980, and 4/1983). In 1982 the Bahrain High Council for Water Resources was established (Legislative Decree 7/1982).

Although Amiri Decree 12/1980 and the subsequent related ordinances and executive frameworks have looked into water management issues from a comprehensive point of view and seemed encouraging, they failed in implementation. The drilling of wells continued, and during the four-year moratorium 1980-1984 the rate of increase in the number of wells drilled exceeded the rate observed during the 1970's. The well participation system, licensing and taxation of water wells has not occurred, and the contribution of the High Council for water Resources, to the solution of the water crisis in Bahrain seemed minimal.

It is obvious from the preceding discussion that the problem is not one of legislation but its enforcement. Experience shows that technical solutions could readily be developed, but until the problems of implementation are solved a formulated water resource strategy is relatively useless with a fate paralleling that of the earlier Amiri decrees.

5. QATAR

The legal situation in State of Qatar parallels that of Kuwait and Bahrain where the State has taken over responsibility for handling groundwater supplies recognizing that ownership of groundwater is fundamental to exercising some control. Under Law 1/1988, the State assigned to the Ministry of Water and Electricity the role of developing policies and procedures, required for all phases of operation. The regulations cover drilling of groundwater wells, the number of wells to be drilled, their depth and the amount of water, which can be extracted. Information has to be provided to the Ministry of the well tests carried out, of the drilling and cleaning techniques used, water analyses and pump installation. The well owner is responsible for the maintenance of the well, the water distribution and conservation as required by the Ministry and the latter has the right to halt operations, which could lead to groundwater pollution. Any unauthorized well is liable to expropriated by the Ministry and penalties applied.

Penalties and/or fines may be exacted for unauthorized drilling, drilling without a permit or with an expired permit, halting operations until the problem is resolved. Fines (up to 10,000 riyals about US$ 2700) can be doubled and equipment confiscated for non-compliance.

The comprehensive nature of these laws is somewhat modified by the addendum of law 39/1998 which exempts from these provisions State owned private enterprises by decree of the Council of Ministers, and which largely effects the hydrocarbon industry.

The laws also make no allowance for pollution which can result from the infiltration of agricultural fertilizers or the uprise of salt water or the injection of brines.

6. OMAN

Because of its position influenced by two climatic systems, and its topography the water situation in Oman is not nearly as severe as in the other Gulf Countries, although water is in limited supply. Historically Oman, as the other Gulf countries, entered a drier phase following the end of the last pluvial, the evidence for which is marked by travertine deposits, the presence of former lakes and relict drainage patterns. The earliest evidence for irrigation dates from about 4000-5000 BP at a time when the southwest monsoon was more extensive and potentially there was surface run-off. As drier conditions developed terrace agriculture and a more nomadic life style developed. The first aflaj on the lower reaches of wadi fans date from about 2700-2800 BP, on the western side of the North Oman Mountains. The establishment of falajes gave rise to stable communities, with water under local community control, which managed distribution for domestic and livestock use as well as irrigation.

Within the falaj community, water rights could be bought and sold, under a number of conditions. For short-term use, water rights could be sold at auction, but sales of water rights and associated land rights, were much less frequent, and are usually well documented, because of the involvement of land, the value of the water is hard to define. Sale with or without land rights, under conditions such that the land could be recovered usually within a ten-years period, by refund of the purchase price is really a form of securing a loan. Outright sale also exists but is less common.

During the last quarter of a century the introduction of powerful pumps, has changed the situation through the extraction of large volumes of groundwater, and has led here, as elsewhere, to the need for some form of legislative control.

Water Regulations

The Water Research Council was formed in 1975 with the task of monitoring and advising on all water related projects. Prior to 1975 the Ministry of Agriculture and Fisheries supervised irrigation water and the Ministry of Electricity and Water had the responsibility for potable water supplies. Since then the organization and management of water has undergone a series of changes culminating in the establishment of the Ministry of Water Resources. It replaced the Public Authority for Water Resources, which had the task of systematic exploration for groundwater, the installation of rain, wadi and groundwater monitoring and the preliminary assessment of the hydrology and saline intrusion into the Al Batinah coastal aquifer.

The Ministerial Decree 13/1995 included articles to cover the definitions, well registrations and permitting procedures and permits. It also covered the registration of contractors, their duties and penalties for non-compliance. Under the Minister there are directorates with specific areas of interest such as restricted areas where well fields provide the water supply, or within the restricted area of a falaj. Areas of shortage are those areas where average extraction, exceeds groundwater recharge, open areas are areas specified for agricultural activity but which has not yet developed.

Well registration with the Ministry is compulsory, and non-registered wells are regarded as illegal and liable to be shut down. Well permits were introduced and cover proposed well deepening, changes in extraction or in pumping machinery. Restrictions are imposed upon any well construction within 3.5 km of the mother well of a falaj system to provide for a well protection zone. The permit must specify whether it is for a new well, increasing the depth of an existing well, and whether there is a change in the use of water, or capacity through the installation of new equipment. The guidelines for permit policy are outline in Tables 9.2, and 9.3 lists some of the reasons for permit non-approval.

The Royal Decree 83/1988 established water resources as a national resource and provided the authority to take any necessary action to protect and conserve underground water and the development of existing supplies. Water rights were vested in the State. Well field protection zones were established to prohibit any action that could affect an established well field. In the late 80's the responsibility for well permits was transferred from the Ministry of Agriculture and Fisheries to the Water Resource Agency along with control of some regional hydrological networks. A programme was also begun to delineate flood prone areas.

Royal Decree 2/1990 required the registration of all existing wells and the issue of new permits, and by decree 374/1992 a national well inventory was established to assess the distribution and quality of groundwater resources, with the data banked in computers. As early as 1989 water resources affairs were separated from environmental matters by the creation of the Ministry of Water Resources, which assumed responsibility for the implementation of decrees 2/1990 and 374/1992.

Water Conservation

a) Recharge and retention dams

With finite water resources retention and recharge dams, have been constructed to augment the water resource. Although their impact is small on the freshwater availability, they enable floodwaters, which would otherwise be lost to the sea or the desert, to be retained for public use and serve a secondary purpose in flood protection and control. It is one of the responsibilities of the Ministry of Water Resources to determine suitable sites, build and maintain and measure the effectiveness of dams. In 1985 about 56 possible recharge sites were identified, of which 15 have subsequently been developed as well as several flood control dams.

Recharge dams are usually located in the upper reaches of wadi fans, where suitable aquifers are available, and where target areas lie within reasonable distances downstream. Such recharge dams are usually only 6-8m high, but up to 10 km in length, and retain broad, shallow reservoirs. Most are to be found in the Al Batinah area, where agriculture, and consequently groundwater extraction, has increased heavily in the last few years. Because of the high storm water runoff the most effective designs integrate the spillway in the dam structure as an overtoppable embankment, protecting the embankment against erosion in a variety of ways.

b) Treated water and brackish water

The development of sewage plants in the major urban areas has made available treated effluent as a water source. Currently, it is used for amenity irrigation, of parks, gardens and roadsides, and its potential use injection as a barrier against saline water intrusion has been considered (Ministry of Water Resources, Oman, 1995).

There is an underused potential for the development of salt tolerant strains of common crops, and of different crops, with judicious selection of soils and irrigation. Within limits yields can be attractive although production decreases, when salinity reaches too high a level.

Recently the government has attempted to relocate some enterprises into areas where water resources are under-utilized. In particular this applies to grass and fodder farms on the Salalah Plain and in the Al Batinah area, which reduces the water demand, in the overdeveloped areas, and encourages growth in areas with higher potential.

c) Domestic and commercial supplies

Metering municipal and domestic supplies with graded charges is one obvious way of conservation through a reduction of demand. It is easiest to apply since the municipal supplies are piped and therefore metering is easiest to apply. Combined with a program of leak prevention can result in significant savings.

d) Agricultural water economy

Although domestic and municipal supplies can be metered, there are fewer means of controlling farm water consumption, which is responsible for 90% of water use. Few meters have been installed on farm wells, and a special study was begun in 1993. The pilot project was launched in southern Al Batinah to evaluate the different meter types, and the current levels of water use, and water efficiency, as the first stage in developing a water conservation policy for farms. The pilot study included a representative sample of farm sizes, methods of irrigation, cropping patterns and water quality.

The installation of modern irrigation as bubble, drip and sprinkler programs provides the main water saving, and the government subsidizes up to 75% of the cost of installation. Some 800 farms have such modern systems, not always with favorable results, but the methods offer the best opportunities for water saving over the long term.

The introduction of storage tanks into the falaj system, permits farmers to fill storage tanks during their periods of access to water. This water can then be metered and released more frequently from the tanks to irrigate vegetables requiring more frequent irrigation than possible under the regular falaj routine. The use of drip irrigation rather than the traditional flooding techniques, has proved successful and helped raise farming income.

e) Conservation campaign

One of the most important factors, in any drive for economy in water use is having an informed public and public support, for conservation methods, to back up any economic measures, such as graded water charges, which may have to be taken. Since the 1980's the Ministry of Water Resources has taken up this role in a multi-pronged approach. Education, as the first step, is undertaken at all levels through schools with increased water resource information inserted into the curriculums, and with competitions and artwork. At a more sophisticated level lectures and discussions can focus on more detail of conservation programs and their importance in the high schools and universities. This is backed up with lectures to a variety of citizens groups. The second line of approach is through the media by means of television by both special prorams and commercials, radio discussions and newspaper articles and advertising. This media approach is supported by the production of calendars with water conservation messages, slogans on envelopes and accompanying bills for water and electricity.

Table 9.2. Guidelines for water permit policy in Oman*.

Reason for Permit	Protection Zone			Areas of Shortage			Open Areas		
	New well	Alternative well	Increase in well depth	New well	Alternative well	Increase in well depth	New well	Alternative well	Increase well depth
Expansion of agricultural/arable land	Y	Y	Y	N	N	N	N	N	N
Preserving existing date or fruit farm	N	Y	Y	N	Y	Y	Y	Y	Y
Preserving, existing, far and other crops	N	N*	N	N	N*	N	Y	Y	Y
Falaj reinforcement within protection zone	Y	Y	Y	Y	Y	Y	Y	Y	Y
Falaj reinforcement outside zone	Y*	Y	Y	Y	Y	Y	Y	Y	Y
Public supply mosque and school	Y	Y	Y	Y	Y	Y	Y	Y	Y
Private supply, domestic and animals	N	Y	Y	N	Y	Y	Y	Y	Y
Private supply outside town or village and distant source	N	Y	Y	Y	Y	Y	Y	Y	Y
Agro-industry (excluding fodder production)	N	Y	Y	N	Y	Y	Y	Y	Y
Plantation and horticulture	N	Y**	Y**	N	Y**	Y**	Y	Y	Y
Industrial areas resolved by the directorate of water resources:									
Private industry	N	N	N	N	Y	Y	Y	Y	Y
Exploration and monitoring	Y	Y	Y	Y	Y	Y	Y	Y	Y
Water sale	N	N	N**	N	N	N**	N	N	N

Y = Yes , N = No
N* = unless due to construction failure.
Y* = with permission from other falajes.
Y** = trees only.
N** = unless no other sources available.
* Modified from Ministry of Water Resources, Oman, 1995 publications.

Table 9.3. Reason for permit refusal water well in Oman*.

1. Location of a well within a restricted areas
2. Location of wells within areas of water shortage
3. Increased pumping which could lead to increased water salinity
4. Extraction would have an adverse effect on adjacent wells
5. Well is not registered
6. The current water supply is sufficient to meet current demands
   The current supply meets the needs for agriculture and tree growth.
   Water is available at other localities in the same general location.

In a similar manner falajes must be registered with approval from the directorate prior to replacing or covering a falaj. The directorate of the ministry has the right and obligation to investigate all sites, and may stop operations in the absence of a permit or until one is obtained. In all cases work must be reported to the appropriate Ministry directorate.
* From Ministry of Water Resources, Oman, 1995 publications.

Chapter 10

TOWARDS THE DEVELOPMENT OF A WATER POLICY MANAGEMENT

INTRODUCTION

Throughout the world the stress upon water resources is increasing. During the 20th century global population tripled, and water withdrawal increased 7 fold, yet it is estimated that one third of the countries of the world, are experiencing medium to high levels of stress on their water resources. Despite water being a basic human necessity, one fifth of the world's population is without access to safe drinking water, and one half is without adequate sanitation (IWRM, 2000). Water serves a dual purpose, as a basic requirement for biotic life and as a sink and transport medium for domestic, agricultural and industrial waste. The deterioration in water quality, a combination of pollution and rising salt levels, due to marine intrusion and the uprise of saline water from deeper horizons, poses a threat to human health, and the health of the ecosystem.

Over the next two decades increased food demands will require an estimated 15-20% increase in irrigation, a problem exacerbated if individual countries strive for food self sufficiency, rather than food security through trade. According to Allan (1995) a supply of 1,000 m^3 of water per ton of grain per year, needed to raise food staples is beyond the capacity of the region's resource endowment. Thus, increased food demand is more related to water availability than to land, yet water governance is still largely sectoral leading to fragmented, and uncoordinated development which results in a parochial outlook on the problem as a whole. A holistic approach to water resource management is required, one which concerns the management of natural systems coordinated with the range of activities which create the demand for water to determine land use and the disposal of water borne waste products.

Water resource planning is thus a participatory decision fully integrating the social and economic sectors. It is a decision making process involving community groups, local government, and community organizations which will permit each to indicate their demands/needs for bulk services and impacts national food policies and energy policy. There is then a possibility for consensus/ common agreement together with the recognition that some sacrifices will be required in some sectors for the common good. Where consensus cannot be achieved recourse will have to be made to an agreed arbitration procedure.

The planners of a water management policy must assess the effects of inflation, balance of payments and land policy, the downstream and external costs, and the benefits or restrictions imposed, upon the natural water system. This includes the effects of deforestation and afforestation, and the urbanization of catchment areas. Policies, which increase water demand, must include arrangements for the removal of waste products. Policies allocating water have to take into account the relative value of water both socially and economically.

In the past the failure to recognize the full value of water stems from the traditional view of water as a free good, or at least one acquired by affordable labor input (Allan, 1995). This has led to water being allocated to projects of low value uses, which provided no incentives to treat water as a limited resource. To extract the maximum benefit from the available resource, the perception of water needs to change and opportunity costs in current allocative patterns recognized. The importance in allocating a scarce resource is to extract the maximum benefit. Charging is applying an economic instrument to affect behavior towards water conservation and effective water use, and helps in ensuring cost recovery and measures the customer willingness to fund additional investment. The goal is full cost recovery for all water uses, with resource management covering operating and maintenance expenditures and capital charges, the opportunity costs arising from alternative water uses, and external factors due to changes in the economic activities of indirectly affected sectors (IWRM, 2000). This cost is not necessarily charged directly to the user, but has then to be borne by society.

Water Resources

In the Arabian Peninsula groundwater accounts for 40-98% of the total freshwater resource. Of the meagre rainfall 1-10% becomes available as soil moisture and ends up nourishing economic crops (Rogers and Lydon, 1994; Abdel Magid, 1995). The major part of surface water arises outside the borders of Arabian Gulf countries (in the Euphrates and Tigris basins). The deficit is made up by drawing on groundwater, during the 20th century the amount has increased sevenfold although the population increase has only increased threefold, currently the wells around Riyadh (Saudi Arabia) pump 2.5 m^3/sec of brackish water from depths around 1,200-1,400m. As the salinity of the water has increased they are now pumped only for mixing with desalinated water. Over exploitation of the Tertiary carbonate aquifers has resulted in a drawdown of 15-45m. The current availability of water in the area per capita is 1,700 m^3, but assuming a 3% rise in population, this will drop to 670 m^3 per capita by the year 2025 according to a World Bank report. Thus, in a region where water is in short supply, it is imperative that a coherent policy be established through a system of permits, regulations and standards to ensure supplies for sustainable development as a basis from which water supplies can be successfully managed. The purpose of such policy is not only to provide an adequate supply of water both in quality and quantity to meet current needs, but one which is adaptable to, and anticipates future needs. It has to be recognized that freshwater is a finite resource as the average yield from the hydrological cycle, a natural capital asset requiring maintenance to sustain the desired services, but one which cannot meet the demands placed upon it. Human activities can reduce the availability and quality through the mining of groundwater, and the pollution of surface and groundwater, changing land use (deforestation, afforestation, urbanization), the temporal and spatial variability of flow and the planned re-use of return flow.

Sustainable development therefore must include a comprehensive and balanced economic system consistent with the needs of the state, and is not just the judicious and rational use and management of water. It has come to be recognized that water is national resource requiring a national authority to set objectives for it is clear that uncontrolled extraction is not in the public interest. The responsibility for water has often come under the control of different ministries depending upon its ultimate use, i.e., one authority controlling potable or commercial water supplies, and yet another responsible for water for agricultural use. The allocation of water resources through permits, laws and regulations according to standards set by competent authorities must reflect the many uses of water. For sustainable development, the frame of reference of the Water Resource Policy provided by the competent authority. It requires the provision of sufficient water to meet current and future needs. A holistic approach is needed to co-ordinate activities at all levels, to ensure that economically, water cost and sustainability are taken into account when making production and consumption choices. Only with the local governments, community organizations which can all indicate their needs for bulk services, can some consensus be obtained, one which will require some sacrifices for the common good. This recognizes some sectors must accept the need for change and the changing value of water. Many past failures which have arisen are attributable to the fact that water has been viewed as a free good or at least there has been a failure to recognize its full value.

The goal should be full cost recovery, and charging is applying an economic instrument to effect behaviour toward water conservation and efficient water use, cost recovery ensuring the customer interest to maintain services. If the full cost is not borne by the user it has to be met from other revenues. Traditionally, the expectation has been that water is free, and sufficient for the irrigation, which provides the food for the population, and there will be a strong resistance to accept that water is to be delivered at cost. There is also unwillingness in the society to recognize the importance of allocation of water for use at all levels, since this generally runs counter to entrenched interests. Yet with water scarcity, water has a delivered cost and may command a variety of prices according to its planned use. Water allocation is therefore controlled by the sector, activity, tract or crop, which will bring a sound return (Allan, 1995).

The relationship between the different segments of water economy are mediated through the political process, but must be based upon sound economic, ecological, and equitable water quality management principles. In the absence of surface water supplies the principal needs are met by the extraction of groundwater. In practice delivery of groundwater has proceeded without clear policy limits, with the result that over-extraction of groundwater in a region already deficient in water, has resulted in falling groundwater tables, and deterioration of water in quality and quantity, with saltwater intrusion near coastal areas, or the uprise of saline waters from deeper horizons.

It is generally recognized that water is a national resource, and therefore requires a national policy to set objectives, but in the past the responsibility for control of water resources in the different Gulf States has been divided among several ministries, such as Agriculture and Fisheries for irrigation and farming, with the Ministry of Water and Electricity responsible for potable supplies for domestic and commercial use, with the involvement of the departments of Commerce and Health to establish safe drinking water standards. The current trend is to have all these responsibilities concentrated in a single ministry as Ministry of the Environment or Ministry of Water Resources.

Less clearly recognized is the trans-boundary movement of subsurface water. Since aquifers cross national boundaries in such cases basin-wide plans for areas such as the Gulf States, where the Tertiary Dammam aquifers serve not only Saudi Arabia but also Kuwait, Bahrain, Qatar and the United Arab Emirates, are essential. Such an issue can be handled by the Gulf Countries Co-operative Council through the formulation of joint strategies. At the international level both the UN Conference on the Environment at Rio de Janeiro in 1992 and the World Commission on Environment and Development (the Bruntland Commission) in 1987 failed to attach much priority to water issues (Stout, 1995). Stout (1995) listed the issues which should be considered to focus attention on water issues, such as the raising and maintenance of a global profile of freshwater, to ensure a sustained commitment to resolve fresh water problems, and to provide advice and recommendations on topics for improving water management, and finally to provide an on-going review of the state of development of resources, and best management practices, that is the optimum allocation of a scarce resource. Membership of any organizing council or management group should include designated representatives of interested groups both national and international, UN agencies, as the World Bank and the development banks (Abdel Magid, 1995).

The International Hydrological programme (IHP) which grew out of the International Hydrological Decade, a UNESCO programme to stimulate research into hydrology and water resources in vulnerable environments (Salih, 1997), selected four major issues for particular consideration. These are the allocation of water, achieving effective implementation, and the environmental and social issues. It also included the land-water linkage.

There is thus a clear recognition of the two main aspects of water policy, and the need for coordinated approach to involve all interested parties. One that is scientific- technological, and the other socio-economic, political and legal. These two aspects are related and there is a constant feedback from the scientific-technical database, which will provide the basis upon which a policy can be formed. Once the policy is formulated, management can operate within the guidelines of established policy. The water policy must not only be designed to meet current needs but also to provide for future development. Water management is therefore a technical issue and operates within the parameters set by the policy. As the water policy affects everyone, it must have general support especially as it involves a change in traditional thought concerning the provision of water and will inevitably involve changes in the pricing structure. However sight must never be lost of the inalienable right of every individual to water, as enshrined in the Shari'ah and the constitutions of the Gulf States.

The principles of water resource planning and management are set out very clearly by Abdel-Magid (1995) using Oman as his example. Water Resource Planning requires a strong institutional and legal framework to support effective management, where management can be defined as the allocation of water resources through permits, laws, regulations and standards set by competent authorities to ensure adequate water supplies for sustainable development within the frame of reference of the Water Resource Policy provided by that authority. It requires the provision of sufficient water to meet current and future needs.

Since any development plan will entail changes affecting the public, it is essential that any plan and its objectives are uderstood by, and has, public support so that ultimately the introduction of a pricing structure, while it may involve discussion over details, will be acceptable in principle. A scaled pricing structure for water according to the amount used represents a major policy change, and most of the Gulf States have a scale of charges (Table 10.1). Initially population centers developed around supplies of water which effectively limited their growth, but as populations expanded in urban centers, and as there were no rivers or streams which could be tapped to augment supplies, recourse was made to the pumping and distribution of groundwater. Initially such water was provided either free or at highly subsidized rates. Since currently most of the domestic supplies of water come from desalination plants, the costs of providing potable water are not negligible the water management decision to recover total costs with a marginal subsidy is a task for the appropriate authorities and is clearly a potent weapon in the armory of social change. Currently water prices are so low that there is little incentive on the part of the user to economize.

Historically, water was perceived as a uniform commodity with the same value in all sectors and the traditional and current view is that access to water is the right of each individual. The Qur'an further lays out that it cannot be denied to anyone, even when in short supply. In the Gulf countries the governments, as representatives of the people, have generally undertaken the task to provide and distribute that supply. Thus, the formulation of a national water management policy is essential, not only with the provision and distribution of water, but to recognize that water has different values in different sectors, and to deal with the complex social, economic and legal issues which can arise from its allocation. The important criterion is to allocate water in those sectors which will bring a sound return. While productive efficiency, i.e. the effective use of unit of water is essential, it is subordinate to high levels of allocative efficency.

Water Policy

The first stage in the development of a water resource management policy, must obviously be the assessment of the available supply, and an investigation of potential new supplies, followed by conservation measures to determine the quantity available.

The second stage is the formulation of a policy, with the assessment of priorities and economic decisions concerning criteria as diverse as allocations and pricing structures. This may prove to be the most difficult phase, since historically water has been free or provided at a minimum cost, with few or no controls on allocation. It will hit the agricultural segment hardest, not surprisingly since agriculture is the biggest user of water (80%), and responsible for the continuing water deficit. There has been a general failure, to recognize the real cost, and the economic value of water services, that is the recognition of water, as an economic good, as well as being a basic necessity and a human right.

Water Policy must also recognize the diversity of water use, and the competing demands of

Table 10.1. Current Water Charges in the Gulf States (sources: Gulf Cooperation Council-Economic Bulletin, XIV, 1999).

Country	Item	Cost Per Cubic Meter	
		Local currency	U.S. Dollar
United Arab Emirates	For all consumers	330 Fils	89.9 Cents
Bahrain	- Industrial consumption	12 Fils	3.2 Cents
	- Residential consumption		
	1st category up to (60 m^2)	25 Fils	6.7 Cents
	2nd category (61-100 m^2)	80 Fils	21.2 Cents
	2nd category (61-100 m^2)	200 Fils	53.2 Cents
	- Minimum monthly fees	1.5 Dinar	4.0 Dollar
	- Non-residential consumption		
	1st category up to (450 m^2)	300 Fils	6.7 Cents
	2nd category from (451 m^2)	400 Fils	1.6 Cents
	- Minimum monthly fees	9.9 Dinar	26.3 Dollar
	- Water irrigation consumption		
	1st category up to (60 m^2)	20 Fils	5.3 Cents
	2nd category (61-100 m^2)	25 Fils	6.7 Cents
	3rd category from (101 m^2)	85 Fils	22.6 Cents
Saudi Arabia	1st category up to (50 m^2)	10 Halala	2.7 Cents
	2nd category (51-100 sm^2)	15 Halala	40 Cents
	3rd category (101-200 m^2)	200 Halala	53.3 Cents
	4th category (201-300 m^2)	400 Halala	1.07 Dollar
	5th category from (301 m^2)	600 Halala	1.6 Dollar
Oman	For any quantity of consumption:		
	Residential and Government buildings	440 Peasa	1.14 Dollar
	Commercial and industrial establishments	660 Peasa	1.72 Dollar
Qatar	One fee for all activities	4.4 Riyals	1.21 Dollars
	Monthly estimated fees for whom supplied by water tankers:		
	For flats and commercial shops	150 Riyals	41.25 Dollars
	For villas and residentials	200 Riyals	55.0 Dollars
Kuwait	Water:		
	Distributed by meter	211 Fils	71.1 Cents
	For industrial establishments at Al-Shuaibah and Abdulla port (industrial areas)	66 Fils	22.4 Cents
	For the tank owners	79 Fils	28.9 Cents
	Desalinated Water:		
	Received from desalinated plants	158.5 Fils	53.9 Cents
	Distributed by meter: (well) water (low salinity)	264 Fils	89.8 cents
	Residential consumption with a fixed price according to the size of the meter	26.42 Fils	9 Cents
	Dairy and commercial farms	5.28 Fils	1.8 Cents
	Government farm	13.21 Fils	4.5 Cents
	Tankers Owners	Free	Free

industry and agriculture, it is needed for recreation and beautification projects, such as the growing of tree and ornamental shrubs, the development of parks and landscapes whose value is not easy to assess. The question of water allocations to meet these competing demands with a supply, which is fundamentally restricted, is a policy decision taken at governmental level, but requires popular support. Further it is possible that some of these decisions may have consequences not readily foreseeable. For example, the construction of retention dams to retain rare storm floodwaters to increase infiltration also retain sediment, which reduces the life of the structure. The retained water may provide a breeding ground for mosquitoes and the hosts of parasites such as Bilharzia. Unfortunately, water is seldom retained for a sufficient length of time to accommodate recreational use.

A comprehensive scheme would require the agency, to have the authority to shut down inefficient wells, and wells yielding water with total dissolved solid values greater than the established norm, and establishing extraction limits on other wells. Yet most systems lack the necessary equipment to measure the amount of water delivered much less the salinity, extraction rates and water levels in all wells. Charges must seen to be fair and efficiently collected with some oversight system to prevent abuse. Under a "public trust" doctrine to protect public rights an allowance must be made for some degree of recharge, which means in effect that extraction should remain below the calculated safe extraction yield. Factored into the control of the costs of well water must be the cost of desalinated and treated water, either as special elements or as part of the overall charge rate.

Water Demands and Supplies

There have been several attempts to assess the past and present water demands of the Gulf States, and project the future demand, which are summarized in a series of tables providing a broad overview of the problems. Table 10.2 looks at the demands over the last thirty years. It demonstrates the continuous growth of demand as the region grew economically and socially. The projected water availability from all sources is summarized in Table 10.3.

The deficit and the exploited water resources for the same year 1992 are shown in Tables (10.4 and 10.5). High demands were placed on the water supply by the demands of agriculture as the region tried to attain food self-sufficiency. In the short term there are three ways in which the deficiency can be met, by increasing the extraction of groundwater, increasing desalination and by the reuse of treated water and by stringent conservation.

Table 10.2. Water demand and projected demand in the Gulf States for the years 1980-2010 (in Mm^3/yr) (compiled from Al-Alawi and Abdulrazzak, 1994).

Country	Year			
	1980	1990	2000	2010
Saudi Arabia	2,362	16,300	23,100	25,300
United Arab Emirates	789	1,490	2,232	2,450
Oman	665	1,236	1,417	1,585
Kuwait	186	383	640	771
Bahrain	138	216	285	315
Qatar	110	194	334	388
Total	**4,250**	**19,819**	**31,629**	**35,395**

Table 10.3. Water availability (Mm^3) for the years 2000 and 2010 (after Al-Alawi and Abdulrazzak, 1994).

Country	Year 2000				Year 2010		
	Treated water (Mm^3)	Desalinated water (Mm^3)	Ground-water (Mm^3)	Surface water (Mm^3)	Treated water (Mm^3)	Desalinated water (Mm^3)	Ground-water (Mm^3)
Saudi Arabia	710	1,289	20,212	900	1,000	1,300	22,100
United Arab Emirates	200	772	1,185	75	250	772	1,359
Oman	50	68	1,072	227	61	68	1,229
Kuwait	80	428	132	—	106	428	237
Bahrain	42	115	93	—	53	141	121
Qatar	43	216	75	—	43	216	129

In the Gulf States the first logical step is in the establishment of a water database, and in the absence of running water this has involved the registration of all well and springs, their yield, water composition and a control of drilling at all phases. For each well, depth is required, well bore, yield at regular intervals to monitor depth to the water table, and the changes in salinity. From these data, maps can be generated to show iso-salinity hydrological gradients, similar control is required over treated water.

In the large states as Saudi Arabia and United Arab Emirates to avoid unwieldy burocracy agricultural or water management districts have been established in which the day to day management rests in the hands of those most concerned. In the agricultural areas where management has rested in the hands of the farmers this technique has proved to be successful, not only in reducing costs but enhancing production. The parent Water Authority can provide a source of funds for the establishment of efficient irrigation systems.

Groundwater provides the principal supply but there appear to be few chances of expanding production and in real terms, in arid environments the recharge is not significant. Although the cost of well head water is low, the main costs lie in the transfer costs, as in pipelines and pumping stations. The principal aquifers are known, as are some of the areal changes in aquifer characteristics due facies changes, which affect the porosity, permeability, and transmissivity of the aquifer. Few areas remain which could significantly supplement the aquifer supply. In most of the Gulf States over-pumping has led to significant declines in water quality and of the groundwater level marked by a rise in the total dissolved solids and increasing salinity. Where "safe yields" have been determined they are less than the extraction rate by a factor of two or more.

In the absence of substantial increases in groundwater, more energy has to be assigned to the increased production of desalinated water, this essentially equates to the construction of new desalination plants. However a doubling the output of the present desalination plants still would fail to meet water requirements even at the current usage rate, much less the projected needs. There are two main processes for the production of desalinated water, the multi-stage flash process (which currently provides about 90% of the potable water supplies in Saudi Arabia) and the reverse osmosis process. Although production costs have decreased in recent years due to a combination of competition and improved technical efficiency, the cost of new plants remains high as does the upgrading of existing facilities. If there were ever a surplus production, the water in excess of requirement could be use in aquifer storage and recovery programs. Tests have shown that the water upon recovery may only need some disinfection, and it has the added advantage of retarding salt-water intrusion.

The only other substantial source of water is from treated wastewater, although there are strong social feelings against its use. However in areas where it has been used in agriculture, the social bias against its use has declined. Treated water costs about a fifth of the cost of desalinated water.

Table 10.4. Renewable water resources (Mm^3/yr) estimated in the Gulf States as of 1992 (compiled from various sources see also Gleick, 1993; Al Alawi and Adbulrazzak, 1994).

Country	Surface water (Mm^3/yr)	Groundwater Recharge (Mm^3/yr)	Current water Exploitation (Mm^3/yr)	Present water deficit (Mm^3/yr)
Saudi Arabia	3,210	2,340	14,430	12,090
United Arab Emirates	150	120	1,000	880
Oman	1,450	475	728	253
Kuwait	—	160	114	—
Bahrain	—	112	140	78
Qatar	—	50	185	135

Table 10.5. Exploited water resources (Mm^3/yr) as of 1992 (compiled from various sources see also Gleick, 1993; Al Alawi and Adbulrazzak, 1994).

Country	Groundwater		Desalinated water		Recycled water		Total
	Mm^3/yr	%	Mm^3/yr	%	Mm^3/yr	%	(Mm^3/yr)
Saudi Arabia	14,430	93.5%	795	5.1%	217	1.4%	15,442
United Arab Emirates	900	69.0%	342	26.2%	62	4.8%	1,304
Oman	645	93.8%	32	4.7%	10.5	1.5	687
Kuwait	80	19.8%	240	59.6%	83	20.6%	403
Bahrain	160	71.4%	56	25.0%	8	3.6%	224
Qatar	144	57.6%	83	33.2%	23	9.2%	250

Currently its main use is in horticultural and agricultural projects, and as a source of water in municipal projects as fountains and green areas. In agriculture its higher organic content may even provide necessary salts and fertilizer. With higher treatment levels it can also be stored in aquifer storage and recovery projects. As urban areas increase in size and the use of treated water becomes more socially acceptable, considerably greater volumes can be expected from this source.

Natural rainfall, although sporadic is often concentrated in heavy storms of short duration. Such storm frequencies are about every five years. The water can be trapped by the construction of retention dams, but the recharge is low and losses from surface evaporation are high, nevertheless 164 dams have been constructed in Saudi Arabia, despite the non-economic construction costs. Other small-scale projects rely on the entrapment of surface rainfall in sediment filled hollows or wadis, even where the rainfall is as low as 50-60 mm/yr. In an analogous manner as rain is collected from runoff on a roof in a rain barrel, rain water runoff from buildings or concrete surfaces can be trapped and stored in underground tanks.

Clay-lined dewponds, which are fed through the condensation of night dew from saturated air, have been in use since Neolithic times. Dew in the near coastal areas is capable of maintaining scrub vegetation a few kilometers from the coast. The construction costs of dewponds are modest, but it is uncertain how much water can be generated in this way, even in a small village community. This water can be used for washing, cleaning and irrigation (see Joudi and Fok, 1999). Mohamed and Al Zubari (1999) describe the collection and recycling of condensate water from the system in operation at Bahrain airport. Condensate is collected from 14 large air conditioning units and stored after filtering in a 27,000 litre tank. The water is treated with anti-corrosion and anti-scale reagents, and is used to supply the toilets and floor washing and the demands of the catering services for non-potable water, all of which represents a saving of desalinated water. The cost of installation when incorporated in the original construction plans and the operating costs are offset by the water economy costs but the profit can only show after long-term use. The method can be applied to the major hotels, apartment complexes and shopping malls, with long-term savings, which could be appreciable.

Within the Gulf States there are no significant alternative water sources. The importation of water has been discussed, with particular reference to the Turkish Peace Project, which involves piping water from Turkey to the Gulf States. The major stumbling block is the cost for the collection and distribution of the water, particularly in times of unrest and political in stability in the countries involved. It is unclear what the effect on the water balance in Turkey would be if that the volume of water imported and whether sufficient to satisfy the exported demands of the Gulf States. Technically the project is feasible, if costly, based upon the experience of the "Great man-made river" in Libya. However, it could be looked upon as a tremendous opening for international co-operation. Other ideas have been proposed from the towing of icebergs from Antarctica to the importing water either in giant tankers or towing submersible plastic containers. Technically these projects are possible, but insufficient attention has been given to their economic feasibility.

The situation remains that, if further deterioration of groundwater is to be prevented, additional measures will have to be taken, and most schemes seem unlikely to provide significant contributions to that end. Clearly, natural groundwater recharge cannot occur under the present climatic regime, and current water usage. Every scheme proposed to increase supply, will require major investment taxing the resources of even the richest of the oil producing countries and point toward the need to establish some of levy on water use.

Water Resource Assessment

In the Gulf region, the water supply cannot be maintained by the uncontrolled extraction of groundwater, even with the use of all other available water resources, such as desalination and treated wastewater. In the absence of an established policy, water shortage will be expected, which can largely be attributed to the agricultural demands, as stated above. Over much of the region, the seasonal rainfall is small in quantity, so that water harvesting is seldom a viable option. In the absence of running water, except on the rare occasions of floods following rainstorms, the assessment of available supplies, requires a basin study, of both the natural discharge and recharge areas. Technically, the geological setting, basin and aquifer extent and their hydrological properties, are required on an "on-going" basis, since some factors may change with time and in space. The basic hydrological data cover precipitation, evaporation, surface runoff drainage patterns, topography and soil studies, groundwater flow and the hydraulic conditions of groundwater flow, transmission storage and leakage. Simulation modeling is useful, and a number of models exist because of the diversity of the parameters ranging from homogeneous isotropic flow to inhomogeneous, anisotropic conditions.

The assessment of water resources can be considered in four phases:

a) The first phase is an inventory of all wells with the relevant technical data as depth, movement of the groundwater, water composition (covering both quality and quantity), and aquifer characteristics. Details of all springs and falajes is needed, their yield and composition, as well as technical details of all desalination plants in operation, their costs and production figures, and the amount and status of all treated water facilities. The effect of dams for water retention and increasing groundwater recharge will be harder to estimate, but cannot be ignored. The presence of standing water must also be evaluated as a health risk for malarial mosquitoes and *Bilharzia*. Even minor sources as dewponds and water from air conditioning units should be included. The potential of increasing of water supply and finding new sources will have to be incorporated in the policy. Since aquifer characteristics are related to aquifer lithology, it is essential to carry out the difficult task of estimation of potential changes in the aquifers themselves, and the consequent changes in water quantity and quality.

b) The second phase will cover conservation techniques, which include control of drilling and extraction rates, the potential of water harvesting, and the construction of underground and surface dams.

c) The third phase will be based on the knowledge derived from the water inventory to determine the cost, and some projection of present needs, and future demand, based upon an accepted growth rate.

d) The final phase is the methods of increasing supply by importing water, which range from towing icebergs or water filled plastic pods to diverting river water long distances. The other means of importing water in the form of importing food, especially grain since a ton of grain requires a thousand tons of water to produce it. Reducing production of crops with heavy water demands would mean sacrificing food self-sufficiency, very much a political issue.

Principal Water Sources

a) Groundwater

Groundwater is the principal source of supply of water, but there appears to be few chances of expanding production without a significant risk of increasing the total dissolved solids (TDS) and a decline in groundwater levels. Although wellhead costs are low, the costs of transfer and distribution are high (such as pipelines and pumping stations). Once the capital costs are met, running expenses are much less a significant item in the cost structure. The principal aquifers are well known, but potential changes in aquifer characteristics as porosity, permeability, transmissivity and composition can arise from changes in aquifer lithology. In the Arabian Gulf region, few areas remain in which significant, new aquifers could be located. For some aquifers, the safe yield has been calculated to determine extraction limits, but even when this has been done the actual extraction has exceeded the safe limit by at least a factor of two. Whilst this should have ruled out excessive increases in groundwater extraction, in the absence of effectively enforced legislation, it seldom seems to have done so.

b) Desalination

Desalination of brackish water or seawater, provides the obvious means of increasing the available supply of potable water. The two desalination technologies commonly practiced in the Middle East are multi-stage flash desalination and reverse osmosis. The first is the traditional leader in the region largely because it was the only method commercially available in the mid-sixties at the time of increasing demand in the Gulf States. Future expansion probably lies with reverse osmosis. It is expected that in the early years of the 21st century the volume of water produced by this technique would equal or exceed that produced by multi-stage flash desalination. Cost reductions have been achieved through improved heat transfer. The successful use of chemicals to reduce scaling, which when combined with thermal processes and improved membrane efficiency will achieve competitive pricing between the two technologies. Moreover membranes can provide an absolute barrier to *Cryptosporidium* and *Giardia* cysts. Microfiltration and ultrafiltration which use porous membranes, remove not only suspended particles, but also bacteria and viruses, are used to pre-treat wastewater in reverse osmosis systems, used in industrial situations.

However a doubling of the present output of desalinated water still fails to meet even current requirements. Although production costs have decreased in recent years, due to improved techniques and competition, the cost of new plants still remains high. For example, a Tampa Bay, Florida, facility using reverse osmosis quotes unit water costs at $1.71 per 1,000 gallons (3.8 m^3). The cost of the water produced in the Gulf States are strongly affected by energy costs, the cost per cubic meter of water can rise from 0.6 cents to 2.5 cents, when natural gas is replaced by oil in multi-stage flash distillation systems (Al Alawi and Abdulrazzak, 1994), which is still less than the 0.45 cents cost of the Tampa plant.

c) Wastewater

The only other substantial source of water is treated wastewater. Although there are strong social feelings against its use, however where it has been used, in urban projects, horticulture and agriculture, the social bias has declined. Wastewater with secondary treatment, i.e. biological oxidation and disinfection, are being used for groundwater recharge of nonpotable aquifers or wetland augmentation. Its value is enhanced in that it may contain a higher than usual organic content, and may thus provide some salts and fertilizers, depending upon the procedures to which it has been subjected. Its attraction is in the cost factor, since it only costs one fifth that of desalinated water. As the urban areas expand greater volumes of wastewater can be expected with a decline in social bias against its use. Treatment is becoming progressively more critical because of the high proportion of untreated domestic and industrial effluents discharged directly into water courses, irrigation canals and drainage ditches. When tertiary treatment is applied appropriate uses of wastewater include landscape and golf course watering, toilet flushing, crop irrigation and indirect potable use, as in groundwater recharge, and surface reservoir augmentation. In coastal areas, injection of wastewater is particularly useful, since it can create a barrier to salt water intrusion.

Conservation on Water Supply

This is the strongest economic factor at work in the provision of water, for the statement that water is in short supply must be qualified. There is no shortage of water, the shortage of water refers to water of a quality suitable for domestic and agricultural use, and if money were no object desalination could provide water for all needs, and desalination techniques are well known, and have been in operation in the Gulf area for decades. The source of the problem lies in the cost of providing water, to supply the rapidly growing population and the expansion of agriculture. This has led to the over-drafting of groundwater, with the consequent deterioration in water quality, and the fall in groundwater level.

The traditional view is that access to water is the right of each individual, and the Qur'an further lays out than it cannot be denied to anyone, even when in short supply. In the Gulf States the governments, as representatives of the people, have generally undertaken the task, to provide and distribute that supply. Thus, the formulation of a national water management policy is essential, not only with the provision and distribution of water, but to deal with the complex social, economic and legal issues which can arise from its allocation. The issues involve not only national policy, but since aquifers cross-national boundaries, international accords are also essential.

In many ways, effective conservation will require a fundamental change in the popular view of water. It remains a vital commodity to which all have access, but not virtually cost free. The economic value of water, has changed since people lined up at the well or spring, as a result of the growth in population, and the demand for improved services. Traditionally, water was provided virtually free, or at a minimum cost, however, the distribution costs, and treatment are high, especially when the costs of desalinated water are factored in, not only the cost of water production and treatment, but the costs of equipment and distribution, so water has an economic value and there is competition for the supply from industries, which can show a greater return on capital than agriculture. The costs to consumers have been recognized, and rate charges introduced by the various states still, only represent a small percentage recovery of the total costs.

In the absence of an increased water supply, the maximization of existing provides an effective way of increasing supply. The range of conservation techniques available range from general economic controls to a variety of practical techniques.

Broad economic controls can be applied to both domestic and industrial supplies, since they are usually piped, and can be monitored at both the supply end and user end. However, the installation and monitoring of domestic and industrial supplies is far from complete, and there is no simple means of mandating domestic use, which would not impinge upon individual rights. The economic cost to the consumer holds the key to water usage, and most States have introduced a scale of charges. However these rates represent only a small proportion of the total cost, and fail to cover provisions for future supplies, the costs of treatment, the costs of desalinated water and salinity treatment. These charges are levied primarily on the domestic and industrial supplies; very few charges are levied directly upon agriculture, which uses the bulk of the available water (Table 10.1). The most practical conservation measures, which can be applied to piped supplies, is in the detection and repair of leaks. Supplies may be enhanced by water harvesting which, by retarding wadi flow following the rare storms by the use of dams and slowly releasing the accumulated water, improves seepage into the underlying aquifer. Desalination is now used in most countries, to provide potable domestic supplies, and waste water treatment provides water mainly for the irrigation of parks, fountains and landscapes in general as well as in agriculture, and as urban centers grow this amount can significantly increase. The sludges from the treatment of wastewater are used as fertilizers or soil conditioners since they still have an appreciable organic content and some soluble salts.

To conserve water supplies, the repair and maintenance of the distribution system can provide significant savings, but perhaps the most important factor is the ability to restrict well drilling, through the issuance of permits, and the restriction of the number of wells and the extraction rates in a given area. This can include shutting in wells where the water quality is low.

The three major users of water in order of magnitude are agriculture, domestic and commercial. Of these the easiest to which to apply conservation methods are the latter two, since the supplies are usually piped, and therefore easier to monitor through the use of flow meters, at both the supply end and the user end. The greatest losses occur through leaks especially older network systems still in use. Loss of water through leakage could amount to 25% or more of the water supplied, and even in some of the newer, and presumed more efficient systems, the losses may still be as high as 15%. Hence detection and repair of leaks becomes a major item in conservation.

There is no practical way in which conservation in domestic and commercial supplies can be mandated, but a graded scale of user charges can provide a powerful stimulus to conserve water, and experience bears this out (Table 10.1). Water has to be recognized as a natural resource, whose value has increased enormously, and one whose cost/price ratio must be adjusted to reflect contemporary values. Economic pricing is not just a question of varying rates but reflects water policy in allocation and involves acceptance of the changed concept of water, from being a virtually free commodity, to one with a value consistent with that of a commodity under stress. It requires pricing, which allows for production and distribution costs, and an element towards recovery of the basic capital costs. The actual value placed upon it is a policy matter according to the country's priorities and will affect the whole state.

The conservation of agricultural water supplies can be affected in a variety of ways, of which the most effective would be the switch to modern irrigation techniques. More than half the water distributed by open channel and flooding techniques fails to benefit the crops as intended, even the traditional falaj channels are covered to reduce evaporation. Much less water is required for drip and sprinkler irrigation which can be timed for those parts of the day when losses from the evaporation are lower, and the water can be directed directly to the crops. The installation of a subsurface network of perforated pipes, which delivers water directly to the roots of individual plants with negligible loss, is 90 to 95% efficient as compared to 50-70% for flood and furrow systems. Where there are figures on its effectiveness it shows a net reduction of water demand from 30-70% and 20-30% over the center pivot system, combined with an increased crop yield of 20-90%. Such systems are costly to install, and only large farmers have used them.

A comprehensive scheme to cover agricultural uses of water is more difficult task, since not only is it concerned with basic hydrological issues, but also with crop policy, whether to sacrifice the objective of food self sufficiency, to one of food security concentrating on crops less demanding of water, and yielding better market prices (Ibnouf and Abdel-Magid, 1994). In Saudi Arabia for example, wheat support prices were reduced with the objective of reducing production to a level meeting national requirements (Dabbagh and Abderrahman, 1997). Wheat and sugarcane production was curtailed or replaced, because of their high water requirements by the production of fodder crops and vegetables for local markets, which also yield a better price, and can be stimulated by the use of subsidies in the initial stages of change.

The most effective cultivation is in a controlled environment where heat light, air and soil temperatures and humidity, can all be monitored, a system particularly suited to growing tomatoes and vegetables for the local market. For open field crops, soil water retention can be increased through mulching, and the use of soil conditioners as bitumen emulsions. Transpiration losses can be reduced through the growth of wind breaks, and there have been a number of experiments to determine the most effective trees. The trees can also provide a supply of timber generally lacking in the Arabian Gulf States.

Other alternatives involve crop selection of more productive varieties, and varieties less sensitive to higher salinities, or even a reduction in the cultivated area. The cost structure applied to agriculture cannot be the full economic cost, because of the effect on the stability of food prices. However it cannot continue at the present low level, which does nothing to stimulate economy.

Water Legislation

The ultimate objective of water policy is to have an economically sound, and equitable system under which the intake covers, not only the running costs of the system, but also provides funds to cover future developments. Legislation provides the framework upon which the policy rests and exists, as the final arbiter to resolve issues, which defy solution any other way. The legislation must therefore be comprehensive, and consistent with the Shari'ah and Islamic Law, as required by the constitutions of the Gulf States. It must cover situations which could not be anticipated at the time Islamic Law was established, and consequently must follow the steps laid out in the medhabs, following the precept that it harms none. It and must be clear and socially accepted.

The water legislation requires a vast code of regulations which can be summarized below:

1. They must permit the operating authority to close down well, where too many exist in a close proximity, and where water quality has deteriorated.
2. There must be a means of assuring that standards are met, and a sequence of penalties for non-compliance, sufficiently severe that compliance is better than noncompliance. It must also provide for water allocations and charging.
3. The laws must cover ownership rights, and the transfer of ownership.
4. They must control the function, and jurisdiction of water and water-related organizations, in both the public and private sectors.
5. There must be a code for the prevention, control and mitigation of harmful effects ranging from flooding to pollution and salinization.
6. They must provide for the establishment of water conservation zones and for research where needed.
7. There must also be laws dealing with the treatment of wastewater and the disposal of brines both from desalination plants and oilfield operations.
8. The law must provide a regulating agency with the authority to close down wells, with low productivity, or with water salinities, outside an acceptable range, to impose penalties for non-compliance, and to establish a scale of water charges.
9. Above all, the legal code must have built-in enforcement schemes, to establish penalties for non-compliance, without ever losing its role as a court of appeal and adjudicator in questions, concerning conflicting interpretations of the law.

The laws must cover ownership rights, and the transfer of ownership. They must control the function and jurisdiction of water, and water-related organizations, in both the public and private sectors. There must be a code for prevention, control and mitigation of harmful effects, ranging from flooding to pollution and salinization. They must provide for the establishment of water conservation zones, and for research where needed. There must also be laws dealing with the treatment of waste water, and the disposal of brines, both from desalinization plants and oil operations. Above all, the legal code must have a built-in enforcement scheme to establish penalties for non-compliance sufficiently severe to make compliance preferable to non-compliance without ever losing its role in a court of appeal and in adjudication on questions involving conflicting interpretations of the law.

The beginnings of a legal system have already started, for during the 1980's decrees regulating groundwater extraction were established in Saudi Arabia. In Bahrain decree 12/1980 called for the regulation of extraction and a reduction in the rate of quality deterioration, followed in 1982 by a further decree regulating extraction and the installation of water meters. In 1991 the Dubai Emirate established environmental protection regulations and in 1993 in the United Arab Emirates the Federal Environmental Agency was created by law 7/1993 which was assigned the role of protecting water resources, and establishing water standards. The following year in Qatar under decree 13 water law no 4 was amended transferring the Environmental Protection Committee to the Ministry of Municipal Affairs and Agriculture. In 1995 in Yemen a law was promulgated which called for a national water resources authority to establish policy, strategy and plans. These activities show increasing concern with water problems although most of the decrees had limited coverage, and relied upon the regulations and requirements of drilling permits as a means of controlling groundwater extraction, in effect there was no comprehensive water code. Metering extraction is not yet widely spread and penalties for non-compliance tend to be too low to be effective deterrents.

Where a Water Authority has been established, to reduce an unwieldy bureaucracy, in the larger countries such as Saudi Arabia, Oman and the United Arab Emirates local agricultural or water management districts have been established. In these districts the day-to-day management rests in the hands of those most concerned. Modeled after the traditional communal, management systems where domestic and agricultural demands are met (Salih, 1997) by village co-operatives based upon inherited and acquired rights, they determine domestic and agricultural requirements. Over the centuries this system has proven its value in the maintenance of the falajes, and the handling of supply problems. Where modern cooperatives have been established, following the traditional system they have proved to be successful in reducing costs and enhancing production calling upon the regional authority to provide funds or subsidies for the installation modern efficient irrigation systems, as sprinkler, drip or bubble techniques. Funds can also be provided to subsidize maintenance of the distribution system, to reduce leaks, which can cause water loss until the system is sufficiently well established, to maintain itself through water charges.

Policy questions arise because of contamination whether from point sources as municipal or industrial wastewater, landfills, storage tanks and from non-point sources as agricultural fetilizers and herbicides, or soil leachate or saline water intrusion. Such contamination is difficult and expensive to remediate, and the effects may prove to be irremedial. Under such circumstances preventive measures have to be developed. A zonal protection scheme is outlined in the UNESCO/ACSAD report (1990).

Zone 1 (The well zone): 10-30m around the well. In this zone no human activity is permitted except that necessary to run the water supply. This includes all traffic, agriculture, application of all fertilizers and pesticides and organic fertilizers.

Zone 2 (The inner Protection Zone): the size of the zone is designed to prevent any pathogenic contamination, and is based upon the flow rate, usually involving a time of 50-60 days. No building or farming, no horticulture or allotments are permitted within the zone; this includes roads, parking, cemeteries, fuel storage, fishponds, carwash or oil changes. Organic fertilizers are permitted only if they are spread immediately

Zone 3 (The outer protection zone): protects the entire catchment area from non-degraded or poorly degraded chemical pollutants, and radioactive pollutants and all activities related to these. Open-air storage of chemicals, waste disposal, and sewage plants are banned from the zone as are all industrial activities where potential pollutants are used or produced.

Projected Energy Conservation (Towards a Partial Solution)

Even when all the possible sources are well managed, and all the conservation methods are emplaced, and the social costs adjusted, it is still clear that the supply is inadequate to the increases which the standard of living will demand much less the question of future growth. Currently, only desalination holds any promise for meeting the shortfall, which means the construction of new plants with the attendant capital costs and running costs. The sources of power within the Gulf area have been the seemingly limitless supplies of natural hydrocarbons, whose continued use spells increased danger to the atmosphere and the Earth's environment.

The way out of this impasse may lie in more fully utilizing the natural source of energy derived from solar radiation. Several methods are possible, among them are:

1. Directing solar energy by parabolic mirrors on to a central collecting tower, where their heating effect can create steam to drive a turbo-electric generator. There are large uninhabited areas, which like most of the rest of the country get an average of nearly 10 hours a day direct sunlight. The mirrors are mounted such that they can track the sun's position in the heavens for maximum effect.
2. Thermo-electric panels of the kind used to power space craft and a myriad other uses, from traffic lights to watches. Despite their reduction in cost, these are still too expensive to be used in the quantity needed.
3. Wind energy, which is finding increased applications in the power generation field from California to Denmark and China. The ancient system initially milling grains and driving pumps to drain the Dutch polders or machinery in the early mill towns has been updated by the development of aerodynamically designed vanes. In Germany and Denmark wind energy supplies more than 10% of the energy needs and the amount is increasing (Flavin, 1981).

In the United Arab Emirates, for example, two areas where a trial could be made of the prospects for generating wind energy for agricultural use are the area around Al Ain, because of its food producing importance for dairy products, and the Al Fujairah Emirate because of its production of vegetables and its compact size. Both are areas with significant water shortage problems. Bakhit and Nairn (1997) calculated that with a 10% reduction in the cultivated area of Al Fujairah Emirate could balance its water budget with the construction of an additional desalination plant. In an Emirate without hydrocarbon resources running costs could be minimized through the used of such natural energy sources. Al Fujairah Emirate and Al Ain city are both are closed to the Oman Mountains, so that there must exist winds of sufficient velocity, the run wind dynamos most of the time. That energy could then drive a desalination plant. One of the first research items will have to be the establishment of climatic platforms to provide continuous records of wind velocities. Such climatic platforms have been designed and successfully used in the USA (Breed and Reheis, 1999). Although the velocities of storm winds are probably well known, the only continuous records of wind velocities in the Gulf area are from airports which are not normally located in areas characterized by high winds. The effect of blown dust and sand will also have to be measured, to determine the potential erosion that the dynamo blades could suffer.

Some order of magnitude figures can be generated for wind energy plants, and an estimate made of the costs involved in generating enough power to drive a desalination unit. In Flavin and Jenssen (1994) initial costs including installation of the older type wind dynamos was between $1,000 and $1,200 per kilowatt with the average generating cost of 7 cent per kilowatt per hour, scaling these numbers to the costs of producing 500 MW gives a cost estimate of $500,000, which will then be used to produce 2,000,000 gallons/day with a value, based upon production costs, of $9,000,000 or a little under a 2% return on investment assuming no "profit" is intended to be added to the basic production cost. With the newer machines entering the market, capable of generating 300 to 750 kilowatt per turbine, the number of turbines to be installed obviously will fall, and generating costs may be reduced to around 5 per kilowatt hour i.e., comparable with electricity generated by more traditional means (Davidson, 1991). The costs of producing fresh water by desalination varies according to the salinity of the feed stock, with brackish water costs are only one quarter the cost of desalinating seawater and the preceding rough estimates are based upon normal seawater desalination.

Wind strengths are notoriously variable, and even in the best sites selected are liable to fail periodically, so for continuous desalination there will necessarily have to be a back-up power source. At the present time that would seem to be oil or gas generated electricity. At a future date it may be possible to set up a heliostat field of parabolic mirrors focusing solar energy of a central tower creating steam to power turbines. Since this is restricted to daylight hours, the energy produced could be fed into a grid system reducing the electrical demand on oil-fired power plants during daylight hours, thus permitting more to be focused upon desalination at nights, when the demand is traditionally low.

Future Conservation Policy and Rational Plans

The major problem, which the Gulf States face in their progress towards self-reliance in agriculture, industry and domestic consumption, is the development of their fresh water resources. Unfortunately the groundwater resources of the Gulf States are very limited and groundwater withdrawal has greatly in exceeded recharge over many years. It is important to stress groundwater management and the acceptance that water is a commodity in short supply, and therefore, has a distinct value and cost and is not something available on demand in any quantity at virtually no cost. The charging of water supplies, even if not at the full economic rate, must be related to the cost of production and distribution and to quantity. The scale of charges may vary according to the use of the water, the current water charges in force in the Gulf States are listed in Table 10.1.

To achieve this end there is a growing need for an awareness program of the problems involved in the conservation and rational use of water, and the protection of groundwater to support what may prove unpopular economic policies. The awareness program should attack all educational levels, from elementary school to adult media programming. It should cover the relevant parameters grouped below:

A. Conservation or resources
   i) Special education programs on the best methods of exploitation.
   ii) Controlled well drilling.
   iii) State established scale or scales of charges for fresh water.
   iv) Modern techniques of water distribution and continuous maintenance.

B. Establishment of agricultural guidelines
   i) Training on modern irrigation techniques, which reduce water demand and enhance production.
   ii) Application of modern drip and sprinkler irrigation techniques.
   iii) Establishment of shade trees resistant to drought and salinity to protect plants and reduce evapotranspiration.
   iv) Encouraging the use of greenhouses.

C. Future studies
   i) Development of a data base for all the Gulf States which preserves all meteorological data, annual surveys of all wells including information of depth, radius, piezometric level, static level, water type and potential aquifer lithology and soil analysis.
   ii) Long term statistical studies for water supply.
   iii) Studies to improve recharge such as introducing dams and water drainage to direct water into catchment areas.

D. Maintaining supplies
   i) Desalination of brackish groundwater and/or seawater at minimal expense through reverse osmosis, or utilizing natural gas or solar energy.
   ii) Recycling sewage water following accepted treatment standards set by the World Health Organization.

Chapter 11

NUMERICAL MODELING OF CERTAIN AQUIFER SYSTEMS IN UNITED ARAB EMIRATES, SAUDI ARABIA AND KUWAIT

INTRODUCTION

Much effort in recent years has been devoted to developing predictive models of groundwater flow and solute transport as a tool in groundwater assessment. The computer program developed by the US Geological Survey (McDonald and Harbaugh, 1988) has been applied in the Wadi al Bih area of the United Arab Emirates. Al Sulaimi and Akbar (1999) described groundwater flow in Kuwait using the VTDN program of the Bureau de Recherches Gèologiques et Minières (Mukhopadhyay et al., 1994) and in Saudi Arabia Yazicigil et al. (1986) looked at the Dammam aquifer using USGS two dimensional groundwater flow program of Trescott et al. (1976). Four situations have been modeled, the Paleozoic and Quaternary aquifers in the United Arab Emirates, the Dammam aquifer system in Saudi Arabia and the Cenozoic aquifer system in Kuwait.

1. Groundwater Flow Model of the Wadi al Bih Aquifer in Northern United Arab Emirates

The importance of the Wadi al Bih aquifer lies in the fact that it represents the only source of water for Ras Al Khaimah city prior to the construction there of a desalination plant in 1998. The Wadi al Bih catchment area covers 475 km^2 and receives an average rainfall of 155 mm (Ministry of Agriculture and Fisheries, 1993), however the amount of rain that can fall in a single storm can exceed the total rainfall of a dry year (35.8 mm recorded in 1984). On December 16th 1995 about 55 mm were measured at the Al Bih station and on March 12th 58.8 mm was recorded at the same station and on December 30th 1979 a flood of 4.5 Mm^3 was recorded at Al Burayrat station downstream in the wadi (Abul Enien, 1996). The estimated annual recharge for the Wadi al Bih aquifer is 9.57 Mm^3. The area includes three water well fields, two located within the Wadi al Bih basin containing 38 wells. The third well field is located in the Al Burayrat area. The daily production from the well fields varies between 300 m^3 and 1,700 m^3/d, with a total production during the period 1990-1995 of 57.8 Mm^3.

The aquifer itself consists of fractured limes tones of the Ru'us al Jibal Group overlain by about 30m of gravel. It contains both old and recent water with a hydraulic conductivity of 50 m/d, a transmissivity of 2,027 m^2/d and a storage coefficient of $5x10^{-3}$ deduced from pumping tests. The hydraulic head decreases from +30m (mean sea level) near the main dam of Wadi al Bih, to -5m at Al Burayrat, the flow being westerly towards the Arabian Gulf (Fig. 11.1).

The groundwater flow model was constructed from data from the two well fields in the eastern part of the basin, and covers the area between the main dam, and the Al Burayrat field (which is not included in the model), in the west. The simulations were carried out using the Waterloo Hydrogeological software Visual Modflow version 2 for the year 1995, which includes the flow package of McDonald and Harbaugh (1988). The objective is to characterize the present groundwater conditions in the Wadi al Bih, and predict the amount, extent and rate of water table, fluctuation under natural and man influence stresses.

Table 11.1. Zone budget output and mass balance calculated for the Wadi al Bih aquifer at the end of the calibration period (1991–1995).

Parameter	Zone budget (m^3/day)		Mass balance (m^3/day)			
	Input	Output	Cumulative volumes in	Cumulative volumes out	Rate in	Rate out
Storage	0	0	0	0	0	0
Constant head	60,446	35,741	60,446	35,741	60,446	35,741
Wells	0	24,705	0	24,705	0	24,705
Recharge	0.1162	0	0.1162	0	0.1162	0
Total	60,446	60,446	60,446	60,446	60,446	60,446

Table 11.2. Field measured hydraulic heads (m) above sea level used in the calibration of the Wadi al Bih groundwater flow model (elevation B1 changed from 103m to 108m, cf Ministry of Agriculture and Fisheries report 1993). Location in Fig. 11.1.

Year	Water-table elevation (in meters relative to sea-level)				
	Well B1	Well B1	Well B2	Wee B3	Well GWR11
January 1991	13.99	18.99	20	-5.08	-4.1
January 1996	14.82	19.82	20.4	-4.89	-5.08

Table 11.3. Field-measured hydraulic heads (in meters relative to sea-level) in Wadi al Bih and Al Burayrat area in April 1996. Location in Fig. 11.1.

Well Number	Hydraulic head (m)	Well number	Hydraulic head (m)
W29	22.61	W5	26.90
W17	22.16	HD	15.36
W10	22.97	RF	21.69

The model assumes a single layer limestone aquifer, overlain by almost dry gravels, which do not contribute to groundwater flow. Groundwater flow is assumed to occur mainly through fractures in the bedrock. The aquifer is assumed to have a constant thickness of about 40m. Groundwater flow is assumed to be isotropic horizontally with conductivity the same in the x, y and z directions. The model was calibrated for five years, from January 1991 to December 1995. The assigned model boundaries were based upon geological and hydrological data for the modeled active cell for which all the climatological and hydrological parameters were calculated. The rest of the basin, and the surrounding mountains, and the alluvial fan, were considered inactive (Fig. 11.2).

Based upon field measurements of water levels, the main Wadi al Bih dam, and the subsidiary dams, were considered to have a constant head of 30m while the cells occupied by the observation well (B3) in the Al Burayrat area, was assigned a constant -5m head (Fig. 11.3). The recharge flux is assumed to pass through the dam, so the cells downstream of the main and secondary dams, are considered to represent recharge boundaries.

Data required by the Visual Modflow package, were prepared in different files. The selected size of each cell in the finite-difference grid was 100m x 100m. The elevation of the modeled aquifer layer varied, between 100m below msl to 120m above msl, to ensure a full representation of all the cells. Well locations were specified by importing a base map in the *.dxf format including the geographic coordinates of the pumping and observation wells, and surrounding features. Each well screen was selected to cover 40m of the well's length, starting from the bottom. The pumping rate for each well during the stress period (1,826 days) was recorded. The water levels in the observation wells at the beginning and end of the stress period were also recorded. Data from the observation wells was used in model calibration.

The rate of recharge flux for the whole area was represented by 10% of the mean annual rainfall, which averages 155 mm/yr and the annual recharge through the main and secondary dams is calculated at 9.57 Mm^3. Based upon the pumping test results, the average hydraulic conductivity (K) of the Wadi al Bih aquifer is estimated at 67 m/d ($7.75x10^{-4}$ m/s), for the x, y and z directions. The specific storage capacity (Ss) is calculated from the relation Ss = S /b, where S is the storativity and b is the aquifer's saturated thickness, and is estimated as Ss = $2.09x10^{-5}$ m^{-1}. The specific yield (Sy) was not measured for Wadi al Bih and the representative value for fractured limestone aquifers was used (Sy = 0.14, Fetter, 1988), with an average porosity of 0.30.

All the input data in files of specific formats were used, to run the Visual Modflow program. At the end of the simulation period (1991-1995), the hydraulic head showed a drop from 35m to 25m, in the eastern part of the modeled area (Fig. 11.4), which directly followed simulating the introduction of the Wadi al Bih well field, and the pumping of its 30 wells. The drop can also be explained in part, by limited recharge the modeled aquifer receives. Westward the head declined until it reached sea level, at the mouth of the wadi. In the Al Burayrat area, the groundwater level is now below sea level (-5m). This situation induces the intrusion of salt water, from the Arabian Gulf into the fresh water in the eastern part of the region. The streamlines drawn by the program show a strong contribution from the secondary dam (Fig. 11.4), which can be explained by the preferential movement of groundwater, through existing subsurface fractures in the area. The zone budget output and mass balance calculated from the model is summarized in Table 11.1.

The constant head values of the main and secondary dams, were adjusted to obtain the best fit between the calculated and observed heads (Tables 11.2 and 11.3). The figures adopted for the cells occupied by the main and secondary dams, were taken at 30m. The measured head values, and program calculated heads are the same in three wells, but differs in one, and even with an elevation correction to within 1m.

The model was used to predict the changes in hydraulic head, as a result of groundwater development in the Wadi al Bih, in the decade

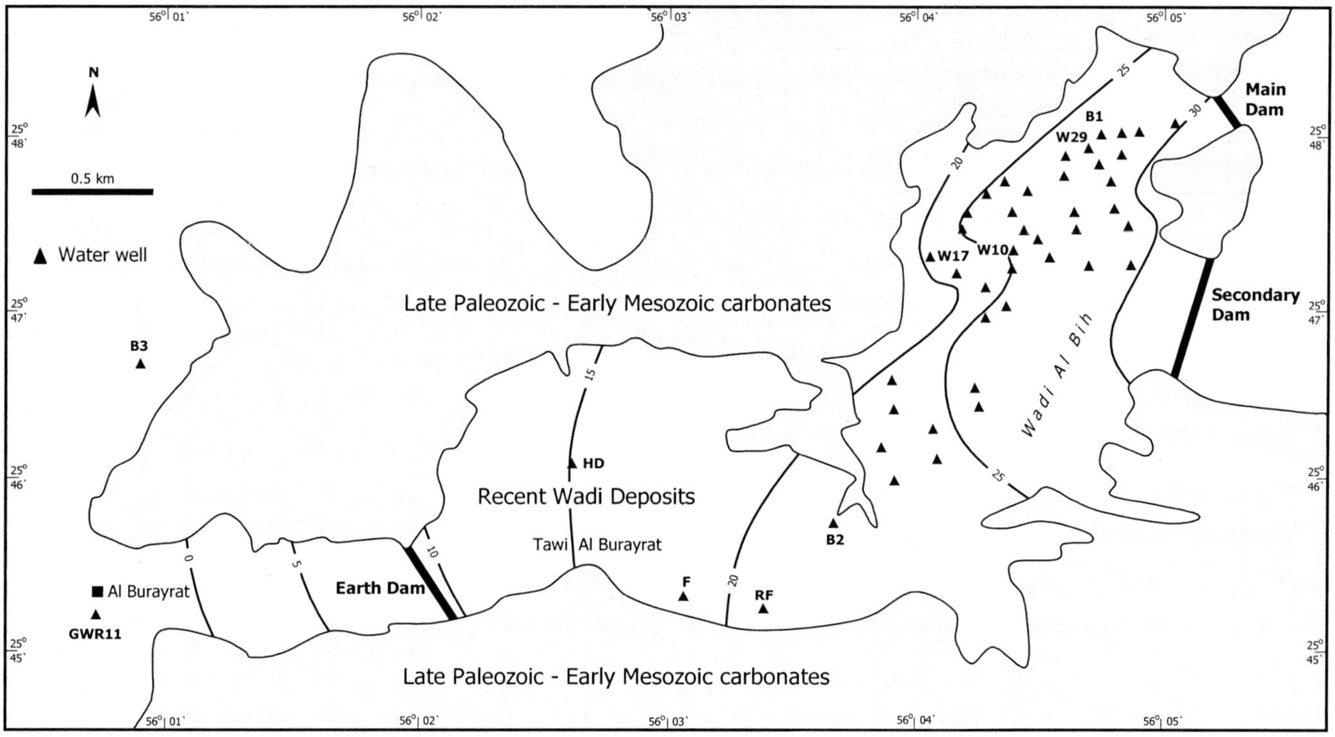

Fig 11.1. Hydraulic head contour map (in meters) relative to sea level of Wadi al Bih limestone aquifer, United Arab Emirates (modified from Al Wahedi, 1997).

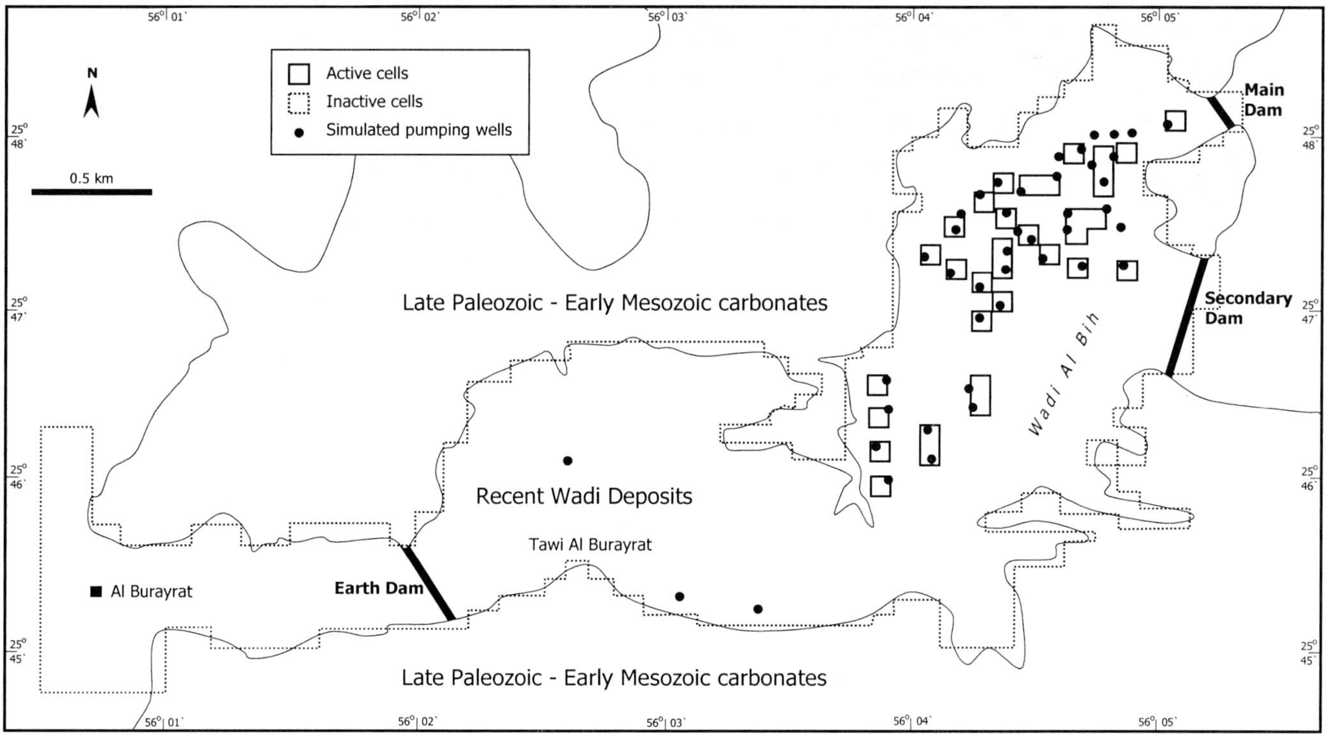

Fig. 11.2. Active and inactive (no flow) cells, with pumping wells in Wadi al Bih groundwater flow model in the United Arab Emirates (modified from Al Wahedi, 1997).

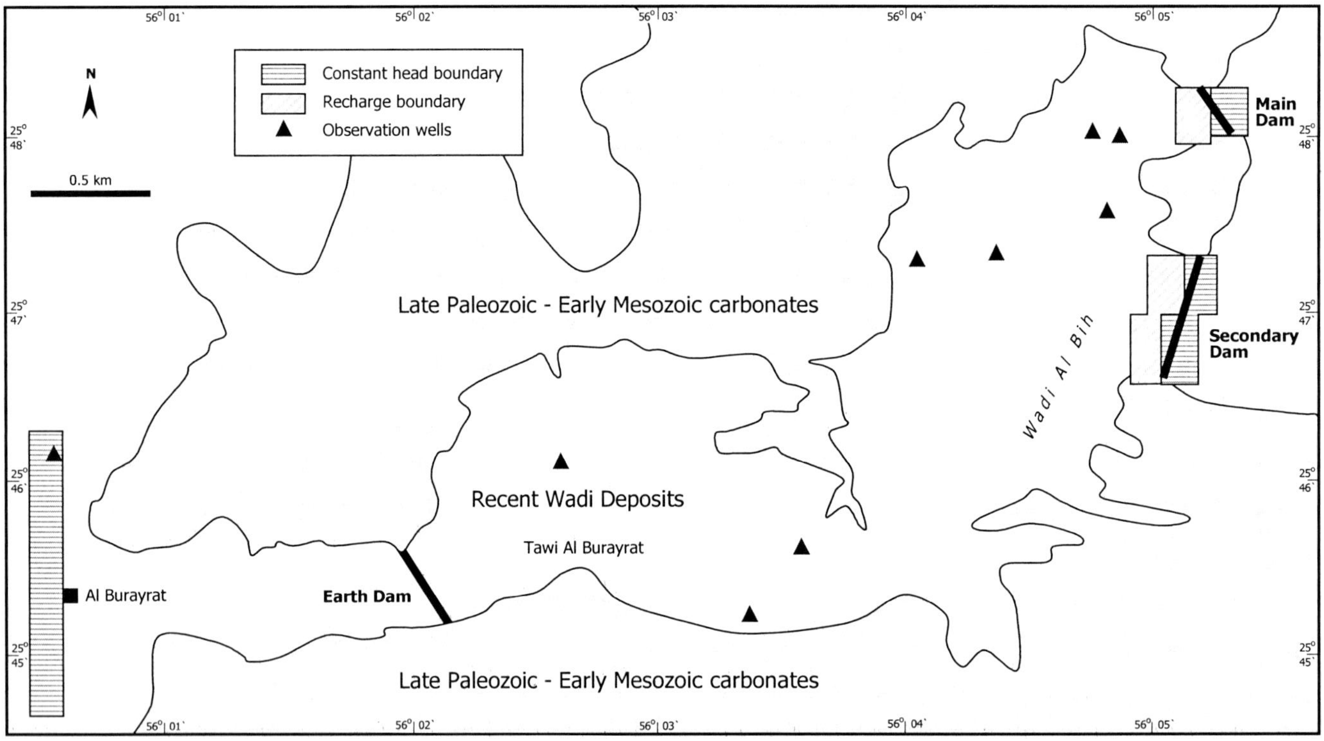

Fig. 11.3. Boundary conditions and observation wells used to calibrate the Wadi al Bih groundwater flow model (modified from Al Wahedi, 1997).

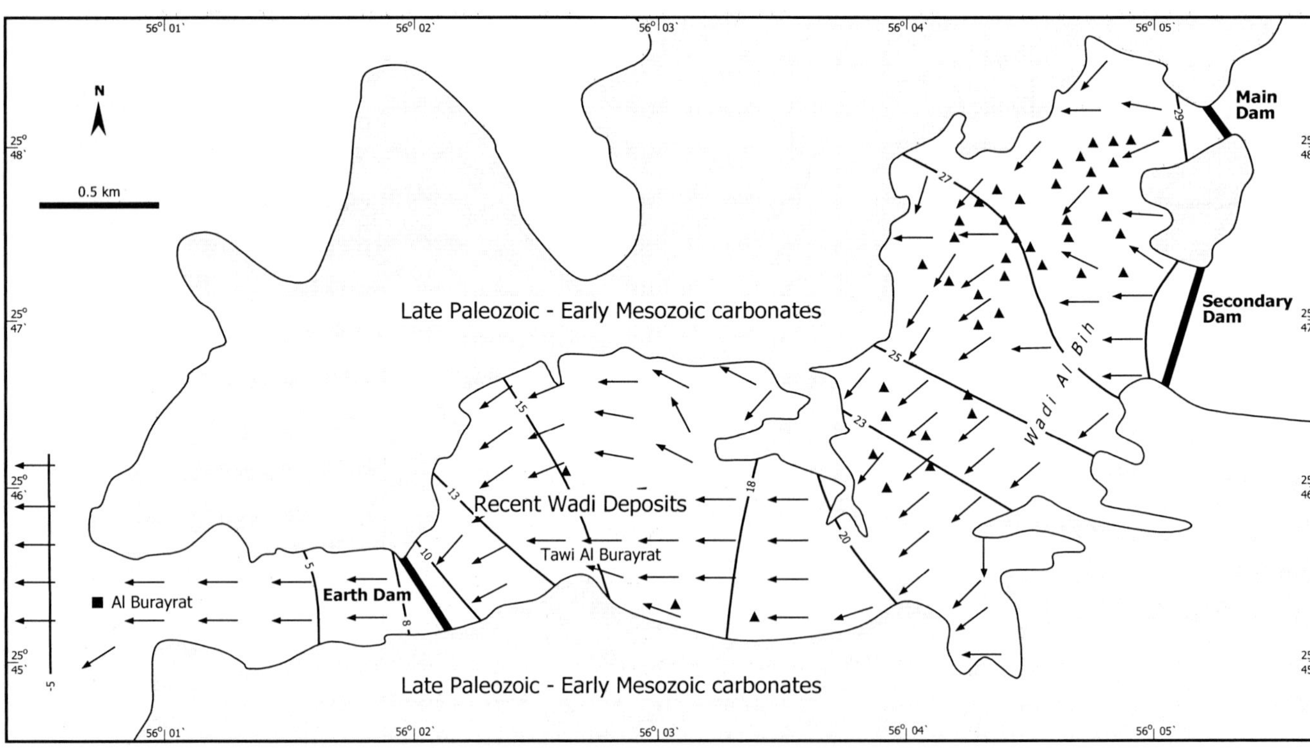

Fig. 11.4. Program calculated hydraulic heads (in meters above sea level) for the Wadi al Bih aquifer in 1995. Arrows indicate the direction of groundwater flow (modified from Al Wahedi, 1997).

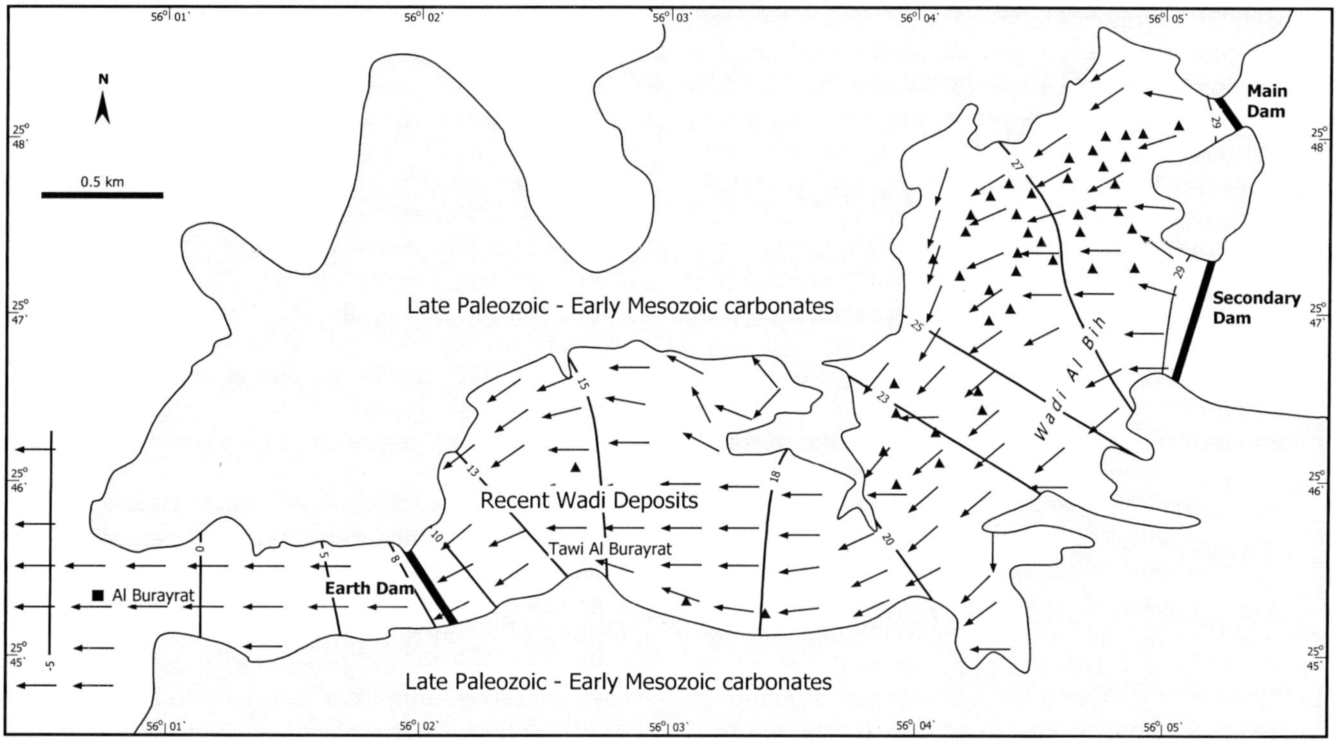

Fig. 11.5. Predicted hydraulic heads (in meters above sea level) in the Wadi al Bih aquifer for the period 1996-2005 at the same pumping and recharge rates as the present (modified from Al Wahedi, 1997).

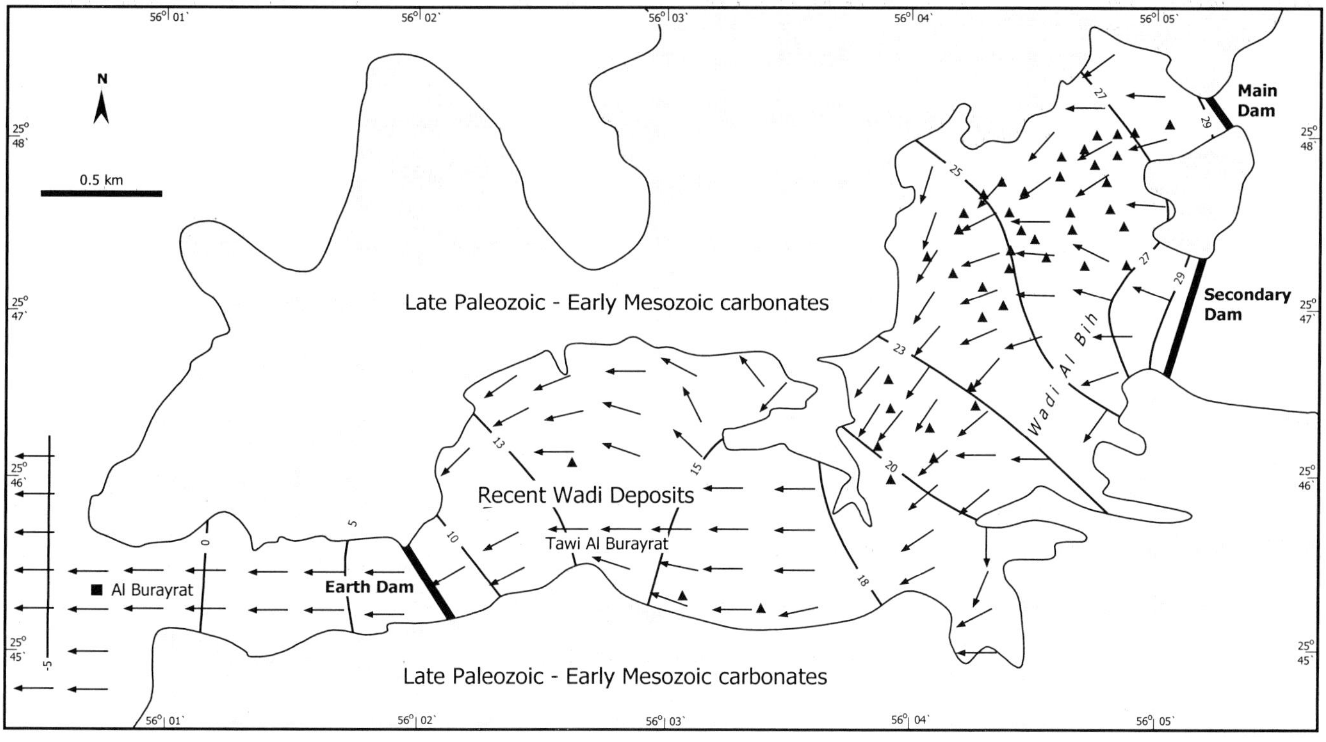

Fig. 11.6. Predicted hydraulic head (in meters above sea level) in Wadi al Bih for the years 1996-2005 at double ppumping nd recharge rates as at present (modified from Al Wahedi, 1997).

1996-2005. Of the unlimited future scenarios only two were tested to show the sensitivity of the model to changes in hydrological conditions. In the two situations tested, one assumed the same recharge and pumping rate, the other assumes the recharge remains constant, but the pumping rate doubled. The second possibility is the more likely, because of the continuous growth in population, agricultural area and industrial activity. The results are figured in Figs. 11.5 and 11.6 and summarized in Tables 11.4 and 11.5.

Comparison of Figures (11.5 and 11.6) shows that, continuing the present groundwater recharge rate, and abstraction from the Wadi al Bih aquifer, will lead to a moderate lowering in the groundwater level. This is reasonable since the average annual extraction (11.23 Mm^3) is greater than the estimated annual recharge (9.57 Mm^3), leading to the decline figured as in Fig. (11.6). On the other hand, if the extraction rate is doubled, the decline in the hydraulic head will be even greater, as illustrated in Fig. 11.6. This also shows a general eastward shift in the iso-potential contours, indicating the continuous lowering of groundwater level, and the increased chances for salt-water intrusion, from the Arabian Gulf, into the fresh water in the eastern part of the Wadi al Bih aquifer.

2. A Geochemical Model of the Wadi al Bih Aquifer, Northern United Arab Emirates

The Wadi al Bih drainage basin lies within, and is surrounded by rocks of Late Permian to Lower Cretaceous age (Hudson and Chatton, 1958). Glennie et al. (1974) introduced the term Hajar Supergroup, which was then subsequently subdivided into the Ru'us al Jibal, Elphinstone and Musandam groups. Within the first group are the Hagil and Bih formations, the former reaches a thickness of 260m and is the main aquifer formation, tapped by most of the water wells in the area. It is fractured, and has karstic features which increase its ability to carry water. Karst features are also found in the underlying Bih limestone. The Ru'us al Jibal mountain range has undergone tectonic folding and faulting. The faults mostly run north-south, except in the region of Jabal Hagab where trends are more westerly or northwesterly. The structural setting influences the hydrogeology, because most of the Wadi al Bih tributaries are located along fault zones, which enhance lateral flow and vertical percolation. The fractures also increase percolation of atmospheric gases, into the open aquifer system.

A geochemical model of the aquifer was made, based upon water samples collected from the Wadi

Table 11.4. Program calculated zone budget and mass balance for Wadi al Bih aquifer at the end of the decade 1996-2005, assuming the same rate of pumping and recharge as at present.

Parameter	Zone budget (m^3/day)		Mass balance			
	Input	Output	Cumulative volumes in	Cumulative volumes out	Rate in	Rate out
Storage	0.000226	0	4.9875	44.204	0.000226	0
Constant head	64,251	35,216	345,090,000	193,910,000	64,251	35,216
Wells	0	29,036	0	151,180,000	0	29,036
Recharge	0.11617	0	636.51	0	0.11617	0
Total	64,252	64,252	345,090,000	345,090,000	64,252	64,252
	0.000059		0		0.003906	

Table 11.5. Program calculated zone budget and mass balance for the Wadi a Bih aquifer at the end of the prediction period 1996–2005, assuming double the present pumping rate but the same recharge rate as present.

Parameter	Zone budget (m^3/day)		Mass balance			
	Input	Output	Cumulative volumes in	Cumulative volumes out	Rate in	Rate out
Storage	0.001388	0	9.2328	44.204	0.001388	0
Constant head	91,055	32,984	443,000,000	185,750,000	91,055	32,984
Wells	0	58,072	0	257,250,000	0	58,072
Recharge	0.11617	0	636.51	0	0.11617	0
Total	91,056	91,056	443,000,000	443,000,000	91,056	91,056
	0.000954		0		0	

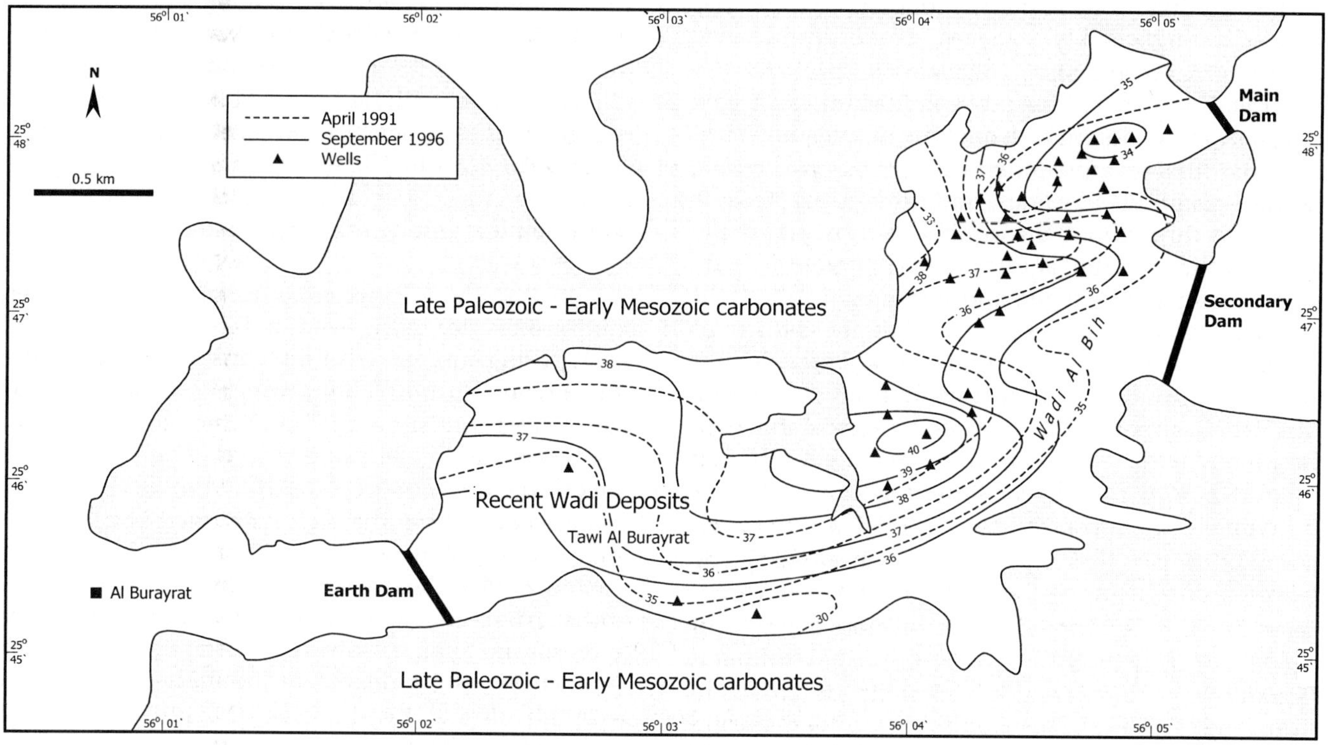

Fig. 11.7. Iso-temperature (°C) contour map of the groundwater in Wadi al Bih limestone aquifer (United Arab Emirates). Data for April 1996 showing in dashed-lines and data for September showing in solid-lines (modified after Al Wahedi, 1997; Al Asam, 1997).

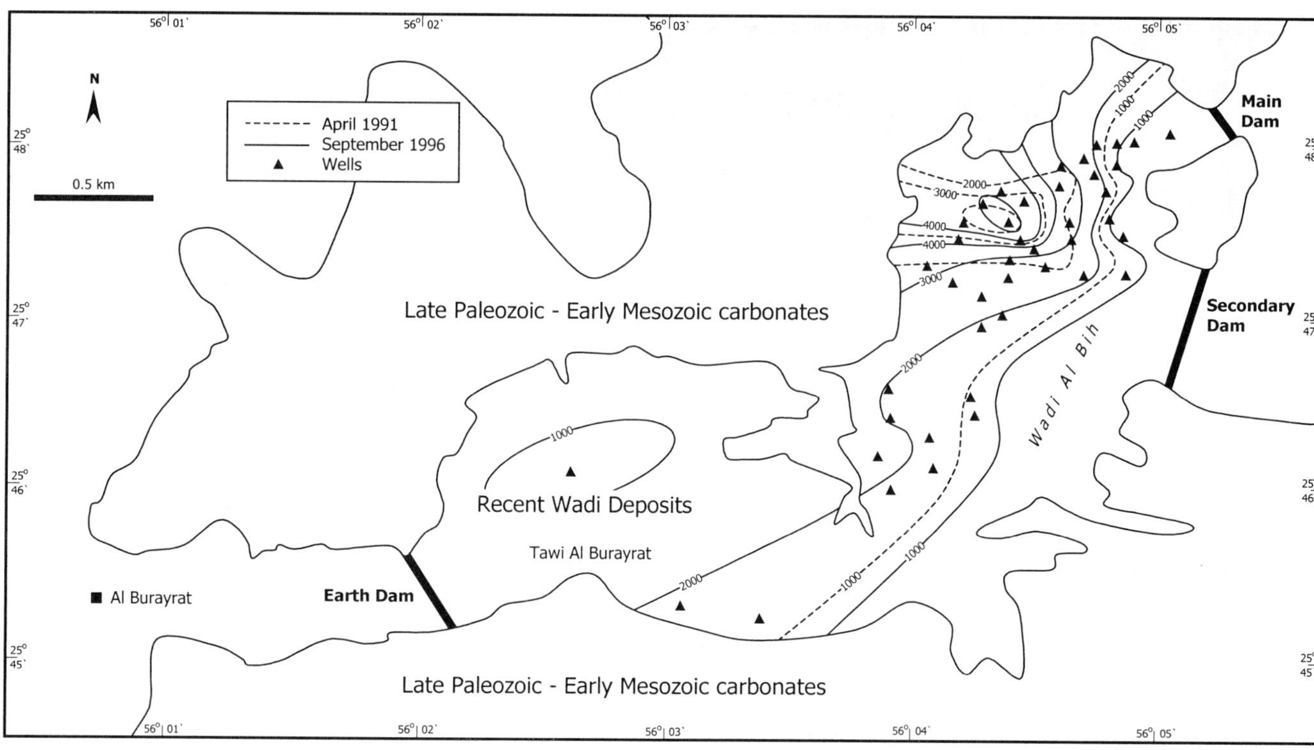

Fig. 11.8. Iso-electrical conductivity contour map (μS/cm) of groundwater in Wadi al Bih limestone aquifer (United Arab Emirates). Data for April 1996 showing in dashed-lines and data for September 1996 showing in solid-lines (modified after Al Wahedi, 1997; Al Asam, 1997).

al Bih aquifer. Geochemical modeling shows that groundwater quickly reaches equilibrium with calcite. Mass transfer calculations predict the precipitation of calcite at equilibrium and the dissolution of dolomite from the Elphinstone and Ru'us al Jibal groups under an open system, which requires high CO_2 values, indicating that CO_2 is available throughout the aquifer, attributed to the karst nature of the aquifer. Interpretation of the stable isotopes of oxygen and hydrogen in the light of chemical modeling suggests that the source of saline water in the Wadi al Bih aquifer is not seawater intrusion, but most likely from the mixing of a deep component, that has undergone water-rock reactions with evaporates, or mixing with an unknown brine from lower stratigraphic units

The electrical conductivity in (μS/cm), pH, and temperature (°C) of samples were measured. The temperature of the groundwater samples varied between 32.8 °C and 43.3 °C, with an average of 36 °C during winter, and 36.3 °C during summer. Groundwater temperatures in wells downstream from the Wadi al Bih dams was lower during September than during April, which may indicate groundwater recharge, during the cooler months is usually of lower temperature, than existing groundwater (Fig. 11.7). The steady increase in groundwater temperature in the Wadi al Bih, may indicate a source of water from a deeper horizon in the aquifer. The pH of the groundwater varies between 7.1 and 7.9, averaging 7.5. The alkalinity of the Wadi al Bih groundwater is attributed to the dissolution of limestone in open conditions. The total dissolved solids in the groundwater varies between 593 and 7,007 mg/l, with an average of 2,122 mg/l, in contrast to the Al Burayrat well field, which ranges from 979 to 6,642 mg/l, with an average of 3,901 mg/l.

Geochemical Interpretation

In 1996 the average total dissolved solid content of groundwater in the Wadi al Bih aquifer increased, 28% during summer over the winter value, related to the higher groundwater pumping, from the well fields during summer, and the lower natural recharge during the same period. These conditions favor the increase of salinity, from deeper saline water especially in the well-field area. The isohaline map shows the increase in total dissolved solid, from east to west in the direction of the Arabian Gulf, and away from the main dam (Fig. 11.8). The steady increase in total dissolved solid towards the west and southwest, is related to the limited recharge of groundwater in these areas, and attributed to an increasing influence of salt water from the Arabian Gulf.

Historical data relating to the Cl^- and the Electrical Conductivity for the period 1980-1994, underlines the dramatic increase in Electrical Conductivity and chloride concentration in the aquifer, during the last fifteen years, which coincides with the expansion of groundwater pumping, for agricultural and domestic purposes (Fig. 11.9).

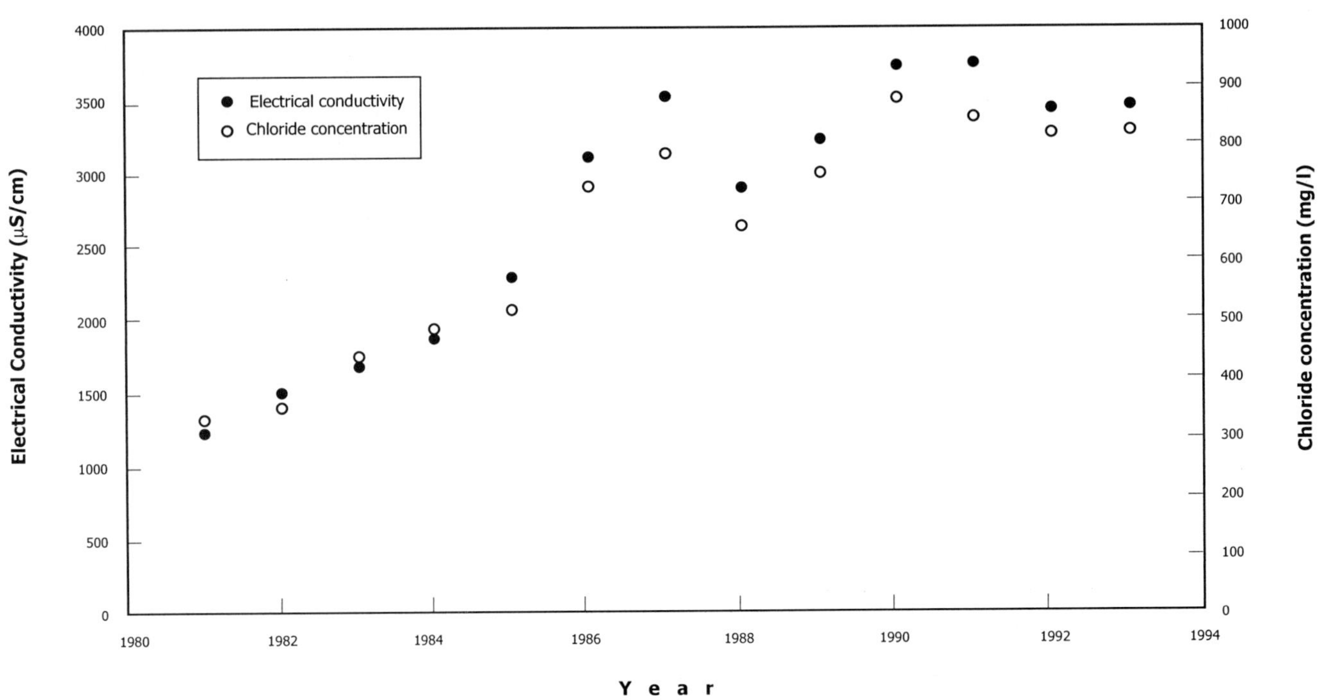

Fig. 11.9. Increasing electrical conductivity and chloride in Wadi al Bih aquifers in the United Arab Emirates between 1981-1994

The main source of Ca^{2+} and Mg^{2+} in the groundwater is due to the weathering of carbonate rocks. The sodium ion Na^+ shows very high concentrations, compared to the calcium and magnesium ions, as a result of the mixing of saline water with groundwater, especially in the Al Burayrat area where the hydraulic head is -5m (msl). The high HCO_3^- indicates the chemical weathering of the limestone and dolomite rocks forming the aquifer. The rain dissolves the carbonate rocks in the catchment area of the wadi, and then recharges the aquifer, with a high concentration of CO_3^{-2} and HCO_3^- during the winter months. This shows in the higher HCO_3^- content of the eastern part of Wadi al Bih, and a low in the western part. Chloride (Cl-) is the dominant anion in the area, and the high concentration (1,400 mg/l) is indicative of a growing salinity problem. A comparison of the anion abundance between the summer and winter months, shows their decrease in the winter months, as a result of dilution by meteoric water. The exceptions to this are the CO_3^{2+} and HCO_3^-, the bicarbonate anion does not seem to obey this rule, because it is a major component of meteoric water. The nitrate concentrations (4.1 mg/l in winter and 5.9 mg/l in summer), are related to the limited agricultural activity in the upstream area.

When the results of water analyses are shown on a Piper diagrams (Fig. 11.10), according to the diagram two water types can be identified (Piper,

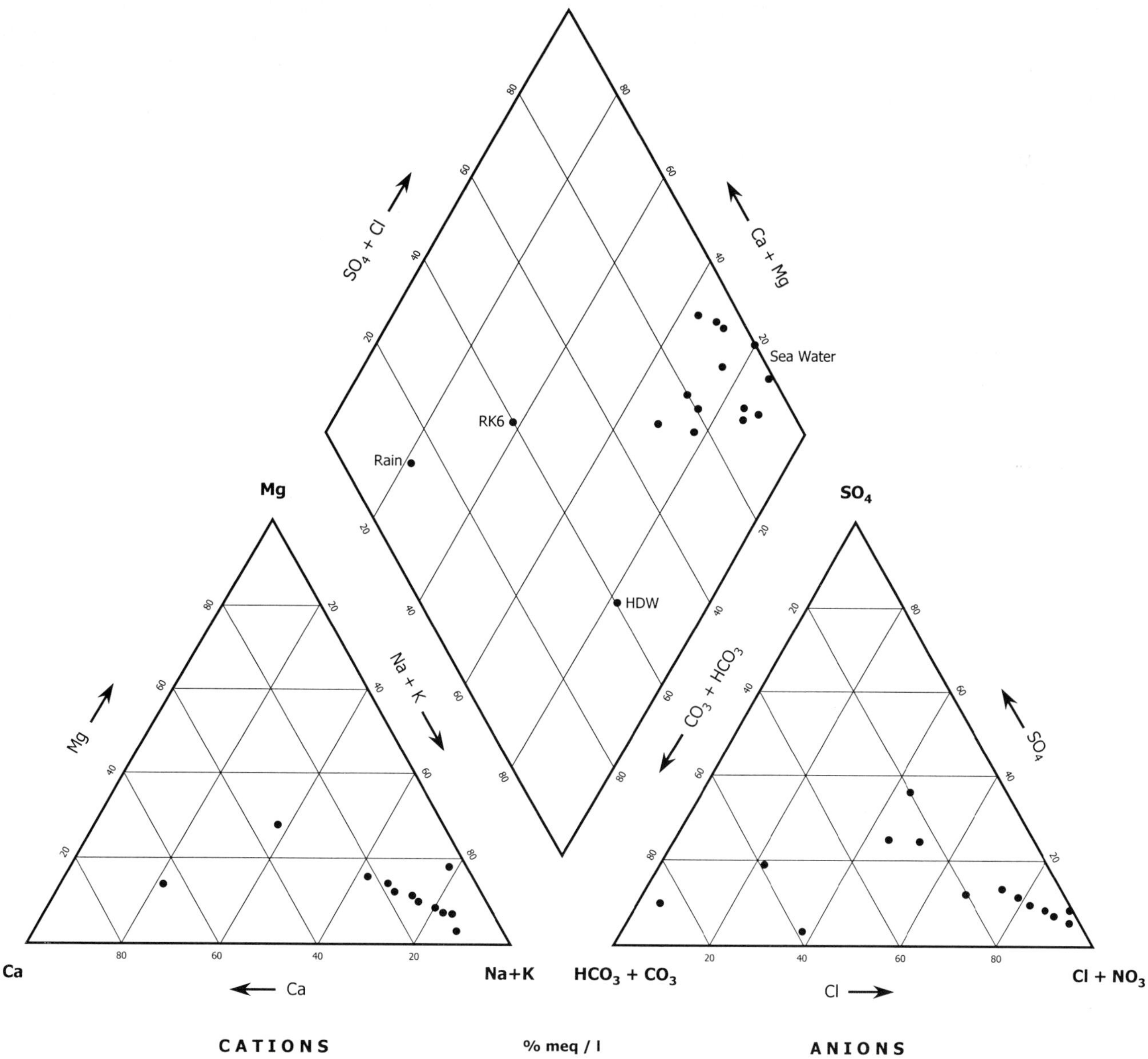

Fig. 11.10. Trilinear plot of groundwater samples from Wadi al Bih limestone Paleozoic aquifer (United Arab Emirates).

1944; Hounslow, 1995) in the Wadi al Bih basin, water rich in Ca^{2+}, Mg^{2+} and HCO_3^- ions and saline rich water with Na^+, K^+, Cl^- and SO_4^{2-}. Except for rainwater samples, all groundwater samples plot on a trend towards the Na^+ apex in the cation triangle, and towards the Cl^- apex in anion triangle, ie., basically indicating NaCl contamination (Hounslow, 1995). The results shown in figure 11.10 could be interpreted as the result of mixing of two end members, seawater and rainfall. However because of the possibility of ionic exchange, and the addition or removal of solid or gas phases, further analysis is required (NETPATH, Plumer et al., 1991).

Absolute ionic concentrations are commonly used to investigate specific water types, but because the water in the Wadi al Bih aquifer is subject to dilution and mixing with other water types, ionic ratios are more revealing. Because chloride is the dominant anion in seawater and bicarbonate is the most abundant anion in groundwater (Todd, 1980), the $Cl^-/HCO_3^- + CO_3^{2-}$ ratio is used to evaluate, the hypothesis of seawater intrusion into the aquifer. The ratio is <1 upstream of the main dam, indicating low salinity groundwater, the ratio everywhere else is >1 suggesting the influence of saline water, but not necessarily sea water particularly, in the western and southwestern parts of the aquifer.

According to Hounslow (1995) the ratio of Cl^- to the total anion content, where >0.8 where the total dissolved solids is <500 mg/l indicates the influence of sea water, whereas a ratio <0.8 indicates weathering of carbonate rocks. The ratio in the aquifer varies between 0.21 and 0.91, showing that both processes are active, carbonate weathering in the upstream area, and salt water intrusion in the western part of the basin. The ratio of HCO_3^-/total anions is >0.8 in all samples from the aquifer, confirming the weathering of carbonate rocks throughout the aquifer.

The $Na^+/(Na^+ + Cl^-)$ molar ratio >0.5, and a total dissolved solids of <500 mg/l, found in the groundwater of the Wadi al Bih area, suggests reverse softening or saline intrusion (Hounslow, 1995). Evidence of potential ion exchange is found in only three cases. As ion depleted rainwater, moves through the carbonate limestone formations of the Wadi al Bih (of the Musandam Group), the Ca^{2+}/Mg^{2+} evolves from a high value to a lower value, as Mg^{2+} is contributed from dolomite dissolution, and Ca^{2+} is lost through calcite precipitation at a later stage. There is thus a steady decrease in the Ca^{2+}/Mg^{2+} molar ratio, from the recharge area in the east, towards the discharge area to the west. The $Ca^{2+}/(Ca^{2+} + Mg^{2+})$ molar ratio in the Wadi al Bih supports the weathering of limestone and dolomite indicated by the Cl^-/sum anions, and the HCO_3^-/sum anion molar ratio, and throws doubt on the influence of sea water as a source of salinity of the aquifer.

It is interesting to note that the slope of the trend in Cl^-/SO_4^{2-} (Fig. 11.11) is represented by a slope greater than 40, well in excess of the slope of 13 expected for a rainwater-seawater mix. It suggests that the end member-mixing source was not modern seawater, but rather a variable of water-rock reactions, with NaCl rich evaporite minerals or brine with a high Cl^-/SO_4^{2-}. Further resolution requires an investigation using stable S isotope analysis.

The degree of saturation of a solution (Saturation Index), with respect to a particular mineral phase or solid can be determined using the WAREQ4F program (Plumer et al., 1976). It was applied for calculation of the saturation indices, of all possible minerals and gas phases, in the Wadi al Bih aquifer. The saturation index of calcite is mainly negative, in the upstream area of the wadi, indicating undersaturation, and the potential for dissolution of limestone rocks dominant in the area. In the middle and downstream of the main dam, the saturation index of calcite becomes slightly positive, evidence of saturation to oversaturation, and precipitation of calcite becomes possible under these conditions. Dolomite is undersaturated in the upper Wadi al Bih basin, and all the dolomite saturation index values are below zero. Further downstream, the dolomite saturation index becomes positive. The calculated Saturation Index values for gypsum are all negative, in both the upstream and downstream of the wadi, indicating undersaturation throughout the aquifer.

The source of the higher salinity water in the Wadi al Bih, can be examined using the NETPATH and WATE4QF models. The results show that, all models using a simple fresh water/seawater mix violate the thermodynamic constraints on the dissolution and precipitation of calcite, dolomite and in some cases carbon dioxide gas. The reason stems from the observed Ca^{2+}/Mg^{2+} in groundwater, for which the solution of mass transfer using NETPATH, requires dolomite as a sink for Mg^{2+} from seawater. Thus mixing of seawater is not a probable source, for the higher saline groundwater, found in the Wadi al Bih aquifer system.

Gonfiantini (1992) in a preliminary analysis of the isotopic data over the period 1984–1992, concluded that salt water intrusion was an on-going process, causing the rise in groundwater salinity. However, a re-investigation and re-evaluation of these data was made to determine the source of the salinity of the water in the al Bih aquifer, and assess the impact of the Wadi al Bih dam on groundwater recharge. It showed that the local meteoric water line (LMWL), for the Wadi al Bih has a lower slope, than the global meteoric water line indicated

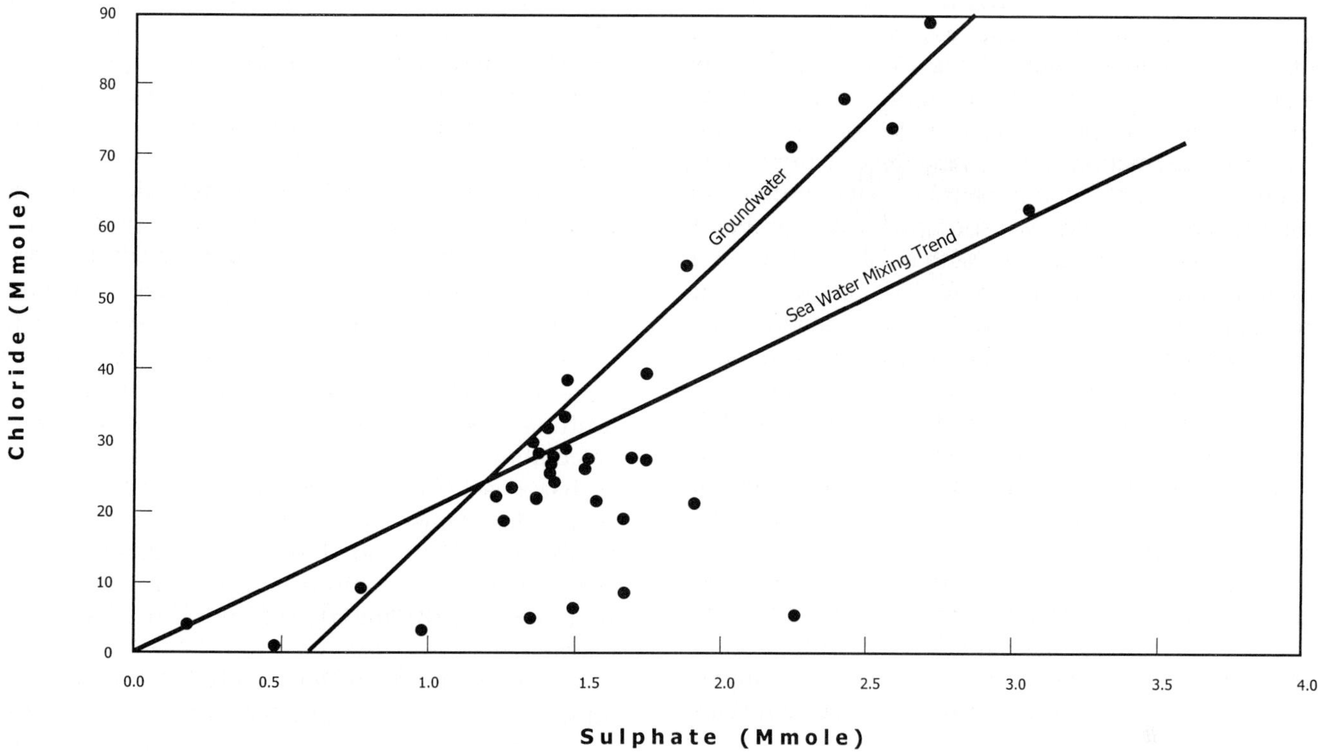

Fig. 11.11. The relationship between sulphate and chloride at Wadi al Bih Paleozoic aquifer in the United Arab Emirates.

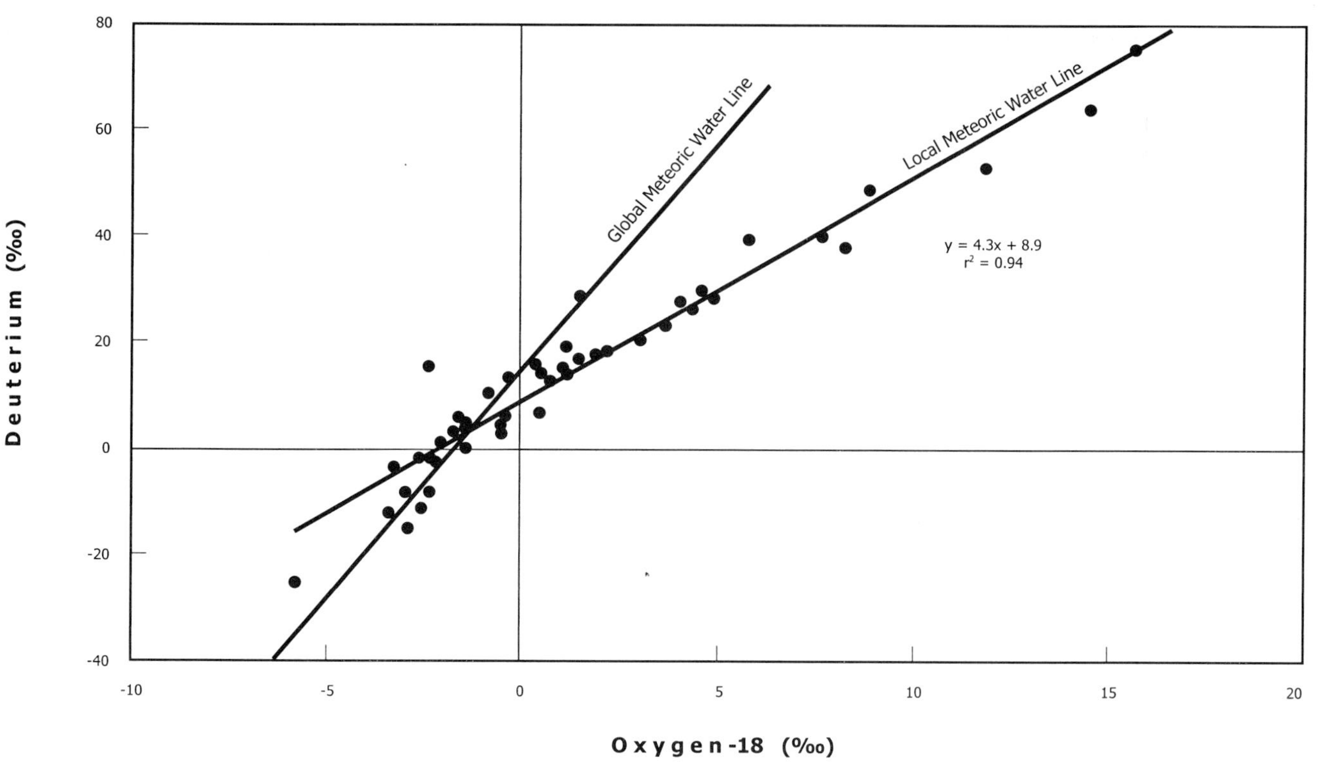

Fig. 11.12. Stable isotope composition of rainwater at Wadi al Bih basin, United Arab Emirates.

evaporation of precipitation (Fig. 11.12), and groundwater samples clustered, in a narrow range of more depleted values (Fig. 11.13). The latter suggests that, neither evaporation nor the mixing process, has a pronounced effect on the stable isotope composition, but a weak best-fit trend does suggest a small component of evaporation or mixing, with an unknown source of water, probably not seawater. As calculated with NETPATH, a range of seawater mixing of up to 35% is needed, to reproduce the observed salinity, but the isotope data does not strongly follow this trend, consequently a highly saline water or brine, may exist in the lower stratigraphic units. A much smaller component of mixing would be required to reproduce the observed change in salinity with a less pronounced effect on the isotopic composition of the groundwater. This unknown component, may only represent a deeper ground flow path for the Wadi al Bih system that undergoes water-rock reaction with evaporite minerals at depth. This hypothesis is supported by the relationship between salinity and higher temperature water in the northeastern section (Figs. 11.7 and 11.8) and the relationship between tritium and salinity (Fig. 11.14). The tritium content (^{3}H) of the Wadi al Bih groundwater is < 4 tritium units (TU) whereas precipitation during the period ranges from 5-10 TU (International Atomic Energy Agency Global precipitation network station Bahrain), and is consistent with a groundwater flow of older (possibly deep flow), and younger (karstic flow) groundwater, both from the same recharge area. The data do not suggest a large component of recharge induced by the Wadi al Bih dam, (which would have a strong evaporative signature).

The depleted isotopic composition of the groundwater, with respect to local precipitation, suggests a recharge zone at higher altitude, which can be estimated if a $\delta^{18}O$ of -3‰ for groundwater, a mean $\delta^{18}O$ of –1.1‰ for measured rainfall, near the Global Meteoric Water Line (GMWL), and an arid zone adiabatic of $\delta^{18}O$ of 0.2‰ per 100m increase in altitude (Erikson, 1983). The result of such a calculation is that, the average elevation of the recharge is 1,050m. As the elevation of the Ru'us al Jibal Mountains surrounding the aquifer varies between 1,050m and 2,090m, it seems logical to conclude that, the main recharge area is in the high mountains, with the runoff infiltrating at higher elevations.

In summary, the water quality in the Wadi al Bih aquifer system, has been affect by water abstracted from the aquifer, so the identification of the recharge zone is of considerable interest, in water quality management. The average temperature of the groundwater indicates that, groundwater recharge from rainfall occurs during the cooler months of the

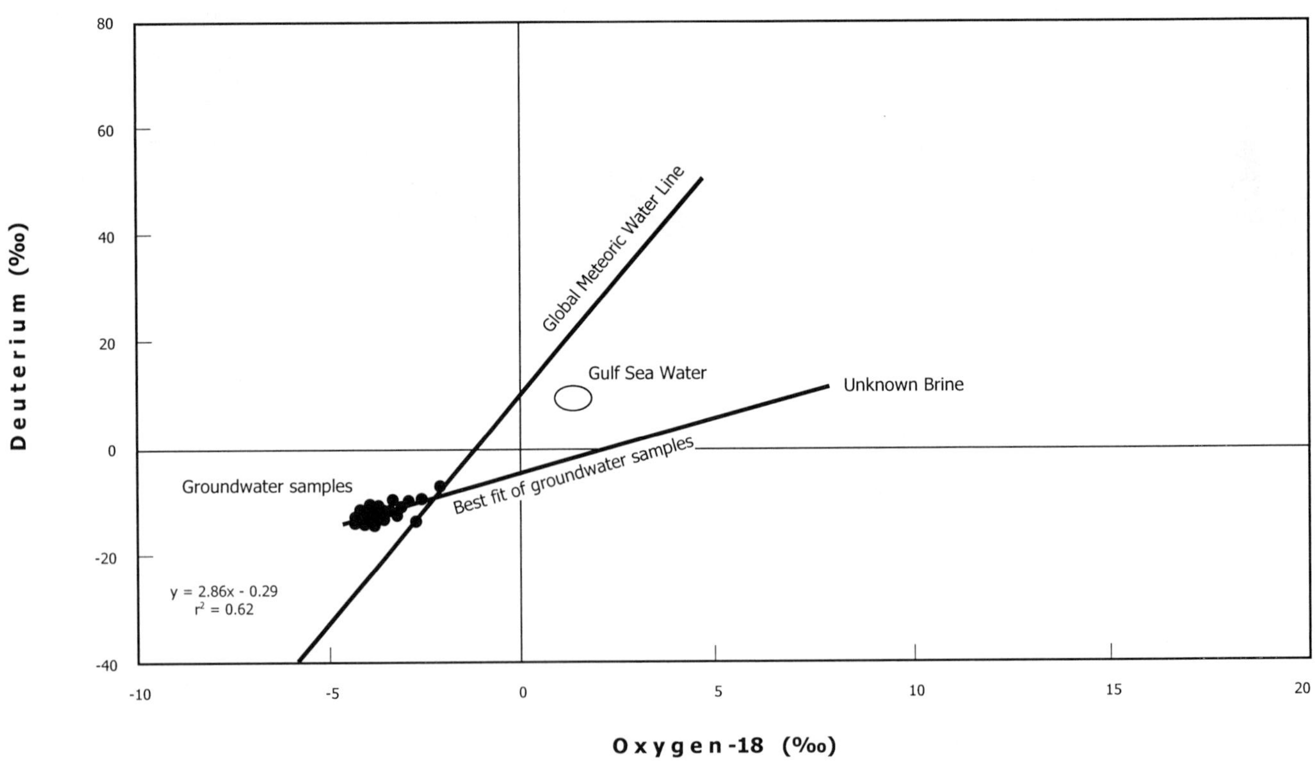

Fig. 11.13. Stable isotope composition of groundwater at Wadi al Bih basin, United Arab Emirates.

year. The steady increase in groundwater temperature northwestwards, may indicate a source of water from deeper horizons entering the aquifer. Evaluation of the chemistry of the water, in particular the Ca^{2+}/Mg^{2+} and the Cl^{-}/SO_4^{2-}, suggests that the end member is not modern seawater, but rather a variable degree of water-rock reactions, with NaCl rich evaporite systems or brine.

Modeling with NETPATH and WATE4QF confirms that, seawater intrusion is not the source of the higher salinity, but suggests a mix of groundwater, with an older and possibly deeper source of water and younger groundwater, both from the same recharge area.

2. Groundwater Flow Model of the Dammam Aquifer in Saudi Arabia

A groundwater flow model for 17,600 km^2 of the Dammam aquifer was designed by Yazicigil et al. (1986) using the USGS two dimensional groundwater flow program of Trescott et al. (1976). The modeled area ranged in elevation from 500m in the west to sea level in the Arabian Gulf to the east. The modeling study provided calibrated values of the hydraulic parameters of the aquifer and responses to different management alternatives.

The basic model hydrogeological and lithological assumptions, and parameters used by Yazicigil et al. (1986) were that the Alat and Khobar members could be treated as a single, Dammam, aquifer with hydrological continuity across the Rus Formation. The Neogene aquifer was not included in their model. Aquifer geometry was derived from data from 100 wells in 12 of which hydraulic heads were monitored. The 1969 Italconsult potentiometric surface was taken to represent predevelopment heads within the modeled area. Groundwater flow is from southwest to northeast with low gradients in the west ($4x10^{-4}$ to $7x10^{-4}$) increasing to the east and northeast to $2.5x10^{-3}$ to $5x10^{-3}$ then decreasing to the values $5x10^{-4}$ to $8x10^{-4}$ along the Arabian Gulf coastal belt consistent with the transmissivity data from pumping test experiments.

The input values of vertical hydraulic conductivity range from $4x10^{-12}$ m/s in synclines, where the Rus Formation is composed of anhydrite and shale to $2x10^{-10}$ m/s in the anticlines where the Rus consists limestone. The storage coefficient ranges from $<10^{-5}$ in the east to $>10^{-2}$ in the west. The applied aquifer transmissivity ranges from zero where the aquifer is missing to 6×10^{-1} m^2/s. The eastern and western boundaries of the modeled area were considered constant head boundaries while the northern and southern boundaries parallel to the streamlines were assumed no-flow boundaries (Fig. 11.15).

Table 11.6. Simulated and observed drawdown data (in meters) used for the verification of the numerical results in the Dammam aquifer system in eastern Arabia (after Yazicigil et al., 1986). Negative values indicate a rise in hydraulic head. [obs.= observed, Sim = Simulated].

Location	Drawdown in meters at end of the year							
	1977		1978		1979		1980	
	Obs.	Sim.	Obs.	Sim.	Obs.	Sim.	Obs.	Sim.
Dhahran	--	--	0.3	0.5	0.5	1.0	1.0	1.6
Al Hulayyah	--	--	--	--	0.0	0.1	-0.1	0.3
Um Al Shaik	0.1	0.2	0.4	0.3	0.5	0.6	0.7	0.8
Al Jubail	--	--	0.3	0.3	-0.9	0.7	-1.6	1.2
Ad Dammam	--	--	--	--	0.4	0.4	0.9	0.8
Al Hofuf-East	1.1	1.2	4.7	3.5	10.1	7.2	15.3	12.5
Al Hofuf-South	1.0	1.0	2.9	2.9	6.2	6.0	10.5	10.4
Wadi Al Miyah	--	--	--	--	--	--	1.1	0.1
Al Atwalah	--	--	0.1	0.3	-0.1	0.7	-0.1	1.0
Al Uqayr	--	--	--	--	-1.1	0.3	-0.1	0.7

Table 11.7. Simulated draw down after 15 years (1980-1995) of water extraction from the Dammam aquifer in eastern Saudi Arabia (after Yazicigil et al., 1986).

Field No.	Area	Pumping rate (m^3/s)	Draw Down (m)
1	Abu Hadriyah	0.27	1.4
2	Al Hinnah	0.07	2.9
3	Jubail	0.86	5.1
4	Jubail	0.58	4.6
5	Abqaiq	0.47	32^{+}
6	Al Hofuf	1.10	145^{+}
7	Al Qatif	0.57	2.5
8	Al Qatif	0.23	2.1
9	Al Qatif	0.18	3.1
10	Al Qatif	1.29	2.9
11	Al Qatif	0.12	3.7
12	Al Qatif	1.83	3.7
13	Al Qatif	0.91	4.3
14	Ras Tanurah	0.24	1.5
15	Ad Dammam	0.35	6.8
16	Ad Dammam	2.01	4.9
17	Ad Dammam	1.34	10.6
18	Khobar	2.70	21.9
19	Tarut Island	0.35	3.4
20	Al Qatif-Aramco	0.38	4.0

In the model calibration, input variables such as aquifer transmissivity, storativity, leakage and boundary conditions are adjusted to obtain a good match between field measured and model calculated hydraulic heads. Figures (11.15a,b) illustrates a good agreement between the models calculated hydraulic heads with the pre-development hydraulic head map of Italconsult (1969). The model calculations show that within the modeled area, the Dammam aquifer receives recharge from the western boundary

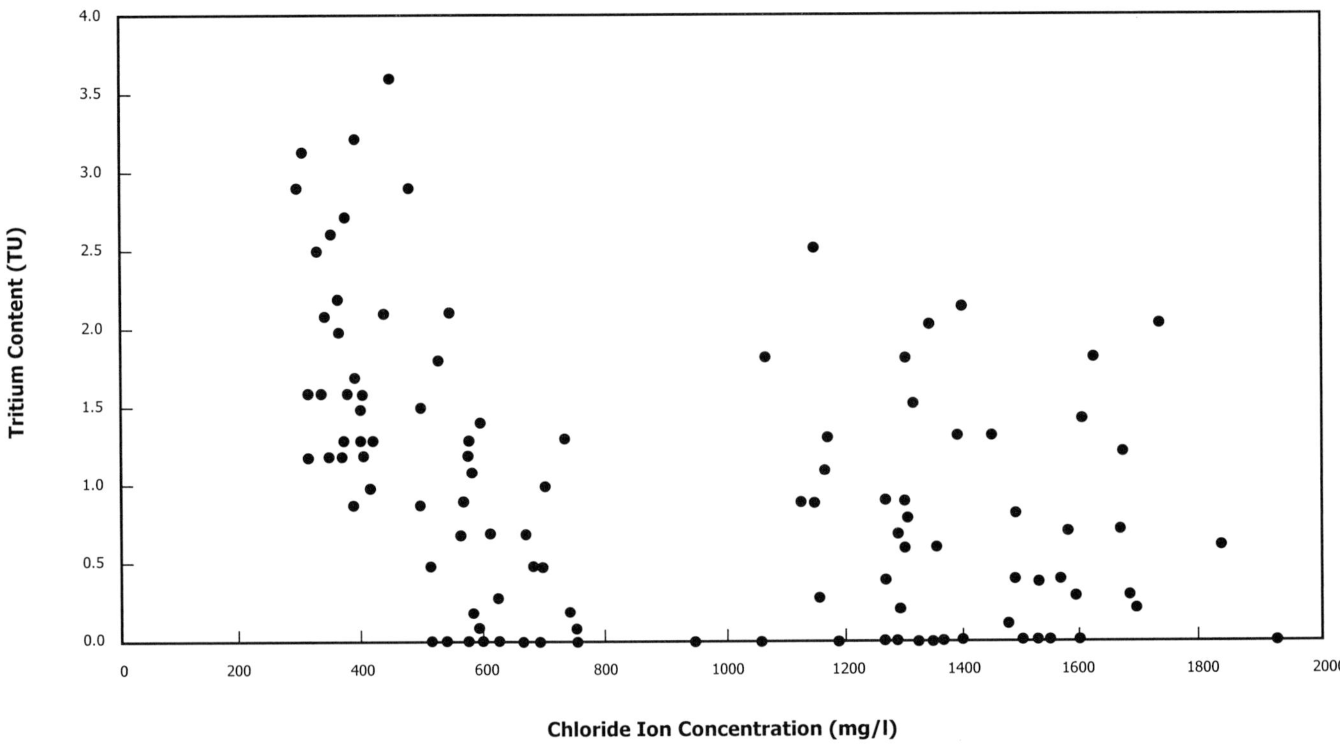

Fig. 11.14. The relationship between chloride ion concentration (mg/l) and tritium content at Wadi al Bih Paleozoic aquifer, United Arab Emirates.

at a rate of 13.4 m^3/s whereas the discharge is through groundwater withdrawal (6.56 m^3/s) and underflow (6 m^3/s) across the eastern boundary. The aquifer also receives 0.24 m^3/s through vertical leakage from the Umm er Radhuma in the Ghawar anticlinal area and loses 0.26 m^3/s to it in other areas. Transient simulations for the period 1971–1980 was based upon pre-development calibration and available data on aquifer utilization during this period. The model generated 1980 piezometric surface map for the Dammam aquifer is illustrated in figure 11.15c.

Model verification was achieved by comparison of simulated drawdowns with field measured data (Table 11.6) for the period 1977–1980, with the differences attributed to the uncertainty of groundwater extraction rates. Model calibration indicates that approximately 14% of the total groundwater extraction within the modeled area comes from aquifer storage.

The calibrated values of transmissivity, storage coefficient and vertical leakage for the aquifer exhibit a wide variation reflecting the heterogeneous nature of the Dammam aquifer system within the modeled area but agree with the results of pumping tests.

At an annual increase in extraction of 19 million m^3/yr from the Dammam aquifer since 1975, the total demand was estimated at 625 million m^3/yr. The model-generated head distribution map (Fig.11.15d) shows three cones of depression at Hofuf, Abqaiq and Al Khobar. Table 11.7 shows the simulated drawdown predicted after 15 years of pumping different well fields. The results indicate that pumping the aquifer at the present rate could dewater the aquifer at Hofuf and Abqaiq. The simulated decline in hydraulic head is 1995 at Hofuf ranged from 40m to 60m compared to the 1980 levels thus a reduced pumping rate is necessary to avoid water depletion in both areas.

3. Groundwater-Flow Model for the Kuwait Aquifer Systems

During the 1990's groundwater yields from the Kuwait aquifers reached about a million cubic metres a day. Most of this water was drawn from aquifer storage for there is only limited recharge from either Iraq or Saudi Arabia. Flow models which incorporate the Eocene Dammam aquifer and the Mio-Pliestocene Kuwait Group aquifers were generated by Mukhopadhyay et al. (1994) using the VTDN computer program of the Bureau de Recherches Gèologiques Minières.

The VTDN software simulates groundwater flow in a multilayered aquifer-aquitard system by a method of finite differences. In Kuwait a three-layer aquifer system (Upper Kuwait Group, Lower Kuwait Group and Dammam aquifers) separated by two semipervious aquitards was simulated. As no

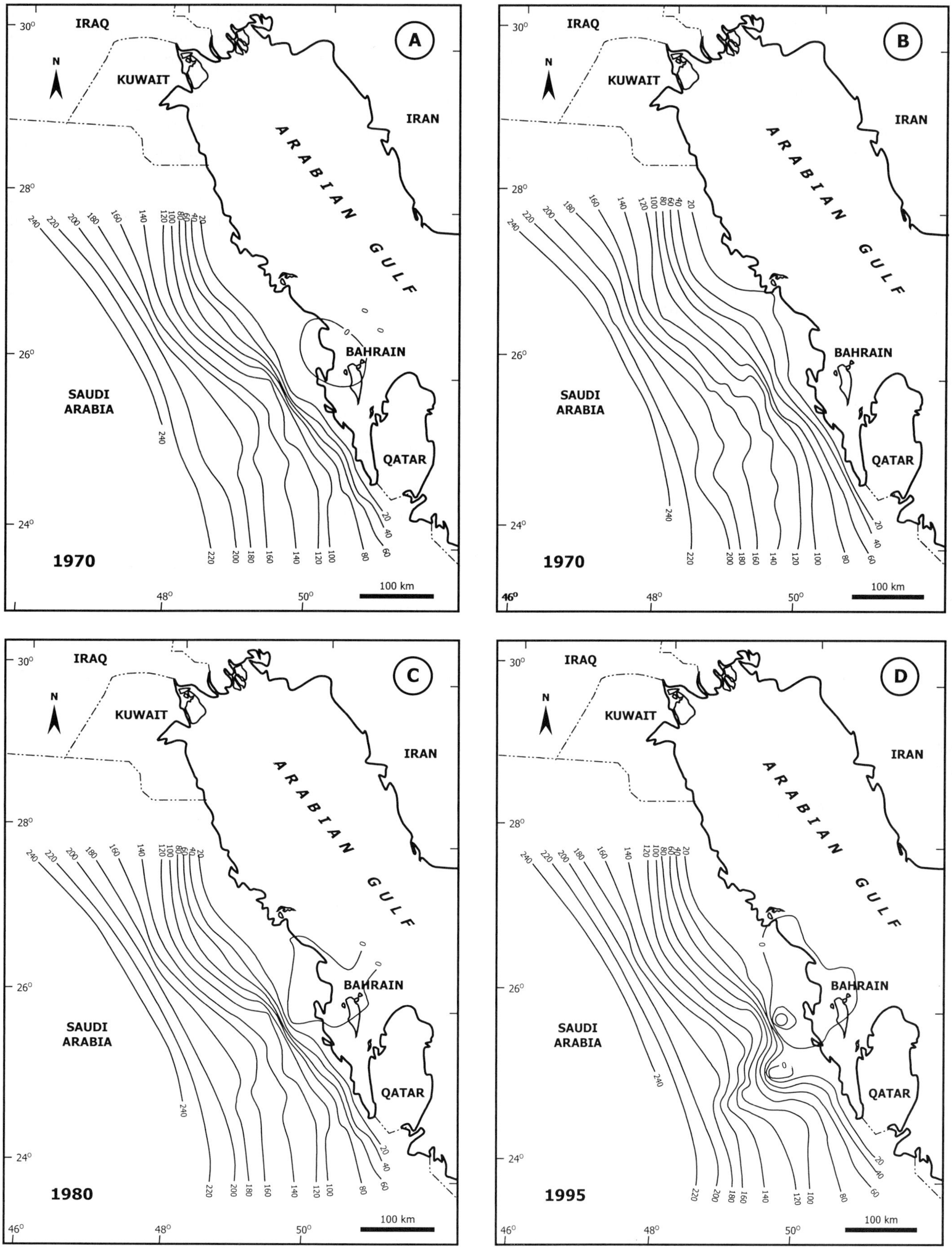

Fig. 11.15. Maps of the Dammam aquifer in eastern Saudi Arabia (compiled and modified from Italconsult, 1969; Bureau de Recherches Gèologiques et Minières, 1977 and Yazicigil et al., 1986). a)Piezometric surface for the year 1970; b) Simulated piezometric surface for the year 1970; c) Simulated piezometric surface for the year 1980; d) Simulated piezometric surface for the year 1995. All elevations measured in meters above sea level.

natural hydrological boundaries exist within Kuwait except the shoreline of the Arabian Gulf, the simulated area extends beyond the boundaries of Kuwait into Iraq and Saudi Arabia. The eastern boundary, where the aquifers are in direct contact with the Arabian Gulf, was set as the constant head boundary. Fixed heads were also assumed for the northwestern and southwestern boundaries where head values were determined from potentiometric surface maps. The input hydraulic parameters of the Kuwait aquifers are listed below:

Table 11.8. Aquifer parameters used in the calibration of the groundwater-flow model of the Kuwait aquifer systems (modified after Mukhopadhyay et al., 1994).

Parameter	Aquifer 1 (Upper Kuwait Grou)	Aquitard 1	Aquifer 2 (Lower Kuwait Group)	Aquitard 2	Aquifer 3 (Dammam Formation)
Transmissivity	1×10^{-4}	--	1×10^{-4}	--	1×10^{-4}
(m^2/s)	2×10^{-2}	--	2×10^{-2}	--	2×10^{-2}
Storage	8×10^{-2}	--	8×10^{-2}	--	1×10^{-4}
Coefficient	--	--	1.8×10^{-2}	--	--
V. Permeability (m/s)	--	5×10^{-9}	--	1.3×10^{-9}	--

Table 11.9. Groundwater balance in the Kuwait Group aquifers and the Dammam Aquifer for the period 1960-1988 (summarized from Al Sulaimi and Akbar, 1999).

Year	Groundwater Budget (m^3/s)				Aquifer
	Recharge	Discharge	Error	Balance	
1960	0.032	-0.032	0	0	Upper Kuwait Group Aquifer
1970	0.648	-0.994	0	-0.346	
1980	0.050	-0.737	0	-0.687	
1988	0.070	-1.061	0.001	-0.990	
1960	1.825	-1.825	0	0	Lower Kuwait Group Aquifer
1970	0.565	-1.562	0.004	-0.993	
1980	0.658	-3.021	0.003	-2.360	
1988	0.715	-5.108	0.001	-4.390	
1960	0.287	-0.288	0.001	0	Dammam Aquifer
1970	1.518	-1.537	0.001	-0.018	
1980	2.482	-2.522	0.002	-0.038	
1988	3.973	-4.046	0.003	-0.070	

The steady state calibration indicates flux rates of 0.9 m^3/s and 1.2 m^3/s from the southwestern and northwestern boundaries respectively. The calibration run of the program simulated hydraulic heads and aquifer budget for the period 1960-1988 and yielded the following results (Figs. 11.16 ; 11.17 and 11.18):

1. The absence of artificial groundwater pumping during the 1960's allowed the natural recharge to balance natural discharges from all aquifers in Kuwait.
2. In the 1970's artificial groundwater pumping in addition to the natural discharge exceeded recharge by 1.4 m^3/s leading to a decline in the hydraulic head in all aquifers.
3. In the 1980's the groundwater discharge exceeded natural recharge by 3 m^3/s leading to further declines in the hydraulic head, exhaustion of aquifer storage and rise in groundwater salinity.
4. In 1988 the negative groundwater balance in all the Kuwait aquifers rose to 6.0 m^3/s, four times the 1970 negative balance and twice the 1980 figure. This has led to aquifer depletion and serious water salinity problems.
5. It is estimated that 80% of the lateral inflow to the Dammam and Kuwait Group aquifers comes from Saudi Arabia, and 20% from Iraq. The lateral outflow is directed eastwards to the Arabian Gulf and northward into Iraq.
4. The direction of intraquifer groundwater flow changed from the Dammam into the Kuwait group aquifers in 1960 was reversed due to the change in the vertical hydraulic gradient.

The groundwater flow model was run to test four different groundwater development scenarios for the period 1988-2010. The main assumptions and results are listed below:

1. Controlled development

Assumptions: The groundwater extraction rate of 4.5 m^3/s of 1988 increases to 5.5 m^3/s in 1995 and then remains constant until 2010 with the current pumping rate from the Kuwait Water Company fields also remaining constant. Increased production capacity of the Ministry of Electricity and Water well fields to 90% efficiency in accord with specific regulations. The predicted water budget of the Kuwait aquifers is listed in Table 11.10 and the predicted cones of depression associated with groundwater pumping are illustrated in figures (11.19-11.21).

2. Intensive production

Assumptions: The groundwater pumping rates from the Kuwait Company well field remains unchanged, but groundwater pumping in the Al Wafrah and Al Abdali agricultural area increases annually by 5%. The production capacity of the Ministry of Electricity and Water to 90% efficiency in accord with specific operation rules and new fields are developed by the Ministry of Electricity and Water with an efficiency in accord with operational rules. The predicted water budgets according to this model are listed in Table 11.11 and the predicted

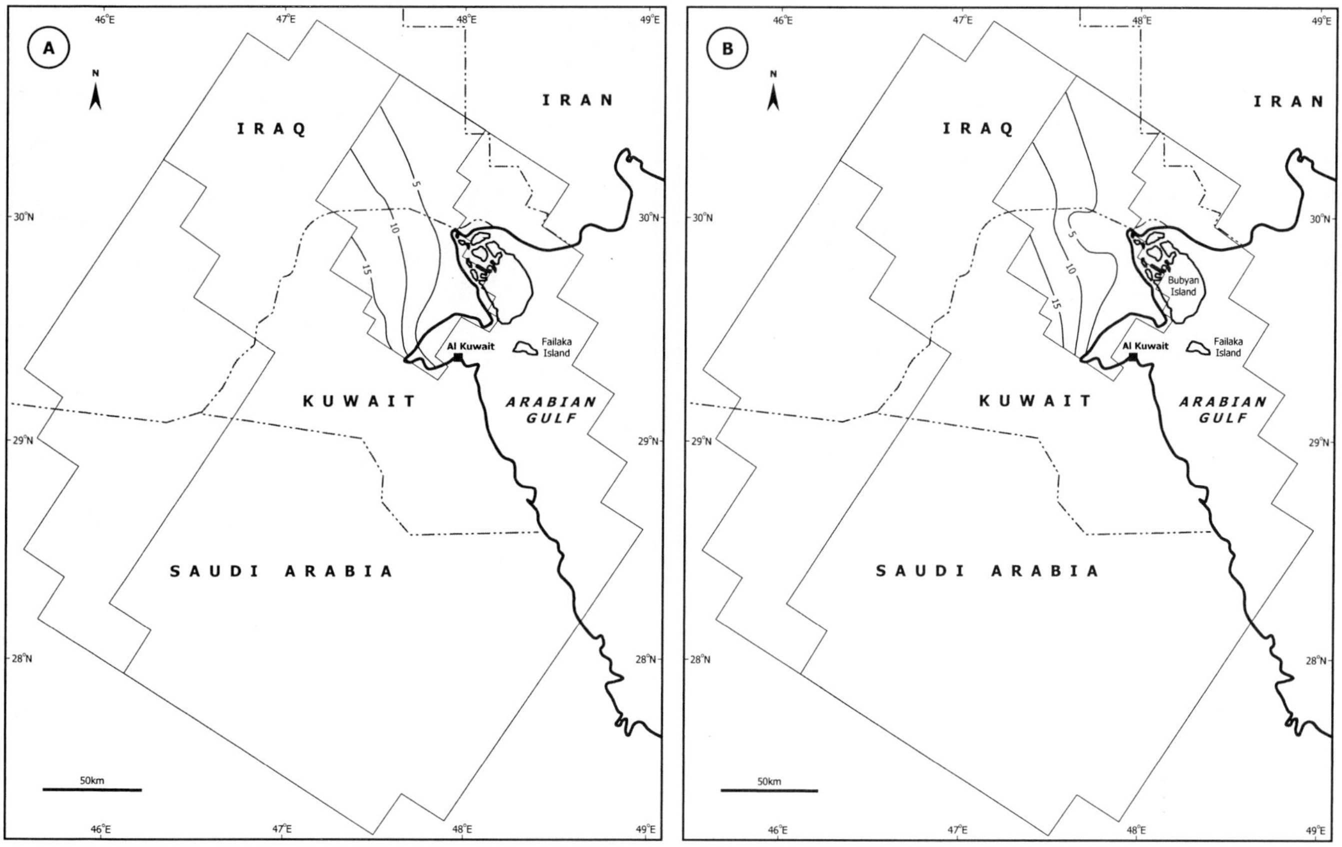

Fig. 11.16. Piezometric surface contour map (in meters above sea level) of the Dibdibba aquifer (Kuwait) in (a) 1960 and (b) 1988 (modified after Al Sulaimi and Akbar, 1999).

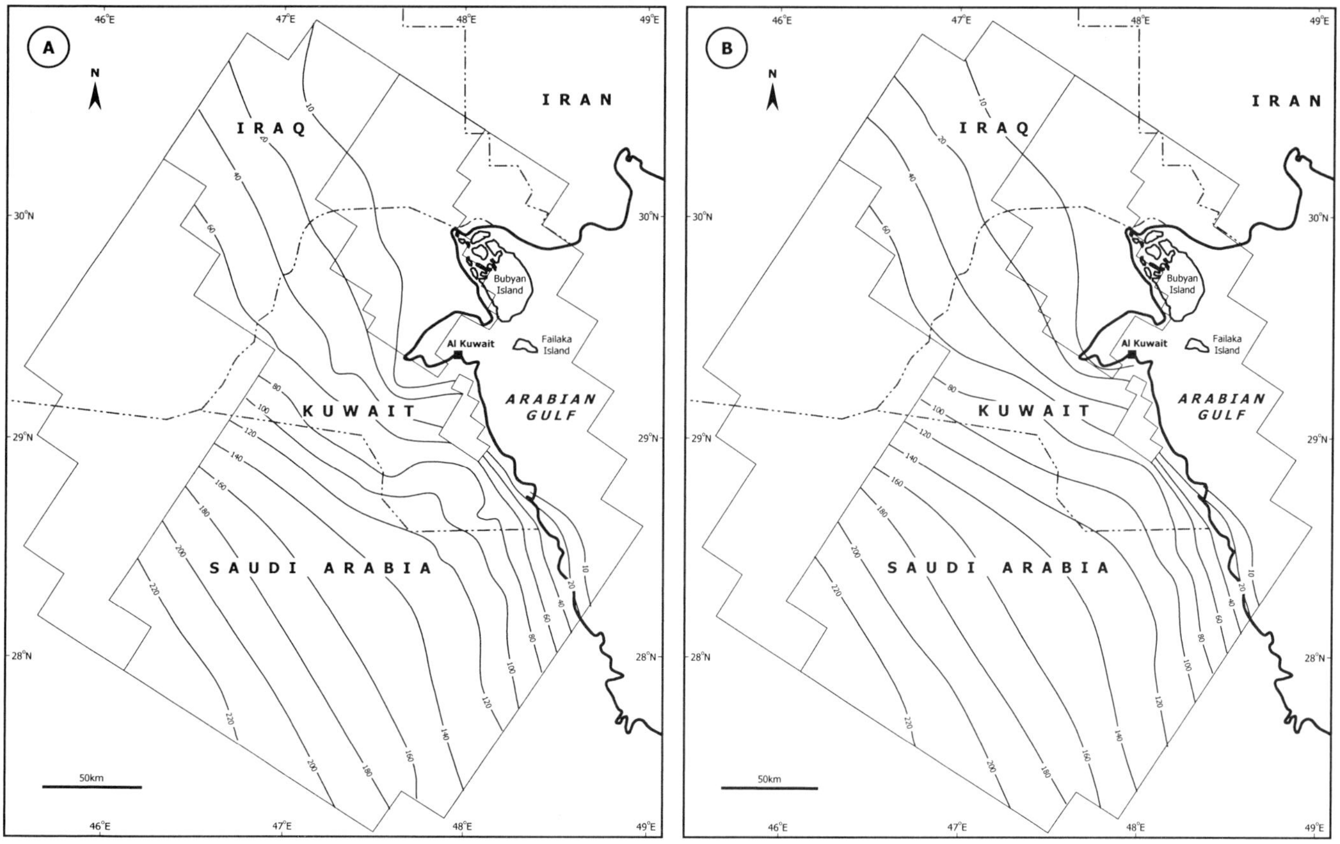

Fig. 11.17. Piezometric surface contour map (in meters above sea level) of the Kuwait Group aquifer (Kuwait) in (a) 1960 and (b) 1988 (modified after Al Sulaimi and Akbar, 1999).

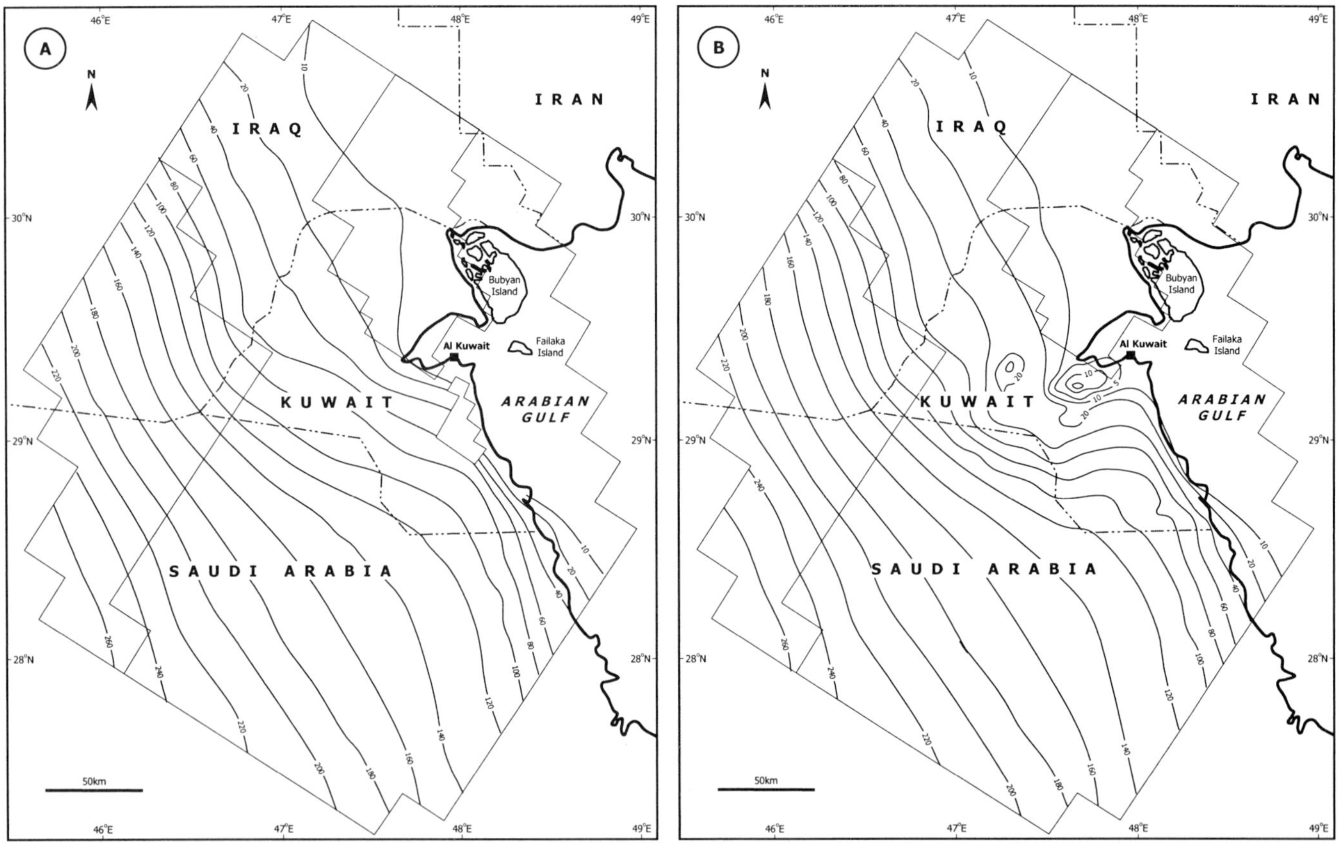

Fig. 11.18. Piezometric surface contour map (in meters above sea level) of the Dammam aquifer (Kuwait) in (a) 1960 and (b) 1988 (modified after Al Sulaimi and Akbar, 1999).

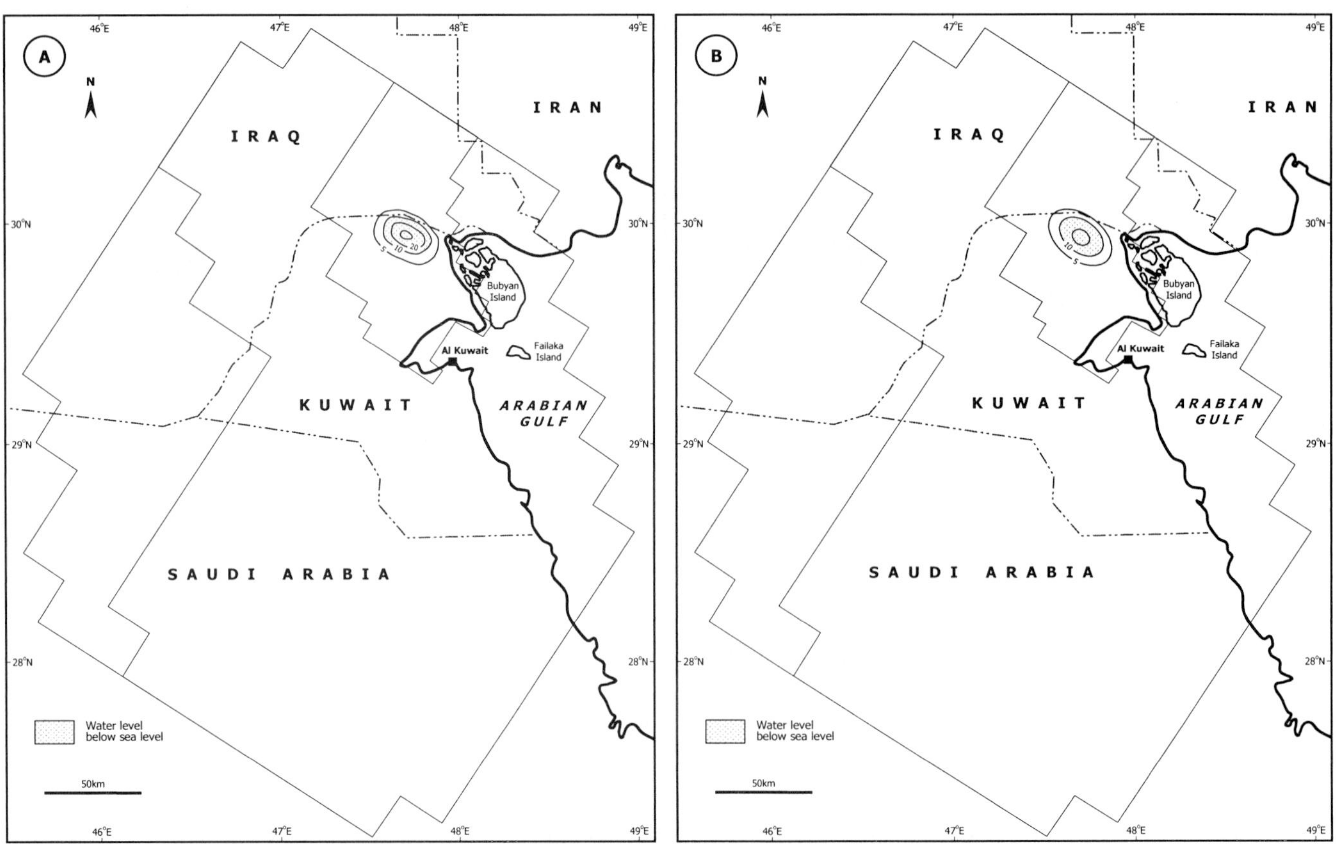

Fig. 11.19. Predicted cone of depression (in meters) in the Dibdibba aquifer (Kuwait) in 2010 according to (a) first scenario and (b) second scenario (modified after Al Sulaimi and Akbar, 1999).

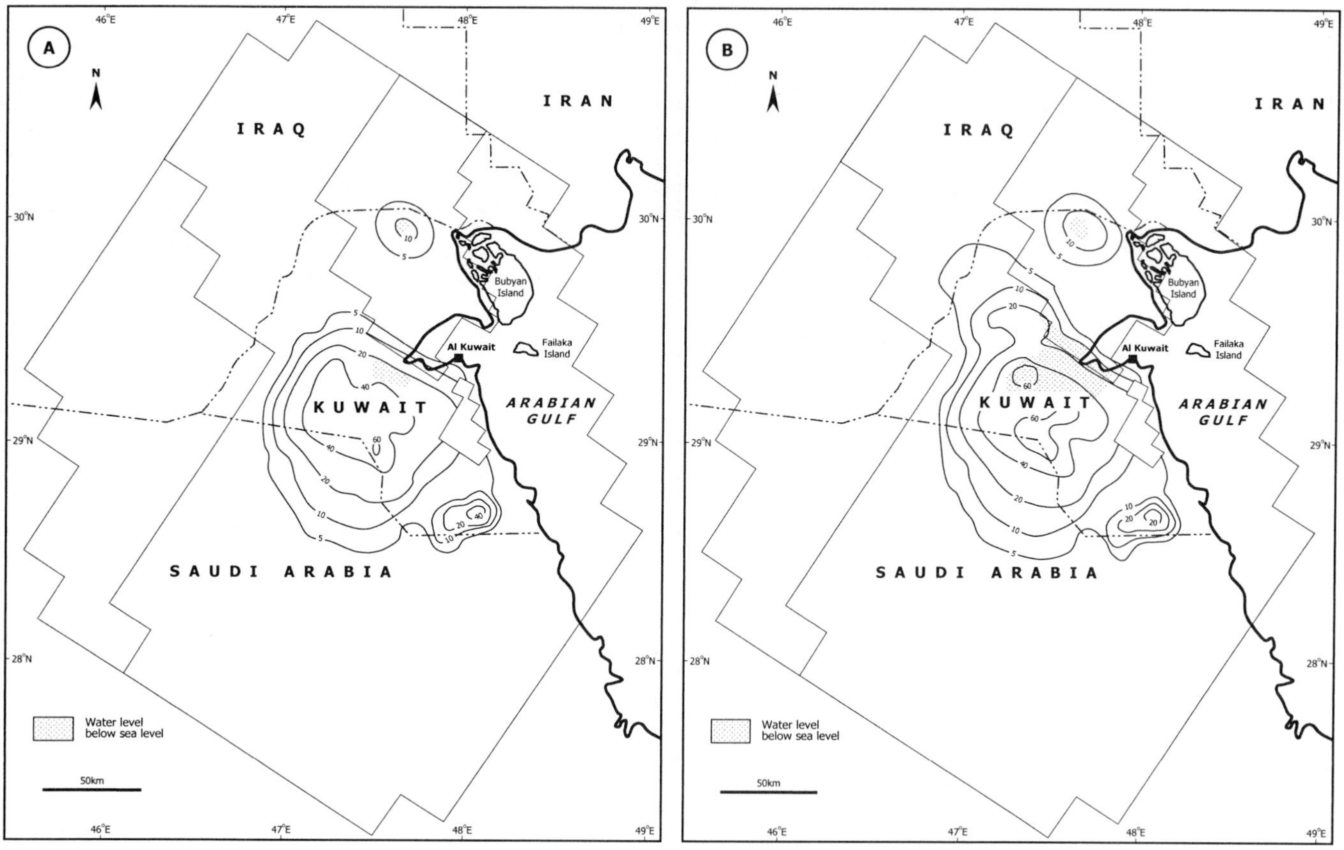

Fig. 11.20. Predicted cone of depression (in meters) in the Kuwait Group aquifer (Kuwait) in 2010 according to (a) first scenario and (b) second scenario (modified after Al Sulaimi and Akbar, 1999).

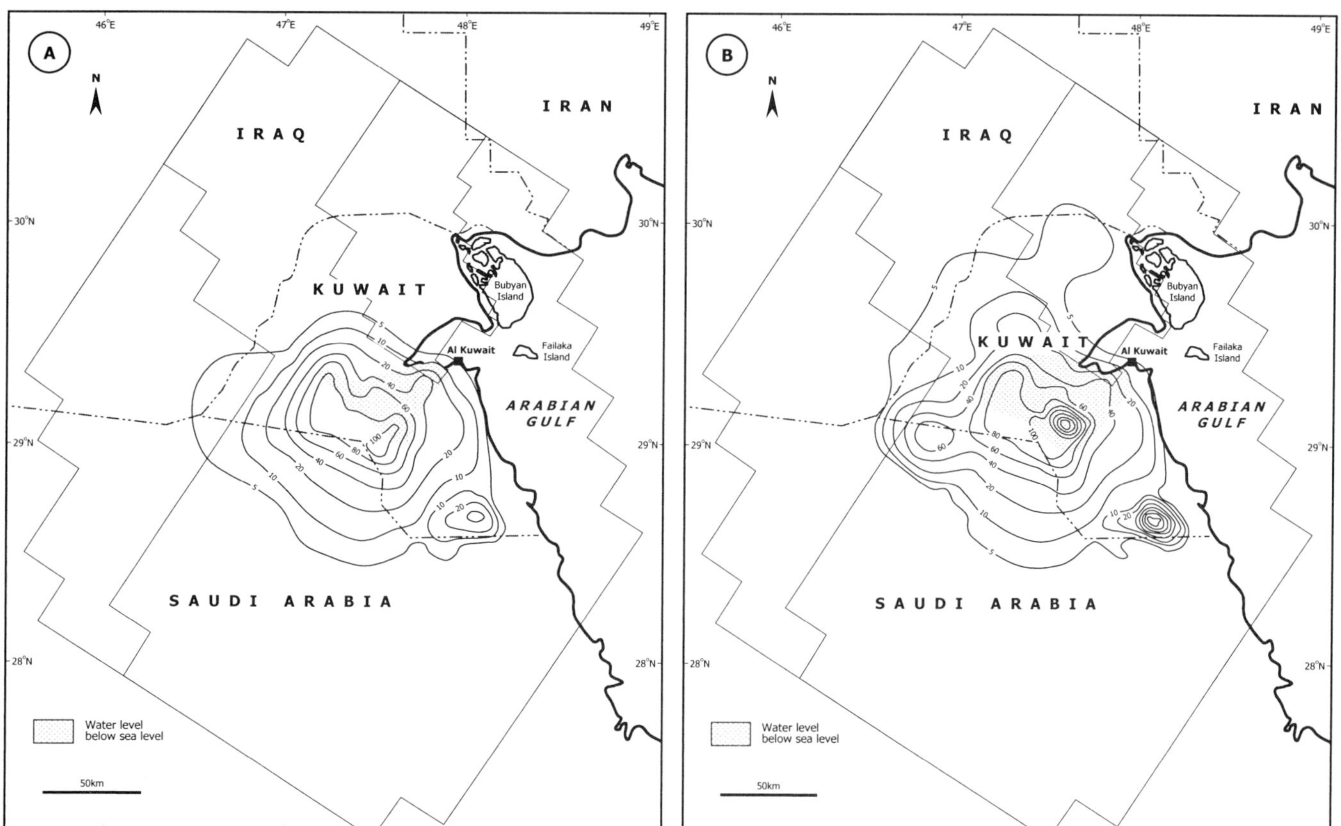

Fig. 11.21. Predicted cone of depression (in meters) in the Dammam aquifer (Kuwait) in 2010 according to (a) first scenario and (b) second scenario (modified after Al Sulaimi and Akbar, 1999).

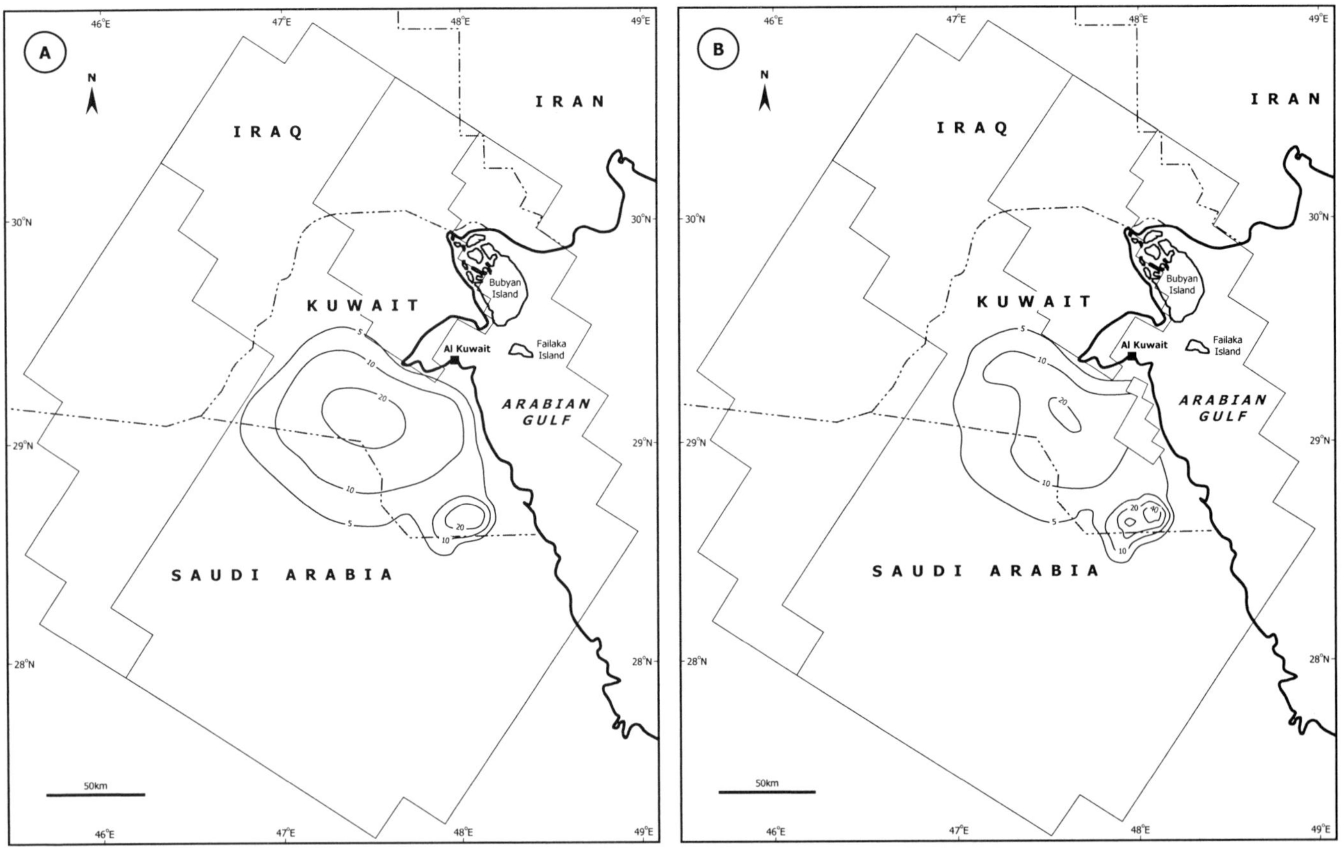

Fig. 11.22. Predicted cone of depression (in meters) in the Kuwait Group aquifer (Kuwait) in 2010 according to (a) third scenario and (b) fourth scenario (modified after Al Sulaimi and Akbar, 1999).

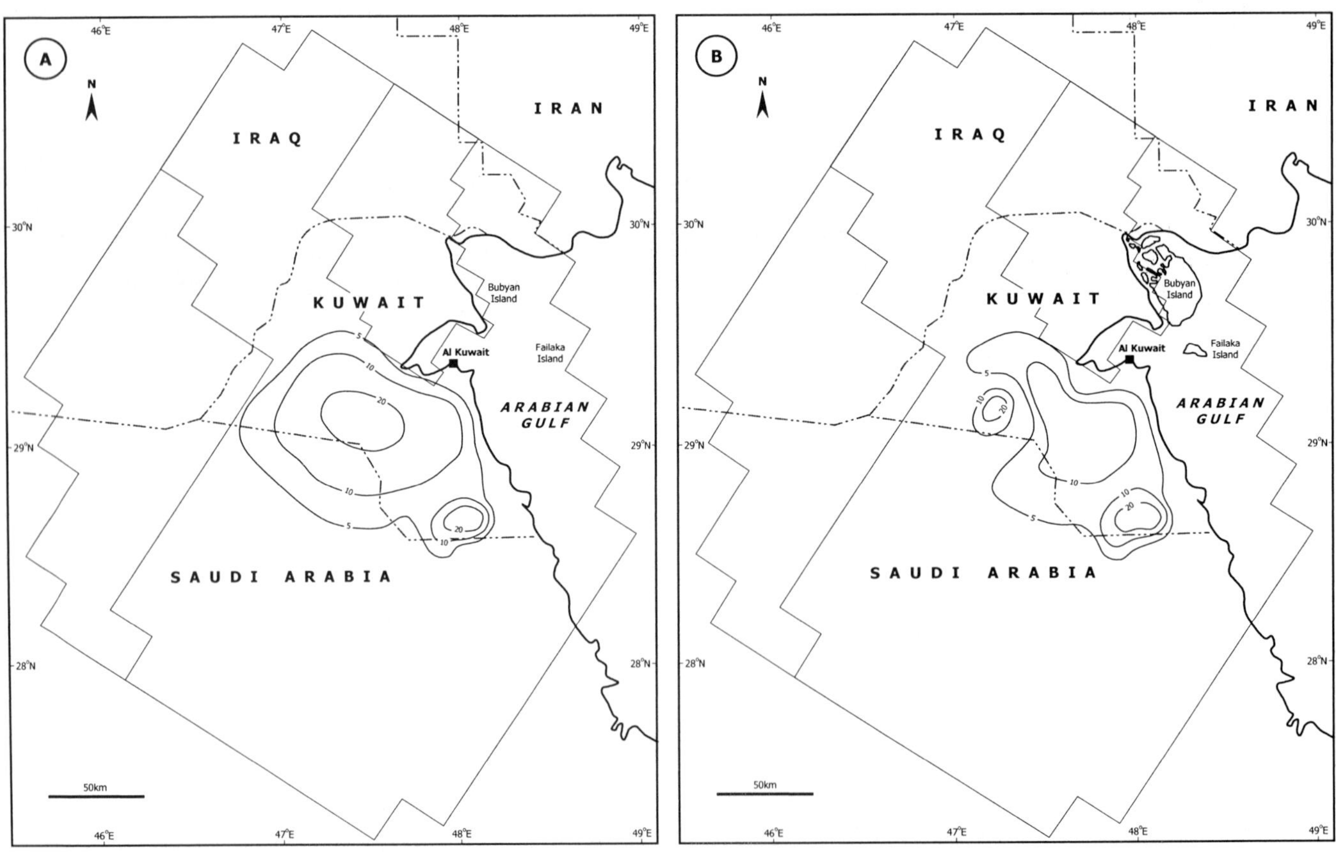

Fig. 11.23. Predicted cone of depression (in meters) in the Dammam aquifer (Kuwait) in 2010 according to (a) third scenario and (b) fourth scenario (modified after Al Sulaimi and Akbar, 1999).

Table 11.10. Groundwater Balance in the Kuwait aquifers for the period 1988–2010 according to the controlled development scenario (summarized from Al Sulaimi and Akbar, 1999).

Year	Groundwater Budget (m^3/s)				Aquifer
	Recharge	Discharge	Error	Balance	
1988	0.532	-0.533	-0.001	1988	Upper Kuwait Group Aquifer
1996	0.472	-0.473	-0.001	1996	
2000	0.445	-0.445	0	2000	
2010	0.359	-0.360	-0.001	2010	
1988	3.042	-3.041	-0.001	1988	Lower Kuwait Group Aquifer
1996	3.729	-3.728	-0.001	1996	
2000	3.719	-3.512	-0.206	2000	
2010	3.714	-2.618	-1.096	2010	
1988	3.042	-3.045	-0.003	1988	Dammam Aquifer
1996	3.730	-3.732	-0.002	1996	
2000	3.719	-3.716	-0.003	2000	
2010	3.714	-3.677	37	2010	

Table 11.11. Groundwater balance in the Kuwait aquifers for the period 1988-2010, according to the intensive pumping scenario of model simulations (summarized from Al Sulaimi and Akbar, 1999).

Year	Groundwater Budget (m^3/s)				Aquifer
	Recharge	Discharge	Error	Balance	
1988	0.532	-1.523	0.001	-0.990	Upper Kuwait Group Aquifer
2000	0.100	-1.884	0.001	-1.873	
2010	0.771	-2.162	0.610	-2.323	
1988	1.180	-6.260	0.001	-5.079	Lower Kuwait Group Aquifer
2000	1.186	-10.157	0.006	-8.965	
2010	1.189	-7.853	0.349	-7.022	
1988	3.973	-4.046	0.003	-0.070	Dammam Aquifer
2000	5.809	-5.896	0.002	-0.085	
2010	4.568	-4.550	-0.054	-0.018	

cones of depression associated with groundwater pumping are shown in figures (11.22 and 11.23).

3. Long-term recovery

Assumptions: In 1995 all well fields cease to produce water and are replaced by alternative sources (desalinated and treated wastewater). The groundwater extraction rate rises from 4.5 m^3/s in 1995 to 8.3 m^3/s and remains at this level to 2010. Groundwater pumping from the well fields of the Kuwait Water Company and Ministry of Electricity and Water stop. But groundwater pumping

Table 11.12. Groundwater balance in the Kuwait aquifers for the period 1988-2010, according to the long-term recovery scenario of model simulations (summarized from Al Sulaimi and Akbar, 1999).

Year	Groundwater Budget (m^3/s)				Aquifer
	Recharge	Discharge	Error	Balance	
1988	0.532	-1.523	0.001	-0.990	Upper Kuwait Group Aquifer
1995	0.527	-1.757	0	-1.320	
2010	0.768	-0.691	-0.002	0.075	
1988	1.180	-6.260	0.001	-5.079	Lower Kuwait Group Aquifer
1995	1.180	-7.531	0.002	-6.349	
2010	1.188	-1.655	0.003	0.200	
1988	3.973	-4.046	0.003	-0.070	Dammam Aquifer
1995	4.548	-4.630	0.002	-0.080	
2010	0.932	-0.931	-0.002	-0.001	

continues in the Al Wafrah agricultural area because of its remoteness from sources of treated wastewater. Predicted water budgets according to this scenario are listed in Table 11.12 and the predicted cones of depression are figured in figures (11.22 and 11.23).

Table 11.13. Groundwater balance in the Kuwait aquifers for the period 1988-2010, according to the intensive pumping scenario of model simulations (summarized from Al Sulaimi and Akbar, 1999).

Year	Groundwater Budget (m^3/s)				Aquifer
	Recharge	Discharge	Error	Balance	
1988	0.532	-1.523	0.001	-0.990	Upper Kuwait Group Aquifer
1995	0.527	-1.757	0	-1.230	
2000	0.752	-0.632	0	0.120	
2010	0.859	-0.750	-0.001	0.110	
1988	1.180	-6.260	0.001	-5.079	Lower Kuwait Group Aquifer
1995	1.180	-7.531	0.002	-6.349	
2000	3.013	-3.225	0.002	-0.214	
2010	3.010	-1.750	-0.014	1.246	
1988	3.973	-4.046	0.003	-0.070	Dammam Aquifer
1995	4.548	-4.630	0.002	-0.080	
2000	2.107	-2.096	-0.003	0.008	
2010	2.107	-2.094	-0.002	0.011	

4. Artificial Recharge

Assumptions: Artificial injection of excess desalinated water and treated wastewater artificially into the Dammam aquifer at specified rates. In 1995 all well fields ceases to produce water and are replaced by alternative sources (desalinated and treated wastewater). The groundwater pumping from the well fields of Kwait Water Company and Ministry of Electricity and Water stop. But groundwater pumping continues in the Al Wafrah agricultural areas because its remoteness from the sources of treated wastewater. Predicted water budgets according to this scenario are listed in Table 11.13 and the predicted cones of depression are figured in figures (11.22 and 11.23).

Groundwater-Flow Models of the Quaterary Aquifer System in United Arab Emirates

The National Drilling Company of United Arab Emirates in cooperation with the U.S. Geological Survey has initiated the Groundwater Research Project for the Emirate of Abu Dhabi in 1988. Since then, the Groundwater Research Project has drilled over 300 wells all over United Arab Emirates for groundwater evaluation. Among the publications of the Groundwater Research Project staff, are two groundwater flow models published for the Quaternary aquifer system by Khalifa (1999) and Silva and Al Noaimi (1999). The following is a brief discussion on each of these models:

A) Al Jaww Plain Model

This model was constructed by Khalifa (1999) with the use of the U.S. Geological Survey groundwater flow model MODFLOW of McDonald and Harbaugh (1988). The model was used to simulate the predevelopment conditions the Quaternary alluvial aquifer and its response to pumping from the Umm Ghafa well field.

The Al Jaww plain is a gently sloping area, which is bounded by the northern Oman Mountains on the east and Jabal Hafit on the west. The Quaternary aquifer is composed mainly of alluvial sands and gravels with silt and clay interbeds, and is underlain by impermeable silt and clay of the Lower Fars Formation. The saturated thickness of the aquifer ranges from 80m in the east and 100m in the west. The groundwater recharge comes mainly from the northern Oman Mountains in the east, which contribute an average daily recharge of 15,900 m^3 (about 90% of aquifer recharge). Jabal Hafit on the other hand accounts from only 10% of aquifer recharge. Between 1982 and 1995, groundwater level in the Quaternary alluvial within the modeled area has declined by 4 to 12 m as a result of groundwater pumping from the Umm Ghafa well field. The hydraulic conductivity of aquifer ranges from 1.6 m/day close to Oman Mountains and 9.6 m/day near Jabal Hafit.

The northern and southern sides of the model area were simulated as no-flow boundaries (parallel to streamlines), while the recharge (east) and discharge (west) borders were simulated as constant-head boundaries. With a difference of ± 5m, the model-calculated groundwater levels agree reasonably with the levels measured in 9 observation wells. The transient simulation period from 1982 to 1995 reproduced water level decline as a result of groundwater pumping.

With the 1995 groundwater abstraction rate (17,900 m^3/day), the groundwater decline in the Quaternary aquifer at the Al Jaww plain by the year 2005 is predicted to vary between 15 at Umm Ghafa and 25m at Mezyad. Figure (11.24) compares the groundwater budgets under predevelopment conditions with no pumping from the Quaternary aquifer within the modeled area and 1995 conditions with pumping.

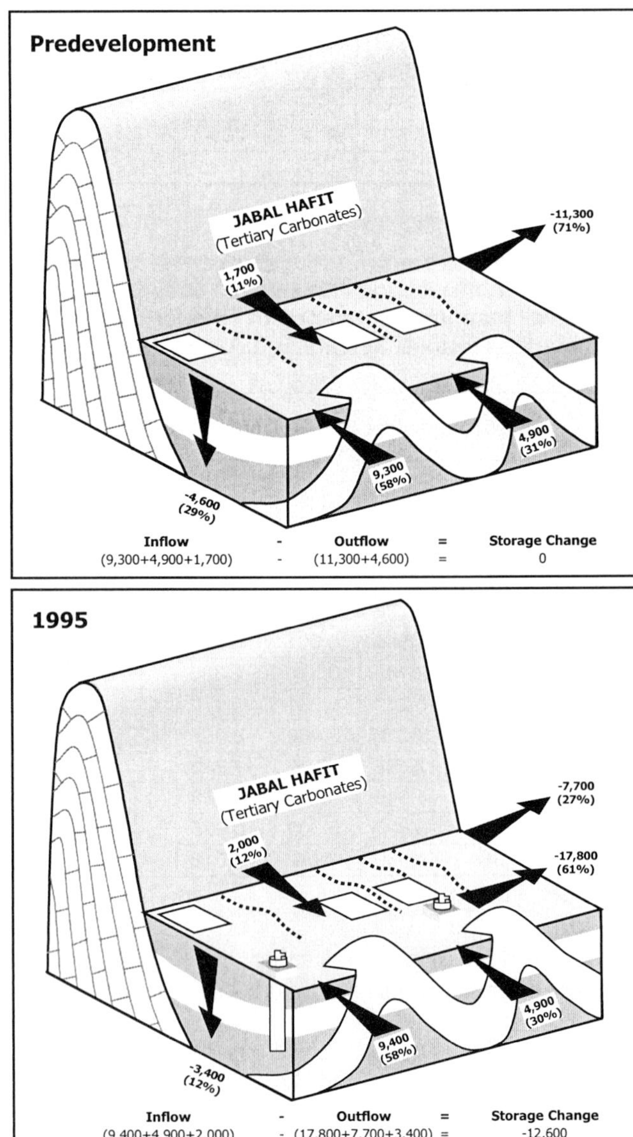

Fig. 11.24. Groundwater budgets under predevelopment and 1995 conditions (in m^3/d) for Quaternary aquifer near Jabal Hafit in the United Arab Emirates (modified after Khalifa, 1997).

Model results indicate that the reduction of groundwater abstraction by 2,000 m^3/day balances the current groundwater consumption taken with the natural recharge stops further water level decline in the Al Jaww plain area.

B) Northeast Abu Dhabi Model

Although numerous wells intercept fresh water of the Quaternary alluvial aquifer in northeastern Abu Dhabi Emirate, about 70,000 m^3/day moves downgradient, reaching brackish groundwater and becomes unfit for drinking or agriculture. A groundwater-flow model was constructed by Silva and Al Noaimi (1999) to evaluate a number of strategies to recover this wasted water.

The main aquifer in the model area is the Quaternary alluvium. Groundwater enters the aquifer at the base of the Oman Mountains as subsurface inflow in buried alluvial channels and as recharge along wadi beds. The aquifer transmissivity ranges from 200 to 1,400 m^2/day and storage coefficient from 0.07 to 0.29 with the high transmissivity zones coinciding with the courses of the alluvial paleochannels. The average annual recharge from rain was estimated to be 3 mm/year. Between 1985 and 1995, the groundwater level declined 1 to 9m. Because of the aquifer saturated thickness rarely exceeds 20m, wells in some areas would become dry unless groundwater pumping is reduced or the aquifer is artificially recharged. In 1995, 31 Mm^3 were pumped for municipal and agricultural uses. Natural groundwater outflow from the model area was estimated at 70,000 m^3/day. This water then mixes with brackish groundwater and becomes unsuitable for most uses.

The simulated period of the model was from 1984 and 2015. The boundaries of the model are the flow boundary at the Oman Mountains front on the east, a constant-head boundary in the west and no-flow boundaries to the north and south. The aquifer was simulated over the period from 1984 to 2015 as a single layer with a free water table.

The simulated hydraulic heads for the 1995-2015 periods indicate that some wells would become dry if the 1995 groundwater-pumping rate continues. Drawdowns of 16 to 20m are calculated in the central portion of the modeled area. Based on the model results strategic placement of wells could recover 35,000 m^3/day now lost through mixing with brackish groundwater and construction of an 80m deep and 17 km long subsurface dam would cause a rise in groundwater levels of 1 to 20m.

REFERENCES

Aba Hussain, A.A., 1992. Sand drift in Bahrain: A case study in Al-Areen wildlife sanctuary. *College of Applied Sciences, Arabian Gulf University Publication*, Bahrain, 255 p.

Abdel Magid, Y., 1995. Planning water resources development in arid zones: An agenda for action in the Arab Region. *Proceedings of the International Conference on Water Resource Management in Arid Countries*, Muscat, Sultanate of Oman, 12-16 March 1995, v. 3, p. 47-54.

Abdel Rahman, H.A. and Omezzine, A., 1996. Aflaj water resources management: Tradable water rights to improve irrigation productivity in Oman. *Water International*, v. 12, p. 70-75.

Abderrahman, W., 1979. The water potential of the Hasa Oasis. *Hofuf Agricultural Research Center*, Saudi Arabia, Publication No. 38, 142 p.

Abderrahman, W.A., 1990. Effect of groundwater use on the chemistry of spring water in Al-Hassa oasis: Journal of King Abdulaziz University Earth Science vol. 3, p. 259-265.

Abu Jaber, N., Kimberley, M. and Cavaroc., V., 1989. Mesozoic-Paleogene basin development within the Eastern Mediterranean borderland. *Jour. Petrol. Geol.*, v. 12, p. 419-436.

Abul Al Enien, H., 1996. Wadi al Bih alluvial fan water resources. *Published by Geographic Department, Kuwait University and Kuwait Geographic Association*, Paper no. 189, 78 p. (in Arabic).

Ahmed, B.Y., 1993. Climate of Saudi Arabia: Geographic Letters. *Published by Kuwait Geographic Society*, 143 p.

Akiti, T., Gonfiantini, R. and Kulaib, A., 1992. Aspects of environmental isotope hydrogeology of the United Arab Emirates. *Proceedings of First Gulf Water Conference*, Dubai, United Arab Emirates, 10-13October 1992, v. 1, 13 p.

Akiti, T.T, Gonfiantini, R. and Mutawa, A., 1988. Aspects of isotope hydrology of the United Arab Emirates. *Paper presented at International Atomic Energy Agency Advisory group meeting, Arid Zone hydrology*, IAEA Vienna, Austria, 50 p.

Al Alawi, J. and Abdulrazzak, M., 1994. Water in the Arabian Peninsula: Problems and perspectives. *In: Peter Rogers and Peter Lydon (ed.), Water in the Arab World: Perspectives and Prognoses*. Published by Harvard University Press, U.S.A., p. 171-202.

Al Asam, M.S., 1997. Application of geophysical and geochemical techniques for the assessment of groundwater recharge from Wadi Al Bih dams, Ras Al Khaimah, United Arab Emirates. *M.Sc. Thesis, United Arab Emirates University*, 154p.

Al Asam, M.S. and Abdul-Rahim, A., 1994. The use of sewage treated water in the United Arab Emirates. *Green Age, Japan*, v. 3, p. 2-5.

Al Awadi, E. and Mukhopadhyay, A., 1995. Hydrogeology of the Dammam Formation of the Gudair Area, Kuwait. *Proceedings of the International Conference on Water Resources Management in Arid Countries*, Muscat, Sultanate of Oman, 12-16 March 1995, v. 2, p. 428-435.

Al Dhowaila, K.H. and Al Mutlak, A.A., 1999. Optimization of sand filtration process for Tertiary treatment. *Proceedings of the 4th Gulf Water Conference,* February 13-17, 1999, Bahrain, v. 2, p. 887-902.

Al Hajari, S.A., 1990. Geology of the Tertiary and its influence on the aquifer system of Qatar and eastern Arabia. *Ph.D. Thesis, Department of Geological Sciences, University of South Carolina*, 246 p.

Al Hawsani, A.H., 1996. Water resources in the United Arab Emirates. *Unpublished Report, Department of Water and Electricity,* Abu Dhabi, UAE, p. 683-768.

Al Khadi, A. and Hancock, P.L., 1980. Structure of the Dhruma-Nisah segment of the Central Arabian Graben System. *Directorate General of Mineral Resources,* Bull. no. 16, Jeddah, Saudi Arabia, 40 p.

Al Laboun, A.A., 1986. Stratigraphy and Hydrocarbon Potential of the Paleozoic Succession in both Tabuk and Widyan Basins, Arabia, *In: M.T. Halbouty (ed.), Future Oil and Gas*

Provinces of the World. Amer. Assoc. Petrol. Geol. Mem., v. 49, p. 373-397.

Al Laboun, A.A., 1987. The Distribution of Permo-Carboniferous silicastic rocks in the greater Arabian Basin, *Amer. Assoc. Petrol. Geol. Bull.*, v. 100, p. 362-373.

Al-Lamki, M.S.S., 1992. Regional Hydrogeology of South Oman. *University College London, M.Sc. Thesis*. 79 p.

Al Lamki, M.S. and Terken, J.M., 1996. The role of hydrogeology in petroleum development, Oman. *GeoArabia*, v. 1, No. 4, p 495-510.

Al Magid, E.H., 1988. Climate of Qatar State. *Ministry of Communication and Transportation, Department of Meteorology*, Doha, Qatar, 76 p.

Al Mutaz, I.S. and Al Sultan, B., 1997. Operation characteristics of Maufouha reserve osmosis plants. *Paper presented at IDA World Congress on Desalination and Water Reuse*, Madrid, Spain, October 6-9, 1997, 25 p.

Al Noaimi, M.A., 1993. Evaluation of Bahrain available water resources, present utilization, and consuming sectors future needs, v. 1: Evaluation of available water resources and utilization patterns. *Report prepared for Bahrain Water Research Committee and published by Bahrain Center for Studies and Research*, Bahrain.

Al Noaimi, M.A., 1999. Evaluation of available water resources and their uses in the State of Bahrain. *Bahrain Center for Studies and Research, Series of Studies and Scientific Research Papers, no. 24*, 131 p. (in Arabic).

Al Ruwaih, F.M., 1985. Hydrochemical classification of the groundwater of Umm Al-Aish, Kuwait. *Journal of University of Kuwait (Science)*, v. 12, p. 287-296.

Al Ruwaih, F.M., 1984. Groundwater chemistry of Dibdibba Formation, North Kuwait. *Groundwater*, v.22 (4), p.412-417.

Al Ruwaih, F.M., 1994. Patterns of groundwater chemistry, Kuwait Group aquifer, Kuwait, *Jour. Env. Hydrol.*, v. 2 (1), p. 24-35.

Al-Ruwaih, F.M., 1995. Assessment and management of a desert basin in Kuwait. *Water International*, v. 20, p. 213-224.

Al Saati, A.J., 1995. Domestic wastewater reuse in the Gulf Corporation Council. *Attawun*, v. 10 (38), p. 33-46. (in Arabic).

Al Sabahi, A.M., 1997. Wastewater treatment and reuse in the Sulatante of Oman. *Proceedings of the 3rd Gulf Water Conference*, 8-13 March 1997, Muscat, Sulatanate of Oman, v. 3, p. 859-888.

Al Sayari, S.S. and Zötl, J.G. (eds.), 1978. Quaternary Period in Saudi Arabia. *Springer-Verlag*, Wein, New York, 334 p.

Al Sulaimi, J. and Akbar, A., 1999. Groundwater in Kuwait (in Arabic). Published by Kuwait Foundation for Advancement of Science, 273 p.

Al Sulaimi, J. and Mukhopadhyay, A., 2000. An overview of the surface and near-surface geology, geomorphology and natural resources of Kuwait. *Earth Science Reviews*, v. 50, p. 227-267.

Al Sulaimi, J., Amer, A., Saleh, N., Salman, A., Ayyash, J. and El-Sayed, M., 1988. Delineation of watersheds in Kuwait and their geomorphological characteristics. Report No. KISR 2856, Kuwait Institute for Scientific Research, 70 p.

Al Sulaimi, J.S. and Al Rabaa, S.M., 1994. Morphological and morphostructural features of Kuwait. *Geomorphology*, v. 11, p. 151-167.

Al Wahedi, A.A., 1997. Application of hydrogeochemistry and groundwater modelling techniques for assessment of water resources of Wadi Al Bih drainage basin, Ras Al Khaimah area, United Arab Emirates. M.Sc. Thesis, United Arab Emirates University. 135 p.

Alavi, M., 1994. Tectonics of the zagros orogenic belt of Iran: New data and interpretation. *Tectonophysics*, v. 229, p. 211-238.

Al Awadhi, N., Puskas, K. and Al-Muzaini, S., 1994. The use of treated wastewater for greenery development in the State of Kuwait. The Joint Saudi-Japanese Workshop, 1994.

Al-Bassam, A.M., Al-Dabbagh, M.E. and Hussein, M.T., 2000. Application of a revised hydrostratigrpahical classification and nomenclature to the Mesozoic and Cenozoic succession of Saudi Arabia. *Jour. Afr. Earth. Sci.*, v. 30 (4), p. 917-927.

Al-Haddad, M., Abu-Jarad, F. and Al-Jabri, I., 1998. Measurements of Radon-222 concentrations in some water samples in Eastern Saudi Arabia using liquid scintillation counter. *Arab. Jour. Sci. Eng.*, v. 23 (1C), p. 145-150.

Al-Hajj, Y.A.M., 1995. Integrated groundwater resource management in Qatar. *Paper presented at Sixth Regional Meeting of the IHP National Committees of the Arab Region,* Amman, Jordan (in Arabic).

Al-Hajji, Y.Y., 1976. A quantitative hydrological study of field "A", South Kuwait. *M.Sc. Thesis, Faculty of the Graduate College of Ohio University,* 124 p.

Allan, J.A., 1995. Water deficits and management options for arid regions with special reference to the Middle East and North Africa. *Proceedings of the International Conference on Water Resources Management in Arid Countries,* Sultanate of Oman, Muscat, 12-16 March 1995, v. 1, p. 1-8.

Al-Marshudi, A., 1995. Operation of the falaj system. In: Traditional agriculture and fishing in the Sultanate of Oman. *College of Agriculture, Sultan Qaboos University,* Muscat, Sultanate of Oman, p. 5-10.

Al-Mogren, A., 1995. Water resources in the Gulf Corporation Council Countries: Natural resources. *Attawun,* v. 10 (38), p. 13-32. (in Arabic).

Al-Mojil, 1963. Report on drilling programme. Unpublished report to Government of Qatar, 14p.

Al-Muzaini, S.M. and Ismail, A.A., 1994. Irrigation water and the national plan for greening Kuwait. *Proceedings of 2nd Gulf Water Conference: Towards an Integrated Water Resources Management,* 5-9 November, Bahrain (in Arabic) p. 89-98.

Alqubaisi, H.A., Zubari, W.K. and Aba Hussain, A.A., 1999. Effect of groundwater depletion on natural springs and surrounding environment in Bahrain. *Proceedings of 4th Gulf Water Conference,* Manama, Bahrain, p. 159-174 (in Arabic).

Al-Saafin, A.K., Bader, T. A., Shhata, W., Hoetzl, H., Wohnlich, S. and Zoetl, J.G., 1989, Groundwater recharge in an arid karst area in Saudi Arabia (Abstract). *Paper Presented at 28th International Geological Congress in Washington, D.C.*

Al-Sagaby, I.A. and Moallim, A.M., 1999. Temporal variation of groundwater chemistry in Saq aquifer, Qassim Region, Saudi Arabia. *Proceedings of 4th Gulf Water Conference,* Bahrain, v.1, p. 243-255.

Alsajwani, T.M.A. and Lawrence, R.J., 1995. Small desalination plants in rural communities in Oman. *Proceedings of the International Conference on Water Resources Management in Arid Countries,* Muscat, Sultanate of Oman, v.2, p. 617-626.

Al-Shaqsi, S.R.S., 1994. Integrated water resources management (Sultanate of Oman). *Proceedings of 2nd Gulf Water Conference: Towards an Integrated Water Resources Management,* Bahrain, 5-9 November, 1994, (in Arabic) vc.1, p. 64-77.

Alsharhan, A.S. and Kendall, C.G.St.C., 1986. Precambrian to Jurassic rocks of the Arabian Gulf and adjacent areas: their facies, depositional setting and hydrocarbon habitat. *Amer. Assoc. Petrol. Geol. Bull.*, v. 70, p. 977-1002.

Alsharhan, A.S. and Nairn, A.E.M., 1986. A review to the Cretaceous formations in the Arabian Peninsula and the Gulf: Part I. Lower Cretaceous (Thamama Group) Stratigraphy and Paleogeography. *Jour. Petrol. Geol.*, v. 9, p. 365-391.

Alsharhan, A.S. and Nairn, A.E.M., 1988. A review of the Cretaceous formations in the Arabian Peninsula and Gulf: Part II, Mid-Cretaceous (Wasia Group), stratigraphy and paleogeography, *Jour. Petrol. Geol.*, v. 11, p. 89-112.

Alsharhan, A.S. and Nairn, A.E.M., 1990. A review of the Cretaceous formations in the Arabian Peninsula and Gulf: Part III, Upper Cretaceous (Aruma Group) stratigraphy and paleogeography. *Jour. Petrol. Geol.*, v.13, p. 247-266.

Alsharhan, A.S. and Nairn, A.E.M., 1994. Geology and hydrocarbon habitat in the Arabian Basin: the Mesozoic of the State of Qatar. *Geologie en Mijnbouw*, v. 72, p. 265-294.

Alsharhan, A.S. and Nairn, A.E.M., 1995. Tertiary of the Arabian Gulf: Sedimentology and hydrocarbon potential. *Palaeogeogr., Palaeoclimatol., Palaeoecol.*, v. 114, p. 369-380.

Alsharhan, A.S. and Nairn, A.E.M., 1997. Sedimentary Basins and Petroleum Geology of the Middle East. *Elsevier Science*, 942 p.

Alsharhan, A.S., Nairn, A.E.M. and Mohammed, A., 1993. Late Palaeozoic glacial sediments of the southern Arabian Peninsula: their lithofacies and hydrocarbon potential. *Mar. Petrol. Geol.*, v. 10, p. 71-78.

Al-Yahyai, R., 2000. The traditional aflaj irrigation system. Published in Internet under the site *www.geocities.com/RainForest/8883/aflaj.html*,p.1-9.

Atkinson, G.W. and Eccleston, B.L., 1986. Water resource and their utilization in the Arab World. Country Report for Qatar, Arab Fund for Economic and Social Development, Kuwait Symposium 12-15 January, 1986.

Atlas of Saudi Arabia, 2000. Published by Saudi Arabian Ministry of Higher Education, 2nd edition, Al Obaikan Publishing Co., 245 p.

Authman, M.N., 1983. Water and Development Process in Saudi Arabia. *Tihama Press, Jeddah*, Saudi Arabia, 302 p.

Bahrain Petroleum Company (BAPCO), 1952. Report on natural springs flow survey, Bahrain Islands, unpublished BAPCO file report.

Bakhit, D.W.M. and Nairn, A.E.M., 1997. A proposed water conservation plan for the Fujairah Emirate. *Journal of Faculty of Science, United Arab Emirates University*, v. 9 (1), p. 171-201.

Bakhit, D.W.M., 1998. Environmental and management problems in the hydrology of the United Arab Emirates. *Unpublished Ph.D. Thesis, University of South Carolina*, Columbia, 407 p.

Bakiewicz, W., Milne, D.M. and Noori, M., 1982. Hydrogeology of the Umm er Radhuma aquifer, Saudi Arabia, with reference to fossil gradients. *Quarter. Jour. Eng. Geol. London*, v. 15, p. 105-126.

Ball, D.M., Al-Hussein, A.H., Al-Jada'an, K. and Stoner, R.F., 1981. Groundwater flow system in Eastern Saudi Arabia. *Presented at the Arab Water Technology Exhibition*, Dubai, 1-5 November, 1981.

Ball, W.J. and Nordstorm, D.K., 1992. WATEQ4F: A program for the calculating speciation of major trace and redox elements in natural waters. *International Groundwater Modelling Center*, Colorado School of Mines, Golden, C0 80401, U.S.A., 189 p.

Ballantyne, W., 1990. A ressertion of the Shari'ah: The jurisprudence of the Gulf States. *In: Heer, N. (ed.), Islamic Law and Jurisprudence, University of Washington Press*, Seattle and London, p. 149-159.

Bates, R.L. and Jackson J.A. (eds), 1987. Glossary of geology (3rd edition). *American Geological Institute*, Alexandria, Virginia, 788 p.

Bazuhair, A.S. and Hussein, M.T., 1990. Springs in Saudi Arabia. *Journal of King Abdulaziz University, Earth Science*, v. 3, p. 251-258.

Behre, S.M., 1986. Geologic and geochronologic constraints on the evolution of the Red Sea - Gulf of Aden and Afar depression. *Jour. Afr. Earth. Sci.*, v. 5, p. 101-117.

Beydoun, Z.R., 1980. Some holocene geomorphological and sedimentological observations from Oman and their Palaeogeological Implications. *Jour. Petrol Geol.*, v. 2, p. 427-437.

Beydoun, Z.R., 1988. The Middle East: Regional geology and petroleum resources. *Scientific Press Ltd., Beaconsfield Bucks*, U.K., 292 p.

Breed, C.S. and Reheis, M.C., 1999. Desert winds: Monitoring wind-related surface processes in Arizona, New Mexico and California, US Geological Survey Prof. Paper 1598, 153 p.

Buday, T. 1980. The regional geology of Iraq. Stratigraphy and paleontology. *State Organization for Minerals Library*, Baghdad, Iraq, 445 p.

Burdon, D.J. and Al-Sharhan, A., 1968. The problem of the paleokarstic Dammam Limestone aquifer in Kuwait. *Journal of Hydrology*, v.6 (4), p. 385-404.

Burdon, D.J., 1966. Report to the Government of States of Kuwait on investigation of the Dammam Limestone aquifer in Kuwait. *Unpublished report prepard by FAO, Rome for the Government of Kuwait.*

Burdon, D.J., 1977. Flow of fossil groundwater. *Quarter. Journ. Eng. Geol. London*, v. 10, p. 97-124.

Burdon, D.J., 1982. Hydrogeological conditions in the Middle East. *Quarter. Journ. Eng. Geol. London*, v. 15, p. 71-82.

Bureau de Recherches Gèologiques et Minières, 1976. Hydrogeological Investigaitons of the Wasia aquifer in the Eastern Province of Saudi Arabia, Regional study final report and appendices. *Ministry of Agriculture and Water,* Riyadh, Saudi Arabia.

Bureau de Recherches Geologiques et Minieres, 1977. Al Hasa Development Project Groundwater Resources Study and Management Program. *Reports prepared for Ministry of Agriculture and Water,* Riyadh, Saudi Arabia, p. 1-136.

Bureau de Recherches Geologiques et Minieres, 1985. Agriculture and soil studies of Saq and overlying aquifers. *Study of Ministry of Agriculture and Water,* Riyadh, Saudi Arabia.

Cavelier, C., Al-Salatt, A. and Huuze, Y., 1970. Geological description of the Qatar Peninsula. *Department of Petroleum Affairs,* Qatar, 39 p.

Chebotarev, I.I., 1955. Mechanisms of natural waters in the crust of weathering. *Geochim. Cosmochim. Acta,* v. 8, p. 22, 137, 198.

Clark, I.D., 1988. Groundwater resources in the Sultanate of Oman: Origin, circulation times, recharge processes and paleoclimatology. Isotopic and geochemical approaches. *Ph.D. Thesis, University Paris XI,* Orsay, 264 p.

Clark, I.D., Fritz, P., Quinn, O.P., Rippon, P.W., Nash, H., 1987. Modern and fossil groundwater in an arid environment: A look at the hydrogeology of Southern Oman. Isotope techniques in water resources development. *Proceedings of an International Symposium held March30-April 3, 1987, International Atomic Energy Agency.* Vienna, Austria, p. 167-187.

Collins, G.A., 1975. Geochemistry of oil field water: Developments in petroleum sciences 1. *Elesevier Scientific Publishing Company,* Amsterdam, Oxford-New York.

Cressey, G.B., 1958. Qanats, karez, and foggras. *Geographical Review,* v. 48, p. 24-44.

Dabbagh, A.E. and Abderrahaman, W.A., 1997. Management of groundwater resources under various irrigation water use scenarios in Saudi Arabia. *Arab. Jour. Sci. Eng.,* v. 22, p. 47-264.

Dabbagh, T., Sadler, P., Al-Saqabi, A. and Sadeqi, M., 1994. Desalination: an emergent option. *In: P. Rogers and P. Lydon (eds.), Water in the Arab World - Perspectives and Prognoses,* Harvard University, Division of Applied Sciences, p. 203-242.

Davidson, R., 1991. Performance up and costs down, Windpower Monthly (November), p. 15.

Davis, H., 1996. Hydrogeologic investigation and simulation of groundwater flow in the Upper Floridan aquifer of north-central Florida and southwestern Georgia and delineation of contributing areas for selected city of Tallahassee, Florida, water-supply wells. U.S. Geological Survey, Water Resources Investigations Report 95-4296, 56 p.

De Jong, E.L., Yazicigil, H. and Al-Layla, R.I., 1989. Scenario Planning for Water Resources. A Saudi Arabian Case Study. *Water International,* v.14, p. 6-12.

Dewdney, J., 1988. The Cambridge Atlas of the Middle East and North Africa. Edited by G. Blake, J. Dewdney and J. Mitchell, *published by Cambridge University Press,* U.K., 74 p.

Dincer, T., 1978. The use of environmental isotopes in arid zone hydrogeology. *International Atomic Energy Agency Symposium on Application of Isotope Techniques to Arid Zone Hydrology,* Vienna, Austria, p. 23-30

Dincer, T., Al-Mugrin, A. and Zimmermann, U., 1974. Study of the infiltration and recharge through the sand dunes in arid zones with special reference to the stable isotopes and thermonuclear Tritium. *Journal of Hydrology,* v. 123, p. 79-109.

Dolatyar, M and Gray, T.S., 2000. Water politics in the Middle East: A context for conflict or cooperation? *Macmillan Press Limited,* U.K., 243p.

Doornkamp, J.C., Brunsden, D. and Jones, D.K.C., 1980. Geology, Geomorphology and Pedology of Bahrain. *Geo Abstracts,* U.K., 443p.

Duce, R.A., Winchester, J.W. and Van Nahl, T.W., 1963. Iodine, bromine and chlorine in the Hawaiian marine atmosphere. *Jour. Geophysic. Res.,* v. 68, p. 3943-3947.

Eccleston, B. and Harhash, I., 1982. The hydrogeology of Qatar. Water Resources and Agricultural Development Project, phase III, 173 p.

Eccleston, B., Pike, I. and Harhash, I., 1981. The water resources of Qatar and development. Technical note No. 5, Water Resources and Agricultural Development Project, 390 p.

Economic Bulletin, 1999. Vol. 14, p. 1-225. Published by Gulf Cooperation Council, Riyadh, Saudi Arabia.

Edgell, H.S., 1987. Geology of studies areas, in groundwater resources evaluation of Saudi Arabia: A long-term strategic plan for fresh groundwater use. *King Fahd University of Petroleum and Minerals Press*, Dhahran, Saudi Arabia, p. 24-77.

Edgell, H.S., 1990. Geological framework of Saudi Arabian groundwater resources. *Journal of King Abdul Aziz University, Jeddah, Earth Sciences*, v.3, p. 267-286.

Edgell, H.S., 1990. Karst in northeastern Saudi Arabia. *Journal of King Abdulaziz University, Jeddah, Earth Sciences*, v. 3, p. 81-94.

Edgell, H.S., 1997. Aquifers of Saudi Arabia and their geological framework. *Arab. Jour. Sci. Eng.*, v. 22, p. 3-31.

El-Khatib, A.B., 1980. Seven Green Spikes-water and agricultural development, 2nd ed., *Ministry of Agriculture and Water*, Riyadh, Saudi Arabia, 362p.

El-Shamy, F., 1990. The hydrochemistry of the spring at Ain bu Sukhnah, U.A.E. *Arab J. Scient. Res.*, v. 8 (1), p. 33-49.

Embabi, N.S., 1991. Dune types and patterns in the United Arab Emirates using Landsat TM-data: *The 24th International Symposium of Remote Sensing of Environment,* Rio de Janeiro, Brazil, May, 1991, 12 p.

Erikson, E., 1983. Stable isotopes and tritium in precipitation. Guidebook on nuclear techniques in hydrology. Technical Report Series No. 91, *International Atomic Energy Agency*, Vienna, Austria, p. 19-33.

Food and Agricultural Organization of the United Nations, 1979. Shared water resources in the Gulf States. United Nations, v. 1, 310 p.

Ferguson, A.D. and Hill, W.G., 1953. Land Springs Survey, Bahrain. *Bahrain Petroleum Company.*

Fetter, C.W., 1988. Applied hydrogeology. *Macmillan Publishing Company,* New York, second edition, 592 p.

Flavin, C. and Jenssen, N., 1994. Power Surge. Norton, New York, 382 p.

Flavin, C., 1981. Windpower: A turning point. World Watch Paper 45.

Fleck, R.J., Greenwood, W.R., Hadley, D.G., Anderson, R.E. and Schmidt, D.L., 1980. Rubidium-strontium geochronology and plate tectonic evolution of the southern part of the Arabian Shield. *U.S. Geol. Survey Prof. Paper, 1131*, 38 p.

Food and Agricultural Organization of the United Nations, 1981. The water resources of Qatar and their development, volume 1.

Freeze, R. A. and Cherry, J.A., 1979. Groundwater. *Prentice-Hall, Inc.*, New Jersey, 604 p.

Garamoon, H.K., 1996. Hydrogeological and geomorphological studies on the Abu Dhabi - Al Ain - Dubai rectangle, United Arab Emirates. *Ph.D. Thesis, Ain Shams University,* Cairo, Egypt, 277 p.

Galloway, D.; Johes, D.R. and Ingebritsen, S.E., (eds.), 1999. Land subsidence in the United States. U.S. Department of the Interior, *U.S. Geological Survey*, Circular 1182, 177p.

German Water Engineering, 1982. Um Ghafa water project, Al-Ain, U.A.E. *Preliminary Consultant Report for Water and Electricity Department*, 11 p.

Gibb, S.A. and Partners, 1970. Water resources survey, supplement to interim report, Subsurface investigations in Al-Ain area. *Department of Development and Public Works*, Abu Dhabi (unpublished report).

Gleick, P.H. (ed.), 1993. Water in Crisis: A guide to the world's freshwater resources. *Published by Oxford University Press,* New York, 473 p.

Glennie, K.W., 1992. Plate tectonics and Oman Mountains. *Tribulus*, v.2 (2), p. 11-21.

Glennie, K.W., 1996. Geology of Abu Dhabi. *In: Osborne, P.E. (ed.) Desert Ecology of Abu Dhabi*, Pisces, Newbury, p. 16-35.

Glennie, K.W., Boeuf, M.G.A., Hughes Clarke, M.W., Moody-Stuart, M., Pilaar, W.F.H. and Reinhardt,

B.M., 1974. Geology of the Oman Mountains: Verhandelingen van het Koninklijk Nederlands geologisch mijnbouwkundig Genootschap. *The Royal Dutch Geological and Mining Society (KNGMG)*, 423 p.

Glennie, K.W., Pugh, J.W. and Goodall, T.M., 1994. Late Quaternary Arabian desert models of Permian Rotliegend Reservoirs. *Shell Exploration Bulletin*, v. 274 (3), p. 1-19.

Gonfiantini, R. and Akiti, T.T., 1985. Isotope investigations as a tool for regional hydrological studies in the United Arab Emirates. *Progress Report for UAE Ministry of Electricity and Water. Prepared by International Atomic Energy Agency*, Vienna.

Gonfiantini, R., 1992. Investigation of ground water resources of the United Arab Emirates by using isotope techniques. *International Atomic Energy Agency Report (UAE/8/92)*, Vienna, Austria, 34 p.

Gorin, G.E., Racz, L.G. and Walter, M.R., 1982. Late Precambrian-Cambrian sediments of Huqf Group, Sultanate of Oman. *Amer. Assoc. Petrol. Geol. Bull.*, v. 66, n. 12, p. 2609-2627.

Grabowski, G.J. Jr. and Norton, I.O., 1995. Tectonic controls on the stratigraphic architecture and hydrocarbon systems of the Arabian Plate. *In: Husseini, M.I. (ed.), Middle East Petroleum Geosciences, Geo'94, Gulf Petrolink*, Bahrain, v. 1, p. 413-430.

Groundwater Development Consultants (International) Ltd, 1979. Umm er Radhuma Study, *Interim Report for the Ministry of Agriculture and Water*, Riyadh, Saudi Arabia.

Groundwater Development Consultants, 1980. Umm er Radhuma study, Bahrain Assignment, v. 3: Groundwater Resources. A report prepared to *Ministry of Agriculture and Water*, Riyadh, Saudi Arabia.

Groundwater Development Consultants, 1983. Project 25/81: Reverse osmosis desalination, aquifer C investigation: A report prepared to *Ministry of Electricity and Water*, State of Bahrain.

Gur, A., 1995. Water reuse options in the Arab region. Paper presented at *Sixth Regional Meeting of the IHP National Committees of the Arab Region*, 3-6 December, Amman, Jordan.

Hamida, I.H. and Yaqubi, A.S., 1979. Hydrogeological and hydrogeochemical aspects of the main groundwater fields in Kuwait, Arabian Gulf. *Proceedings First Geological Congress of the Middle East (GEOCOME-1)*, Ankara, Turkey, p. 271-296.

Hamilton, N.J., 1965, Water Resources of Bahrain Island, Bahrain. *Bahrain Petroleum Company*.

Harhash, I.E. and Nasser, M.H., 1982. The hydrological setting of the Peninsula of Qatar (in Arabic). Proceedings of the First Symposium of the Future Development of Water Resources in the Gulf and Arabian Peninsula, Kuwait, 3-5 March 1981. *Journal of the Gulf and Arabian Peninsula Studies, Kuwait University*, v. 2, p. 151-202.

Harris, P.M., Frost, S.H., Seiglie, G.A. and Schneiderman, N., 1984. Regional unconformities and depositional cycles, Cretaceous of the Arabian Peninsula. *In: Schlee, J.S. (ed.), Interegional unconformities and hydrocarbon accumulation. Amer. Assoc. Petrol. Geol. Mem.*, v. 36, p. 67-79.

Hassan, A.A. and Al-Aidarous, A., 1985. Regional aquifer geology – onshore Abu Dhabi. *Geology Department, ADCO Project Report 1584-50*, Abu Dhabi, U.A.E., 28 p.

Hassan, M.H. and Cagatay, M.N., 1994. Hydrogeochemistry of Khobar aquifer in eastern Saudi Arabia: *Proceedings of 2nd Gulf Water Conference*, Bahrain, v. 2, p. 163-178.

Hassan, M.H., 1998. Major distribution in Khobar aquifer, eastern Saudi Arabia. *Arab. Jour. Sci. Eng.*, v. 23 (1C) p. 67-88.

Henson, F.R.S., 1951. Observation on the geology and petroleum occurrences of the Middle East. *Proceedings of 3rd World Petroleum Congress*, The Hague, v. 1, p. 118-140.

Hill, W.G., 1953. Submarine Springs Survey. Bahrain, *Bahrain Petroleum Company*.

Holm, D.A., 1960. Desert Geomorphology in the Arabian Peninsula. *Science*, v. 132 (3437), p. 1369-1379.

Hötzl, H., Job, C., Moser, H., Rauert, W., Sticher, W. and Zötl, J.G., 1980. Isotope methods as a tool for Quaternary studies in Saudi Arabia. *In: Arid Zone Hydrology Investigations with Isotope Techniques. International Atomic Energy Agency*, Vienna, Austria, p.215-235.

Hötzl, H., Maurin, V. and Zötl, J.G., 1978. Geologic history of the Al Hasa area since the Pliocene. *In: Al-Sayari, S.S. and Zötl, J.G. (ed), Quaternary Period in Saudi Arabia, Springer-Verlag,* Vienna, New York, p. 58-77.

Hounslow, A.W., 1995. Water Quality Data, Analysis and Interpretation. *CRC Press, Inc., Florida,* 397 p.

Hudson, R.G.S. and Chatton, M., 1959. The Musandam limestone (Jurassic to Lower Cretaceous) of Oman Arabia. *Museum National D'histoire Naturelle,* Paris, v. VII, p. 69-91.

Hydroconsult, 1978. Reconnaissance report and development proposals, Abu Dhabi, U.A.E., Eastern Region water resources. *Government of Abu Dhabi, Ministry of Petroleum and Mineral Resources Report,* 126 p.

Ibnouf, M.A.O. and Abdel-Magid, I.M., 1994. Oman water resources: Management, problems and policy alternatives. *Proceedings of the 2nd Gulf Water Conference, 5-9 November, 1994,* Bahrain, v. 2, p. 19-31.

Ibrahim, M.W., Khan, M.S. and Khatib, H., 1981. Structural evolution of Harmaliyah Oil Field, eastern Saudi Arabia. *Amer. Assoc. Petrol Geol. Bull.,* v. 65, n. 11, p. 2403-2416.

International Atomic Energy Agency, 1984. Isotope Hydrology 1983. *Proceedings of Vienna Symposium,* Vienna, Austria.

Ismail, N., 1995. Strategic projection for planning and management of water resources in the Gulf Corporation Council countries. *Attawun,* v. 10 938), p. 47-62 (in Arabic).

Italconsult, 1969. Water and Agriculture Development Studies for Area IV, Final Report, Geohydrology. *Ministry of Agriculture and Water,* Riyadh, Saudi Arabia, 80 p.

Italconsult, 1971. Water and Agricultural Studies in Bahrain. Volume 1, Bahrain Ministry of Commerce and Agriculture, Saudi Arabia.

IWRM (Integrated Water Resources Management), 2000. TAC background Notes, Global Water partnership, Stockholm, 68 p.

Jensen, J., 1993. Safe Drinking Water Act. *In: Arbuckle, J.G., Brownell, F.W., Case, D.R., Halbleib, W.T., Jensen, L.J., Landfair, S.W., Lee, R.T., Miller, M.L., Nardi, K.J., Olney, A.P., Sarvadi, D.G., Spensly, J.W., Steinway, D.M., Sullivan, T.F.P. (eds.), Environmental Law Handbook* (Chapter 8), Government Institutes, Inc. Rockville, MD, p. 251-266.

Jöb, C., 1978. Hydrochemical investigations in the areas of Al Qatif and Al Hasa with some remarks on water samples from Wadi Al Miyah and Wadi Al Sah'ba near Haradh. *In: Al-Sayari, S.S. and Zötl, J.G. (eds.). Quaternary Period in Saudi Arabia,* Springer Verlag, New York, p. 93-134.

Joudi, D.N. and Fok, Y., 1999. Rainwater harvesting in arid regions. *Proceedings of 4th Gulf Water Conference,* 13-17 February, 1999, Bahrain, v. 2, p. 471-489.

Khalifa, A.A., 1995. Surface water and groundwater resources in U.A.E. Culture and Science Society, Meeting on Water Balance in U.A.E., Dubai, 12p.

Khalifa, A.M., 1999. Simulated impacts of pumping from well fields near Umm Ghafa, Eastern Abu Dhabi Emirate. *Proceedings of 4th Gulf Water Conference,* 13-17 February, 1999, Bahrain, p. 317-329.

Khalifa, A.M., 1997. Hydrogeology of the geothermal fractured-rock well field at Jabal Hafit, Abu Dhabi Emirate. *Proceedings of 3rd Gulf Water Conference,* Muscat, Sultanate of Oman, v. 1, p. 125-140.

Khalifa, M.A. and Rizk, Z.S., 1994. Subsurface geology and hydrogeology of Al-Mustawi area, Al-Qasim province, Saudi Arabia. *Symposium on Desert Studies in the Kingdom of Saudi Arabia 'Extant and Implementation', Center for Desert Studies, King Saud University,* Saudi Arabia, v. 3, p. 237-267.

Kharin, N., Tateishi, R. and Harhash, H., 1999. Degradation of the drylands of Asia. *Chiba University Press,* Chiba, Japan, 90 p.

Khordagui, H., 1997. Environmental consideration of brine-water disposal from desalination plants. *Proceedings of 3rd Gulf Water Conference,* Muscat, Oman, p. 957-976.

Khouri, J., 1982. Hydrogeology of the Syrian Steppe and adjoining arid areas. *Quart. Journ. Eng. Geol., London,* v. 15, p. 135-154.

King Fahd University Staff, 1987. Groundwater resources evaluation in Saudi Arabia and long-term strategic plan for fresh groundwater use.

King Fahd University Petroleum and Minerals, *KFUPM Press*, Dhahran, Saudi Arabia, 168 p.

Kinsman, D.J.J., 1969. Models of formation, sedimentary associations, and diagenetic features of shallow water and supratidal evaporites. *Amer. Assoc. Petrol. Geol. Bull.*, v. 52, p. 830-840.

Lloyd, J.W. and Farag, M.H., 1978. Fossil groundwater gradients in arid regional sedimentary basins. *Groundwater*, v. 16, p. 388-393.

Macumber, P.G., 1995. Freshwater lenses in the hyper-arid region of Central Oman. *The Sultanate of Oman International Conference on Water Management in Arid Countries,* Muscat, Sultanate of Oman, v. 2, p. 480-487.

Macumber, P.G., Al Abri, R. and Al Akhzami, S., 1998. Hydrochemical facies in the groundwater of central and southern Oman. *In: A.S. Alsharhan, K.W. Glennie, G.L. Whittle and C.G.ST.C. Kendall (eds.), Quaternary Deserts and Climatic Change,* Balkema, Rotterdam p. 511-520.

Macumber, P.G., Al-Said, S.B.G., Kew, G.A. and Tennakoon, T.B., 1994. Hydrogeologic implications of a cyclonic rainfall event in Central Oman. *In: H. Nash and McCall, G.J.H. (ed.), Groundwater Quality*, Chapman & Hall, p. 87-97.

Macumber, P.G., Niwas, J.H., Abadi, A. and Seneviratne, R., 1997. A new isotopic water line for northern Oman. *Proceedings of the Third Gulf Water Conference,* Muscat, Sultanate of Oman, p. 141-152.

Madani, I.M., Al-Shiryan, A.I., Lori, I. and Al-Khalifa, H., 1992. Public awareness and attitudes toward various uses of renovated water. *Environment International*, v. 18, p. 489-495.

Makdisi, J., 1990. An inquiry into Islmaic influences during the formative period of the common law; *In: Heer, N. (ed.), Islamic Law and Jurisprudence, University of Washington Press*, Seattle & London, p. 303-307.

Mandeel, M.A., 1992. Desalination and the environment. Department of Chemical Engineering, University of Kuwait (in Arabic).

Mar Del Plata Action Plan, 1977. United Nations Water Conference – Mar Del Plata, Argentina, 14-25 March, 1977.

Mayboom, P., 1966. Groundwater studies in the Assiniboine River drainage basin part I, The Evolution of a Flow System in South Central Saskatchewan. *Geol. Surv. Can. Bull.*, v. 139, 65p.

McClure, H.A., 1978. Early Paleozoic Glaciation in Arabia, *Palaeogeogr., Palaeoclimatol., Palaeoecol.*, v. 25, p. 315-326.

McDonald, M.G. and Harbaugh, A.W., 1988. A modular three-dimensional finite-difference groundwater flow model. *U.S. Geological Survey Report 83-875.*

Meinzer, O.E., 1923. The occurrence of groundwater in the United States, with discussion of principles. U. S. Geological Survey, Water Supply Paper, 489 p.

Ministry of Agriculture and Fisheries, United Arab Emirates, 1993. Hydrology. Water and Soil Department, v. 3, 294 p.

Ministry of Electricity and Water, Qatar, 1987. Development and utilization of water resources in the Middle East. Country Report for Qatar, 21p.

Ministry of Water Resources, Oman, 1995. Water Resources of the Sultanate of Oman: An Introductory Guide. *Published by Ministry of Water Resources,* Sultanate of Oman, 78 p.

Mohamed, A.E. and Al Zubari, W.K., 1999. Bas experience in condensate water utilization. *Proceedings of the 4th Gulf Water Conference, 13-17 February, 1999,* State of Bahrain, v. 1, p. 151-159.

Möser, E.P. Rauert, W., Sticher, W. and Zötl, J.G., 1978. *In: Al-Sayari, S.S. and Zötl, J.G. (eds.), Quaternary Period in Saudi Arabia.* Springer Verlag, New York, p. 153-162.

Moshrif, M.A. and Kelling, G., 1984. Stratigraphy and sedimentary history of Upper-Lower and Middle Cretaceous rocks, central Saudi Arabia. *Deputy Ministry for Mineral Resources, Jeddah, Saudi Arabia, Mineral Resources Bulletin,* 28, 28p.

Mukhopadhyay, A., Al Sulaimi, J. and Barat, J.M., 1994. Numerical modeling of groundwater resource management options in Kuwait. *Groundwater*, v. 32(6), p. 917-928.

Mukhopadhyay, A., Al Sulaimi, J., Al Awadi, E. and Al Ruwaih, F., 1996. An overview of the Tertiary Geology and hydrogeology of the northern part of the Arabian Gulf region with special reference to Kuwait. *Earth-Science Reviews*, v. 40 (3-4), p. 259-295.

Murris, R.J., 1981. Middle East stratigraphic evolution and oil habitat. *Geologie Mijnbouw*, v. 60, p. 467-486.

Musayab, R., 1988. Water resources and development in the State of Bahrain. Water Supply Directorate, Ministry of Works, Power and Water, Bahrain.

Naimi, A.I., 1965. The groundwater of northeastern Arabia. *Fifth Arab Petroleum Congress,* Cairo, Egypt, 24 p.

Nardi, K.J., 1993. Underground Storage Tanks, *In: Arbuckle, J.G., Brownell, F.W., Case, D.R., Halbleib, W.T., Jensen, L.J., Landfair, S.W., Lee, R.T. and Miller, M.L. (eds.), Environmental Law Handbook*, p. 541-571.

Omar, S.A., Al-Yaqubi, A. and Senay, Y., 1981. Geology and groundwater hydrology of the State of Kuwait. *Journal of the Gulf and the Arabian Peninsula Studies*, v. 1, p. 5-67.

Owen, R.M.S. and Nasr, S.N., 1958. Stratigraphy of the Kuwait-Basrah areas; In: L.G. Weeks (ed). Habitat of Oil, a Symposium. *Amer. Assoc. Petrol. Geol. Mem.*, p. 1252-1278.

Parker, D.H., 1985. The hydrogeology of Cainozonic aquifers in the PDO concession area, Sultanate of Oman, PDO unpublished report.

Pike, J.G., 1978. The Agro-climatology of Qatar. FAO Project Technical Note No. 1, Doha, Qatar.

Piper, A.M., 1944. A graphic procedure in the geochemical interpretation of water analysis. *Trans. Amer. Geophysical Union*, v. 25, p. 914-928.

Plumer, L.N., Prestemon, E.C. and Parkhurst, D.L., 1991. An interactive code (NETPATH) for modeling net geochemical reactions along a flow path. US Geological Survey Water Resources Investigations Report 91-4078.

Porritt, P.D., 1953. Artesian well survey, Bahrain. *Bahrain Petroleum Company.*

Powers, R.W., 1968. Saudi Arabia (excluding Arabian Shield). *Lexique Stratigraphique International, Asie, Vol. III, Fascicule 10 b1, C.N.R.S.*, Paris, 177 p.

Powers, R.W., Ramirez, L.F., Redmond, C.D. and Elberg, E.L. Jr., 1966. Geology of the Arabian Peninsula, Sedimentary Geology of Saudi Arabia. *U.S. Geol. Survey, Prof. Paper, 560-D*, 127 p.

Quinn, O., 1986. Regional hydrological evaluation of the Najd aquifers. Sultanate of Oman (Report OFR-48-86). *Public Authority for Water Resources*, Oman.

Raveendran, E. and Madany, I.M., 1991. The quality of groundwater in Bahrain. *Science of the Total Environment*, v. 103 (2-3), p. 177-183.

Ries, A.C. and Shackleton, R.M., 1990. Structures in the Huqf-Haushi Uplift, east central Oman. *In: Robertson, AH.F., Searle, M.P., and Ries, A.C., eds., The geology and tectonics of the Oman region. Geol. Soc. London.*, Spec. Publ., No. 49, p. 653-664.

Ritter, W., 1981. Did Arabian oases run dry? *In: Meckelein, W. (ed.), Desertification in Extremely Arid Environments. Stuttgarter Geographische Studien, Bd. 95*, Stuttgart, p. 73-92.

Rizk, Z.S. and Alsharhan A.S., 1999. Application of natural isotopes for hydrogeologic investigations in United Arab Emirates. *Proceedings of the 4th Gulf Water Conference*, Bahrain, 13-17 February, 1999, v. 1, p. 197-228.

Rizk, Z.S. and El-Etr, H.A., 1997. Hydrogeology and hydrogeochemistry of some springs in the United Arab Emirates. *Arab. Jour. Sci. Eng.*, v. 22, p. 95-111.

Rizk, Z.S., 1996. Falajes of United Arab Emirates: Geological setting and hydrogeological characteristics. *Arab. Jour. Sci. Eng.*, v. 23, p. 3.

Rizk, Z.S., Alsharhan, A.S. and Shino, S., 1997. Evaluation of groundwater resources of United Arab Emirates. *Proceedings of 3rd Gulf Water Conference*, Muscat, Sultanate of Oman, v. 1, p. 95-122.

Rizk, Z.S., Garamoon, H.K. and El-Etr, H., 1998. Contribution to hydrogeochemistry of the Quaternary aquifer in the Al Ain area, United Arab Emirates. *In: A.S. Alsharhan, K.W. Glennie, G.L. Whittle and C.G.ST.C. Kendall (eds.), Quaternary Deserts and Climatic Change*, Published by Balkema, Rotterdam, p. 439-454.

Robinson, B.W. and Al Ruwaih, F., 1985. The stable isotopic composition of water and sulfate from the Raudhatain and Umm al Aish freshwater fields, Kuwait. *Chemical Geology*, v.58 (1-2), p. 129-136.

Rogers, P. and Lydon, P. (eds.), 1994. *Water in the Arab World: Perspectives and Prognoses*, Published by the Division of Applied Sciences, Harward University, Harward University Press, 369 p.

Saint Marc, P., 1978. Arabian Peninsula. *In: Moullade, M. and Nairn, A.E.M. (eds). The Phanerozoic Geology of the World, II. The Mesozoic. Elsevier, Amsterdam*, p. 435-462.

Sakr, I.M., 1989. An introduction to hydrogeology of Gulf Cooperation Council (in Arabic). Al Ain Establishment for Advertizing Distribution and Publishing, 445 p.

Salih, A.M.A., 1997. UNESCO's international hydrological programme and sustainable water resources management in the Arab Region. *Proceedings of the 3rd Gulf Water Conference, 8-13 March 1997*, Muscat, Sultanate of Oman, v. 1, p. 81-94.

Saudi Arabia Ministry of Agriculture and Water, 1984. *Water Atlas of Saudi Arabia; Water Resources Development*. Riyadh, Saudi Arabia, 112 p.

Schoeller, H., 1962. Les eaux souterraines, Messan etc cie., editeurs, 120 boulevard St. German, Paris VI.

Schyfsma, E., 1978. Climate. *In: Al-Sayari, S.S. and Zötl, J.G., ed., Quaternary Period in Saudi Arabia*, Springer Verlag, New York, p. 31-44.

Sen, Z. and Al-Dakheel, A., 1986. Hydrochemical facies evaluation in Umm er Radhuma limestone, eastern Saudi Arabia. *Ground Water*, v. 24(5), p. 626-635.

Senay, Y., Hamdan, L. and Yaqubi, A., 1987. Study of the subsurface water rise in the residential areas of Kuwait. vol. 3: Hydrogeology, Kuwait Institute for Scientific Research Report.

Sharaf, A.T., 1980. Climate of Kuwait: University Cultural Foundation. *First Edition, Alexandria*, 237 p.

Sharaf, M.A. and Hussein, M.T., 1996. Groundwater quality in Saq aquifer, Saudi Arabia. *Hydrogeol. Sci. Jour.*, v. 41 (5), p. 683-670.

Sharief, F.A.M., 1982. Lithofacies distribution of the Permian-Triassic rocks in the Middle East. *Jour. Petrol. Geol.*, v. 5, p. 203-206.

Sharief, F.A.M., 1983. Permian and Triassic geological history and tectonics of the Middle East. *Jour. Petrol. Geol.*, v. 6, p. 95-102.

Shuval, H.I., 1969. Health factors in the reuse of wastewater for agricultural, industrial and municipal purposes, in problems in community water management. *Public Health Papers, No. 38, WHO*, Geneva.

Shuval, H.I., Adin, A., Fattal, B., Ravitz, E. and Yekutiel, P., 1985. Wastewater irrigation in developing countries – Health effects and technical solutions. *World Bank Technical Paper No. 51*, The World Bank, Washington DC, 324 p.

Silva, E. and Al Noaimi, F., 1999. Digital simulation of groundwater salvage in northeastern Abu Dhabi Emirate. *Proceedings of 4th Gulf Water Conference*, 13-17 February, 1999, Bahrain, p. 303-316.

Songreah, 1967. Report on increasing groundwater supplies for the city of Doha, Qatar. Unpublished report, Water Department, Government of Qatar (cited in FAO, 1981).

South Carolina Department of Health and Environmental Control, 1990. Undergraduate Storage Tank Control Regulations. *Groundwater Protection Division*, South Carolina, U.S.A., 45 p.

South Carolina Department of Health and Environmental Control, 1993. Underground Injection Control Regulations, *Grounwater Protection Division*, South Carolina, U.S.A., 40 p.

South Carolina Department of Health and Environmental Control, 1994. Wellhead Protection: Technical Guidance Document for South Carolina. *Local Groundwater Protection, Groundwater Protection Division*, South Carolina, U.S.A., 40 p.

Starr, J.R. and Stoll, D.C., 1988. The politics of scarcity. Water in the Middle East, Boulder, Colorado, Westview.

Statistical Abstract, 1990/1991, General Statistics Organization, Directorate of Statistics, Bahrain.

Statistical Data, 1991. Water supply demand and consumption statistical data. Bahrain: *Water Supply Directorate, Ministry of Electricity and Water, Bahrain.*

Stoesser, D.B. and Camp, V.E., 1985. Pan-African microplate accretion of the Arabian Shield. *Geol. Soc. Amer. Bull.*, v. 96, p. 817-826.

Stout, G.E., 1995. Global water management issues. *Proceedings of the International Conference on Water Resource,*12-16 March, 1995, Muscat, Oman, v. 1, p. 55-58.

Sulin, V.A., 1946. Petroleum formation in system of natural waters. *Costoptekhizdat*, Moscow, 96 p.

Sutcliff, J.W., 1966, Groundwater Extraction of Bahrain Island and Coastal Hasa. *Ministry of Commerce and Agriculture*, Bahrain.

Sutton, S., 1984. The falaj - a traditional cooperative system of water management. *Waterlines*, v. 2 (3), p. 8-11.

Terratest Ltd, 1975. Abu Dhabi mineral survey - Stage II. Detailed investigation of promising areas, Final report. Phase II. Ain Bu Sukhnah (unbuplished report).

Thornthwaite, C.W., 1948. An approach towards a rational classification of climate. *Geographical Review*, v. 38, p. 55-94.

Todd, D. K., 1980. Groundwater hydrology. *John Wiley and Sons, New York (second edition)*, 535 p.

Toth, J., 1963. A theoretical analysis of groundwater flow in small drainage basins. *Jour. Geophysic. Res.*, v. 68, p. 4795-4812.

Trescott, P.C., Pinder, G.F. and Larson, S.P., 1976. Model for aquifer simulation in two dimesnsions with results of numerical experiments. *U.S. Geological Survey, Techniques of Water Resources Investigations*, Book 7, Chapter C1, 116 p.

U.S. Salinity Laboratory, 1954. Diagnosis and improvement of saline and alkaline soils. *U.S. Department of Agriculture Handbook*, No. 60, 160 p.

UNESCO/ACSAD, 1990, Water resources in the Arab region. The Arab Center for the Studies Zones and Dry Lands, Damascus, Syria, 166 p.

United Arab Emirates National Atlas, 1993. Remote Sensing Centre, U.A.E. University, 188 p.

United Nations, 1982. Groundwater in the eastern Mediterranean and western Asia. *Department of Technical Co-operation for Development, Natural Resources, Water Series No. 9*, p. 195-206.

US Geological Survey and United Arab Emirates National Drilling Company, 1996. Groundwater Resources of Al Ain Area, Abu Dhabi Emirate, United Arab Emirates. Unpublished report, National Drilling Company, Abu Dhabi, UAE.

Vail, P.R., Mitchum, R.M. and Thompson, J.R., 1977. Seismic stratigraphy and global changes of sea-level, part 4: Global cycles of relative changes of sea-level, in Payton, C.E., ed., Seismic stratigraphy application to hydrocarbon exploration. *Amer. Assoc. Petrol. Geol. Mem.*, v. 26, p. 83-97.

Wallender, W.W., 1989. Economic surface irrigation within environmental constraints. *International Symposium on Agriculture and Fisheries Development in Oman*, Sultan Qaboos University, Muscat, 15-19 October, 1989.

Walton, W.C., 1962. Selected analytical methods for well and aquifer evaluation. *Illinois State Water Survey Bulletin*, v. 49, 81 p.

Wender, L.E., Bryant, J.W., Dickens, M.F., Neville, A.S. and Al-Moqbel, A.M., 1988. Paleozoic (Pre-Khuff) hydrocarbon geology of the Ghawar area, Eastern Saudi Arabia. *GeoArabia*, v. 3, p. 273-302.

White, W.B., 1977. Conceptual models for carbonate aquifers: Revisited. *In: Dilamarter, P.R. and Csallany, S.C. (ed.), Hydrologic problems in karst regions*, Bowling Green, Kentucky, Western Kentucky University, U.S.A., p. 176-187.

Whittle, G.L. and Alsharhan, A.S., 1994. Dolomitization and chertification of the Early Eocene Rus Formation in Abu Dhabi, United Arab Emirates. *Sed. Geol.*, v. 92, p. 272-285.

Whittle, G.L., Alsharhan, A.S. and El Deeb, W.M.Z., 1995. Bio-lithofacies diagenesis in the Early-Middle Oligocene of Abu Dhabi, United Arab Emirates. *Carbonates and Evaporites*, v. 10 (1), p. 54-64.

Whybrow, P.J. and McClure, H.A., 1981. Fossil mangrove roots and Palaeoenvironments of the Miocene of the eastern Arabian Peninsula. *Palaeogeogr., Palaeoclimatol., Palaeoecol.*, v.32, p.213-225.

Wilson, H.H., 1969. Late Cretaceous and eugeosynclinal sedimentation, gravity tectonics, and ophiolite emplacement in Oman Mountains, southeast Arabia. *Amer. Assoc. Petrol. Geol. Bull.*, v. 53, n. 3, p. 626-671.

Wilson, J.L. and Peterson, J.A., 1986. Petroleum stratigraphy of the northeast Africa – Middle East region. *Symposium on the hydrocarbon potential of the intense thrust zone, sponsored by Organization of Arab Petroleum Exporting Countries (OAPEC), Kuwait, and Ministry of Petroleum and Mineral Resources,* U.A.E., Abu Dhabi, 14-18 December 1986, p. 227-330.

Woodward, D.G., Menges, C.M. and Khalifa, M.A., 1992. Hydrologic exploration in Eastern Abu Dhabi emirate using uphole data from petroleum seismic surveys. *Proceedings of 1st Gulf Water Conference,* Dubai, United Arab Emirates, v. 1, 19 p.

World Health Organization (WHO), 1971. International standards of drinking water. 3rd edition, Geneva, Switzerland.

World Health Organization (WHO), 1989. Health guidelines for the use of wastewater in agriculture and aquaculture. Technical Report Series No. 778, Geneva, Switzerland.

Wright, E.P.W., 1967. Report on groundwater resources of Bahrain Islands: A report submitted to the Ministry of Works and Agriculture, State of Bahrain, by the British Embassy in Bahrain.

Wright, V. P., Ries, A. C. and Munn, S. G., 1990. Intraplatformal basin-fill deposits from the Infracambrian Huqf Group, east central Oman *In: Robertson, A.H.F., Searle, M.P., and Ries, A.C., eds., The geology and tectonics of the Oman region. Geol. Soc. London,* Spec. Publ., No. 49, p. 601-616.

Wushiki, H., 1991. $^{18}O/^{16}O$ and D/H of the meteoric waters in South Arabia. *Mass Spectroscopy*, v. 39 (5), p. 239-250.

Yazicigil, H., Al-Layla, R.I. and de Jong, R.L., 1986. Numerical Modeling of the Dammam Aquifer in Eastern Saudi Arabia. *Arab. Jour. Sci. Eng.*, v.11(4), p. 349-362.

Yurtsever, Y., 1992. Preliminary evaluation of isotope results from reconnaissance samples collected in Kuwait: International Atomic Energy Agency Technical Cooperation project. *Isotope Hydrology of the Middle East*, 12 p.

Yurtsever, Y. and Payne, B.R., 1979. Application of environmental isotopes to groundwater investigations in Qatar, isotope hydrology 1978. *Proceedings of Symposium*, Vienna, v. 2, p. 465-490.

Yurtsever, Y., 1999. An overview of nuclear science and technology in groundwater assessment/ management and International Atomic Energy Agency activities in the Gulf Region. *Proceedings of the 4th Gulf Water Conference,* 13-17 February, 1999, State of Bahrain, v. 1, p. 161-173.

Zekai, S. and Al-Dakheel, A., 1986. Hydrochemical facies evaluation in Umm er Radhuma limestone, Eastern Saudi Arabia. *Groundwater,* v. 24 (5), p. 626-635.

Zubari, W.K., 1987. A numerical three-dimensional flow model of the Dammam aquifer system, Bahrain and Eastern Saudi Arabia. *M.Sc. Thesis, Department of Earth Resources, Ohio University,* Athens, Oh, USA, 409 p.

Zubari, W.K., 1997. Towards the establishment of a total water cycle management and re-use program in the GCC Countries. *Proceedings of the 3rd Water Conference,* Muscat, Sultanate of Oman, 8-3 March, 1997, v. 1, p. 1-16.

Zubari, W.K., 1999. The Dammam aquifer in Bahrain hydrochemical characterization and alternative for management of groundwater quality. *Hydrogeology Journal*, v. 7, p. 197-208.

Zubari, W.K. and Lori, E.J., 1991. Report on Water Resources Studies: Evaluation, Exploitation, and Legislation, in Bahrain in the period 1924-1991. Published by *Bahrain Center for Studies and Serearch* (in Arabic).

Zubari, W.K. and Madany, I.M., 1992. Effect of Development on Groundwater Resources in Bahrain: A Quantitative and Qualitative Study in the Period 1925-1991. *Proceedings of the Development and Environment Conference, 15-17 November 1992, The Secretariat General, Cooperation Council for the Arab States of the Gulf,* Riyadh, Saudi Arabia (in Arabic).

Zubari, W.K., Al-Junaid, S.S. and Alaa El-Din, M.N., 1998. Impacts of deteriorating groundwater quality on desertification of soils in Bahrain (1956-1992), a case study. *Proceedings of the International Conference on desert development in the Arab Gulf countries*: Substainable development in arid zone. In: Omar, S.A.S.; Misak, R., Al-Ajami, D. and Al Awadi, N. (eds.), v. 2, Management and improvement of desert resources, A.A. Balkema, Rotterdam, p. 581-594.

Zubari, W.K., Khater, A.R., Al Noaimi, M.A. and Al Junaid, S.S., 1997. Spatial and temporal trends in groundwater salinity in Bahrain. *Arab. Jour. Sci. Eng.*, v. 22(1C), p. 82-94.

Zubari, W.K., Mubarak, M.A. and Madany, I.M., 1993. Development impacts on groundwater resources in Bahrain. *Water Resources Development*, v. 9 (3), p. 263-279.

Zubari,W.K. and Al-Junaid, S.S., 1992. Groundwater situation in Bahrain - A quantitative and qualitative study 1980-1990. *Proceedings of the First Gulf Water Conference*, v. 1, Water Science and Technology Association, Dubai, 10-14 October, 1992, (in Arabic).

SUBJECT INDEX

Y

Z

Appendix A

GLOSSARY OF TERMS AND LOCAL NAMES USED IN WATER RESOURCES STUDIES IN ARABIAN GULF REGION

(A)

Ab	water, river.
Ab-rah	water course, aqueduct.
Abu	father, possessor of.
Abyad	white.
Ain (pl. ayoon)	a spring (springs) or surface seepage of groundwater.
Ainy	type of falaj which originates from a motherwell.
Akhdar	green.
Al	the (definite article).
'Ala'	height.
A'la'	higher, upper.
Al-Kharsa'ah	sinkhole with most of its roof still standing.
Al-Khurays	wide depression with thin sheet of water.
Al-Sunú	man-made holes for the preservation of water.
Ahmar	red.
Ard	ground, earth, land.
'Arq	a long linear dune (cf. 'irq and seif).
Asfar	yellow.
Aswad	black.
Ayn	spring, well.
Azraq	blue.

(B)

Bab	door, gate, narrow strait.
Bahr (or Behar)	sea (plural: seas).
Barchan	originally a Turkish word adopted by the Arabs and subsequently European Geomorphologists to describe a crescent-shaped sand dune, opening down- wind (North Africa also spelt barkan or barkhan).
Batin (or batan)	a desert depression or coastal embayment.
Baydà	white.
Bir (or Ber)	water well.
Bowhairah	lake or flooded area.
Bu	abbreviation of Abu.
Burj (or burq)	literally means "tower" but it is occasionally used to describe an isolated hill.
Burga	peak.

(C)

Cistern	man-made underground cement-lined chamber to store flood water.

(D)

Dahana (or dhanna)	sandy area with many water wells (a high water-table).
Dahl	sinkhole formed by lime-stone dissolution.
Dahlat	ephemeral river bed (wadi).
Dahr (or dahar)	plateau or flat topped hill.

Daia (or daya) an oasis situated in a depression on a sandstone or limestone plateau.

Dakaka (or Dakakah) a region of good camel-grazing pastures, in the southern Rub' al Khali, consisting of well-vegetated, undulating hard-packed red sands (cf. dikaka and zibar).

Dalu a bucket of camel or goat skin, used to draw water at a well.

Dawhat bay.

Dawudi type of falaj which originates from a motherwell.

Dhow Arabian style of boat, usually single-masted.

Dikaka accumulations of dune sand.

(E)

Erg vein, linear dune.

(F)

Falaj long underground aqueduct used for irrigation; cf. qanat.

Fasht rocky reef.

Faydat depression, salt flat or ephemeral lake.

Feidj sand-free interdune corridor in area of linear dunes.

Foggara synonmy: falaj.

(G)

Galal desert stream.

Ghabbat (or khabbat) a generally sloping depression, sometimes occupied by a shallow ephemeral lake.

Ghadir a temporary pond or marsh left after heavy rain.

Ghaith rain.

Ghàr cave.

Gharraqat marsh or lake.

Gharb west.

Gharbi western.

Ghaur depression, lowland or plain.

Ghayli type of falaj in which water is derived from wadi baseblow.

Ghor dry mudflat or dry non-saline lake.

Ghubbat deep-water bay or inlet.

Golban wells.

(H)

Hadd sandspit, low sandy point.

Hajj one who has made the pilgrimage to Mecca.

Hajr hump-backed tract of sand.

Hàlat dry or drying sandbank.

Halaq small ephemeral stream.

Hamadah a bare rock surface or rocky (or hamada) desert plateau in a desert region.

Hamrà' red.

Harra lava flow or lava field.

Hasa pebbles.

Hatiyah depression, oasis or valley.

Hawa cave, depression or sinkhole.

Hawd basin.

Hawr shallow marsh (fresh or brackish).

Hayr pearl bank.

Hidbat a well-vegetated, low lying area.

(or hadhbat)	The term can be used more specifically to apply to a vegetated depression bounded by mountains or hills.
Hisar	fort.
Huta	a local eastern Arabian name for a flat interdune area, generally with vegetation.

(I)

Ibn	son, descendant.
Iddi	type of Falalj which originates from mother well
Imam	religious leader, holy man.
Imàrah	seat of a governor.
'Irg ma'kuf	linear mega-dunes with a hooked end, like a shepherd's crook (also known as 'irg muta' 'akkif or 'irg muhayyar) drawn from the dunes resemblance to the curled tail of the saluki dog).

(J)

Jau (jaw, jaub or jawf)	valley, hollow or depression (pl. jiban).
Jebel	hill, mountain.
Jazàir	island, peninsula.
Jazireh	island, peninsula.
Janub	south.
Janubi	southern.
Jifr	a hole or a water well dug by hand.
Jiri	low and where water accumulates and forms a pool.

(K)

Kabir (or kabeer)	large, great.
Kahf	cave.
Kahlij	gulf, bay estuary.
Kandi (or *kani*)	stream of ephemeral channel (wadi).
Karkur	ephemeral channel or valley.
Karez	equivalent term of *falaj*
Khabra	a depression or dry lake without salts or evaporates (plural: khabari).
Khadrà'	green.
Khafs	pond in a natural depression.
Kharat	mudflat or rainpool.
Kharif	the season autumn, generally thought as the time of the year when the rains first fall.
Kharimah (or kharimat)	the terminus of an ephemeral river system often forming a gravel interdune flat, between mega-dunes (plural: khara'im).
Khashm	literally means "nose" but it is also used to refer to a cliff promontory or mountain spur.
Khawr	arm of the sea, inlet, channel.
Khor (or khawr)	literally means brackish, salt-water which is only suitable for watering camels, (khowr) can also be used to refer to a marine estuary (plural Khor: khiran); e.g. Khor Dubai.
Kúl	fort, town.

(M)

Ma'a (or mai)	water.
Madhma	literally "land of thirst" used to refer to a large area of desert with few wells.
Mahdar	a village, usually without permanent buildings, which is only occupied for a part of the year.

Mahader interdune areas where water table is shallow and agricultural way of life exist.

Majra the course of an ephemeral river channel (plural: majari).

Mallahat (or mallahet) salt marsh, salt flat or salt lake.

Mamlahan inland sabkhas that have been excavated for salt.

Maraggat soft shoal.

Marsá anchorage.

Masila dry, sandy ephemeral channel bed.

Mashriq east.

Mayhrib west.

Meydan square, field.

Minà' port, anchorage.

(N)

Nad (or Naddood) hill of sand.

Nahal ephemeral channel (*wadi*).

Nahr river, stream, canal.

Najwat steep-to shoal, pearl bank, coral reef.

Naqà large sand dune.

Naqqazah river terrace.

Nukda The tributary of an ephemeral channel (or *wadi*) emptying into the sands.

(Q)

Qa' literally used to refer to a depression of low-lying area, which might be occupied by a plain, marsh, salt-flat or dry mudflat.

Qaba dome.

Qal'at castle, fort.

Qalayyib salt pan or salty well.

Qalt well.

Qanat equivalent Iranian term of falaj

Qarar (or qararah) depression, ephemeral channel or valley.

Qarhoud a pyramid-shaped sand dune, formed by the intersection of other dunes (plural: qaraheid); cf. star dune.

Qarn isolated hill.

Qaryeh village.

Qasàr above-water rock, rocky islet.

Qasr castle, fort.

Qit'at rock, patch of rocks.

(R)

Ramlah meaning sand dune, or a sand (or ramlat) or dune-covered region (plural: rimal); e.g. Ramlat Wahiba literally means the dune-covered region partly occupied by the Wahiba family (or tribe). English speakers know the region as the Wahiba Sands.

Ra's cape, point, peak, headland, hill, ridge or spur.

Rawd ephemeral channel (wadi).

Rawda lowland used for agriculture.

Reg algerian synonym for "Serir".

Rud stream.

Ruqq reef, shoal, sandbank.

Ru'ús plural of Ra's.

(S)

Sabkha (or sabkhah) flat marshy depression with salty gypsiferous clayey algal sandy crust (plural: sibakh).

Sadd obstruction, barrier, dam.

Saghir small.

Sahara the great desert (see sahra).

Sahl desert plain (plural: suhool).

Saniyah well or small traditional farm (plural: sawani).

Sayyid descendant of the Prophet, lord chief.

Seif seasonal rainfall in eastern Oman brought by extratropical circulation systems, Laterally: summer.

Seyl rain or the run-off of rainfall into ephemeral channels.

Serir deflation lag of pebbles on the desert surface. Libyan synonym of "Reg".

Saruq (or sarug) a flat interdune depression generally with vegetation (Thomas, 1932).

Shaggat ephemeral channel (cf. wadi).

Shamal literal translation of an Arabic word meaning North, but is used locally in Arabia for the strong northerly wind that blows down the Arabian Gulf.

Shanna fresh or sweet water.

Shaqqat ephemeral channel (cf. wadi).

Sharq east.

Shatib ephemeral channel or valley.

Shatt beach or shore (sometimes including coastal salt-flats or ephemeral lakes; cf. shati).

Shaydh chief of a tribe, elder, religious leader.

Shu'ayb stream or ephemeral channel (cf. *wadi*).

Shib rocky shoal, inshore reef.

Shutban ephemeral channel (cf. wadi).

Sif (Sifut) coast, sandy beach.

Simum (or simoom) a hot wind.

Súq (Souk) market place.

Súr wall, rampart.

(T)

Tall hill.

Taw (or tawii) water well.

Tmad shallow well or water-hole.

(U)

Umm mother of.

'Uqlat well or rainpool.

Uyun (or uymm) spring.

(W)

Wadi watercourse, river bed, valley (pl. wadian).

Waha oasis.

Wàli local governor.

Washkah palm tree.

Wasmi autumn rain.

Wilayat an administrative area within the Sultanate of Oman

(Z)

Zajarah type of well found on Al Batinah region in Oman.

Zargà' blue.

Zawr bend, twist, turn.

Zumul sand dune.

Appendix B*

GLOSSARY OF SCIENTIFIC & TECHNICAL TERMS RELATED TO WATER RESOURCES

(A)

Actual evapotranspiration: The evapotranspiration that actually occurs under given climatic and soil-moisture conditions.

Aeolianite: A consolidated sedimentary rock formed of wind-deposited sand; commonly but not necessarily rich in carbonate grains.

Alluvial: Pertaining to or composed of alluvium or deposited by a stream or running water.

Alluvial fan: An outspread deposit of detrital material, with generally a low outward slope.

Alluvial terrace: A terrace formed when a river incises into its own valley fill.

Alluvium: A general term for clay, silt, sand, gravel, or similar unconsolidated material deposited during comparatively recent geologic time by a stream or other body of running water as a sorted or semisorted sediment in the bed of the stream or on its floodplain or delta, or as a cone or fan at the base of a mountain slope.

Anhydrite: Calcium surface, $CaSO_4$; orthorhombic; transparent to translucent; Mohs' hardness 3 to 3.5; specific gravity 2.93. A source of cement, sulfuric acid, and plaster

Anion: A negatively charged ion that migrates to an anode, in electrolysis.

Anion exchange: Ion exchange process in which anions in solution are exchanged for other anions from an ion exchanger.

Anisotropy: The condition under which one or more of the hydraulic properties of an aquifer vary according to the direction of flow.

Annulus: The space between the drill string or casing and the wall of the borehole or outer casing.

Anticlinal ridge: A mountain range developed along the axis of an anticline.

Applied Stress: The downward stress imposed on a specified horizontal plane within an aquifer system. At any given level in the aquifer system, the applied stress is the force or weight (per unit area) of sediments and moisture above the water table, plus the submerged weight (per unit area), accounting for buoyancy of the saturated sediments overlying the specified plane at that level, plus or minus the net seepage stress generated by flow through the specified plane in the aquifer system.

Aquiclude: Rock having a low hydraulic conductivity. A saturated but relatively impermeable material that does not yield appreciable quantities of water; clay is an example.

Aquifer: A layer of rock which holds water and allows water to percolate through it.

1. Rock or sediment in a formation, group of formations, or part of a formation which is saturated and sufficiently permeable to transmit economic quantities of water to wells and springs.
2. A highly permeable layer of rock or soil that holds or can transmit groundwater.
3. Rock having a high hydraulic conductivity.
4. Groundwater occurs in many types of geologic formations, those known as aquifers are most important. An aquifer may be defined as a formation that contains sufficient saturated permeable material to yield significant quantities of water to wells and springs. This implies an ability to store and to transmit water; unconsolidated sands and gravels are a typical example.

Aquifer System: A heterogeneous body of interbedded permeable and poorly permeable geologic units that function as a water-yielding hydraulic unit at a regional scale. The aquifer system may comprise one or more aquifers within which aquitards are interspersed. Confining units may separate the aquifers and impede the vertical exchange of ground water between aquifers within the aquifer system.

Aquifer test: A test involving the withdrawal of measured quantities of water from or addition of water to, a well and the measurement of resulting changes in head in the aquifer both during and after the period of discharge or addition.

Aquifuge: A relatively impermeable formation neither containing nor transmitting water; solid granite belongs to this category.

Aquitard: A saturated but poorly permeable stratum that impedes ground-water movement and does not yield water freely to wells, but that may transmit appreciable water to or from adjacent aquifers and, where sufficiently thick, may constitute an important groundwater storage zone; sandy clay is an example.

Arenite: A detrital rock whose grains are mostly 0.06-2.0 mm in diameter. Synonym of "sandstone".

Argillaceous: A term applied to all rocks or substances composed of clay minerals or having a notable proportion of clay in their composition. Referring to a rock having an appreciable percentage of clay minerals. Synonym of "clayey".

Argillite: A nonfissile mudrock that is very highly indurated, perhaps weakly metamorphosed, and lacking salty cleavage

Arkose: A detrital rock that contains an appreciable percentage of feldspar grains, typically at least 20%.

Artesian: An adjective referring to confined aquifers. Sometimes the term artesian is used to denote a portion of a confined aquifer where the altitudes of the potentiometric surface are above land surface. But more generally the term indicates that the altitudes of the potentiometric surface are above the altitude of the base of the confining unit.

Artesian flow: Groundwater under sufficient head to rise above the aquifer.

Artesian well: A free-flowing well tapped into a pressurized aquifer

Artificial recharge: The process by which water can be injected or added to an aquifer. Dug basins, drilled wells, or simply the spread of water across the land surface are all means of artificial recharge. An increase in or development of new groundwater storage through an induced flow of surface water into aquifers or by direct recharge via basins, ditches or boreholes, etc.

Avalanche slope: The slope that forms when wind-blown sand from the windward side of a dune passes into the air of the leeward side. The sand will start to slip if further deposition would result in the maximum, angle or repose for dry sand of 34° to be exceeded.

(B)

Bajada: Coalescing alluvial fans, forming a continuous waste slope.

Barchan (or barchan dune): Crescent-shaped sand dune, which migrates downwind in the direction of its horns. It has a gentle windward slope and a steeper slipface on its lee slope. Barchans sometimes unite laterally to form rather irregular barchanoid dunes.

Barchanoid dunes: Cross between a barchan and a transverse dunes.

Barrier boundary: An aquifer-system boundary represented by a rock mass that is not a source of water.

Baseflow: That part of stream discharge derived from groundwater seeping into the stream.

Baseflow recession: The declining rate of discharge of a stream fed only by baseflow for an extended period. Typically, a baseflow recession will be exponential.

Baseflow recession hydrograph: A hydrograph that shows a baseflow-recession curve.

Base exchange: The displacement of a cation bound to a site on the surface of a solid, as in silica-alumina clay-mineral packets, by a cation in solution.

Base levels: Theoretical downward limits of stream erosion.

Beach: Gently sloping shore of a body of water.

Bed: A layer of rock distinguishable in the field; a subdivision of a formation, thicker than 1 cm.

Bedrock: Solid rock underlying soil and unconsolidated sediment.

Belt of soil moisture: The top layer of soil from which plant roots obtain water.

Bentonite: A colloidal clay, largely made up of the mineral sodium mont-morillonite, a hydrated aluminium silicate.

Bird's-eye structure: A distinctive pattern of small (typically 1-3 mm) voids in limestones; may be filled with minerals, commonly calcite or anhydrite.

Bit: The cutting tool attached to the bottom of the drill stem.

Borehole geophysics: The general field of geophysics developed around the lowering of various probes into a well.

Boundary layer: The lower contact of a body of moving water.

Braiding: The interweaving of fluvial channels to produce a pattern like a plait. Forms in flooded wadi channels when the water: sediment ratio falls, causing newly deposited sediment to choke the channel; also when wind-blown sand blocks a channel. During the next flood, the water cuts another route through the obstructions, thereby increasing the degree of braiding.

Brine: Surface or subsurface water containing more than 35 ppt. of dissolved substances; more saline than normal seawater. Synonym of "Hypersaline".

Buried valley: A depression in an ancient land surface or in bedrock now covered by younger deposits.

(C)

Calcareous: Containing a substantial proportion of calcium carbonate; generally applied to rocks and soils.

Calcareous rocks: Sedimentary rocks dominantly made of carbonate minerals.

Calcrete: A limestone precipitated as surface or near-surface crusts and nodules by the evaporation of soil moisture in semi-arid climates. Synonym of "caliche"; a variety of "Duricrust". Horizon of hard, dense calcium carbonate or sediment cemented by calcium carbonate, found in soils of hot, arid and semi-arid regions.

Caliche: Surface sediments in arid or semi-arid areas that become cemented by evaporation of lime-rich ground-water.

Capillarity: The property of surface tension seen in the rise of liquids in tubes of very small bore or the action of blotting paper. Important as the mechanism whereby groundwater is brought to the surface in sabkhas, where precipitation of evaporite minerals such as halite takes place.

Capillary: Resembling a hair; of very small bore. If a tube of very small bore is immersed in water, the water will rise up within the tube as a result of capillary attraction.

Capillary forces: The forces acting on soil moisture in unsaturated zone, attributable to molecular attraction between soil particles and water.

Capillary fringe: Soil zone above the water table in which soil pores are partially water filled.

Carbonate rocks: Rocks composed mostly of carbonate minerals.

Carbonation: Weathering process gene-rating carbonate minerals.

Cation: An ion having a positive charge and, in electrolytes, characteristically moving to a negative electrode.

Cation exchange: Ion exchange process in which cations in solution are exchanged for other cations from an ion exchanger.

Cave: A natural underground open space or series of open spaces and passages large enough for a person to enter, generally with a connection to the surface; often formed by solution of limestone.

Caverns: Large underground openings (caves), usually formed by the dissolution of carbonate rocks.

Cementation: The process of precipitating a binding material around rock-forming grains. Solidification of sediment, usually caused by the introduction of pore-filling material or by recrystallization of fine-grained materials. Synonym of "Lithification".

Chalcedony: A rock composed of micro-crystalline quartz crystals of fibrous habit, as seen with a polarizing micro-scope.

Chalk: A soft, pure micritic limestone composed of the shells of planktonic foraminifera and/or coccoliths.

Chert: A rock composed of micro-crystalline or cryptocrystalline inter-locking quartz crystals, typically with diameters less than 20 μm.

Clastic: Referring to a rock or sediment composed of fragments of preexisting materials.

Clay: Used to indicate (1) fragmental material less than 4 μm in size, and also (2) a group of minerals of phyllosilicate structure and particular composition; i.e. clay size versus clay minerals. The clay-size fraction of most silicate sediments is composed mostly of clay minerals.

Clay membrane: An assemblage of clay-mineral flakes that act as a selective filter for dissolved substances in water passing through them.

Claystone: A lithified, nonfissile rock composed mostly of clay-size silicate materials.

Climate: The weather conditions of an area averaged over a period of time. Also, extremes in weather behaviour.

Climatic destabilization: Climatic change that involves an increase in the frequency of extremes in weather behaviour (record temperatures and precipitation, for example).

Climatic optimum: The state of an ideal climate; inferred for existing desert regions to have had sufficient annual rainfall to render the area fit for habitation. Such conditions are thought to have prevailed over much of the Saharan and Arabian deserts between about 9,000 and 6,500 years ago.

Coastal plain: A low, level plain composed of horizontal or of gently sloping strata of clastic material. One of its margins is the coast. It may represent a portion of the sea floor that recently emerged, and it borders the preexisting land which was uplifted with it. or advancing deposition forming a low lying area bordering a sea/margin.

Coefficient of runoff: The ratio of surface water runoff to the original rainfall.

Coefficient of storage: The volume of water an aquifer releases from or takes into storage per unit surface area of the aquifer per unit change in head.

Coefficient of thermal expansion: The expansion of a unit length or volume of a material per degree rise in temperature. Every minerals has its own particular coefficient leading to differential expansion or contraction in rocks of complex mineralogy.

Common ion effect: The decrease in the solubility of a salt dissolved in water already containing some of the ions of the salt.

Cone of depression: Dewatered volume around a pumping well.

Confined aquifer: An aquifer that is overlain by a confining bed. The confining bed has a significantly lower hydraulic conductivity than the aquifer.

Confined groundwater: The water contained in a confined aquifer. Pore-water pressure is greater than atmospheric at the top of the confined aquifer

Confining bed: A body of material of low hydraulic conductivity that is stratigraphically adjacent to one or more aquifers. It may lie above or below the aquifer.

Confining Unit: A saturated, relatively low-permeability geologic unit that is areally extensive and serves to confine an adjacent artesian aquifer or aquifers. Leaky confining units may transmit appreciable water to and from adjacent aquifers.

Connate water: The water entrapped in sedimentary rock at the time of its deposition. Water trapped in the pore space of sediment at the time of deposition.

Consolidation: In soil mechanics, consolidation is the adjustment of a saturated soil in response to increased load, involving the squeezing of water from the pores and a decrease in void ratio or porosity of the soil.

Contaminant: Any solute that enters the hydrologic cycle through human action.

Convection: The transfer of heat from one part of a fluid or gas to another by flow of the fluid or gas from the hotter parts of the colder. A fluid will rise if heated from below because, through expansion, it becomes less dense than the cold.

Coppice dune: A thicket of small shrubs on dunes.

Coriolis Force: the force created by the Earth's rotation that causes any moving object on the Earth's surface (e.g. water, gas) to curve to the right in the Northern Hemisphere and to the left in the Southern Hemisphere.

Cross beds: Layers within a stratified unit that are oriented that an angle to the dominant stratification.

Cuesta: A more or less lobed ridge, showing a steep descent on one side and gentle one on the other side.

(D)

Darcy's law: An equation derived for the flow of fluids on the assumption that the flow is laminar and that inertia can be neglected.

Deflation: the blowing away of dry fine-grained rock material (sand and dust), by the wind. A form of aeolian erosion at work chiefly in deserts.

Deluge: Heavy fall of rain or flood; synonym for the biblical 'Noah's Flood'.

Density: The mass or quantity of a substance per unit volume. Units are kilograms per cubic meter or grams per cubic centimeter.

Dentritic: Resembling a branching pattern of a tree.

Depression storage: Water from precipitation which collects in puddles at the land surface.

Desalination: The process whereby the salt content of water is greatly reduced so that saline water can be used for human consumption or irrigation.

Desert: An almost barren tract of land in which precipitation is so scanty and spasmodic that it will not adequately support vegetation, and where the potential rate of evaporation far exceeds precipitation.

Desert pavement: A desert surface comprising closely packed pebble or cobble-sized clasts, which commonly are angular.

Desert varnish: Patina of manganese or iron oxide on the surface pebble or cobble-sized clasts, which commonly are angular.

Detrital: Referring to sediment compo-sed of pieces of preexisting materials derived from outside the depositional basin.

Diagenesis: The chemical and physical changes occurring in sediments before consolidation or while in the environment of deposition. All of the chemical, physical and biological changes in the characteristics of a sediment accumulation from the time the grains are deposited until they are metamorphosed or melted, excluding weathering.

Dikakah: Sand accumulations in and around scrub vegetation. Named after Arabic word for such partly stabilised dune sand. See Coppice dune.

Dip slope: The angle which a stratum makes with a horizontal plane.

Discharge: The volume of water flowing past a fixed point in a river, stream, or pipe per unit of time. The volume of water per unit time passing through a given cross section.

Discharge area: An area in which there are upward components of hydraulic head in the aquifer. Groundwater is flowing toward the surface in a discharge area and may escape as a spring, seep, or baseflow, or by evaporation and transpiration.

Discharge velocity: An apparent velocity, calculated from Darcy's law, which represents the flow rate at which water would move through an aquifer if the aquifer was an open conduit. Also called specific discharge

Distillation: The process whereby a liquid is evaporated and recondensed as a purified liquid.

Dolomicrite: Mud-size dolomite crystals in a carbonate rock.

Dolomite: A mineral with formula $CaMg\ (CO_3)_2$ but also used to refer to the sedimentary rock composed largely or entirely of the mineral.

Dolostone: A term used by some petrologists to designate a sedimentary rock composed largely or entirely of the mineral dolomite.

Dorag dolomitization: Replacement of limestone by dolomite through the mechanism of a mixture of freshwater (about 95%) and seawater (about 5%), presumably along coastlines in subtropical and tropical areas.

Downwarp: Down fold.

Draa: A large dune with a height generally in excess of about 60m and a width in the order of 500m or more; a megadune. Believed by some workers to form a special class of dune. The term seems to be applied to those dunes that were formed, or modified form earlier types, during the last glaciation.

Drainage basin (Catchment basin): A geographical region drained by a river or stream.

- The land area from which surface runoff drains into a stream system.
- A region bounded by a drainage divide and occupied by a stream system; the provenance of sediment in the lower part of the trunk stream.

Drainage divide: A boundary line along a topographically high area which separates two adjacent drainage basins.

Drawdown: A lowering of the water table of an unconfined aquifer or the potentiometric surface of a confined aquifer caused by pumping of groundwater from wells.

Drill-pipe: Special pipe used to transmit rotation from the rotating mechanism to the bit. The pipe also transmits weight to the bit and conveys air or fluid which removes cuttings from the hole and cools the bit.

Drilling fluid: A water- or air-based fluid used in the water-well drilling operation to remove cuttings from the hole, to clean and cool the bit, to reduce friction between the drill string and the sides of the hole, and to seal the borehole.

Dune: Accumulation of wind-blown sand that possesses one or more slipfaces. Its size is dependent on the availability of sand and the ability of the wind to carry sand to the top without removing it again. The finest sand grains are usually found at the crest.

(E)

Earth Geoid: The sea-level equipotential surface or figure of the Earth. If the Earth were completely covered by a shallow sea, the surface of this sea would conform to the geoid shaped by the hydrodynamic equilibrium of the water subject to gravitational and rotational forces. Mountains and valleys are departures from this reference geoid.

Effective porosity: The amount of interconnected pore space through which fluids can pass, expressed as a percent of bulk volume. Part of the total porosity will be occupied by static fluid being held to the mineral surface by surface tension, so effective porosity will be less than total porosity. The percentage of a rock volume that consists of interconnected void spaces.

Effective Stress: Stress (pressure) that is borne by and transmitted through the grain-to-grain contacts of a deposit, and thus affects its porosity and other physical properties. In one-dimensional compression, effective stress is the average grain-to-grain load per unit area in a stress is the weight (per unit area) of sediments and moisture above the water table, plus the submerged weight (per unit area) of sediments between the water table and the specified depth, plus or minus the seepage stress (hydrodynamic drag) produced by downward or upward components, respectively, of water movement through the saturated sediments above the specified depth. Effective stress may also be defined as the difference between the geostatic stress and fluid pressure at a given depth in a saturated deposit, and represents that portion of the applied stress which becomes effective as intergranular stress.

Electrical conductivity (EC): It is also called "specific conductance". It refers to the reciprocal of electrical resistance of a substance when it is placed in between two electrodes of 1 cm^2 in area each placed face to face at a distance of 1 cm.

Electrical conductance: A measure of the ease with which a conducting current can be caused to flow through a material under the influence of an applied electric field. It is the reciprocal of resistivity and is measured in mhos per meter.

Electrical resistance model: An analog model of groundwater flow based upon the flow of electricity through a circuit containing resistors and capacitors.

Electrical resistivity: The property of a material which resists the flow of electrical current measured per unit length through a unit cross-sectional area.

Eluviation: The downward movement of material in a soil by groundwater percolation.

Eolianite: All consolidated sedimentary rocks deposited by the wind. More commonly used in connection with calcite-cemented dune sands of coastal regions.

Equipotential line: A line in a two-dimensional groundwater flow field such that the total hydraulic head is the same for all points along the line.

Equipotential surface: A surface in a two-dimensional groundwater flow field such that the total hydraulic head is the same everywhere on the surface.

Erosion: The breakdown of materials of the earth's crust by various physical and chemical processes and the transporting of the particles by wind, moving water, or moving ice.

Escarpment: A cliff or a relatively steep slope separating level or gently sloping tracts.

Eustatic change in sea level: World-wide change in sea level caused by changes in the volume of sea water, as by the melting of glaciers.

Evaporation: The process by which liquid is transformed to the vapor state. The process by which water passes from the liquid to the vapor state.

Evaporative cooling tower: A mecha-nical device that relies on the evaporation of water for the cooling of heated water effluents.

Evapotranspiration: The sum of evaporation plus transpiration.

Evaporite: A sedimentary deposit composed of minerals more soluble than dolomite, such as gypsum, halite, or sylvite.

Exfoliation: the weathering process whereby thin layers peel off the surface of a rock.

(F)

Facies: A characteristic or group of characteristics of a rock that distinguish it from other rocks, e.g. redbed facies, salt facies, sandy facies.

Fault: A fracture or a zone of fractures along which there has been displacement of the sides relative to one another parallel to the fracture.

Filter pack: Sand or gravel that is smooth, uniform, clean, well-rounded and siliceous. It is placed in the annul of the well between the borehole wall and the well screen to prevent formation material from entering the screen.

Flood plain: The portion of a river valley built of sediments and covered with water at flood stage.

Flow net: The set of intersecting equipotential lines and flow lines representing two-dimensional steady flow thro-ugh porous media.

Formation: An extensive, mappable lithologic unit in a sedimentary sequence.

Formation waters: Waters present in the pores of a sedimentary unit.

Forset slope: The inclined bedding that face down-current in both aeolian and aquatically deposited sediments; may be used to deduce the direction of depositing currents.

Fracture trace: The surface representation of a fracture zone. It may be a characteristic line of vegetation or linear soil-moisture pattern or a topographic sag.

Fully penetrating well: A well drilled to the bottom of an aquifer, constructed in such a way that it withdraws water from the entire thickness of the aquifer.

(G)

Gamma-gamma radiation log: A bore-hole log in which a source of gamma radiation as well as a detector are lowered into the borehole. This log measures bulk density of the formation and fluids.

Geomorphology: The description of natural phenomena and the investigation of the history of geologic changes through the interpretation of topographic forms.

Geostatic (Lithostatic) Stress: The total weight (per unit area) of sediments and water above some plane of reference. Geostatic stress normal to any horizontal plane of reference in a saturated deposit may also be defined as the sum of the effective stress and the fluid pressure at that depth.

Granule: A sediment particle with a diameter of 2-4 mm.

Granule ripple: wind-formed ripple composed largely of granules.

Groundwater conservation: A system of measures to be taken for prevention from and liquidation of water pollution and depletion.

Groundwater control: A system of measures intended to obtain regular information on groundwater pollution and depletion, as well as to predict these events, and to provide the observance of the established order in groundwater protection against pollution and depletion by all consumers.

Groundwater pollution: Man-made changes in water quality (physical, chemical and biological properties) as compared to a natural water state and maximum permissible quality standards.

Groundwater table: The surface between the zone of saturation and the zone of aeration; the surface of an unconfined aquifer.

Ground subsidence: The sinking of the earth's surface that occurs when subsurface mineral deposits or ground- water are withdrawn.

Groundwater: The subsurface water in the zone of saturation beneath the water table

- The water contained in inter-connected pores located in a confined aquifer
- The part of the subsurface water that is below the groundwater table, in the zone of saturation. Synonym of "phreatic water".

Groundwater basin: A rather vague designation pertaining to a groundwater reservoir which is more or less separate from neighboring groundwater reservoirs. A ground-water basin could be separated from adjacent basins by geologic boundaries or by hydrologic boundaries.

Groundwater flow: The movement of water through openings in sediment and rock which occurs in the zone of saturation.

Groundwater mining: The practice of withdrawing groundwater at rates in excess of the natural recharge.

Groundwater reservoir: Extractable underground fresh water.

Grout: A fluid mixture of cement and water (neat cement) of a consistency that can be forced through a pipe and placed as required. Various additives, such as sand, bentonite, and hydrated lime, may be included in the mixture to meet certain requirements. Bentonite and water are sometimes used for grout.

Grouting: The operation by which grout is placed between the casing and the sides of the well bore to a predetermined height above the bottom of the well. This secures the casing in place and excludes water and other fluids from the well bore.

Gypsum: An evaporite mineral Calcium sulphate ($CaSO_4.2H_2O$), typically found just below the surface of coastal and inland sabkhas alters to anhydrite when it loses its water of crystallisation.

(H)

Halite: An evaporite mineral common salt (NaCl), which forms a thin crust over the surface of both coastal and inland sabkhas; also precipitated as bedded halite on the floor of some deep evaporite basins (e.g. beneath the salt domes of Iran, the Arabian Gulf and SE Oman).

Hardness: A property of water causing formation of an insoluble residue when the water is used with soap. It is primarily caused by calcium and magnesium ions.

Head: Fluid pressure due to differences in height in two water surfaces.

Heterogeneous: Pertaining to a substance having different characteristics in different locations. A synonym is nonuniform.

High-magnesian calcite: A calcite containing more than 4% $MgCO_3$ in solid solution.

Homogeneous: Uniform in structure or composition throughout.

Horse Latitudes: The sub-tropical belts of high atmospheric pressure at about 30° north and south of the equator.

Hydraulic conductivity (Permeability coefficient): The ratio of flow velocity to driving force for water flow in porous media.

- A coefficient or proportionality describing the rate at which water can move through a permeable medium. The density and kinematic viscosity of the water must be considered in determining hydraulic conductivity.
- For practical work in groundwater hydrology, where water is the prevailing fluid, permeability coefficient (K) is employed. A medium has a unit permeability coefficient if it will transport in unit time a unit volume of groundwater at the prevailing kinematic viscosity through a cross section of unit area, measured at right angles to the direction of flow, under a unit hydraulic gradient, having units of velocity.

Hydraulic gradient: The change in total head with a change in distance in a given direction. The direction is that which yields a maximum rate of decrease in head.

Hydraulic Head: A measure of the potential for fluid flow. The height of the free surface of a body of water above a given subsurface point.

Hydrochemical facies: Bodies of water with separate but distinct chemical compositions contained in an aquifer.

Hydrocompaction: The process of volume decrease and density increase that occurs when certain moisture-deficient deposits compact as they are wetted for the first time since burial. The vertical downward movement of the land surface that results from this process has also been termed "shallow subsidence" and "near-surface subsidence."

Hydrogeologic: Those factors that deal with surface waters and related geologic aspects of surface waters.

Hydrologic cycle: The ceaseless flow of water from one reservoir to another. The cycle of phenomena through which water passes above, on, and under the earth's surface.

Hydrostratigraphic unit: A formation, part of a formation, or a group of formations in which there are similar hydrologic characteristics allowing for grouping into aquifers or confining layers.

Hygroscopic: The ability of minerals like halite to absorb moisture readily without becoming a liquid.

Hypersaline: Referring to water contains more than 35 ppt of dissolved solids. Synonym of "Brine".

(I)

Igneous rocks: That solidified from molten or partly molten material, that is from a magma.

Induration: A process whereby soft sediment becomes hard rock. The term originally implied baking in proximity to an igneous rock, but is now used to include hardening by the pressure of overlying sediments or by cementation.

Infiltration: Seepage of surface water into soil and rock layers. The flow of water downward from the land surface into and through the upper soil layers.

Infiltration basins: Area-type engineering structures for an artificial recharge of fresh groundwater storage through an infiltration of surface water filling these structures (basins, canals) into aquifers.

Infiltration capacity: The maximum rate at which infiltration can occur under specific conditions of soil moisture. For a given soil, the infiltration capacity is a function of the water content.

Infiltration water intake: Engineering structure (well, borehole, underground gallery or adit) used for extraction of groundwater and surface water percolating from streams and water bodies. Syn: longshore water intake; water intake in river valleys.

Insolation: Exposure to radiation from the sun. More particularly, the term implies the wide diurnal temperature range found in desert areas, and the temperature difference encountered between the hot interior of a rock exposed to the sun's rays and its cooler exterior just after sunset, which can cause a rock to split.

Interdune area: The area between dunes. In the ideal case it is devoid of dune sand. Commonly, however, it comprises lightly vegetated areas of coppice dunes (dikakah), which may cover an older sequence of dune sands, interdune sabkha sediments or the deposits of active or formerly active wadis. (see also Saruq). In some areas, sand dunes are so closely packed spaced that there is no true interdune area between them.

Interference: The condition occurring when the area of influence of a water well comes into contact with or overlaps that of a neighbouring well, as when two wells are pumping from the aquifer or are located near each other.

Intermediate zone: That part of the unsaturated zone below the root zone and above the capillary fringe.

Interstitial: Referring to location in pore spaces, between the detrital grains.

Intrastratal solution: Dissolution of minerals after they have been deposited in a body of sediment and buried.

Intrinsic permeability: Pertaining to the relative ease with which a porous medium can transmit a liquid under a hydraulic or potential gradient. It is a property of the porous medium and is independent of the nature of the liquid or the potential field.

Intrusive rocks: Those igneous rocks formed from magma injected beneath the earth's surface. Generally these rocks have large crystals caused by slow cooling.

Ion exchange: A process by which an ion in a mineral lattice is replaced by another ion which was present in an aqueous solution.

Isocon: A line drawn on a map to indicate equal concentrations of a solute in groundwater.

Isohyetal line: A line drawn on a map, all points along which receive equal amounts of precipitation.

Isostatic subsidence: Crustal subsidence adjusting to a newly acquired load, usually of sediment, water or ice. The crust rises if that load is removed.

Isotropy: The condition in which hydraulic properties of the aquifer are equal in all directions.

(J)

Juvenile water: Water derived directly from the magma and that is coming to the earth's surface for the first time.

(K)

Karst: A type of topography that is formed on limestone, dolomite, gypsum and other rocks, primarily by dissolution, and that is characterized by sinkholes, caves, and subterranean drainage.

Karstification: Action by water, mainly chemical but also mechanical, that produces features of a karst topography.

Karst topography: It is developed on calcareous rocks and characterized by sink holes, disappearing streams, and caverns.

Kinetic energy: the energy possessed by a moving body by virtue of its motion ($1/2\ mv^2$). See surface creep.

(L)

Lag deposit: A relatively coarse sediment (coarse sand, pebbles, boulders etc.) that was left behind when other finer grains were removed by the transporting medium or air or water.

Lamina: A layer of sediment or sedimentary rock less than 1 cm thick.

Laminar flow: Water flow in which the stream lines remain distinct and in which the flow direction at every point remains unchanged with time. It is characteristic of the movement of groundwater.

Lamination: Stratification on a scale less than 1 cm.

Leaching: The dissolving, transporting, and redepositing of materials by water seeping downward. The selective removal of soluble constituents from minerals and rocks by percolating water; used mostly in reference to surficial processes.

Leaky confining layer: A low-permeability layer that can transmit water at sufficient rates to furnish some recharge to a well pumping from an underlying aquifer. Also called "Aquitard".

Limeclast: A carbonate rock fragment of clastic or detrital origin contained within a carbonate rock; the fragment may be intrabasinal (intraclastic) or extrabasinal (terrigenous).

Limestone: A sedimentary rock consisting largely or entirely of calcium carbonate.

Linear dune: Any straight or gently arced dune; its name does not imply its mode of origin. Often used as a loose synonym for longitudinal dune.

Lithification: The conversion of unconsolidated sediment into a coherent aggregate. Synonym of "Cementation" or "Induration".

Lithostratigraphy: Identification of rocks by their lithological character.

Longitudinal dune: dune whose long axis parallels the prevailing dune-forming wind; it grows by extending downwind. Avalanche slopes, where present, are almost parallel to the axis of the dune and can face toward either flank. May occur as a swarm of parallel dunes as in the Rub al Khali or the Wahiba.

Lost circulation: The result of drilling fluid escaping from the borehole into the formation by way of crevices or porous media.

Low-energy environment: A depositional environment characterized by relatively low kinetic energies, e.g. lakes.

Low-magnesian calcite: A calcite containing less than 4% $MgCO_3$ in solid solution.

(M)

Marl: A friable mixture of subequal amounts of micrite and clay minerals.

Matrix: The fine-grained material in a sediment with a conspicuous range in grain size, e.g., the clay in a texturally immature sandstone.

Megadune: Any large dune whose height exceeds about 60 m and has a crestal spacing of 500m or more. Most are thought to have formed during the last major glaciation.

Member: A body of rock, not necessarily mappable, that is a subdivision of a formation, may consist of one or more beds.

Metamorphic rocks: Any rock derived from pre-existing rocks by mineralogical, chemical, and/or structural changes, essentially in the solid state, in response to marked changes in temperature, pressure, shearing stress, and chemical environment, generally at depth in the earth's crust.

Meteoric water: Water of recent atmospheric origin.

Micrite: Microcrystalline calcite or aragonite mud, normally of clay size, in a carbonate rock.

Miliolite: A cemented, carbonate-rich dune sand (aeolianite), which was first described form the northwest coast of India where it typically contains ooids and miliolid foraminifera. These latter are not always present, however.

Monsoon: A monsoon is the type of wind system in which there is a complete or almost complete reversal of prevailing direction from season to season. It is especially prominent within the tropics on the eastern sides of great landmasses. The Southwest Trade Winds in the southern Indian Ocean are warm and saturated with moisture. Because of intense heating of the land in summer, a low-pressure area develops over northwestern Indian and West Pakistan which deflects the Southeast Trades northwards across the equator. Part of these deflected trade winds travel parallel to the east coast of Africa and then swing parallel to the South Arabian coast (as the Southwest Monsoon) where the hill slopes facing the sea receive some rain. These moist winds bring torrential rainfall to much of central and northern India. The summer monsoon lasts between April and September, although its duration at any one place depends upon its geographical location. The winter monsoon of reversed direction usually produces only light winds in northern India and Arabia because of the protection afforded by the mountain masses to the north. Where the Southwest Monsoon blows (north) across the Wahiba Sands, the elevation of the sands above sea level is too low to induce rainfall over this hot area. The winds are, however, strong. Similarly over the Rajasthan Desert, the South-west Monsoon brings little rainfall because of the desert's relatively low elevation.

Mud: Sediments whose particles are less than 0.06 mm in size.

Mudcracks: Desiccation cracks in mud usually outlining polygonal blocks.

Mudrock: Detrital rock whose grains are mostly less than 0.06 mm in size.

Mudstone: Nonfissile mudrock containing subequal amounts of silt- and clay-size silicate detritus; also a mud-supported limestone that contains less than 10% sand- or grave-size particles.

(N)

Nappe: A complex rock mass of large dimensions brought forward over rocks by recumbent folding or by thrusting.

Naturally developed well: A well in which the screen is placed in direct contact with the aquifer materials; no filter pack is used.

Natural gamma radiation log: A borehole log that measures the natural gamma radiation emitted by the formation rocks. It can be used to delineate subsurface rock types.

Nebkha dune: Sand accumulations in and around scrub vegetation. Named after Arabic word for such partly stabilized dune sand.

Neutron log: A borehole log obtained by lowering a radioactive element, which is a source of neutrons, and a neutron detector into the well. The neutron log measures the amount of water present; hence, the porosity of the formation.

Nodular anhydrite: Synonym of "chicken-wire structure".

Nodule: A small, hard, round, irregularly shaped body of a different composition from the rock in which it is located, e.g. a phosphate nodule in a shale.

Northeast Monsoon: Sometimes called the Little Monsoon, brings relatively light, cold dry dusty winds to eastern Arabia from northern India and Pakistan.

(O)

Observation well: A nonpumping well used to observe the elevation of the water table or the potentiometric surface. An observation well is generally of larger diameter than a piezometer and typically is screened or slotted throughout the thickness of the aquifer.

Optimal Yield: An optimal amount of ground water, by virtue of its use, that should be withdrawn from an aquifer system or ground-water basin each year. It is a dynamic quantity that must be determined from a set of alternative ground-water management decisions subject to goals, objectives, and constraints of the management plan.

Orogeny: The process of relatively rapid mountain building by folding, faulting and upheaval of specific portions of the earth's crust.

Osmosis: In aquifers, osmosis permits the diffusion of dissolved salts from areas of high concentration to those of low concentration.

Overburden: The loose soil, silt, sand, gravel, or other unconsolidated material overlying bedrock, either transported or formed in place; regolith.

Overthrust: An upthrust with low angle thrust plane and large net slip.

(P)

Packstone: A clastic limestone in which the grains are resting on each other (grain supported) but some micrite is present.

Paleogeography: The study and description of the physical geography of the geologic past, e.g. topographic relief.

Paleogeology: The study and description of the geology of the geologic past, e.g. its tectonics and outcrop patterns of rocks.

Paleokarst: A karstified area that has been buried by later deposition of sediments.

Partial penetration: When the intake portion of the well is less than the full thickness of the aquifer.

Partially penetrating well: A well constructed in such a way that it draws water directly from a fractional part of the total thickness of the aquifer. The fractional part may be located at the top of the bottom or anywhere in between the aquifer.

Parts per million (ppm): The unit measure of the concentration of a component substance; for example, a 1 ppm concentration of arsenic is 1 part of arsenic to 999,999 parts of other material. It is the unit to express the ratio of a trace amount. 1 ppm is one to a million (0.0001%).

Perennial Yield: The amount of usable water from an aquifer that can be economically consumed each year for an indefinite period of time. It is a specified amount that is commonly specified equal to the mean annual recharge to the aquifer system, which thereby limits the amount of ground water that can be pumped for beneficial use.

Perched water: Unconfined groundwater separated from an underlying main body of groundwater by an unsaturated zone.

Permeability: The capability of soil or rock to transmit water or air. The capacity of a porous material to transmit a fluid; depends on the characteristics of the material, the fluid, and the physical conditions of the flow.

pH: The negative logarithm to the base 10 of the hydrogen ion activity of a solution. This hydrogen ion concentration (pH) is an index to show alkalinity. The neutral point is pH 7.0, and as the value becomes smaller than this, acidity becomes stronger; and as the value approaches 14.0, alkalinity becomes stronger. With the exception of special cases, surface water of river is in the neighborhood of pH 7.0, while seawater is normally slightly alkaline at around pH 8.2.

Phreatic water: Water in the zone of saturation. Synonym of "Groundwater"

Piezometer: A nonpumping well, generally of small diameter, which is used to measure the elevation of the water table or potentiometric surface. A piezometer generally has a short well screen through which water can enter.

Piezometer nest: A set of two or more piezometers set close to each other but screened to different depths.

Playas: Smooth, flat, often saline plains.

Plutonic rocks: Igneous rocks occurring in general in large rock masses formed at considerable depths within the earth's crust.

Porosity: The volume of open space in a soil or rock layer.

- The ratio of volume of interstices in a rock to its total volume.
- The ratio of the volume of void spaces in a rock or sediment to the total volume of the rock or sediment.
- The percentage of rock volume that consists of void space.

Potential evapotranspiration: The evapotranspiration that would occur under given climatic conditions if there were unlimited soil mixture

Potentiometric surface (Piezometric surface): The surface to which water in an aquifer will rise under hydrostatic pressure. A surface that represents the level to which water will rise in tightly cased wells. If the head varies significantly with depth in the aquifer, then there may be more than one potentiometric surface. The water table is a particular potentiometric surface for an unconfined aquifer.

Precipitation: Liquid or solid forms of water falling to earth from clouds.

Pumping cone: The area around a discharging well where the hydraulic head in the aquifer has been lowered by pumping. Also called "cone of depression".

Pumping test (Aquifer test): A test made by pumping a well for a period of time and observing the change in hydraulic head in the aquifer. A pumping test may be used to determine the capacity of the well and the hydraulic characteristics of the aquifer. Also called "Aquifer test". The test is accomplished by measuring the static water level, during the continuous pumping from well or trench to find the hydraulic characteristics of aquifer.

(R)

Radius of influence: The radial distance from the center of a well bore to the point where there is no lowering of the water table or potentiometric surface (the edge of its cone of depression).

Rain shadow: An area of little or no rain that is sheltered from the prevailing rain-bearing winds by a range of hills or mountains, over which rainfall may have been heavy as moist air was forced to rise to cooler altitudes. It is the lee side of a range.

Recharge: The addition of water to the groundwater zone of saturation, naturally by precipitation or runoff, or artificially by spreading or injection.

Recharge area: An area in which there are downward components of hydraulic head in the aquifer. Infiltration moves downward into the deeper parts of an aquifer in a recharge area.

Recharge basin: A basin or pit excavated to provide a means of allowing water to soak into the ground at rates exceeding those that would occur naturally.

Recharge boundary: An aquifer system boundary that adds water to the aquifer. Streams and lakes are typical recharge boundaries.

Relative humidity: The percent of water vapor present in air compared with the maximum amount of water vapor in saturated air at a specified temperature.

Residual drawdown: The difference between the original static water level and the depth to water at a given instant during the recovery period.

Resistivity log: A borehole log made by lowering two current electrodes into the borehole and measuring the resistivity between two additional electrodes. It measures the electrical resistivity of the formation and contained fluids near the probe.

Reverse osmosis method (RO): It is also referred to as the reverse osmosis membrane method. It is a method to obtain clean water by applying pressure against osmotic pressure. For example, when a vessel is partitioned into compartments by a semipermeable membrane (acetic cellulose membrane), and salt water is placed in one compartment and water in the other up to the same level, the water infiltrates through the semipermeable membrane and migrates into the thick salt water and generates pressure by the difference in level. This difference in pressure is called osmotic pressure, and when greater pressure is applied to the salt water to counter this osmotic pressure, only water migrates to the water through the semipermeable membrane. As the osmotic pressure of sea water for example is 24.8 kg/cm^2, if a larger pressure than this is applied, clear water is obtainable. By applying this theory, the reverse osmosis method is used for treatment.

Ripple: A surface undulation, generally of unconsolidated sand, whose wavelength depends on wind strength and is constant with time. The ripple axis is always transverse to the wind. The coarest grains are found at the crest. The ripple height depends on the range of grainsize present and the wind strength.

Ripple index: The ratio between the wavelength of a ripple and its height. For aeolian sands commonly between 15 and 20.

Runoff: The total amount of water flowing in a stream. It includes overland flow, return flow, interflow and baseflow.

(S)

Sabkha: A flat area of clay, silt or sand, commonly with crusts of salt. Sub-divided into: (1) Coastal sabkha: a coastal flat at or just above the level of normal high tide. Its sediments consist of sand, silt or clay and its surface is often covered with a salt crust formed by the evaporation of water drawn to the surface by capillary action of from occasional marine inundations. The coastal sabkha is characterised by the presence of algal mats and the occurrence of gypsum and anhydrite within its sediment. It is subjected to deflation down to the water table; (2) Inland sabkha: a flat area of clay, silt or sand, commonly with saline encrustations, that is typical of desert areas of inland drainage, and in some interdune areas. Their salts may be formed by evaporation of surface water, or of water drawn to the surface from the water table by capillary action.

Safe Yield: The amount of ground water that can be safely withdrawn from a ground-water basin annually, without producing an undesirable result. Undesirable results include but are not limited to depletion of ground-water storage, the intrusion of water of undesirable quality, the contraventions of existing water rights, the deterioration of the economic advantages of pumping (such as excessively lowered water levels and the attendant increased pumping lifts and associated energy costs), excessive depletion of streamflow by induced infiltration, and land subsidence.

Saltation: The repeated forward movement of wind- (or water-) driven sand grains along low trajectories.

Salt plug: A body of rock salt which resembles in form, and in its relation to enclosing rocks, an intrusive igneous plug.

Sand drift: A low mound of aeolian sand lacking a slip-face, which can form as the result of a fall in the velocity of a sand-laden wind. Commonly found on the leeward side of wind-breaks such as vegetation, boulders, outcrops or other irregular surface. sometimes referred to as a stringer of sand.

Sand sea: Any large area covered by sand dunes that are so close together that there is no real interdune area.

Saruq: Arabic for 'easy going'. The name given to some interdune areas, especially in central Abu Dhabi, that tend parallel to the axes of adjacent mega-giant dunes. As the Arabic implies, it gives easy passage to camels, and provided the coppice dunes are not too closely spaces, also to 4-wheel-drive vehicles.

Seepage Stress: Force (per unit area) transferred from the water to the medium by viscous friction when water flows through a porous medium. The force transferred to the medium is equal to the loss of hydraulic head and is termed the seepage force exerted in the direction of flow.

Sedimentary rocks: Rocks resulting from the consolidation of loose sediment that has accumulated in layers.

Seif dune: Gently sinuous form of linear dune. Believed to develop downwind as the resultant of oblique seasonal winds from the two flanks. Usually occur singly or in groups of two or three.

Sinkhole: A depression in a karst area. At land surface its shape is generally circular and its size measured in meters to tens of meters; underground it is commonly funnel-shaped and associated with subterranean drainage.

Shamal: Arabic word for north; applied to north or northwest wind that blows down the Arabian Gulf and clockwise across the Rub al Khali.

Shelf: Zone extending from the low-water line to a depth at which there is a marked increase in slope towards great depth.

Slip-face: The slope that forms when wind-blown sand from the windward side of a dune passes into the air of the leeward side. The sand will start to slip if further deposition would result in the maximum, angle or repose for dry sand of 34° to be exceeded. Also known as avalanche slope.

Southwest Monsoon: Because of intense heating of the land in summer, a low-pressure area develops over northwestern India and Pakistan, which deflects the Southeast Tradewinds of he southern Indian Ocean northwards across the equator. These winds swing parallel to the coast of southern Arabia. Southwest Monsoon bringing some rain to the Hadhramaut coast between April and September.

Specific capacity: The rate of discharge of a water well per unit of drawdown, commonly expressed in m^3/hr/m or m^3/day/m. It varies with duration of discharge.

Specific Storage: The volume of water that an aquifer system releases or takes into storage per unit volume per unit change in head. The specific storage is equivalent to the Storage Coefficient divided by the thickness of the aquifer system.

Specific heat: The quantity of heat necessary to raise the temperature of one gram of a substance by one degree celsius. The specific heat of water is much greater than that of rock. Thus rock heats up and loses heat much more readily than water. This is why marine areas have a much lower annual temperature range than localities far inland, a factor that controls monsoon.

Specific yield: The ratio of the volume of water that a given mass of saturated rock or soil will yield by gravity to the volume of that mass. This ratio is stated as a percentage.

Spring: Any natural discharge of water from rock or soil onto the land surface or into a surface-water body.

Star dune: A roughly star-shaped pyramidal dune with three or more radiating arms with slip faces. Thought to form where seasonal winds are strongly oblique to each other. May result also by modification of older transverse or longitudinal dunes.

Stratigraphy: The study of rock strata, especially of their distribution, deposition and age.

Static water level: The level of water in a well that is not being affected by withdrawal of groundwater.

Storage: The capacity of an aquifer, aquitard, or aquifer system to release or accept water into ground-water storage, per unit change in hydraulic head.

Storage Coefficient: The volume of water that an aquifer system releases or takes into storage per unit surface area per unit change in head.

Stress: In a solid body, the force (per unit area) acting on any surface within it; also refers to the applied force (per unit area) that creates the internal force. Stress is variously expressed in units of pressure, such as pounds per square inch, kilograms per square meter, or Pascals.

Subsidence: Sinking or settlement of the land surface, due to any of several processes. As commonly used, the term relates to the vertical downward movement of natural surfaces although small-scale horizontal components may be present. The term does not include landslides, which have large-scale horizontal displacements, or settlements of artificial fills.

Surface creep: The slow movement of large grains under the impact of smaller grains in saltation, by means of the transfer of kinetic energy from one to the other.

(T)

Terrestrial: Continental.

Threshold velocity: the wind velocity needed to set those desert sediments in subtropical areas of land into motion.

Total dissolved solids (TDS): A term that expresses the quantity of dissolved material in a sample of water, either the residue on evaporation, dried at 180 Degrees Celcius, or, for many waters that contain more than about 1,000 mg/l, the sum of the chemical constituents.

Trade winds: The winds that blow from the subtropical belts of high pressure towards the equatorial region of low pressure. They blow from NE in the northern hemisphere and from SE in the southern hemisphere. The term refers to the subtropical ocean winds that could be depended on by sail-driven trading ships. Sometimes extended to trade wind deserts.

Trade wind desert: A term sometimes applied to those deserts in subtropical areas of land that controls monsoons.

Transmissivity: The rate at which water is transmitted through a unit width of an aquifer under a unit hydraulic gradient. Transmissivity values are given in m^3 per day through a vertical section of an aquifer one meter wide and extending the full saturated height of an aquifer under a hydraulic gradient of 1.

Transpiration: The process by which water absorbed by plants, usually through the roots, is evaporated into the atmosphere from the plant surface.

Transverse dune: A dune whose long axis is at right angles to the prevailing dune forming wind. Likely to break up into barchanoid and then barchan dunes if the supply of sand is not maintained.

(U)

Ultra basic rocks: Igneous rocks with a silica content less than 45 per cent.

Unconfined aquifer: An aquifer where the water table is exposed to the atmosphere through openings in the overlying materials.

Up-warp: Up fold.

(V)

Vug: A small cavity or chamber in rock that may be lined with crystals.

(W)

Wadi: Desert watercourse, dry except after rain.

Water of crystallation: The water pre-sent in hydrated compounds such as gypsum ($CaSO_4.2H_2O$). If the temperature of the gypsum crystal is raised above about 50 °C, either by deep burial or by near-surface heating in a desert, it loses its water of crystallation ($2H_2O$) and becomes the anhydrous mineral anhydrite.

Water table: The surface between the vadose zone and the ground water; that surface of a body of unconfined groundwater at which the pressure is equal to that of the atmosphere.

Well screen: A filtering device used to keep sediment from entering a water well.

Well yield: The volume of water discharged from a well in cubic meters per day.

---

* These definitions are based on Bates and Jakson (1987) and Galloway et al. (1999).